Students and External Readers	Staff & Research Students
DATE DUE FOR RETURN	
UNIVE	

Methods in Enzymology

Volume 88
BIOMEMBRANES
Part I
Visual Pigments and Purple Membranes, II

METHODS IN ENZYMOLOGY

EDITORS-IN-CHIEF

Sidney P. Colowick   Nathan O. Kaplan

Methods in Enzymology

Volume 88

Biomembranes

Part I

Visual Pigments and Purple Membranes, II

EDITED BY

Lester Packer

MEMBRANE BIOENERGETICS GROUP
UNIVERSITY OF CALIFORNIA
BERKELEY, CALIFORNIA

Editorial Advisory Board

1982

ACADEMIC PRESS

A Subsidiary of Harcourt Brace Jovanovich, Publishers

New York   London
Paris   San Diego   San Francisco   São Paulo   Sydney   Tokyo   Toronto

ACADEMIC PRESS, INC.
111 Fifth Avenue, New York, New York 10003

United Kingdom Edition published by
ACADEMIC PRESS, INC. (LONDON) LTD.
24/28 Oval Road, London NW1 7DX

Library of Congress Cataloging in Publication Data
Main entry under title

Visual pigments and purple membranes.

   (Biomembranes ; pt. I) (Methods in enzymology ; v. 88
   Includes bibliographical references and index.
   1. Visual pigments. 2. Visual purple. 3. Cell
membranes. I. Packer, Lester. II. Series. III. Series:
Methods in enzymology ; v. 88. [DNLM: 1. Cell membrane.
2. Membranes--Enzymology. W1 ME9615K v. 31, etc.]
QP601.M49   vol. 88   [QP671.V5]   574.19   82-1736
ISBN 0-12-181988-4   [591.1'823]   AACR2

PRINTED IN THE UNITED STATES OF AMERICA

82 83 84 85   9 8 7 6 5 4 3 2 1

Table of Contents

Section I. Bacteriorhodopsin

A. Purple Membrane Preparations and Protein Structure

B. Reconstituted Systems

C. Molecular Structure of Purple Membranes

D. Chemistry, Spectroscopy, and Photochemistry

E. Specialized Physical Techniques

F. Ion Transport and Physiology

G. Biogenesis, Genetics, and Microorganisms

H. Light-Dependent Behavioral Responses of the Intact Organism

I. Other Retinal Proteins

Section II. General Methods for Retinal Proteins

A. Bacteriorodopsin and Rhodopsin Molecular Structure

B. Model Chromophores

C. Physical and Chemical Methods

Contributors to Volume 88

Article numbers are in parentheses following the names of contributors.
Affiliations listed are current.

N. G. ABDULAEV (86), *Department of Protein Chemistry, Shemyakin Institute of Bioorganic Chemistry, USSR Academy of Sciences, Moscow, USSR*

PATRICK L. AHL (88), *Department of Physics, Boston University, Boston, Massachusetts 02215*

PRAMOD V. ARGADE (76), *Departments of Physics and Physiology, Boston University, Boston, Massachusetts 02215*

ALFRED E. ASATO (65), *Department of Chemistry, University of Hawaii at Manoa, Honolulu, Hawaii 96822*

EVERT P. BAKKER (5), *Fachbereich Biologie/Chemie, Fachgebiet Mikrobiologie, Universität Osnabrück, D-4500 Osnabrück, Federal Republic of Germany*

VALERIA BALOGH-NAIR (64), *Department of Chemistry, Columbia University, New York, New York 10027*

HAGAN BAYLEY (10), *Department of Biochemistry, College of Physicians and Surgeons, Columbia University, New York, New York, 10032*

BRIAN BECHER (33), *School of Basic Medical Sciences, University of Illinois, Urbana, Illinois 61801*

JURGEN BERGMEYER (12), *Max-Planck-Institut für Biochemie, Abteilung Membranchemie, Munich, Federal Republic of Germany*

ROBERT R. BIRGE (67), *Department of Chemistry, University of California, Riverside, California 92521*

RICHARD A. BLATCHLY (63), *Department of Chemistry, Columbia University, New York, New York 10027*

A. E. BLAUROCK (16), *Laboratories for Cell Biology, Department of Anatomy, University of North Carolina, Chapel Hill, North Carolina 27514*

ROBERTO A. BOGOMOLNI (50, 58), *Cardiovascular Research Institute, University of California, San Francisco, California 94143*

MARK BRAIMAN (77), *Department of Chemistry, University of California, Berkeley, California 94720*

DAVID S. CAFISO (83), *Department of Chemistry, Stanford University, Stanford, California 93405*

ROBERT CALLENDER (74), *Physics Department, City College of the City University of New York, New York, New York 10031*

PEDRO CANDAU (89), *Membrane Bioenergetics Group and, Department of Physiology, University of California, Berkeley, California 94720*

S. ROY CAPLAN (5), *Department of Membrane Research, Weizmann Institute of Science, Rehovot 76100, Israel*

RICHARD J. CHERRY (31), *Department of Chemistry, University of Essex, Wivenhoe Park, Colchester, Essex CO4 3SQ, England*

NOEL A. CLARK (42, 84), *Department of Physics, University of Colorado, Boulder, Colorado 80309*

RICHARD A. CONE (88), *Department of Biophysics, The John Hopkins University, Baltimore, Maryland 21218*

ALAN COOPER (79), *Chemistry Department, Glasgow University, Glasgow G12 8QQ, Scotland*

HENRY L. CRESPI (1), *Chemistry Division, Argonne National Laboratory, Argonne, Illinois 60439*

ROSALIE K. CROUCH (20), *Departments of Ophthalmology and Biochemistry, Medical University of South Carolina, Charleston, South Carolina 29425*

NORBERT A. DENCHER (2, 6, 19, 56), *Biophysics Group, Department of Physics, Freie Universität Berlin, D-1000 Berlin 33, Federal Republic of Germany*

NANCY W. DOWNER (80, 82), *Department of Biochemistry, University of Arizona, Tucson, Arizona 85721*

IAN DUNDAS (47), *Departments of Microbiology and Plant Physiology, University of Bergen, Bergen, Norway*

THOMAS G. EBREY (61, 66), *Department of Physiology and Biophysics, University of Illinois at Urbana-Champaign, Urbana, Illinois 61801*

LAURA EISENSTEIN (39), *Department of Physics, University of Illinois at Urbana-Champaign, Urbana, Illinois 61801*

M. A. EL-SAYED (73), *Department of Chemistry, University of California, Los Angeles, California 90024*

D. M. ENGELMAN (11), *Department of Molecular Biophysics and Biochemistry, Yale University, New Haven, Connecticut 06511*

JOAN J. ENGLANDER (80), *Department of Biochemistry and Biophysics, University of Pennsylvania School of Medicine, Philadelphia, Pennsylvania 19104*

KNUTE A. FISHER (28), *Cardiovascular Research Institute, Department of Biochemistry and Biophysics, University of California, San Francisco, California 94143*

JAMES M. FORSYTH (36), *College of Engineering, Laboratory for Laser Energetics, Rochester, New York 14623*

ROBERT D. FRANKEL (36), *College of Engineering, Laboratory for Laser Energetics, Rochester, New York 14623*

JOSEPH M. FUKUMOTO (40), *Department of Chemistry, University of California, Los Angeles, California 90024*

WOLFGANG GÄRTNER (70), *Institut für Biochemie der Universität Würzburg, Röntgenring 11, Federal Republic of Germany*

GERHARD E. GERBER (9), *Department of Biochemistry, McMaster University, Hamilton, Ontario L8N 3Z5, Canada*

WERNER GOEBEL (49), *Institut für Genetik und Mikrobiologie, Universität Würzburg, 8700 Würzburg, Federal Republic of Germany*

A. GOLDMAN (11), *Department of Molecular Biophysics and Biochemistry, Yale University, New Haven, Connecticut 06511*

S. GRUNER (37), *Department of Physics, Princeton University, Princeton, New Jersey 08540*

BO HÖJEBERG (10), *Departments of Biology and Chemistry, Massachusetts Institute of Technology, Cambridge, Massachusetts 02139*

TOSHIAKI HAMANAKA (34), *Department of Biophysical Engineering, Faculty of Engineering Science, Osaka University, Toyonaka, Osaka 560, Japan*

BENNO HESS (22, 23, 32), *Max-Planck-Institut für Ernährungsphysiologie, 4600 Dortmund 1, Federal Republic of Germany*

MAARTEN P. HEYN (2, 6), *Biozentrum der Universität Basel, Abt. Biophysikalische Chemie, Basel, Switzerland*

EILO HILDEBRAND (56), *Institute of Neurobiology, Nuclear Research Center, Jülich, Federal Republic of Germany*

KENJI HIRAKI (34), *Department of Biophysical Engineering, Faculty of Engineering Science, Osaka University, Toyonaka, Osaka 560, Japan*

BARRY HONIG (61), *Department of Physiology and Biophysics, University of Illinois at Urbana-Champaign, Urbana, Illinois 61801*

WILLIAM D. HOPEWELL (40), *IBM Instruments, Inc, San Jose, California 95110*

PHILLIP N. HOWLES (57), *Roswell Park Memorial Institute, Buffalo, New York 14263*

KUO-SEN HUANG (10), *Departments of Biology and Chemistry, Massachusetts Institute of Technology, Cambridge, Massachusetts 02139*

WAYNE L. HUBBELL (83), *Chemistry Department, University of California, Berkeley, California 94720*

TATSUO IWASA (18), *Department of Physics, Faculty of Science, Tohoku University, Aobayama, Sendai 980, Japan*

NAOKI KAMO (46), *Faculty of Pharmaceutical Sciences, Hokkaido University, Sapporo 060, Japan*

M. KATES (13), *Department of Biochemistry, University of Ottawa, Ottawa, Ontario K1N 9B4, Canada*

LAJOS KESZTHELYI (38), *Institute of Biophysics, Biological Research Center, 6701 Szeged, Hungary*

H. GOBIND KHORANA (9, 10), *Departments of Chemistry and Biology, Massachusetts Institute of Technology, Cambridge, Massachusetts 02139*

GLEN I. KING (30), *Department of Physiology and Biophysics, University of California, Irvine, California 92717*

ROBERT A. KINSEY (41), *School of Chemical Sciences, University of Illinois at Urbana-Champaign, Urbana, Illinois 61801*

AUGUSTIN KINTANAR (41), *School of Chemical Sciences, University of Illinois at Urbana-Champaign, Urbana, Illinois 61801*

DAVID S. KLIGER (68), *Division of Natural Sciences, University of California, Santa Cruz, California 95064*

TETSUYA KONISHI (24), *Department of Radiochemistry–Biology, Niigata College of Pharmacy, 5829 Kamishin'ei, Niigata 950-21, Japan*

JUAN I. KORENBROT (8), *Department of Physiology, University of California School of Medicine, San Francisco, California 94143*

RAFI KORENSTEIN (22, 23), *Department of Membrane Research, Weizmann Institute of Science, Rehovot 76100, Israel*

W. KREUTZ (87), *Albert-Ludwigs-Universität, Institut für Biophysik und Strahlenbiologie, Frieiburg, Federal Republic of Germany*

DIETRICH KUSCHMITZ (32), *Max-Planck-Institute für Ernährungsphysiologie, 4600 Dortmund 1, Federal Republic of Germany*

S. C. KUSHWAHA (13), *Department of Biochemistry, University of Ottawa, Ottawa, Ontario K1N 9B4, Canada*

JACK R. LANCASTER, JR. (54), *Departments of Chemistry and Biochemistry, Utah State University, Logan, Utah 84322*

THOMAS A. LANGWORTHY (52), *Department of Microbiology, School of Medicine, University of South Dakota, Vermillion, South Dakota 57069*

JANOS K. LANYI (59), *Max-Planck-Institut für Biochemie, Abteilung Membranchemie, Munich, Federal Republic of Germany*

TERRANCE LEIGHTON (48), *Department of Microbiology and Immunology, University of California, Berkeley, California 94720*

HORST-DIETER LEMKE (12), *Max-Planck-Institut für Biochemie, Abteilung Membranbiochemie, D-8033 Martinsried, Munich, Federal Republic of Germany*

AARON LEWIS (27, 72, 78), *School of Applied and Engineering Physics, Cornell University, Ithaca, New York 14850*

MEI-JUNE LIAO (10), *Departments of Biology and Chemistry, Massachusetts Institute of Technology, Cambridge, Massachusetts 02139*

MOW LIN (69), *Chemistry Department, Brookhaven National Laboratory, Upton, New York 11973*

CHRISTINA LIND (10), *Departments of Biology and Chemistry, Massachusetts Institute of Technology, Cambridge, Massachusetts 02139*

EDWARD V. LINDLEY (57), *Section of Biochemistry, Molecular and Cell Biology, College of Agriculture and Life Sciences, Cornell University, Ithaca, New York 14850*

ROBERT S. H. LIU (65), *Department of Chemistry, University of Hawaii at Manoa, Honolulu, Hawaii 96822*

ERWIN LONDON (10), *Departments of Biology and Chemistry, Massachusetts Institute of Technology, Cambridge, Massachusetts 02139*

RICHARD H. LOZIER (17, 37), *Cardiovascular Research Institute and Department of Biochemistry and Biophysics, University of California, San Francisco, California 941431*

WERNER MÄNTELE (87), *Albert-Ludwigs-Universität, Institut für Biophysik und Strahlenbiologie, Freiburg, Federal Republic of Germany*

RUSSELL E. MACDONALD (57), *Section of Biochemistry, Molecular and Cell Biology, College of Agriculture and Life Sciences, Cornell University, Ithaca, New York 14853*

ROBERT MACELROY (91), *Extraterrestrial Research Division, NASA, AMES Research Center, Moffett, California 94035*

DEREK MARSH (90), *Max-Planck-Institut für Biophysikalische Chemie, Abteilung Spektroskopie, D-3400 Göttingen, Federal Republic of Germany*

RICHARD MATHIES (75, 77), *Department of Chemistry, University of California, Berkeley, California 94720*

ROLF J. MEHLHORN (43, 89, 91), *Membrane Bioenergetics Group and Department of Physiology, University of California, Berkeley, California 94720*

HARTMUT MICHEL (14), *Membranbiochemie, Max-Planck-Institut für Biochemie, D-8033 Martinsried, Munich, Federal Republic of Germany*

TOSHIO MITSUI (34), *Department of Biophysical Engineering, Faculty of Engineering Science, Osaka University, Toyonaka, Osaka 560, Japan*

MICHAEL G. MOTTO (21), *Department of Chemistry, Columbia University, New York, New York 10027*

YASUO MUKOHATA (53), *Department of Biology, Faculty of Science, Osaka University, Toyonaka, Osaka 560, Japan*

ELIANE NABEDRYK-VIALA (81), *Groupe de Biophysique des Systemes Membranaires, Service de Biophysique, Departement de Biologie, Centre d'Etudes Nuclaires de Saclay, 91191 Gif sur Yvette, France*

KOJI NAKANISHI (63, 64), *Department of Chemistry, Columbia University, New York, New York 10027*

DOROTHEA-CH. NEUGEBAUER (29), *Zoolo-gisches Institute der Universität, D-4400 Münster, Federal Republic of Germany*

DIETER OESTERHELT (3, 12, 14, 45, 55), *Membranbiochemie, Max-Planck-Institut für Biochemie, D-8033 Martinsried, Munich, Federal Republic of Germany*

ERIC OLDFIELD (41), *School of Chemical Sciences, University of Illinois at Urbana-Champaign, Urbana, Illinois 61801*

HOWARD BEVERLEY OSBORNE (81, 82), *Laboratoire de Biologie Moléculaire et Cellulaire, Département de Recherche Fondamentale, Centre d'Etudes Nuclaires, 38041 Grenoble, France*

MICHAEL OTTOLENGHI (62), *Department of Physical Chemistry, The Hebrew University of Jerusalem, Jerusalem 91904, Israel*

YU. A. OVCHINNIKOV (86), *Department of Protein Chemistry, Shemyakin Institute of Bioorganic Chemistry, USSR Academy of Sciences, Moscow, USSR*

LESTER PACKER (46, 89, 91), *Membrane Bioenergetics Group, University of California, Berkeley, California 94720*

WILLIAM W. PARSON (35), *Department of Biochemistry, University of Washington, Seattle, Washington 98195*

G. J. PERREAULT (27), *School of Applied and Engineering Physics, Cornell University, Ithaca, New York 14850*

FELICITAS PFEIFER (49), *Institut für Genetik und Mikrobiologie, Universität Würzburg, 8700 Würzburg, Federal Republic of Germany*

I. PROBST (43), *Physiologisch-Chemisches Institut der Universität Göttingen, Göttingen, Federal Republic of Germany*

ALEXANDRE QUINTANILHA (83), *Department of Physiology-Anatomy, University of California, Berkeley, California 94720*

TONI RACANELLI (46), *Membrane Bioenergetics Division, University of California, Berkeley, California 94720*

CHARLES N. RAFFERTY (19), *Laboratory of Vision Research, National Eye Institute,*

National Institute of Health, Bethesda, Maryland 20205

KENNETH J. Rothschild (42, 76, 84), Departments of Physics and Physiology, Boston University, Boston, Massachusetts 02215

ROSEMARY SANCHES (84), Instituto de Fisica de São Carlos, 13560 São Carlos, SP, Brazil

SYAM SARMA (48), Department of Microbiology and Immunology, University of California, Berkeley, California 94720

BENNO P. SCHOENBORN (30), Department of Biology, Brookhaven National Laboratory, Upton, New York 11973

FRITZ SEIBERT (87), Albert-Ludwigs-Universität, Institut fur Biophysik und Strahlenbiologie, Freiburg, Federal Republic of Germany

STANLEY SELTZER (69), Chemistry Department, Brookhaven National Laboratory, Upton, New York 11973

H. SIGRIST (25), Institute of Biochemistry, University of Bern, CH-3012 Bern, Switzerland

V. P. SKULACHEV (7), Laboratory of Molecular Biology and Bioorganic Chemistry, Moscow State University, Moscow 117234, USSR

WALTER SPERLING (19), Institut für Neurobiologie, KFA Jülich, 5170 Jülich, Federal Republic of Germany

G. D. SPROTT (13), Division of Biological Sciences, National Research Council of Canada, Ottawa, Ontario K1A 0R6, Canada

ELENA NEGRI SPUDICH (26), Department of Anatomy, Albert Einstein College of Medicine, Bronx, New York 10461

JOHN LEE SPUDICH (26), Department of Anatomy, Albert Einstein College of Medicine, Bronx, New York 10461

J. STAMATOFF (37), Department of Analytical Chemistry, Celanese Research Company, Summit, New Jersey 07901

T. A. STEITZ (11), Department of Molecular Biophysics and Biochemistry, Yale University, New Haven, Connecticut 06511

YASUO SUGIYAMA (53), Department of Biology, Faculty of Science, Osaka University, Toyonaka, Osaka 560, Japan

MANFRED SUMPER (51), Institut für Biochemie, Genetik und Mikrobiologie, Universität Regensburg, 8400 Regensburg, Federal Republic of Germany

FUMIO TOKUNAGA (18), Department of Physics, Faculty of Science, Tohoku University, Aobayama, Sendai 980, Japan

TERJE TORSVIK (47), Departments of Microbiology and Plant Physiology, University of Bergen, Bergen, Norway

PAUL TOWNER (70), Institut für Biochemie der Universität Würzburg, Röntgenring 11, Federal Republic of Germany

MOTOYUKI TSUDA (71, 85), Department of Physics, Sapporo Medical College, Sapporo 060, Japan

JIRO USUKURA (15), Department of Anatomy, Faculty of Medicine, University of Tokyo, Hongo 7-3-1, Bunkyo-ku, Tokyo 113, Japan

GOTTFRIED WAGNER (44), Institut für Allgemeine Botanik und Pflanzenphysiologie, Justus Liebig-Universität Geissen, D-6300 Giessen, Federal Republic of Germany

B. A. WALLACE (60), Department of Biochemistry, Columbia University, New York, New York 10032

ANTHONY WATTS (90), Department of Biochemistry, University of Oxford, Oxford OX1 3QU, England

HANS JURGEN WEBER (48, 50, 58), Cardiovascular Research Institute, University of California, San Francisco, California 94143

GOTTFRIED WEIDINGER (49), Institut für Genetik und Mikrobiologie, Universität Würzburg, 8700 Würzburg, Federal Republic of Germany

EICHI YAMADA (15), Department of Anatomy, Faculty of Medicine, University of Tokyo, Hongo 7-3-1, Bunkyo-ku, Tokyo 113, Japan

P. ZAHLER (25), *Institute of Biochemistry, University of Bern, CH-3012 Bern, Switzerland*

HAYWARD ZWERLING (91), *Lawrence Berkeley Laboratory and Physiology-Anatomy Department, University of California, Berkeley, California 94720*

K. VAN DAM (4), *Laboratory of Biochemistry, B.C.P. Jansen Institute, University of Amsterdam, 1000 HD Amsterdam, The Netherlands*

P. W. M. VAN DIJCK (4), *Gist Brocades, R and D Division of Microbiology, Department of Cellular Biochemistry, 2600 MA Delft, The Netherlands*

Preface

The methods described in this volume are those which particularly pertain to the investigation of the retinal pigments discovered in halobacteria about a decade ago. In addition, the volume includes many of the most current and useful methods common to investigations of all retinal protein systems, particularly the physical and chemical methods. It will be an indispensible reference for both new and established investigators in this exciting field of biological research.

We wish to thank the advisory board for their counsel in selecting the topics and the contributors. A special thanks is due to John Hazlett for invaluable editorial and administrative assistance in the compilation and organization of this volume.

Lester Packer

METHODS IN ENZYMOLOGY

EDITED BY

Sidney P. Colowick and Nathan O. Kaplan

VANDERBILT UNIVERSITY
SCHOOL OF MEDICINE
NASHVILLE, TENNESSEE

DEPARTMENT OF CHEMISTRY
UNIVERSITY OF CALIFORNIA
AT SAN DIEGO
LA JOLLA, CALIFORNIA

I. Preparation and Assay of Enzymes
II. Preparation and Assay of Enzymes
III. Preparation and Assay of Substrates
IV. Special Techniques for the Enzymologist
V. Preparation and Assay of Enzymes
VI. Preparation and Assay of Enzymes (*Continued*)
    Preparation and Assay of Substrates
    Special Techniques
VII. Cumulative Subject Index

METHODS IN ENZYMOLOGY

EDITORS-IN-CHIEF

Sidney P. Colowick   Nathan O. Kaplan

VOLUME VIII. Complex Carbohydrates
Edited by ELIZABETH F. NEUFELD AND VICTOR GINSBURG

VOLUME IX. Carbohydrate Metabolism
Edited by WILLIS A. WOOD

VOLUME X. Oxidation and Phosphorylation
Edited by RONALD W. ESTABROOK AND MAYNARD E. PULLMAN

VOLUME XI. Enzyme Structure
Edited by C. H. W. HIRS

VOLUME XII. Nucleic Acids (Parts A and B)
Edited by LAWRENCE GROSSMAN AND KIVIE MOLDAVE

VOLUME XIII. Citric Acid Cycle
Edited by J. M. LOWENSTEIN

VOLUME XIV. Lipids
Edited by J. M. LOWENSTEIN

VOLUME XV. Steroids and Terpenoids
Edited by RAYMOND B. CLAYTON

VOLUME XVI. Fast Reactions
Edited by KENNETH KUSTIN

VOLUME XVII. Metabolism of Amino Acids and Amines (Parts A and B)
Edited by HERBERT TABOR AND CELIA WHITE TABOR

VOLUME XLV. Proteolytic Enzymes (Part B)
Edited by LASZLO LORAND

VOLUME XLVI. Affinity Labeling
Edited by WILLIAM B. JAKOBY AND MEIR WILCHEK

VOLUME XLVII. Enzyme Structure (Part E)
Edited by C. H. W. HIRS AND SERGE N. TIMASHEFF

VOLUME XLVIII. Enzyme Structure (Part F)
Edited by C. H. W. HIRS AND SERGE N. TIMASHEFF

VOLUME XLIX. Enzyme Structure (Part G)
Edited by C. H. W. HIRS AND SERGE N. TIMASHEFF

VOLUME L. Complex Carbohydrates (Part C)
Edited by VICTOR GINSBURG

VOLUME LI. Purine and Pyrimidine Nucleotide Metabolism
Edited by PATRICIA A. HOFFEE AND MARY ELLEN JONES

VOLUME LII, Biomembranes (Part C: Biological Oxidations)
Edited by SIDNEY FLEISCHER AND LESTER PACKER

VOLUME LIII. Biomembranes (Part D: Biological Oxidations)
Edited by SIDNEY FLEISCHER AND LESTER PACKER

VOLUME LIV. Biomembranes (Part E: Biological Oxidations)
Edited by SIDNEY FLEISCHER AND LESTER PACKER

VOLUME LV. Biomembranes (Part F: Bioenergetics)
Edited by SIDNEY FLEISCHER AND LESTER PACKER

VOLUME LVI. Biomembranes (Part G: Bioenergetics)
Edited by SIDNEY FLEISCHER AND LESTER PACKER

VOLUME LVII. Bioluminescence and Chemiluminescence
Edited by MARLENE A. DELUCA

VOLUME LVIII. Cell Culture
Edited by WILLIAM B. JAKOBY AND IRA H. PASTAN

VOLUME LIX. Nucleic Acids and Protein Synthesis (Part G)
Edited by KIVIE MOLDAVE AND LAWRENCE GROSSMAN

VOLUME LX. Nucleic Acids and Protein Synthesis (Part H)
Edited by KIVIE MOLDAVE AND LAWRENCE GROSSMAN

VOLUME 61. Enzyme Structure (Part H)
Edited by C. H. W. HIRS AND SERGE N. TIMASHEFF

VOLUME 62. Vitamins and Coenzymes (Part D)
Edited by DONALD B. MCCORMICK AND LEMUEL D. WRIGHT

VOLUME 63. Enzyme Kinetics and Mechanism (Part A: Initial Rate and
Inhibitor Methods)
Edited by DANIEL L. PURICH

VOLUME 64. Enzyme Kinetics and Mechanism (Part B: Isotopic Probes
and Complex Enzyme Systems)
Edited by DANIEL L. PURICH

VOLUME 65. Nucleic Acids (Part I)
Edited by LAWRENCE GROSSMAN AND KIVIE MOLDAVE

VOLUME 66. Vitamins and Coenzymes (Part E)
Edited by DONALD B. MCCORMICK AND LEMUEL D. WRIGHT

VOLUME 67. Vitamins and Coenzymes (Part F)
Edited by DONALD B. MCCORMICK AND LEMUEL D. WRIGHT

VOLUME 68. Recombinant DNA
Edited by RAY WU

VOLUME 69. Photosynthesis and Nitrogen Fixation (Part C)
Edited by ANTHONY SAN PIETRO

VOLUME 70. Immunochemical Techniques (Part A)
Edited by HELEN VAN VUNAKIS AND JOHN J. LANGONE

VOLUME 71. Lipids (Part C)
Edited by JOHN M. LOWENSTEIN

Section I

Bacteriorhodopsin

A. Purple Membrane Preparations and Protein Structure
Articles 1 through 3

B. Reconstituted Systems
Articles 4 through 8

C. Molecular Structure of Purple Membranes
Articles 9 through 16

D. Chemistry, Spectroscopy, and Photochemistry
Articles 17 through 27

E. Specialized Physical Techniques
Articles 28 through 42

F. Ion Transport and Physiology
Articles 43 through 46

G. Biogenesis, Genetics, and Microorganisms
Articles 47 through 55

H. Light-Dependent Behavioral Responses of the Intact Organism
Article 56

I. Other Retinal Proteins
Articles 57 through 59

[1] The Isolation of Deuterated Bacteriorhodopsin from Fully Deuterated *Halobacterium halobium*[*]

By HENRY L. CRESPI

With the development of methods for the mass, autotrophic cultivation of algae in heavy water containing 99.8 atom percent $D(^2H)$,[1–4] it is now possible to devise fully deuterated culture media for heterotrophic organisms that are of practical and general utility. Fully deuterated algae are a source of carbohydrates, amino acids, and growth factors[5–7] that make it possible to culture a wide variety of fully deuterated microorganisms. These deuterated microorganisms and compounds obtained from them greatly extend the power of several instrumental techniques—nuclear magnetic and electron spin resonance spectroscopy, neutron scattering analysis, and infrared and Raman spectroscopy.[8]

Preparation of Algal Hydrolysate

Typically, 500 g (dry weight) of whole cells of deuterated green or blue-green algae (*Scenedesmus obliquus, Chlorella vulgaris*, or *Synechococcus lividus*[9]), or 500 g of a mixture of these algae, are refluxed gently for 24 hr with approximately 4 liters of 1 N DCl in D_2O. The reflux condenser is fitted with a T outlet through which a gentle steam of nitrogen gas is flowed. The flask should contain a surfeit of boiling stones and should be brought slowly to a boil, as there will be some foaming initially. After 24 hr let cool, add 20–30 g of activated charcoal, let stand several minutes, and vacuum filter (No. 4 paper) and wash. The filtrate is treated again with activated charcoal, filtered, and then evaporated (45–50°, ro-

[*] Work performed under the auspices of the Office of Basic Energy Sciences, Division of Chemical Sciences, U.S. Department of Energy.

[1] H. L. Crespi, S. M. Conrad, R. A. Uphaus, and J. J. Katz, *Ann. N.Y. Acad. Sci.* **84**, 648 (1960).
[2] J. J. Katz and H. L. Crespi, *Science* **151**, 1187 (1966).
[3] H. F. DaBoll, H. L. Crespi, and J. J. Katz, *Biotechnol. Bioeng.* **4**, 281 (1962).
[4] R. G. Taecker, H. L. Crespi, H. F. DaBoll, and J. J. Katz, *Biotechnol. Bioeng.* **13**, 779 (1971).
[5] M. I. Blake, H. L. Crespi, V. Mohan, and J. J. Katz, *J. Pharm. Sci.* **50**, 425 (1961).
[6] H. L. Crespi and J. J. Katz, this series, Vol. 26, Part C, p. 627.
[7] H. L. Crespi, J. Marmur, and J. J. Katz, *J. Am. Chem. Soc.* **84**, 3489 (1962).
[8] H. L. Crespi, *in* "Stable Isotopes in the Life Sciences," STI/PUB/442, p. 111. Vienna, 1977.
[9] Obtainable in ^{1}H culture from Culture Collection of Algae, Department of Botany, University of Texas, Austin.

tary, vacuum evaporator) to a very thick syrup to remove a good portion of the DCl and so minimize the amount of NaOD needed to neutralize. The syrup is then made up to about 1.8 liters and brought to "pH" 6.5[10] with NaOD, or NaOH dissolved in D_2O if isotopic purity is not critical. The voluminous precipitate that forms often filters slowly. Filter and wash with D_2O.

Treat the filtrate once again with activated charcoal, 10–20 g. The final filtrate should be yellow to light tan. Evaporate (max. 40°) to a thick, granular syrup and store in the freezer. About 500 ml of syrup are obtained, containing about 350 g of deuterated organic material and considerable salt.

Culture Conditions

The nutrient medium contains in 1 liter: NaCl, 250 g; anhydrous $MgSO_4$, 10.0 g; trisodium citrate·$2H_2O$, 3.0 g; KCl, 2.0 g; anhydrous $CaCl_2$, O.2 g; $FeSO_4·7H_2O$, 6 mg; microelement solution, 0.5 ml[11]; and syrupy hydrolysate, 25 ml. The pH is adjusted to 7.2 ± 0.2 with NaOD. The typical protocol for culture in D_2O is as follows: The initial culture is made in 60% D_2O (3 vol D_2O to 2 vol ^1H_2O) nutrient medium containing 5.0 g/liter yeast extract (Difco Certified) and 7.5 g/l casamino acids (Difco Technical), rather than algal hydrolysate. About 55 ml of this medium is autoclaved in a 125-ml conical flask capped with aluminum foil.[12] The flask is inoculated from a (^1H) slant and growth allowed to take place at 35–37° on a reciprocal shaker under a bank of two 40-W fluorescent lights. After a day or two, growth begins, and on the third or fourth day this flask is used to inoculate a flask of fully deuterated nutrient medium.

This second culture is done with 100 ml of sterile fully deuterated nutrient medium in a 250-ml conical flask. When the fully deuterated nutrient medium (about 110 ml) is autoclaved,[13] a precipitate develops. The tilted flask is allowed to stand for a day, and the clear supernatant is decanted into an empty sterile flask. The precipitate is discarded. This flask is then inoculated with about 5 ml of the 60% D_2O culture and placed on the reciprocal shaker. After growth (but generally no pigmentation) this flask is used to inoculate two 1-liter cultures. The 1-liter cultures, contained in either 2- or 3-liter flasks, are placed on a rotary shaker under 160 W of fluorescent lighting. All lights are from 30 to 60 cm from culture

---

[10] A meter reading of 6.5. All references to pH indicate the actual meter reading.

[11] The microelement solution contains per liter: H_3BO_2, 2.8 g; $MnCl_2·4H_2O$, 0.9 g; $CuSO_4·5H_2O$, 0.08 g; Na_2MoO_4, 0.13 g; $ZnSO_4·7H_2O$, 0.22 g.

[12] All culture flasks are capped with aluminum foil. There is no aeration.

[13] It would probably be satisfactory to sterilize by filtration, but this technique has not been tested.

surfaces. From the second through the fourth day, an additional 500 W of incandescent light is added. Pigmentation develops on the fourth day and cells are harvested on the seventh or eighth day. It is not necessary to sterilize the large flasks, but if, as is likely to happen, the laboratory becomes contaminated with a halophilic organism, one must take care to harvest just at full pigmentation and not let the culture overgrow.

Isolation of the Purple Membrane

All centrifugations except the first are at 4°. The cell suspension is centrifuged at 600 g for 2 min to remove coarse material. The cells are then pelleted by centrifuging 20 min at 5000 g. The pellet is slurried in residual supernatant fluid, a small fraction of a milligram (for each liter of cells) of DNase are added to each centrifuge bottle, the viscous slurry is suspended in D_2O, and the contents of all the bottles are poured into a conical flask. Final volume should be about 50 ml/liter of harvest. The suspension is stirred vigorously (magnetic stirring bar) for 2–3 hr at room temperature. The suspension is then centrifuged at 40,000 g for 15 min. The supernatant is discarded and the purple sediment resuspended in D_2O. Hard, gummy pelleted material should not be resuspended. The cycle of stirring and centrifuging (40,000 g for 30 min) is repeated two or three times (stir 1–2 hr)[14]. If the final pure suspension is to be stored for any length of time, add a little salt to assure stability. The yield will be about 15 mg/liter of culture.

Purple membrane labeled with ^{13}C or ^{15}N can be generated in a similar manner using suitably labeled algae.

[14] One can intersperse with a washing with EDTA-D_2O buffer to minimize trace metal impurities [(D. B. Kell and A. M. Griffiths, *Photobiochem. Photobiophys.* **2**, 105 (1981))].

[2] Preparation and Properties of Monomeric Bacteriorhodopsin

By Norbert A. Dencher and Maarten P. Heyn

The bacteriorhodopsin (BR) molecules in the purple membrane (PM) are organized into a two-dimensional hexagonal lattice of trimers that are surrounded by about 30 lipid molecules. In order to study the structure and function of isolated BR molecules it is advantageous to solubilize the PM into BR monomers.[1-4] These BR monomers can be reconstituted with

[1] M. P. Heyn, P. J. Bauer, and N. A. Dencher, *Biochem. Biophys. Res. Commun.* **67**, 897 (1975).

exogenous phospholipids to form vesicles[5,6] and planar bilayer membranes.[7] They can also be reassembled into two- and three-dimensional crystals.[8,9] To date Triton X-100 and octyl-β-D-glucoside are the only detergents that fulfil the essential criteria for this purpose: They solubilize the PM to the state of monomers without significantly affecting the spectral and functional properties and can be removed again if desired.[2,3] Other detergents tested—Ammonyx LO,[10,11] cetyltrimethylammonium bromide,[10] cholate,[12] deoxycholate,[12] digitonin,[10] dodecyl trimethylammonium bromide,[13] Emulphogene,[10] and Tween 80[10]—either denature BR or do not yield monomers.

Preparation of BR Monomers

A PM sample is considered to be solubilized if it fails to sediment when subjected to 200,000 g for 45 min. The percentage solubilization is determined by dividing the absorbance of the supernatant after centrifugation by the absorbance before centrifugation. The rate and extent of BR monomer formation during solubilization can be measured by monitoring the disappearance of the exciton coupling effects in the visible circular dichroism spectrum.[1,2,14] For Triton X-100 and octyl glucoside these two parameters are not only a function of the detergent to BR ratio but are also strongly dependent on pH, ionic strength, and temperature. Whereas at pH 6.9 (25 mM phosphate buffer) complete solubilization is reached in about 20 hr at 20° with a Triton X-100 to BR ratio of 6.2 (w/w), at pH 5.0 (100 mM acetate buffer) it takes 48 hr to reach approximately 60%.[2] At pH 5.0 in the same buffer and 25° complete solubilization could be ob-

[2] N. A. Dencher and M. P. Heyn, *FEBS Lett.* **96**, 322 (1978).

[3] N. A. Dencher and M. P. Heyn, in "Energetics and Structure of Halophilic Microorganisms" (S. R. Caplan and M. Ginzburg, eds.), p. 233. Elsevier/North-Holland Biomedical Press, Amsterdam, 1978.

[4] R. Casadio, H. Gutowitz, P. Mowery, M. Taylor, and W. Stoeckenius, *Biochim. Biophys. Acta* **590**, 13 (1980).

[5] R. J. Cherry, U. Müller, R. Henderson, and M. P. Heyn, *J. Mol. Biol.* **121**, 283 (1978).

[6] N. A. Dencher and M. P. Heyn, *FEBS Lett.* **108**, 307 (1979).

[7] E. Bamberg, N. A. Dencher, A. Fahr, and M. P. Heyn, *Proc. Natl. Acad. Sci. U.S.A.* **78**, 7502 (1981).

[8] H. Michel, D. Oesterhelt, and R. Henderson, *Proc. Natl. Acad. Sci. U.S.A.* **77**, 338 (1980).

[9] H. Michel and D. Oesterhelt, *Proc. Natl. Acad. Sci. U.S.A.* **77**, 1283 (1980).

[10] Norbert A. Dencher and M. P. Heyn, unpublished observation.

[11] B. Becher, F. Tokunaga, and T. G. Ebrey, *Biochemistry* **17**, 2293 (1978).

[12] S. -B. Hwang and W. Stoeckenius, *J. Memb. Biol.* **33**, 325 (1977).

[13] M. Happe and P. Overath, *Biochem. Biophys. Res. Commun.* **72**, 1504 (1976).

[14] B. Becher and T. G. Ebrey, *Biochem. Biophys. Res. Commun.* **69**, 1 (1976).

tained within 48 hr by increasing the detergent-to-BR ratio to 35.[4] A similar pH dependence is observed for octyl glucoside. With this detergent (41 mM) in 25 mM phosphate buffer, pH 6.9, no solubilization occurred during 48 hr in samples containing in addition sodium chloride in a concentration higher than 140 mM. Even in the presence of 15 mM NaCl the solubilization rate is approximately three times smaller than that in the pure buffer system.[2] In general, the solubilization rate is similar for both detergents. In the case of octyl glucoside, however, differences in the rate and extent can be observed, depending on the quality of this detergent.

Solubilization of BR in Triton X-100. PM suspended in 25 mM phosphate buffer, pH 6.9, is mixed with a 10% (w/w) solution of Triton X-100 (available from Packard Instrument Co., Inc., Sigma) in the same buffer to give a detergent-to-BR ratio by weight of about 4. The buffer volume is chosen in such a way that the Triton X-100 concentration lies between 0.2 and 0.5% (w/w). After sonication for 20 sec in a water bath sonifier the sample is kept for about 30 hr in the dark at room temperature (approx. 20°). Subsequently the sample is centrifuged at 200,000 g for 45 min to remove any nonsolubilized material (usually less than 2%). Prior to this step the buffer may be changed by dialysis for 12 hr (e.g., 100 mM sodium acetate, pH 5.0, containing 0.2% Triton X-100).

The detergent concentration should be above the critical micelle concentration (0.017% [15]) and can be increased up to 2%. Titration experiments show that a minimum ratio of 1.8—2.0 mg Triton X-100/mg PM is required for successful solubilization.[5] In order to increase the solubilization rate, higher detergent-to-BR ratios can be used.

Solubilization of BR in Octyl-β-D-glucoside. Octyl glucoside can be synthesized following published procedures[16–19] or purchased from various companies (e.g., Calbiochem-Behring Corp., Riedel-de-Haen). Optimum conditions for the solubilization of BR are obtained with 40 mM octyl glucoside ($\hat{=}$ 1.2% (w/w), MW = 294.4) and a detergent-to-protein ratio by weight of 20 in low ionic strength buffer of neutral pH (e.g., 25 mM phosphate buffer, pH 6.9). Following 20-sec sonication, the sample is allowed to stand in the dark at room temperature for 20–30 hr. After this time 85–95% of the PM is solubilized to BR monomers.

Depending on the quality of the detergent used it is sometimes necessary to increase the octyl glucoside concentration up to 100 mM or possible to decrease the detergent-to-BR ratio below 10 at a final concentration

[15] P. W. Holloway and J. T. Katz, *Biochemistry* **11**, 3689 (1972).
[16] C. R. Noller and W. C. Rockwell, *J. Am. Chem. Soc.* **60**, 2076 (1938).
[17] C. Baron and T. E. Thompson, *Biochim. Biophys. Acta* **382**, 276 (1975).
[18] J. F. W. Keana and R. B. Roman, *Memb. Biochem.* **1**, 323 (1978).
[19] J. T. Lin, S. Riedel, and R. Kinne, *Biochim. Biophys. Acta* **557**, 179 (1979).

of 40 mM. Furthermore, it is occasionally advantageous to terminate the solubilization after a shorter period (e.g., after 12 hr) and to use only the supernatant obtained after centrifugation for subsequent experiments. This avoids possible contamination of the sample with denatured BR.

Properties of Solubilized BR Monomers

Various experimental approaches show that BR is solubilized in both detergents to the state of monomers. The exciton coupling effects in the visible CD spectrum disappear and are replaced in the case of Triton X-100 by a small positive CD monomer band centered at about 560 nm.[1,2,14] Gel filtration experiments give values of 28 ± 5 Å^2 and approximately 50 Å^4 for the Stokes radius of the BR–lipid–octyl glucoside and BR–lipid–Triton X-100 complex, respectively, excluding the possibility that the micelles contain more than one BR molecule. This conclusion is confirmed by the determination of the molecular weight of BR in the mixed Triton–lipid–BR micelles.[20]

The Triton X-100-solubilized BR preparations are stable for several days when stored in the dark at room temperature and show no evidence for chromophore loss. BR solubilized in octyl glucoside can be stored at 4° for some days; at room temperature, however, progressive chromophore loss occurs after about 2 days. It is advantageous to store the solubilized samples at low pH, e.g., at pH 5.0.

In terms of a number of criteria, the properties of BR in the solubilized form differ only slightly from those in the native membrane. The secondary structure of Triton X-100 and octyl glucoside-solubilized BR, determined from the circular dichroism at 208 and 222 nm, contains about 70% α-helix and is thus quite similar to that of BR in the PM.[2,20] Apart from a decrease in the extinction coefficient and a small blue shift of the visible absorption band the absorption spectrum of monomeric BR is unchanged. The phenomenon of light–dark adaptation can still be observed in the solubilized state; however, the extent of light adaptation is less than that in PM.[2,4] The absorption maximum occurs at 552 nm in the light-adapted state, shifting to 546 nm upon dark adaptation (0.2% Triton X-100, pH 6.9, 20°). The photocycle of Triton X-100-solubilized BR is qualitatively the same as in the PM. Whereas the half-time of the decay of the M_{412} intermediate remains unchanged, the formation is about three times faster. In octyl glucoside-solubilized BR, however, the cycle seems to be slightly different.[2] Recently it was shown that illumination of Triton X-100-solubilized BR results in a reversible hydrolysis of the retinal aldi-

[20] J. A. Reynolds and W. Stoeckenius, *Proc. Natl. Acad. Sci. U.S.A.* **74**, 2803 (1977).

mine.[21] Furthermore, reconstitution experiments prove that BR monomers can effectively pump protons.[6]

Removal of Endogenous Phospholipids

The action of Triton X-100 and octyl glucoside leads to partial delipidation of BR. Sucrose density gradient centrifugation and gel filtration in the presence of these detergents allow a separation of detergent–lipid and detergent–lipid–protein micelles. When PM solubilized in 5% Triton X-100 or 5% octyl glucoside is chromatographed on BioGel A-0.5 m in buffer containing 1% octyl glucoside and 25 mM sodium phosphate at pH 6.9, about 90% of the polar lipids are removed from BR.[22] Solubilization of the PM with Triton X-100 followed by gel filtration in deoxycholate solution results in a removal of at least 99% of endogenous lipids.[22]

Removal of Detergent

The detergents can be removed from the solution containing solubilized BR by prolonged dialysis. The low critical micelle concentration of Triton X-100 is a disadvantage in this procedure. Since the critical micelle concentration of octyl glucoside is about 100 times higher (25 mM [23]), this detergent can be removed more rapidly. Using the return of the amplitude of the excitation CD effect as a signal to monitor the detergent removal, equilibrium was reached in 40 hr with octyl glucoside as opposed to about 7 days with Triton X-100 (4°, 100 mM acetate buffer, pH 5.0). Most of the effects of solubilization are reversible upon slow dialysis; e.g., the absorption maximum of BR shifts back to its original position and the exciton CD bands reappear with the same specific ellipticity as in the native PM.[2] During dialysis, reassembly occurs and large hexagonal crystalline domains are formed again.[5] A small amount of detergent, however, remains bound to BR.[5]

Centrifugation using a detergent-free sucrose gradient and gel filtration provide alternative faster methods to remove the detergents. A rapid removal of the detergents from solubilized BR can be achieved in the time course of minutes by absorption to resins, e.g., Bio-Beads SM-2 (Bio-Rad Labs)[19,24] or Amberlite XAD-2 (Rohm and Haas Ltd.).[25] For octyl gluco-

[21] A. M. Shkrob, A. V. Rodionov-and Yu. A. Ovchinnikov, *Bioorg. Khim.* **4**, 354 (1978).
[22] K.-S. Huang, H. Bayley, and H. G. Khorana, *Proc. Natl. Acad. Sci. U.S.A.* **77**, 323 (1980).
[23] K. Shinoda, T. Yamaguchi, and R. Hori, *Bull. Chem. Soc. Jpn.* **34**, 237 (1961).
[24] P. W. Holloway, *Anal. Biochem.* **53**, 304 (1973).
[25] P. S. J. Cheetham, *Anal. Biochem.* **92**, 447 (1979).

side a final detergent concentration of 0.018% can be reached with Bio-Beads SM-2,[19] i.e., much lower than its critical micelle concentration of 0.74%.

Choice of Detergent

Both Triton X-100 and octyl glucoside can be considered to be suitable detergents to prepare BR monomers, since no gross structural and functional alterations occur during solubilization. In contrast to Triton X-100, octyl-β-D-glucoside is a chemically well-defined compound. For spectroscopic investigations octyl glucoside is superior to Triton X-100. The far UV absorbance is much higher for Triton X-100 and its strong fluorescence in the near UV makes fluorescence work in this range exceedingly difficult. The higher critical micelle concentration of octyl glucoside allows more rapid detergent removal and makes it more suitable for reconstitution work. Also its smaller micelle size and weight can be of advantage. Asolectin, diphytanoylphosphatidylcholine, egg lecithin, and phosphatidylserine vesicles can only be prepared with BR solubilized in octyl glucoside.[6,26] On the other hand, BR is more stable in Triton X-100. And last but not least, Triton X-100 is much cheaper than octyl glucoside.

[26] M. P. Heyn and N. A. Dencher, this volume, Article [6].

[3] Reconstitution of the Retinal Proteins Bacteriorhodopsin and Halorhodopsin

By Dieter Oesterhelt

Reconstitution of the retinal proteins bacteriorhodopsin (BR) and halorhodopsin (HR) from the corresponding apoproteins BO and HO with retinal and retinal analogue compounds is a valuable tool for elucidation of structure–function relationships. This article describes the experimental methods for the preparation of retinal-free membranes and membrane vesicles containing the apoproteins BO or HO, the methods for the reconstitution of BR and HR in membranes, cell vesicles, and cells, and an assay for both chromoproteins.

Strains

Growth of halobacterial cells is described in Chapter [45] of this volume. The strains of *Halobacterium halobium* suited for the experimental

	Phenotypic properties			
Strain	BO	HO	Retinal	Source
S9	+	(+)	+	Obtained by L. Jan
M1	+	(+)	+	Isolated in the author's laboratory
M2	(+)^a	+	+	Isolated in the author's laboratory
JW5	+	(+)	—	By J. Weber
L-33	—	+	+	By J. Lanyi
L-07	—	+	—	By J. Lanyi
W296	+	(+)	+	Obtained by G. Weidinger

^{a} (+) means reduced amounts compared to +.

procedures described are summarized in the table. Included are strain M1 or S9 (BO$^+$, HO$^+$, Re$^+$), containing bacteriorhodopsin but very small amounts of halorhodopsin under standard growth conditions; strain M2 (BO$^+$, HO$^+$, Re$^+$), which under the same conditions produces smaller amounts of bacteriorhodopsin than M1 or S9 but enhanced amounts of halorhodopsin; strain W296 and strain JW5 (BO$^+$, HO$^+$, Re$^-$), which are retinal$^-$ mutant strains; strain L-33 (BO$^-$, HO$^+$, Re$^+$), lacking bacteriorhodopsin but containing more halorhodopsin than S9 under the standard growth conditions; and strain L-O7 (BO$^-$, HO$^+$, Re$^-$), which is derived from L-33 and cannot synthesize retinal.

Preparation of Cell Vesicles

Cells from a 10-liter fermenter are harvested by centrifugation (15,000 g for 15 min) and washed with 300 ml of 4 M NaCl. The cells are resuspended in 200 ml 4 M NaCl and 20 ml portions are sonicated in a centrifuge tube 5 × 10 sec with 10-sec intervals at 4° with the macrotip of a Branson sonifer (level 6) at optimal energy output. Cell rupture is indicated by a decrease in turbidity and is checked by microscopic inspection. Intact cells are removed by centrifugation for 15 min at 15,000 g and the vesicles in the supernatant are spun down for 45 min at 200,000 g and resuspended in 100 ml 4 M NaCl. The centrifugation for 15 min at 15,000 g is repeated and the supernatant centrifuged again for 45 min at 200,000 g. Finally, the vesicles are resuspended in 12 ml 4 M NaCl and the protein content is determined by the Lowry method with serum albumin as a standard.[1] Typical values are between 25 and 50 mg protein per milliliter.

[1] O. H. Lowry, N. J. Rosebrough, A. L. Fair, and R. J. Randall, *J. Biol. Chem.* **193**, 265 (1951).

Isolation of Purple and White Membranes

Purple membranes are isolated from strains M1 or S9 as described in Vol. 31 of this series.[2] Typical yields are 300–500 mg of membranes per 100 liters of a culture. White membranes are isolated from W296 or JW5 cells using the same procedure with a slight modification: a 20–45% (w/w) gradient is used instead of a 30–50% (w/w) sucrose density gradient. In this gradient a broad single band of yellowish color is seen. In the lower part of this band membranes containing only BO (white membranes) are enriched as revealed by SDS–PAGE (R. Schoeningh, unpublished results). These fractions are pooled and washed by centrifugation as described for the purple membranes.

Preparation of Apomembranes from Purple Membranes

Preparation of apomembranes is based on the reaction of purple membranes with hydroxylamine in light, which yields a colorless membrane and retinal oxime.[3] The kinetics of the reaction depend on light intensity, temperature, pH, and hydroxylamine concentration. The temperature effect is especially pronounced: a fourfold increase in velocity occurs per 10° change in temperature. No reaction was observed below pH 5; between pH 6 and 7.5 the velocity is nearly constant and above pH 9 the reaction starts to proceed also in the dark. This might be due to an increasing amount of the 500-nm chromophore which equilibrates with the purple complex and is known to react with hydroxylamine in the dark.[4]

For preparative purposes a thermostated vessel containing 400 ml of a purple membrane suspension with an optical density at 560 nm of 1–3 is placed in front of a 900-W xenon lamp. A 2 M hydroxylamine solution, pH 7, is prepared by mixing equal amounts of 4 M solutions of hydroxylamine hydrochloride and 4 N sodium hydroxide, and 40 ml of this mixture is added to the purple membrane suspension. The temperature is kept constant at 20° and light filtered through 20 cm of water and a cutoff filter [(orange glass (OG) filter 515, Schott)] is focused onto the stirred sample. The suspension becomes colorless after about 15 hr and the membranes are centrifuged and washed carefully with water (about five times) to remove any residual traces of hydroxylamine that could interfere with the reconstitution reaction (see below). The membranes are finally resuspended in about 50 ml of water, divided into aliquots, and stored frozen.

[2] D. Oesterhelt and W. Stoeckenius, this series, Vol. 31, Part A, p. 667.
[3] D. Oesterhelt, L. Schuhmann, and H. Gruber, *FEBS Lett.* **44**, 257 (1974).
[4] U. Fischer and D. Oesterhelt, *Biophys. J.* **28**, 211 (1979).

Assay for Bacteriorhodopsin and Halorhodopsin in Cells and
Vesicles

Principle. The assay for bacteriorhodopsin is based on the measurement of the initial velocity of proton efflux from cells or vesicles upon illumination. Without addition of a lipophilic cation such as tetraphenylphosphonium (TPP⁺) compensating ion fluxes including that of protons will give rise to complex pH changes of the medium as known for a long time (for discussion of these phenomena see Hartmann and Oesterhelt,[5] Bogomolni *et al.*,[6] and Greene and Lanyi.[7]) Low concentrations (about 1 μM) of TPP⁺ can be used as a membrane potential indicator, but higher concentrations (about 1 mM) of the same substance will quench the potential.[8] Under these conditions the proton efflux through BR is electrically compensated by the influx of TPP⁺ and a large drop in the pH value of the medium is observed. The proton efflux is linear in time for the first 15–30 sec and is calibrated by addition of alkaline or acid solutions. BR activity is expressed either in grams of protons per minute or in grams of protons per minute per molecule of bacteriorhodopsin if the cellular content of BR is known (for measurement of BR content see Chap 45). At constant irradiance and TPP⁺ concentration the initial rate of proton efflux is proportional to total amount of BR present in the sample and, at constant cell concentration, proportional to cellular BR content. Typical turnover numbers are 270 protons per minute per molecule BR at 20° and under conditions of light saturation. This value is clearly below the value one would expect from the turnover number of the photochemical cycle of BR (about 100 per second at room temperature). A plausible explanation of this fact is that TPP⁺ at a concentration of 250 μM used in the assay system saturates the aqueous basal salt solution but does not quench completely the membrane potential.[8] In spite of this the assay system described allows us to compare quantitatively the activity of BR in different halobacterial strains and the effectiveness of retinal analog compounds in reconstituted chromoproteins.

Measurement of HR activity in intact cells or vesicles is based on a similar principle as described earlier for BR. The uncoupler 3-(carbonyl cyanide 3-chlorophenylhydrazone) (CCCP) is used to compensate for the

[5] R. Hartmann and D. Oesterhelt, *Eur. J. Biochem.* **77**, 325 (1977).
[6] R. A. Bogomolni, R. A. Baker, R. H. Lozier, and W. Stoeckenius, *Biochim. Biophys. Acta* **440**, 68 (1976).
[7] R. V. Greene and J. K. Lanyi, *J. Biol. Chem.* **254**, 10986 (1979).
[8] H. Michel and D. Oesterhelt, *FEBS Lett.* **65**, 275 (1976).

efflux of sodium ions, thereby quenching membrane potential.[9] The initial velocity of proton influx mediated by CCCP is used as a measure of halorhodopsin activity. Typical values found for L-33 cells are 380 protons per minute per molecule HR at 20°. The halorhodopsin content of the cells is determined by difference spectroscopy of a sample bleached by 0.2 M hydroxylamine and a reference without hydroxylamine. HR content is calculated using an absorption coefficient of 48,000 cm^{-1}mol^{-1}.[10]

Assay System. Because uncouplers short-circuit the proton flow through BR and lipophilic cations compensate for the Na$^+$ flow through HR, both retinal proteins when present in the same cell can be tested separately by addition of either uncoupler (HR) or TPP$^+$ (BR). *Halobacterium halobium* S9 cells are grown until the stationary phase is reached. The cells are sedimented and resuspended in basal salt to an O.D. of 4 corresponding to 2 mg protein per milliliter.[5] Then 10–500 μl of this suspension are diluted to 8 ml with basal salt and tetraphenyl phosphonium bromide (100 mM in ethanol) is added to a final concentration of 250 μM. The pH is adjusted to 6.8–6.9 and the sample kept in the dark for 5 min. Light from a 150-W projector (OG 515 filter, Schott) is used for illumination, giving an irradiance of about 100 mW/cm^2 at the place of the sample, which is stirred in a cyclindrical and thermostated vessel at 20°. The initial rate of proton extrusion is measured with a standard pH meter connected to a strip chart recorder and the pH changes are calibrated by addition of known amounts of acid and alkaline solutions. In a control experiment an acid jump is applied by fast addition of hydrochloric acid from a syringe. The rate recorded must exceed the rate of acidification seen on illumination to assume that the monitoring system is not the limiting step. Halorhodopsin is assayed in exactly the same way as BR but with 50 μM CCCP added instead of 250 μM TPP. BR and HR activities in vesicles of S9, M1, M2 and L-33 cells are tested as described earlier with 50–200 μl of vesicles (25–50 mg protein/ml) in the sample volume of 8 ml.

Reconstitution of BR and HR in Intact Cells and Cell Vesicles

Cells. W296, JW5, or L-O7 cells from a stationary state culture are resuspended in basal salt to an optical density of 4 and an excess of retinal (1 μl/ml cell suspension) in methanolic solution (10 mM) is added. Alternatively, retinal is added stepwise to titrate the BO content of the cells. Complete reconstitution takes place in the range of 2–5 hr at room tem-

[9] G. Wagner, D. Oesterhelt, G. Krippahl, and J. K. Lanyi, *FEBS Lett.* **131,** 341 (1981).
[10] J. K. Lanyi and H. J. Weber, *J. Biol. Chem.* **255,** 243 (1980).

perature. The BO concentration of the white mutant cells is around 1 nmol/ml cell suspension. The assay for BR is carried out as described for wild-type cells on page 14. Measurement of photophosphorylation reconstituted in these cells is described in Chap. 14.

For analysis of retinal analog membranes an alternative procedure can be used. Strains W296 and JW5 are grown under standard conditions in 2-liter Erlenmeyer flasks (700 ml medium) at 40° and 105 rotations per minute. On days 2, 3, and 4 a total of 600 nmol retinal analog compound in 200 nmol aliquots is added to the culture, and after the end of growth the cells are centrifuged and membranes isolated as described for purple membranes. These membranes have the same lipid-to-BR ratio as purple membranes and are very useful for photochemical experiments.

Vesicles. To 100–200 μl of a vesicle suspensions (about 30 mg/ml) 0–10 nmol of all-*trans*-retinal in methanol (1 μmol/ml) are added and the samples allowed to stand in the dark at room temperature for 90 min (W296 vesicles) or 5–10 hr (L-O7 vesicles). The reconstitution reaction of the chromoproteins is followed by difference spectroscopy with a vesicle suspension without retinal but with the same amount of solvent serving as the reference. The samples are diluted with 4 M NaCl to a final volume of 8 ml and transferred to a cylindrical vessel (see above) for measurement of activity. Halorhodopsin is measured in the presence of 50 μM CCCP and bacteriorhodopsin in the presence of 250 μM TPP as described for intact cells. Retinal analogs can be tested for their activity under the same conditions of incubation. Furthermore, the reconstitution reaction of HO with retinal is very useful for identification and isolation of the retinyl binding moiety of HR and is described here.

Preparation of [³H]Retinal

A vial containing 100 mCi (12.7 μmol) sodium [³H] borohydride of high specific activity (e.g., 7.8 Ci/mmol) is opened and 50 μmol all-*trans*-retinal in 0.7 ml isopropanol added. After addition of 0.20 ml water the reaction mixture is stirred under nitrogen for 90 min at room temperature. TLC [solvent mixture petrolether (40–60°)–acetone = 10:1] showed that appreciable but not total reduction to retinol had occurred. The solvent was then evaporated by a stream of nitrogen with the vial placed in a warm water bath. After addition of 142 mg of manganese dioxide in 3 ml of petrolether (40–60° boiling point), the mixture was stirred under nitrogen at room temperature and reoxidation of retinol to retinal followed by TLC. The time required for complete oxidation depends on the quality of the manganese dioxide used. The solution is then filtered and the solvent evaporated by a stream of nitrogen. The residue dissolved in methanol is pure all-*trans*-[³H]retinal with traces of the 13-*cis* isomer. Its specific radioactivity is about 2 Ci/mmol.

Reconstitution and Reduction of HR

To L-O7 vesicles in 4 M sodium chloride (10–50 mg/ml protein) [³H]-retinal is added in methanolic solution until no further increase in the absorption at 588 nm is observed (about 1.5 nmol halorhodopsin per 10 mg of protein). After 5–10 hr an equal volume of 3 M sodium chloride plus 1 M sodium acetate, pH 5.0, containing 2% cyanoborohydride is added. After addition of a saturating amount of diethyl ether the sample is incubated in the dark at room temperature for 3 hr. This treatment produces a covalent linkage between the retinal moiety and HO and the product can be identified on SDS–gel by fluorography.[11]

Reconstitution of BR in Apomembranes and White Membranes

The reconstitution reaction of BR in apomembranes has been analyzed in great detail.[12–15] It is useful for studies on retinal(ol)–protein interaction and for the incorporation of retinal analog compounds into BO. However, retinal oxime cannot be removed by washing procedures during apomembrane preparation, and organic solvents or detergents will extract retinal oxime but usually damage the membrane structure. While retinal oxime does not seem to interfere with the reconstitution reaction itself, it might prevent correct determination of the activity of retinal analog compounds in BR in liposomal systems because retinal is produced measurably from retinal oxime by light and even in the dark on prolonged storage of the apomembranes. Additionally, reconstituted samples of apomembranes cannot be stored in the frozen state because they become discolored. Therefore the use of white membranes from retinal⁻ mutant cells is sometime preferable. On the other hand, BO or reconstituted BR in white membranes has a different lipid-to-protein ratio compared with purple membranes and apomembranes (R. Schoeningh, unpublished results) and therefore might behave physicochemically different compared to BR in purple membranes. To eliminate this problem mutant cells can be

[11] J. K. Lanyi and D. Oesterhelt, *J. Biol. Chem.* **257**, 2674 (1982).
[12] D. Oesterhelt and V. Christoffel, *Biochem. Soc. Trans.* **4**, 556 (1976).
[13] F. Tokunaga and T. G. Ebrey, *Biochemistry* **17**, 1915 (1978).
[14] T. Schreckenbach, B. Walckhoff, and D. Oesterhelt, *Eur. J. Biochem.* **76**, 499 (1977).
[15] T. Schreckenbach, B. Walckhoff, and D. Oesterhelt, *Biochemistry* **17**, 5353 (1978).
[16] D. Oesterhelt and B. Hess, *Eur. J. Biochem.* **37**, 316 (1973).

grown in the presence of retinals. The isolation of purple membranes or analog membranes from such cultures is described earlier. Concentrations of bacterio-opsin in apomembranes and white membranes are calculated from the increase in absorbance at 568 nm upon addition of all-*trans*-retinal using an absorption coefficient of 63,000 cm/mol.[16] For a reconstitution experiment membranes are suspended in 0.1 M potassium phosphate buffer of the desired pH to a BO concentration of about 10 μM and 1 mM solutions of retinal analog compounds in isopropanol or methanol are added in microliter aliquots. The chromophore formation is followed spectroscopically. To study inhibitory action reaction of retinal analogue compounds 10 μl of isopropanol containing various concentrations of the inhibitor are mixed into the membrane suspension (1 ml) and equilibrated in a refrigerated bath at 5° for 5 min. Then 20 μl of isopropanol containing 16 nmol all-*trans*-retinal is quickly stirred into the suspension and the increase in absorbance at 570 nm monitored. A detailed investigation of inhibitory properties of retinal analog compounds is found in Towner *et al.*[17]

[17] P. Towner, W. Gaertner, B. Walckhoff, D. Oesterhelt, and H. Hopf, *Eur. J. Biochem.* **117**, 353 (1981).

[4] Bacteriorhodopsin in Phospholipid Vesicles

By PIET W. M. VAN DIJCK, and KAREL VAN DAM

Introduction

Bacteriorhodopsin, an integral chromoprotein in the plasma membrane of *Halobacterium halobium,* is a light-driven proton pump. It is the only protein present in the so-called purple membrane patches of the bacterial membrane and is organized in a hexagonal pattern of trimers. These patches can cover up to 50% of the plasma membrane under anaerobic conditions. The protein can be purified by a simple washing scheme, based on the fact that the purple membrane is the only plasma membrane substructure that does not disintegrate at low ionic strength. It is one of the very few integral membrane proteins of which both the primary and three-dimensional structures have been elucidated. Furthermore, the protein is remarkably stable. These properties have rendered bacteriorhodopsin one of the favorite membrane proteins to be studied in model systems.[1-4]

[1] W. Stoeckenius, R. H. Lozier, and R. A. Bogomolni, *Biochim. Biophys. Acta* **505**, 215 (1979).

Various methods for incorporating bacteriorhodopsin into lipid vesicles have been published. It is the purpose of this contribution to describe those methods with which we have positive experience[5,6] and to compare the different preparations with respect to vesicle size, homogeneity, and proton pump activity.

Materials

Bacteriorhodopsin was isolated from *Halobacterium halobium* strain R1 or S9 according to the procedure described by Oesterhelt and Stoeckenius.[7] Both strains, gifts of Drs. D. Oesterhelt and R. Peters, respectively, were grown as described before.[8] The S9 strain, which is rather unstable, was plated after two consecutive batches on 2% agar in growth medium. The bright purple colonies, easily distinguishable from the orange-red revertants, were used for new inoculations.

Egg phosphatidylcholine (type V-E), octylglucoside, cholate, deoxycholate and Amberlite XAD-2 were obtained from Sigma (St. Louis) and Triton X-100 from Serva (Heidelberg). Cholate and deoxycholate were purified and stored as detailed by Hartzell *et al.*[9]

Bio-beads SM-2 (20–50 mesh) was obtained from Bio-Rad Labs (Richmond); [³H]Triton X-100 and [³H]deoxycholate were from NEN Chemicals (Frankfurt). All other chemicals were of analar grade quality.

All experiments have been carried out in 150 mM KCl, 2 mM EDTA, pH 6.9 (this will be referred to as salt in the next sections), unless stated otherwise.

A. Reconstitution Methods Not Involving Detergents

1. Sonication

The method is in essence that of Racker.[10] Egg phosphatidylcholine (40 mg) is dried down from a chloroform–methanol solution (100 mg/ml) by rotary evaporation as a thin lipid film in a 100-ml round-bottom flask and washed twice by 10 ml of diethyl ether or absolute ethanol. The lipid is dispersed[11] in 2 ml of salt and 2 mg of purple membrane by shaking with

[2] Yu. A. Ovchinnikov, N. G. Abdulaev, M. Yu. Feigina, A. V. Kiselev, and N. A. Lobanov, *FEBS Lett.* **100**, 219 (1979).

[3] H. G. Khorana, G. E. Gerber, W. C. Herlihy, C. P. Gray, R. J. Anderegg, K. Nihei, and K. Biemann, *Proc. Natl. Acad. Sci. U.S.A.* **76**, 5046 (1979).

[4] D. M. Engelman, R. Henderson, A. D. McLachlan, and B. A. Wallace, *Proc. Natl. Acad. Sci. U.S.A.* **77**, 2023 (1979).

[5] P. W. M. van Dijck and K. van Dam, unpublished data (1980).

some glass beads in the flask. The multilayered liposomes are transferred to a sonication vessel (MSE sonifier, flat bottom tip, 4-μm amplitude) and sonicated 60 times 15 sec with 45-sec intervals. The vessel is flushed with argon and cooled by a basin filled with tapwater (about 10°). The resulting preparation consists of unilamellar vesicles (SUV) of variable size (mean average diameter 240 Å). Analysis on a density gradient shows that the protein is very heterogeneously distributed over the SUVs. Up to 50% of the vesicles does not contain protein at all, whereas at lipid-to-protein ratios of less than 10 an increasing proportion of the protein is not incorporated into the SUVs.[5]

In agreement with results obtained previously,[12,13] carboxypeptidase digestion shows that the protein is incorporated in the lipid bilayers for more than 90% in the inside-out orientation.[5]

The proton pump activity (measured by the alkalinization of the outside medium on illumination) of the SUVs is on the order of 3 H$^+$/bacteriorhodopsin. Below a lipid-to-protein ratio of 10 (w/w) the activity decreases, presumably as the result of enhanced passive proton permeability of the bilayers and the appearance of nonincorporated protein.

2. French Pressure Cell Vesicles (FPV)

SUVs are stored at $-70°$ for 8 hr and thawed at room temperature. This leads to aggregation and fusion of the SUVs to multilamellar liposomes. Also residual purple membrane that had not been incorporated into the SUVs is now present in the bilayers of the liposomes.

To disrupt the large liposomes we have used the gentle method of repeated passage through a French pressure cell as outlined by Barenholz *et al.*[14] The dispersion is extruded four times at room temperature from a 1-in. diameter cell operated at 20,000 psi (126 $\times$ 10^6 pascals) (Aminco

[6] P. W. M. van Dijck, K. Nicolay, J. Leunissen-Bijvelt, K. van Dam, and R. Kaptein, *Eur. J. Biochem.* **117**, 639 (1981).

[7] D. Oesterhelt and W. Stoeckenius, this series, Vol. 31, Part A, p. 667.

[8] A. Danon and W. Stoeckenius, *Proc. Natl. Acad. Sci. U.S.A.* **71**, 1234 (1974).

[9] C. R. Hartzell, H. Beinert, B. F. van Gelder, and T. E. King, this series, Vol. 53, Part D, p. 54.

[10] E. Racker, *Biochem. Biophys. Res. Commun.* **55**, 224 (1973).

[11] A. D. Bangham, J. de Gier, and G. D. Greville, *Chem. Phys. Lipids* **1**, 225 (1967).

[12] G. E. Gerber, C. P. Gray, D. Wildenauer, and H. G. Khorana, *Proc. Natl. Acad. Sci. U.S.A.* **74**, 5426 (1977).

[13] K. J. Hellingwerf, Ph.D. Thesis, University of Amsterdam, Offsetdrukkerij Veenstra-Visser, Groningen (1979).

[14] Y. Barenholz, S. Anselm, and D. Lichtenberg, *FEBS Lett.* **99**, 210 (1979).

French Pressure Cell Press, Travenol Labs, Inc., Silver Spring). The pressure difference instantly disrupts the multilamellar liposomes into unilamellar FPVs. FPVs have a mean average diameter of 380 Å. Even at a low lipid-to-protein ratio of 4 (w/w) all the protein is incorporated. Furthermore, protein-free vesicles are virtually absent; even at a lipid-to-protein ratio of 40 the protein-free vesicle population is less than 20%. FPV preparations also have the bacteriorhodopsin present in the inside-out mode for more than 90%.[5] The proton pump activity, which is the resultant of bacteriorhodopsin-induced proton accumulation inside the vesicles, counteracted by passive proton permeability and inhibition of proton uptake, due to a developing membrane potential, is larger in FPVs than in SUVs (5 and 3 H^+/bacteriorhodopsin, respectively). This can be explained by the larger internal volume of the FPVs and the smaller effect of the protein on the passive proton permeability because of the lesser curvature of FPV membrane.

3. Filtration

Fused multilamellar liposomes can in prinicple be broken by other methods, one of which is the filtration technique as decribed by Olson et al.[15] In this procedure multilamellar dispersions are extruded through a series of Nucleopore polycarbonate filters (Nucleopore Inc.) of decreasing pore diameter (3 to 0.2 μm). Vesicles are pinched off from the large liposomes in the extrusion process. The average diameter of the extruded vesicles, which are not per se unilamellar, is determined by the pore size of the filter. The recovery after sequential extrusion is at least 80%.

4. Dimethyl Sulfoxide

Bacteriorhodopsin can withstand concentrations of DMSO up to 60% (v/v).[16] Lipid (40 mg of egg phosphatidylcholine) is dissolved in 2 ml of DMSO at 37°; this is mixed dropwise with 2 mg of purple membrane in 2 ml of salt at 37°. The DMSO concentration is then lowered to 20% (v/v) by adding 6 ml of medium (also dropwise to avoid a rise in temperature from the heat of mixing). In this step the spectrum of the bacteriorhodopsin returns to the normal spectrum.[16] Analysis of the reconstituted material, which is recovered by centrifugation at 70,000 g for 1 hr, reveals that associates have been formed with a low lipid-to-protein ratio [about 1 (w/w)] and a proton pump activity of about 1 H^+/bacteriorhodopsin.

[15] F. Olson, C. A. Hunt, F. C. Szoka, W. J. Vail, and D. Papahadjopoulos, *Biochim. Biophys. Acta* **557**, 9 (1979).
[16] D. Oesterhelt, M. Meentzen, and L. Schuhmann, *Eur. J. Biochem.* **40**, 453 (1973).

5. Hydration of Lipid–Protein Complexes

Darszon et al.[17] have recently introduced a technique leading to the formation of very large vesicles ("gigantosomes"). For the reconstitution of bacteriorhodopsin we have used the following protocol. Purple membrane, 5 mg of protein in 1 ml of water, is added to 8 ml of hexane, which contains 100 mg of egg phosphatidylcholine. This mixture is sonicated (as detailed in Sec. A,1), after which the two-phase system is allowed to separate. The hexane phase, containing lipid–bacteriorhodopsin complexes,[17] is transferred to four flat-bottom glass vials of 25-mm diameter. Hexane is evaporated by passing a stream of argon slowly over the vials. After addition of 2 ml of salt they are placed in the dark at room temperature for at least 48 hr. Then the hydrated material is collected, layered on 1% sucrose, and centrifuged for 10 min at 1500 g. Only the "gigantosomes" remain in the top layer. Electron microscopy reveals that the vesicles are in the range of 0.2–3 μm and that they often are multilayered. A large fraction does not show any protein particles; protein-free vesicles can also be demonstrated in density gradients.

The activity of the gigantosomes obtained in this way is very low (1 H$^+$/bacteriorhodopsin) suggesting an almost random orientation of the protein. This is supported by the fact that the protein particles are randomly distributed over both fracture faces. The procedure has not yet been evaluated further.

B. Detergent-Mediated Reconstitution Methods

For the use of detergents in reconstitution and the properties of detergents in general the reader is referred to Helenius et al.[18-20] In detergent-mediated reconstitution both relative and absolute concentrations of lipid, protein, and detergent determine the equilibria between the several phases and therefore the effectivity and reproducibility of the reconstitution. Therefore, all experiments have been performed in the same volume and salt medium, and with the same amounts of lipid and protein.

An appropriate amount of detergent (solid or as a solution in water) is mixed with 40 mg of egg phosphatidylcholine in chloroform–methanol (1 : 1, v/v) and taken to dryness. The mixed lipid–detergent film is washed twice with diethyl ether. After addition of salt either purple membrane

[17] A. Darszon, C. A. Vandenberg, M. Schönfeld, M. H. Ellisman, N. C. Spitzer, and M. Montal, Proc. Natl. Acad. Sci. U.S.A. 77, 239 (1980).
[18] A. Helenius and K. Simons, Biochim. Biophys. Acta 415, 29 (1975).
[19] C. Tanford and J. A. Reynolds, Biochim. Biophys. Acta 457, 133 (1976).
[20] A. Helenius, D. R. McCaslin, E. Fries, and C. Tanford, this series, Vol. 56, Part G, p. 734.

sheets or deoxycholate-purified and delipidated bacteriorhodopsin aggregates are added (final volume 2 ml). After incubation of the mixed micellar solutions at room temperature they are transferred to pretreated dialysis bags,[21] knotted tightly to stretch the membrane pores. Dialysis is performed for 3–4 days against 1 liter of salt, refreshed every morning and evening, and kept at a temperature between 15 and 20°. To the last change of dialysis solution 2 g of hydrated and washed Amberlite XAD-2 is added.[22]

1. Reconstitution Methods with Purple Membrane Sheets

a. *Triton X-100.* Triton X-100, a nonionic detergent, is often used in reconstitution studies since it solubilizes membrane components quite effectively. It is, however, difficult to remove by dialysis, since the critical micellar concentration is about 0.03% (w/v). We have used the following protocol.

To a dried film of 40 mg egg phosphatidylcholine and 100 mg Triton X-100 is added salt and 2 mg of purple membrane. The detergent is removed from the mixed micellar solution by the method of Holloway,[23] a batchwise incubation with 2 g of washed moist Bio-beads SM-2 for 3 hr. Since Bio-Beads bind Triton X-100 in the form of micelles, the final level of Triton, as is verified by using [³H]Triton X-100, will never be lower than the critical micellar concentration. In addition also some 30% of the protein, probably bound in the form of mixed protein–detergent micelles, is lost in this procedure.

The Triton concentration is further reduced by dialysis and reaches values of less than 1 molecule of Triton per 1000 lipid molecules.

The result of this reconstitution is a preparation of multilamellar vesicles with diameters ranging from 5000 to 10,000 Å. The proton pump activities vary from five to eight H⁺/bacteriorhodopsin. The lower activities are probably a reflection of protein that is not incorporated into the vesicles. The vesicles are found in one band on density gradients; protein-free vesicles are absent.

When Triton X-100 is removed by extensive dialysis only, no pump activity is found, but these vesicles are predominantly unilamellar.[22,24]

b. *Octyl Glucoside.* To a mixed film of 40 mg of egg phosphatidylcholine and 25 mg of octyl glucoside, salt and 2 mg of purple membrane are added. After an incubation period the detergent is removed by dialysis.

[21] P. McPhie, this series, Vol. 22, p. 23.
[22] S. B. Hwang and W. Stoeckenius, *J. Membr. Biol.* **33**, 325 (1977).
[23] P. W. Holloway, *Anal. Biochem.* **53**, 304 (1973).
[24] N. A. Dencher and M. P. Heyn, *FEBS Lett.* **108**, 307 (1979).

The resulting reconstituted preparation has a viscous and inhomogeneous appearance. Analysis on density gradients shows a top band of empty lipid vesicles, a protein-containing band, and nonincorporated protein material extending as sticky threadlike material in the center of the tube throughout the gradient. Proton pump activities are in the range of 4–12 H^+/bacteriorhodopsin, in agreement with values obtained by Dencher and Heyn.[24] Reconstitution is not achieved by complete solubilization in the form of mixed micelles, a process occurring between 3 and 4% octyl glucoside, but by integration of protein in a destabilized membrane. Conditions seem to be optimal around 1.25% octyl glucoside. Reconstitution with higher concentrations of octyl glucoside yields a comparable suspension with respect to viscosity and appearance, but proton pump activities are consistently lower (in agreement with the data of Racker et al.[25]) This suggests that at higher octyl glucoside concentrations the protein is either irreversibly inactivated or, more likely, more randomly oriented in the reconstituted vesicles.

c. Cholate. This procedure is adapted from Racker and Stoeckenius.[26] Egg phosphatidyl choline (40 mg) is mixed with 0.2 ml 20% (w/v) sodium cholate and dried in a mixed film. Salt is added before 2 mg of purple membrane. (Otherwise the bacteriorhodopsin instantly denatures as a consequence of the high pH of the mixed film.) Cholate is removed by dialysis. The resulting preparation consists of protein-free vesicles, protein-rich vesicles, and nonincorporated protein. Proton pump activities are of the order of 5–10 H^+/bacteriorhodopsin.

Hwang and Stoeckenius[22] reported on vesicles that had been obtained in a slightly different manner (1.3% cholate, short sonication), which were rather heterogenous in size and showed a tendency to incorporate the protein in a random fashion. Activities that are found are of the order of 2–3 H^+/bacteriorhodopsin.

2. Reconstitution Methods Employing Deoxycholate Purified Bacteriorhodopsin

Bacteriorhodopsin (5–10 mg in 1 ml) is solubilized in 10% sodium deoxycholate for 24 hr at room temperature. It is delipidated and purified from traces of red membrane by layering the material on a 10–60% (w/w) continuous sucrose gradient (volume 15 ml) and centrifuging for 15 hr at 75,000 g (TST.28.38 rotor of Kontron-TGA-65 ultracentrifuge: Kontron, Zürich). The purple band is collected and used for further reconstitutions.

[25] E. Racker, B. Violand, S. O. Neal, M. Alfonzo, and J. Telford, Arch. Biochem. Biophys. **198**, 470 (1979).
[26] E. Racker and W. Stoeckenius, J. Biol. Chem. **249**, 662 (1974).

The amount of DOC still associated with the protein is determined by using [³H]deoxycholate and corresponds to 7 mol of detergent per mole of bacteriorhodopsin. Endogenous lipid is removed for 80% in this procedure.[22]

a. Deoxycholate. Egg phosphatidylcholine (40 mg) is mixed with 0.4 ml 10% (w/v) sodium deoxycholate solution pH 8.0 and taken to dryness. Salt and 2 mg of deoxycholate-purified bacteriorhodopsin are added. After 1 hr incubation the deoxycholate is removed by dialysis. The reconstituted vesicles migrate in one band on a density gradient. They have not been analyzed with respect to size and show a pump activity of about 4 H⁺/bacteriorhodopsin.

b. Triton X-100. Besides the absence of nonincorporated protein there are no differences in behavior on a density gradient between vesicles reconstituted either with purple membrane or with deoxycholate-purified bacteriorhodopsin. Also proton pump activities are found to be of the same order of magnitude.

c. Octyl Glucoside. Vesicles reconstituted with purified bacteriorhodopsin are visually more homogeneous than those prepared from purple membranes. Only a small fraction on nonincorporated protein is present, whereas most of the vesicles contain protein. The diameter of the vesicles is 1000–3000 Å. The protein particles are distributed over both fracture planes, which is in apparent contradiction to the high pumping activity. Furthermore, some larger particle aggregates appear and these are sometimes found to be clustered. The proton pump activity ranges from 16 to 22 H⁺/bacteriorhodopsin.

d. Cholate. As already noted by Hwang and Stoeckenius[22] deoxycholate-purified bacteriorhodopsin reconstituted in lipid vesicles by cholate display high proton pump activities. We have found values in the range of 18–38 H⁺/bacteriorhodopsin. All the vesicles migrate on a density gradient in one band with no indications of protein-free vesicles or nonincorporated protein. The vesicles have a limited size distribution and an average diameter of about 800 Å. Almost all protein particles are located on the concave fracture plane. Our analyses are fully consistent with those of Hwang and Stoeckenius.[22]

Very recently, Khorana and co-workers[27] have extended the deoxycholate-cholate procedure by solubilization of purple membrane fragments in Triton X-100, after which the bacteriorhodopsin is delipidated for more than 99% and brought into a deoxycholate environment by column chromatography. The proton pump activities are high. In vesicles of

---

[27] K.-S. Huang, H. Bayley, and H. G. Khorana, *Proc. Natl. Acad. Sci. U.S.A.* **77**, 323 (1980).

Preparation	Size (Å)	Maximal extent of H$^+$ uptake (H$^+$/BRh)	Comments
SUV (A,1)	260	3	U
FPV (A,2)	380	5	U
DMSO (A,4)	n.d.[a]	1	—
"Gigantosomes" (A,5)	(2–25) × 10^3	1	M(+U),PR
Triton X-100 (B,1,a)	(5–10) × 10^3	5–8	M,PR
Octyl glucoside (B,1,b)	(2–10) × 10^3	4–12	U(+M),PR
Cholate (B,1,c)	n.d.	5–10	—
Deoxycholate (B,2,a)	n.d.	4	—
DOC/Triton X-100 (B,2,b)	n.d.	5–8	—
DOC/Octyl glucoside (B,2,c)	(1–3) × 10^3	16–22	U,PR
DOC/Cholate (B,2,d)	800	18–38	U,PC

[a] n.d. = not determined, U = predominantly unilamellar, M = predominantly multilamellar, M(+U) = more than 50% multilamellar, U(+M) = more than 50% unilamellar, PR = protein particles on both fracture planes, PC = protein particles preferentially on concave fracture planes.

phosphatidyl choline and bacteriorhodopsin (50:1, w/w) a value of 24 H$^+$/bacteriorhodopsin is reported. Carboxypeptidase digestion indicates a uniform inside-out orientation of the bacteriorhodopsin.

Concluding Remarks

The data on the different preparations are listed in the table. It is clear that the methods reported here for the reconstitution of bacteriorhodopsin in model vesicles are far from exhaustive. The mechanism that triggers the incorporation of bacteriorhodopsin into the lipid bilayer in the correct orientation is still far from clear. However, one can tentatively conclude that the protein will find its proper position in a region of detergent concentrations where there is a transition from a mixed micellar solution to a destabilized bilayer. The curvature of the bilayer will play a role in the orientation of the protein. Optimal procedures aim at minimizing the degree of protein randomization. Furthermore, large vesicles are bound to be superior in pumping activity because of their large internal volume.

Acknowledgments

The authors thank José Leunissen-Bijvelt for the freeze-fracture electron microscopy contributions. P. W. M. van Dijck is supported by a postdoctoral fellowship from the Netherlands Organization for the Advancement of Pure Research (ZWO) under auspices of the Netherlands Foundation for Chemical Research (SON)

[5] Phospholipid Substitution of the Purple Membrane

By EVERT P. BAKKER and S. ROY CAPLAN

Bacteriorhodopsin is an ideal protein for studies on protein–lipid interactions in artificial membranes: its properties are well defined, its enzymatic activity is easy to assay, and it can easily be reconstituted with various phospholipids (reviewed by Stoeckenius et al.[1]). A complication is, however, the fact that the purple membrane sheets used for the reconstitution experiments contain not only bacteriorhodopsin, but also endogenous lipids (mainly diphytanylether analogs of phospho- and glycolipids[2,3]). Consequently, these endogenous lipids will be coreconstituted with bacteriorhodopsin, and under some conditions these lipids may remain specifically associated with the protein (see later). For that reason it is a prerequisite for studies on the interaction between bacteriorhodopsin and lipids to remove the endogenous purple membrane lipids from the protein before it is reconstituted. This can be achieved in two different ways. First, bacteriorhodopsin can be delipidated with detergents.[4,5] It has been reported that 99% of the endogenous lipids have been removed by this method.[5] The second, somewhat less laborious method is based on exchange of endogenous for exogenous lipids, mediated by the detergent cholate.[6] This method, which leads to a replacement of 90–95% of the purple membrane lipids,[7] is described here.

Principle

Purple membrane sheets are incubated with the detergent cholate in the presence of excess exchange phospholipid. After an incubation period, during which phospholipid exchange takes place rather than membrane fusion, the mixture is layered on a sucrose gradient and centrifuged. The bacteriorhodopsin-containing membranes band at higher density than phospholipid and detergent not associated with the protein.

[1] W. Stoeckenius, R. H. Lozier, and R. A. Bogolmoni, Biochim. Biophys. Acta 505, 215(1979).

[2] M. Kates, S. C. Kushwaha, and G. D. Sprott, Chapter 13, this volume.

[3] S. C. Kushwaha, M. Kates, and W. G. Martin, Can. J. Biochem. 53, 284(1975).

[4] M. Happe and P. Overath, Biochem. Biophys. Res. Commun. 72, 1504(1976).

[5] K. S. Huang, H. Bayley, and H. G. Khorana, Proc. Natl. Acad. Sci. U.S.A. 77, 323(1980).

[6] G. B. Warren, P. A. Toon, N. J. M. Birdsall, A. G. Lee, and J. C. Metcalfe, Proc. Natl. Acad. Sci. U.S.A. 71, 622(1974).

[7] E. P. Bakker and S. R. Caplan, Biochim. Biophys. Acta 503, 362(1978).

METHODS IN ENZYMOLOGY, VOL. 88

Reagents

Sodium cholate. A solution of 20% (w/v) is brought to pH = 7.2 with HCl. If this solution is not colorless, the cholate salt should be pretreated with active charcoal.[8] Sodium cholate from Merck ("for biochemical purposes") does not require this treatment.

Phospholipids. L-Dilauroylphosphatidylcholine (DLPC), L-dimyristoylphosphatidylcholine (DMPC), L-dipalmitoylphosphatidylcholine (DPPC) from Calbiochem. Egg phosphatidylcholine (EPC) either from Sigma (type VE) or from Lipid Products (London). Alternatively, EPC can be isolated from egg yolk.[9] Aliquots of 25 mg phospholipid are dissolved in 2.5 ml chloroform–methanol = 2:1 (by volume).

Purple membrane isolated according to Oesterhelt and Stoeckenius.[10] The sheets are suspended at 5 mg protein per milliliter in 150 mM KCl, 100 mM Tris–HCl, pH = 6.5. The concentration of bacteriorhodopsin is then about 200 μM and that of the endogenous lipids about 1.7 mg/ml.[10]

Thin-layer chromatography glass plates from Merck (Silicagel 60 without fluorescence indicator).

Cupric phosphoric acid spray.[11] 18 g of phosphoric acid and 8 g $CuSO_4 \cdot 5H_2O$ per 100 ml H_2O.

Procedure

The phospholipid solution is taken to dryness in a 50 ml round bottom flask connected to a rotary evaporator. To the dry film is added 2.0 ml of a solution containing 100 mM Tris–HCl, 150 mM KCl, and 30 mg/ml sodium cholate. The pH of this solution should be 6.5 (adjusted with HCl). The solution is gently shaken until all phospholipid has dissolved. With DPPC the temperature should be raised to 45° in order to achieve solubilization of this compound. A brief sonication step may help to accelerate the solubilization of the phospholipids. To the solution is added 1.0 ml of the purple membrane suspension, and the clear solution is incubated for 1 hr in the dark at room temperature. With DPPC the suspension is heated several times to 45°, and subsequently left to cool. After 1 hr of incubation the solution is layered in a centrifuge tube on top of a 30-ml linear sucrose gradient (20–60% (w/v)) and centrifuged at 4° for 16 hr at

[8] Y. Kagawa and E. Racker, *J. Biol. Chem.* **246**, 5477(1971).
[9] W. S. Singleton, M. S. Gray, M. L. Brown, and J. L. White, *J. Am. Oil Chem. Soc.* **42**, 53(1965).
[10] D. Oesterhelt and W. Stoeckenius, this series, Vol. 31, Part A, p. 667.
[11] E. H. Coch, G. Kessler, and J. S. Meyer, *Clin. Chem. (Winston-Salem, N.C.)* **20**, 1368(1974).

100,000–150,000 g. Depending on the lipid composition, the purple membrane fraction will band between 45 and 50% sucrose. By contrast, excess phospholipid and detergent will remain at the top of the centrifuge tube. The purple fractions are collected and washed once with 25 ml of a 150 mM KCl solution, and the exchange procedure is repeated exactly as described above, except that the sodium cholate concentration is reduced by one-half. The purple membrane fractions are collected from the sucrose gradient, washed once with 150 mM KCl, and dialyzed for a total of 24 hr against 2 × 1 liter of 150 mM KCl. The membranes are collected, suspended at 2–4 mg protein per milliliter in 150 mM KCl and stored at 4°.

Note. The incubation is carried out at pH = 6.5 because bacteriorhodopsin has a tendency to become red, and to denature subsequently, when brought together with the detergent cholate and phospholipids with unsaturated fatty acids (e.g., EPC) at alkaline pH. Even during the incubation with EPC at pH = 6.5 bacteriorhodopsin changes its color from purple to deep red. The purple color returns, however, after removal of the detergent.

Extent of Lipid Substitution

The polar lipids of the various purple membrane fractions are separated by thin-layer chromatography. The extent of lipid substitution is calculated from the intensities of the spots of the endogenous lipids of the various fractions. For this purpose, equal amounts of native and lipid-substituted purple membrane (equivalent to about 10 nmol of bacteriorhodopsin) are delipidated.[12] The lipid extract is dried with a stream of nitrogen and dissolved in 100 μl chloroform–methanol = 2 : 1 (by volume). The following samples are applied to TLC plates: 50 μl for the lipid-substituted purple membrane and aliquots of 5, 10, 20, and 50 μl for the native purple membrane. The polar lipids of the various fractions are separated with a solvent mixture of chloroform–methanol–water = 65 : 35 : 4 (by volume.[7,13] The plate is sprayed with the cupric phosphoric acid spray and heated for 10–20 min at 180°. The lipids appear as grey-brown spots on a light background. (*Note.* It takes saturated lipids longer to become visible than it does unsaturated lipids.) The extent of lipid substitution is calculated by comparing the relative intensities of the spots on the chromatogram of the two major endogenous polar lipids in the lipid-substituted purple membrane with those of the native purple membrane series.[7] The

---

[12] E. G. Bligh and W. J. Dyer, *Can. J. Biochem. Physiol.* **37**, 911(1959).
[13] H. Wagner, L. Horhammer, and P. Wolff, *Biochem. Z.* **334**, 175(1961).

TABLE

THE EXTENT OF LIPID SUBSTITUTION OF THE ENDOGENOUS POLAR LIPIDS OF THE PURPLE MEMBRANE BY PHOSPHATIDYLCHOLINE SPECIES[a]

Preparation	Percent exchange of endogenous lipids	Lipid–phosphorus content (mol/mol of bacteriorhodopsin)[b]
Native purple membrane	—	10
DLPC-purple membrane	70	9
DMPC-purple membrane	80	10
DPPC-purple membrane	90	10
EPC-purple membrane	95	11

[a] Adapted with permission from E. P. Bakker and S. R. Caplan, Biochim. Biophys. Acta 503, 362 (1978).[7]

[b] Lipid phosphorus was determined according to P. S. Chen, T. Y. Toribara, and H. Warner, Anal. Chem. 28, 1756 (1956). The concentration of bacteriorhodopsin was determined spectroscopically.

error is of the order of 5–10%. An alternative method for the estimation of the extent of lipid substitution is to grow cells on [^{32}P]phosphate and to determine the amount of ^{32}P-labeled polar lipid associated with bacteriorhodopsin.[5]

The extent to which the polar lipids of the purple membrane were substituted by the various species of phosphatidylcholine (PC) is given in the table. It is remarkable that the extent of substitution increases with the length of the fatty acid chains of the PC species used for the substitution. This may indicate that the lipid bilayer formed by the shorter-chain phospholipids is too thin to supply bacteriorhodopsin with the apolar environment it requires.[7,14] Support for this notion comes from the observation that delipidated bacteriorhodopsin reconstitutes poorly with DMPC.[5] The longer-chain PC species, DPPC and EPC, replace the polar lipids of the purple membrane to the extent of 90 and 95%, respectively (see the table). Similar extents of substitution have been obtained with egg phosphatidylethanolamine or total soybean phospholipids as exogenous lipids. In all these preparations the amount of endogenous lipid left is equal to 1 mol or less per mole of bacteriorhodopsin present.

Properties of Lipid-Substituted Purple Membrane

The photochemical properties of bacteriorhodopsin in the lipid-substituted purple membranes are very similar to those of the native membrane,

[14] R. Henderson and P. N. T. Unwin, Nature (London) 257, 28(1975).

except that the O_{660} intermediate[15,16] is not observed.[7,17] This may be because the bacteriorhodopsin photocycle is slower in the lipid-substituted membranes than it is in the native membrane,[7,17] and it is known that under such conditions O_{660} disappears.[18] Lipid-substituted purple membranes reconstitute well with excess added phospholipids. These observations indicate that the endogenous purple membrane lipids do not have a special function in the mechanism of action of bacteriorhodopsin. Apparently, the protein has only a general requirement for phospholipids in order to be active.

Physical studies indicate that in EPC-purple membrane bacteriorhodopsin forms small protein clusters with a large rotational mobility.[17,19,20] The same is true for DPPC-purple membrane above 45°. However, both the rotational mobility[20] and the rate of the photocycle[17] decrease dramatically when DPPC-purple membrane is cooled below 40°. Since DPPC is known to undergo a phase transitioning in this range of temperature,[21] this may indicate that the activity of bacteriorhodopsin is influenced directly by the physical state of the phospholipids surrounding it. It is remarkable that such a profound effect of the lipid environment on the activity of bacteriorhodopsin has not been observed with native purple membrane sheets[22] or with monomeric bacteriorhodopsin[23,24] reconstituted with DMPC. At temperatures below that of the phase transition of DMPC, bacteriorhodopsin is, in these systems, excluded from the DMPC phase. It becomes surrounded by its own endogenous lipids, in which it forms a hexagonal lattice as in the native membrane.[23,24] It is therefore to be expected that a dramatic effect of temperature on the photocycle will be observed only when all endogenous lipids have been replaced by DMPC.

[15] M. A. Slifkin and S. R. Caplan, *Nature (London)* **253**, 56(1975).

[16] R. H. Lozier, R. A. Bogomolni, and W. Stoeckenius, *Biophys. J.* **15**, 955(1975).

[17] W. V. Sherman and S. R. Caplan, *Biochim. Biophys. Acta* **502**, 222(1978).

[18] N. A. Dencher and M. Wilms, *Biophys. Struct. Mech.* **1**, 259(1975).

[19] E. P. Bakker, M. Eisenbach, H. Garty, C. Pasternak, and S. R. Caplan, in "Molecular Aspects of Membrane Transport" (D. Oxender and C. F. Fox, eds.), p. 553. Alan R. Liss, Inc., New York, 1978.

[20] C. Pasternak and M. Shimizky, in "Energetics and Structure of Halophilic Microorganisms" (S. R. Caplan and M. Ginzburg, eds.), p. 309. Elsevier/North-Holland Biomedical Press, Amsterdam, 1978.

[21] B. D. Ladbrooke, R. M. Williams, and D. Chapman, *Biochim. Biophys. Acta* **150**, 333(1968).

[22] E. Racker and P. C. Hinkle, *J. Membr. Biol.* **17**, 181(1974).

[23] R. J. Cherry, U. Müller, R. Henderson, and M. P. Heyn, *J. Mol. Biol.* **121**, 283 (1978).

[24] N. A. Dencher and M. P. Heyn, *FEBS Lett.* **108**, 307 (1979).

[6] Reconstitution of Monomeric Bacteriorhodopsin into Phospholipid Vesicles

By MAARTEN P. HEYN and NORBERT A. DENCHER

Introduction

Monomeric bacteriorhodopsin was reconstituted into phospholipid vesicles by mixing lipid and detergent-solubilized protein together and removing the detergent by dialysis. Such vesicles allow an investigation of the spectroscopic and proton transport properties of bacteriorhodopsin monomers and provide a useful model system for the study of lipid–protein interactions.

Preparation of Bacteriorhodopsin-Phospholipid Vesicles

A summary of the vesicle preparation method has appeared elsewhere.[1] Considerable progress has been made since. Some of these improvements have been described recently.[2]

Bacteriorhodopsin is solubilized to the state of monomers in the nonionic detergents Triton X-100 or octyl glucoside (octyl-β-D-glucopyranoside).[3,4] For dimyristoyl- and dipalmitoylphosphatidylcholine (DMPC and DPPC) the procedure works best with Triton X-100. Asolectin, diphytanoylphosphatidylcholine, egg lecithin, and phosphatidylserine vesicles can be prepared only with bacteriorhodopsin solubilized in octyl glucoside. The lipids (10–20 mg in 1 ml chloroform) are spread homogeneously on the bottom of a round-bottom flask. The solvent is evaporated by a stream of nitrogen while the flask is being gently shaken. To avoid undesirable cooling the flask is kept at about 30° in a waterbath. When the lipid film has formed, the flask is connected for 20 min to high vacuum to remove any residual solvent (20°). Using a water bath, both the lipid flask and the solution of solubilized bacteriorhodopsin (in 100 mM acetate buffer, pH 5, and 0.2–0.5% (w/w) Triton X-100 or 40 mM octyl glucoside) are separately kept at a temperature above the lipid phase transition (e.g., 29° for DMPC and 44° for DPPC). The solution of solubilized bacteriorhodopsin (at the concentration required to achieve the desired lipid-to-protein ratio after mixing) is poured into the round-bottom flask and clean glass beads are added (diameter 2 mm). After shaking two or three times by hand, the

[1] R. J. Cherry, U. Müller, R. Henderson, and M. P. Heyn, *J. Mol. Biol.* **121**, 283 (1978).
[2] N. A. Dencher and M. P. Heyn, *FEBS Lett.* **108**, 307 (1979).
[3] N. A. Dencher and M. P. Heyn, *FEBS Lett.* **96**, 322 (1978).
[4] N. A. Dencher and M. P. Heyn, this volume, Article [2].

lipid film comes off the bottom of the flask. The mixture is kept in the water bath at the required temperature for 10 min more with occasional shaking. Since bacteriorhodopsin solubilized in octyl glucoside is light sensitive,[3] the flask is protected from light by aluminum foil. The suspension is allowed to stand another 10 min in the water bath. The mixture is poured into a dialysis bag and dialyzed for 3–4 hr against 1 liter 100 mM sodium acetate buffer (pH 5), which is at 29° for all lipids. A temperature above the phase transition is deemed too risky for DPPC for such a long time. The dialysis is carried out under constant stirring in a 1-liter glass cylinder kept at 29° and protected from the light by aluminum foil. After the first buffer change with buffer of room temperature, the dialysis is continued in the cold room (4°). The buffer is changed three times on the first day. On the following days one change is sufficient. The reconstituted lipid–protein vesicles are formed during the detergent dialysis. Octyl glucoside has a critical micelle concentration (CMC) of 25 mM, which facilitates its rapid removal by dialysis. For vesicles made with bacteriorhodopsin solubilized in this detergent, a minimal dialysis period of 2 days is required. For vesicles made with bacteriorhodopsin solubilized in Triton (CMC about 0.017%) about 8 days of dialysis are required. At the end of the dialysis period the vesicle suspensions are clear, unless the molar lipid-to-protein ratio is above 350. Pure lipid vesicles, which are often required as a control, are prepared exactly as above, except that the solution added to the lipids consists of buffer with the same detergent concentration as used in the presence of bacteriorhodopsin.

The vesicles are further purified by isopycnic sucrose density centrifugation. The dialysate is layered on a 5–40% (w/w) sucrose gradient and centrifuged at 250,000 g for 5 hr at 8° (Beckman SW 41-Ti rotor). The bacteriorhodopsin vesicle band is easily identifiable by its color. Bands are

THE EXTENT OF LIPID SUBSTITUTION OF THE ENDOGENOUS POLAR LIPIDS OF THE PURPLE MEMBRANE BY PHOSPHATIDYLCHOLINE SPECIES[a]

Preparation	Percentage exchange of endogenous lipids	Lipid–phosphorus content[b] (mol/mol of bacteriorhodopsin)
Native purple membrane	—	10
DLPC-purple membrane	70	9
DMPC-purple membrane	80	10
DPPC-purple membrane	90	10
EPC-purple membrane	95	11

[a] Adapted with permission from E. P. Bakker and S. R. Caplan, *Biochim. Biophys. Acta* **503**, 362 (1978).

[b] Lipid phosphorus was determined according to P. S. Chen, T. Y. Toribara, and H. Warner, *Anal. Chem.* **28**, 1756 (1956). The concentration of bacteriorhodopsin was determined spectroscopically.

collected from the top. In terms of band width, the best results are obtained with DMPC-bacteriorhodopsin vesicles. With this system one very sharp purple band is observed. With decreasing lipid-to-protein ratio this band is located at a higher density in the gradient. Below this purple band one or two weak flaky white bands occur. The pure DMPC vesicles give rise to one extremely sharp pancake-like band. With DPPC and Triton-solubilized bacteriorhodopsin the band is broader, and occasionally two separate bands of different density and lipid-to-protein ratio are obtained. With diphytanoylphosphatidylcholine, dipalmitoylphosphatidylserine, egg lecithin and asolectin vesicles, a single broad band occurs. The purity of the lipids is important. Good results are obtained with DMPC and DPPC from Fluka (Buchs, Switzerland), show one spot in thin-layer chromatography. After collection of the bands from the gradient, the sugar is removed by dialysis against 100 mM sodium acetate buffer (pH 5). The phospholipid concentration of the vesicles is determined by the method of Ames and Dubin.[5] The bacteriorhodopsin concentration is determined spectroscopically[6] with bleached vesicles as reference and by the Hartree–Lowry method[7] with corrections.[6] The agreement between these two methods is excellent. Vesicles were prepared with molar phospholipid to bacteriorhodopsin ratios ranging from 40 to 500. For low lipid-to-protein ratios the resulting ratios are somewhat higher than the planned ones (e.g., 92 versus 73). For the high ratios the converse is the case. Radioactive Triton was used to determine the amount of detergent remaining after dialysis. About one Triton per two bacteriorhodopsin remained.[1] For reconstituted vesicles prepared with bacteriorhodopsin solubilized in octyl glucoside, the amount of residual detergent is expected to be even less on the basis of the very high CMC of this detergent. Experiments with radioactive octyl glucoside in similar detergent-dialysis procedures show that more than 99.9% of the octyl glucoside is removed in three days.[8–10]

Properties and Applications of Bacteriorhodopsin-Phospholipid Vesicles

The reconstituted vesicles were examined in the electron microscope by negative staining and were found to be predominantly unilamellar with

[5] B. N. Ames and D. T. Dubin, *J. Biol. Chem.* **235**, 769 (1960).
[6] M. Rehorek and M. P. Heyn, *Biochemistry* **18**, 4977 (1979).
[7] E. F. Hartree, *Anal. Biochem.* **48**, 422 (1972).
[8] W. A. Petri, Jr. and R. R. Wagner, *J. Biol. Chem.* **254**, 4313 (1979).
[9] V. H. Engelhard, B. C. Guild, A. Helenius, C. Terhorst, and J. L. Strominger, *Proc. Natl. Acad. Sci. U.S.A.* **75**, 3230 (1978).
[10] A. Helenius, E. Fries, and J. Kartenbeck, *J. Cell Biol.* **75**, 866 (1977).

radii varying between 100 and 300 nm.[1] Similar values for the average radius were obtained from quasi-elastic light scattering.[11] The size distribution appeared to be narrowest for vesicles with the lowest lipid-to-protein ratios. The vesicle suspensions were clear and suitable for spectroscopy (except for molar lipid-to-protein ratios above 350). Illumination of a suspension of vesicles resulted in a reversible alkalization of the external medium, i.e., in a light-induced proton translocation opposite to that observed in intact halobacteria.[2] The fact that these vesicles pump protons shows that a net orientation of bacteriorhodopsin molecules exists. What percentage of bacteriorhodopsin molecules are oriented in one direction is at present not known. Whereas protons were pumped inward with all the lipids tested, the extent of the steady-state pH change varied with the type of lipid. A maximal net inward translocation of 12 protons per bacteriorhodopsin was observed with asolectin and egg lecithin.[2] Perhaps the most useful property of these vesicles is that they show lipid–protein segregation. Based on X-ray diffraction, electron microscopy, circular dichroism, and transient dichroism, it could be shown that above the phase transition of the lipids the bacteriorhodopsin molecules are monomeric (provided the molar lipid-to-protein ratio is above 40), whereas below the transition they are aggregated in the same hexagonal lattice as in the purple membrane.[1] Qualitatively, the photocycle of bacteriorhodopsin is the same in the aggregated and monomeric form.[2,12] In comparison with the purple membrane, the cycle is about seven times slower in DMPC vesicles.[12] The Arrhenius plot for the decay rate of the M intermediate has no break at the phase transition of DMPC.[12] The value of this rate constant depends, however, on the type of lipid used in the reconstitution.[12] The phenomenon of light–dark adaptation can still be observed with these vesicles, although its extent is reduced in the monomeric state.[1,12] The long-term stability of these vesicles is excellent. After 1 year of storage at 4°, the proton pumping activity is unaffected and the spectral properties are unchanged.

This vesicle system allows a direct comparison between the properties of monomeric and hexagonally crystallized bacteriorhodopsin in the same lipid environment. It permitted us to show that bacteriorhodopsin monomers carry out a photocycle and pump protons.[2] Protein aggregation does not seem to be required for these functions. The effects of the type of lipid and its physical state on the proton pump and on the kinetics of the photocycle can also be conveniently studied.[12] In previous work on the reconstitution of bacteriorhodopsin in lipid vesicles, whole purple membranes were incorporated. Such systems are clearly not suitable for

[11] M. P. Heyn, R. J. Cherry, and N. A. Dencher, *Biochemistry* (in press).
[12] N. A. Dencher, M. P. Heyn, and K.-D. Kohl, submitted for publication.

studying lipid–protein interactions. The present vesicles, on the other hand, are quite useful for such purposes. Moreover, they can be prepared with a variety of lipids and with a wide range of lipid-to-protein ratios. The effect of bacteriorhodopson on the order and mobility of the lipids, on the membrane viscosity, and on the lipid phase transition was recently reported.[11] Finally, these vesicles with their net bacteriorhodopsin orientation have been used to incorporate bacteriorhodopsin transmembranously into planar lipid bilayers.[13] On illumination, steady photocurrents were observed.[13]

[13] E. Bamberg, N. A. Dencher, A. Fahr, and M. P. Heyn, submitted for publication.

[7] A Single Turnover Study of Photoelectric Current-Generating Proteins

By V. P. Skulachev

Several years ago we introduced a direct method for measuring the electrogenic activity of membrane proteins.[1-6] Natural membranes (open sheets or vesicles) or reconstituted membranes (proteoliposomes) were adsorbed onto one of the two surfaces of a planar phospholipid membrane. Addition of an energy source was found to induce translocation of charges by the studied current-generating proteins plugged through the natural or proteoliposomal membrane. In the case of, e.g., bacteriorhodopsin, the translocated charge is H^+ ion which is pumped from the bulk solution into a water-containing cavity between the bacteriorhodopsin membrane and the planar phospholipid membrane. Then H^+ ion (or any other cation) diffuses through the planar membrane to the bulk solution on the opposite side of the planar membrane. Thus a transmembrane continuous current and a potential difference arise. They can be measured

[1] L. A. Drachev, A. A. Jasaitis, A. D., Kaulen, A. A. Kondrashin, E. A. Liberman, I. B. Nemecek, S. A. Ostroumov, A. Yu. Semenov, and V. P. Skulachev, *Nature (London)* **249**, 321 (1974).
[2] E. L. Barsky, L. A. Drachev, A. D. Kaulen, A. A. Kondrashin, E. A. Liberman, S. A. Ostroumov, V. D. Samuilov, A. Yu. Semenov, and V. P. Skulachev, *Bioorg. Khim.* **1**, 113 (1975).
[3] L. A. Drachev, V. N. Frolov, A. D. Kaulen, E. A. Liberman, S. A. Ostroumov, V. G. Plakunova, A. Yu. Semenov, and V. P. Skulachev, *J. Biol. Chem.* **251**, 7059 (1976).
[4] V. P. Skulachev, *FEBS Lett.* **64**, 23 (1976).
[5] L. A. Drachev, A. D. Kaulen, A. Yu. Semenov, I. I. Severina, and V. P. Skulachev, *Anal. Biochem.* **96**, 250 (1979).
[6] V. P. Skulachev, this series, Vol. 55, pp. 586 and 751.

METHODS IN ENZYMOLOGY, VOL. 88

with two electrodes immersed in the bulk solutions on both sides of the planar membrane. The results obtained with such a method were recently confirmed in other laboratories.[7-12]

The same technique can be used to measure charge displacement inside a molecule of a current-generating protein if the trajectory of the charge does not parallel the membrane plane. This creates a unique possibility for resolving the intermediate stages of electrogenesis if all the generators can be simultaneously actuated. This is the case for photoelectric generators.

Comparing different kinds of supporting (planar) membranes and films, we found that a black phospholipid membrane and a collodion film impregnated with a decane solution of phospholipids have the best time resolution (better than 50 nsec [5,13-17]). This means that charge translocation processes can be measured in such a fast time scale. The collodion film proved to be more convenient than the black membrane, since the former is much more stable.

Collodion Film

A collodion film was formed on the water surface by adding a 1% amyl acetate solution of nitrocellulose. After amyl acetate evaporation, the film was transferred to air by means of a ring made of a Millipore filter. The film was dried for at least an hour, impregnated with a decane solution of lecithin (70 mg/ml), and clamped in a dismountable Teflon chamber so that it separated the two electrolyte-containing compartments.

Measurement of Electric Response

The electric potential difference between the two compartments was measured with platinum or silver chloride electrodes connected with an

---

[7] Th. Schreckenbach, *in* "Photosynthesis in Relation to Model Systems" (J. Barber, ed.), 189. Elsevier/North-Holland Biomedical Press, Amsterdam, 1979.

[8] T. R. Herrman and G. W. Rayfield, *Biochim. Biophys. Acta* **443**, 623 (1976).

[9] P. Shieh and L. Packer, *Biochem. Biophys. Res. Commun.* **71**, 603 (1976).

[10] M. C. Blok, K. J. Hellingwerf, and K. Van Dam, *FEBS Lett.* **76**, 45 (1977).

[11] M. C. Blok and K. Van Dam, *Biochim. Biophys. Acta* **550**, 527 (**1979**).

[12] Z. Dancshazy and B. Karvaly. *FEBS Lett.* **72**, 136 (1976).

[13] L. A. Drachev, A. D. Kaulen, and V. P. Skulachev, *Mol. Biol.* (*Moscow*) **11**, 1377 (1977).

[14] L. A. Drachev, A. D. Kaulen, and V. P. Skulachev, *FEBS Lett.* **87**, 161 (1978).

[15] Z. Dancshazy, L.A. Drachev, P. Ormos, K. Nagy, and V. P. Skulachev, *FEBS Lett.* **96**, 59 (1978).

[16] V. P. Skulachev, *in* "Photosynthesis in Relation to Model Systems" (J. Barber, ed.), p. 176. Elsevier/North-Holland Biomedical Press, Amsterdam, 1979.

[17] V. P. Skulachev, *Sov. Sci. Rev., Sect. D* **1**, 83(1980).

operational amplifier Analog Devices 48K. To prevent light-induced artifacts, the electrodes were screened with black polyethylene. From the amplifier an electric signal was transmitted to a Data Lab transient recorder DL-905 or 922 and then to computer Nova 3D or to recorder KSP4.

Procedure of Incorporation

Incorporation of bacteriorhodopsin into the collodion film was carried out by the following procedure.

Suspension of bacteriorhodopsin sheets was sonicated for 3 min in disintegrator Braun-sonic 1515 and added to one of the Teflon chamber compartments to achieve a final concentration of bacteriorhodopsin of about 50 μg/ml. At pH 6.0, spontaneous incorporation of the sheets into the phospholipid-impregnated film took place if the NaCl concentration in the bathing solution was sufficiently high (0.1 M). In this case a decane solution of lecithin (70 mg/ml) was used to impregnate the collodion film. If the salt concentration was low, the impregnation mixture was supplemented with octadecylamine (1 mg/ml).[12] In both cases the decane solution of lecithin should be stored for several days at 0° or incubated for several min at 70°. Incorporation of bacteriorhodopsin proteoliposomes into the lecithin-impregnated collodion film was found to occur at pH 3.5–4.5. Proteoliposomes were added to one of the film-separated compartments, the final protein concentration being 10–50 μg/ml. After 2 hr incubation, pH of the bathing solution was increased by adding KOH. So, neither of these procedures required Ca^{2+}, which was previously used to activate the incorporation.[1-6] Applying these procedures, we obtained high values of the photoelectric responses to continuous illumination, reaching 300 mV with the bacteriorhodopsin-free compartment being found positively charged. A laser flash ($t_{1/2}$ = 15 nsec, λ = 530 nm, light impulse energy 5 mJ) inducing a single turnover of bacteriorhodopsin was shown to generate potential of the same direction, reaching 65 mV.

A similar method was used to study the photoelectrogenic responses of animal rhodopsin and bacterial photosynthetic reaction center complexes. In the former case photoreceptor disks and in the latter chromatophores or the reaction center-containing proteoliposomers were studied.

Spectral Measurements

To correlate fast electric events with spectral transitions of current-generating proteins, light transmission measurements in the same experimental chamber can be done. In this case, responses of an excess of nonincorporated sheets, disks, or chromatophores added into one of the two film-separated compartments were monitored by means of a flash-photolysis apparatus with two monochromators. Monitoring light, having passed through a monochromator, illuminated the experimental chamber. Then the light passed through the second monochromator to reach a photoelectric multiplier. From the photomultiplier, a signal was transmitted to an amplifier, then to Data Lab 905 (or 922), and finally to a computer, Nova 3D. We mostly used an average value of computer-stored signals induced by 50–100 flashes.

Both the electric and the spectral measurements were limited by the time resolution of Data Lab (50 nsec).

To analyze the kinetics of the electric and spectral responses, we used a computer Nova 3D program for curve expansion decomposition according to the method of least squares.

Natural Membranes

Three kinds of natural membranes containing biological photoelectric generators have been studied by means of the preceding techniques, namely, bacteriorhodopsin sheets, photoreceptor disks, and bacterial chromatophores.

Membrane fragments with bacteriorhodopsin as the only protein (sheets) were isolated from *Halobacterium halobium* R_1 (for method, see Oesterhelt and Stoekenius[18]). Bacteriorhodopsin proteoliposomes were reconstituted as described previously.[6]

Photoreceptor disks were obtained from the outer segments of rods of cattle retina,[19] chromatophores from *Rhodospirillum rubrum*.[20]

Photoelectric Responses of Bacterial and Animal Rhodopsins and Photosynthetic Reaction Centers

Figure 1 shows the results of measurement of electric potential generation by bacteriorhodopsin sheets incorporated into a phospholipid-impregnated collodion film (curve $\triangle V$), as well as of light transmittance changes at 570 nm [curve $(\triangle T/T)_{570}$] and at 412 nm [curve $(\triangle T/T)_{412}$] in a suspension of the sheets.[21]

[18] D. Oesterhelt and W. Stoeckenius, this series, Vol. 31, Part A, p. 667.
[19] V. I. Bolshakov, G. R. Kalamkarov, and M. A. Ostrovsky, in "Frontiers of Bioorganic Chemistry and Molecular Biology" (S. N. Ananchenko, ed.), p. 433. Pergamon, Oxford, 1980.
[20] A. A. Kondrashin, V. G. Remennikov, V. D. Samuilov, and V. P. Skulachev, *Eur. J. Biochem.* **113**, 219 (1980).
[21] L. A. Drachev, A. D. Kaulen, L. V. Khitrina, and V. P. Skulachev, *Eur. J. Biochem.* **117**, 461 (1981).

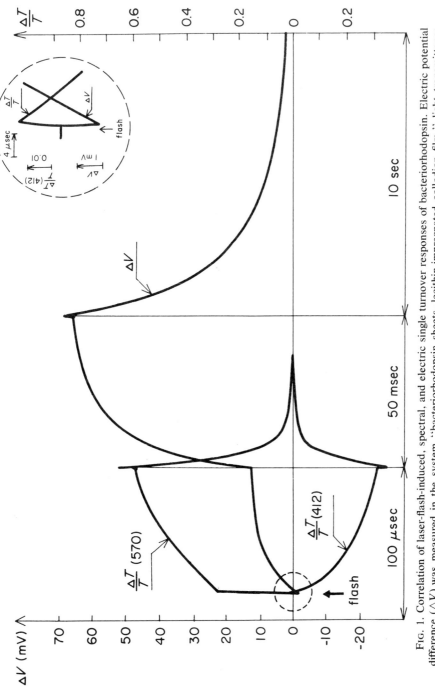

FIG. 1. Correlation of laser-flash-induced, spectral, and electric single turnover responses of bacteriorhodopsin. Electric potential difference (ΔV) was measured in the system "bacteriorhodopsin sheets–lecithin-impregnated collodion film," light transmittance changes at 570 nm, $(\Delta T/T)_{570}$, and at 412 nm, $(\Delta T/T)_{412}$, were measured in the sheet suspension. Bathing solution for both electric and spectral measurements contained 0.1 M NaCl, 5 mM MES buffer, 3 mM citrate (pH 6.0).

To visualize the main stages of the electrogenic effect, four different time scales were used, from microseconds to seconds. The first electrogenic stage of a laser flash-induced response is faster than the time resolution of our apparatus. It is oppositely directed compared to the slower stages of the bacteriorhodopsin electrogenesis. There are two slower stages, one developing in the microsecond scale and the other taking several milliseconds. The decay of the produced electric potential difference occurs within seconds.

To measure the spectral response of bacteriorhodopsin to the laser flash, we used suspensions of bacteriorhodopsin sheets, since the amount of the pigment incorporated into the collodion film is too small for such a measurement to be made.

It is seen (Fig. 1) that the earliest spectral change is an increase in the 570-nm light transmittance, which, like the first (opposite) phase of electrogenes, proves to be faster than the time resolution of the device. This effect is due to the conversion of bacteriorhodopsin to its bathointermediate.[22] Then a slow phase of the 570-nm transmittance increase appears which roughly correlates with the 412-nm transmittance decrease. Apparently, the slow 570-nm change is associated with the conversion of the bathointermediate to the 412-nm intermediate as soon as the bathointermediate has some absorption at 570 nm. The decays of both 570 and 412-nm changes occur in the millisecond scale.

Thus one can conclude that three main electrogenic phases of bacteriorhodopsin electrogenesis are associated with the following three main stages of the photocycle:

The first (opposite) phase: 570-nm bacteriorhodopsin → bathointermediate
The microsecond phase: bathointermediate → 412-nm intermediate
The millisecond phase: 412-nm intermediate → 570-nm bacteriorhodopsin

Figure 2 shows the data of an experiment with a black membrane used instead of the collodion film. Bacteriorhodopsin proteoliposomes were incorporated into a black planar membrane made of a decane solution of asolectin (100 mg/ml). The incubation medium contained 0.05 M Tris–HCl (pH 7.4) and 0.03 M $CaCl_2$. One can see that in the black membrane and collodion film systems, the kinetics of the microsecond and millisecond electrogenic phases are similar. (The time scales used in Fig. 2 do not allow one to observe the opposite electrogenic phase.) This and other observations (not shown) suggest that there is no essential difference between the black membrane and collodion film in electric measurements of the preceding type. At the same time, the collodion film, being very stable, is much more convenient.

[22] W. Stoeckenius, R. H. Lozier, and R. A. Bogomolni, *Biochim. Biophys. Acta* **505,** 215 (1979).

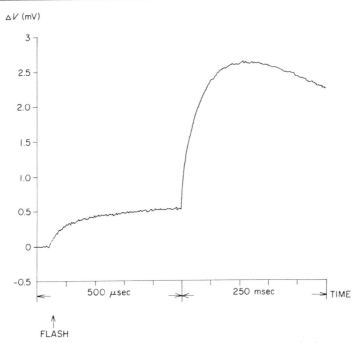

FIG. 2. Single turnover photoelectric response of bacteriorhodopsin in the system "bacteriorhodopsin proteoliposomes–black asolectin membrane." Bathing solution contained 0.05 M Tris–HCl (pH 7.1) and 0.03 M CaCl$_2$.

In Fig. 3 the photoelectric response of animal rhodopsin is shown. Photoreceptor disks were incorporated into the collodion film in the dark. Then the laser flash was given. One can see the formation of a photopotential that closely resembles that of bacteriorhodopsin.[23,24] Again, the first electrogenic phase is small and oppositely directed, as compared with the slow phase. As in bacteriorhodospin, the computer analysis revealed two phases contributing to the main photoelectric response, one developing in the microsecond and another in the millisecond time scale. However, in animal rhodopsin the microsecond phase was about 10 times slower than in bacteriorhodopsin. In both rhodopsins the amplitudes of the phases increase from the first to the third phase. The direction of the photopotential proves to be the same. It corresponds to the transfer of a positive charge from the membrane surface where the C-end of the rhodopsin polypeptide chain was exposed, to the surface where its N-end is

[23] V. I. Bolshakov, A. L. Drachev, L. A. Drachev, G. R. Kalamkarov, A. D. Kaulen, M. A. Ostrovsky, and V. P. Skulachev, *Dokl. Akad. Nauk SSSR* **249**, 1462 (1979).
[24] L. A. Drachev, G. R. Kalamkarov, A. D. Kaulen, M. A. Ostrovsky, and V. P. Skulachev, *FEBS Lett.* **119**, 125 (1980).

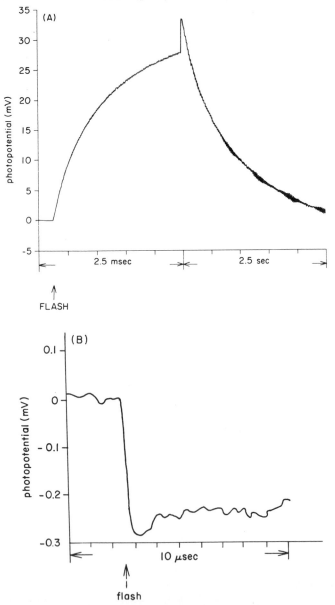

FIG. 3. Single turnover photoelectric response of animal rhodopsin in the system "photoreceptor discs–phospholipid impregnated collodion film." Bathing solution: 0.1 M NaCl and 5 mM MES–KOH (pH 6.0).

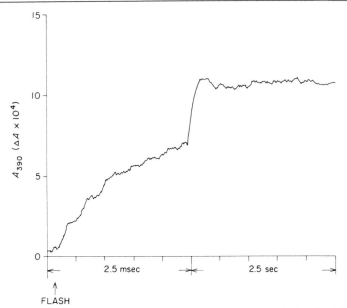

FIG. 4. Spectral response of photoreceptor discs to 15-nsec laser flash. The 390-nm light absorbance was measured. Incubation mixture: 0.1 M NaCl and 5 mM MES–KOH (pH 6.0).

found. Discharge of the photopotential in both cases takes seconds if the systems were exposed to the first flash. Repeated flashes induce strong acceleration of the decay and decrease in the amplitude of the photopotential (not shown; see Bolshakov et al.[23] and Drachev et al.[24]).

In Fig. 4 a spectral response of animal rhodopsin is shown. Comparison of electrical and optical measurements indicates that there is a correlation between the main electrogenic phases and rhodopsin bleaching (metarhodopsin I → metharhodopsin II transition).

Figure 5 demonstrates the photoelectric response of Rh. rubrum chromatophores incorporated into a collodion film. Again, as in rhodopsins, a single turnover of the photogenerator (in this case, photosynthetic reaction center complexes) was found to result in formation of significant electric potential. However, the parameters of the photoresponse proved to be different as compared with rhodopsins. One can see that the first electrogenic phase is the major one, while the second (τ about 200 μsec) is rather small.[16] One more small phase (τ about 20 μsec) can be revealed when we use Chromatium minutissimum chromatophores retaining their cytochrome c pool (not shown; see Drachev et al.[25]). All the phases were unidirectional, which indicated the positive charging of the chromato-

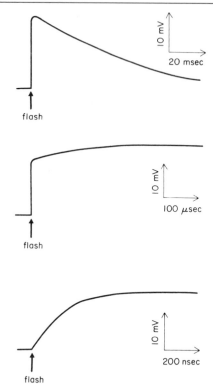

Fig. 5. Single turnover photoelectric response of photosynthetic reaction centers in the system "*Rh. rubrum* chromatophores–phospholipid-impregnated collodion film." Bathing solution: 0.05 M Tris–Hcl (pH 7.3), 0.04 M MgSO$_4$, 2 × 10^{-4} M phenazine methosulfate, 4 × 10^{-4} M vitamin K$_3$, 4 mM sodium ascorbate. Asolectin was used to impregnate the film.

phore-free side of the film. The opposite direction of the electric field was found in the experiments with rhodopsins. The decay of the chromato-phore-generated photopotential was of complex kinetics, the main portion developing in the millisecond time scale.[25] Spectral measurement showed that the fast electrogenic phase is due to electron transfer from bac-teriochlorophyll to the Fe-Q component in photosynthetic reaction center complexes. The fast phase of the photopotential decay is associated with the return of electrons from Fe-Q to bacteriochlorophyll. The 20-μsec phase corresponds to the reduction of oxidized bacteriochlorophyll by cy-tochrome c_2, and the 200-μsec phase, with the reduction and/or protona-tion of the secondary quinone ("free" CoQ) (not shown; see Drachev *et al.*[25]).

[25] L. A. Drachev, A. Yu. Semenov, V. P. Skulachev, I. A. Smirnova, S. K. Chamorovsky, A. A. Kononenko, A. B. Rubin, and N. Ya. Uspenskaya, *Eur. J. Biochem.* **117**, 483 (1981).

The similarity of fast electrogenic responses of two rhodopsins and their apparent difference from that of photosynthetic reaction centers, were regarded as indicating the bacteriorhodopsin and animal rhodopsin can have a common biological function. This may be a light-dependent charge separation. However, the electric field produced by the animal rhodopsin was assumed to induce a local electric breakdown of the membrane to release Ca^{2+} from the disk interior, whereas the bacteriorhodopsin-generated electric field is used to support various $\Delta\bar{\mu}H$-dependent types of work.[23,24]

In conclusion, the use of the preceding fast techniques allows essential parameters of change translocation in photoelectric generators to be monitored. This gives important information about the mechanism and function of such generators.

[8] The Assembly of Bacteriorhodopsin-Containing Planar Membranes by the Sequential Transfer of Air–Water Interface Films

By JUAN I. KORENBROT

Introduction

Planar lipid bilayer membranes accessible to electrical measurements and to control of the ionic milieu bathing them can now be formed with relative ease.[1-3] These membranes have provided a unique experimental tool to understand the mechanisms of ion membrane transport mediated by model ionophores.[4-6]

To understand the mechanisms of ion transport in biological membranes, techniques have been developed to form similar planar membranes consisting not only of lipid but also of proteins that are presumed to be involved in transport and are obtained from the biological membranes.[7,8] In addition, formation of such model membranes allows direct experimental study of the quantitative features of membrane transport

[1] P. Mueller, D. O. Rudin, H. T. Tien, and W. C. Wescott, *Nature* (*London*) **194**, 979 (1962).
[2] A. Finkelstein, this series. Vol 32, Part B, p. 489.
[3] M. Montal, this series, Vol. 32, Part B, p. 545.
[4] D. Haydon and S. Hladky, *Q. Rev. Biophys.* **5**, 187 (1972).
[5] G. Eisenman, "Membranes, A Series of Advances." Dekker, New York, 1974.
[6] S. A. McLaughlin and M. Eisenberg, *Annu. Rev. Biophys. Bioeng.* **4**, 335 (1975).
[7] M. Montal, *Annu. Rev. Biophys. Bioeng.* **5**, 119 (1976).
[8] J. I. Korenbrot, *Annu. Rev. Physiol.* **39**, 19 (1977).

METHODS IN ENZYMOLOGY, VOL. 88

and of the possible regulatory effects of transmembrane chemical and electrical gradients. Bacteriorhodopsin is one of the membrane-bound proteins that have been successfully incorporated into planar model membranes. This success arises in part from the fact that the model membranes have been formed by the incorporation not of molecularly dispersed chromoproteins, but rather of purple membrane fragments. Planar lipid films containing purple membrane fragments were first formed by Drachev et al.,[9] who "painted" thick films by applying onto a hydrophobic aperture a suspension of purple membrane fragments and lipid in decane. Drachev et al.[10] found that these films could be more efficiently formed by adsorbing onto a preformed thick lipid film either isolated purple membrane fragments or liposomes into which purple membrane fragments were first incorporated (proteoliposomes). Shieh and Packer[11] further improved the method by reinforcing the preformed thick lipid film with a polymer, thus prolonging its lifetime. The ultimate in stability has been obtained by adsorbing purple membrane fragments onto thick lipid films formed by lipid impregnation of solid supports such as millipore filter membranes[12] or collodion membranes.[13] Dancshazy and Karvaly[14] first described the adsorption of purple membrane fragments onto thin lipid films. All these model systems have been extremely useful and have provided important information on the function of bacteriorhodopsin; they have, however, important limitations in their analytical use. The available methods do not allow control of the amount or orientation of bacteriorhodopsin in the model membrane. In addition, adsorption of purple membrane fragments or proteoliposomes onto a preformed bilayer results in a physical system best described as two membranes, purple and lipid bilayer, in series.[10] The two membranes form a sandwich-like structure in which each maintains its characteristic impedance.[15,16] Thus, the purple membrane is accessible to the aqueous phase on only one of its surfaces. An alternative experimental method that permits the formation of planar films of large and variable area containing known amounts of oriented, crystalline bacteriorhodopsin is described here. These films are

[9] L. A. Drachev, A. D. Kaulen, S. A. Ostrumov, and V. P. Skulachev, FEBS Lett. 39, 43 (1974).

[10] L. A. Drachev, V. N. Frolov, A. D. Kaulen, E. A. Liberman, S. A. Ostrumov, U. G. Plakunova, A. Y. Semenou, and V. P. Skulachev, J. Biol. Chem. 251, 7059 (1976).

[11] P. Shieh and L. Packer, Biochem. Biophys. Res. Commun. 71, 603 (1976).

[12] M. C. Blok and K. Van Dam, Biochim. Biophys. Acta 507, 48 (1978).

[13] L. A. Drachev, A. D. Kaulen, and V. P. Skulachev, FEBS Lett. 87, 161 (1978).

[14] Z. Dancshazy and B. Karvaly, FEBS Lett. 72, 136 (1976).

[15] T. R. Herrman and G. W. Rayfield, Biophys. J. 21, 111 (1978).

[16] E. Bamberg, H.-J. Apell, N. Dencher, W. Sperling, H. Stieve, and P. Lauger, Biophys. Struct. Mech. 5, 277 (1979).

easily accessible to electrical and spectrophotometric measurements and to the complete control of the ionic milieu on either side of the membrane. The method described here consists of the assembly of a membrane through the sequential transfer of two air–water interface films onto a hydrophilic, electrically conductive solid support. The first interface film consists of oriented, single-sheet fragments of purple membrane randomly distributed over the water surface and separated by a lipid monolayer. The second film consists only of lipids. The following sections include first a description of the formation of the interface film and next a description of the procedure by which the interface films are assembled into planar membranes.

Formation of Air–Water Interface Films

Instrumentation. The physical and electrical properties of insoluble films formed at the air–water interface are classically studied in instruments generally referred to as surface balances. These balances consist of a shallow trough, which contains the aqueous phase, and a surface barrier. The surface barrier rests across the edges of the trough and is used to compress the surface film by reducing the effective aqueous surface. Excellent monographs describing experimental techniques in surface chemistry have been published by Gaines[17] and by Kuhn et al.[18] and should be consulted for complete details. In our instrument we use a rectangular trough (outside dimensions 40 × 12.5 × 2 cm, inside dimensions 36.5 × 10.5 × 0.8 cm), milled from a solid piece of Teflon (polytetrafluoroethylene), and a tight-fitting barrier (Fig. 1). At one end of the trough there is a cylindrical well (radius 2.25 cm, depth 3 cm). To permit illumination, a glass sealed port hole (radius 0.75 cm) is machined on the wall of the well. The tight-fitting barrier is held in a carriage and is driven along the trough by the motion of the carriage on a threaded rod extending the length of the trough. The rod is connected to a digitally controlled servomotor. The servo system (M. Fong Electronics, Oakland, CA) is designed so that the position of the surface barrier on the trough is controlled by the surface pressure. Thus, the surface barrier moves quickly (time constant of response <100 msec) and automatically as required to maintain the surface pressure of an interface film within ±0.05 mN/m of a predesignated value. Although constant surface pressure must be maintained during

[17] G. L. Gaines, "Insoluble Monolayers at Liquid–Gas Interfaces." Wiley (Interscience), New York, 1966.
[18] K. Kuhn, D. Mobius, and H. Bucher, *in* "Physical Methods of Chemistry" (A. Weissberger and B. W. Rossiter, eds.), Vol. 1 Part 3B, p. 577. Wiley (Interscience), New York, 1972.

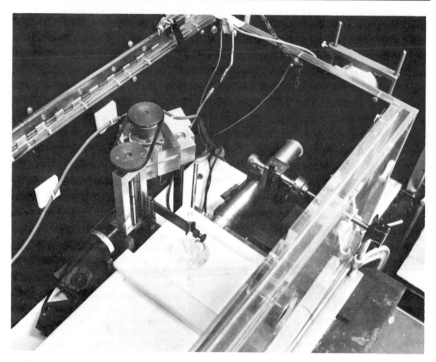

Fig. 1. Photograph of the surface balance within the electrically shielded enclosure. The surface barrier is in place. The support glass chamber is positioned within the well in the trough. On the wall of this well is a glass-sealed porthole that allows illumination of the assembled membranes. The glass chamber is held vertically by means of an arm in a device that can smoothly raise or lower the chamber.

transfer of the film onto a solid support, the instrumentation need not be as elaborate as that described here. Constant surface pressure can also be maintained in simple surface balances with the use of piston oils and flexible surface barriers.[19,20]

The surface balance is completely enclosed and is supported on a mechanical stage designed for vibration isolation. This is very important to obtain stable membranes. The enclosure provides electrical shielding and minimizes contamination of the aqueous surface. In this method, as in all procedures in surface sciences, attention must be paid to maintaining scrupulous cleanliness. The trough and all glassware must be frequently cleaned with strong oxidizing agents. All organic solvents must be redis-

[19] N. K. Adam, *Proc. R. Soc. London, Ser. B* **122,** 134 (1937).
[20] C. Y. C. Pak and J. D. Arnold, *J. Colloid Sci.* **16,** 513 (1961).

tilled and the water used must be of very high purity, as obtained, for example, by double distillation in the presence of oxidizing agents. To measure surface pressure, we have used the Wilhelmy method.[17] The electronic instrumentation necessary to record photocurrents and photovoltages generated across the membranes has been described in other volumes of this series[2] and will not be discussed here.

Spreading the Interface Films. The interface films consisting of purple membrane fragments and lipid are formed by spreading on a clean water surface a suspension of purple membrane fragments in a solution of purified soya phosphatidylcholine (soya PC) in hexane. Purple membrane fragments are isolated and purified from the plasma membrane of *Halobacterium halobium* R_1[21] and stored at 4° in basal salt solution (4.3 M NaCl 0.0275 M KCl, 0.080 M MgSO$_4$, 0.009 M Na citrate, pH 7.0). Soya PC can be obtained from commercial sources and should be purified.[22] The hexane suspension of purple membrane is prepared as follows: Purple membrane fragments in the concentration range of 0.16–0.8 mg/ml of bacteriorhodopsin (0.38–1.93 O.D. at 570 mm taking ϵ = 63,000 and MW = 26,000) are suspended in 2 ml basal salt and sonicated for 30 sec in a bath sonicator (Model G-80-80 Laboratory Supply Co., Hicksville, N.Y.). Then 0.5 ml of a solution of 0.28 mg/ml of soya PC in freshly redistilled hexane is layered on top of the sonicated suspension. This two-phase system is vigorously mixed with a vortex mixer (10 sec) and is immediately centrifuged at 1500 g for 10 min. After centrifugation, three phases are distinguished: a lower, clear aqueous phase; a middle, narrow turbid phase containing the purple membrane fragments and probably some water suspended in hexane; and an upper, clear hexane phase. The aqueous phase is carefully and completely removed with a Pasteur pipette. To the remaining phases, 2.2 ml of hexane are added. The production of the membrane suspension as described offers no particular difficulty and is extremely reproducible. The suspension can be prepared up to 2 hr before forming the interface film, and a fresh suspension should be prepared for each interface film formed. Immediately before forming the interface film, the suspension of purple membrane in hexane is sonicated for 60 sec in the bath sonicator. To form the film, 100 μl of the sonicated suspension is applied with a microsyringe onto the clean water surface. The weight ratio of bacteriorhodopsin to PC in the hexane phase determines the physical properties of the interface film. In the range of weight ratios prescribed here, the interface films consist of nonoverlapping, single-sheet fragments of purple membrane separated by a lipid monolayer.

[21] D. Oesterhelt and W. Stoeckenius, this series, Vol. 31, Part A, p. 667.
[22] D. J. Hanahan, J. C. Dittmer, and E. Warashina, *J. Biol. Chem.* **228,** 685 (1957).

The membrane fragments are typically 0.1 μm in diameter and are randomly distributed over the aqueous surface.[23] In films formed from a suspension of 7:1 bacteriorhodopsin: soya PC (weight ratio), the concentration most frequently used, the fragments occupy about 35% of the aqueous surface at collapse surface pressure (47 mN/m), and the optical absorbance of the film at 570 nm is about 4 × 10⁻⁴ O.D. Approximately 85% of the fragments are oriented in the same direction.[23] The bacteriorhodopsin molecules in the interface films are spectrophotometrically intact and fully functional.[24,25] The second interface film (lipid film) is spread from a solution in hexane containing soya PC (0.5 mg/ml) and purified hexadecane (32:1 mole ratio hexadecane: soya PC).

Assembly of Planar Membrane from the Interface Films

Preparation of Support Glass Chambers. To assemble planar membranes the interface films are transferred from the water surface onto an electrically conductive, hydrophilic support. This support consists of a thin film casted from nitrocellulose. The nitrocellulose film is itself mechanically supported on a polyester film rigidly attached to a glass chamber. The glass chambers used are illustrated in Fig. 2.(They can be obtained from the glassblowing shop of VWR, San Francisco, Ca.) The chambers consist of an open cylinder (1.5 cm I.D. × 1.7 cm O.D., 2.5 cm long) with flared ends (outside diameter of the flare 2.5 cm). Three glass tubes (3 mm I.D. × 5 mm O.D.) are fused to the cylinder, a central one (7 cm long) and two lateral ones (2 cm), each forming an angle of approximately 45° with the central one. To ready the chambers for the assembly of membranes, they are cleaned by successively immersing them in a solution of detergent and in a warm solution of chromic acid. They are then exhaustively rinsed with water and dried in an oven. A disk of polyester film (diameter 13 mm, 127 μm thick, obtained from Transylwrap Corp., San Francisco, CA) is then glued onto one end of the glass chamber with the use of a quick-setting epoxy mixture ("5 Minute" epoxy from Devcon Corp., Danvers, MA). This polyester disk has a hole in its center. Two size holes have been used: 2.3 ± 0.1 × 10⁻² cm² made by slowly applying to the polyester film an electrically heated platinum tip (diameter at the tip, 0.3 mm) or 0.7 cm² made with the use of a metal punch. The polyester disk is cleaned with detergent and rinsed extensively with water before it is attached to the glass chamber. The epoxy is allowed to dry for 2–4 hr.

[23] S.-B. Hwang, J. I. Korenbrot, and W. Stoeckenius, *J. Membr. Biol.* **36**, 115 (1977).
[24] S.-B. Hwang, J. I. Korenbrot, and W. Stoeckenius, *J. Membr. Biol.* **36**, 137 (1977).
[25] S.-B. Hwang, J. I. Korenbrot, and W. Stoeckenius, *Biochim. Biophys. Acta* **509**, 300 (1978).

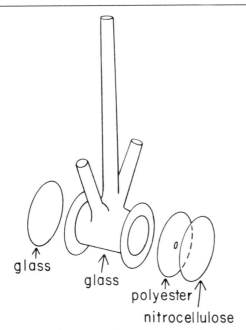

glass

glass

polyester

nitrocellulose

FIG. 2. Schematic drawing of the glass chamber used in the membrane assembly. The glass chamber consists of an open cylindrical body. One end of the cylinder is closed by a glass disk, while the other end sustains a polyester disk that has an aperture in its center. The polyester disk serves as mechanical support of the thin nitrocellulose film onto which the model membranes are assembled. The open vertical arms of the glass chamber accommodate electrodes and allow the exchange of the solution contained in the chamber.

A thin nitrocellulose film is then deposited on the face of the polyester film across the hole (see below). After 1–3 hr drying time, a thoroughly cleaned glass disk (2.5 cm diameter, 0.18–0.22 mm thick, Thomas Scientific Co., Philadelphia, PA) is glued onto the remaining open end of the glass chamber. All these preparative procedures must be carried out with particular attention to preserve the cleanliness of the glass chambers. Gloves should be worn and the chambers should be kept in closed containers during the various waiting periods. This is important, because interface films will not transfer onto contaminated surfaces.

Casting and Deposition of Nitrocellulose Film. Nitrocellulose is a nitrated derivative of cellulose that contains 10.5–12% nitrogen (approximately two nitrates per glucose) (purified Parlodion strips, Mallinkrodt Chem. Co.). A 1% (w/v) solution of purified nitrocellulose in isoamyl acetate is prepared by stirring overnight at room temperature in a closed glass container. This solution should be used within 1 week of its preparation and then discarded. Thin films are prepared by either of the two following procedures, depending on the electrical features desired in the film. The first procedure is based on the method originally described by Carnell and Cassidy:[26,27] A glass disk (2.5 cm diameter × 0.18–0.22 mm thick) is thoroughly cleaned, rinsed, and dried under a stream of nitrogen. The solution of nitrocellulose (50 ml) is placed in a clean glass vessel and the glass slide, being held perfectly vertical, is mechanically dipped into the solution at a rate of 0.16 cm/sec. The slide is held under the solution for 2 min and is then withdrawn at a rate of 0.077 cm/sec. The polymer film is allowed to dry, away from the vapors of the polymer solution, for precisely 90 sec. At the end of that time, the nitrocellulose film is floated off the glass slide by manually dipping the slide, while holding it at about 45°, into a glass vessel filled to the brim with water. The nitrocellulose film now floating on the water surface is transferred across the face of the polyester fixed on the glass chamber by manually inserting the chamber into the aqueous bulk and then withdrawing it under the film at about a 60° angle. One to 2 hr later, any uncovered polyester surface is covered with a thin coat of polymer solution applied with a disposable glass capillary. The second film casting procedure is based on the method of Lackshminarayanaiah and Shanes[28,29]: The glass chambers are immersed in 400 ml of a 0.1 N NaCl aqueous solution contained in a glass vessel. The glass vessel is covered, and through an aperture in the cover, sufficient volume of 1% polymer solution is slowly applied onto the aqueous surface to produce a 360-Å-thick film (density of nitrocellulose is 1.66 g/cm³). Thirty minutes is allowed for solvent disappearance and formation of the polymeric film. At the end of that time the glass chamber is manually withdrawn across the aqueous surface at a 60° angle. Any polymer film covering nonpolyester surfaces is carefully wiped off. Again, any uncovered polyester surface is covered 1–2 hr later. These procedures are carried out at room temperature (19–21°), in an air-conditioned room and in the absence of any air current. Some of the physical features of the nitrocellulose films thus produced are listed in the accompanying table. These films can be obtained with great reproducibility; however, to do so the method of production described must be strictly observed. There are a large number of experimental parameters that can define and alter the physical characteristics of the nitrocellulose films.[27] We have explored many different protocols to produce the nitrocellulose films. The procedure described is the one we have found to be the most reliable and consistent. Membrane

[26] P. H. Carnell, *J. Appl. Polym. Sci.* **9**, 1863 (1965).
[27] P. H. Carnell and H. G. Cassidy, *J. Polym. Sci.* **155**, 233 (1961).
[28] N. Lackshminarayanaiah and A. M. Shanes, *J. Appl. Polym. Sci.* **9**, 689 (1965).
[29] N. Lackshminarayanaiah, *J. Appl. Polym. Sci.* **9**, 689 (1965).

PHYSICAL PROPERTIES OF NITROCELLULOSE SUPPORT FILM

Property	Value
Thickness	360 ± 10 Å
Capacitance	$0.156\ \mu F/cm^2$
Dielectric constant	6.7
Resistance (in 50 mM NaCl)	
Carnell–Cassidy	$2 \pm 3 \times 10^3\ \Omega\ cm^2$
Lackshminarayanaiah–Shanes	$8 \pm 5 \times 10^2\ \Omega\ cm^2$

Transfer numbers	(in nitrocellulose films)	(in aqueous solution)
NaCl:		
t_{Na}	0.409 ± 0.028	0.391
t_{Cl}	0.590 ± 0.029	0.609
KCl:		
t_K	0.498 ± 0.002	0.609
t_{Cl}	0.501 ± 0.002	0.510
HCl:		
t_H	0.817 ± 0.013	
t_{Cl}	0.182 ± 0.013	

assembly is equally successful on polymer films casted by either of the techniques described. However, the films casted by the Lacksminarayan-aiah–Shanes technique are somewhat more porous and should be used when the effects of varying salt concentration inside the glass chamber are to be tested. The completely assembled glass chambers are used no sooner than 12 hr and no later than 3 days after being completed. Assembled glass chambers can be used only once, and must then be thoroughly cleaned and reassembled. The epoxy glue can be removed by immersion in boiling glycerol.

Membrane Assembly. All procedures are carried out at room temperature (19–21°). Immediately before use, the glass chambers are cleaned by gently dripping redistilled hexane over their surface. Hexane is allowed to evaporate (10 min) and the chambers, with their inside dry, are then immersed under a clean surface in the aqueous subphase contained in the well of the surface balance. The chambers are held vertically in the well. The long, central glass stem of the chamber is clamped onto a mechanical device that can smoothly raise and lower the chamber across the aqueous surface. Immediately after immersing the glass chamber, the purple membrane–lipid interface film is spread by applying onto the clean water surface 100 μl of the purple membrane lipid suspension in hexane. The film is brought to a surface pressure of 30 mN/m. While holding that pressure with the use of the automatic servomechanism, the glass chamber is withdrawn across the aqueous surface at a rate of 0.015 cm/min. As the surface film transfers onto the surface of the glass chamber, the surface barrier advances to maintain a constant surface pressure and the surface area is, therefore, reduced. The efficiency of film transfer is measured by the loss of surface area. The transfer ratio is defined as the ratio of the area of surface film lost to the geometrical surface area of the substrate onto which the film is transferred. The transfer ratio of the purple membrane interface film onto the glass chamber must be about 1 for successful membrane assembly. If the transfer ratio is not 1, the procedure must be stopped and started again with a different chamber and a fresh membrane suspension. The most common cause of failure in transfer is contamination: to correct it, the Teflon trough should be completely cleaned and a fresh aqueous subphase prepared. After the glass chamber is withdrawn, it remains suspended above the aqueous surface while the surface is thoroughly cleaned. These operations take 4–6 min. The surface barrier is returned to its starting point and 40 μl of the lipid solution in hexane is then spread. The lipid interface film is immediately brought to and maintained at a surface pressure of 30 mN/m. Following a 9- to 10-min period, sufficient time to allow excess hexadecane to disappear from the interface in order to obtain a mixed lipid film,[17,30,31] the glass chamber is transported downward through the interface film. Film transfer is carried out at the same speed as the upward motion and is again monitored through the loss of surface area.

The motion of the glass chamber is stopped when the hole in the polyester film is lined up with the glass porthole used for illumination. The transfer ratio should be 0.6 ± 0.1. The lower ratio most probably reflects the fact that about 35% of the surface in the first film consists of purple membrane fragment and most, therefore, by hydrophilic and not expected to interact with the hydrophobic surface of the second film. Again, if this transfer is not adequate the membrane, assembly will fail. In our experience the second transfer is inadequate in approximately 5% of the experiments. Failure in the transfer is corrected by carefully cleaning the trough and starting anew. After the motion of the glass chamber has been stopped, the chamber is gently filled with aqueous solution through one of the diagonal side arms while an electrode is inserted in the other one. The membrane is now assembled and the electrical measurements of light-induced membrane currents and voltages can commence.[31]

The stability of the transferred interface films must depend not only on the maintenance of strong lateral interactions between the molecular components of the film, but also on the strength of the bonds between these

[30] G. L. Gaines, *J. Phys. Chem.* **65**, 382 (1961).
[31] J. I. Korenbrot and S.-B. Hwang, *J. Gen. Physiol.* **76**, 649–682 (1980).

components and the surface of the solid.[18] Since the surface of the purple membrane–lipid interface film contains negative charges, as does the nitrocellulose, the aqueous subphase used in the membrane assembly procedures in our experiments typically contains $4 \times 10^{-4}\ M$ $CdCl_2$. The monovalent salt composition of the subphase may be varied as desired to accommodate different experimental protocols. However, the effects of the subphase composition on the transfer ratio must be tested, and experiments should be done only when the transfer ratio meets their expected values. In our experience so far, we have tested subphases containing salt only up to 50 mM ionic strength.

The assembled membranes are stable for several hours (4–6). They are, however, easily destroyed by mechanical shock. Therefore, careful attention must be paid to proper vibration isolation of the surface balance instrumentation, and manipulations of the glass chamber after completing membrane assembly should be minimized.

The assembled membranes are accessible to the aqueous phase on both of its surfaces, since the polymeric thin support is sufficiently porous to allow the rapid and complete exchange of solutions. Nonetheless, a limitation of the assembly method is that the solid support is not simply an aqueous phase. Therefore, any quantitative analysis of the data obtained in the experimental membranes must consider the electrical features of the support itself.[31] A special advantage of the assembly of large and stable membranes containing bacteriorhodopsin is that it becomes possible to correlate the measurement of electrical current with a measurement of the flux of specific ions. In addition, it is also possible to correlate the transport mechanism of bacteriorhodopsin with its photoinduced spectroscopic changes.

Acknowledgments

Juan I. Korenbrot is an Alfred P. Sloan Research Fellow. This research was supported by NIH Grants EY01586 and EY00050 and a grant from the Office of Water Research and Technology, Department of the Interior.

[9] Primary Structure of Bacteriorhodopsin: Sequencing Methods for Membrane Proteins

By Gerhard E. Gerber* and H. Gobind Khorana†

Methodology for the determination of amino acid sequences of integral membrane proteins is only now beginning to be developed. The straightforward application of the procedures and methods established for soluble proteins is not generally feasible for work with membrane proteins because of their predominantly hydrophobic character. This latter fact, which causes aggregation and water insolubility, renders difficult or impractical the fragmentation by proteolysis and separation of the fragments. Until recently, the coat protein (59 amino acids) of the filamentous virus M-13 and the sialoglycoprotein, glycophorin (131 amino acids), both of which contain only very short (15–20 amino acids) hydrophobic regions, were the only two proteins that had been sequenced. More recently, progress in sequencing is increasing and the sequences of cytochrome b_5, of a number of the dicyclohexylcarbodiimide-binding subunits of mitochondrial ATPases, and those of a number of subunits of cytochrome oxidase have been reported. Very recently, the total amino acid sequence of the purple membrane protein, bacteriorhodopsin, of *Halobacterium halobium* has been determined in two laboratories,[1,2] while partial sequences of this protein have been reported from additional laboratories.[3-5] The present article is based on the primary structure work reported by Gerber and Khorana and their colleagues.[2,6] During this work new methods were developed for separation and sequencing, and these are discussed in relation to the general problem of sequencing integral membrane proteins.

General Principles: Isolation of Bacteriorhodopsin Fragments and Their Sequencing

The fact that bacteriorhodopsin is largely embedded in the bilayer[7] provides an opportunity to look for specific cleavages by proteolytic en-

[1] Yu. A. Ovchinnikov, N. G. Abdulaev, M. Yu. Feigina, A. V. Kiselev, and N. A. Lobanov, *FEBS Lett.* **100**, 219 (1979).

[2] H. G. Khorana, G. E. Gerber, W. C. Herlihy, C. P. Gray, R. J. Anderegg, K. Nihei, and K. Biemann, *Proc. Natl. Acad. Sci. U.S.A.* **76**, 5046 (1979).

[3] J. E. Walker, A. F. Carne, and H. W. Schmitt, *Nature (London)* **278**, 653 (1979).

[4] L. M. Keefer and R. A. Bradshaw, *Fed. Proc., Fed. Am. Soc. Exp. Biol.* **36**, 1799 (1977).

[5] J. Bridgen and I. D. Walker, *Biochemistry* **15**, 792 (1976).

METHODS IN ENZYMOLOGY, VOL. 88

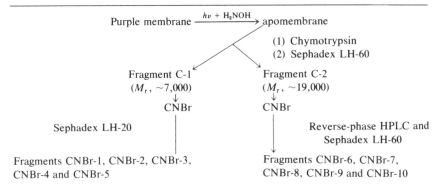

SCHEME 1. Enzymatic and chemical steps used in the degradation of bacteriorhodopsin and methods for the separation of the fragments.

zymes in the "loops" protruding out of the bilayer. However, the purple membrane sheets undergo only a single cleavage near the carboxyl terminus.[8] A second cleavage, which normally is very sluggish, or nonexistent, is facilitated by prior cleavage of the retinal Schiff base by illumination in the presence of hydroxylamine.[9] The resulting apomembrane on digestion with chymotrypsin yielded two clean fragments, C-1 and C-2 (Scheme 1). C-1, the large fragment, was readily shown to contain the original carboxyl terminus, while C-2 originated from the amino terminus. Degradation of the fragments with cyanogen bromide at methionine sites yielded a total of 11 fragments, consistent with the number of methionine residues present in the protein. Separation of the short-sized fragments that resulted from C-2 was achieved by gel permeation in organic solvents using Sephadex LH-20. The corresponding problem of the separation of C-1 fragments containing larger and more hydrophobic fragments was solved by a combination of reverse-phase HPLC[10] and gel permeation using Sephadex LH-60 in organic solvents. The two fragments C-1 and C-2 were overlapped by HPLC isolation of the appropriate CNBr frag-

[6] G. E. Gerber, R. J. Anderegg, W. C. Herlihy, C. P. Gray, K. Biemann, and H. G. Khorana, *Proc. Natl. Acad. Sci. U.S.A.* **76**, 227 (1979).
[7] W. Stoeckenius, R. H. Lozier, and R. A. Bogomolni, *Biochim. Biophys. Acta* **505**, 215 (1979).
[8] G. E. Gerber, C. P. Gray, D. Wildenauer, and H. G. Khorana, *Proc. Natl. Acad. Sci. U.S.A.* **74**, 5426 (1977).
[9] D. Oesterhelt, L. Schuhmann, and H. Gruber, *FEBS Lett.* **44**, 257 (1974).
[10] *Abbreviations:* GCMS, gas chromatographic mass spectrometry; HPLC, high-pressure liquid chromatography; NBS, *N*-bromosuccinimide; SPITC, 4-sulfophenylisothiocyanate; IPTAP, TETA, and AEAP glass, isothiocyanatophenylthiocarbamoylaminopropyl, triethylenetetramino, and aminoethylaminopropyl glass, respectively.

ment from the whole protein. Further, degradation with N-bromosuccinimide was used to obtain additional fragments from the protein.

Amino acid sequencing of the fragments was performed by a combination of stepwise (Edman) degradation methods and gas chromatographic mass spectrometry (GCMS). Edman degradations were carried out both in solution phase (Beckman sequencer) and in solid phase after attachment of the peptides to a derivatized glass support. Some fragments could not be completely sequenced directly and further enzymatic cleavages were required. These fragments were derivatized with hydrophilic reagents such as SPITC to render them soluble under the appropriate aqueous conditions.

GCMS provided an independent method, and this was carried out in collaboration with Drs. K. Biemann, R. Anderegg, and W. Herlihy, and the methods used were as previously developed by this group.[11]

Peptide mixtures for GCMS sequencing were generated by partial acidic hydrolysis or digestion with relatively nonspecific enzymes. Acid hydrolysis is well suited for membrane protein fragments because it can be used for both soluble and insoluble fragments. The data, in general, were abundant for hydrophobic regions, with tetra- and pentapeptides being produced in high yield. Enzymatic digestion generally gave a set of peptides that were complementary to those obtained by acid hydrolysis. Because the CNBr fragments were insoluble in aqueous buffers, digestions were restricted to enzymes that can hydrolyze peptides in aqueous suspension, such as elastase, thermolysin, and proteinase K. The combined use of stepwise sequencer techniques and GCMS, two independent methods, ensured confidence in the results. Further, the use of GCMS simplified the problems of ordering of the various fragments, because it provided direct identification of some of the methionine-containing peptides overlapping the CNBr fragments.

Materials

Trypsin treated with L-(tosylamide-2-phenyl)ethyl chloromethyl ketone, clostripain, chymotrypsin, and elastase were from Worthington Biochemicals. Pyroglutamate aminopeptidase was from Boehringer Mannheim.

N-Bromosuccinimide (NBS), fluorescamine, and isothiocyanatophenylthiocarbamoylaminopropyl (IPTAP) glass were purchased from Pierce. Triethylenetetramino (TETA) and aminoethylaminopropyl (AEAP) glass were from Sequemat or Pierce. $LiAl^2H_4$ and $B_2{}^2H_6$ from

[11] H. Nau and K. Biemann, Anal. Biochem. 73, 139 (1976).

Alfa-Ventron (Danvers, MA), Sephadex LH-20 and LH-60 from Pharmacia, and μBondapak C_{18} columns from Waters Associates (Milford, MA).

Methods

General Methods

Preparation of Purple Membrane and Apomembrane. Halobacterium *halobium* cells were grown and the purple membrane was isolated by sucrose density gradient according to the published procedure.[12] The purple membrane was suspended (2 mg/ml) in 4.0 M NaCl containing 1.0 M $NH_2OH \cdot HCl$ (pH 7.0) and the stirred suspension was irradiated at 25° with a 500-W quartz halogen lamp with a Schott 530 filter until the purple color had completely disappeared. The apomembrane was collected by centrifugation, washed twice with distilled water, and used immediately for proteolysis.

Amino Acid Analysis. Amino acid compositions were determined with a Beckman 119C amino acid analyzer. Hydrolysis was performed for 24 to 96 hr in sealed, evacuated tubes in either 6 M HCl or 3 M mercaptoethanesulfonic acid at 110°. The sequences reported were all consistent with the amino acid compositions found for the fragments.

Electrophoresis of Polypeptide Fragments. Fragmentation was followed by polyacrylamide gel electrophoresis as described by Laemmli[13] and Swank and Munkres.[14] Aliquots of the peptides were dried, residues were suspended in 20 μl of 0.2 M sodium borate buffer at pH 9.0, and 20 μl of fluorescamine solution (1 mg/ml of acetone) was added. The samples were subjected to electrophoresis on a 12% acrylamide gel and the bands were visualized under ultraviolet light.

Fragmentation of the Protein

Chymotryptic Digestion of Apomembrane. The apomembrane was suspended at a concentration of 2 mg/ml in 50 mM Tris-HCl (pH 8) containing 5 mM $CaCl_2$ and incubated at 37° with chymotrypsin (wt ratio of enzyme to apomembrane, 1:100) for 5 hr. The membrane was collected by centrifugation (45 min, 50,000 g), washed twice with distilled water, and lyophilized. The apomembrane was cleaved by chymotrypsin to form two fragments, C-1 and C-2. C-1 (M_r 19,000) contained the COOH terminus of bacteriorhodopsin, while C-2 (M_r 6900) contained the blocked

---

[12] D. Oesterhelt and W. Stoeckenius, this series, Vol. 31, Part A, p. 667.
[13] U. K. Laemmli, *Nature* (*London*) **227**, 580 (1970).
[14] H. N. Rydon and P. W. G. Smith, *Nature* (*London*) **169**, 922 (1952).

NH$_2$ terminus. The fragments (C-1 and C-2) were separated on Sephadex LH-60 as described later.

CNBr Cleavage of C-1 and C-2. The fragment (2 μmol) was dissolved in 2 ml of 88% formic acid and diluted with water to an acid concentration of 70%. CNBr (50-fold molar excess over the methionine content of the fragment) was added in 70% formic acid. The solution was kept in the dark at room temperature for 24 hr. Excess reagents were removed under reduced pressure. Fragments formed from C-2 were separated by gel permeation, while the CNBr fragments of C-1 were isolated by gel filtration or reverse phase HPLC as described below. CNBr-1 and CNBr-2 could also be further resolved, if necessary, by HPLC. In some experiments bacteriorhodopsin obtained after delipidation of the purple membrane by gel permeation chromatography (see the following discussion) was directly subjected to fragmentation by CNBr. In this case the fragment CNBr-9 was replaced by CNBr-9a, which contained the additional NH$_2$-terminal amino acid sequence Val-Pro-Phe.

NBS Cleavage of C-1 The fragment (1.5 μmol) in 200 μl of 88% (v/v) formic acid was mixed with 8.0 ml of 8 M urea in 0.3 M acetic acid (pH 3.5), and to it was added a solution of NBS (133.5 mg; 100-fold excess over tryptophan) in 200 μl of the above urea–acetic acid solvent. After incubation at room temperature for 30 min, the fragments were separated by gel permeation chromatography and HPLC. For analysis, aliquots of column effluents were first derivatized with fluorescamine and then subjected to polyacrylamide gel electrophoresis.

Cleavage with Clostripain. 4-Sulfophenylisothiocyanate (SPITC) derivatives of CNBr-10 and CNBr-10a were prepared as discussed below for CNBr-2 and purified by chromatography on Sephadex LH-20. These fragments were also derivatized with succinic anhydride[15] to give water-soluble products. The derivatized fragments (100–200 nmol) were dissolved in 2.0 ml of 20 mM triethylamine, and 2.0 ml of 200 mM N-Tris(hydroxymethyl)methyl-2-amino-ethanesulfonic acid buffer (pH 7.6) containing 2 mM dithiothreitol and 2 mM CaCl$_2$ was added. Clostripain was added at an enzyme-to-substrate ratio of 1:100, and the mixture was incubated at 37° for 18 hr. The digest was then evaporated to dryness, the residue was dissolved in 88% formic acid, and the solution was chromatographed on Sephadex LH-20. The material emerging at the void volume was used for sequence determination.

Tryptic Digestion of CNBr-6. The peptide (200 nmol) was dissolved in 0.4 ml of 50 mM Tris-HCl, pH 8.0, containing 5 mM CaCl$_2$ and incubated with trypsin (enzyme/substrate, 1:100 by weight) at 37° for 5 hr. After

[15] I. M. Klotz, this series, Vol. 11, p. 576.

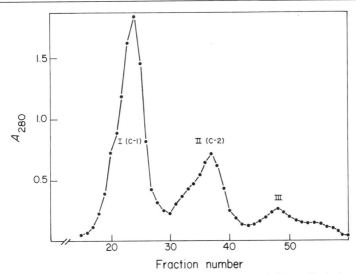

Fraction number

FIG. 1. Chromatography of chymotryptic fragments C-1 and C-2 on Sephadex LH-60. The lyophilized chymotryptic digest (see text) was dissolved in 3.0 ml of 88% formic acid, 7.0 ml of ethanol were added, and the solution was chromatographed on Sephadex LH-60 (details in text).

electrophoresis at pH 2.0, the peptide containing the COOH terminus of CNBr-6 was used for sequencer analysis.

Isolation of Protein Fragments

The membrane or its fragments were not soluble, even in 88% formic acid; however, after lyophilization the residue readily dissolved in this solvent. The addition of ethanol to a final concentration of 70% gave homogeneous solutions, while the addition of water to the formic acid solution resulted in precipitation. This solvent system was, therefore, used for the isolation of fragments by gel permeation as well as by reverse phase HPLC.

Gel Permeation Chromatography. This chromatography was carried out on columns (2.5 × 80 cm) of Sephadex LH-20 or LH-60 equilibrated in 88% HCOOH/ethanol (30:70). Lyophilized membrane or fragments were dissolved in 88% (v/v) formic acid, ethanol was added to an ethanol concentration of 70% by volume, and the solution was used for chromatography.

An excellent separation and recovery of C-1 and C-2 was obtained by using Sephadex LH-60 in 88% formic acid/ethanol (30:70) (Fig. 1). The third peak (III) had an amino acid composition identical to that of C-2 and

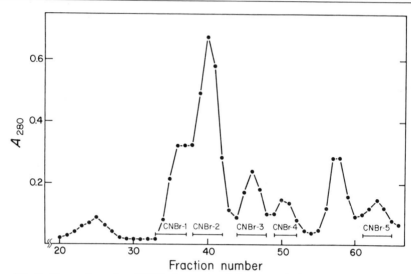

FIG. 2. Chromatography of CNBr fragments of chymotryptic fragment C-2 on Sephadex LH-20. The CNBr digest of C-2 (2 μmol) was dried and dissolved in 3.0 ml of 88% formic acid, and the solution was diluted with 7.0 ml of ethanol and chromatographed on Sephadex LH-20, as in text.

was obtained due to acidic cleavage of the Asp-36 to Pro-37 bond. Similarly, the CNBr fragments of C-2 all separated on Sephadex LH-20 (Fig. 2).

Although the separation was primarily on the basis of size, in one case aggregation was observed. Application of the total CNBr digest of C-1 on a Sephadex LH-60 column (2.5 × 80 cm) gave the pattern shown in Fig. 3A. Sodium dodecyl sulfate/polyacrylamide gel electrophoresis of the material in each of the peaks (Fig. 3B) showed that pool a (void volume) contained mainly a peptide traveling faster than the other CNBr fragments (CNBr-7, CNBr-8, and CNBr-9). From its composition this peptide proved to be CNBr-10, the fifth and smallest CNBr fragment. Its purification in monomeric form was achieved as follows: Pool a was dissolved in trifluoroacetic acid (0.5 ml) at 0°, the solution was diluted with 7.0 ml of ethanol, and the final volume was adjusted to 10.0 ml with 88% formic acid. This was immediately applied to the same LH-60 column. Polyacrylamide gel analysis (Fig. 3C) showed that the different fragments now eluted essentially according to their size. CNBr-10 thus prepared was used in sequence work.

Reverse-Phase HPLC of Fragments. The solvent formic acid/ethanol/water used for gel permeation chromatography was also useful for reverse-phase HPLC on μBondapak C_{18}. The system adopted utilized 5%

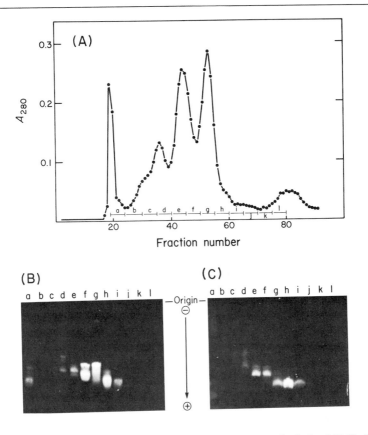

FIG. 3. (A) Chromatography of the CNBr fragments of C-1 on Sephadex LH-60. CNBr cleavage and chromatography were carried out as described in text. (B) Samples were pooled as shown (a–l) and were analyzed by polyacrylamide gel electrophoresis. (C) Pool was dried, redissolved in 0.5 ml of anhydrous trifluoroacetic acid, and diluted with 7.0 ml of ethanol and 2.5 ml of 88% formic acid. The sample was chromatographed on Sephadex LH-60, and pools were subjected to polyacrylamide gel electrophoresis.

formic acid in water as solvent A and 5% formic acid in ethanol as solvent B. Injection of the sample in 88% formic acid avoided problems of fragment solubility, provided the concentration of ethanol was sufficient at the beginning of the program (40%). The use of lower initial concentrations of ethanol at the start of the gradient resulted in poor recoveries of the hydrophobic fragments. This system was also able to separate large and very hydrophobic fragments, C-1 and C-2. The CNBr fragments derived from C-1 were completely separated by this system (Fig. 4). In

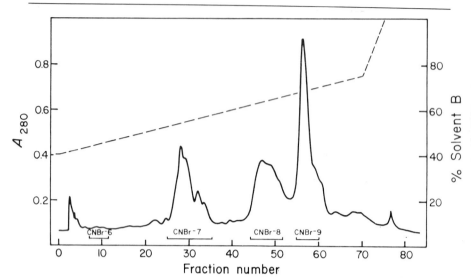

FIG. 4. HPLC of CNBr fragments of chymotryptic fragment C-1. The CNBr digest of C-1 was dried and dissolved in 1.5 ml of 88% formic acid, and the solution was injected into a column (0.8 × 30 cm) of μBondapak C_{18}. The column was developed as shown.

addition, the CNBr fragments from C-2 that were not completely resolved by Sephadex LH-20 chromatography (CNBr-1 and CNBr-2) were easily resolved. The solvent was easily removed, the peptides were conveniently monitored spectrophotometrically, and the total separation time was 1 hr.

The recovery of fragments from the column was generally good (75–85%). A serious exception to this was CNBr-10, which was not identified in any of the fractions obtained from the HPLC column. Presumably the aggregation discussed earlier is responsible for this effect.

Sequencing Methods

All fragments were subjected to analysis by automated Edman degradation and GCMS. The combination of these complementary methods provided the complete sequence of most fragments directly. Only four of the fragments required additional fragmentation for completion of their sequence.

Automated Edman sequencing. A Beckman 890C sequencer was used with 50–400 nmol of peptide. Large fragments (e.g., C-1) were run by using essentially the standard Beckman fast 1.0 *M* Quadrol program but including 0.2% sodium dodecyl sulfate in the Quadrol and with 2.0 mg of sodium dodecyl sulfate added to the sample.[16] Smaller fragments and peptides were applied in the presence of 2–4 mg of Polybrene[17] with a modified dilute Quadrol program using 0.2 M Quadrol and reduced separate benzene and ethyl acetate washes. The thiazolinones were converted either manually in 1 M HCl or with a Sequemat P-6 autoconverter in 1.5 M HCl in methanol. Phenylthiohydantoin amino acids were identified both by HPLC on μBondapak C_{18} in a methanol gradient (14–35%, v/v) in 0.01 M sodium acetate (pH 4.1)[18] and by amino acid analysis after hydrolysis (20 hr at 150° in 6 M HCl or 24–36 hr in 55% HI at 110°).

Attachment of CNBr Fragments to Glass Supports. (i) *CNBr-10 to IPTAP Glass:* The fragment (50 nmol) was lactonized and allowed to react with ethylenediamine.[6] Excess reagent was removed under reduced pressure and the residue was dissolved in 200 μl of dimethylformamide. To the solution was added 50 mg of IPTAP glass, and the mixture was shaken at 45° for 18 hr.[19] The IPTAP glass was washed three times with 1 ml of dimethylformamide and finally with 1 ml of 88% formic acid before the start of sequencer analysis. (ii) *CNBr-9a to TETA and AEAP Glass:* The fragment CNBr-9a was derivatized with [^{14}C]succinic anhydride and then treated with trifluoroacetic acid to form the homoserine lactone derivative. The latter was coupled to TETA and AEAP glass by following the published procedures.[20,21] Coupling of the peptide to the glass was monitored by measuring the radioactivity remaining in solution and in successive washes of the glass with methanol, ether, and trifluoroacetic acid.

The glass-coupled 3-carboxypropionylated CNBr-9a was cleaved with trypsin in 50 mM ammonium bicarbonate buffer (pH 7.8) at 37° for 4 hr using an enzyme-to-substrate ratio of 1 : 100. The sequence of the amino acids in the resulting 36 amino acid long peptide (83–118) remaining on glass was determined by the solid-phase sequencer program of Horn and Bonner.[22]

Partial Acid Hydrolysis of Peptides for GCMS Sequencing. Chymotryptic and CNBr fragments (100–400 nmol per experiment) were subjected to partial acid hydrolysis in 6 M HCl at 110° for 20–35 min or longer for very large or hydrophobic peptides. Because the peptides were

[16] G. S. Bailey, D. Gilbert, D. F. Hill, and G. B. Peterson *J. Biol. Chem.* **252**, 2218 (1977).
[17] D. G. Klapper, C. E. Wilde, and J. D. Capra, *Anal. Biochem.* **85**, 126 (1978).
[18] P. J. Bridgen, G. A. M. Cross, and J. Bridgen, *Nature (London)* **263**, 613 (1976).
[19] R. A. Laursen, M. J. Horn, and A. G. Bonner, *FEBS Lett.* **21**, 67 (1972).
[20] M. J. Horn and R. A. Laursen, *FEBS Lett.* **253**, 285 (1973).
[21] E. Wachter and R. Werhahn, *Anal. Biochem.* **97**, 56 (1979).
[22] M. J. Horn and A. G. Bonner, *in* "Solid Phase Methods in Protein Sequence Analysis" (A. Previero and M.-A. Coletti-Previero, eds.), p. 163. Elsevier/North-Holland, New York, 1977.

insoluble in 6 M HCl, they were first dissolved in a minimum volume of 88% formic acid or 50% (v/v) trifluoroacetic acid.

Enzymic Digestion of CNBr Fragments for GCMS Sequencing. Peptides (100–400 nmol) were suspended by sonication in 1–2 ml of ammonium acetate buffer (pH 8.5, 0.17 M in acetate) containing 2 mM $CaCl_2$, and the mixture was incubated 24–72 hr with chymotrypsin, elastase, thermolysin, or proteinase K. Portions of fresh enzyme were added periodically, and the digestion was followed by monitoring the absorbance at 280 nm of the supernatant solution after centrifugation. Final enzyme-to-substrate weight ratios ranged from 1:100 to 1:15. Digestions were carried out at 37° and at 55° for thermolysin and proteinase K.

Conversion of Peptide Mixtures to Polyamino Alcohols and Mass Spectrometry. The sequence of reactions for the derivatization of a mixture of peptides to polyamino alcohols has been described in detail elsewhere.[23] The peptides were methylated, trifluoroacetylated, and then reduced with either $LiAl^2H_4$[11] or $B_2{}^2H_6$.[24] The polyamino alcohols were silylated and injected directly into the gas chromatograph. The GCMS system and the high-resolution mass spectrometric techniques employed have been described elsewhere.[25] A computer program[26] assisted in the interpretation of the mass spectra of the peptide derivatives.

Results

Sequence of CNBr Fragments 1-5. C-2 contains five methionine residues; therefore, six fragments were expected on CNBr treatment. As described earlier, five distinct peptide fractions were obtained on Sephadex LH-20 (Fig. 2). A GCMS experiment performed directly on fraction CNBr-5 showed it to consist of two small peptides: CNBr-5a (Tyr-Leu-Ser-Hse) and CNBr-5b (Val-Pro-Phe). The latter did not terminate in homoserine and, therefore, it was the COOH terminus of C-2. Automated Edman degradation of fragments CNBr-1, CNBr-3, and CNBr-4 gave the NH_2-terminal sequences shown in Fig. 5; GCMS data on acid hydrolysates and enzyme digests supported these results and completed the COOH-terminal sequences of these fragments.[6]

CNBr-2: The Blocked N-Terminus. CNBr-2 failed to release any phenylthiohydantoin amino acid in an Edman experiment, and it was con-

[23] J. A. Kelley, N. Nau, H.-J. Forster, and K. Biemann, *Biomed. Mass Spectrom.* **2**, 313 (1975).

[24] H. Frank, H. J. Chaves das Neves, and E. Bayer, *J. Chromatogr.* **152**, 357 (1978).

[25] K. Biemann, *in* "Biochemical Applications of Mass Spectrometry" (G. R. Waller, ed.), p. 96. Wiley (Interscience), New York, 1972.

[26] J. B. Biller, W. C. Herlihy, and K. Biemann, *ACS Symp. Ser.* **54**, 18 (1977).

cluded to contain the blocked NH_2 terminus of the protein. Analysis of partial acid hydrolysates of C-2 and CNBr-2 by GCMS allowed the reconstruction of the partial sequence Glx-Ala-Glx-Ile-Thr-Gly (Fig. 5). An elastase (or proteinase K) digest contained a soluble pentapeptide (Glu_2-Ala_1-Thr_1-Ile_1) that failed to adsorb on Dowex 50 $\times 2$ (H^+) and, therefore, lacked a free NH_2 terminus. The high-resolution mass spectrum of this peptide showed ions corresponding to $<$Glu-Ala-. . . .[6] Digestion of the pentapeptide with pyroglutamyl-peptide hydrolase released pyroglutamic acid, which was characterized by the identity of its electrophoretic mobility (pH 3.1 and pH 6.5) with that of authentic pyroglutamic acid. The remaining peptide contained one less glutamic acid, and dansylation[27] identified the alanine as its new NH_2 terminus.

For further sequence work, CNBr-2 (200 nmol), which was entirely insoluble in aqueous buffers, was solubilized by reaction with 4-SPITC as follows. The COOH-terminal homoserine was lactonized by treatment with 1 ml of anhydrous trifluoroacetic acid at 20° for 1 hr. After evaporation, the residue was dissolved in 1 ml of 10% (v/v) ethylenediamine in dimethylformamide and kept at room temperature for 3 hr. Excess reagent was removed under reduced pressure, and the residue dissolved in 100 μl of 88% formic acid, 1 ml of pyridine, 500 μl of triethylamine, and 100 μl of water. After addition of 0.1 mmol (25 mg) of 4-SPITC, the solution was kept at 45° for 1 hr and dried under reduced pressure; the excess reagent was then removed by passage through LH-20. The CNBr-2 derivative, detected by A_{280}, was dried, and the residue was dissolved in 1.0 ml of N-methylmorpholine, dried again, and dissolved in 3.0 ml of water. After centrifugation, the supernate contained 80% of the CNBr-2 derivative. The solubilized peptide (100 nmol) on digestion with 0.5 unit of pyroglutamyl-peptide hydrolase as mentioned earlier released the NH_2-terminal pyroglutamic acid. The remaining peptide was repurified by chromatography on Sephadex LH-20 and then subjected to sequencer analysis. This allowed the assignment of positions 2–15, which overlapped with the sequence established by GCMS (positions 11–20) (Fig. 5).

The COOH Terminus of Bacteriorhodopsin: Amino Acid Sequence of CNBr-6. Automated Edman degradation of fragment CNBr-6 and of a tryptic peptide containing its COOH terminus together with the GCMS data[6] completed the sequence of CNBr-6. It was identified as the COOH terminus of bacteriorhodopsin because it ended in a sequence corresponding to that reported earlier[8] for bacteriorhodopsin.

Sequence of CNBr-7. The sequence deduced and shown in Fig. 6 for

[27] A. M. Weiner, T. Platt, and K. Weber, *J. Biol. Chem.* **247**, 3242 (1972).
[28] A. Szewczuk and J. Kwiatkowska, *Eur. J. Biochem.* **15**, 92 (1970).

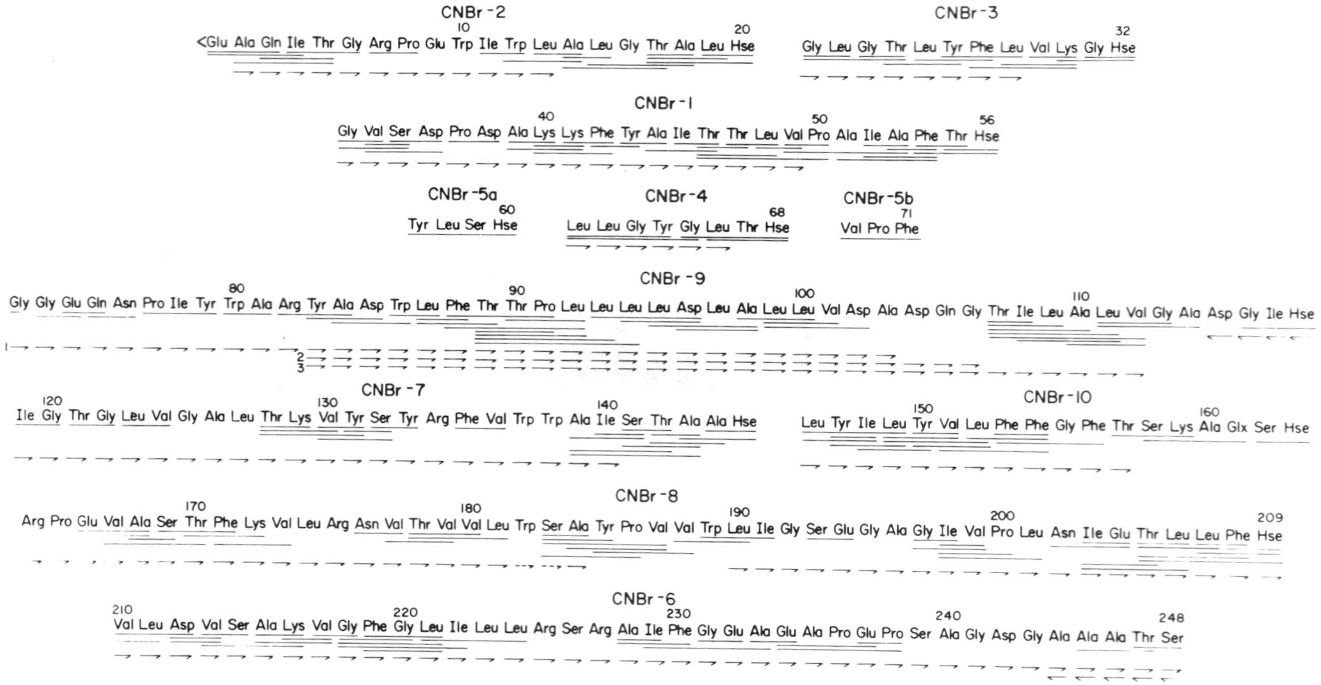

CNBr-2

10   20
<Glu Glu Gln Ile Thr Gly Arg Pro Glu Trp Ile Trp Leu Ala Leu Gly Thr Ala Leu Hse

CNBr-1

             40   50
Gly Val Ser Asp Pro Asp Ala Lys Phe Tyr Ala Ile Thr Thr Leu Val Pro Ala Ile Ala Phe Thr Hse

CNBr-3

                  32
Gly Leu Gly Thr Leu Tyr Phe Leu Val Lys Gly Hse

CNBr-5a 60
Tyr Leu Ser Hse

CNBr-4   68
Leu Leu Gly Tyr Gly Leu Thr Hse

CNBr-5b 71
Val Pro Phe

CNBr-9

80   90   100   110
Gly Gly Gln Asn Pro Ile Tyr Trp Ala Arg Tyr Ala Asp Trp Leu Phe Thr Thr Pro Leu Leu Leu Leu Asp Leu Ala Leu Leu Val Asp Ala Asp Gln Gly Thr Ile Leu Ala Leu Val Gly Ala Asp Gly Ile Hse

CNBr-7

                   120   130   140   150   160
Ile Gly Thr Gly Leu Val Gly Ala Leu Thr Lys Val Tyr Ser Tyr Arg Phe Val Trp Trp Ala Ile Ser Thr Ala Ala Hse Leu Tyr Ile Leu Tyr Val Leu Phe Phe Gly Phe Thr Ser Lys Ala Glx Ser Hse

CNBr-8

             170   180   190   200   209
Arg Pro Glu Val Ala Ser Thr Phe Lys Val Leu Arg Asn Val Thr Val Val Leu Trp Ser Ala Tyr Pro Val Val Trp Leu Ile Gly Ser Glu Gly Ala Gly Ile Val Pro Leu Asn Ile Glu Thr Leu Leu Phe Hse

CNBr-6

210   220   230   240   248
Val Leu Asp Val Ser Ala Lys Val Gly Phe Gly Leu Ile Leu Leu Arg Ser Arg Ala Ile Phe Gly Glu Ala Glu Ala Pro Glu Pro Ser Ala Gly Asp Gly Ala Ala Ala Thr Ser

CNBr-10

Fig. 5. Amino acid sequences of CNBr fragments formed from bacteriorhodopsin. Underlinings indicate peptides identified by GCMS; the arrows indicate data obtained by automated Edman sequencing. The three sets of arrows under CNBr-9 indicate three sequencer experiments as described in text.

GLU ALA GLN ILE THR GLY ARG PRO GLU TRP ILE TRP LEU ALA LEU GLY THR ALA LEU MET GLY LEU GLY THR LEU
5   10   15   20   25

TYR PHE LEU VAL LYS GLY MET GLY VAL SER ASP PRO ASP ALA LYS PHE TYR ALA ILE THR THR LEU VAL PRO
30   35   40   45   50

ALA ILE ALA PHE THR MET TYR LEU SER MET LEU LEU GLY TYR GLY LEU THR MET VAL PRO PHE GLY GLY GLN
55   60   65   70   75

ASN PRO ILE TYR TRP ALA ARG TYR ALA ASP TRP LEU PHE THR THR PRO LEU LEU LEU ASP LEU ALA LEU LEU
80   85   90   95   100

VAL ASP ALA ASP GLY THR ILE LEU ALA ILE VAL GLY ALA ASP GLY LEU MET ILE GLY THR GLY LEU VAL GLY
105   110   115   120   125

ALA LEU THR LYS VAL TYR SER TYR ARG PHE VAL TRP TRP ALA ILE SER THR ALA ALA MET LEU TYR ILE LEU TYR
130   135   140   145   150

VAL LEU PHE PHE GLY PHE THR SER LYS ALA GLU SER MET ARG PRO GLU VAL ALA SER THR PHE LYS VAL LEU ARG
155   160   165   170   175

ASN VAL THR VAL VAL LEU TRP SER ALA TYR PRO VAL VAL TRP LEU ILE GLY SER GLU GLY ALA GLY ILE VAL PRO
180   185   190   195   200

LEU ASN ILE GLU THR LEU LEU PHE MET VAL LEU ASP VAL SER ALA LYS VAL GLY PHE GLY LEU ILE LEU LEU ARG
205   210   215   220   225

SER ARG ALA ILE PHE GLY GLU ALA GLU ALA PRO GLU PRO SER ALA GLY ASP GLY ALA ALA ALA THR SER
230   235   240   245

FIG. 6. Complete amino acid sequence of bacteriorhodopsin.

this fragment differs from that reported by Ovchinnikov and co-workers,[1] there being two contiguous tryptophan residues (positions 137 and 138) instead of one. The molar extinction of CNBr-7 at 280 nm was consistent with two tryptophan residues. Sequencer analysis confirmed this conclusion and placed the second tryptophan at position 138.

Sequence of CNBr-8. A sequencer run on CNBr-8 gave unambiguous results as far as Leu-181. Sequencer analysis of the COOH-terminal NBS fragment of C-1 that started at Leu-190 gave unequivocal assignments up to position 219. The experiment also provided an overlap of CNBr-8 and CNBr-6. The only outstanding question pertained to position 182. Amino acid composition of CNBr-8 showed two tryptophan residues. From the preceding sequencer runs and the GCMS data, the only unassigned position is 182, and this must be occupied by the second tryptophan.

Sequence of CNBr-9. This fragment formed the NH_2 terminus of C-1, and sequencer analysis of the latter gave the sequence up to Ala-103 (sequence 1 in CNBr-9, Fig. 5). Digestion of the SPITC derivative with clostripain yielded a fragment consisting of the sequence Tyr-83 to Hse-118. Automated Edman degradation of this fragment established the sequence up to Gly-106 (sequence 2 in CNBr-9 in Fig. 5). The amino acid at position 105 was identified as glutamine by HPLC analysis of the phenylthiohydantoin. In further experiments, the fragment Tyr-83 to Hse-118 was coupled to TETA or AEAP glass and subjected to automated Edman degradation. The sequence up to Val-112 was thus obtained (sequence 3 in CNBr-9 in Fig. 5).

Derivatization of the COOH terminus in CNBr-9 with [14C]ethylenediamine followed by degradation with elastase yielded a radioactive fragment that contained Ile and had the composition expected for the COOH-terminal pentapeptide. This result supports the assignment of Ile at position 117. Further, carboxypeptidase Y treatment of N-(3-carboxypropionyl)-CNBr-9 gave evidence for the assignment of Asp at position 115. These results, together with the GCMS data, completed the sequence of CNBr-9 as shown in Fig. 5.

Sequence of CNBr-10. GCMS on acid hydrolysates of total C-1 and the unpurified aggregate (pool a of Fig. 4A) gave a partial sequence for this fragment. Amino acid composition of the purified CNBr-10 and Edman degradation of the latter after attachment to IPTAP glass enabled derivation of the complete sequence.

The Total Sequence of Bacteriorhodopsin. The ordering of the preceding CNBr fragments so as to derive the sequence of C-2 was accomplished by the identification of short Met-containing peptides that formed overlaps between the CNBr fragments. Thus, a GCMS experiment on a partial acid hydrolysate of C-2 allowed the identification of 12 Met-containing peptides, which included Thr-Ala-Leu-Met-Gly, Thr-Met-Try-Leu, Ser-Met-Leu-Leu-Gly, and Thr-Met-Val-Pro-Phe. Therefore, the CNBr fragments must be linked as follows: . . . (CNBr-1)-(CNBr-5a)-(CNBr-4)-(CNBr-5b)-OH. Because CNBr-2 contains the blocked NH_2 terminus, the only possible arrangement of the fragments is (CNBr-2)-(CNBr-3)-(CNBr-1). The resulting sequence of C-2 is included in Fig. 6.

Sequence analysis of C-1 established CNBr-9 to be at its NH_2 terminus while carboxypeptidase analysis showed CNBr-6 to be the COOH terminus of bacteriorhodopsin and of C-1. GCMS analysis of a partial acid hydrolysate of C-1 yielded several methionine-containing peptides, including Leu-Phe-Met-Val (207–210) and Phe-Met-Val-Leu (208–211). The latter peptides and the sequence of the NBS fragment, Leu-190 to Ser-248, described earlier, established the overlap and order of CNBr-8 and CNBr-6. The order (CNBr-7)-(CNBr-10) was established for these two fragments by the identification of the overlap peptide Ala-Met-Leu by GCMS.

The order (CNBr-9)-(CNBr-7) was further supported by (i) the amino acid composition of the NBS fragment, Leu-87 to Trp-137 and (ii) the identification of the tripeptide Leu/Ile-Met-Leu/Ile by GCMS. Thus, the order of the CNBr fragments in C-1 is (CNBr-9)-(CNBr-7)-(CNBr-10)-(CNBr-8)-(CNBr-6).

The necessary overlap between C-2 and C-1 was provided by the following experiments: The whole protein was cleaved with CNBr, and the fragments were chromatographed by HPLC. A peptide with an amino acid composition corresponding to CNBr-9 but having, in addition, the amino acids Val, Pro, and Phe was isolated by HPLC. Further, the peptide Val-Pro-Phe-Gly was identified by GCMS in a partial acid hydrolysate of this fragment. Finally, a sequencer experiment gave the sequence Val-Pro-Phe-Gly-Gly-Glu-Gln-Asn-Pro-Ile-. . . .

The amino acid sequence described here differs from that reported by Ovchinnikov and co-workers[1] with respect to an additional tryptophan (position 138) and amino acid assignments at positions 105, 111, 117, 146, and 206. Of the total 248 amino acids present in bacteriorhodopsin, 70% are hydrophobic, and there is marked clustering of the hydrophobic as well as of the hydrophilic amino acids. Thus, 80% of hydrophobic amino acids occur in clusters of four or more, the longest region containing 14 contiguous hydrophobic amino acids (143–156). The presence of tracts of hydrophilic and hydrophobic amino acids may be a general property of membrane proteins. The membrane-associated segments of a number of integral membrane proteins[29–31] seem to acquire hydrophobicity in this

---

[29] M. Tomita and V. T. Marchesi, *Proc. Natl. Acad. Sci. U.S.A.* **72**, 2964 (1975).
[30] W. Wickner, *Proc. Natl. Acad. Sci. U.S.A.* **73**, 1159 (1976).

way,[32] although the overall compositions of these proteins are not markedly hydrophobic.

Calculations were performed by the method of Chou and Fasman[33] for the occurrence of β-turns in bacteriorhodopsin, although this empirical method may not be applicable to membrane proteins. Indications were obtained for (1) the presence of some N-terminal amino acid residues out of the bilayer, (2) a β-turn at Arg-7 to Trp-10, (3) a turn at sequence Ser-35 to Lys-40, and (4) a turn near the Phe-71 residue, where enzymic cleavages occur (chymotrypsin and papain). If, indeed, there is a turn at about the sequence 36–39 (see also Walker *et al.*[3]), this is very probably on the cytoplasmic side. Nothing is known about the location of the remaining turns except that a papain cleavage has been observed at residue 162.[1] The electron diffraction data of Unwin and Henderson[34] indicate that bacteriorhodopsin probably consists of seven helices that traverse the membrane and are largely embedded in it. More recently, an attempt has been made to fit the amino acid sequence to the three-dimensional density map of the molecule. A single model[35] was deduced as the most probable. An important feature of the model is that the location of the charged amino acids, which must of necessity be placed within the bilayer, is such that inter- or intrahelix salt bridges can be formed to neutralize the charges.

The Sequencing of Membrane Proteins: General Comments

Solubilization, selective degradation, and isolation of required fragments in pure form constitute the first challenging phase of work on primary structure determination of membrane proteins. The methods introduced in the present work for separation of the very hydrophobic peptides undoubtedly will undergo further development so as to give greater flexibility and scope. Thus, gel permeation, which has been enormously useful in separation of soluble proteins and peptides, can be modified as described in the present work so as to facilitate separation of hydrophobic polypeptides as well. Similarly, the scope and versatility of HLPC are rapidly increasing as a wide range of functionalized solid supports become available, and HPLC promises to be at least as useful for work with hydrophobic proteins and peptides as with soluble protein fragments.

[31] J. Ozols and C. Gerard, *J. Biol. Chem.* **252,** 8549 (1977).
[32] J. P. Segrest and R. J. Feldman, *J. Mol. Biol.* **87,** 853 (1974).
[33] P. Y. Chou and G. D. Fasman, *Adv. Enzymol.* **47,** 45 (1978).
[34] P. N. T. Unwin and R. Henderson, *J. Mol. Biol.* **94,** 425 (1975).
[35] D. M. Engelman, R. Henderson, A. D. McLachlan, and B. A. Wallace, *Proc. Natl. Acad. Sci. U.S.A.* **77,** 2023 (1980).

For proteolytic cleavages it is often desirable or necessary to solubilize the proteins or large polypeptide fragments in aqueous buffers. Some of the procedures used in the present work serve as examples. Thus, the use of isocyanato benzene sulfonic acids or succinic and related anhydrides for derivatizing available amino groups is often helpful and may, in fact, provide specific handles for separation and characterization of subfragments following further cleavages.

Stepwise sequencing in solution phase by using the spinning cup techniques frequently suffers from losses of material during solvent extractions. The seriousness of the problem increases with increasing hydrophobicity of the peptide fragment. A number of modifications in the automated Edman degradation programs have been introduced. While certain modifications may successfully minimize losses, especially in conjunction with derivatization that increases hydrophilicity, solid phase sequencing is an attractive alternative method for sequencing very hydrophobic peptides. This technique originally developed by Laursen[36,37] has been used with great success for sequencing of the dicyclohexylcarbodiimide-binding subunits of mitochondrial ATPases. A variety of methods for attachment of the polypeptides to derivatized glass supports (see earlier discussion) have been recommended, and although the yields at the attachment step may not be completely satisfactory, the important compensation is the high yield of the PTH-amino acids sustained at the repetitive steps. Furthermore, solid phase sequencers are far simpler and more economical in operation than the spinning cup liquid phase sequencers.

Sequencing by stepwise degradation and GCMS were used in conjunction in the present work, and the two methods complement each other effectively. Their combined use greatly facilitates analysis of the fragments, and the results are much more reliable than when either technique is used alone. The sequencer removes one amino acid at a time from the NH_2 terminus of a fragment, and this is identified by various procedures; the GCMS technique, on the other hand, is based on a partial acid or enzymatic hydrolysis of the peptide followed (after suitable derivatization) by separation of the small peptides by gas chromatography and structural analysis by mass spectrometry. Thus, while the sequence becomes progressively more difficult to identify by the sequencer method with increasing distance from the NH_2 terminus, the GCMS technique is equally effective in all positions within the molecule. However, some specific sequences (e.g., those that are particularly acid-labile) may be difficult to observe by GCMS. The alignment of partial sequences identified by

[36] R. A. Laursen, *Eur. J. Biochem.* **20**, 89 (1971).
[37] R. A. Laursen and W. Machleidt, *Methods Biochem. Anal.* **26**, 201 (1980).

GCMS can often be achieved more rapidly by sequencer data than by working out different acid or enzymatic hydrolysis conditions. The GCMS data are particularly useful in the construction of the COOH-terminal sequences of CNBr peptides, because all but one of them end in homoserine, which can be uniquely detected by this method. Overlapping peptides extend this sequence until it connects with the Edman data. This complementarity of the two methods is clearly illustrated in Fig. 5, which shows the abundance of GCMS information in the COOH-terminal regions. Furthermore, the use of the two independent methods was of particular importance for those residues for which our sequence differs from the sequences of other groups.[1,3]

It should, however, be emphasized that the accessibility of a GCMS facility is by no means critical to the determination of membrane protein structures. In particular, as methods for fragmentation and separation of fragments become more and more efficient and experience with the sequencing methods as reviewed here increases, standard degradative methods should provide completely reliable sequence data.

Acknowledgments

This work was supported by grants from the Institute of Allergy and Infectious Disease (AI11479), the National Cancer Institute (CA11981), the National Institute of General Medical Sciences (GM05472 and RR00317), and the National Science Foundation (PCM78-13713) and by funds made available to the Massachusetts Institute of Technology by the Sloan Foundation. G.E.G. was a Medical Research Council of Canada Postdoctoral Fellow (1975–1978).

[10] Delipidation, Renaturation, and Reconstitution of Bacteriorhodopsin

By Hagan Bayley, Bo Höjeberg, Kuo-Sen Huang, H. Gobind Khorana, Mei-June Liao, Christina Lind, and Erwin London

We describe here the complete displacement of the purple membrane lipids by detergents and the subsequent reconstitution of the solubilized bacteriorhodopsin with exogenous or endogenous lipids to form func-

Abbreviations used: BR, bacteriorhodopsin; BO, bacterio-opsin; dBR, delipidated bacteriorhodopsin; dBO, delipidated bacterio-opsin; DOC, deoxycholate; SDS, sodium dodecyl sulfate; SBL, soybean phospholipids; HHPL, *Halobacterium halobium* phospholipids; dBR/DOC, delipidated bacteriorhodopsin in buffer containing deoxycholate (and similarly dBO/DOC and dBO/SDS).

tional vesicles. In these vesicles all the protein molecules are oriented with their COOH-termini exposed at the external surface; the opposite orientation is found in the intact cell membrane of *Halobacterium halobium*. In addition, a method is given for the complete denaturation by organic solvents of lipid- and detergent-free bacterio-opsin. The denatured protein can be refolded and becomes fully active when reconstituted into vesicles.

The effects of different lipids and retinal analogs on proton translocating activity can be tested using vesicles made by the reconstitution procedures.[1,2] The vesicles formed from the endogenous lipids have extraordinary properties, including high stability (more than 4 months at 4°), optical clarity, and such high impermeability to ions that extensive proton translocation into the vesicles is observed only in the presence of K^+ and valinomycin, which together prevent formation of a membrane potential.[3] Our studies of denaturation and renaturation bear on the problem of membrane biogenesis.[4]

A. General Methods

Materials. Triton X-100 was from Rohm & Haas. Cholic acid (Sigma) was recrystallized from acetone and deoxycholic acid (Sigma) from ethanol and water, after treatment of the hot solutions with charcoal. All-*trans*-retinal (Sigma) was stored in ethanol at −20°, and the purity was checked periodically by HPLC.

Soybean Phospholipids. Soybean phospholipids were partially purified by a slight modification of the method of Kagawa and Racker.[5] Asolectin (50 g, from Associated Concentrates) was stirred, under Ar for 3 days, with acetone (1 liter) that had been freshly distilled from anhydrous K_2CO_3. The insoluble material was collected by filtration and mixed with anhydrous ether (100 ml, Mallinckrodt). A little dark-colored solid was removed by centrifugation (12,000 g) at 4° for 15 min. The solvent was evaporated from the supernatant solution, and the residue was dissolved in $CHCl_3$ and stored in small vials at −20°. The concentration of the lipid

[1] K.-S. Huang, H. Bayley, and H. G. Khorana, *Proc. Natl. Acad. Sci. U.S.A.* **77**, 323 (1980).
[2] H. Bayley, R. Radhakrishnan, K.-S. Huang, and H. G. Khorana, *J. Biol. Chem.* **256**, 3797 (1981).
[3] C. Lind, B. Höjeberg, and H. G. Khorana, *J. Biol. Chem.* **256**, 8298 (1981).
[4] K.-S. Huang, H. Bayley, M.-J. Liao, E. London, and H. G. Khorana, *J. Biol. Chem.* **256**, 3802 (1981).
[5] Y. Kagawa and E. Racker, *J. Biol. Chem.* **246**, 5477 (1971).

was calculated after a phosphorus assay[6] assuming an average MW of 800.

When the lipid was required for reconstitution experiments, most of the solvent was removed in a stream of N_2, the residue was dissolved in dry ether, and the solvent was again removed. The treatment with ether was repeated, and then the lipid was dried in a vacuum (<50 μm Hg) for 12 hr before use.

Polar Lipids from H. halobium. The polar lipids from *H. halobium* (HHPL) were isolated essentially as described by Kates and co-workers.[7] Cells from a 10-liter culture were harvested and washed twice with 4 *M* NaCl (200 ml). The washed cells (100 g net weight) were suspended in 4 *M* NaCl (100 ml) and extracted with methanol (500 ml) and $CHCl_3$ (250 ml) overnight. The mixture was centrifuged at 1000 *g* for 5 min, and the pellet was extracted for 2 hr with half the volume of solvents used the first time. To the combined supernatant solutions were added $CHCl_3$ and water (400 ml each), and the two phases were allowed to separate. The $CHCl_3$ phase, containing both nonpolar and polar lipids, was taken to dryness under reduced pressure and dissolved in a small volume of $CHCl_3$. The polar lipids were precipitated by adding 10 volumes of ice-cold acetone supplemented with 0.03% (w/v) $MgCl_2$. After 1 hr at 4°, the precipitate was collected by centrifugation (1000 *g*, 5 min) and dissolved in $CHCl_3$, and the precipitation step with acetone was repeated. The isolated polar lipids, in $CHCl_3$, were stored at $-20°$. The yield was usually 6–10 mg lipids per gram of bacteria (wet weight). Analysis of the isolated polar lipids for phosphate content[6] and hexose content[8] (using sucrose as a standard) revealed a lipid composition similar to that reported previously,[9] i.e., phosphatidylglycerol phosphate, 50–55%; glycolipid sulfate, 35–40%; phosphatidylglycerol plus phosphatidylglycerolsulfate, 8–10%.

Proton Translocation Assay. Proton translocation was assayed by the method of Racker and Stoeckenius[10] in which the pH of the medium containing the vesicles is monitored with a glass electrode during illumination. We used as a light source a Kodak type 800 slide projector, fitted with a Schott OG 530 filter and, to contain the assay solution, we used a water-jacketed glass vessel maintained at 30°. The pH electrode (Beckman type 39505) was connected to a Corning Digital 112 pH meter, the

[6] B. N. Ames, this series, Vol. 8, p. 115.
[7] C. N. Joo, T. Shier, and M. Kates, *J. Lipid Res.* **9**, 782 (1968).
[8] M. Dubois, K. A. Gilles, J. K. Hamilton, P. A. Rebers, and F. Smith, *Anal. Chem.* **28**, 350 (1956).
[9] A. J. Hancock and M. Kates, *J. Lipid Res.* **14**, 422 (1973).
[10] E. Racker and W. Stoeckenius, *J. Biol. Chem.* **249**, 662 (1974).

output of which was connected to a Esterline Angus Speed Servo recorder capable of offsetting up to 100 mV. Portions of the vesicle preparation (10–50 μl) were assayed in 1–2 ml of unbuffered salt solution (see following section). A fresh solution is prepared each day, as old solutions can cause the baseline to drift. Ten to fifteen minutes after the electrode had been inserted into the assay mixture, a flat or almost flat baseline was established, and the initial pH was adjusted to the desired initial value (usually between 6.2 and 6.4) using 0.001 M HCl or 0.001 M NaOH. Proton uptake was initiated by illumination, which was continued until a plateau was reached. After extinguishing the light and waiting for a flat baseline to be reestablished, the electrode was calibrated by injecting aliquots (5 μl) of 0.001 M HCl (prepared on the day of use by diluting 1 M HCl) into the assay solution. The initial rate and final extent of translocation were then calculated.

B. Delipidated Bacteriorhodopsin

Bacteriorhodopsin in deoxycholate (dBR/DOC). A pellet of purple membrane (50 mg of bacteriorhodopsin) was homogenized with 5% (v/v) Triton X-100, 0.1 M Na acetate, pH 5.0 (2.5 ml) using a glass rod.[11] After 2 days at 25° in the dark,[12] with occasional agitation, most of the membrane had become solubilized. The supernatant from a brief centrifugation (8000 g, 1 min), to remove any aggregated material, was applied at 4° in the dark to a column of BioGel A-0.5 m (2.5 × 170 cm), which was eluted with 0.25% (w/v) DOC,[13] 10 mM Tris, 0.15 M NaCl, 0.025% (w/v) NaN$_3$ adjusted to pH 8.0 with NaOH. Crimson-colored fractions (12.5 ml each) containing delipidated protein and no Triton X-100 appeared after a small purple peak of incompletely solubilized material and before a peak containing lipid and Triton X-100.[1] An elution profile from an analytical column is shown in Fig. 1. Using lipid biosynthetically labeled with ^{32}P (or ^{14}C) and [^{3}H]Triton X-100, we realized a goal that had been long sought, the complete removal of both the endogenous lipid and the detergent used to solubilize the purple membrane.[1] The pooled fractions ($A_{280}/A_{540} \leqslant 2.0$) from the preparative column were concentrated using an Amicon CF25 Centriflo ultrafiltration cone. A four-fold increase in the concentration of protein gave only a 1.25-fold increase in the concentration of DOC, but more DOC was retained at higher protein concentrations. The yield of

---

[11] R. Henderson, *Annu. Rev. Biophys. Bioeng.* **6**, 87 (1977).
[12] The membrane is solubilized more rapidly at high pH. We have recently used 5% Triton X-100, 0.1 M Tris-HCl, pH 7.0, for 12 hr and obtained similar results.
[13] 1.0% (w/v) cholate can be used in place of DOC. This detergent is more easily removed by dialysis.

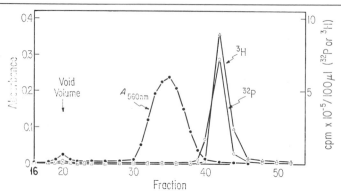

FIG. 1. Removal of endogenous lipid from bacteriorhodopsin. ^{32}P-labeled purple membrane (2 mg; 3.0 × 10^7 cpm) was solubilized in buffer (0.1 ml) containing ^{3}H-labeled Triton X-100 (3.8 × 10^7 cpm), and the solution was applied to a column of BioGel A-0.5m (1.0 × 100 cm), which was eluted with the buffer containing 0.25% DOC (see text). Fractions (1.2 ml) were collected at a flow rate of 7 ml/hr. From Huang et al.[1]

protein was ~80%. Delipidated bacteriorhodopsin (dBR/DOC; λ_{max} 538) (ϵ = 40,000); 280 (ϵ = 76,000[14]) is stable for more than 3 months when stored at −20° after quick-freezing, but ~20% of the sample bleaches per week at 4° in the dark.

Bacterio-opsin in Sodium Dodecyl Sulfate (dBO/SDS). dBR/DOC (50 mg protein) was concentrated to 5 ml, mixed with 1 M NH$_2$OH, 1 M NaCl, pH 8.0 (1 ml), and irradiated with the 500-W quartz halogen lamp of a Kodak type 800 slide projector using an orange filter (Schott OG 530). After 15 min the solution was pink and a further portion of NH$_2$OH solution (1 ml) was added. After a further 45 min of irradiation at 25°, the entire sample was chromatographed on BioGel A-0.5 m in 0.25% DOC as described for the preparation of dBR/DOC (Fig. 2). The A_{280} and A_{350} of the fractions revealed a peak of aggregated dBO/DOC at the void volume followed by dispersed protein well separated from a peak of retinal oxime. The protein containing fractions (numbers 25–38, Fig. 2) were pooled, concentrated to 20 ml, and dialyzed over 4 days at room temperature against 1% SDS (Bio-Rad electrophoresis grade), 0.01 M NaP$_i$, pH 8.0 (5 × 1.8 liters). The recovery of protein was over 90%. λ_{max} 280 nm (ϵ = 65,000). dBO/SDS was stored at room temperature.

C. Reconstitution

Reconstitution of Bacteriorhodopsin in Deoxycholate (dBR/DOC) with Soybean Lipids (SBL). SBL (120 mg) from which the storage solvent

[14] The previously reported value of ϵ_{538} = 58,000 was based on the Lowry assay for protein, which we have recently found to be inaccurate for dBR/DOC.

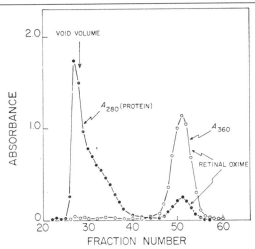

FIG. 2. Removal of retinal oxime from delipidated bacterio-opsin in deoxycholate. The profile is for a smaller preparation than that described in the text. Bleached dBR/DOC (8.5 mg protein in 0.7 ml) was applied to a column of Bio-Gel A-0.5m (1.0 × 100 cm), which was eluted at 7 ml/hr as described in the text. Fractions of 1.15 ml were collected. From Huang *et al.*[4]

had been carefully removed was dissolved by sonication in 2% sodium cholate, 0.15 M KCl, pH 8.0 (2.5 ml). The solution was mixed at 4° with dBR/DOC (2.5 ml; $A_{540} = 1.2$) and then dialyzed at 25° against 10 mM Tris-HCl, 0.15 M NaCl, 0.025% (w/v) NaN$_3$, pH 8.0 (4 × 2 liters) for 2 days and against 0.15 M NaCl, 0.025% (w/v) NaN$_3$ (2 × 2 liters) for one more day. Proton uptake on illumination was assayed in 2 M NaCl, pH 6.2, at 30° by the method of Racker and Stoeckenius described earlier. The initial rate was usually around 1.0 H$^+$/BR/sec and the final extent of translocation approximately 45 H$^+$/BR.[1]

Reconstitution of dBR/DOC with Polar Lipids from H. halobium (*HHPL*). Reconstitution of dBR/DOC with HHPL was performed as described earlier for SBL except that NaCl, in the dialysis buffer, was replaced by KCl, in order to load the vesicles with potassium. Maximal proton translocation was observed in vesicle preparations where the lipid-to-protein ratio was between 40:1 and 80:1 (w/v), and in this range analytical sucrose density centrifugation revealed that all the lipid was associated with protein. In contrast, the vesicles made from SBL comprised a large fraction of free lipid and several different fractions of protein-containing vesicles.[3,15]

BR-containing HHPL vesicles of very low turbidity can be obtained

---

[15] B. Höjeberg, C. Lind, and H. Bayley, unpublished work.

by gel filtration on Sepharose CL4B (Fig. 3). Vesicles (0.3 ml), concentrated to 0.8–1.0 mg BR/ml by ultrafiltration with an Amicon CF25 Centriflo cone, were applied to a Sepharose CL4B column (0.5 × 110 cm) equilibrated with 5 mM Tris-Cl, pH 7.0, 0.15 M KCl, and 0.025% (w/v) NaN$_3$. The sample was eluted at room temperature in thè dark with the equilibrating buffer, and fractions of 0.25 ml were collected. Two peaks containing BR were found. The first peak, in the void volume, contained a vesicle population with high turbidity and low proton translocation activity, whereas the majority of the vesicles exhibiting low turbidity and high translocating activity were recovered in the later eluting peak. The amount of turbid material varied from one preparation to another (though it never exceeded 20% of the vesicle preparation) and was found to be drastically reduced if freshly prepared dBR/DOC was used for reconstitution rather than protein that had been stored frozen.

Proton uptake by BR-containing HHPL vesicles (0.15–0.25 nmol BR) was assayed in KCl or NaCl (2 ml) with the initial pH adjusted to 6.2–6.3, ıs described earlier. Generally, proton translocation was found to be higher when measured in NaCl than in KCl with an optimum salt concentration of 0.5 M in both cases. An initial rate of 3–6 H$^+$/BR/sec and an extent of 35–55 H$^+$/BR was observed when the vesicles were assayed in the presence of valinomycin (1–3 μM); in its absence these values were reduced by factors of 3–5 and 5–10, respectively.

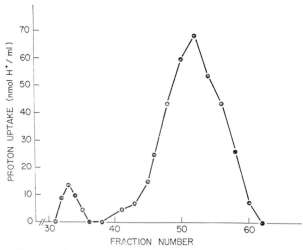

FIG. 3. Gel filtration of BR-containing HHPL vesicles on Sepharose CL4B. The column was run as described in the text. Fractions (0.24 ml) were assayed for proton translocation activity and two pools, fractions 31–35 and 45–60, were collected and concentrated for further analysis From Lind et al.[3]

Reconstitution of Bacterio-opsin in Sodium Dodecyl Sulfate (dBO/SDS) with Soybean Lipids. SBL (15 mg/ml in 2% sodium cholate, 0.15 M KCl, pH 8.0; 0.6 ml) and dBO/SDS (A_{280} = 2.12; 0.25 ml) were mixed. All-*trans*-retinal (2 equivalents in 5 μl ethanol) was added and the chromophore reappeared over $\sim$20 min at 25° in the dark. After an incubation of 3.5 hr, the sample was dialyzed against 10 mM NaP$_i$, 0.15 M NaCl, 0.025% (w/v) NaN$_3$, pH 8.0 (4 × 1 liter) for 2 days and then against 0.15 M NaCl, 0.025% (w/v) NaN$_3$ (2 × 1 liter) for 1 day. The vesicles so formed take up protons on illumination: $\sim$1.0 H$^+$/BR/sec, $\sim$45 H$^+$/BR.[4]

D. Denaturation and Refolding of Bacterio-opsin

Lyophilized apomembrane[2,16] was dissolved in 88% formic acid–ethanol (3:7, the formic acid is added first) and passed through a column of Sephadex LH-60 in the same solvent to remove the lipids. The solvent was evaporated from a portion of the void volume fractions that contained the protein, and the residue (100 μg) was dissolved in anhydrous CF$_3$COOH (100 μl), under N$_2$, in the dark.[17] To renature the protein the solution was diluted with ethanol (500 μl), taken to pH 5.6 with concentrated NH$_4$OH, and dialyzed against 1% SDS, 0.01 M NaP$_i$, pH 8.0, for 2 days (2 × 1 liter). A protein precipitate that soon redissolves appears during the solvent exchange. The final solution of dBO/SDS can be reconstituted, as described earlier, into fully active vesicles.[18]

[16] G. E. Gerber, C. P. Gray, D. Wildenauer, and H. G. Khorana, *Proc. Natl. Acad. Sci. U.S.A.* **74**, 5426 (1977).
[17] It was shown by ^{1}H that the protein in CF$_3$COOH is a random coil (note 4 and M.-J. Liao, unpublished).
[18] We have also shown that two chymotryptic fragments of bacteriorhodopsin (residues 1–71 and 72–248) can each be transferred into SDS solution by a similar method. When the fragments are combined and reconstituted into vesicles, full proton pumping activity is obtained.[4]

[11] The Identification of Helical Segments in the Polypeptide Chain of Bacteriorhodopsin

By D. M. ENGELMAN, A. GOLDMAN, and T. A. STEITZ

Introduction

Understanding of the structure of bacteriorhodopsin has been advanced considerably by information of two kinds. First, application of image reconstruction techniques using the electron microscope led Henderson and Unwin[1] to the finding that the molecule contains seven rods of density that traverse the membrane lipid bilayer region. These rods of density have dimensions and packing appropriate for alpha helices. Second, Khorana et al.[2] and Ovchinnikov et al.[3] have determined the complete sequence of amino acids in the protein in independent investigations. Their sequences agree in most respects, and the sequence obtained by Khorana and his group has now been largely confirmed by DNA sequencing procedures (H. G. Khorana, personal communication). Juxtaposition of these two kinds of information suggests immediately that it is important to understand which regions of the polypeptide chain correspond to the helical regions that span the membrane bilayer. Assignment of these regions would then serve as the basis for a number of structural investigations that include labeling studies in the electron microscope, neutron diffraction, chemical modification, cross-linking, etc.

In this brief review we summarize the kinds of information that can be used to identify the helical regions and attempt to improve the assignments based on several approaches.

Structural Constraints

One of the major constraints in identifying the portions of the sequence that form the seven helices is their length. Unfortunately, the resolution of the electron microscope structure in the direction perpendicular to the membrane sheet is only about 14 Å, and the lengths of the helices cannot be established with confidence from the electron density map.[1] The X-ray diffraction studies of Blaurock[4] and Henderson[5] provide important information that can be used to place limits on the range of lengths to be expected for the helices. Since the helices traverse the membrane at angles between 0 and 20° from the perpendicular, their lengths are constrained by the minimum packing of membrane sheets in a dehydrated specimen (48 Å) and the thickness of the hydrophobic region of the lipid bilayer (30 Å). The latter number is derived from the lipid profile obtained by Blaurock[4] using extracted lipids from *Halobacterium halobium*. Since the helices must connect to other portions of the polypeptide chain, their

[1] R. Henderson and P. N. T. Unwin, *Nature (London)* **257**, (1975).
[2] H. G. Khorana, G. E. Gerber, W. C. Herlihy, C. P. Gray, R. J. Anderegg, K. Nihei, and K. Biemann, *Proc. Natl. Acad. Sci. U.S.A.* **76**, 5046 (1979).
[3] Yu. Ovchinnikov, N. Abdulaev, M. Fergira, A. Kiselev, and N. Lobanov, *FEBS Lett.* **100**, 219 (1979).
[4] A. E. Blaurock, *J. Mol. Biol.* **93**, 139 (1975).
[5] R. Henderson, *J. Mol. Biol.* **93**, 123 (1975).

ends are unlikely to lie exactly at the 48-Å limit and probably can be presumed to be less than 45 Å with confidence. The observed rise per residue in the helices is 1.5 Å per amino acid.[4,5] Consequently, the range of possible lengths for the helices implies a range of 20–30 amino acids as the length of a helix. Since the sequence contains 248 amino acids, this range corresponds to 54–85% helix. The range obtained from circular dichroism measurements is 70–80% helix,[6] which implies a range from 25 to 28 amino acids per helix. The true helix lengths appear to be greater than the minimum, so the actual structure may contain some variation in length. For the present treatment, 25 amino acid helices are considered on the basis that longer helices do not permit sufficient link regions between helices (see below).

Proteolytic Cleavage

An important source of information is provided by the action of proteolytic enzymes on the bacteriorhodopsin structure. The underlying assumption in such studies is that the soluble enzymes will act to cleave only polypeptide links that are exposed to the aqueous environment. Consequently, one expects that such cleavage sites will be in connecting loops between helices and can, therefore, be used to identify the spacer regions between helices. The proteolytic cleavage sites reported to date are shown in the Table I. They fall in four regions of the polypeptide chain, the first being in the N-terminal region, up to the bond between Arg-7 and Pro-8, the second being from glycine 65 to valine 76, the third is a single site between Ser-161 and Met-162, and the fourth is a cleavage that results in liberation of 21 amino acids from the C terminus (Arg-227-Ala-228). These data from the laboratories of Ovchinnikov and Walker provide important clues concerning the link regions.

Since proteolytic cleavage results in the liberation of many residues from the C terminus, it may be asked whether this portion of the molecule might be one or a part of one of the helices. Electron microscopic studies of membranes from which 10 or 21 residue pieces of the carboxy terminus are removed show that the density map in projection is almost unaltered by such a modification.[7] Thus, one can exclude the 21 residues of the carboxy terminus as part of any helix.

[6] M. M. Long, D. W. Urry, and W. Stoeckenius, *Biochem. Biophys. Res. Commun.* **75**, 725 (1977).

[7] B. A. Wallace and R. Henderson, *in* "Electron Microscopy of Molecular Dimensions" (W. Baumeister and W. Vogell, eds.), p. 57, Springer-Verlag, Berlin and New York, 1980.

TABLE I
PROTEOLYTIC CLEAVAGE SITES

Enzyme	Site
Papain	Gly-Ile $(3-4)^a$
Papain	Gly-Gly $(72-73)^a$
Papain	Gly-Leu $(65-66)^a$
Papain	Ser-Met $(162-163)^a$
Chymotrypsin	Phe-Gly $(71-72)^a$
Trypsin	Arg-Ala $(227-228)^b$
Pronase + proteinase K	Arg-Pro $(7-8)^b$
Pronase + proteinase K	Phe-Gly $(71-72)^b$
Pronase + proteinase K	Gln-Asn $(75-76)^b$

[a] Yu. Ovchinnikov, N. Abdulaev, M. Fergira, A. Kiselev, and N. Lobanov, *FEBS Lett.* **100,** 219 (1979).
[b] J. E. Walker, A. F. Carne, and H. Schmitt, *Nature (London)* **278,** 653 (1979).

Other Modification Studies

In addition to proteolysis, information from other kinds of modification may be useful. One important result comes from the application of lactoperoxidase-catalyzed iodination. N. Katre and R. Stroud (personal communication) have used an immobilized lactoperoxidase molecule to modify purple membrane fragments. Subsequent analysis showed 75% of the radioactive iodine to be located in the cyanogen bromide fragment from amino acid 118 to 145. In this region only two tyrosines are present, tyrosine 131 and 133. Consequently, the expectation is that one or both of these tyrosines must be exposed at the aqueous surface of the bacteriorhodopsin molecule.

Another modification of interest is the derivitization of the membrane by a biotinyl reagent that reacts with lysine amino groups.[8] In this study it was thought that a single lysine was modified by the reagent in such a way as to permit binding of an avidin–ferritin complex. The use of electron microscopy resulted in a determination that only one side of the membrane, that part oriented toward the outside of the plasma membrane of the *Halobacterium*, was labeled. Although the location of the lysine in the sequence was not determined, it is evident that the modified group must

[8] R. Henderson, J. S. Jubb, and S. Whytock, *J. Mol. Biol.* **123,** 259 (1978).

be near the surface. Only lysine 129 is likely to be exposed on the outside surface of the protein, implicating it as the likely binding site for the biotinyl reagent.

Additional evidence of this kind comes from the work of Wallace using the tetrakisacetoxymethylmercury (TAMM) reagent. Again, the reagent is thought to interact with the ε amino group of lysine and binds with a one-to-one molar stoichiometry to bacteriorhodopsin. It labels the same surface as the biotin reagent, and it has been shown that only one peptide in the cyanogen bromide cleavage pattern is labeled.[7] As in the case of the biotin reagent, only lysine 129 appears to be a candidate for such modification. Although the biotin and TAMM modification studies cannot be taken as strong evidence, they provide a useful element of support for the choice of assignment that is presented later.

Energy Calculations

If the regions of the polypeptide that pass through the lipid bilayer are examined in terms of the free energy cost of moving them from an aqueous to a nonaqueous environment, a striking correspondence between the assignments dictated by modification studies and the free energy calculation is found. Steitz et al.[9] have explored the progressive insertion of the sequence of bacteriorhodopsin into a lipid bilayer. The sequence is taken as a single continuous alpha helix, which is moved progressively, one residue at a time, into and through the membrane barrier. The helix axis is maintained perpendicular to the barrier, and the transfer free energy for each successive segment is computed. A region of arbitrary width may be selected for analysis, but the appropriate choice for bacteriorhodopsin is 30 Å. The circumstances are illustrated in Fig. 1. The insertion free energy is divided into two parts, that due to the transfer of surface area from the aqueous to the nonaqueous environment, and that due to the presence of polar groups.[9] In Fig. 2 the free energy of burying successive alpha helical regions in the nonpolar regions is shown for bacteriorhodopsin. Each successive segment is identified by its initial amino acid, starting at the N terminus. It is clear that there are many negative excursions that occur along the amino acid sequence. These minima may correspond to the positions in the sequence of the seven alpha helical regions where they traverse the nonpolar region of the lipid bilayer, since they would be in their most stable location under that circumstance.

Based on this analysis, choices of the initial amino acid in a 19-residue sequence spanning the hydrophobic portion of the lipid bilayer can be

[9] T. A. Steitz, A. Goldman, and D. M. Engelman, *Biophys. J.* **37,** 124 (1981).

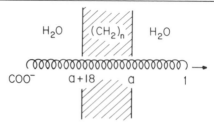

FIG. 1. In order to evaluate the free energy of insertion of the polypeptide chain into the lipid bilayer, the polypeptide is viewed as a continuous alpha helix of 248 amino acids. The calculation is made by a conceptual insertion of this helix in progressive steps of one amino acid into a nonpolar region as shown. The total insertion free energy for successive 19 amino acid segments of the chain is calculated based on an evaluation of the hydrophobic and hydrophilic components of the transfer free energy.[10]

evaluated. A 19-residue region is taken, since the amino acid side chains at the ends of the 21-residue stretch required to span the hydrophobic portion of the bilayer are presumed to be accessible to solvent. Thus the use of transfer free energy calculations permits an independent assessment of possible locations of the helical segments. Seven choices of helical segments are defined by the analysis and are indicated by arrows in Fig. 2. The link regions, which are also defined by this analysis, include all sites of modification by proteolytic cleavage and by chemical derivitization that have been described above. This gives substantial support to the idea

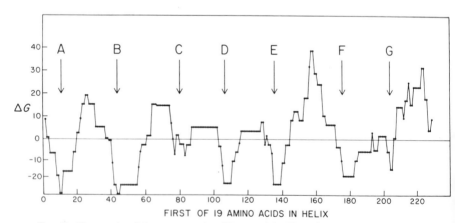

FIG. 2. The result of the computation illustrated in Fig. 1 is shown. The water-to-oil transfer free energy for successive 19 amino acid segments is plotted as a function of the first amino acid in a given segment. The curve shows a number of negative excursions, corresponding to favorable partitioning of a segment of the chain into the nonaqueous phase of a lipid bilayer. The positions that are thought to represent the starts of helices of bacteriorhodopsin are labeled A through F.

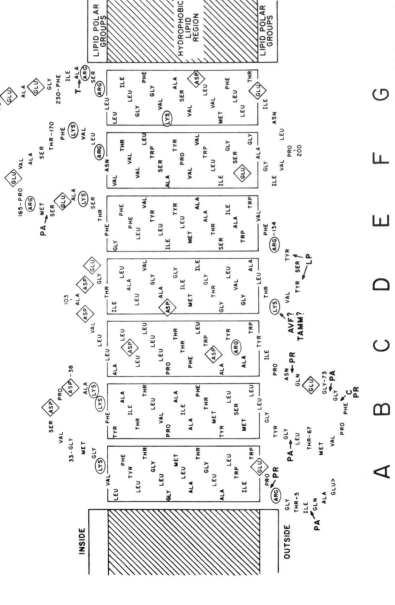

Fig. 3. Combining evidence from modification studies and free-energy calculations leads to the model shown. The helical regions are indicated as they may be related to the lipid bilayer structure. The nonpolar portions are indicated by the boxed areas. Sites of modification in the nonhelical portions are shown. (See Table I and text.) PR, pronase; PA, papain; TR, trypsin; C, chymotrypsin; AVF, avidin–ferritin bound to biotin; TAMM, tetrakisacetoxymethylmercury; LPL, lactoperoxidase iodination.

that the hydrophobic character of the helical regions can be used as a principle of identification.

A Choice of Model

In previous efforts to define the regions of sequence that correspond to segments, detailed analyses of the energetics of insertion were not considered.[3,10] If one presumes that the helices are longer than the 21 amino acids required for a minimal spanning of the nonaqueous portion of the lipid bilayer, it is difficult to decide what the appropriate choice of length should be. For present purposes we take a length of 25 amino acids as representing the reasonable minimum choice defined by the structural constraints. Using this length, together with the assignments implied by the modification and energetic considerations, we obtain the model shown in Fig. 3. We propose this model as the basis for further tests of ideas concerning the location of helices and hope that it will suggest approaches to tests of its validity.

In a recent study, Huang et al.[11] have used a photoactive retinal analogue to cross-link the retinal binding site, a Schiff base to lysine 216, to a region of the polypeptide near serine 193. If the chromaphore extends at an angle of about 20° from the plane of the membrane[12,13] and the photoactive group on the B-ionone ring located at about 15 Å from the axis of helix G, the site on helix F would be appropriately positioned for such derivatization in the assignment shown in Fig. 3.

Thus, a number of lines of evidence combine to provide a strong suggestion for the assignment of helical regions in the bacteriorhodopsin polypeptide. While the ends of the helices may vary by a few amino acids, large departures from the present assignment seem unlikely. A similar set of approaches may be useful in defining secondary structural features in a number of other globular membrane proteins, including rhodopsin, erythrocyte band III protein, and the calcium ATPase from sarcoplasmic reticulum.

[10] D. M. Engelman, R. Henderson, A. D. McLachlan, and B. A. Wallace, Proc. Natl. Acad. Sci. U.S.A. 77, 2023 (1980).
[11] K. S. Huang, H. Bayley, and H. G. Khorana, Fed. Proc., Fed. Am. Soc. Exp. Biol. 40, 1659 (1981).
[12] T. G. Ebrey, B. Becher, B. Mao, P. Kilbridge, and B. Honig, J. Mol. Biol. 112, 377 (1977).
[13] A. N. Kriebel and A. C. Albrecht, J. Chem. Phys. 65, 4576 (1976).

[12] Determination of Modified Positions in the Polypeptide Chain of Bacteriorhodopsin

By HORST-DIETER LEMKE, JURGEN BERGMEYER, and DIETER OESTERHELT

Introduction

The primary structure of bacteriorhodopsin (BR)* in the purple membrane (PM) is known[1,2] (also Chapters [9] and [86]) and the binding site of the retinal moiety has been reported to be Lys-276.[3,3a,3b] Chemical modification is an approach for identification of amino acids participating in chromophoric structure and proton translocation. Reports on the modification of the purple membrane by tetranitromethane and by other amino acid reagents have appeared.[4-8] Although specific effects on the chromophore and its photochemical cycle have been observed, no determination of the position of the modified residues in the polypeptide chain has been carried out and therefore the conclusions about their participation in the chromoprotein's function has remained speculative and very general.

This article describes high-performance liquid chromatography (HPLC) analysis of dipeptides derived from modified bacteriorhodopsin by subtilisin digestion. The structure of the peptides is established by coelution with authentic compounds. Some of the 247 possible dipeptide sequences are unique in bacteriorhodopsin and thus can be used to identify unequivocally the modified positions. In cases where a dipeptide sequence occurs several times in the chain, cyanogen bromide cleavage of the protein with subsequent separation of the fragments on a HPLC col-

Abbreviations used: BR, bacteriorhodopsin; BO, bacterio-opsin; PM, purple membrane; TLC, thin-layer chromatography; HPLC, high-performance liquid chromatography; Tyr(NO₂), 3-nitrotyrosine; CNBr, cyanogen bromide.

[1] Yu. A. Ovchinnikov, N. G. Abdulaev, M. Yu. Feigina, A. V. Kiselev, and N. A. Lobanov, *FEBS Lett.* **100**, 219 (1979).
[2] H. G. Khorana, G. E. Gerber, W. C. Herlihy, C. P. Gray, R. J. Anderegg, K. Nihei, and K. Biemann, *Proc. Natl. Acad. Sci. U.S.A.* **76**, 5046 (1979).
[3] H. Bayley, K.-S. Huang, R. Radhahnishaw, A. H. Ross, Y. Takagaki, and H. G. Khorana, *Proc. Natl. Acad. Sci. USA* **78**, 2225 (1981).
[3a] H.-D. Lemke, and D. Oesterhelt, *FEBS Lett.* **128**, 255 (1981).
[3b] E. Mullen, A. H. Johnson, and M. Akhtar, *FEBS Lett.* **130**, 187 (1981).
[4] T. Konishi and L. Packer, *FEBS Lett.* **92**, 1 (1978).
[5] M. C. Campos-Cavieres, T. A. Moore, and R. N. Perham, *Biochem. J.* **179**, 233 (1979).
[6] L. Packer, S. Tristham, J. M. Herz, C. Russell, and C. L. Borders, *FEBS Lett.* **108**, 243 (1979).
[7] R. Renthal, G. I. Harris, and G. I. Parrish, *Biochim. Biophys. Acta* **547**, 258 (1979).
[8] G. Harris, R. Renthal, I. Tuley, and N. Robinson, *Biochim. Biophys. Res. Commun.* **91**, 926 (1979).

METHODS IN ENZYMOLOGY, VOL. 88

umn can be used to determine the location of the modified positions. In some cases mass spectroscopy and sequence analysis of larger fragments facilitates the quantitation of extent of modification at various positions. As examples for the application of this method we describe the nitration of tyrosines[9] and an azo coupling reaction of tyrosine 64.[10] In both cases specific absorptions can be used for detection of modified peptides. Alternatively, the use of radioactivly labeled agents for modification allows the specific detection of these peptides.

Methods

Nitration of PM.[9] A solution of tetranitromethane (10% in ethanol, 0.2 ml) is added to 17 ml PM suspension (100 μM, $\epsilon = 63,000$ cm^{-1} M^{-1}) in 0.1 M Tris–HCl buffer, pH 8.0, prepared as described in Oesterhelt and Stoeckenius.[11] After 30 min at dim room light and 25° the reaction is terminated by pouring the mixture into a 10-fold volume of an aqueous solution of 2-mercaptoethanol (1% v/v). The membranes are washed four times with water and collected by centrifugation (150,000 g, 45 min). Extent of modification is determined by amino acid analysis.

Amino Acid Analysis Standard amino acid analysis is carried out with a commercial amino acid analyzer using the one-column procedure. Tyr(NO$_2$) elutes immediately after phenylalanine. The protein or the peptides are hydrolyzed in evacuated, nitrogen-flushed vials with constant boiling HCl containing 0.1% phenol for 24 hr at 115° or 20 hr for peptides.

Preparation of Diazotized Sulfanilic Acid. A solution of NaNO$_2$ (0.75 mmol dissolved in 7.5 ml H$_2$O) is added at 0° to 0.75 mmol sulfanilic acid dissolved in 10 ml 0.3 N HCl. After 30 min of reaction the solution is adjusted to pH 5 with 1 N NaOH and then diluted with water to a final volume of 25 ml.

Coupling of PM with Diazotized Sulfanilic Acid. Two milliliters of the diazotized sulfanilic acid solution are added dropwise at 0° to 6 ml of a sonified PM suspension (260 μM, 50% of maximal output of a Branson sonifier microtip for 1 min in pulse mode) in 0.05 M sodium borate buffer, pH 10, containing 1 M NaCl. After 3 hr the reaction is terminated by addition of 1 ml of a lysine solution (0.2 M). The reaction is allowed to

[9] H. D. Lemke and D. Oesterhelt, *Eur. J. Biochem.* **115**, 595 (1981).
[10] J. Bergmeyer, J. Straub, and D. Oesterhelt, *in* "High Pressure Liquid Chromatography in Protein and Peptide Chemistry" (F. Lottspeich, A. Henschen, and K.-P. Hupe, eds), p. 315(-324) de Gruyter, Berlin, 1981.
[11] D. Oesterhelt and W. Stoeckenius, this series, Vol. 31, Part A, p. 667.

proceed for another 15 min. The membranes are then washed three times with the reaction buffer and three times with water (150,000 g, 30 min).

Delipidization of the Modified Membrane. One μmole of modified PM in 3 ml of water is mixed with 10 ml of acetone–aqueous ammonia (5:1, v/v), and 6 M trichloroacetic acid is then added until the protein precipitates (about 100 μl). The precipitate is collected by centrifugation and washed twice with acetone by homogenization and centrifugation in order to remove residual lipids.

Proteolytic Cleavage of BO by Subtilisin. One μmole of modified bacterio-opsin (BO) is suspended in 5.3 ml of N-methylmorpholine/acetate buffer (0.1 M, pH 7.5) by sonification (see earlier) and digested with subtilisin (5% w/w, Sigma bacterial protease VII) at 37° for 8 hr and then centrifuged for 5 min in a bench centrifuge. The enzyme obtained from different suppliers may produce different peptide patterns and even different batches may have the same effect. Compared with the chromatogram in Fig. 5 of footnote 9, which shows an appreciable amount of Tyr(NO$_2$) as peak 1, our present batch of subtilisin under the same conditions does not yield Tyr(NO$_2$) but larger peptides (peptides 6 and 7, Fig. 1 and the Table).

Cyanogen Bromide Cleavage of BO. One μmole delipidated BO (see earlier) is first dissolved in 9 ml of formic acid and then diluted with water to a final acid concentration of 70%. Cyanogen bromide (180 mg) in 1.3 ml of 70% formic acid is added and the mixture is flushed with nitrogen for 2 min and incubated for 24 hr in the dark at 22°. It is then diluted with 50 ml of water and lyophilized.

HPLC of Subtilisin Digests. The supernatant of the subtilisin digest is lyophilized and redissolved in formic acid. Peptides are separated on a reversed-phase column (Shandon ODS-Hypersil; 5 μm, 4.5 × 300 mm, Bischoff, Stuttgart) using a Knauer Gradient-System. Gradient elution was performed with 5% aqueous formic acid with a linearly increasing content of acetonitrile (0.5%/min). Elution rate is 1.5 ml/min and fractions are collected in 1 ml portions for further analysis. Nitrated peptides are detected by their specific absorption at 360 nm, other peptides by their absorption at 280 nm or 215 nm. On a column of this diameter maximally 13 mg material are applied. However, using a preparative column (16 × 250 mm) of the same type, up to 40 mg in a single run can be separated without noticeable loss of resolution. For further purification thin-layer chromatography is used. Samples (50 nmol) are placed on precoated silica gel plates (Merck, Darmstadt) and developed with isopropanol–acetone–25% aqueous ammonia (90:90:20). Peptides are detected with 10% ammonia in acetone (nitrated peptides) or fluorescamine (1 mg in 100 ml ace-

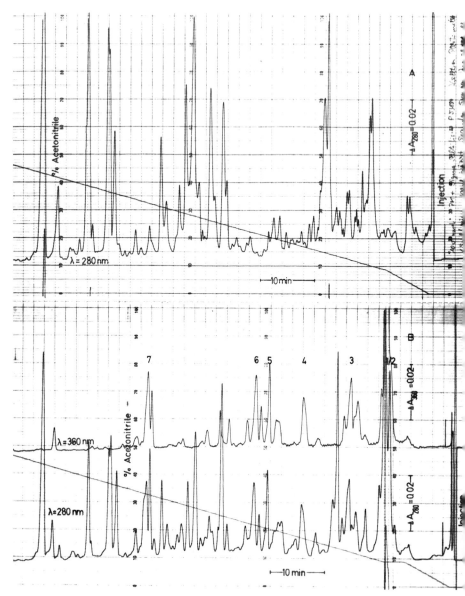

FIG. 1. Separation of peptides derived from subtilisin digested PM by reversed-phase HPLC. (A) Chromatogram of unmodified PM. Aromatic amino-acid-containing peptides are detected by their absorption at 280 nm. (B) Chromatogram of nitrated PM. Nitrated peptides 1–7 are detected by their specific absorption at 360 nm. This trace delays behind the trace at 280 nm as seen by the two marker lines before injection. For better separation of peak 1 and 2 the gradient is interrupted for 4 min. Therefore elution conditions for A and B are identical only in the second part of the profile. B was run four weeks later than A, using a different batch of reversed-phase material and freshly prepared elution buffers, indicating the high reproducibility of that procedure.

TABLE

AMINO ACID ANALYSIS OF NITRATED PEPTIDES AFTER CHROMATOGRAPHY ON REVERSED-PHASE AND ANION EXCHANGE HPLC[a]

	Peptide no.						
	1	2	3	4	5	6	7
Threonine	—	1.0	—	—	—	0.8	—
Serine	0.8	0.7	—	—	0.6	—	0.2
Glycine	—	1.2	0.1	0.1	0.2	3.0	—
Alanine	—	—	—	0.1	—	—	—
Valine	—	—	1.0	0.1	—	—	0.1
Leucine	—	—	—	1.0	—	1.4	1.1
Tyrosine	—	0.3	—	—	—	—	—
3-Nitrotyrosine	1.0	1.0	1.0	1.0	1.0	1.0	1.0
Phenylalanine	—	0.6	—	—	1.1	—	1.0
	Ser-Tyr$_{133}$	Gly-Tyr$_{64}$	Val-Tyr$_{131}$	Leu-Tyr$_{133}$	Tyr$_{26}$-Phe	Gly,Tyr$_{64}$, Gly,Leu,Thr	Thy$_{26}$,Phe,Leu

[a] Peptide 2 is not obtained in a pure state. Its structure was derived from coelution and by mass spectroscopy.[9] Peptide 6 was purified with TLC as the second step. The glycine content has to be corrected by the amount of glycine in the plate (control sample, about 1 mol).

93

tone containing 1 ml pyridine). Yellow peptides are eluted with 1.5 M aqueous ammonia and after lyophilization subjected to further analysis. Alternatively, separation of all peptides can be achieved if fractions derived from the reversed phase column are injected onto anionic exchange columns[12] (see below).

HPLC of CNBr-Peptides. The lyophilized peptide mixture is dissolved in formic acid and 50–500 nmol aliquots are injected into a Latek-HPLC gradient system equipped with a Lichrosorb RP 18 column (4.5 × 300 mm, 10 μm, Knauer, Berlin). Gradient elution is performed with 5% aqueous formic acid with a linearly increasing content of acetonitrile (from 0 to 100% within 1 hr). Elution rate is 2 ml/min and 1 ml fractions are collected. Further purification of the isolated modified CNBr-peptide V (amino acid residues 61–68) is achieved by HPLC on an anion exchange column (Varian MicroPak AX-5, 4 × 300 mm). Gradient elution is performed using acetonitrile with a linearly increasing content of 10 mM triethylammonium acetate, pH 4.3 (from 0 to 50% in 100 min).[12] Elution rate is 1.5 ml/min and 1-ml fractions are collected.

Results

Treatment of the purple membrane with tetranitromethane under conditions described earlier leads to the nitration of 3 moles tyrosine per mole bacteriorhodopsin. Figure 1 compares the chromatogram of the subtilisin digest of an unmodified membrane (Fig. 1A) with that of a nitrated sample (Fig. 1B). Nitrated peptides are detected by their specific absorption at 360 nm and further purified either by TLC or by HPLC using an anionic exchange column.

An example of such a chromatogram is given in Fig. 2. The table summarizes the amino acid analysis of the purified peptides. The stoichiometry found allows the deduction of the peptide's net amino acid composition. The structures of the dipeptides (peaks 1–5) were derived from coelution experiments with authentic compounds.[9] The resolution power of the reversed-phase column allows the separation of the most closely related peptides, such as Ile-Tyr(NO₂) and Leu-Tyr(NO₂) or Phe-Tyr(NO₂) and Tyr(NO₂)-Phe. This means that the structure of dipeptides can be analyzed without prior purification.

The amino acid composition of the larger peptides 6 and 7 (Fig. 1 and the table) unambiguously identifies the tyrosines 64 and 26 as carrying the nitro group. Also peptides 1–3 and 5 can only be derived from positions 26, 64, 131, or 133. The only ambiguity concerns peptide 4 (Leu-

---

[12] M. Dizdaroglu and M. G. Simic, J. Chromatogr. **195**, 119 (1980).

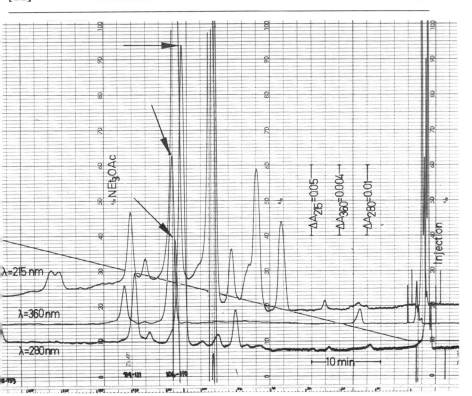

FIG. 2. Rechromatography of peak 5 from the chromatogram in Fig. 1B on an anion exchange column. Peptides are detected by their absorption at 215, 280, or 360 nm. The shift of the three traces can be seen from the last eluted peak absorbing at all three wavelengths. NEt₃OAc indicates the percentage of 10 mM triethylammonium acetate pH 4.3, in acetonitrile; $T = 40°$. The main nitropeptide marked by three arrows, is absorbing at 360 nm (indicated by the arrow in the middle) and is well separated from the main impurity only detectable at 215 nm (peptide bond absorption).

Tyr(NO₂)), a sequence that occurs at positions 25–26, 146–147, and 149–150. The CNBr cleavage of nitrated BO and subsequent HPLC as described under Methods and shown in Fig. 3 for unnitrated BO reveal that fragment II (positions 21–32) but not fragment VIII (positions 146–163) contain Tyr(NO₂).

As a result of quantitative analysis,[9] tyrosine 26 and 64 are shown to be fully nitrated, whereas tyrosine 131 and 133 share the third mole of nitro groups. After reduction of the nitrated membrane with sodium dithionite peptide, analysis as described earlier demonstrates that only position 26 remains nitrated, whereas the other Tyr(NO₂)'s are reduced by this water-soluble reagent.

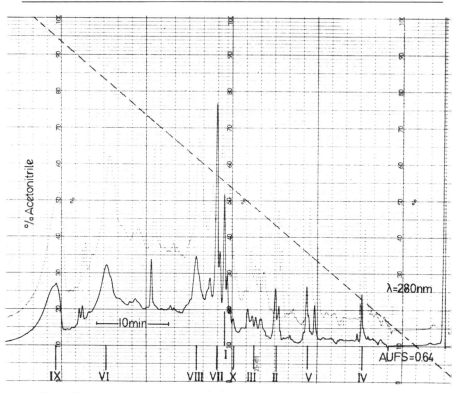

Fig. 3. Separation of all CNBr fragments (I-X) of BO by reversed-phase HPLC. I: positions 1–20; II: positions 21–32; III: positions 33–56; IV: positions 57–60; V: positions 61–68; VI: positions 69–118; VII: positions 119–145; VIII: positions 146–163; IX: positions 164–209; X: positions 210–248. Satellite peaks are presumably due to the homoserine–homoserine lactone equilibrium in solution. The low absorption at 280 nm of fragment X is due to the absence of tryptophane and tyrosine residues.

This confirms structural models of bacteriorhodopsin that place the tyrosine 64, 131, and 133 at the outer extracytoplasmic surface of the purple membrane.[1,13]

The location of tyrosine 64 at the outer surface is confirmed by modification using diazo coupling with diazotized sulfanilic acid. Under the conditions described in the section on methods, only tyrosine 64 reacts. The chromatogram in Fig. 4A shows only one peak with an absorption at 330 nm characteristic for the sulfanilic acid moiety. This peak corresponds to fragment V (position 61–68) of the chromatogram in Fig. 3. It

[13] D. M. Engelman, R. Henderson, A. D. McLachlan, and B. A. Wallace, Proc. Natl. Acad. Sci. U.S.A. 77, 2023 (1980).

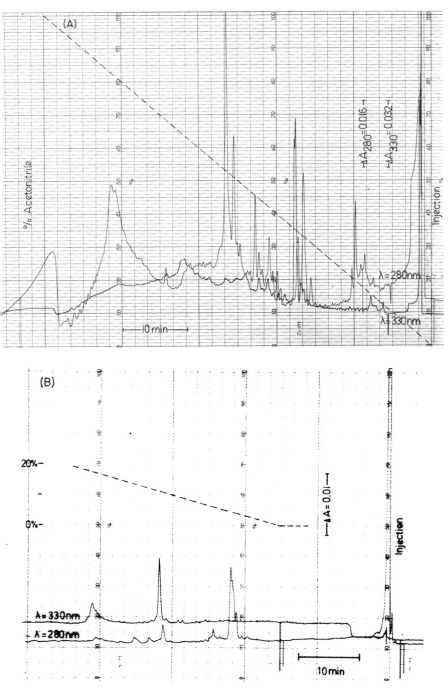

FIG. 4. (A) Chromatogram of CNBr fragments of diazo-coupled PM. Only fragment V shows absorption at 330 nm. (B) Rechromatography of fragment V on an anion exchange column. The broken line shows the gradient (0–20%) of 10 mM triethylammonium acetate, pH 4.3, in acetonitrile.

can be further purified by chromatography on an anionic exchange column (Fig. 4B). After rechromatography all other fragments also show amino acid compositions in accordance with the theoretical one and therefore justify the assignment in Fig. 3 (compare Bergmeyer *et al.*[10]).

In summary, the method described here for identification of modified positions in BR is simple, reliable, and at best requires a two-step HPLC procedure.

[13] Lipids of Purple Membrane from Extreme Halophiles and of Methanogenic Bacteria

By M. KATES, S. C. KUSHWAHA, and G. D. SPROTT

A. Purple Membrane (PM)[1]

1. Preparation of PM[1,2]

Reagents

Growth medium: dissolve NaCl (250 g), KCl (2.0 g), sodium citrate (3.0 g), $MgSO_4$ (9.8 g), and $FeSO_4$ (50 mg) in 1 liter dist. H_2O; autoclave this solution and add oxoid peptone (10.0 g).

Basal salt solution: NaCl (250 g), KCl (2.0 g), and $MgSO_4$ (9.8 g) in 1 liter of dist. H_2O; adjust final pH to 6.5.

Sucrose, 1.3 M and 1.5 M.

Deoxyribonuclease (DNase), DN-100 from beef pancreas (Sigma Chemical Co.).

Procedure. Cells of *Halobacterium cutirubrum*[2] or *Halobacterium halobium*[1,3] are first grown aerobically at 37° in 2-liter batches of the above growth medium (volume of inoculum: volume of growth medium, 1:4) in 4-liter Erlenmeyer shaker-flasks in an incubator shaker (New Brunswick Scienctific Co., Inc.) at a shaking rate of 180 rpm for 72 hr. Aeration is then reduced by lowering the shaking rate to 90 rpm and incubation is continued under fluorescent light (a bank of six 20-in. "cool white" fluorescent tubes) for 72 hr, all other conditions remaining the same. Cells from six 2-liter batches are harvested by centrifugation at 10,000 *g* for 20 min, washed twice with 200-ml basal salt solution, and treated with 10 mg

[1] D. Oesterhelt and W. Stoeckenius, *Nature (London)*, New Biol. **233**, 149 (1971).
[2] S. C. Kushwaha, M. Kates, and W. G. Martin, *Can. J. Biochem.* **53**, 284 (1975).
[3] S. C. Kushwaha, M. Kates, and W. Stoeckenius, *Biochim. Biophys. Acta* **426**, 703 (1976).

of DNase for 30 min at 22° with stirring. The cell suspension is then dialyzed at 4° against 6 liters of distilled water for 16 hr with at least two changes of water. The dialysate is centrifuged at 10,000 g for 20 min to remove cell debris, and the supernatant centrifuged at 50,000 g for 1.5–2.0 hr. The pellet is suspended in 200–300 ml of dist. water, and the suspension centrifuged at 10,000 g for 20 min to further remove debris and then at 50,000 g for 1.5 hr to yield the "crude" purple membrane. This is suspended in about 18 ml of dist. water and purified by centrifugation on a sucrose density gradient (6 ml of 1.3 M sucrose layered on top of 3 ml of 1.5 M sucrose at 4°) at 260,000 g for 18 hr in a SW-27 Ti swinging bucket rotor in a L₂C-65B Beckman ultracentrifuge at 5°. The PM band appears at the interface of the 1.5 M and 1.3 M sucrose layers (the red band appears at the top of the 1.3 M sucrose layer) and is collected by displacement, dialyzed against distilled water to remove sucrose, and recentrifuged on the same sucrose gradient, as described earlier. The PM, which now forms a single sharp band, is removed, stored at 4° overnight if necessary, and dialyzed against deionized distilled water just before use. Yield of PM, 2–10 mg per liter of culture. For long-term storage, the dialyzed PM suspension is centrifuged at 48,000 g 1 hr and the PM pellet resuspended in basal salt solution (10–15 ml per 100 mg PM).

Protein is determined by the method of Lowry et al.[4] using bovine serum albumin as standard. Dry weights of PM are determined after drying suitable aliquots of a dialyzed PM suspension in a weighing bottle over KOH in vacuo (<1 mm Hg) at room temperature.

The PM preparations had λ_{max} 565 and 275 nm with molar extinction coefficient $\epsilon_{565} = 4.8 \times 10^4$ and ratio $A_{565}/A_{275} = 2.05$, essentially as reported by Oesterhelt and Stoeckenius.[1] The PM preparations also showed a single band on SDS-gel electrophoresis.[2,3]

2. Preparation of Lipid-Depleted PM

Lipid-depleted PM was prepared by treatment of PM (55.9 mg) with deoxycholate/cholate according to the procedure of Hwang and Stoeckenius.[5] The deoxycholate/cholate treatment was repeated three times. The delipidated PM was finally dialyzed against 75 mM K₂SO₄ to remove sucrose and analyzed for protein,[4] lipid-P by the modified Bartlett procedure,[6] and vitamin-MK.[2]

[4] O. H. Lowry, N. J. Rosenbrough, A. L. Farr, and R. J. Randall, J. Biol. Chem. **193**, 265 (1951).

[5] S. B. Hwang and W. Stoeckenius, J. Membr. Biol. **33**, 325 (1977).

[6] M. Kates, "Techniques of Lipidology," pp. 365, 436, 438. Am. Elsevier, New York, 1972.

3. *Total Lipid Extraction*

Whole cells were extracted as described previously.[6] The sucrose-free PM was extracted essentially by the method of Bligh and Dyer[7] as described previously,[2] modified as follows: A suspension of 100 mg PM in 24 ml dist. H_2O is diluted with 60 ml of methanol and stirred in the dark under N_2 for 45 min; 30 ml of $CHCl_3$ is added and the stirring is continued for 30 min. The mixture is centrifuged, and the supernatant is collected; the pellet is extracted with 42 ml of methanol–H_2O (2.5:1) for 10 min with stirring, 15 ml of $CHCl_3$ is added, and stirring is continued for 10 min. After centrifugation the combined supernatants are diluted with 45 ml of $CHCl_3$ and 45 ml of 0.2 M KCl with gentle mixing and the phases are allowed to separate in a separating funnel for several hours. The lower chloroform phase is removed, diluted with benzene and ethanol, and brought to dryness on a rotary evaporator at 30°. The residual total lipids are dissolved in chloroform to a known volume and subjected to separation and analysis as described in the following discussion. The total lipids have a slight yellow color because of the retinal present (yield from 100 mg PM, 20 mg). Polar and nonpolar lipids were separated by acetone precipitation.[2]

4. *Separation of Lipid Components*

Nonpolar lipid components of PM were separated by one-dimensional TLC on silica gel G (0.5-mm-thick layers) in the following solvent systems: A, 0.3% ethyl ether in hexane for squalene, dihydrosqualene, tetrahydrosqualene, and β-carotene; B, 6% ethyl ether in hexane for vitamin MK-8; and C, 1% ethyl ether in chloroform for retinal. Phospholipids and glycolipids were separated by two-dimensional TLC on silica gel H (0.5-mm-thick layers) using solvent E, chloroform–methanol conc. NH_4OH (65:35:5, v/v) in the first direction, and solvent F, $CHCl_3$ – 90% HAc-MeOH (30:20:4, v/v/v) in the second direction. Before use, TLC plates are washed by ascending development with $CHCl_3$-MeOH (1:1, v/v), air dried, and activated at 110° for 12 hr. Retinal and carotenoids are detected by their visual colors and squalenes and vitamin MK-8 by iodine vapor; the spots are scraped off the plate, eluted with acetone, and quantitated spectrophotometrically using the following $E_{1cm}^{1\%}$ values: retinal, 1510 at 383 nm in ethanol; β-carotene, 2505 at 450 nm in petroleum ether; vitamin MK-8, 268 at 248 nm in petroleum ether; squalenes were determined gravimetrically. Retinal was also determined by the thiobarbituric acid method of Futterman and Saslaw[8] directly on the total lipid extract. Phos-

[7] E. G. Bligh and W. J. Dyer, *Can. J. Biochem. Physiol.* **37**, 911 (1959).
[8] S. Futterman and L. D. Saslaw, *J. Biol. Chem.* **236**, 1652 (1961).

pholipids are detected by the phosphate spray reagent[6] and glycolipids by the α-naphthol spray reagent.[6] For quantitative purposes, the TLC plate was stained with I_2 vapor, and the individual phospholipid and glycolipid spots were scraped directly into digestion tubes and analyzed for phosphorus[6] or sugar,[6] respectively.

For preparative purposes, polar lipids are isolated by TLC on preparative silica gel H plates (1.0–0.75 mm thick; 25–30 mg lipids per plate) developed twice in solvent system E, as described elsewhere.[9,10] The bands corresponding to the lipid components were detected by visualization with Rhodamine 6G or dichlorofluorescein under ultraviolet light and eluted[10] from the silica with chloroform–methanol–0.1 N HCl (1:2:0.8, v/v); the eluates are converted to a two-phase system by addition of 1 vol each of chloroform and water and the chloroform layer is immediately neutralized with 0.2 N methanolic NH_4OH, diluted with benzene and brought to dryness *in vacuo*. The residual ammonium salts are precipitated from chloroform with 20 vol of acetone in the cold, collected by centrifugation and dried *in vacuo*.

5. *Lipid Composition of Whole Cells and PM*

Total lipids of extreme halophiles account for about 3–4% of cell dry weight. Polar lipids constitute about 90% of the total lipids and have been shown[11] to be derivatives of a glycerol diether, 2,3-di-*O*-phytanyl-*sn*-glycerol as follows: (1) phosphatidylglycerophosphate (PGP), (2) phosphatidylglycerol (PG), (3) phosphatidylglycerosulfate (PGS), and (4) a glycolipid sulfate (GLS), the structures of which are shown in Fig. 1. The composition of these polar lipids in whole cells (as well as in PM) is of unusual simplicity (Table I), only two major components being present, PGP (70 mol %) and GLS (20 mol %), two other components, PGS and PG being present in low amounts (4 and 5 mol %, respectively).

Recently, a minor glycolipid sulfate has been identified[12] in *H. cutirubrum* as a α-galactofuranosyl derivative of GLS; it accounts for about 1–2 mol % of the total polar lipids.

Neutral lipids,[11] accounting for about 10% of total lipids, consist of squalenes (mostly squalene and smaller amounts of di- and tetrahydrosqualene), vitamin MK-8, C_{40}-carotenoids, geranylgeraniol, diphytanyl glycerol ether, bacterioruberins (C_{50} red pigments), and retinal, which is of course present in the form of bacteriorhodopsin; the composition of the

[9] A. J. Hancock and M. Kates, *J. Lipid Res.* **14**, 422 (1973).
[10] M. Kates and P. Deroo, *J. Lipid Res.* **14**, 438 (1973).
[11] M. Kates, *Prog. Chem. Fats Other Lipids* **15**, 301 (1978).
[12] B. W. Smallbone and M. Kates, *Biochim. Biophys. Acta* **665**, 551 (1981).

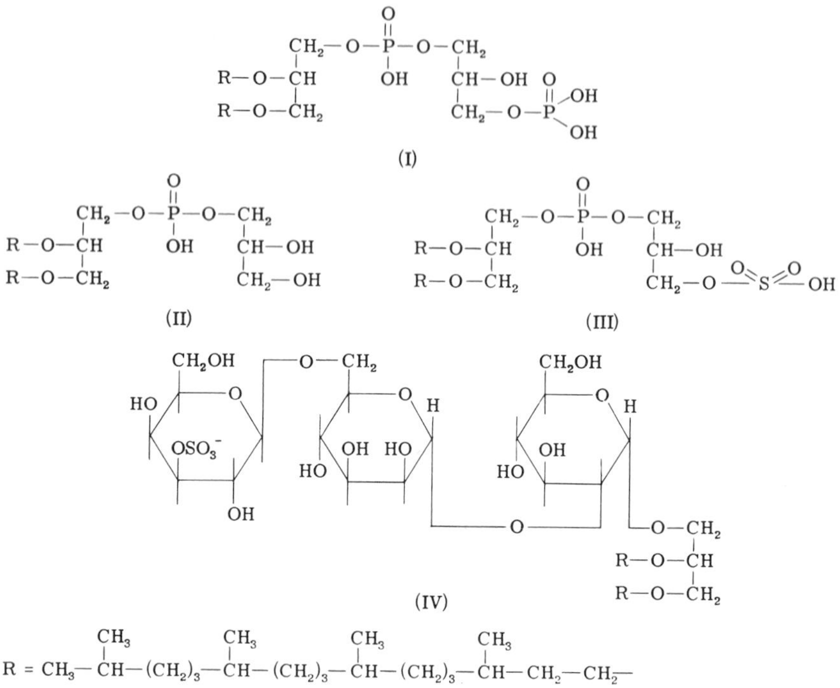

FIG. 1. Structures of major polar lipids in *Halobacterium cutirubrum:* I, phosphatidyl-glycerophosphate (PGP); II, phosphatidylglycerol (PG); III, phosphatidyl glycerosulfate (PGS); IV, glycolipid sulfate (GLS). In all compounds the alkyl group R is 3*R*, 7*R*, 11*R*, 15-tetramethylhexadecyl.

neutral lipids is given in Table I. Note that squalenes constitute a considerable proportion of the total lipids on a mole basis (10 mol %).

Analyses of the PM[2,11] show that it contains about 20% by weight of lipid and 77% by weight of protein, giving a lipid-to-protein weight ratio of 0.25:1. The mole ratio of lipid to protein (exclusive of retinal) was about 7:1 or about 21 molecules per unit cell (Table I); the polar lipids alone amount to 6 molecules per bacterio-opsin molecule (or 18 molecules per unit cell). These values are somewhat lower than those calculated by Blaurock and Stoeckenius[13] (40 molecules lipid per unit cell) or by Blaurock[14] (33 molecules lipid per unit cell), but higher than that calculated by Stoeckenius *et al.*[15] (12–14 molecules per unit cell). The composition of the polar lipid is similar to that of whole cells (Table I) but the

[13] A. E. Blaurock and W. Stoeckenius, *Nature (London), New Biol.* **233**, (1971).
[14] A. E. Blaurock, *J. Mol. Biol.* **93**, 139 (1975).
[15] W. Stoeckenius, R. A. Bogomolni, and R. H. Lozier, *in* "Molecular Aspects of Membrane Phenomena" (H. R. Kaback, H. Neurath, G. K. Radda, R. Schwyzer, and W. R. Wiley, eds.), p. 306. Springer-Verlag, Berlin and New York, 1975.

proportions of PGP and GLS are appreciably lower and higher, respectively, than in whole cells. The high content of the triglycosyl diether is due to hydrolytic breakdown of the glycolipid sulfate (GLS → TGD + H_2SO_4) during preparation of the PM; the values for TGD are therefore combined with those of GLS.

Note again that squalenes constitute about 10 mol % of total lipids in PM, amounting to about two molecules per unit cell (Table I). The squalenes appear to aid in stabilization of the halophile membrane structure by facilitating interaction of divalent ions with the acidic groups of the polar lipids.[16]

The very low content of vitamin MK-8 (0.5 molecule per unit cell) suggests that this quinone may not be a component of the PM but may have arisen from contamination with red membrane that presumably contains the vitamin MK-8 as part of its electron transport system. However, if this is true, the fact that the content of vitamin MK-8 in PM (0.32%) is similar to that of red membrane (0.44%) (see Kushwaha et al.[2]) would suggest that the PM is contaminated with about 70% of red membrane, which is clearly not the case. We have found that repeated delipidation with deoxycholate according to the procedure of Hwang and Stoeckenius[5] results in the removal of about 90% each of the phospholipids and of vitamin MK-8. The residual amount of quinone (0.05 mol/unit cell) may be integrally bound to the protein, but the function of such a small amount of vitamin MK is difficult to envisage.

Our previous analyses for retinal in PM,[2,3] involving extraction with cetyl pyridinium bromide in the presence of hydroxylamine, gave low retinal contents (0.59–0.61 mol/mol protein of MW 26,000). More recent analyses (B. Smallbone and M. Kates, unpublished data) utilizing extraction with 90% acetone, estimation of retinal by the thiobarbituric acid procedure,[8] and determination of protein by amino acid analysis gave retinal: protein mol ratios of 0.85 ± 0.05 (Table I) approaching a value of 1, as previously reported.[1,17] The low values for retinal content may be due to incomplete cleavage of the protonated Schiff-base linkage of retinal to a lysyl residue in bacteriorhodopsin and/or reformation of new, more stable Schiff-base linkage(s) between the extracted free retinal and other lysyl residues made available in the denatured protein. Either process would result in incomplete extraction of retinal and hence in low values for the retinal content.

Analyses for total-P and lipid-P in PM[2] suggested the presence of 1 or 2 mol phosphate bound to the protein. However, the question whether bacterio-opsin is a phosphoprotein can only be settled by direct analysis

[16] J. K. Lanyi, W. Z. Plachy, and M. Kates, *Biochemistry* **13**, 4914 (1974).
[17] G. K. Papadopoulis, T. L. Hsiao, and J. Y. Cassim, *Biochem. Biophys. Res. Commun.* **81**, 127 (1978).

TABLE I
LIPID COMPOSITION OF EXTREMELY HALOPHILIC BACTERIA AND OF PURPLE MEMBRANE

Lipid component	MW	H. cutirubrum[2,11] Whole cells[a]		H. cutirubrum[2,11] Purple membrane[a]			H. marismortui[18] Whole cells	
		Mol % total lipids	Mol % polar lipids	Mol % total lipids	Mol % polar lipids	Mol per unit cell	Mol % total lipids	Mol % polar lipids
Neutral lipids								
Squalenes	410 (ave)	9	—	10	—	2	22	—
Vitamin MK-8	716	1	—	2	—	0.5	4	—
Retinal	284	0.3[b]	—	12[b]	—	2.6[b]	0.1	—
C$_{50}$ bacterioruberins	740	2	—	0	—	0	2	—
Polar lipids								
Phosphatidylglycerol	830	4	5	5	6	1	10	11
Phosphatidylglycerophosphate	930	58	70	49	61	11	56	61
Phosphatidylglycerosulfate	920	4	4	4	5	1	15	17
Glycolipid sulfate	1240	17	21	7 ⎫ 22	28	5	0	0
Triglycosyl diether[c]	1140	trace	trace	15 ⎭			10	11

[a] Lipids of whole cells also contain C$_{40}$ carotenoids (0.1 mol %), geranylgeraniol (3.5 mol %) and diphytanylglycerol (1.3 mol %). These components are absent in purple membrane.

[b] Recent results using the thiobarbituric acid procedure[8] on 90% acetone extracts of PM (B. Smallbone and M. Kates, unpublished results).

[c] Structure of triglycosyl diether is galactose–mannose–glucose diether in H. cutirubrum and glucose–mannose–glucose diether in H. marismortui.

of the lipid-free opsin. Bacterio-opsin is probably not a glycoprotein since all of the hexose found in PM is accounted for by the glycolipids.[2]

6. *Lipid Composition of Other Halophilic Bacteria*

The halophilic bacteria examined so far, *H. cutirubrum* and *H. halobium*, which grow naturally in salt flats, appear to have the same lipid composition.[3] However, a recent study[18] of the lipids of a Dead Sea halophile *Halobacterium marisimortui* has shown this organism to have a simple but different polar lipid composition to that of the halophiles previously studied[11] (Table I). The Dead Sea organism, while also having a high content (62 mol %) of PGP, had no detectable glycolipid sulfate component but made up for the deficit in sulfate by having an increased content of PGS (17 mol %) (Table I). By also having an increased content of PG (11 mol %) it appeared to maintain the number of negative charges per mole ionic lipid at a value of about 2, as in *H. cutirubrum*. Another difference in the lipids of the Dead Sea halophile is the presence of a high proportion (11 mol %) of a triglycosyl diether that is similar to the one in *H. cutirubrum* but differs in having a terminal glucose instead of galactose. Since it has been shown[19,20] that the glycolipid sulfate is necessary for formation of stable bilayers of *H. cutirubrum* polar lipids, probably as a result of its large polar head, it may be that the triglycosyl diether in *H. marismortui* has a similar membrane stabilizing function. It would be of interest to examine the lipids of PM in *H. marismortui*, but so far synthesis of PM in this organism could not be demonstrated.[18]

B. Lipids of *Methanospirillum hungatei* GPI

1. *Culture of Organism*

Reagents[21,22]
MINERAL SOLUTION 1: Dissolve 3.9 g K_2HPO_4 in 1 liter deionized water.
MINERAL SOLUTION 2: Dissolve the following in 1 liter deionized water: KH_2PO_4 (2.4 g), $(NH_4)_2SO_4$ (6.0 g), NaCl (0.59 g), $MgSO_4 \cdot 7H_2O$ (1.2 g), and $CaCl_2 \cdot 2H_2O$ (0.79 g).
TRACE MINERAL SOLUTION: Dissolve nitrilotriacetic acid (1.5 g) by adjusting the pH to 6.5 with KOH, then dissolve the following in 1 liter of

[18] R. W. Evans, S. C. Kushwaha, and M. Kates, *Biochim. Biophys. Acta* **619**, 533 (1980).
[19] J. S. Chen, P. G. Barton, D. Brown, and M. Kates, *Biochim. Biophys. Acta* **352**, 202 (1974).
[20] B. W. Smallbone, Ph.D. thesis, University of Ottawa, Ottawa, Canada (1982).
[21] G. B. Patel and L. A. Roth, *Can. J. Microbiol.* **23**, 893 (1977).
[22] E. A. Wolin, M. J. Wolin, and R. S. Wolfe, *J. Biol. Chem.* **238**, 2882 (1963).

the resulting solution, $MgSO_4 \cdot 7H_2O$ (3.0 g), $MnSO_4 \cdot H_2O$ (0.5 g), $CoCl_2 \cdot 6H_2O$ (0.1 g), $CaCl_2 \cdot 2H_2O$ (0.1 g), $ZnSO_4 \cdot 7H_2O$ (0.1 g), $CuSO_4 \cdot 5H_2O$ (0.1 g), $AlK(SO_4)_2 \cdot 12H_2O$ (0.01 g), H_3BO_3 (0.01 g), $Na_2MoO_4 \cdot 2H_2O$ (0.024 g) and $NiCl_2 \cdot 6H_2O$ (0.12 g).

TRACE VITAMIN MIXTURE: Dissolve the following in 1 liter deionized water: biotin (2 mg), folic acid (2 mg), pyridoxine·HCl (10 mg), thiamine·HCl (5 mg), riboflavin (5 mg), nicotinic acid (5 mg), p-aminobenzoic acid (5 mg), lipoic acid (5 mg), and vitamin B_{12} (0.5 mg).

$FeSO_4$ SOLUTION: $FeSO_4 \cdot 7H_2O$ is dissolved in 100 ml deionized water acidified with 3 drops conc. HCl.

RESAZURIN SOLUTION: Resazurin (0.1 g) is dissolved in 100 ml deionized water.

NA_2CO_3 SOLUTION: Na_2CO_3 (8.0 g) is dissolved in 100 ml deionized water and brought to a boil under H_2-CO_2 (4:1, v/v) in a 500-ml boiling flask. After bubbling the gas mixture through the solution for 30 min, 10-ml aliquots are dispensed anaerobically by syringe into 120-ml serum bottles.

CYSTEINE-SULFIDE REDUCING SOLUTION: Cysteine·HCl (2.5 g) is dissolved in 50 ml deionized water and 3 N NaOH is added to pH 10. The solution is placed in a 500-ml boiling flask and flushed with N_2, $Na_2S \cdot 9H_2O$ (2.5 g) is added, and the volume is increased to 200 ml. The solution is taken to a boil under N_2, transferred anaerobically to serum bottles, and autoclaved.

Procedure. Synthetic acetate broth (SA) is prepared by adding the following reagents in the order given to 800 ml of deionized water: mineral solution 1 (75 ml), mineral solution 2 (75 ml), $FeSO_4$ solution (10 ml), sodium acetate (2.5 g), trace mineral solution (10 ml), trace vitamin mixture (10 ml), and resazurin solution (1.0 ml). The medium is taken to a boil under H_2-CO_2 (4:1, v/v) and allowed to cool for 15 min. Na_2CO_3 solution (5.0 ml) is added anaerobically and the medium is reduced according to Hungate[23] with cysteine–sulfide solution (12 ml). Aliquots are dispensed anaerobically by syringe to either 120-ml serum vials[24] or to 1-liter storage bottles modified according to Balch and Wolfe.[25] Bottles containing the medium with H_2-CO_2 (4:1, v/v) in the head space at atmospheric pressure are autoclaved for 15 min at 15 psi. *Methanospirillum hungatei* GPI (isolated by Patel *et al.*[26] and deposited with the German Collection of Microorganisms, Gottingen, as catalog No. DSM 1101) is grown at 35° under an atmosphere of H_2-CO_2 (4:1, v/v) in the synthetic broth (SA)

[23] R. E. Hungate, *Bacteriol. Rev.* **14**, 1 (1950).
[24] T. L. Miller and M. J. Wolin, *Appl. Microbiol.* **27**, 985 (1974).
[25] W. E. Balch and R. S. Wolfe, *J. Bacteriol.* **137**, 264, (1978).
[26] G. B. Patel, L. A. Roth, L. Van Den Berg, and D. S. Clark, *Can. J. Microbiol.* **22**, 1404 (1976).

described earlier. Cultures are maintained by weekly transfer into the 120-ml serum bottles (10 ml broth). Five-day-old broth cultures grown in serum bottles served as inoculum for 100-ml aliquots of medium (5%, v/v). Cultures were shaken at 150 rpm and supplied daily with H_2-CO_2 (4:1, v/v) to 10 psi.

Fementor cultures of 6 liters are prepared as described elsewhere[27] except for the concentrations of cysteine (0.51 mM) and Na_2S (0.33 mM) used to reduce the SA medium. As growth proceeds, the stirring rate is increased to 400 rpm. Further additions of cysteine (0.066 mM) and Na_2S (0.043 mM) are made every second day during growth. The initial inoculum consists of several 100-ml aliquots of cells grown in 1-liter bottles. Thereafter, 1 liter of a 7-day-old fermentor culture is blown anaerobically into a second fermentor containing fresh SA medium. Cultures are harvested in late logarithmic phase after about 5 days of growth and washed once with water under an atmosphere of H_2.

2. Extraction of Lipids[28]

Total lipids are extracted from harvested cells by the method of Bligh and Dyer[7] modified as follows: A suspension (total vol, 80 ml) of about 30 g (wet wt) of cells in 4 M NaCl is diluted with 200 ml of methanol and 100 ml of chloroform and the mixture stirred at room temperature for 1 hr. The mixture is then centrifuged and the supernatant is collected. The pellet is suspended in 80 ml of 4 M NaCl and extracted twice more as described earlier. The combined supernatants are diluted with 300 ml each of chloroform and distilled water in a 4-liter separatory funnel, mixed gently, and left at room temperature for several hours to allow the phases to separate. The chloroform phase is then removed, diluted with benzene and 99% ethanol (to aid in removal of traces of water), and brought to dryness on a rotary evaporator at 30–35°. The residual lipids are dissolved in chloroform and stored at −20°. Yield of lipids, 210 mg from 30 g wet cells (5.5% of dry cells).

3. Separation of Lipid Components[28]

Total lipids (210 mg) are fractionated on a column (18 × 3 cm) of silicic acid (50 g, Bio-sil A, 100–200 mesh, activated at 125° for 4 hr) using the following elution sequence: chloroform, 200 ml (fraction 1, neutral

[27] G. D. Sprott, R. C. McKellar, K. M. Shaw, J. Giroux, and W. G. Martin, *Can. J. Microbiol.* **25**, 192 (1979).
[28] S. C. Kushwaha, M. Kates, G. D. Sprott, and I. C. P. Smith, *Biochim. Biophys. Acta* **664**, 156 (1981).

lipids); acetone, 200 ml (fraction 2, glycolipids and phosphatidyl glycerol); chloroform-methanol (3:2, v/v), 250 ml (fraction 3, phosphoglycolipids); and methanol, 750 ml (fraction 4, phosphoglycolipids).

Phosphoglycolipids. The two main phosphoglycolipids (PGL-I and
PGL-II) are isolated from fraction 3 and 4 by preparative TLC on silica
gel G (25–30 mg lipid/plate; 0.75-mm-thick layer) in chloroform–methanol–diethylamine–H_2O (110:50:8:3.5, v/v); PGL-I has R_f 0.46 ± 0.05
and PGL-II has R_f 0.34 ± 0.05 in this solvent system. The lipid bands are
detected with Rhodamine 6G spray[6] and eluted from the silica gel with
chloroform–methanol–H_2O (1:2:0.8, v/v), as described earlier for the
polar lipids of PM. The PGL-I and PGL-II thus obtained are precipitated
from a minimum amount of chloroform (ca. 1 ml) by addition of 30 vol of
acetone in the cold. Each precipitate is collected by centrifugation, repre-
cipitated twice more by the same procedure, and dried *in vacuo;* yield,
PGL-I, 115 mg; PGL-II, 32 mg.

These TLC pure lipids are converted to the ammonium salt form as
follows: a solution of PGL-I or PGL-II (30–50 mg) in 19 ml of chloro-
forom–methanol-0.2 N HCl (1:2:0.8, v/v) is immediately diluted with
5 ml of chloroform and 5 ml of 0.2 *N* HCl. The diphasic mixture is immediately centrifuged and the separated chloroform phase neutralized with
0.2 *N* methanolic NH_4OH to pH 8–9, diluted with benzene, and brought
to dryness on a rotary evaporator. The residual ammonium salt form of
PGL-I or PGL-II is dissolved to a known volume in chloroform and the
solution is used for analysis of lipid-P[6] and sugar.[6,29]

Glycolipids (DGD-I and DGD-II) and Phosphatidylglycerol (PG).
Glycolipids and phosphatidylglycerol are isolated from Fraction 2 by preparative TLC on silica gel G (0.75 mm thick layer) using double development in chloroform–methanol–acetic acid–H_2O (85:15:10:3, v/v); R_f
values are DGD-I, 0.48; DGD-II, 0.57, and PG, 0.43. Traces of glycolipids
DGT-I (R_f, 0.53) and DGT-II (R_f, 0.63) are also present. Each component
is detected and eluted from the silica gel as described earlier. The PG
component was converted to the ammonium salt as described for PGL-I
and PGL-II. The glycolipids and PG were analyzed for sugar[29] and lipid-P,[6]
respectively.

4. *Lipid Components of M. hungatii*[28] (Fig. 2)

Total lipids extracted from whole cells accounted for 5.5% of cell dry
weight. Neutral lipids constituted about 6% of the total lipids, the remain-

[29] S. C. Kushwaha and M. Kates, *Lipids* 16, 372 (1981).

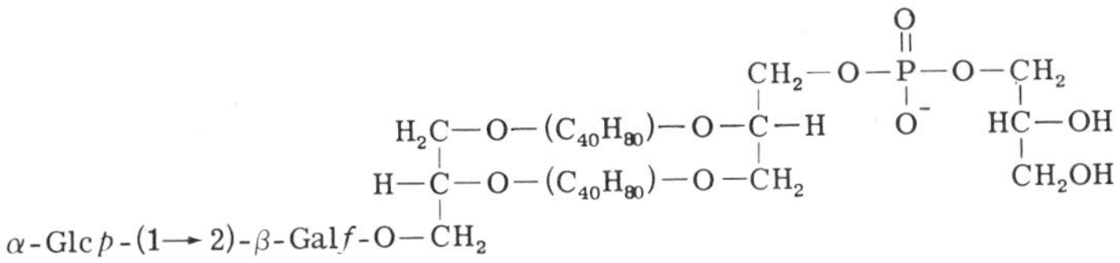

PGL-I

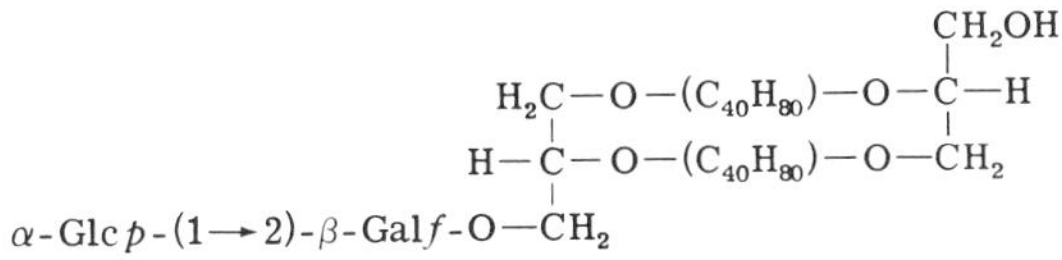

$$
\begin{array}{c}
\qquad\qquad\qquad\qquad\overset{\displaystyle O}{\underset{\displaystyle |}{\underset{}{\parallel}}} \\
\qquad\qquad CH_2-O-P-O-CH_2 \\
H_2C-O-(C_{40}H_{80})-O-\overset{|}{C}-H \qquad O^- \qquad H\overset{|}{C}-OH \\
H-\overset{|}{C}-O-(C_{40}H_{80})-O-\overset{|}{C}H_2 \qquad\qquad CH_2OH \\
\alpha\text{-Glc}p\text{-}(1\rightarrow 2)\text{-}\beta\text{-Gal}f\text{-O}-\overset{|}{C}H_2
\end{array}
$$

PGL-I

DGT-I

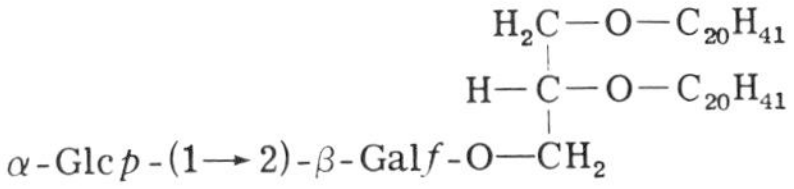

$$
\begin{array}{c}
\qquad\qquad\qquad CH_2OH \\
H_2C-O-(C_{40}H_{80})-O-\overset{|}{C}-H \\
H-\overset{|}{C}-O-(C_{40}H_{80})-O-\overset{|}{C}H_2 \\
\alpha\text{-Glc}p\text{-}(1\rightarrow 2)\text{-}\beta\text{-Gal}f\text{-O}-\overset{|}{C}H_2
\end{array}
$$

DGT-I

DGD-I

$$
\begin{array}{c}
H_2C-O-C_{20}H_{41} \\
H-\overset{|}{C}-O-C_{20}H_{41} \\
\alpha\text{-Glc}p\text{-}(1\rightarrow 2)\text{-}\beta\text{-Gal}f\text{-O}-\overset{|}{C}H_2
\end{array}
$$

DGD-I

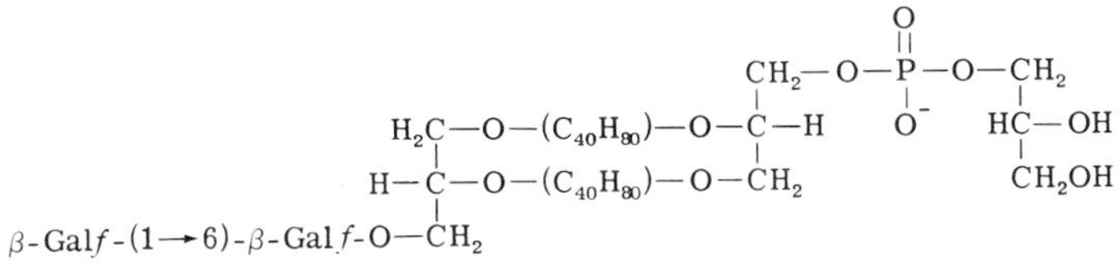

PGL-II

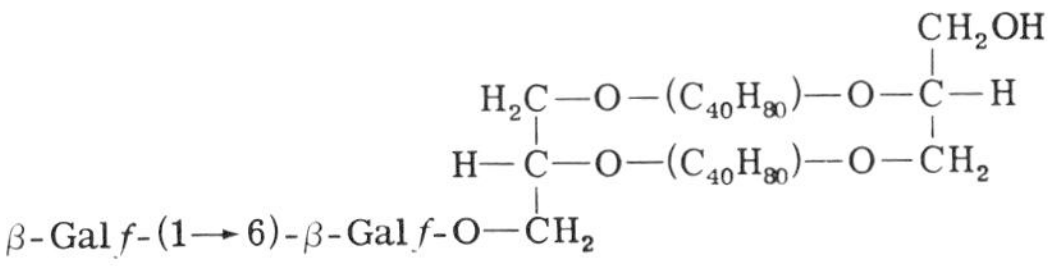

$$
\begin{array}{c}
\qquad\qquad\qquad\qquad\overset{\displaystyle O}{\parallel} \\
\qquad\qquad CH_2-O-P-O-CH_2 \\
H_2C-O-(C_{40}H_{80})-O-\overset{|}{C}-H \qquad O^- \qquad H\overset{|}{C}-OH \\
H-\overset{|}{C}-O-(C_{40}H_{80})-O-\overset{|}{C}H_2 \qquad\qquad CH_2OH \\
\beta\text{-Gal}f\text{-}(1\rightarrow 6)\text{-}\beta\text{-Gal}f\text{-O}-\overset{|}{C}H_2
\end{array}
$$

PGL-II

DGT-II

$$
\begin{array}{c}
\qquad\qquad\qquad CH_2OH \\
H_2C-O-(C_{40}H_{80})-O-\overset{|}{C}-H \\
H-\overset{|}{C}-O-(C_{40}H_{80})-O-\overset{|}{C}H_2 \\
\beta\text{-Gal}f\text{-}(1\rightarrow 6)\text{-}\beta\text{-Gal}f\text{-O}-\overset{|}{C}H_2
\end{array}
$$

DGT-II

$$
\begin{array}{ll}
H_2C-O-C_{20}H_{41} & C_{20}H_{11}-O-\overset{|}{C}-H \qquad\qquad CH_2-O-\overset{\displaystyle O}{\overset{\parallel}{P}}-O-CH_2 \\
H-\overset{|}{C}-O-C_{20}H_{41} & C_{20}H_{41}-O-\overset{|}{C}H_2 \qquad\qquad O^- \quad HO-\overset{|}{C}H \\
\beta\text{-Gal}f\text{-}(1\rightarrow 6)\text{-}\beta\text{-Gal}f\text{-O}-\overset{|}{C}H_2 & \qquad\qquad\qquad\qquad CH_2OH
\end{array}
$$

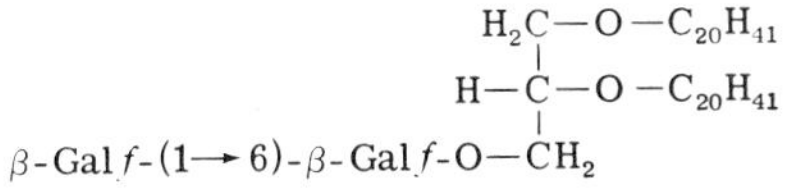

DGD-II 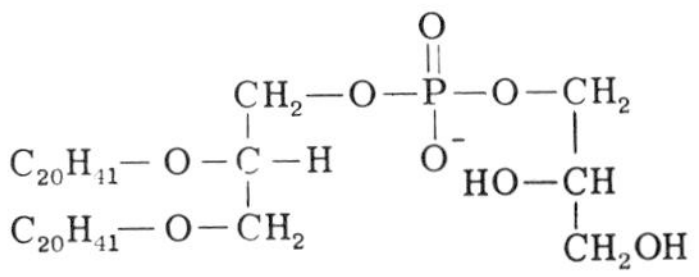   PG

FIG. 2. Structures of complex lipids of *Methanospirillum hungatei*. PGL-I, phosphogly-colipid-I; DGT-I, diglycosyltetraether-I; DGD-I, diglycosyldiether-I; PGL-II, phosphogly-colipid-II; DGT-II, diglycosyltetraether-II; DGD-II, diglycosyldiether-II; PG, phosphati-dylglycerol (diether type).

ing 94% being polar lipids. The polar lipids have been shown previously[30,31] to be derived exclusively from biphytanyl diglycerol tetraether (C_{40}-tetraether) as well as from the 2,3-di-O-phytanyl-sn-glycerol[11] (C_{20}-diether) present in extremely halophilic bacteria.

The major phosphoglycolipids, PGL-I and PGL-II, have been shown[28] to be derived from the C_{40}-tetraether; they both have asymmetric structures with sugar residues on one side and a phosphoglycerol group on the other side of the C_{40}-tetraether moiety (see Fig. 2). They differ only in the type and linkage of the sugar residues, PGL-I having an α-glucopyranose group joined by a 1–2 glycosidic link to a β-galactofuranose group, PGL-II having two β-galactofuranose groups joined by a 1–6 glycosidic link (Fig. 2). Two minor glycolipids (DGT-I and DTG-II) containing the C_{40}-tetraether are present in trace amounts and these are derived from PGL-I and PGL-II, respectively, by removal of the phosphoglycerol group.

The other glycolipids, DGD-I and DGD-II, are derived from the C_{20}-diether and have the same sugar residues as PGL-I and PGL-II, respectively (Fig. 2).[28] The phosphatidylglycerol component is not identical[28] to the 2,3-di-O-phytanyl-sn-1-glycerophosphoryl-sn-3′–glycerol found in extreme halophiles,[11] but is its diastereomer, 2,3-di-O-phytanyl-sn-1-glycerophosphoryl-sn-1′-glycerol.

The polar lipids of $M.$ $hungatii$ have almost as simple a composition as those of the extreme halophiles insofar as there are only three main components, PGL-I (45 mol %), PGL-II (13 mol %), and DGD-I (28 mol %) and two minor components, PG (10 mol %) and DGD-II (3 mol %) (Table II).

Examination of the structures of the $M.$ $hungatii$ lipids (Fig. 2) shows that they bear a similarity to those of extreme halophiles in that PG, DGD-I, and DGD-II are derived from the diphytanyl glycerol ether. The major components PGL-I and PGL-II, however, appear to have covalently bonded asymmetric dimeric structures that would be expected to form lipid monolayers with the properties of covalently bonded lipid bilayers. Such a lipid structure would impart rigidity and stability to the membranes of this methanogen not possible with conventional lipid bilayer structures. Another interesting point about the "dimeric" lipids PGL-I and PGL-II is that they appear to be biosynthesized by head-to-head condensation of 1 mol of PG with 1 mol of DGD-I or DGD-II, respectively. Thus there would appear to be a phylogenetic relationship between the lipids of methanogens and extreme halophiles as well as certain thermoacidophiles.[32] Furthermore, lipid biosynthesis appears to have undergone

[30] R. A. Makula and M. E. Singer, $Biochem.$ $Biophys.$ $Res.$ $Commun.$ **82**, 716 (1978).
[31] T. G. Tornabene and T. A. Langworthy, $Science$ **203**, 51 (1979).
[32] T. A. Langworthy, in "The Mycoplasmas" (M. F. Barile, S. Razin, J. G. Tully, and R. F. Whitcomb, eds.), Vol. 1, p. 495. Academic Press, New York, 1979.

TABLE II

LIPID COMPOSITION OF A METHANOGENIC BACTERIUM

Methanospirillum hungatii[a]

Lipid component	MW[b]	R_f[c]	Total lipids (wt %)	Polar lipids (mol %)
Phosphoglycolipids				
PGL-I	1813	0.34	50	45
PGL-II	1813	0.46	14	13
Phospholipid				
Phosphatidylglycerol	835	0.61	5	10
Glycolipids				
DGD-I	978	0.68	17	28
DGT-I	1624	0.73	0.5	0.3
DGD-II	978	0.78	2	3
DGT-II	1624	0.83	0.2	0.2
Neutral lipids	—	0.91	6	—

[a] From S. C. Kushwaha, M. Kates, G. D. Sprott, and I. C. P. Smith, *Biochem Biophys. Acta*, **664,** 156 (1981).

[b] Molecular weights of acidic lipids are given for the ammonium salts.

[c] On silica gel G in chloroform–methanol–diethylamine–water (110:50:8:3.5, v/v).

further elaboration in methanogens and thermoacidophiles compared to the extreme halophiles. Studies of the biosynthetic pathways of the C_{40}-tetraether and C_{20}-diether lipids in these organisms should prove to be extremely interesting.

Acknowledgments

This work was supported by the Medical Research Council of Canada (Grant MA-4103). The assistance of Mr. F. Cooper of N.R.C. Canada with the GC–Mass Spectral analyses is gratefully acknowledged.

[14] Preparation of New Two- and Three-Dimensional Crystal Forms of Bacteriorhodopsin

By HARTMUT MICHEL and DIETER OESTERHELT

Introduction

Bacteriorhodopsin, as isolated from Halobacteria,[1] is arranged in a highly stable, two-dimensional crystalline, hexagonal lattice of p3 sym-

[1] D. Oesterhelt and W. Stoeckenius, this series, Vol. 31, Part A, p. 667.

metry. The dimension of the unit cell is 62–63 Å[2,3] but can be reduced to 59–60 Å on partial delipidization[4,6] while retaining the p3 symmetry. Recent crystallization attempts led to the formation of a new, stable two-dimensional crystal form of purple membrane with an orthorhombic lattice having $p22_12_1$ symmetry,[5,6] as well as to the formation of three-dimensional crystals of bacteriorhodopsin[7] and a three-dimensional crystalline arrangement of the original purple membrane.[8] These new crystal forms should be of help in structure elucidation of bacteriorhodopsin. Additionally, the orthorhombic form of purple membrane, which possesses two identical sides, is a valuable membrane for various comparative purposes.

Preparation of Purple Membrane with Orthorhombic Lattice

Originally, three methods (A–C, cf. 5,6) were developed for the preparation of purple membrane with orthorhombic lattice (o-PM). At present we use mainly methods A and C with minor modifications. We recommend use of method A when large sheets of o-PM (e.g., for electron microscopy) are needed and method C when the size of the rolls is of minor importance.

o-PM is formed as sheets preferentially, but also tubes and closed vesicles. The sheets of o-PM have a strong tendency to roll up, and the rolls are clearly visible under the light microscope. Formation of o-PM therefore can be controlled by light microscope inspection. Additionally, appearance of an intense silky sheen accompanies formation of o-PM.

Method A. 4.5 ml of 20 mM N,N-dodecyltrimethylammonium chloride (DTAC, obtained from Serva, Heidelberg, West Germany; the corresponding bromide salt, obtained from Sigma, cannot be used) in 150 mM potassium acetate, pH 3, is added to 0.5 ml purple membrane suspension containing 240 nmol bacteriorhodopsin.* Then the suspension is sonicated two times at 25-W output, for 30 sec each, with the microtip of a Branson sonifier (model B15). This is possibly a critical step in the preparation of o-PM using method A. According to our experience the real output of different sonifiers of the same type varies. Too intense sonication would lead to destruction and formation of small purple membrane

[2] A. E. Blaurock and W. Stoeckenius, *Nature (London), New Biol.* **233**, 152 (1971).
[3] R. Henderson, *J. Mol. Biol.* **93**, 123 (1975).
[4] R. Henderson, J. S. Jubb, and M. G. Rossmann, *J. Mol. Biol.* **154**, 501 (1982).
[5] H. Michel, D. Oesterhelt, and R. Henderson, *Proc. Natl. Acad. Sci. U.S.A.* **77**, 338 (1980).
[6] H. Michel, D. Oesterhelt, and R. Henderson in Electron Microscopy at Molecular Dimensions (W. Baumeister and W. Vogell eds.), Springer Verlag, 61 (1980).
[7] H. Michel and D. Oesterhelt, *Proc. Natl. Acad. Sci. U.S.A.* **77**, 1283 (1980).
[8] R. Henderson and D. Shotton, *J. Mol. Biol.* **139**, 99 (1980).
* The amounts of bacteriorhodopsin given are based on a molar extinction coefficient of 63,000 in the light-adapted state and on a molar weight of 26,000.

pieces, which would not be spun down in the following centrifugation step. The sonicated suspension is centrifuged at 70,000 g for 20 min in an ultracentrifuge. Approximately 90% of the membranes is sedimented at this stage. The slightly purple supernatant is discarded. The pellet is then resuspended in 5 ml 10 mM Hepes buffer, pH 7, and sonicated twice for 10 sec at 25 W. After centrifugation as earlier, the pellet is resuspended in 2 ml of a 10 mM Hepes buffer, containing 0.1% NaN$_3$, 0.2% Triton X-100 at pH 7, and is stirred for 1 hr in the dark at room temperature. It is then dialyzed for 16 hr against 250 ml dialysis buffer (0.1 M potassium phosphate buffer, preferably pH 5.6, 0.1% NaN$_3$, 0.2% Triton X-100) at room temperature. During dialysis an amorphous precipitate and a few rolled-up sheets of o-PM are formed. The contents of the dialysis bag are then transferred into reagent tubes and shaken on a rotary shaker (Minishaker MSR, Kühner, Basel, Switzerland) at 60 rpm. The temperature should be 27–28°. At higher temperatures large multilamellar vesicles rather than rolled-up sheets are obtained. At lower temperatures the velocity of the formation of o-PM is slowed down. At 4° no o-PM is formed.

The pH of the dialysis buffer can be varied between 4.5 and 6. At the higher pH values (5.6–5.8) the rolled up sheets of o-PM are considerably larger. The rolls reach lengths up to 0.2 mm, but the velocity of formation and the overall yield is smaller. Shaking should be continued for at least 3 weeks. At pH 4.5 the rolls are shorter, but within 3 days 60–80% of the bacteriorhodopsin present is found as rolls of o-PM. The rolls of o-PM are purified as described under method C.

Method C. This method involves simply addition of low amounts of long-chain amines and Triton X-100 to purple membrane, leading to a conversion of the hexagonal form into the orthorhombic one. In contrast to our previous observations using different batches of purple membrane, we now often obtain o-PM at molar ratios of DTAC to bacteriorhodopsin higher than 2. The optimal ratio varies from purple membrane preparation to purple membrane preparation. Therefore we recommend small-scale preparations of o-PM at different DTAC to bacteriorhodopsin ratios at first, in order to find the optimal DTAC to bacteriorhodopsin ratio for the batch of purple membrane used. Convenient small scales are 40 nmol of bacteriorhodopsin in 0.5 ml 0.1 M potassium phosphate, 0.2% Triton X-100, 0.1% NaN$_3$, pH 5.2, containing different amounts of DTAC to achieve molar ratios of DTAC:bacteriorhodopsin of 1–12. For this purpose 10 μl 10% Triton X-100, 50 μl 1 M KH$_2$PO$_4$ solution, 5 μl NaN$_3$ solution (10%), 80 μl 0.5 mM bacteriorhodopsin (purple membrane suspension), 4–48 μl 10 mM DTAC solution are mixed in a small reagent tube and water is added to a final volume of 0.5 ml.

After 2 weeks of shaking (60 rpm) at 27–28°, the samples should be inspected by the light microscope, and the optimal ratio, usually about 6, should be used to set up a large-scale preparation of o-PM. A lot of different long-chain amines can be used. With cetyltrimethylammonium bromide the optimal cetyltrimethylammonium bromide to bacteriorhodopsin ratio is slightly below that for DTAC.

The use of purple membranes isolated from *Halobacterium halobium* strain R_1M_1 instead of strain S9 is to be preferred for methods A and C, because the velocity of o-PM formation as well as the yield is higher. More than 70% conversion to o-PM is normally observed within 3 weeks.

Purification of o-PM. o-PM can easily be purified on sucrose density gradients, since its density is lower than that of the native hexagonal purple membrane. One milliliter of the suspension is layered on top of a linear gradient from 32.5 to 42.5% (weight per weight) sucrose, using the SW-41, SW-40 rotors (Beckman) or TST 41 (Kontron). Figure 1 shows such a gradient after 13 hr centrifugation at 25,000 rpm. The rolls of o-PM in the middle of the gradients are clearly separated from amorphous material at the bottom, and vesicular purple membrane and solubilized bacteriorhodopsin at the top of the gradient. The fractions containing the rolls of o-PM are collected, spun down after dilution with water, and freed from the sucrose by at least five washings with water and centrifugation (15 min, 9000 rpm, Sorvall SS-34 rotor). It should be mentioned that the o-PM then still contains traces of Triton X-100, which cannot be removed by further washing and even sonication. However, the Triton X-100 content is low; only 0.1–0.3 mol per mol of bacteriorhodopsin is found.

Differences between Orthorhombic and Hexagonal Purple Membrane. At low pH (around 3) o-PM is not precipitated. It therefore can be used for spectroscopy at low pH without further treatment. o-PM is not stable in basal salt–ether systems. Aggregates are obtained, which quickly bleach, and the needlelike appearance is lost.

Preparation of Three-Dimensional Crystals of Bacteriorhodopsin

The recent crystallization of bacteriorhodopsin gives some hope that standard X-ray methods can be applied to membrane proteins (Fig. 2). Nevertheless the crystallization procedure needs to be improved before final success becomes possible. Therefore we will restrict ourselves to the basic procedure.

Solubilization of Bacteriorhodopsin. It is not easy to obtain a solution of bacteriorhodopsin in octyl glucoside, the detergent used in the crystallization procedure. At pH 7 and room temperature bacteriorhodopsin is soluble in octyl glucoside (20 mM phosphate buffer) but not stable ($t_{1/2} < 10$ hr); on the other hand, at pH 5, octyl glucoside-solubilized bac-

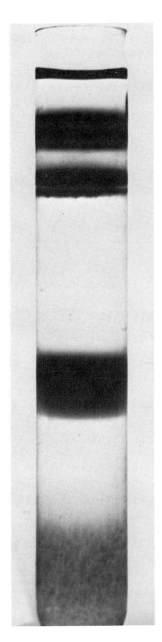

FIG. 1. o-PM after centrifugation on a sucrose density gradient (see text).

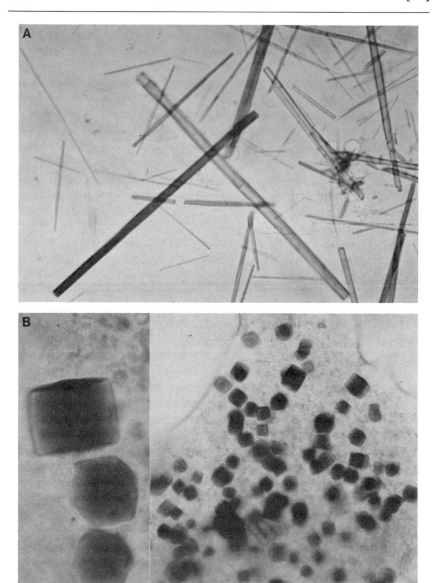

FIG. 2. Three-dimensional crystals of bacteriorhodopsin. (A) Needlelike crystals. The picture was taken with light polarized parallel to the darker (purple) needle ($\approx \times 1200$). (B) Cubelike crystals (left, $\approx \times 6000$; right, $\times 1200$).

teriorhodopsin is reasonably stable, but the velocity of solubilization is too slow. Therefore, as a compromise, purple membranes (corresponding to 60 nmol of bacteriorhodopsin per milliliter) are stirred in 1.3 ml 1.2% β-D-octyl glucoside (Riedel-de-Haen, Hannover, West Germany) containing sodium phosphate buffer (20 mM, pH 5.5) at 30°. After 3 days of stirring, the nonsolubilized bacteriorhodopsin is spun down at 150,000 g for 45 min. The supernatant has a blue-shifted absorption maximum at 552 nm and an optical density of more than 1.7. The amount of denatured bacteriorhodopsin is less than 15%, as estimated from the appearance of the 380-nm absorption which is due to the chromophore destruction.

Crystallization. The bacteriorhodopsin solution is then brought to a concentration of 0.7 M ammonium sulfate or sodium phosphate, pH 5–5.5, by addition of saturated solutions and concentrated threefold over phosphorus pentoxide in an evacuated desiccator. Crystallization is achieved by vapor diffusion of the respective solutions against 2.5 M ammonium sulfate (leading to the cubelike crystal form B) or against 2.5 M sodium phosphate, pH 4.7 (leading to the needlelike crystal form B). Crystallization is observed in the pH range of 6.2–3.8, where slightly different concentrations of phosphate are required. The crystals, which probably still contain most of the lipids of purple membrane, become disordered on removal of octyl glucoside, as seen from the loss of the strict chromophore orientation parallel to the long needle axis.

Problems. Just above the salt concentration necessary for crystallization a phase separation occurs. Solubilized bacteriorhodopsin is enriched in the octyl glucoside-containing phase and subject to more rapid denaturation than in the aqueous phase. The cubelike crystals lose their sharp edges in the octylglucoside phase and do not diffract X-rays any longer. Phase separation therefore should be avoided.

An even more important problem is that the three-dimensional crystals are not the most stable form. The bacteriorhodopsin is converted into membraneous, octyl glucoside-containing stacks of threadlike appearance. This material may constitute a new two-dimensional crystal form of bacteriorhodopsin, but in contrast to hexagonal and orthorhombic purple membrane it is stable only at high salt. It is formed preferentially in ammonium sulfate solutions, but also in sodium phosphate solutions, especially at lower temperatures.

Note added in proof: Substantial improvement of crystal quality is achieved by addition of heptane-1,2,3-triol (1%) or cyclooctane-1,2,3,4-tetrol (1%) obtainable from Oxyl, Bobingen, West Germany, prior to crystallization.

[15] Freeze-Substitution and Freeze-Etching Method for Studying the Ultrastructure of Photoreceptive Membrane

By JIRO USUKURA and EICHI YAMADA

Freeze-fracture and etching methods which were initiated by Steere[1] reveal the three-dimensional structure of biological membrane in spite of the limited resolution due to platinum shadowing. However, conventional chemical fixation followed by glycerol treatment prior to freezing reduces the potential merits of this method. For example, the etching that enables us to expose cytoplasmic organization and true surfaces of the membrane system by sublimation of ice is impossible, in a strict sense, after immersion in glycerol. Therefore, the freeze-etching replica method combined with rapid freezing as physical fixation instead of chemical fixation is desirable, and in this way, we can expect to reveal the structure close to the native state. Van Harreveld and Crowell[2] accomplished the rapid freezing by contact with the metal cooled at liquid nitrogen temperature. However, rapid freezing at liquid nitrogen temperature has been exclusively utilized for the freeze-substitution method and not for the replication.[3] The reason for this may be due partly to the fact that the well-preserved zone of tissues by rapid freezing at liquid nitrogen temperature is very narrow and limited in depth, and it is rather difficult to get the replica from this particular area. More recently, Heuser et al.[4] attempted freezing by metal contact at liquid helium temperature in order to increase the well-preserved zone. However, cooling by liquid helium may not always be practical because of handling difficulty and cost. This article introduces the freeze-etching as well as freeze-substitution method with simple rapid freezing at liquid nitrogen temperature, focusing on the structure of photoreceptive membranes.

Preparation

The halophilic bacterium *Halobacterium halobium* R_1 and retina of the frog *Rana catesbeiana* were used as materials. The harvested bacteria are

[1] R. L. Steere, *J. Biophys. Biochem. Cytol.* **3**, 45 (1957).
[2] A. Van Harreveld and J. Crowell, *Anat. Rec.* **149**, 381 (1964).
[3] A. Van Harreveld, J. Crowell, and S. K. Malhotra, *J. Cell Biol.* **25**, 117 (1965).
[4] J. E. Heuser, T. S. Reese, M. J. Dennis, Y. Jan, L. Jan, and L. Evans, *J. Cell Biol.* **81**, 275 (1979).

washed once with "basal salt' solution,[5] and the resultant pellet is used as the sample. The posterior half of frog eyeball adapted in the dark for 2 hr is cut into small pieces about 3 mm square, and then neural retina is removed carefully from the pigment epithelium. All dissections and preparations are carried out just before freezing.

Freezing Method

The extremely halophilic bacteria were well preserved even by simple plunging into liquid Freon 22 or 13 cooled with liquid nitrogen.[6] However, such plunging into coolants cannot prevent ice crystal formation in the retina. Based on the theory of Coulter and Terracio,[7] a simple apparatus for rapid freezing was constructed (Fig. 1). This apparatus consists of a pure copper block placed at the center of a 2-liter ice chest and a device that bring the specimen to the metal block and make it touch with its surface in parallel. The copper block is 7 cm high by 5 cm in diameter with a

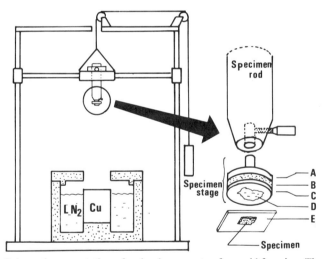

FIG. 1. Schematic presentation of a simple apparatus for rapid freezing. The scheme to the right is a detail of the specimen stage and the specimen rod shown in the left figure. The specimen stage consists of plastic stage (A), sponge layer (B), and Teflon disk (C). The specimen placed on square Teflon sheet (E) is attached by body fluid or glycerol (D).

[5] D. Oesterhelt and W. Stoeckenius, this series, Vol. 31, Part A, p. 667.

[6] J. Usukura, E. Yamada, F. Tokunaga, and T. Yoshizawa, *J. Ultrastruct. Res.* **70,** 204 (1980).

[7] H. D. Coulter and L. Terracio, *Anat. Rec.* **187,** 477 (1977).

polished upper surface. The cover of the ice chest has a round window 6 cm in diameter, which is situated about 3 cm above the copper block. To cool the copper block, liquid nitrogen is poured through the window until the block reaches that temperature. Dissected neural retina is placed on the Teflon sheet of 1 cm² (0.1 mm thickness), setting the outer segment upside. Subsequently, the square Teflon sheet, together with the specimen, is attached to the Teflon disk of the specimen stage with body fluid or glycerol as shown in Fig. 1, and then the specimen stage is fixed by a screw to the top of specimen rod. Freezing is accomplished by making the specimen directly contact the polished copper surface by pushing down the specimen rod. After holding it down firmly for several seconds, the specimen rod is lifted up again and is immediately soaked in liquid nitrogen in another styrofoam cup. In liquid nitrogen, a square Teflon sheet with tissue is removed from a Teflon disk by using cold forceps and is stored in liquid nitrogen. By displacing the chest a little, it is possible to freeze eight or more slices of tissue before repolishing the metal surface.

Frozen tissues resemble tiny, thin paper, because of compression by contact. However, the narrow zone of tissue, about 8 μm from the contacting surface, where it is initially frozen, showed good preservation without compression and ice crystal formation. (See examples in Figs. 3,- 4.)

Freeze-Fracture and Etching Replication

The specimen stage with carrier for fracturing and rotary shadowing is precooled with liquid nitrogen in the styrofoam tray (15 × 10 × 3 cm). Frozen tissues transferred from the styrofoam cup are removed carefully in the liquid nitrogen from a square Teflon sheet. Thereafter, frozen tissues are put into three deep grooves on the specimen stage (Fig. 2).[8] It is important to put the tissues firmly into a groove by adjusting the number of tissue pieces. Otherwise, tissues will be pulled out with knife on fracturing. Usually we are able to put three frozen tissues into one groove. Frozen tissues mounted on the specimen stage together with the carrier are brought into a freeze-fracture apparatus, FD-2S (Eiko Engineering Co.).[9] Immediately, specimens are cleaved perpendicular to the setting position at 3-5 × 10^{-7}mm Hg (Fig. 2). For etching, the cleaved specimen remains in a high vacuum for about 15 min, while specimen temperature is raised from $-160°$ to $-100°$. Cleaved or etched plane of tissue is

[8] J. Usukura and E. Yamada, *J. Electron Microsc.* **29,** 376 (1980).
[9] H. Akahori and T. Watahiki, *J. Electron Microsc.* **26,** 61 (1977).

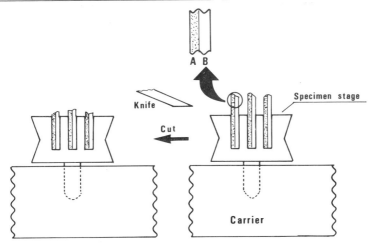

FIG. 2. Schematic demonstration of how to put the frozen tissues into grooves on the specimen stage and cleave them. (A) Well-preserved region of frozen tissue; (B) the region of ice crystal formation.

shadowed at an angle of 60° by evaporation of platinum and carbon, while it is rotating at 0.5 rps. The shadowed samples are floated on the detergent, including NaClO, and the removed replicas are then washed several times with distilled water. Although this type of freeze-fracturing yields narrow replica, the narrow zone along one of margins always corresponds to the well-preserved region.

On the other hand, in the case of halophilic bacteria, the pellet directly mounted on the specimen stage is frozen in the conventional way, as described earlier. Frozen pellet is cleaved and shadowed following the previously mentioned condition. At first, however, the shadowed sample is floated on distilled water and then soaked in the detergent. After several washings with distilled water, replicas are mounted on the grid.

Freeze-substitution

Freeze-substitution[2,3,6] is suitable for the fixation of deformable tissues or organisms sensitive to change in surrounding conditions such as *H. halobium*. Any movement or efflux of substances comprised in the samples is reduced as compared with usual chemical fixation, since the specimen is maintained in a frozen state throughout the process of substitution.

Frozen specimen is immersed in cold absolute acetone containing 4% OsO_4 at -80 to $-90°$ and kept at the same temperature for a few days in

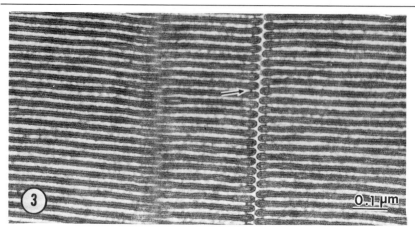

FIG. 3. Thin section electron micrograph prepared from freeze-substituted frog rod outer segment. The disks are arranged at regular interval of about 30 nm, linking each other with subtle substance (arrow). The luminal space of the disk is almost obliterated except at the margin.

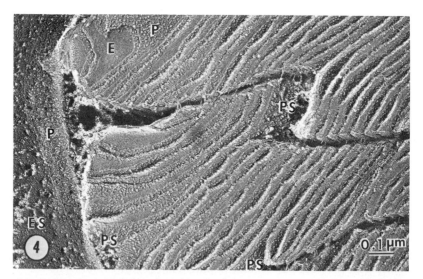

FIG. 4. Electron micrograph of frog rod outer segment freeze-deep-etched. Exposed true surface (ES) of plasma membrane enveloping the stack of disks contains various kind of fine particles, while true cytoplasmic surface (PS) of disk membrane shows many particles presumed to be peripheral membrane proteins or cytoplasmic heads of intrinsic membrane proteins. Membrane particles assumed to be rhodopsins are recognized on the P fracture face of disk membrane. P: P face; E: E face.

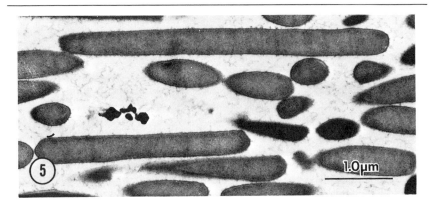

FIG. 5. Thin section of *H. halobium* by freeze-substitution method. Cytoplasmic substances are densely packed homogeneously in the rod-shaped bacterium.

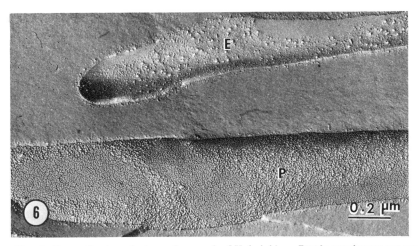

FIG. 6. Freeze-fracture electron micrograph of *H. halobium*. Purple membranes are recognized as patches in plasma membrane. P: P face of plasma membrane; E: E face of plasma membrane.

the freezer or in a dry-ice–acetone mixture. The substituted specimen is gradually warmed up to the room temperature and then washed several times with fresh pure acetone. Specimens are embedded directly in epoxy resin via a resin–acetone mixture.

As the examples obtained in our replication and substitution methods, photoreceptive membranes are illustrated in Figs. 3, 4, 5, 6. Compared to the conventional method, it is clear that the present method is an excellent one to display structure close to the native state.

[16] Analysis of Bacteriorhodopsin Structure by X-Ray Diffraction

By A. E. BLAUROCK

The purple membrane has two properties that make it a particularly good specimen for studying by physical methods. First, it is isolated from *Halobacterium halobium* as more or less flat sheets rather than as closed vesicles.[1] Second, it retains its structure even under vacuum[2] rather than distintegrating on drying as do the nerve myelin membrane[3,4] and the red blood cell membrane.[5] It is the combination of these two unusual properties that led early in my work in Dr. W. Stoeckenius' laboratory to the discovery of a third, most interesting property, that there is a crystalline array extending into the plane of the membrane.

Stoeckenius and Rowen[1] observed flat sheets in electron microscopic views of the purple membrane. Embedded and sectioned, the membrane viewed edge-on appeared as a familiar "unit-membrane,"[6] but with a definite, abrupt ending rather than closing on itself to form a vesicle. In negative-stain views, fragments of the membrane had an oval outline. In addition there was a suggestion of a regular array in the negative-stain image.[1]

The discovery that the membrane could be dried and still retain its structure was by serendipity. My first X-ray diffraction exposure of the purple membrane was of a moderately concentrated suspension prepared by Dr. W. Wober. The unoriented diffraction pattern contained several sharp, concentric rings, which clearly indicated crystallinity of some sort. Since the prevailing dogma was that natural membranes were rather fluid structures, one immediately suspected that the preparation had been contaminated with a crystalline substance, possibly a salt. Accordingly it was with considerable scepticism, but with Dr. Wober's and Dr. Stoeckenius' doubts in mind that there could be salt crystals, that I tried a method similar to one used by Levine and Wilkins[7] for lipids. Thus some of the suspension of purple membrane was dried down on a cylindrical surface and exposed to an X-ray beam tangent to the surface (Fig. 1A). The specimen

[1] W. Stoeckenius and R. Rowen, *J. Cell Biol.* **34,** 365 (1967).
[2] A. E. Blaurock and W. Stoeckenius, *Nature (London), New Biol.* **233,** 152 (1971).
[3] J. B. Finean, *J. Biophys. Biochem. Cytol.* **8,** 13 (1960).
[4] J. B. Finean, *J. Biophys. Biochem. Cytol.* **8,** 31 (1960).
[5] S. Knutton, J. B. Finean, R. Coleman, and A. R. Limbrick, *J. Cell Sci.* **7,** 357 (1970).
[6] J. D. Robertson, *in* "Cellular Membranes in Development" (M. Locke, ed.), p. 1. Academic Press, New York, 1964.
[7] Y. K. Levine and M. H. F. Wilkins, *Nature (London), New Biol.* **230,** 69 (1971).

METHODS IN ENZYMOLOGY, VOL. 88

was exposed under vacuum (1.10^{-2} Torr; details of the X-ray diffraction camera are given elsewhere[8]). The resulting pattern was as clear in its implications as it was unexpected: a well-oriented diffraction pattern indicating a two-dimensional crystalline array extending in the plane of the purple membrane. Thus, as shown schematically in Fig. 1A, a simple series of equally spaced, sharp reflections on the axis perpendicular to the drying surface, the profile axis, indicated that the membranes were stacked with their planes parallel to the surface. A second series of reflections centered on the axis parallel to the surface, the in-plane axis, indicated a crystalline packing of molecules side by side in the plane of the membrane. From the size of the unit cell, about 60 Å in diameter, it seemed likely that the membrane protein, at that time still not characterized, was the basis for the crystalline array. Subsequent biochemical analysis confirmed this view by demonstrating a single species of protein,[9] which was named bacteriorhodopsin for its striking resemblance to the visual pigments.

The significance of the sharp diffraction rings from the suspension was immediately clear: The rings were the summation of the in-plane reflections from a very large number of membrane fragments having all possible orientations in the suspension. The stacking reflections would, of course, be absent in this case. Thus much the same crystalline array was present in the suspended membrane as in the vacuum-dried membrane.

The geometry of the crystalline array was identified by bearing in mind the following considerations. For an X-ray beam incident perpendicular to the plane, the diffraction pattern from a single fragment of a two dimensionally crystalline membrane would be a two-dimensional array of diffraction spots. Laboratory X-ray beams are, however, much too weak to demonstrate the array of diffraction spots from a single fragment of the purple membrane; instead, one must work with very large numbers of fragments. Now, while the planar fragments had by good fortune oriented parallel to one another, the presence of a large number of reflections on the in-plane axis strongly suggested that the crystalline arrays in all the fragments had not come into alignment. Thus, the principal axes of the arrays of molecules in the planes of the membranes were not all parallel to one another but instead pointed at random in the plane of the drying surface. Accordingly, one would assume that all the possible reflections in the two-dimensional array would be seen on the in-plane axis, rather than only those that lay on some arbitrary straight line passing through the center of the array. In that case one has only to consider the distances from the center of the diffraction pattern to all the possible reflections. It is,

[8] A. E. Blaurock, *Biophys. J.* **13**, 290 (1973).
[9] D. Oesterhelt and W. Stoeckenius, *Nature (London), New Biol.* **233**, 149 (1971).

then, a simple matter to derive a formula for the distances once the geometry of the array is specified.

I soon determined that the array of reflections is hexagonal. One naturally would begin by considering simple geometries for the array, e.g., hexagonal, square, rectangular. The hexagonal array was readily shown to account for all the observed distances from the center of the diffraction pattern to the in-plane reflections. A very few of the predicted reflections were not visible, presumably because the intensities were too weak. The formula for the distances is

$$(\text{constant}) \cdot (h^2 + h\,k + k^2)^{1/2} \tag{1}$$

where h and k are integers and the constant takes in parameters of the size of the array itself and of the experimental apparatus.[10,11] A square array was eliminated since the corresponding formula is

$$(\text{constant}) \cdot (h^2 + k^2)^{1/2} \tag{2}$$

The distances predicted by Eq. (2) clearly are not the same as those by Eq. (1). A rectangular array could be eliminated for the same reason.

The hexagonal array of reflections from a single fragment of the purple membrane has since been demonstrated in the electron microscope by Unwin and Henderson.[12] Their observations of electron diffraction confirmed the predicted crystalline packing of molecules in single fragments of the purple membrane.

Having demonstrated that the fragments of purple membrane could be profitably oriented, it was desirable to look for methods to orient the fragments other than by drying. Although it has been shown that complete drying does not greatly affect the structure,[2,10,11] methods were nonetheless sought for orienting the fragments while keeping them fully hydrated in order to detect and avoid any possible deleterious effects of the drying, as well as for other reasons (see below). Clearly the basis for orienting the fragments is the large anisometry: The average fragment is some 5000 Å in the longest dimension[1] but only some 50 Å thick,[2,11] in the direction perpendicular to the plane of the fragment, for an axial ratio of 100:1. Evidently one would want to crowd the fragments together in order to restrict the range of available orientations. By doing this in a gradual, orderly way one could hope to obtain a specimen in which the fragments are permanently and uniformly oriented with their planes parallel to one another. Among the methods tried were (1) a limited and gradual dehydration of a moderately concentrated suspension of membranes inside a thin-wall

[10] A. E. Blaurock, *J. Mol. Biol.* **93**, 139 (1975).
[11] R. Henderson, *J. Mol. Biol.* **93**, 123 (1975).
[12] P. N. T. Unwin and R. Henderson, *J. Mol. Biol.* **94**, 425 (1975).

glass X-ray capillary, the capillary being left unsealed to allow the water to evaporate (Fig. 1B, 1); (2) centrifuging some of the suspension to a firm pellet and sealing some of the pelleted, but still moist, material in an X-ray capillary[2]; (3) sealing some of the suspension in an X-ray capillary and centrifuging with the capillary axis parallel to the large g field, the capillary being supported in cold butter; (4) electrophoresing onto a plug of polyacrylamide in an X-ray capillary[13] (Fig. 1B, 2); (5) drying some of the suspension onto aluminum foil and breaking off flakes of the dried material for exposure free of any substrate; and (6) orienting the fragments, suspended in water, by means of a magnetic field.

Given the large axial ratio, it is in fact difficult to avoid having the fragments orient once they are crowded together. (Local stacking probably accounts for the opalescence of a concentrated suspension.) Thus most of the preceding methods were likely to succeed to some degree, and more or less good results were obtained by methods 1–5.

It is only in the case of (6) that the membranes have been oriented while still in a comparatively dilute suspension.[14] In this case the degree of orientation is not nearly as good as that obtained by the other methods. This is so because the orienting force will be very small for a fragment slightly out of perpendicular to the magnetic field and the force evidently only becomes appreciable when the fragment rotates to being nearly parallel to the field even at 10,000–15,000 Gauss. In addition, the orientation relaxes rapidly when the field is turned off.

Method 1 produced particularly useful specimens. Thus in one case the planes of the fragments were parallel to the meniscus to within $\pm 5°$.[15] Despite the fairly good orientation, however, the fragments remained widely spaced from one another: Just three stacking reflections were observed, indicating an average stacking distance of 180 Å, and only slowly varying—so-called continuous—intensity was observed at larger diffraction angles. These observations indicated that the distances between fragments were highly variable, with a standard deviation of some 50 Å. In consequence, one was seeing the diffraction as though from single, isolated, but oriented fragments. The diffraction from oriented but effectively isolated structures is the most informative (see the following discussion). A specimen of this kind was examined for birefringence and dichroism; both were readily observable.[2]

Orientation by partial or complete drying probably begins at the air–water interface where the membrane fragments will become more concentrated because of the loss of water, and then continues more and more

[13] A. E. Blaurock and D. Oesterhelt, unpublished (1970).
[14] D.-Ch. Neugebauer, A. E. Blaurock, and D. L. Worcester, FEBS Lett. 78, 31 (1977).
[15] A. E. Blaurock, unpublished (1970).

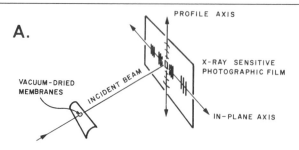

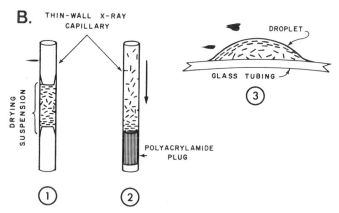

FIG. 1. (A) Schematic diagram of the X-ray diffraction experiment. A droplet of a suspension of the purple membrane is first dried on a cylindrically curve surface. The surface is then aligned with the X-ray beam tangent to the dried material on the surface, and the diffraction pattern is recorded on X-ray film. The dried membranes form multilayer stacks, giving rise to the series of equally spaced reflections on the profile axis. The crystalline arrays of protein molecules extending in the planes of the membrane fragments give rise to the series of reflections centered on the in-plane axis. The reflections are line-like, extending parallel to the profile axis, because the single membrane is not crystalline perpendicular to its plane. The small rectangle at the center of the film is the shadow of the beam stop (not shown). (B) Some methods of preparing oriented specimens. Membrane fragments are represented schematically by the dashes. (1) A suspension of the purple membrane placed in an X-ray capillary with the ends open to allow some of the water to evaporate. (2) Electrophoresis of fragments of the purple membrane onto the surface of a plug of polyacrylamide. The direction of motion of the membranes is indicated by the arrow. The ends of the capillary are connected to the voltage source via small-bore plastic tubing. (3) Cross-sectional view of a droplet of suspension drying on a cylindrically curved surface, in this case a piece of glass tubing, for exposure as in part A.

deeply into the suspension (Fig. 1B). A very small amount of sucrose in the suspension ($\sim 1\%$) may improve the orientation somewhat.

The oriented specimens also were useful in looking for α-helix in the bacteriorhodopsin. I noted an indication of α-helical structure in the original oriented diffraction pattern from the vacuum-dried membranes: a fairly intense band of intensity centered on the profile axis at about 5 Å and extending off the axis some distance.[15] An alternative explanation occurred to me as well: The branched methyl groups on the dihydrophytyl chains of the membrane lipids[16] would, if the lipid fatty chains were oriented perpendicular to the plane of membrane, give rise to a reflection on the profile axis with a Bragg spacing equal to the average distance between those groups. Based on the structure of the chain backbone, the expected reflection would be at about 5 Å, as observed. This consideration dampened my curiosity for a while.

In 1973 my interest renewed itself and I had a cylindrical film cassette made in order to look for the 1.5-Å reflection,[17] which would most convincingly demonstrate the presence of α-helix. My intuition was to look at a fully hydrated specimen since it was evident that complete drying caused some disordering of the crystalline array[2,10] and since the effects of disorder generally are more severe the larger the diffraction angle, i.e., the smaller the distance of interest. A rather broad reflection was recorded at 1.5 Å, which occasionally appeared to split into two reflections.[10]

Subsequently Dr. R. Henderson showed me the well-defined reflection he had recorded at 1.5 Å from fully dried material, indicating α-helix in the bacteriorhodopsin.[11] I therefore tried a flake of material dried on Al-foil and was rewarded with a fairly narrow reflection at 1.5 Å, confirming his observation.[10] With hindsight it appears that one-half of the occasionally split reflection that I had observed from fully hydrated, oriented fragments probably is the reflection from α-helix in the bacteriorhodopsin. The origin of the other half of the split reflection is not clear to me. The flake of membrane fragments also rewarded me with the finest small-angle pattern I have ever recorded (Plate I in Blaurock[10]).

The 1.5-Å reflection is seen as an arc centered on the axis perpendicular to the axis of the in-plane reflections.[10,11] This observation shows that the α-helix runs more or less perpendicularly to the plane of the membrane. The arcing suggests that more than one segment of α-helix is present.[10] The presence on the in-plane axis of intense reflections with Bragg spacings near 10 Å is consistent with segments of α-helix running more or

[16] M. Kates, B. Palameta, C. N. Joo, D. J. Kushner, and N. E. Gibbons, *Biochemistry* **5**, 4092 (1966).

[17] M. F. Perutz, *Nature* (*London*) **167**, 1053 (1951).

less perpendicularly to the plane of the membrane.[10,11] Henderson analyzed the diffracted intensity and suggested that a large proportion of the bacteriorhodopsin molecule consists of α-helix.[11] This suggestion has been borne out by the electron microscopic low-dose image.

The small-angle diffraction pattern from the material sealed inside an X-ray capillary after pelleting at high g forces (method 2, $\sim$400,000 g) showed that the membrane fragments had oriented in the pellet and that the orientation survived the transfer to the X-ray capillary. The continuous intensity recorded on the profile axis from such specimens clearly was consistent with bilayer structure of the lipids in the purple membrane.[2,23] Changing the medium in which the membrane fragments are pelleted (method 2) from water to concentrated salt solution produced changes in the small-angle pattern[23] which confirmed the bilayer hypothesis and provided an absolute scale of electron density.[18]

Continuous intensity on the profile axis is considerably more informative than a series of isolated Bragg reflections. Thus without additional data all conceivable choices of phase angles for the Bragg reflections are equally possible, leading to an infinite number of distinct profiles. That this is not the case for the continuous profile intensity has been shown in a seminal theoretical treatment by Dr. G. I. King. His conclusion is that only a limited number of choices of phase angles are possible. His theory stimulated me to examine the possible profiles for the purple membrane.

My results quickly confirmed King's[19] theory: I was able to find only two possible profiles (Fig. 2), one of which King had calculated by independent means.[20] Overall, the two profiles both indicate that the membrane lipids are in a bilayer. The high density at the middle of the membrane, compared to the middle of the bilayer of extracted lipids (dashed curve in Fig. 2) indicates a high proportion of protein in the middle of the membrane bilayer. Both profiles in Fig. 2 are somewhat asymmetric, but the basis for the asymmetry is X-ray data that are to a small degree uncertain.[20] Thus the asymmetry is regarded as an upper limit to the actual, physical asymmetry of the bacteriorhodopsin molecule. I note that the same two profiles have since been derived by other, independent means.[21,22] Stroud and Agard[22] also were stimulated by King's[19] theoretical development.

The conclusions from the X-ray work are as follows. First, the lipids in

[18] A. E. Blaurock, *Chem. Phys. Lipids* **8**, 285 (1972).
[19] G. I. King, *Acta Crystallogr.* **A31**, 130 (1975).
[20] A. E. Blaurock and G. I. King, *Science* **196**, 1101 (1977).
[21] T. Mitsui, *Adv. Biophys.* **10**, 97 (1978).
[22] R. M. Stroud and D. A. Agard, *Biophys. J* **25**, 495 (1979).
[23] A. E. Blaurock, *Biochim. Biophys. Acta* (Rev. Biomembranes) **650** (1981).

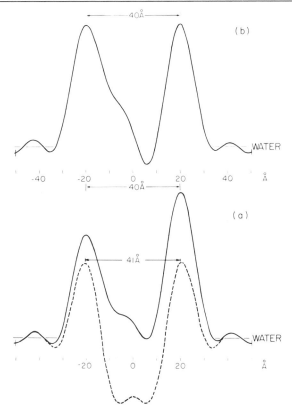

FIG. 2. Electron-density profile of the purple membrane. The electron density, averaged over many unit cells of the crystalline array, is plotted as a function of distance from the middle of the membrane, in the direction perpendicular to the sheet. These are the only two profiles consistent with the continuous profile diffraction.[20] Hence what they share in common—two narrow peaks, each indicating a layer of lipid headgroups at the surface of the membrane; an average density near the middle of the bilayer similar to the density of water; and an overall thickness no greater than 60 Å—must be characteristic of the membrane. The profile of the extracted lipids (dashed curve) shows that the two layers of headgroups are much the same distance apart as in the membrane. The density near the middle of the lipid bilayer (core) is well below that of water, as expected for a core of dihydrophytyl chains.[10] The comparison indicates a high proportion of protein in the core of the purple membrane.[10,20] The asymmetry would suggest either that the protein is more electron dense, or that there is a greater volume of protein, and consequently less lipid, in one-half of the membrane than in the other; it remains, however, to confirm the asymmetry.[20] [Appeared originally as Fig. 3 in ref. 20. Copyright 1977 by the American Association for the Advancement of Science.]

the purple membrane are organized in a bilayer-like structure with much the same thickness as that of the isolated lipids. Second, each molecule of bacteriorhodopsin extends through the bilayer from one surface to the other. Third, there are three identical molecules of bacteriorhodopsin in

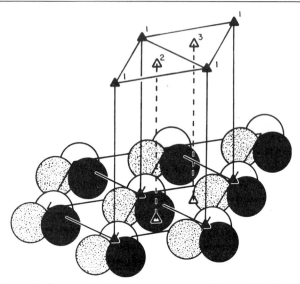

FIG. 3. Schematic diagram showing the arrangement of bacteriorhodopsin molecules in the purple membrane. All molecules are identical, but the different shadings show that the molecules are viewed from three different directions. The molecules are related by three different threefold axes. The molecules have been placed nearer one of the threefold axes (▲) for clarity. Eight unit cells are shown. [Appeared originally as Fig. 2 in ref. 11.]

the unit cell. The three are related by a threefold axis at the center of the three molecules; the axis is perpendicular to the membrane. The unit cells are arranged in a hexagonal array (Fig. 3). Fourth, the lipid molecules occur in patches between bacteriorhodopsin molecules. Fifth, the bacteriorhodopsin molecules do not project far outside the bilayer. Sixth, each of these contains segments of α-helix which run more or less perpendicularly to the plane of the membrane. Seventh, the membrane shows, at most, only a small asymmetry in terms of mass although there is an important asymmetry in terms of function. Finally, the in-plane reflections could not easily be phased by an X-ray diffraction experiment. Instead, this has been done by the elegant means of low-dose imaging in the electron microscope. This work is fulfilling the promise offered by the crystalline array in the purple membrane: a detailed knowledge of the structure of a membrane protein having an interesting function.

Acknowledgments

I thank Ms. M. Blackburn for drafting Fig. 1, and Mr. F. W. King for photographic work.

[17] Rapid Kinetic Optical Absorption Spectroscopy of Bacteriorhodopsin Photocycles

By RICHARD H. LOZIER

Introduction

Optical absorption spectroscopy is a useful tool in the study of many biological pigments; it is particularly appropriate for the investigation of bacteriorhodopsin, a pigment whose biological function is the utilization of light energy to pump protons out of the halobacterial cell in whose membrane the pigment is incorporated. The availability of pulsed lasers and rapid kinetic optical absorption measuring equipment allows the investigation of this process at very high time resolution.[1] In addition to measuring absorption changes of the bacteriorhodopsin chromophore, which are undoubtedly relevant to the mechanism of proton translocation, it is possible to measure proton release, uptake, and, in vesicular systems, translocation, in the same samples and under identical conditions, using pH-indicating dyes. Using polarized actinic and measuring light, information on chromophore motion occurring during the photochemical cycle of the pigment can be obtained. Thus a wealth of structural and functional information can be gained. Although detailed molecular interpretation of visible absorption spectra may not be possible—as compared with resonance Raman, UV and IR absorption, nmr, and fluorescence spectroscopy, and electron, neutron, and X-ray diffraction—the relative ease of visible optical absorption measurements and their high signal-to-noise and time resolution makes them vital to the design and interpretation of experiments using these other techniques. In addition, much molecular information can be inferred from the effects of environmental factors (temperature, pH of suspending medium, etc.) and specific biochemical or genetic modifications of bacteriorhodopsin on the light-induced absorption changes. Finally, optical absorption spectroscopy may be used to monitor the function of bacteriorhodopsin *in vivo* and in various model systems such as cell ghosts, cell envelope vesicles, and proteoliposomes.

---

[1] Although time resolution of less than 1 psec has been achieved in other laboratories, techniques discussed here will cover times greater than 50 nsec.

METHODS IN ENZYMOLOGY, VOL. 88

Important Parameters to Be Considered When Making Optical
Absorption Measurements on Bacteriorhodopsin

Bacteriorhodopsin in purple membrane isolated from halobacteria by
standard procedures is a readily available and unusually stable biological
material. Because of this, very high resolution measurements can be
made and many subtle effects observed. The following parameters are or
may be important and thus should be controlled if these effects are to be
reproducible and interpretable.

Source. The bacterial strain and its growth conditions (e.g., light, temperature, medium) may affect the quality of bacteriorhodopsin in isolated
purple membrane. Although no definite differences in the lipid and protein
composition of purple membrane from *H. halobium* and *H. cutirubrum*
have been found,[2] purple membrane from different strains or the same
strain under different growth conditions may differ in membrane patch
size and shape, carotenoid impurity, etc.

Isolation Procedure. Several purple membrane isolation procedures
have been published (e.g., Oesterhelt and Stoeckenius[3] and Becher and
Cassim[4]). The principal difference noted is the amount of contamination
by the red carotenoid pigment, but this parameter varies considerably
even when the same preparatory procedure is used. A trade-off must of
course be made between yield and purity, but any deviation from a standard procedure (e.g., omission of the sucrose density gradient centrifugation or the use of sonication to attempt to disoldge red membrane fragments from the purple membrane) should be noted. The time and tedium
involved in the sucrose density gradient step, as well as some concern
about possible effects of sucrose on the purple membrane, have prompted
us[5] to search for an alternative. Washing alone has not proved effective in
removing carotenoid impurity. Recently, we have tried a Percoll (Pharmacia) gradient in lieu of the sucrose gradient. Preliminary comparison of the
kinetics of the light-induced optical absorbance changes showed no obvious differences. However, quantitative removal of the Percoll may be difficult. Removal of sucrose by dialysis versus centrifugation may be relevant if the procedures result in different particle size distribution, age of
preparation at completion of isolation, etc.

Purity. Purity of purple membrane is often assayed by the amount of
red pigment present. This is obviously an unreliable procedure insomuch

[2] S. C. Kushwaha, M. Kates, and W. Stoeckenius, *Biochim. Biophys. Acta* **426,** 703 (1976).
[3] D. Oesterhelt and W. Stoeckenius, this series, Vol. 31, Part A, p. 667.
[4] B. M. Becher and J. Y. Cassim, *Prep. Biochem.* **5,** 161 (1975).
[5] Use of first person plural refers to the author and colleagues within the group of
W. Stoeckenius.

as the quantity of red pigment is influenced by the bacterial strain and growth conditions, and the red pigment is under some conditions less stable than bacteriorhodopsin [it is photo- and autobleachable (oxidizable?)]. The presence of contaminating pigments (cytochromes and flavoproteins from the "red" membrane may be present as well as carotenoids) can be assayed by the visible and UV absorption spectrum. The ratio of the 280-nm protein band to the visible absorption is a good indicator but must be used with caution because it is sensitive to light scattering (which may be instrument and sample dependent) and the light adaptation state of the bacteriorhodopsin, which in turn is temperature, pH, and illumination history dependent.

Age. The stability of bacteriorhodopsin in isolated purple membrane is legendary; you can store it unprotected from microbes and under ambient temperature and illumination conditions for months, or take it briefly to pH -1, pH 12, or 100° and still have a purple sample. However, it now appears that purple membrane may "age" rapidly[6]; it is therefore prudent to use fresh membrane as much as possible.

Storage. Until recently, purple membrane has been stored in our laboratory in the refrigerator, usually suspended in distilled water containing 0.01–0.2% sodium azide. (This practice was never proved safe or effective; bacteria are often seen in such samples). Sometimes purple membrane is stored at 4° in basal salt solution; this practice has the disadvantage that the purple membrane aggregates and precipitates, making accurate aliquots difficult to take. In either case, I wash the purple membrane with distilled water, using a Thomas Teflon pestle homogenizer if necessary to disaggregate the membrane and reisolate the membrane by centrifugation at 18,000 rpm for 20 min using a Sorvall SS-34 rotor in a Sorvall RC2-B centrifuge (39,100 g). Freezing membrane suspensions in dilute aqueous media is not recommended because the membrane precipitates out of suspension. A recent idea in this laboratory is to store purple membrane still in sucrose (from the sucrose density gradient centrifugation) in a deep freeze, thawing aliquots and washing them free from sucrose immediately prior to use. We hope that under these conditions bacterial contamination, aggregation, and aging will not occur.

pH. The kinetics, and to a lesser extent the static absorption, of bacteriorhodopsin in isolated purple membrane are sensitive to pH. The pH dependence is convoluted with salt and temperature dependences (see below). I have done many early studies with purple membrane in distilled water, in which the purple membrane is very stable in time. The pH can be measured with a pH meter and adjusted by small additions of acid or

[6] R. Govindjee, T. G. Ebrey, and A. R. Crofts, *Biophys. J.* **30**, 231 (1980).

alkali. For work near neutral pH, I have used 0.01 M potassium phosphate buffer for pH control. It has been reported[7] that purple membrane suspended in 50% dimethyl sulfoxide behaves differently in phosphate buffer than in several cationic buffers at the same pH and concentration. Ideally, a standard set of buffers should be adopted after appropriate tests, but this has not yet been done and is not straightforward since any specific effects of buffers may depend on other factors such as pH, concentration, salts present, etc. pH is also important in that it influences the rate of dark adaptation; at low pH the presence of 13-cis pigment may be a complicating factor (*vide infra*). At extreme pH (less than 5 or greater than 10) the static (light-adapted but in the dark) spectrum may be influenced. The safe pH range is probably influenced by temperature and salt.

Salt. Halobacteria grow in 4 M NaCl plus other salts, and most halophilic enzymes require high salt for stability and function; it would thus seem reasonable to study purple membrane suspended in concentrated salt solutions. This is not practical, however, since, as mentioned earlier, the purple membrane aggregates and precipitates in high salt. Fortunately, purple membrane does not require high salt for stability or function. However, the kinetic behavior, and as mentioned earlier, the pH dependence of the kinetic behavior, is influenced by salts. A full systematic investigation of the effects of various cations and anions on the static absorbance and light-induced absorbance changes has not yet been done. However, some studies on salt-induced light-scattering changes (presumably due to aggregation changes induced by surface charge screening) and safranine-O binding have been done (P. C. Mowery, S. L. Helgerson, and W. Stoeckenius, unpublished). In the future, the influence of anions and cations on photocycle kinetics may be studied with purple membrane incorporated in polyacrylamide gels to prevent membrane aggregation.[8,9] Differential effects of anions and cations (including protons) on the extracellular and cytoplasmic sides of the purple membrane may be of interest, since *in vivo* the ionic composition of the aqueous phases bathing the external and cytoplasmic surfaces are quite different. It should be possible

---

[7] D. Oesterhelt, M. Meentzen, and L. Schuhmann, *Eur. J. Biochem.* **40**, 453 (1973).

[8] The high-temperature stability of bacteriorhodopsin in purple membrane incorporated into a polyacrylamide gel may be decreased (S. -B. Hwang, personal communication). This suggests that some damage to the purple membrane may have occurred during the preparation of the gel or that the gel presents a hostile environment to the membrane at higher temperatures. Controls should be done at low ionic strength (where the gel is not required to prevent aggregation) to determine whether the incorporation of purple membrane into the gel modifies the kinetic behavior of the bacteriorhodopsin.

[9] P. C. Mowery, R. H. Lozier, Q. Chae, Y.-W. Tseng, M. Taylor, and W. Stoeckenius, *Biochemistry* **18**, 4100 (1979).

to study these effects using purple membrane incorporated into planar lipid bilayers, liposomes, cell envelope vesicles, or cell ghosts. Salts may influence photocycle kinetics by altering the pK values, and thus the protonation states, of groups involved in the proton translocation.

Temperature. Chemical kinetics are generally very sensitive to temperature, and the kinetics of bacteriorhodopsin absorbance changes are no exception. Rate constants may change by more than 10% for a 1° change. For critical experiments, temperature control better than 0.1° may be required. Secondary effects of temperature may include changes in the static absorption spectrum (and thus the fraction of light absorbed by actinic and measuring beams), changes in protonation of groups on the membrane (due to temperature dependence of pH and pK values of exogenous buffers and buffering groups on the protein and lipid of the purple membrane), and changes in equilibrium between different forms (e.g., the ratio of 13-*cis* to all-*trans* chromophore in dark-adapted membrane). Temperature effects are, of course, convoluted with pH, salt, etc, effects. At low temperatures, effects of freezing of the suspending medium and at high temperatures denaturation must be considered. It is prudent to measure the absorption spectrum of a sample at a control temperature (e.g., 20°) before and after experiments at extreme temperatures to test for irreversible effects. The reproducibility of the light-induced absorption changes may be an even more sensitive indicator of damage to the sample.

Deuterium Isotope Effects. Deuterium isotope effects on the kinetics of light-induced absorbance changes of bacteriorhodopsin are easily demonstrated. The effects are apparently due predominantly to exchangeable protons,[10] and may well be relevant to the proton translocation mechanism. No effects on the static visible absorption spectrum have been noted for either exchangeable or covalently bound protons versus deuterons. The kinetics of light-induced absorption changes in deuterium oxide are certainly expected to also depend on temperature, salts, pD, etc. Other effects may arise from pK differences for protons versus deuterons.

Solvents/Suspending Media. The remarkable stability of bacteriorhodopsin in the purple membrane is apparently dependent on protein–protein interaction in the two-dimensional crystalline lattice that the protein forms in the plane of the membrane. Dissolution of the lattice with detergents[11] and exposure to organic solvents[7] cause significant changes in static absorption and kinetics and generally reduce the stability (both

[10] R. H. Lozier and W. Niederberger, *Fed. Proc., Fed. Am. Soc. Exp. Biol.* **36**, 1805 (1977).
[11] J. A. Reynolds and W. Stoeckenius, *Proc. Natl. Acad. Sci. U.S.A.* **74**, 2803 (1977).

thermal and photochemical) of bacteriorhodopsin. Addition of glycerol, sucrose, Ficoll, ethylene glycol, etc., as antifreeze for low-temperature studies or to increase viscosity to diminish membrane tumbling (*vide infra*) may have secondary effects. These include changes in the kinetics of the light-induced absorbance changes and in the static absorption spectrum. The former may be due to solvent (different dielectric environment) or viscosity effects. Changes in absorption spectra may be due to chemical solvent effects or physical light-scattering changes due to changes in index of refraction of the medium versus the membrane.

Membrane Potentials. The absorption spectra and kinetics of bacteriorhodopsin in purple membrane may be influenced by static charges on the membrane (which in turn are influenced by salt, pH, temperature, etc.) and by transmembrane potentials. The latter can of course only be observed in cells, closed vesicles, or planar membranes separating compartments, etc., that contain bacteriorhodopsin. In such systems effects of absolute pH and salt concentrations on each side of the membrane must of course be considered. Studies on planar membranes and vesicles containing oriented purple membrane patches may eventually allow the calibration of bacteriorhodopsin kinetics to serve as a probe of the same parameters *in vivo*.

Light. The illumination history of a purple membrane sample can be significant; although bacteriorhodopsin is very stable to light under normal conditions, it may become photolabile under extremes of pH, temperature, etc. Illumination may, under certain conditions, preferentially bleach carotenoids, as noted earlier, and thus may serve to remove unwanted absorption from a purple membrane sample. In addition to irreversible changes, the dark–light adaptation state depends on illumination history in addition to temperature, pH, etc. Deliberate (actinic), inadvertent (during sample handling), and unavoidable (measuring) illumination should be controlled in color, intensity, duration, and polarization as much as possible.

Concentration/Optical Path Length/Optical Density. The penetration of actinic light into a purple membrane-containing sample depends of course on the concentration of the pigment, the optical path length, and the wavelength of the actinic light. A trade-off must be made between making a sample optically thin enough to avoid complications due to actinic light intensity variation across the sample and yet thick enough to give adequate signal-to-noise in the optical transmission measurements.

Additional parameters that may be important under certain circumstances include degree of hydration,[12] pressure (Tsuda and Ebrey[13] and

[12] R. Korenstein and B. Hess, *Nature (London)* **270**, 184 (1977).
[13] M. Tsuda and T. G. Ebrey, *Biophys. J.* **30**, 149 (1980).

L. Eisenstein and P. Ormos, personal communication), and sample anisotropy (from partial drying or orientation in a magnetic[14] or electric[15] field).

Equipment and Procedures

In the next section I discuss in considerable detail the instrumentation and procedures that I have found to work for me. Other instrumentation may be found that works as well or better, but this discussion addresses specific technical problems that will occur, to greater or lesser degree, with any instrumentation. It is hoped the discussion of the specific instrumentation with which I am familiar will be useful in evaluating the performance of instruments available to the reader.

Scanning Spectrophotometers

In our laboratory, static absorption spectra are measured with a Cary 14 spectrophotometer, an Aminco DW-2a spectrophotometer, or a single-beam integrating-sphere spectrophotometer.[16] The Cary 14 is equipped with a 0–0.1, 0.1–0.2/0–1, 1–2 absorbance slidewire and a Cary scattered transmission accessory which has been modified by the installation of an EMI 9659QA photomultiplier, selected for uniform cathode sensitivity, and by thermostating the cuvette holder to provide temperature control of the cuvette contents to better than 0.5°C near 20°. The relative merits of the three instruments are summarized in Table I. The Cary is the best compromise for most purposes. I typically prepare a stock solution

TABLE I
SCANNING SPECTROPHOTOMETERS

	Characteristic[a]		
Instrument	Sensitivity	Scattered light collection	Ease of operation
Cary 14 with scattered transmission accessory	2	2	1
Aminco DW-2a	1	3	2
Single beam with integrating sphere	3	1	3

[a] 1 = best, 3 = worst.

[14] D.-Ch. Neugebauer, A. E. Blaurock, and D. L. Worcester, *FEBS Lett.* **78**, 31 (1977).
[15] L. Keszthelyi and P. Ormos, *FEBS Lett.* **109**, 189 (1980).
[16] R. A. Bogomolni, R. A. Baker, R. H. Lozier, and W. Stoeckenius, *Biochemistry* **19**, 2152 (1980).

of purple membrane in water that is diluted with stock solutions to give the desired final composition in 3.0 ml in a 1 × 1-cm fluorescence-type cuvette just prior to use. As mentioned above, purple membrane forms very stable suspensions in water, but it will partially precipitate from even 10 mM potassium phosphate buffer in a few days time. After running a baseline of suspending medium versus suspending medium, the sample cuvette (or a matched cuvette), carefully washed and dried, is used to contain the purple membrane suspension. At neutral pH and 20°, the purple membrane can be fully light-adapted by placing the cuvette a few centimeters from a cool white fluorescent lamp for several minutes. (An unfiltered tungsten lamp should not be used since it will heat the sample, thus accelerating dark adaptation.) The Cary spectrophotometer is typically started at 750 or 800 nm (this is made possible by the installation of the red-sensitive EMI9659 photomultiplier) and no balance adjustment is made to the absorbance level. The apparent optical density between 750 and 800 nm is due mainly to light scattering, and is generally less than 0.01 absorbance units for a sample having 0.5 absorbance in the 570-nm band. By scanning the entire visible and through the near UV to ~240 nm (switching to the UV lamp at ~350 nm) much can be told about the condition of the sample, including evidence of aggregation (apparent absorption above 750 nm) and degree of carotenoid and protein impurity. Carotenoid impurity can be assessed by the residual absorption bands near 500 nm after bleaching the sample with detergent (1.5% Ammonyx LO in purple membrane suspension, in 5 mM carbonate buffer, pH 10) or by strong alkali (pH > 12.5) plus light. Protein impurity can be assessed by the 280-nm/470-nm absorbance ratio, which should be significantly less than 2. Fortunately, the spectrum of dark-adapted samples can be measured without significant light adaptation in the measuring beam of the Cary, and the spectrum of a light-adapted sample can be measured without significant dark adaptation, at neutral pH and 20°. These statements can be verified by measuring twice the absorption spectrum. The Aminco DW-2a is preferable to the Cary only when very low (< ~0.05) absorbance samples must be used, since its signal-to-noise ratio and sensitivity are greater, but its scattered light collection and stray light properties are not as good as the Cary. The integrating sphere is usually justified only for highly scattering samples (e.g., whole cells). Sufficiently slow absorption changes, such as the dark adaptation at neutral pH and 20°, can be measured by repeatedly scanning the absorption spectrum. It is prudent to protect the sample from the measuring beam between sweeps to minimize the possibility of slow actinic effects of the measuring beam. Absorption changes requiring several tens of seconds or more can be measured by continuously monitoring at a wavelength of interest. With appropriate means for cross-illumination, much faster light-induced absorption changes (those occurring in times longer than ~ 0.01 sec) can be measured with the Aminco DW-2a. Even faster absorption measurements require the use of a rapid kinetic spectrophotometer.

Flash Kinetic Spectrophotometer

Our kinetic spectrophotometer has been developed and improved over several years, using for the most part commercially available components. A block diagram of the instrument is shown in Fig. 1 and its basic elements are discussed below.

Measuring Beam Source. The ideal measuring lamp would have small size, uniformity in time and space, high intensity, and a broad, smooth emission spectrum. The best compromise we have found for most of our studies is a compact filament 45-W, 6.6-A tungsten halogen lamp (G.E. or Sylvania 6.6A/T 2 1/2 Q/CL), which we have used successfully from ~ 350 to 750 nm and to ~ 100 nsec time resolution. Some other tungsten halogen lamps (sometimes referred to as "quartz halogen" lamps), such as Phillips Rallye car lamps, are suitable. Tungsten halogen lamps are superior to regular tungsten lamps in that they have higher color temperature and are therefore better sources in the blue and near UV region. Although more powerful tungsten halogen lamps are available, e.g., 100-, 200-, and 1000-W versions, these do not have higher color temperatures, but rather larger (and less compact) filaments. Some work in this laboratory has been done below 350 nm using a deuterium arc lamp (R. Bogomolni, personal communication). We have little experience using xenon arc lamps as measuring sources; although their color temperature is high, their stability is inferior to the tungsten halogen lamps in our limited experience. Pulsed xenon lamps are often used for nanosecond studies in other laboratories. We have not attempted studies faster than the 50 nsec/pt sampling time of our instrument (see below) and have been able to use the dc tungsten halogen lamp to this time resolution. Our tungsten halogen lamp is powered by a stable dc power supply (e.g., Lambda LE103 FM or New Jersey Electronics SVC40-10). Although the lamp is nominally a 45-W 1000-hr lamp, it is advantageous to overpower the lamp (up to ~ 80 W) for better intensity in the blue and near UV spectral regions. Of course, lamp life is significantly decreased when the lamp is overpowered. The image of the lamp filament (filament size of our lamp is nominally 1.5×3 mm) is focused with a 25-mm diameter 75-mm focal length lens onto the entrance slit (2 mm wide by 5 mm high) of a Jarrell-Ash 82–410 $\frac{1}{4}$ m f:3.2 monochromator having a 1180 groove/mm 500-nm blaze grating and 3.3 nm/mm optical bandwidth. The image of the filament at the circu-

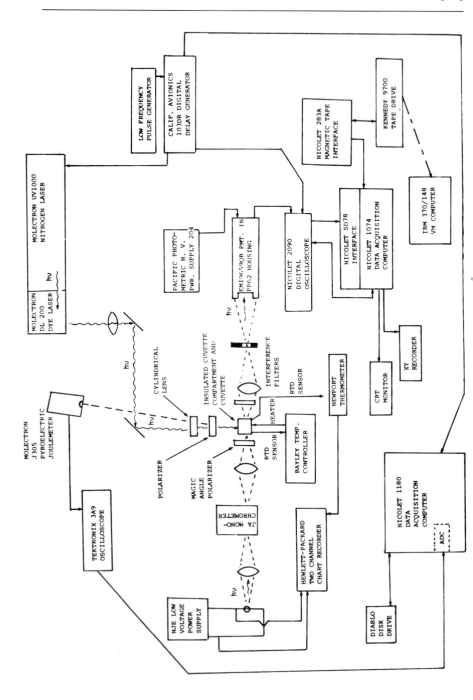

lar 2-mm diameter exit is refocused with a second f : 3.0 lens into the center of the cuvette. The intensity and polarization of the measuring beam depends on the lamp operating wattage, the monochromator efficiency, and the selected measuring wavelength. Light from the source is slightly polarized because of the asymmetry of the lamp filament and envelope. In addition, grating monochromators introduced a wavelength-dependent polarization that results in vertical polarization below the blaze wavelength and horizontal polarization above the blaze wavelength. To eliminate absorption changes due to motion of the chromophores (either within the membrane or caused by tumbling of membranes in suspension) a "magic-angle" polarizer (see Table II) can be used before (or possibly after) the sample. Placing the polarizer before the sample has the advantage of decreasing the measuring-beam intensity on the sample (thus reducing the possibility of actinic effects of the measuring beam) and establishing a known constant polarization for all measuring wavelengths. Placing the polarizer after the sample has the advantage of blocking a fraction of the stray actinic light from reaching the detector. However, data interpretation is complicated if the light incident on the sample is partially polarized, as it is in our case. For the magic-angle polarizer we use polarizing sheet (e.g., Edmond Scientific catalog No. 60490) that provides good polarization ($>1000 : 1$) with transmission >0.2 (0.5 is theoretical maximum) across the visible spectrum.

Cuvette. The type of sample being investigated may dictate the type of cuvette used. I find 1×1-cm fluorescence cuvettes advantageous for most work with purple membrane suspensions. These require a moderate sample volume (~ 3 ml) and allow for measurements to be made in the

TABLE II

POLARIZATION OF MEASURING AND ACTINIC BEAMS TO EXCLUDE
SIGNALS FROM CHROMOPHORE MOTIONS

Measuring geometry	Magic angle polarization of measuring beam
Actinic beam perpendicular to measuring beam and unpolarized	35.3° from normal to plane of actinic and measuring beams
Actinic beam perpendicular to measuring beam and polarized normal to plane of measuring and actinic beams	54.7° from normal to plane of actinic and measuring beams
Actinic beam perpendicular to measuring beam and partially polarized with relative intensities I_v and I_h	$\Theta = \arccos\left[\left(\frac{1}{3}\frac{I_v + I_h}{I_v}\right)^{1/2}\right]$
Measuring and actinic beams coaxial, both polarized	54.7° between planes of polarization

Cary or Aminco in the same cuvette. pH can be measured in these cuvettes. (I use a Beckman 39030 electrode with a Corning 110 pH meter.) For anaerobic work (sometimes required for pH dye experiments using unbuffered suspensions; see the appendix) the stopperable type 111 cuvette (Quaracell) is advantageous, since it permits insertion of an efficient magnetic stirring bar ($\frac{1}{4}$-in. Bel-Art Spinfin) through the top opening. Although stirring during optical transmission measurements adds much noise (due to Schieren effect of flow-oriented purple membrane), by placing a stirring bar inside the cuvette below the measuring beam and placing a motor-driven magnet below the cuvette, the sample can be stirred after additions through a penetrable stopper. (I use silicon rubber cut with a cork borer.) In earlier experiments at variable temperature, we used jacketed cuvettes with a 1.0-cm light path in the measuring beam and a 4-mm width. These cuvettes have the advantages of small sample volume (~ 1 ml), ease of temperature control, and low optical density in the direction of the actinic beam (when cross-illumination is used), but they do not work well in the Cary (its measuring beam is too wide) and they are too small to admit a stirring bar or most pH electrodes. For highly light-sensitive samples, flow cells may be advantageous. We have used a flow cell to allow convenient sample changes between actinic flashes when studying dark-adapted purple membrane. However, the sample must be stationary during the measurement so that the part of the sample illuminated by the actinic beam remains in the measuring beam.[17] Short path length cuvettes (e.g., 1 mm or less) may be adventageous for highly concentrated samples, but they require either positioning at 45° from the measuring and actinic beams, which are at 90° with respect to each other, or coaxial measuring and actinic beams; the first alternative leads to inadvertant polarizations by reflection, the latter to some difficulties with stray laser light reaching the detector and with achieving "magic-angle" polarization. For experiments requiring high actinic intensity, I have used a 2 × 2-mm cuvette to decrease the diameter of the actinic beam required fully to illuminate the volume of sample exposed to the measuring beam. The cuvette was masked in the direction of the measuring beam using black tape to ensure that none of the measuring beam avoids passing through the sample.

Cuvette Temperature Control. The cuvette temperature can be controlled in the range 0–100° using a constant temperature water circulating

---

[17] It is assumed that with a stationary sample diffusion of membrane not subjected to the actinic beam into the measuring beam during the measurement of the light-induced absorbance changes does not occur. To ensure that this assumption is valid, it is prudent to make the part of the sample exposed to the actinic beam extend beyond the part of the sample exposed to the measuring beam.

bath, or more precisely with a flowing nitrogen system described below. Either system can be used with the jacketed cuvette or with a 1 × 1-cm cuvette. The jacketed cuvette is mounted on a thermally insulating block (e.g., acrylic plastic), whereas the 1 × 1-cm cuvette is held by a machined copper block provided with apertures for the actinic and measuring beams and channeled for the circulation of the thermostating fluid. This block is insulated on the outside with pipe insulating tape (Armstrong Armaflex); for work far from ambient temperature it can have "storm windows" made from microscope cover glasses on the measuring and actinic beam apertures. For lower temperatures (to 77°K), we could use an Oxford 704 DeWar; so far we have used this DeWar only for static spectra and slow kinetics on the Aminco DW-2a.

The cuvette and temperature block assembly is contained in a 4 × 4 × 4-in. aluminum box insulated on the inside with $\frac{1}{8}$ in. or more of Armaflex insulation tape or sheet, except for 2-cm apertures for the entering and exiting measuring beam and for the actinic beam entrance, all of which can be closed by 2 × 2-in. projection slide cover glasses (Kodak). Sample temperature is controlled using a Bayley model 121 Precision Temperature Controller. Nitrogen gas is flowed at ~6 liters/min through a copper heat exchanger (in a DeWar containing powdered dry ice for temperatures of 20° or lower), through a glass tube containing a 330-ohm heater[18] connected to the output of the controller, and then to the cuvette block via rubber tubing. The glass and rubber tubing from the heat exchanger to sample box is insulated with neoprene insulation (Armaflex tubing). The nitrogen gas leaving the cuvette block bathes the platinum resistance thermometer (RTD) of the controller. The exiting nitrogen purges the sample box, which serves to prevent moisture from condensing on the cuvette, which might otherwise occur when operating below ambient temperature. The sample temperature is monitored by a small stainless steal sheathed RTD placed in the sample a few millimeters above the intersection of the measuring and actinic beams. The resistance of the RTD is monitored and converted to a digital temperature reading by a Newport model 267B digital pyrometer with 0.01° resolution and 0.04 accuracy. No effect of the measuring or actinic beams on the temperature of the sample has been found. However, the measured temperature of the sample is sensitive to the position of the RTD and to stirring of the sample with a magnetic stirrer (a few tenths of a degree at 65°). A problem to be considered is evapo-

---

[18] The heater was constructed by wrapping fine (e.g., #40) heater wire on mica cut to fit inside the glass tube ($\frac{13}{16}$-in I.D. × 1-in. O.D. × 8 in. long). Small cuts in the mica hold the wire to prevent shorting. Electrical connection is made via small brass bolts that extend through the silicon rubber stoppers used to seal the ends of the glass tube. Small glass connecting tubes (Kimax $\frac{3}{16}$ in.) also penetrate the stoppers to provide for the gas flow.

ration of the sample, which is enhanced by the dry nitrogen atmosphere, especially at elevated temperatures. This problem may be minimized by loosely stoppering the cuvette, with the RTD penetrating the stopper, and minimizing the time spent at elevated temperature.

Cuvette-to-Detector Optics. After passing the sample compartment, the measuring beam must be passed through optical filters (and/or a monochromator) and to the detector. In our present configuration, the measuring beam is refocused by a third f: 3.0 lens onto an aperture and then diverges onto the photocathode of the detector. A combination of 2 × 2-in interference (Baird Atomic, 10-nm half-bandwidth) and colored glass (Corning) filters, chosen to optimize transmission of the measuring beam while minimizing the transmission of scattered actinic light, are placed between the sample and the final lens, as close as possible to the lens. This position of the filters is an improvement over our previous placement of filters in a holder on the front of the photomultiplier housing, presumably because of decreased collection of filter fluorescence. Alternatively, a second monochromator can be positioned with the transmitted measuring beam focused on its entrance slit and the detector positioned after this monochromator. The monochromator has the advantage of more flexible choice of measuring wavelength and ease of changing wavelength. However, in many situations filters in addition to the monochromator are required to reject sufficiently scattered actinic light, whereas the filters alone are generally sufficient.

Actinic Light Sources. Several pulsed actinic light sources have been used in this laboratory (see Table III); they include xenon flashlamps (e.g., United States Scientific Instruments Strobrite, General Radio Strobotac, and camera strobes), a flashlamp-pumped dye laser (Phase-R

TABLE III
ACTINIC LIGHT SOURCES

Source	Wavelength	Energy	Pulse duration
Xenon flashlamp (USSI type 3015 Strobrite)	White light	$3 \times 10^8 - 2 \times 10^7$ beam candles[a]	1.3–7.0 μsec (taken at $\frac{1}{3}$ intensity
Flashlamp-pumped dye laser (Phase R-DL 1200 V)	Tunable across visible by changing dye	150 mJ[b]	250 nsec[b]
Nitrogen-laser-pumped dye laser (Molectron UV1000/DL200)	Tunable across visible by changing dye and tuning grating	0.75 mJ[c]	10 nsec

[a] Intensity rather than energy specification.
[b] With Rhodamine-6G dye (588-nm output).
[c] With Rhodamine-6G dye (580-nm output).

DL-1200V), and a nitrogen-laser-pumped dye laser (Molectron UV1000/DL200). Desirable features in an actinic source include high energy at the desired wavelength, lack of energy at other wavelengths (narrow bandwidth and low stray light), short pulse duration, wavelength tunability, reproducibility (in time and energy), variable repetition rate, and focusability. Camera strobes are relatively powerful and inexpensive but their pulse widths (generally ~ 1 msec) are too long for most work. Pulse widths of ~ 2 μsec are available from the Strobrite and Strobotac but at much reduced intensity. In any event, the white light output of the xenon flashlamps requires filtering to remove light near the measuring wavelength, which is often a difficult or impossible task, and low f number optics[19] must be used to focus the xenon flash onto the sample coincident with the measuring beam. The flashlamp-pumped dye laser has the highest power in a narrow bandwidth and a moderately short pulse duration. The nitrogen-laser-pumped dye laser has lower energy but higher time resolution (~ 10 nsec pulse width) and greater ease of wavelength tunability. The collimated beams from the lasers are an advantage for beam steering and focusing. Our solution for beam steering and focusing is shown schematically in Fig. 1. Two mirrors allow adjustment of the laser beam height and direction so that the beam can be made incident normal to a cuvette face that is parallel to the measuring beam axis and at the same height as the measuring beam. A weak lens (e.g., 2-m focal length) and a cylindrical lens (0.25-m focal length) allows the laser beam to be focused to the proper size and shape to overlap the measuring-beam path through the cuvette. The flashlamp-pumped dye laser (not shown in Fig. 1) has been steered in a similar manner and can be combined with the first laser beam using a beam combining (partially silvered) mirror, or even a plane glass since the power of the flashlamp-pumped laser is so much greater than that of the nitrogen-laser-pumped dye laser. This arrangement does not allow for independent choice of polarization for the two lasers. In future experiments I will probably bring the two laser beams from opposite sides of the cuvette so that the polarization of each can be fixed separately. For a current study of the light intensity dependence of the kinetics, the flashlamp-pumped dye laser is aimed directly toward the cuvette with no beam steering optics.

Several considerations apply to the polarization of the actinic beam. We have sometimes used a polarizing sheet to polarize the actinic beam vertically, and sometimes measured the natural polarization of the laser and adjusted the angle of the measuring-beam polarizer accordingly (see

---

[19] The correct magic angle may depend on the f number of the optics. The values for the magic angle given in Table II assume collimated beams.

Table II and Nagle *et al.*[20]). The latter method has the advantage of not attenuating the actinic intensity, but with our nitrogen-laser-pumped dye laser the polarization is wavelength dependent so the measuring beam polarizer must be reset if the actinic wavelength is changed. In the future I will try using an intracavity crystal polarizer in this laser to decrease the loss of light while achieving good polarization. Unfortunately, this polarizer must be retuned when the actinic wavelength is changed. Another alternative is to deliberately depolarize the actinic beam with a crystal depolarizer and use the appropriate angle for the measuring beam polarizer. Until recently (see Nagle *et al.*[20]) we did not use well-defined polarization conditions in our flash experiments except when deliberately looking for chromophore motion,[10] a shortcoming I think is shared by many others. Gillbro[21] is an exception in that he explicitly considered the problem of the mixing of chemical and physical rate constants; however, he concluded that the problem was not significant under most of his conditions. However, such effects are expected and are empirically demonstrable in our hands.[20] Under conditions of extremely high actinic intensities, the anisotropy introduced in the sample by the actinic flash disappears and polarization conditions become unimportant. However, the limited availability of powerful actinic sources and concerns about cooperative phenomena[22] may make high actinic intensities not the solution of choice.

Detectors. Photomultipliers are excellent detectors in that they provide high and distortion-free gain. To optimize signal to noise, one wants to use the highest measuring light intensity consistent with the sample being essentially in its relaxed state; i.e., the measuring beam should be as intense as possible without being significantly actinic. For most of our work we have used a 2-in.-diameter end-on photomultiplier with extended S-20 photocathode and quartz window (EMI9659QB). This detector is usable across the visible and into the near UV and IR. It has nominally a 10-nsec rise time; however, its high gain capability (11 stages of amplification, up to 10 million gain) makes it somewhat too sensitive for our usual measuring light levels. To avoid exceeding the allowable photomultiplier anode current, we operate at reduced gain ($\sim 400-700$ V, $\sim 50-1000$ gain) with a consequent increase in rise time. We have just purchased a lower gain, faster (and cheaper) photomultiplier (EMI9846QB) that should be more appropriate. The alternative of decreasing the measuring light intensity and increasing photomultiplier gain would result in higher photon noise level. The photomultiplier is contained in a Pacific Photometric model 62 housing fitted with a high-current voltage divider (260 kohms

[20] J. F. Nagle, L. A. Parodi, and R. H. Lozier, *Biophys. J.*, **38**, 161 (1982).
[21] T. Gillbro, *Biochim. Biophys. Acta* **504**, 175 (1978).
[22] R. Korenstein, B. Hess, and M. Markus, *FEBS Lett.* **102**, 155 (1979).

overall resistance with 150-V Zener diode between cathode and first dynode). The high-current dynode chain is used to improve linearity of photomultiplier response to light level at the higher anode currents sometimes used (up to $\sim 100 \mu A$). A highly stable high voltage dc power supply (e.g., Power Designs 2K-10 or Pacific Photometric model 204 2KVDC-10 mA) is used to bias the photomultiplier. For work in the ultraviolet, a photomultiplier having low sensitivity in the visible would be desirable to minimize the effect of stray visible radiation in the measuring beam and to decrease sensitivity to stray actinic light. Such photomultipliers are available from several manufacturers, but I have no experience in their use. Recently, we purchased an EMI housing (for the EMI9659 photomultiplier) equipped with a gating circuit (EMIGB1001A) that allows brief reversing of the potential on the first dynode during the laser flash to reject the signal from scattered laser light. This is useful only at intermediate time resolution, since the photomultiplier is turned off for at least 5 μsecs. The gated photomultiplier may prove particularly valuable for studies on highly scattering samples (e.g., whole cells), since such measurements are doubly complicated by loss of measuring light by scattering and increased scattering of laser light toward the detector.

Photocurrent Measuring Device. The time course of the photomultiplier anode current (and thus the level of light transmitted through the sample) is measured by an analog or digital oscilloscope or signal averaging computer. Digital signal processing is nearly essential if large amounts of data are to be acquired and analyzed; in such cases the expense of film would soon make taking pictures of analog oscilloscope traces uneconomical, to say nothing of the advantage of having digital data for analysis. I have used several systems including (1) a Nicolet 1074 Instrument Computer with SD-71B 12-bit 50 kHz $+/-0.25$ V full-scale analog-to-digital converter, (2) a Nicolet 2090 Explorer III digital oscilloscope with model 204 8-bit 7 MHz $+/-100$ mV analog-to-digital converter, and a Nicolet 1180 computer with 12-bit 3-μsec digitizer. The Nicolet 1074 is the simplest to operate and has the best signal-to-noise ratio, but its analog bandwidth does not allow monitoring of the K and L intermediates of the photocycle unless the sample temperature is lowered substantially below ambient. The photocurrent from the EMI9659 is terminated with a 24-kohm load resistor to ground to give an RC time constant ($C = \sim 200$ pF) of ~ 5 μsec, allowing the full bandwidth of the SD-71B to be utilized. The dark current level[23] of the photomultiplier (with measur-

[23] The dark current level is generally so small as to be indistinguishable from no signal, as seen by turning the photomultiplier high voltage off. If a significant difference is seen, it usually means that stray light is reaching the detector and that steps should be taken to eliminate any light leaks.

ing beam blocked) is set to $+0.25$ V using the DC level of the SD-71B; the initial transmission level (prior to the actinic flash) is set to 0 V by adjusting the measuring light intensity and photomultiplier gain.[24] The resulting photocurrent is then $I = E/R = 0.25$ V/24 kohm $= \sim 10$ μA, which is near optimal for photomultiplier stability. If higher time resolution is required, the Nicolet 2090 oscilloscope is used instead of the SD-71B. A 1-kohm load resistor is then used, which gives RC $= 200$ nsec and $I = 0.1/1000 = 100$ μA. This represents a compromise in that both the RC time and the photomultiplier anode current are higher than optimal, but a satisfactory intermediate amplifier has not been found. For measurements requiring intermediate time resolution, the Nicolet 1180 is sometimes used. It has the disadvantage that an intermediate amplifier must be used (we use a Tektronix 3A9 vertical amplifier in a Tektronix 564 oscilloscope) at some sacrifice in simplicity and stability.

Sweep and Flash Timing. Synchronization of the triggering of the computer time base sweep and the laser flash is provided by a California Avionics 103DR digital delay generator. A leading baseline of approximately one-tenth of the sweep time is measured prior to the laser flash. The digital delay generator is triggered at a rate consistent with essentially complete relaxation of the sample between flashes (e.g., 2 Hz) for signal averaging. Data are stored in blocks of 1024, 2048, or 4096 (usually 1024) points per trace in the 1074 memory and transferred to digital magnetic tape for permanent storage. These tapes are read into a large general-purpose computer (IBM 370/148 at UCSF) for detailed analysis. Analog traces of the data can be obtained from the 1074 or 1180 on an X-Y recorder. Data can be transferred from the 1074 to the 1180 memory for certain data reduction functions not requiring the IBM 370. The 1180 is equipped with a disk for storage of data acquired by the 1180 or transferred to the 1180 from the 1074. When it is necessary to investigate absorption changes over many orders of magnitude in time, data are collected at several sweep speeds and the resulting curves are combined in the IBM 370. When stray laser light influences the measured signals, a flash artifact trace is measured with the measuring beam blocked and used to correct the data trace. A rule of thumb states that the laser artifact must remain on scale (within the range of the analog-to-digital converter) or data should not be taken to avoid damage to the photomultiplier and severe distortion of the data. If the signal being measured is small, a more stringent requirement may need to be applied.

Actinic Light Intensity Monitor. For most experiments I have not used

---

[24] Adjusting lamp wattage rather than photomultiplier gain has the advantage of maintaining a constant noise level for various measuring wavelengths. However, in the near UV, additional photomultiplier gain is needed even at the maximum permissible lamp wattage.

an actinic light intensity monitor. All of the actinic sources listed in Table III, when properly tuned, are reproducible in amplitude to better than $+/-5\%$. Since in most applications signal averaging is used, the actinic light intensity fluctuations also average out. In recent experiments where data were taken at many wavelengths and temperatures over many hours, a laser intensity monitor was used to look for long-term stability. (The laser output drops slowly because of dye degradation.) For this purpose a Molectron J3-05 pyroelectric joulemeter was used. The joulemeter was positioned to receive the reflection from the first (planar) surface of the cylindrical lens used for laser beam focusing. The output of the joulemeter was averaged using the Nicolet 1180 computer. I am presently using the same joulemeter with the flashlamp-pumped dye laser for a study of the light intensity dependence of the kinetics. The joulemeter is now placed after the sample looking directly at the laser beam; it is protected from excessive radiation using Schott glass neutral density filters (Rolyn Optical), since metal film neutral density filters are damaged by the laser beam.

An ingenious instrument for following optical transmission changes over many orders of magnitude in time has been described by Austin *et al.*[25] This device has a logarithmic time base that allows fast and slow processes to be measured with optimum signal-to-noise in a single sweep. We are currently building a logarithmic time base for our flash instrument and hope to incorporate the signal-to-noise optimization feature in the future.

The flash kinetic spectrophotometer described earlier measures transmission at a single wavelength versus time; to get complete spectral information the experiment must be repeated at many measuring wavelengths. Another approach is to use an instrument that measures many wavelengths simultaneously and to repeat the experiment with different delays between the actinic flash and the measurement. Commercially available instruments can be adapted for this purpose (e.g., Tektronix RSS or PAR OMA). The price of the greater wavelength resolution is decreased photometric accuracy (due to stray light and detector dynamic range limitations) and poorer coverage in time. To my knowledge a device having broad and smooth coverage in both time and wavelength has not yet been devised.

Data Reduction and Analysis

It is usual to assume that all spectral states of bacteriorhodopsin obey Beer's law, so that the absorbance of a sample at any wavelength λ and

[25] R. H. Austin, K. W. Beeson, S. S. Chan, P. G. DeBrunner, R. Downing, L. Eisenstein, H. Frauenfelder, and T. M. Norlund, *Rev. Sci. Instrum.* **47**, 445 (1976).

time t, $A(\lambda,t)$, is equal to the sum of the absorbances of the states present

$$A(\lambda,t) = 1 \sum_{i=1}^{n} \epsilon_i(\lambda)c_i(t) \tag{1}$$

where $\epsilon_i(\lambda)$ is the extinction coefficient of state i at wavelength λ, $c_i(t)$ is the concentration of state i at time t, and 1 is the path length of the measuring beam through the sample. This assumption holds reasonably well for BR_{568}^{LA} and BR_{558}^{DA},[26] for which it can be tested, but is subject to some error, since bacteriorhodopsin is normally studied in purple membrane suspensions, which results in some error from light-scattering effects.[27] Also, chromophore–chromophore interactions within the purple membrane lattice may cause deviations from Beer's law, but these effects are probably small.[28]

It is generally agreed that a light-adapted sample of purple membrane at room temperature and neutral pH contains essentially only one spectral state BR_{568}^{LA} or BR_{568}^{trans}, although chromophore extraction and analysis suggests that a few percent of the bacteriorhodopsin molecules contain the 13-*cis* chromophore rather than the dominant all-*trans* isomer.[9] Chromophore extraction and analysis show that BR_{558}^{DA} contains ~1:1 14-*cis* and all-*trans* isomers, referred to as BR^{cis} and BR^{trans}. If a light-adapted suspension is cooled to 77°K, the absorption maximum shifts to ~575 nm and a photoequilibrium between BR^{LA} [29] and its photoproduct K[30] can be demonstrated. The rather sharp isosbestic between bR and K suggests that only two photoactive states are present[10]; in contrast, when bR^{DA} is cooled to 77°K and illuminated with various wavelengths of light, a sharp isosbestic is not found, suggesting that both BR^{cis} and BR^{trans} undergo photoequilibration with their unique products K^{cis} and K^{trans}.[33,34] The spectra

---

[26] We introduced the abbreviation "BR" to represent the chromoprotein bacteriorhodopsin, with superscripts LA and DA to indicate the light- and dark-adapted states. Subscripts are used to indicate the wavelength (in nanometers) of maximum extinction in the visible.

[27] The scanning spectrophotometers described earlier are designed to collect as much scattered light as possible so as to minimize apparent absorbance due to light scattering. In the flash kinetic spectrophotometer, the f number of the cuvette-to-detector optics is deliberately made higher to reject scattered actinic light. I have assumed that the light-scattering properties of the sample do not change on the time scale of the absorbance changes, but I have not explicitly tested this assumption.

[28] M. Rehorek and M. P. Heyn, *Biochemistry* **18**, 4977 (1979).

[29] The wavelength subscript is omitted here because of its temperature dependence.

[30] Intermediates of the photocycle of bR have been designated as K through O.[31]

[31] R. H. Lozier, R. A. Bogomolni, and W. Stoeckenius, *Biophys. J.* **15**, 955 (1975).

[32] When the superscript is omitted the sample should be assumed to be in the light-adapted state.

[33] F. Tokunaga, I. Iwasa, and T. Yoshizawa, *FEBS Lett.* **72**, 33 (1976).

[34] Here the superscripts cis and trans refer to the chromophore configuration of the parent pigments BR^{cis} and BR^{trans}.

for the K's can be calculated from the measured absorption spectra of the mixtures if the fraction present as each species is known. Unfortunately, the fraction is not well known, and the calculated spectra for K from BR^{LA} have absorption maxima ranging from 610[10] to 628 nm.[35] Determinations of the fraction present as K, x_K, and the spectrum of K, $\epsilon_K(\lambda)$ have been deduced from quantum efficiency measurements and the assumption of a unidirectional unbranched cycle from L to M, but the reasoning is circular to the extent that the quantum efficiency measurements require knowledge of the spectra. Recent work in this laboratory[36] may provide an independent way to estimate x_K from spectra measured using linearly polarized actinic and measuring light. On warming, K relaxes via several intermediates to BR.[31] Spectra of intermediates L and M have been calculated from such data.[10,35] At room temperature, at least one additional intermediate O can be seen,[31] and there is evidence for more intermediates N, P, etc.[21,31,37,38] Therefore, one of the first questions to be asked is, under a particular set of conditions, how many intermediates must be assumed to explain the measured light-induced absorption changes? The strategy and progress to date for interpreting the data from the flash spectrophotometer and thus answering this question are outlined in the following paragraphs.

The first step in the data reduction is the conversion of the measured transmission changes to absorption changes, $\Delta A_\lambda (t)$, via Eq. (2)

$$\Delta A_\lambda(t) = -\log \frac{I_\lambda(t)}{I_\lambda{}^0} \qquad (2)$$

where $I_\lambda (t)$ is the measuring light intensity signal at wavelength λ and time t relative to the signal at the same wavelength prior to the actinic flash $I_\lambda{}^0$. The absolute optical absorbance at any wavelength and time $A(\lambda,t)$ is known from the static optical absorption spectrum (measured with a wavelength scanning spectrophotometer), $A_s(\lambda)$, plus the change in optical absorbance $\Delta A_\lambda(t)$.

If magic-angle polarization is used (see Table II), the observed absorption changes should be due to chemical kinetics only (no changes due to physical processes such as membrane tumbling or chromophore motion within the membrane).[39] If it is further assumed that the chemical kinetics

[35] B. Becher, F. Tokunaga, and T. G. Ebrey, *Biochemistry* **17**, 2293 (1978).

[36] J. F. Nagle, S. M. Bhattacharjee, L. A. Parodi, and R. H. Lozier, in preparation.

[37] M. A. Marcus and A. Lewis, *Biochemistry* **17**, 4722 (1978).

[38] M. Stockburger, W. Klusmann, H. Gattermann, G. Massig, and R. Peters, *Biochemistry* **18**, 4886 (1979).

[39] Polarization of the measuring beam in the wavelength scanning spectrophotometer is not important as long as the measuring beam is not actinic and the sample is isotropic.

are first (or pseudofirst) order (as is usually assumed), then the data should be fit by

$$\Delta A_\lambda(t) = \sum_{i=1}^{NL} b_i e^{-k_i t} \tag{3}$$

where NL represents the number of exponential terms (and thus the number of intermediates predicted) of the fit, k_i are functions of the microscopic chemical rate constants, and b_i are functions of the k_i and the extinction coefficients of the intermediates. A simple method for determining the k_i and b_i is graphical analysis, where log $\Delta A_\lambda(t)$ versus t is plotted. If the k_i are well separated, this method may be satisfactory; however, it is known that the photocycle of light-adapted BR contains at least one pair of nearly degenerate rate constants. A more powerful method to determine the b_i and k_i is a nonlinear least squares computer program such as DISCRETE.[40] DISCRETE automatically determines the best fit of the data at one λ versus t with one up to nine exponential terms and outputs the k_i and b_i and their standard deviations. It selects the most likely solution based on statistical criteria. A limitation of DISCRETE is that it can only deal with one kinetic trace at a time (see below). After determination of the k_i from the data, it is common to plot log k_i vs $1/T$ (Arrhenius plot). If these plots are linear for all k_i, it is evidence in favor of a simple unbranched unidirectional cycle. However, we find these plots to be functions of λ,[41] which is physically unreasonable. Therefore, we have developed a version of a computer program (VARPRO)[42,43] that can force the same set of k_i to satisfy all λ, i.e.,

$$\Delta A(\lambda,t) = \sum_{i=1}^{NL} b_i(\lambda) e^{-k_i t} \tag{4}$$

The primary output of VARPRO is the $b_i(\lambda)$ and k_i, the norm of the residuals, $RN(\lambda)$, for each wavelength

$$[RN(\lambda)]^2 = \sum_{i=1}^{N} [A_b(\lambda,t_i) - A_m(\lambda,t_i)]^2$$

and the grand norm of the residuals RN

$$[RN]^2 = \sum_{\lambda=1}^{S} [RN(\lambda)]^2$$

[40] S. W. Provencher, *Biophys. J.* **16**, 27 (1976).
[41] W. Niederberger and R. H. Lozier, unpublished results.
[42] R. H. Lozier, R. LeVeque, and G. Golub, *Am. Soc. Photobiol., Program Abstr.* **7**, 101 (1979).
[43] G. H. Golub and R. J. LeVeque, *Proc. Army Numer. Anal. Comp. Conf., 1979* ARO Report 79-3 (1979).

where N is the number of times, S is the number of measuring wavelengths, and the subscripts b and m refer, respectively, to the best fit and the measured absorbances. Unfortunately, VARPRO does not yet provide the standard deviations for the $b_i(\lambda)$ and k_i automatically, although some sophisticated error analysis has been done by J. F. Nagle.[20]

With the further reasonable assumption that the appropriate photocycle model does not change rapidly with temperature, we analyzed data[20] from many wavelengths and times at a few temperatures using VARPRO. Because of systematic error, or faulty assumptions, we did not find a clear result when the k_i were plotted versus $1/T$; that is, it is difficult to establish unambiguously the appropriate value for NL and the temperature dependence of the k_i. However, the results seem to preclude an unbranched unidirectional cycle, and support the involvement of intermediates such as N and P. Further experiments are planned to try to obtain more certainty in the value of NL, and hence the $k_i(T)$ and $b_i(\lambda,T)$.

Given a model for the photocycle (i.e., the number of intermediates and the reaction pathways between intermediates) and an estimate of the fraction of the bacteriorhodopsin cycling, the output of VARPRO can be used to calculate spectra of the intermediates. Unfortunately, the number of possible models is very large and many are mathematically equally good. Therefore physical criteria must be used to select a reasonable model. We have tested two simple classes of models (unidirectional unbranched and unidirectional simply branched from any intermediate back to BR) using a program we call EXTINCT.[20] No model from these two classes satisfies the physical criteria that we have assumed should be observed. These criteria include, in addition to the validity of Eqs. (1) and (4),

(i) The extinction coefficients of the intermediates are nonnegative at all λ.

(ii) The wavelength dependence of the extinctions is smooth.

(iii) The temperature dependences of the spectra should be small and smooth.

(iv) The fraction cycling cannot exceed unity and should have no (or a small smooth) temperature dependence.

In its present form, EXTINCT deals only with the two classes of models mentioned earlier, which have analytical solutions. A more general program to calculate spectra from a kinetic model and the output of VARPRO has been written by D. Beece and L. Eisenstein (personal communication), but no completely satisfactory model has been found using this program either. The main difficulty is the large number of possible models and the number of free parameters in the more complicated models. Our intention is to continue to approach the problem of establishing a viable model from two directions simultaneously: (1) to eliminate simple models or classes of models, then add one additional free parameter and test this next more complicated model or class of models, and (2) to try specific models that we or others feel may possibly work, using the more general program. If no model can be found that fulfills the criteria we have established, we will have to reexamine the criteria to see whether they should be changed or relaxed.

One possible complication that may be responsible for the difficulty in finding a simple model that accounts for the presently available data may be the presence of a significant amount of a second photoactive pigment, e.g., the 13-*cis* isomer present in BR^{LA}.[9] This question can be addressed by comparing the $b_i(\lambda)$ and k_i obtained with different actinic wavelengths. If k_i are found that belong to two or more photoactive pigments, the assignment of the k_i to different pigments can be made by the actinic wavelength dependence of the $b_i(\lambda)$, assuming that the action spectra of the photoactive pigments are not identical. The fractions of BR_{cis} and BR_{trans} in a light-adapted sample may depend on pH and temperature, and this should be reflected in the results of flash experiments employing two or more actinic wavelengths.

Another potential problem is a light-intensity dependence of the kinetics. Although we have not established a light intensity effect, it has been reported by others.[22] Before attributing such effects to cooperativity between chromophores, it should be established that proper magic-angle conditions are used, that the spectrophotometer has no significant systematic error, and that photolysis of intermediates is not a complicating factor. The latter problem is minimized by using a short actinic flash and temperatures low enough that only the early photoproducts are present during the flash.

We[44] and others[45,46] have used deliberate illumination of intermediates to try to gain additional insights. Unfortunately, the chemical relaxation of the photoproducts of the intermediates may be as complex as the primary photocycle itself.

As implied repeatedly earlier, failure to use proper polarization conditions can lead to erroneous interpretation of kinetic data because the rates of physical motions of the chromophore are convoluted with the chemical kinetics. Conversely, if measurements are made not at the magic angle

[44] R. H. Lozier, W. Niederberger, M. Ottolenghi, G. Sivorinovsky, and W. Stoeckenius, *in* "Energetics and Structure of Halophilic Microorganisms" (S. R. Caplan and M. Ginzburg, eds.), p 123. Elsevier/North-Holland Biomedical Press, Amsterdam, 1978.

[45] Zs. Dancshazy, L. A. Drachev, P. Ormos, K. Nagy, and V. P. Skulachev, *FEBS Lett.* **96,** 59 (1978).

[46] O. Kalinsky, U. Lachish, and M. Ottolenghi, *Photochem. Photobiol.* **28,** 261 (1978).

but at angles both parallel and perpendicular to a plane polarized actinic beam, both the chemical and physical rate constants and amplitudes can be extracted from the data. The chemical rate constants can be obtained because

$$\Delta A_m = \frac{\Delta A_v + 2\Delta A_h}{3}$$

where the subscripts m, v, and h refer to magic angle, vertical, and horizontal polarization of the measuring beam (with the actinic beam polarized vertically). The physical rate constants are obtained from the time dependence of the polarization anisotropy

$$r(t) = \frac{\Delta A_v(t) - \Delta A_h(t)}{\Delta A_v(t) + 2\Delta A_h(t)}$$

If $r(t)$ were analyzed using VARPRO, the resulting b_i and k_i could be interpreted in terms of rates and amplitudes of chromophore motion, although we have not attempted this yet. If a motion is inherent in the chemical conversion from one intermediate to another, the physical k_i may match the chemical k_i. If chromophore motion occurs by motion of bacteriorhodopsin within the membrane, e.g., by rotation of the bacteriorhodopsin molecule about an axis normal to the plane of the membrane, information can be gained about such motion.[47] Finally, tumbling of the purple membrane fragments in suspension will give rise to additional physical rate constants. The number of tumbling rate constants may be very large because a group of identical assymetric particles can have up to five rate constants[48] and these five rate constants would be expected to be further complicated by the fact that a purple membrane suspension is heterogeneous with respect to particle size and shape. It may be possible to exclude the relatively uninteresting tumbling motions of the membrane by casting the purple membrane into polyacrylamide gels so that motions related to the proton pump cycle can be seen more clearly. If similar experiments are performed on oriented immobilized purple membrane samples, the angular motions of the chromophore in and out of the plane of the membrane can be separated.

Finally, I hope the optical data can be extended to other spectral regions and techniques. For example, the involvement of aromatic amino acid residues may be investigated by similar experiments in the ultraviolet, and the kinetic behavior of various resonance Raman lines may be in-

[47] M. P. Heyn, R. J. Cherry, and U. Muller, *J. Mol. Biol.* **117**, 607 (1977).
[48] G. G. Belford, R. L. Belford, and G. Weber, *Proc. Natl. Acad. Sci. U.S.A.* **69**, 1392 (1972).

vestigated and compared with the rate constants seen using visible absorption spectroscopy. Kinetics of proton transfer might be investigated by applying similar analysis to data obtained using D^+ versus H^+,[10] pH indicator dyes,[49] or flash calorimetry.[50] Light-induced electrical signals from oriented membranes[15] and light-induced structural changes detected by X-ray diffraction (Article [37] by J. Stamatoff *et al.*, this volume) could also be subjected to VARPRO analysis. Ultimately, the structure and function of bacteriorhodopsin will be known at high resolution. Although different investigative techniques may be most sensitive to different aspects of the molecular mechanism, the rates and amplitudes seen by any of these techniques must come from the master set which actually applies to this fascinating molecular machine.

Appendix

I have spent considerable effort trying to optimize conditions for measuring the kinetics and stoichiometry of proton release, uptake, and translocation using optical absorption spectroscopy with pH-indicating dyes (see Lozier *et al.*[49]). I prefer the use of absorption over fluorescence spectroscopy for this purpose for the following reasons: (1) The absorption signals of bacteriorhodopsin and the dye should be additive, whereas a fluorescence signal is subject to absorption within the sample, and this absorption may be modulated by absorption changes of the bacteriorhodopsin. (Minimization of self-absorption would require a sample optically thinner than optimal for measuring absorption changes of bacteriorhodopsin.) (2) The fluorescence excitation light may be actinic to the bacteriorhodopsin and excitation light absorbed by the dye would be modulated by bacteriorhodopsin absorption changes. (It is dangerous to use the rule of thumb that fluorescence is more sensitive than absorption spectroscopy when studying light-sensitive samples!) (3) Absorption spectroscopy uses the same beam to monitor both pigments (bacteriorhodopsin and the dye) so that there can be no problem of improper beam overlap. A disadvantage of absorption (or fluorescence) spectroscopy is that is cannot be used in well-buffered samples, as pointed out by Ort and Parson, but their procedure for measuring the kinetics and stoichiometry using flash calorimetry[50,51] is quite complicated, requiring measurements using different buffers and temperatures to sort out volume changes due to several different causes. A third approach, measuring photoconductivity, can be com-

[49] R. H. Lozier, W. Niederberger, R. A. Bogomolni, S.-B. Hwang, and W. Stoeckenius, *Biochim. Biophys. Acta* **440**, 545 (1976).
[50] D. R. Ort and W. W. Parson, *J. Biol. Chem.* **253**, 6158 (1978).
[51] D. R. Ort and W. W. Parson, *Biophys. J.* **25**, 341 (1979).

plicated by motions of charges other than protons. The latter two methods also do not permit simultaneous measurement of the bacteriorhodopsin kinetics, as does absorption spectroscopy. However, I think it should be a goal to resolve differences in conclusions reached from data obtained by the three methods, rather than to try to rely solely on absorption data.

The first step in performing a proton kinetics/stoichiometry experiment is the choice of the indicator dye. Desirable properties include the following:(1) the dye should be highly soluble in the aqueous medium and have low affinity for the purple membrane (for this, a neutral/negatively charged dye avoids attraction of the dye to the negatively charged membrane), (2) the dye should have a pK near the pH at which the measurement is to be made, (3) the dye should have a large differential extinction coefficient between its protonated and unprotonated forms at a wavelength accessible to the spectrophotometer, and preferably at a wavelength where absorption changes of bacteriorhodopsin are small, (4) the dye should have low extinction at the wavelength of the actinic light, and (5) the dye should be nonfluorescent. (This feature is not mandatory, but is convenient, especially when working with scattering samples where geometrical separation of fluoresced and transmitted light is difficult.) Dyes that we have used successfully include brilliant yellow[31] and umbelliferone and p-nitrophenol.[49] Brilliant yellow (p$K \sim 8$) suffers from its significant extinction in the visible and umbelliferone (p$K \sim 8$) suffers from its fluorescence. p-Nitrophenol (p$K \sim 7$) has been the dye of choice (see also Govindjee, Ebrey, and Crofts[6]), but it would of course be desirable to have a larger selection of usable dyes.

After the dye has been chosen, the next step is to decide between two possible protocols: (1) comparing the absorption changes of two samples, with and without dye, or (2) comparing two samples, both having dye but with and without (colorless) buffer. (A difference measurement is required because I know of no suitable pH indicator that has a large differential extinction coefficient at a wavelength where bacteriorhodopsin has no significant absorption changes.) We used protocol 1[31,49] because adding buffer changed the kinetics of bacteriorhodopsin. However, I now favor protocol 2 because the pH of both samples can then be conveniently set using the Cary spectrophotometer to monitor the dye protonation state (rather than using a pH meter); the problem of the buffer sensitivity of the bacteriorhodopsin kinetics is solved by matching ionic strength by adding the appropriate amount of salt to the unbuffered sample. However, protocol 2 requires that a buffer be choosen that, like the dye, does not interact with the purple membrane and, of course, does not interact with the dye except by equilibration of protons.

The next step is to decide on the concentrations of purple membrane, dye, buffer, etc., to use. The following equation may be useful for this purpose:

$$\Delta A_{\text{dye}}^{\lambda\text{meas}} = s\phi I \frac{A_{\text{BR}}^{\lambda\text{act}}}{A_{\text{BR}}^{\lambda\text{act}} + A_{\text{dye}}^{\lambda\text{act}}} [1 - 10^{-(A_{\text{bR}}^{\lambda\text{act}} + A_{\text{dye}}^{\lambda\text{act}})}]$$

$$\Delta\epsilon_{\text{dye}}^{\lambda\text{meas}} \frac{B_{\text{dye}}}{B_{\text{dye}} + B_{\text{medium}} + B_{\text{PM}}} \quad \text{(i)}$$

where $\Delta A_{\text{dye}}^{\lambda\text{meas}}$ = absorption change due to dye

s = stoichiometry of protons per bacteriorhodopsin molecule cycling

ϕ = quantum efficiency for cycling

I = actinic light "concentration" (einsteins/liter)

$A_{\text{bR}}^{\lambda\text{act}}$ = absorbance of the bacteriorhodopsin at the actinic wavelength

$A_{\text{dye}}^{\lambda\text{act}}$ = absorbance of the dye at the actinic wavelength

$\Delta\epsilon_{\text{dye}}^{\lambda\text{meas}}$ = differential extinction coefficient between protonated and unprotonated forms of dye at measuring wavelength

B_{dye} = buffering due to dye

= 0.576 c_{dye} for pH = pK_{dye}

= 2.303 $\dfrac{K_{\text{dye}}c_{\text{dye}}[\text{H}^+]}{(K_{\text{dye}} + [\text{H}^+])^2}$ for pH $\neq$ pK_{dye}

where K_{dye} = antilog($-$pK_{dye})

B_{medium} = 2.303([H$^+$] + [OH$^-$] for CO$_2$ free water (to be determined empirically for other media)

B_{PM} = buffering due to purple membrane (to be determined empirically)

Equation (i) simplifies to Eq. (ii) if a dye having no extinction at the actinic wavelength is used:

$$\Delta A_{\text{dye}}^{\lambda\text{meas}} = s\phi I(1 - 10^{-A_{\text{BR}}^{\lambda\text{act}}}) \Delta\epsilon_{\text{dye}}^{\lambda\text{meas}} \frac{B_{\text{dye}}}{B_{\text{dye}} + B_{\text{medium}} + B_{\text{PM}}} \quad \text{(ii)}$$

It is obvious that such a dye should be used if possible. Equations (i) and (ii) also show that the volume should be minimized to the point where B_{medium} is negligible if possible. The optimal signal-to-noise ratio for an absorbance measurement occurs when the absolute absorbance is 0.43, although there is a nearly constant minimum error for absorbances between 0.7 and 0.19. The absorbance of the sample can be set using Eq. (iii):

$$A_{\text{sample}}^{\lambda\text{meas}} = A_{\text{BR}}^{\lambda\text{meas}} + A_{\text{dye}}^{\lambda\text{meas}} = (\epsilon_{\text{BR}}^{\lambda\text{meas}} c_{\text{BR}} + \epsilon_{\text{dye}}^{*\lambda\text{meas}} c_{\text{dye}})1 = 0.43 \quad \text{(iii)}$$

where

$$\epsilon_{\text{dye}}^{*\lambda\text{meas}} = \frac{\epsilon_{\text{AH}}^{\lambda\text{meas}} 10^{pK_{\text{dye}}-pH} + \epsilon_{A-}^{\lambda\text{meas}}}{1 + 10^{pK_{\text{dye}}-pH}}$$

is the pH-dependent extinction coefficient of the dye where $\epsilon_{\text{AH}}^{\lambda\text{meas}}$ and $\epsilon_{A-}^{\lambda\text{meas}}$ are the extinction coefficient of the protonated and unprotonated forms of the dye at the measuring wavelength. I have typically used 1×1-cm cuvettes for convenience and have generally found $B_{\text{medium}} \ll B_{\text{PM}}$, but this could conceivably change at high or low pH, although B_{PM} also increases at extreme pH and I have not yet perfected conditions to the point where I have gotten usable data beyond the pH 6–8 range. It should be noted that at high I, $A_{\text{BR}}^{\lambda\text{act}}$ becomes a function of I, so that use of very high light intensity brings diminishing returns; in addition, it has been suggested that s may be a function of the fraction of the bacteriorhodopsin cycling.[52] If protocol 2 is used, the buffer concentration should be at least 100 times that of the dye, or more if the pH of the medium is closer to pK_{dye} than pK_{buffer}.

Samples for proton kinetic/stoichiometry experiments must be very carefully prepared because the dye signal is taken as the difference between the samples $+/-$ dye (or $+/-$ buffer) and is often small or comparable to the signal from bacteriorhodopsin. To ensure sample match, the samples are compared on the Cary. When both samples contain dye, the dye absorbance can be matched by additions of acid or alkali to either or both cuvettes as necessary. (When no dye is used in one sample, a pH meter must be used to match the pH of the two samples.) The samples must be protected from air to minimize pH drift of the unbuffered sample(s). Of course, temperature, actinic light intensity, etc., must be the same for both samples. A complete kinetic experiment requires measuring the absorption changes of both samples (1) at a wavelength not absorbed by the dye (or at an isosbestic of the dye), but at which bacteriorhodopsin has a significant absorption change (to ensure that the samples are matched), and (2) at a wavelength where the dye has a large absorption change. In the future I hope to do such experiments and analyze the kinetics of the dye and the bacteriorhodopsin by the procedures described under DATA REDUCTION AND ANALYSIS, in order to determine whether k_i associate with protonation changes can be correlated with the k_i of the photocycle. To complete a stoichiometry experiment, the absorption changes of the dye induced by known additions of acid and/or alkali

[52] B. Hess, R. Korenstein, and D. Kuschmitz, in "Energetics and Structure of Halophilic Microorganisms" (S. R. Caplan and M. Ginzburg, eds.), p. 89. Elsevier/North-Holland Biomedical Press, Amsterdam, 1978.

(beware of atmospheric CO_2 changing the titer of the alkali!) must be done. I have used both the Cary and the flash instrument for this purpose; the Cary has the advantage of showing that the measured absorption changes are due to proton concentration changes only, as indicated by a constant isosbestic point for the dye, the flash instrument has the advantage of higher time resolution. (Time resolution is limited by the injection and mixing time.) I like to use both acid and alkali pulses to control for pH drift associated with CO_2 introduced by additions and mixing. Also, the linearity of the dye absorption change with the amount of acid or alkali added must be determined to obtain an accurate stoichiometry. In my past experiments, proper control of polarization was not employed; future experiments will obviously include this consideration.

The preceding discussion refers principally to measurements on purple membrane fragments that are not incorporated into closed vesicles; obviously, additional considerations pertain to cells, cell envelopes, and proteoliposomes in which proton translocation can be measured.[6,49] These considerations include (1) the internal and external aqueous volumes, (2) the permeability of the vesicles to the dye (and protons!), (3) the buffering capacity of the internal aqueous phase and membrane surface, and (4) the presence of membrane potentials that might actively accumulate a permeable dye. Another consideration that comes up is whether the membrane–aqueous interface forms a kinetically distinct compartment. (This consideration applies also to open membrane fragments.) If such a compartment exists, the affinity of the dye for this compartment and the exchange of dye and protons between the aqueous phase(s) and this compartment are also relevant. Obviously, much work will need to be done to sort this all out.

Acknowledgments

I am grateful to Professor W. Stoeckenius for introducing me to work on the purple membrane and his long-term support of my search for an adequate photocycle model. I am also indebted to many colleagues who have helped me in my pursuit. These include R. Bogomolni, W. Niederberger, D. Tow, M. Ottolenghi, L. Eisenstein, G. Golub, R. LeVeque, J. Nagle, L. Parodi, and others. This work was supported by NIH grant GM 27057.

[18] The Photoreaction Cycle of Bacteriorhodopsin: Low-Temperature Spectrophotometry

By FUMIO TOKUNAGA and TATSUO IWASA

Photochemical reactions of bacteriorhodopsin are photic reactions followed by a series of thermal transformations between intermediates. As the intermediates decay rapidly at physiological conditions, to study them it is often desirable to slow down the rates of their thermal reactions. The intermediates can be accumulated by lowering the reaction temperature, and then absorption spectra of the samples containing intermediates can be measured by a conventional spectrophotometer. Although pulsed laser and rapid measurement systems have been developed recently, low-temperature spectrophotometry is still a powerful technique for measuring accurate absorption spectra of the intermediates.

Equipment

Cryostats specially designed to fit with each particular spectrophotometer should be used for the low-temperature spectrophotometry. In the case of visual pigments, only at liquid helium temperatures can the intermediate hypsorhodopsin be stabilized,[1] while for bacteriorhodopsin (BR), such an intermediate has never been observed at liquid helium temperatures,[2] so cryostats for temperatures down to that of liquid nitrogen probably are sufficient. Examples of such cryostats are given in another chapter in this volume.[3]

Sample Preparation

Purified purple membrane or purple membrane analogs are suspended in an adequate medium such as phosphate buffer (10 mM, pH 7), carbonate buffer (10 mM, pH 10), or distilled water. The suspension is mixed with glycerol at a final concentration of 67 or 75%, in order to have a transparent sample on cooling. For measuring absorption spectra at temperatures between -30 and $-70°$, 75% is recommended, because micro-

---

[1] T. Yoshizawa, in "Handbook of Sensory Physiology" (H. J. A. Dartnall, ed.), Vol. 7, Part 1, p. 146. Springer-Verlag, Berlin and New York, 1972.

[2] T. Iwasa, F. Tokunaga, and T. Yoshizawa, FEBS Lett. 101, 121 (1979).

[3] T. Yoshizawa and Y. Schichida, this volume, Article [49].

METHODS IN ENZYMOLOGY, VOL. 88

crystals of water sometimes appear in 67% glycerol in this temperature range; microcrystals change the scattering rather badly and the baseline declines.

The optical cell usually consists of a transparent glass in front of a rubber spacer ring and then opal glass in the back, which unifies the scattering. A piece of opal glass is also placed on the reference beam to compensate for the scattering effect; the light path length of the sample is usually 1 or 2 mm.

There are two species of bacteriorhodopsin; one is *trans*-BR, which has all-*trans*-retinal as its chromophore and the other is 13-*cis*-BR, which has 13-*cis*. In the dark at room temperatures these two species equilibrate with each other. Irradiation with visible light causes the conversion of 13-*cis*-BR to *trans*-BR. For experiments with *trans*-BR, the sample should be light-adapted by irradiation near 0° where dark adaptation cannot be observed for at least an hour.[4]

13-*cis*-BR can be formed by incubating 13-*cis*-retinal with bacterioopsin but it is easily converted to *trans*-BR either thermally or by irradiation, so it is difficult to prepare pure 13-*cis*-BR. The rate of the dark adaptation is lowered by cooling, so nearly 100% 13-*cis*-BR can be prepared by regeneration in an ice bath in the dark.[5] Pure 13-*cis*-BR cannot be prepared from native purple membrane.

Photoreaction Cycle Intermediates

Batho-products. Below − 120° and even at liquid helium temperatures (9 K) a bathochromically shifted product is formed by irradiation of *trans*-BR.[2] The product is termed batho-*trans*-BR, because it is produced from *trans*-BR. Batho-*trans*-BR does not mean that this contains all-*trans*-retinal as the chromophore. Batho-*trans*-BR and the original *trans*-BR are interconvertible by light; the batho-product is completely reverted with light of wavelengths longer than 680 nm. However, since batho-*trans*-BR absorbs in the wavelength region shorter than its λ_{max}, pure batho-*trans*-BR cannot be obtained. The irradiated sample at liquid nitrogen temperatures is a mixture of *trans*-BR and batho-*trans*-BR. The molar fraction of batho-*trans*-BR in the mixture depends on the irradiation wavelength. The molar fraction of batho-*trans*-BR in the mixture irradiated at 500 nm at − 190° is close to 30%.[6,7]

[4] F. Tokunaga, T. Iwasa, and T. Yoshizawa, *FEBS Lett.* **72**, 33 (1976).
[5] A. Maeda, T. Iwasa, and T. Yoshizawa, *J. Biochem.* (*Tokyo*) **82**, 1599 (1977).
[6] T. Iwasa, F. Tokunaga, and T. Yoshizawa, *Biophys. Struct. Mech.* **6**, 253 (1980).
[7] T. Iwasa, F. Tokunaga, and T. Yoshizawa, *Photochem. Photobiol.* **33**, 539 (1981).

Batho-13-*cis*-BR is produced by irradiation of 13-*cis*-BR with green light (500 nm) below − 150°.[7]

If dark-adapted bacteriorhodopsin is irradiated with green light (500 nm), the mixture produced contains both batho-*trans*-BR and batho-13-*cis*-BR.

Lumi-product. Lumi-*trans*-BR is produced by warming batho-*trans*-BR above − 120° or by irradiating *trans*-BR, pH ≈ 7.0, at a temperature between − 120 and − 90°. At − 75° (dry ice–acetone temperature) irradiation of *trans*-BR produces lumi-*trans*-BR, but it is not stable at this temperature. A lumi-product has not yet been observed from 13-*cis*-BR.

Meta-product. For measuring the absorption spectrum of meta-*trans*-BR (Meta-bR^t in Figs, M), the high salt (e.g., 1 M NaCl) and high pH (e.g., pH 10) preparation is better, because in such a medium meta-*trans*-BR can be stabilized.[8] Meta-*trans*-BR can be observed by warming lumi-*trans*-BR or by irradiating *trans*-BR at around − 50°. The amount of meta-*trans*-BR formed depends on the pH and salt concentration. With high salt (1 M NaCl) and high pH (pH 10, 10 mM carbonate buffer), meta-*trans*-BR is stable below − 50°, but with low salt and neutral pH (10 mM phosphate buffer, pH 7) meta-*trans*-BR is not stable even at − 50°. A meta-product has not been observed from 13-*cis*-BR.

The intermediates N and O, which have been reported on the basis of the flash photolysis experiments, have not been observed at low temperature spectrophotometry yet. In the case of visual pigments, the same intermediates have been observed by low-temperature spectrophotometry as by flash photolysis. The inability to see the later bacteriorhodopsin intermediates using low-temperature spectrophotometry is probably due to their activation energies; that is, in the case of visual pigments, the later intermediates have the larger activation energies, while for bacteriorhodopsin, the activation energies of the N and O intermediates may be smaller than that of meta-*trans*-BR.

Branching Pathway. On warming, batho-*trans*-BR is converted only to lumi-*trans*-BR, but lumi-*trans*-BR is converted to meta-*trans*-BR and also back to the original pigment, trans-BR.[6] The ratio of lumi-*trans*-BR which is converted to meta-*trans*-BR to that converted directly to *trans*-BR depends on the temperature. At − 90° all lumi-*trans*-BR is converted directly to *trans*-BR, but at 0° all lumi-*trans*-BR is converted to meta-*trans*-BR.[9]

Figure 1 shows the photocycles of 13-*cis*-BR and *trans*-BR as observed by low-temperature spectrophotometry.

[8] B. Becher and T. G. Ebrey, *Biophys. J.* **17**, 185 (1977).

[9] O. Kalisky, M. Ottolenghi, B. Honig and R. Korenstein, *Biochemistry* **20**, 649 (1980).

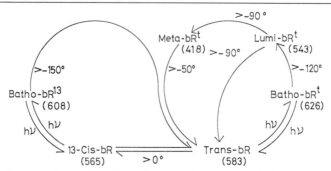

FIG. 1. Photoreaction cycles of bacteriorhodopsin. Photochemical reactions are designated by hν and thermal reaction, by the transition temperatures of the intermediates.

Absorption Spectra of the Photocycle Intermediates

The photocycle intermediates can be converted back to the original pigment by light; so in general the irradiated product is a mixture of the original pigment and at least one intermediate. To calculate the adsorption spectrum of an intermediate, the molar fraction of the intermediate in the mixture must be known. As meta-*trans*-BR does not have absorbances at wavelengths longer than 480 nm, the absorbance decrease in this region

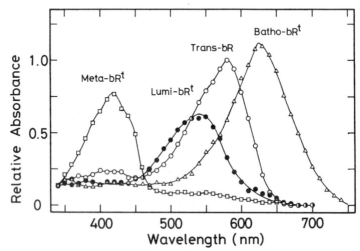

FIG. 2. The absorption spectra of *trans*-BR and its intermediates. (○) *Trans*-BR (λ_{max}: 583 nm, −190°). (△) Batho-*trans* BR (λ_{max}: 626 nm, −190°C). (●) Lumi-*trans*-BR (λ_{max}: 543 nm, −90°). (□) Meta-*trans*-BR (λ_{max}: 418 nm, −90°). The spectrum of each intermediate is normalized by the maximal absorbance of *trans*-BR at the same temperature.

has been used to calculate the molar fraction of this intermediate. The molar fraction of the intermediates preceding meta-*trans*-BR can also be estimated by converting the intermediate to meta-*trans*-BR. However, lumi-*trans*-BR can be converted directly to *trans*-BR, so this branching pathway can make an error in the estimation of the amount of the early intermediates. Another method is to estimate by using the difference spectra among *trans*-BR, lumi-*trans*-BR and meta-*trans*-BR. The detailed method is found in Iwasa *et al.*[6]

The calculated absorption spectra of the photocycle intermediates of *trans*-BR are shown in Fig. 2.

[19] Photochemistry and Isomer Determination of 13-*cis*-and *trans*-Bacteriorhodopsin

By NORBERT A. DENCHER, CHARLES N. RAFFERTY, and WALTER SPERLING

Bacteriorhodopsin (BR) is arranged in the purple membrane (PM) of *Halobacterium halobium* in a two-dimensional hexagonal lattice. Its chromophoric group is retinal bound via a Schiff base in a 1:1 ratio to the protein.[1] The C-13 double bond of the retinylidene moiety has the alternative of being either in the *trans* or in the *cis* configuration. The corresponding BR isomers are called BR^{568}_{trans} (*trans*-BR) and BR^{548}_{13-cis} (13-*cis*-BR, the superscript indicating the λ_{max} of the absorption spectra. In the dark BR is an isomerase isomerizing specifically the C-13 double bond of the retinylidene moiety. The photocycle of *trans*-BR could be characterized by flash and low-temperature spectroscopy,[2,3] since this BR isomer was available in pure form in the light-adapted PM.[4] 13-*cis* BR, however, exists under natural conditions only together with the *trans* isomer. Mixtures containing both BR isomers are more difficult to study than samples containing only one single isomer. Consequently, the investigation of the two separate BR isomers is desirable. Therefore we prepared both BR^{548}_{13-cis} and BR^{568}_{trans} by incubating the retinal-free protein, bacterioopsin (BO), with the respective isomers, 13-*cis*-and *trans*-retinal. With these two BR isomers photometric experiments can be performed that clarify their photochem-

[1] D. Oesterhelt and W. Stoeckenius, *Nature (London), New Biol.* **233**, 149 (1971).

[2] R. H. Lozier, R. A. Bogomolni, and W. Stoeckenius, *Biophys. J.* **15**, 955 (1975).

[3] N. Dencher and M. Wilms, *Biophys. Struct. Mech.* **1**, 259 (1975).

[4] D. Oesterhelt, M. Meentzen, and L. Schuhmann, *Eur. J. Biochem.* **40**, 453 (1973).

METHODS IN ENZYMOLOGY, VOL. 88

istry and interconversion.[5-7] To determine the ratio of the two isomers we developed an analytic method based on their distinct photoreactions. The method is nondestructive to BR and can be applied to living bacteria, isolated PM, and regenerated BR.

Preparation of Bacteriorhodopsin, BR_{trans}^{568}, and BR_{13-cis}^{548}

The chromophore-free protein BO is prepared by illuminating aqueous suspensions of PM in the presence of hydroxylamine. Concentration of BR: $1.3-1.7 \times 10^{-5} M$, of hydroxylamine: $1.0-1.3 M$, pH 6.9. The light on a halogen tungsten lamp is used for illumination ($\lambda > 515$ nm, 10 mW/cm^2). After 4–6 hr at 20° BR is completely bleached as indicated by the absorption spectrum. The reaction rate is faster at alkaline pH and if high light intensities are applied. As in the case of rhodopsin, the chromophore-free bacteriorhodopsin (BO) is more sensitive to the denaturing effects of sonication, temperature, and detergent solubilization than BR. The purity of BO obtained by this method is about 98/2, measured by the ratio BO/BR (Fig. 1). More than 95% of the original pigment concentration can be regenerated in these samples by the addition of retinal.

Prior to regeneration, the excess hydroxylamine is removed by washing the sample with distilled water or by dialysis overnight against the appropriate buffer at 4°. The retinal oxime, however, is not removed but remains attached to the protein. Photoreactions arising from these retinal oximes can be ignored in spectroscopic experiments if excitation and measuring light are outside the absorption band of retinal oxime. Light absorbed by retinal oxime is capable of reconverting BO to BR.[7] Hence it may be advantageous or necessary to use retinal oxime-free BO. The retinal oxime can be extracted from the sample by organic solvents[8] or destroyed by intense irradiation (wavelengths below 340 nm should be omitted). Retinal oxime-free BO can also be obtained from halobacteria grown in the presence of $1-1.5$ mM nicotine[9,10] or from mutant strains that are deficient in retinal synthesis. Regeneration of BR_{13-cis}^{548} and BR_{trans}^{568} is achieved by incubating a buffered aqueous BO suspension—for example,

[5] N. A. Dencher, C. N. Rafferty, and W. Sperling, *Ber. Kernforschungsanlage Juelich* **Juel-1374**, 1–42 (1976).
[6] W. Sperling, P. Carl, C. N. Rafferty, and N. A. Dencher, *Biophys. Struct. Mech.* **3**, 79 (1977).
[7] W. Sperling, C. N. Rafferty, K.-D. Kohl, and N. A. Dencher, *FEBS Lett.* **97**, 129 (1979).
[8] F. Tokunaga and T. Ebrey, *Biochemistry* **17**, 1915 (1978).
[9] M. Sumper, H. Reitmeier, and D. Oesterhelt, *Angew. Chem., Int. Ed. Engl.* **15**, 187 (1976).
[10] N. A. Dencher and E. Hildebrand, *Z. Naturforsch. C. Biosci.* **34C**, 841 (1979).

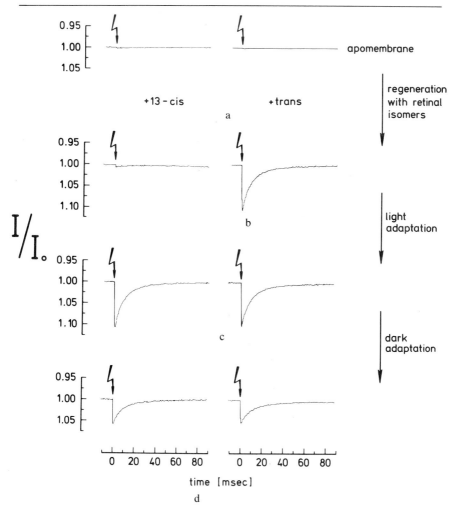

FIG. 1. Time course of transmission changes at 545 nm of apomembrane and regenerated bacteriorhodopsin. a, c, and d horizontal line described in the figure; b, horizontal line shows transmission changes of regenerated BR. The samples were flashed 3 min (BR_{13-cis}^{548}) and 10 min (BR_{trans}^{568}) after addition of the retinal solution to the apomembrane. $\lambda_{flash} > 570$ nm, Schott filter OG 570; temperature 20°; pH 6.88. I_0 = intensity of the measuring light incident on the photomultiplier at the time before the flash; I = intensity of the measuring light incident on the photomultiplier at the time given on the abscissa. (From Dencher et al.[5])

at pH 7—with the respective ethanolic retinal isomer solution: 0.5 ml of BO suspension is mixed in the dark with 2 μl retinal solution. Retinal is usually added in a three- to sixfold molar excess. The final absorbance of the main visible absorption band of the regenerated BR is about 0.5 in a 1-cm pathlength cuvette. For measurements below 0°, ethanolic retinal solutions are added to BO suspended in a buffered glycerol/water mixture (2/1, v/v) and after 2 min at 20° the samples are cooled to the desired temperature. The retinal isomers are purified by high-pressure liquid chromatography[6] and are 99% pure. Because most of the spectroscopic experiments are done in the visible spectral range, the excess free retinal usually does not interfere with the measurements. Special care has to be taken so that, on the one hand, not too much isomerization occurs during regeneration but that, on the other hand, most BR is regenerated at the time of the flash. Equilibration between BR_{13-cis}^{548} and BR_{trans}^{568} is mainly an isomerization of the chromophore and a first-order reaction in both directions. Regeneration of BR is a second-order reaction. Because first- and second-order reactions are involved, this goal can be easily attained. We found that for most purposes a three- to sixfold excess of retinal to BO and a reaction time of 2–10 min are sufficient to obtain optimal conditions at room temperature. It should be kept in mind that regeneration with 13-*cis*-retinal is about three times faster than with the *trans* isomer.[11]

Flash Spectroscopy of BR_{13-cis}^{548} and BR_{trans}^{568}

Freshly prepared samples of 13-*cis* and *trans*-BR were separately investigated by flash spectroscopy. Figure 1 illustrates some results. The first two recordings (first horizontal line) show the absorption changes at 545 nm of the light-adapted BO after photoexcitation. The tiny absorption change after the flash indicates that the BO is contaminated with less than 2% BR. The BO is then regenerated with the respective retinal isomers, 13-*cis* and *trans*-retinal, to the bacteriorhodopsin isomers, BR_{13-cis}^{548} and BR_{trans}^{568} (second horizontal line). The second horizontal line demonstrates that BR_{13-cis}^{548} and BR_{trans}^{568} form different photoproducts and different subsequent transients. The small absorption change in the case of BR_{13-cis}^{548} is due to the remaining BR in the apomembrane preparation (see first horizontal line), to a small amount of BR_{trans}^{568} isomerized from BR_{13-cis}^{548}, and possibly to a trace of regenerated BR_{trans}^{568} originating from a small contamination of *trans*-retinal in the 13-*cis*-retinal solution used for regeneration. The third and fourth horizontal lines show what happens if the two samples are sub-

[11] M. E. Keen and N. A. Dencher, *EMBO Workshop Transduction Mech. Photoreceptors*, p. 64 (1976).

jected to a flash after light adaptation and after subsequent dark adaptation. Light adaptation is achieved by illuminating the sample for about 1 min with light of an intensity on the order of 10 mW/cm², dark adaptation by leaving the sample in the dark for about 5 hr at room temperature. At other wavelengths of the measuring beam and other resolution times analogous recordings are obtained. Figure 2, for example, represents the appearance and decay of ^{610}C, an intermediate of the 13-*cis* cycle. Figure 3 shows difference spectra plotted from data like those of Fig. 2, where a and b indicate the differences used in Fig. 3. From curves like those depicted in Figs. 1, 2, and 3, the reaction scheme of the whole system can be derived. The traces shown in Figs. 1 and 2 represent transmission changes of the sample on exposure to a single weak light flash. Generally not more than 5–8% of the BR molecules are bleached by a single flash; therefore double-photon hits are less likely. In the case of 13-*cis*-BR, averaging of subsequent experiments is possible only in a very restricted way because a portion of the excited 13-*cis*-BR molecules is converted to *trans*-BR. On the other hand, *trans*-BR, if measured at room temperature, is an ideal system for the averaging technique, since it always reverts to its ground state within a few milliseconds. In both cases, however, the slow thermal equilibration between the two BR isomers has to be taken into account. To avoid undesirable photoreactions, the measuring beam should be as weak as possible and the time the measuring beam is impinging on the sample prior to the flash should be as short as possible. In our flash apparatus a shutter opens the measuring beam 5 msec prior to the flash. (The

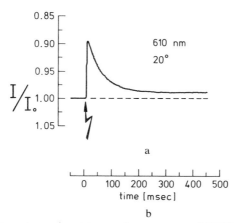

FIG. 2. Transmission changes at 610 nm after exposure of BR^{548}_{13-cis} to a flash ($\lambda >$ 455 nm, Schott GG 455 cutoff filter). Temperature 20°. (a) I/I_0 before the flash minus I/I_0 1 msec after the flash. (b) I/I_0 before the flash minus I/I_0 300 msec after the flash. (From Dencher *et al.*[5])

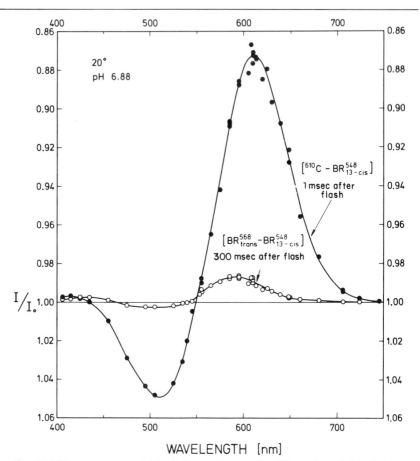

WAVELENGTH [nm]

Fig. 3. Difference spectra obtained after excitation of BR_{13-cis}^{548} at 20°. ph 6.88. The spectra were plotted from data like those of Fig. 2. The difference spectrum 1 msec after the flash (●) indicates the difference between the ^{610}C and the BR_{13-cis}^{548} spectrum, the difference spectrum 300 msec after the flash (○) the difference between the BR_{trans}^{568} and the BR_{13-cis}^{548} spectrum. (From Dencher et al.[5])

technique of studying fast reactions in photochemical systems has been described in numerous reviews, e.g., DeVault.)[12]

The experiments with regenerated BR show that the photocycle originating from BR_{trans}^{568} is in principle identical with that previously described for light-adapted PM.[2,3] We confirmed the fast appearance of ^{630}T (K) during the flash, its decay to ^{550}T (L), the further reaction to ^{411}T (M), and the formation and decay of ^{646}T (O).[5,6]

[12] D. DeVault, this series, Vol. 54, p. 32.

Photoexcitation of BR_{13-cis}^{548} yields three subsequent intermediates ^xC, ^yC, ^{610}C with red-shifted absorption maxima. In contrast to the BR_{trans}^{568} photocycle no intermediates with a short wavelength absorption spectrum like ^{411}T could be observed. Photoisomerization of 13-*cis* BR to *trans*-BR proceeds via these three intermediates. The longest-lived transient ^{610}C (Fig. 3) decays with a half-lifetime of 37 msec (Fig. 2) at 20°. At one intermediate of the 13-*cis* cycle the reaction pathway splits, part of the molecules proceeding to 13-*cis* BR, part of the molecules going to *trans*-BR.[5,6]

Assay for BR_{13-cis}^{548} and BR_{trans}^{568}

The observed differences in the photoreactions of BR_{13-cis}^{548} and BR_{trans}^{568} can be utilized to determine the ratio of BR_{trans}^{568} to BR_{13-cis}^{548} in various preparations of BR under conditions of dark and of light adaptation. The long-lived transient ^{411}T of the BR_{trans}^{568} cycle turned out to be a convenient quantity to measure the amount of BR_{trans}^{568} in a given sample by flash photometry. To use ^{411}T as a measure for BR_{trans}^{568} the following requirements have to be met:

(1) The exciting light (analytic flash) should be weak and short enough to avoid secondary photon hits during the photocycle of BR_{trans}^{568}. Under these conditions the concentration of ^{411}T after the exciting flash is proportional to the concentration of BR_{trans}^{568} present before the flash.

(2) In samples containing both BR_{13-cis}^{548} and BR_{trans}^{568}, absorption changes of BR_{13-cis}^{548} and its transients have to be taken into account.

The interpretation of the measurements can be simplified by choosing a wavelength of the measuring beam where the BR_{13-cis}^{548} path is "spectroscopically decoupled," i.e., no absorption change occurs. Wavelengths where no absorption difference between ^{610}C and BR_{13-cis}^{548} exists are their isosbestic points. At the time after the analytic flash when ^{610}C and BR_{13-cis}^{548} prevail in the BR_{13-cis}^{548} photoreaction pathway, ^{411}T is predominant in the BR_{trans}^{568} cycle. If the appropriate time after the analytic flash (see Fig. 1) and the right wavelength (isosbestic point) are chosen, absorption changes represent only changes within the BR_{trans}^{568} cycle, i.e., the appearance of ^{411}T (disappearance of BR_{trans}^{568}) and disappearance of ^{411}T (appearance of BR_{trans}^{568}). The wavelength we choose for the measuring light is 545 nm, which is close to the isosbestic point $^{610}C/BR_{13-cis}^{548}$ (see Fig. 3). Actually the absorption decrease at 545 nm represents mainly the disappearance of BR_{trans}^{568}, which is, of course, proportional to the appearance of ^{411}T. Test measurements at other wavelengths (e.g., 411 nm) confirmed the consistency of the whole system. The analytic flashes used are so weak and the turnover of the BR so small that the light of the flash practically does not change the ratio of BR_{trans}^{568} to BR_{13-cis}^{548}.

Summarizing, the ratio $[BR_{trans}^{568}]/[BR_{13-cis}^{548}]$ is determined as follows:

(1) The first analytic flash determines the (relative) amount of BR_{trans}^{568} (Fig. 1d, horizontal line).

(2) The sample is light-adapted to transfer practically all BR into the BR_{trans}^{568} form.

(3) A second analytic flash determines the (relative) amount of BR_{trans}^{568} present after light adaptation (Fig. 1c, horizontal line). This corresponds to the sum of BR_{13-cis}^{548} and BR_{trans}^{568} present in the original sample. The difference between the BR_{trans}^{568} measured with the second analytic flash and that measured with the first analytic flash is the original amount of BR_{13-cis}^{548}. After transforming the I/I_0 values into absorbance, which is directly proportional to the concentration, the ratio

$$K = \frac{[BR_{trans}^{568}]}{[BR_{13-cis}^{548}]}$$

is calculated.

With this nondestructive analytical method it was found that in the dark-adapted state the purple membrane of living bacteria, isolated purple membrane, and regenerated bacteriorhodopsin all contain a mixture of BR_{13-cis}^{548} and BR_{trans}^{568} in a ratio of about 1:1. Under conditions of light adaptation, only BR_{trans}^{568} is present.[5,6,13] Another technique to determine the isomeric ratio of BR consists in extracting the retinal with organic solvents and measuring the composition of the retinal isomers by HPLC. We and others applied the extraction method and obtained results in quantitative agreement with the flash spectroscopic data.[5,14-16] Disadvantages of this method are, besides the fact that the BR is destroyed, that isomerization of retinal during the extraction procedure or incomplete extraction may lead to wrong results.

[13] K. Ohno, Y. Takeuchi, and M. Yoshida, *Biochim. Biophys. Acta* **462,** 575 (1977).
[14] N. A. Dencher, P. Carl, and W. Sperling, *Proc. Int. Congr. Biochem., 10th, 1976* 05-1-317, p. 270 (1976).
[15] M. J. Pettei, A. P. Yudd, K. Nakanishi, R. Henselman, and W. Stoeckenius, *Biochemistry* **16,** 1955 (1977).
[16] A. Maeda, T. Iwasa, and T. Yoshizawa, *J. Biochem. (Tokyo)* **82,** 1599 (1977).

[20] Spin Labeling of Bacteriorhodopsin

By ROSALIE K. CROUCH

Spin labels are sensitive monitors of protein environment and thus are useful probes for studying the visual and purple membrane pigments. The spin label technique has been used for rhodopsin by labeling sulfhydryl groups on the protein[1] and by incorporating spin-labeled fatty acids[2] in the membrane. This latter technique has also been used with bacteriorhodopsin,[3] but as this protein contains no sulfhydryl groups, the sulfhydryl derivatization method is not applicable.

If the chromophore itself contains the spin label, then there is assurance that the label is in the binding site. Retinal derivatives can be incorporated into the binding site of both opsin and bleached purple membrane and have been successfully used to probe the binding site and photochemistry of these pigments.[4] The preparation of the spin-labeled derivative, I, 4-(2,2,5,5-tetramethyl-1-pyrrolidinyloxy)retinal, and the formation

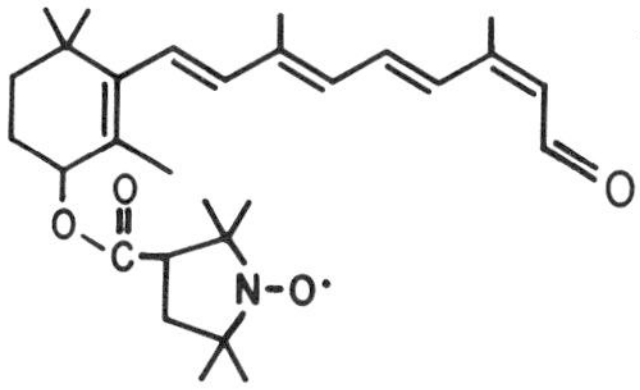

I (13 −*cis* isomer)

of pigment with this derivative and bleached purple membrane are described in this article.

Preparation of Spin-Labeled Retinal

All-*trans*-retinal (60 mg, 0.2 mmol) dissolved in 10 ml carbon tetrachloride is added to a stirred suspension of stoichiometric amounts of recrystallized *N*-bromosuccinimide. 2,2,5,5-Tetramethyl-1-pyrrolidinyloxy-3-carboxylic acid in 15 ml carbon tetrachloride is added with stirring under nitrogen for several minutes to initiate bromination and stirred for

[1] E. Fujimori, *Vision Res.* **15**, 63 (1965); A. Barion, D. Thomas, B. Osborne, and P. Devaux, *Biochem. Biophys. Res. Commun.* **78**, 442 (1977); M. Delmelle and N. Vermaux, *Biochim. Biophys. Acta* **464**, 370 (1977).

[2] K. Hong and W. L. Hubbell, *Proc. Natl. Acad. Sci. U.S.A.* **69**, 2617 (1972).

[3] C. Chignell and D. Chignell, *Biochem. Biophys. Res. Commun.* **62**, 136 (1975).

[4] T. G. Ebrey, Chapter 66, this volume.

2 hr at room temperature. An excess of N,N-dimethylaniline is added and the reaction mixture stirred for 18 hr. After filtering, the solution is washed with cold 5% sulfuric acid, water, 2% sodium bicarbonate, and again with water. After drying over Na_2SO_4, the solvent is evaporated. The spin-labeled retinal (I) is isolated in about 25% yield by thin-layer chromatography with 20% ethyl acetate in hexane or silica gel GF. The product can be further purified by high-pressure liquid chromatography (HPLC) with a μBondapak CN column employing 10% ether in hexane as solvent. The product may be characterized by infrared (C = 0, 1725 cm^{-1}), absorption in ethanol (λ_{max} = 380 nm), mass ([70 eV] m/e, [m$^+$] − 307, 299, and 161) and esr spectrometry (Fig. 1a).

Preparation of Spin-Labeled Bacteriorhodopsin

Bleached purple membrane is prepared as described by Ebrey.[4] A solution of the SLR in ethanol is added by 2-μl increments to bleached membrane in sodium acetate buffer (0.05 M, pH 5.5), shaken, and the absorption spectra recorded using bleached purple membrane as the reference. Spin-labeled retinal is added until the absorption at the λ_{max} (480 nm) shows no further change and free chromophore absorption just begins to appear.

Properties of Spin-Labeled Bacteriorhodopsin

The absorption spectrum of the pigment shows a λ_{max} at 480 nm, considerably blue-shifted from the native pigment. The electron spin resonance spectrum (Fig. 1b) shows an anisotropic signal as expected from a spin label immobilized by attachment to slowly rotating membrane sheets.

The spin-labeled pigment is unstable, hydrolyzing at the ester linkage to the free-spin-labeled carboxylic acid and the 4-hydroxy pigment[5] (λ_{max} of 535 nm). This reaction is somewhat slower in acidic buffers, but at room temperature and pH 5.5, the hydrolysis is complete within 24 hr. Therefore any studies on this spin-labeled pigment derivative must be initiated immediately on pigment regeneration.

A second spin-labeled retinal containing the nitroxide free radical has now been prepared.[6] In this analog, the nitroxide nitrogen is incorporated

[5] D. Oesterhelt and U. Christoffel, *Biochem. Soc. Trans.* **4**, 556 (1976).
[6] R. Crouch, T. G. Ebrey, and R. Govindjee, *J. Amer. Chem. Soc.* **103**, 7364 (1981).

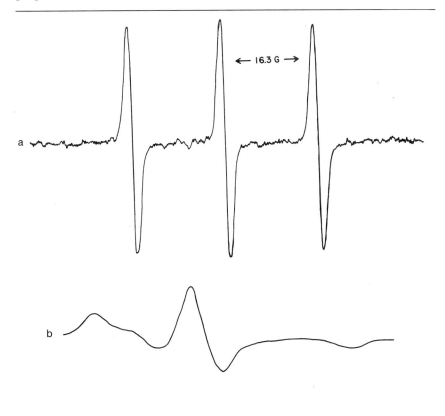

FIG. 1. Electron spin resonance spectra. (a) Spin-labeled retinal in 1% EtOH/sodium acetate buffer, pH 5.5, 0.5 M; 4 mw time scan; 1.0 sec time constant; modulation amplitude 1.6 G; gain = 1.0×10^4; $g_0 - 2.0071$. (b) Spin-labeled pigment 1 min after formation in 0.5 M sodium acetate buffer, pH 5.5; 1-min time scan; 0.3 sec time constant; modulation amplitude 1.6 G; gain = 2.0×10^4.

into the ring of the retinal so that the hydrolysis difficulties reported here are avoided. The ESR spectra of the resulting pigment showed an aniso-tropic signal characteristic of a near-rigid limit nitroxide spectrum (corre-lation time $\sim 10^{-7}$ sec). These results indicate that the environment of the chromophore ring has a high degree of orientation and is rigid within the membrane.

[21] Heavy Atom Labeling of Retinal in Bacteriorhodopsin

By MICHAEL G. MOTTO and KOJI NAKANISHI

Introduction

Bacteriorhodopsin, because of its two-dimensional crystal lattice in the membrane, lends itself to study by various diffraction techniques. Using electron diffraction Unwin and Henderson[1,2] were able to determine the three-dimensional structure of bacteriorhodopsin. By the use of similar techniques with retinals containing heavy atoms, it might be possible to determine the location of the retinal moiety within the protein.

Two compounds were synthesized for this purpose, the 9-demethyl-9-bromoretinal **1** and the 13-demethyl-13-bromoretinal **2**, in which either the 9-or the 13-methyl is replaced with a bromine atom. It was thought that a bromine in these positions would minimize any extra steric interactions that might impede the binding of the chromophore to the protein, and also provide a heavy label that could be used in the diffraction studies.

Synthesis

The 9-demethyl-9-bromoretinal was synthesized as follows: Ethyl 4-oxocrotonate (**3**)[3] was treated with bromine in CCl_4 and then worked up with aqueous potassium carbonate to yield 65% of ethyl 3-bromo-4-oxocrotonate (**4**). The aldehyde was then condensed with Wittig salt (**5**)[4] to give the ester (**6**) in 63% yield. Dibal reduction and MnO_2 oxidation gave the aldehyde (**7**) in quantitative yield. Emmons reaction of aldehyde (**7**) with triethylphosphonosenecioate yielded 60% of the retinyl ester (**8**), which was reduced (Dibal) and oxidized (MnO_2) to give 70% of the 9-demethyl-9-bromoretinal as a mixture of 13-*cis* and all-*trans* isomers. These isomers were then separated by HPLC, Lichrosorb, 10% ether in hexane to obtain the pure all-*trans* chromophore λ_{max} (MeOH) 372 nm, $\epsilon = 39,300$; and λ_{max} (hexane) 363 nm, $\epsilon = 45,000$.

The 13-demethyl-13-bromoretinal was synthesized as follows: 2-butyne-1,4-diol was reacted with *tert*-butyldimethyl silyl chloride[5] to yield

[1] R. Henderson and P. N. T. Unwin, *Nature (London)* **252**, 28 (1975).
[2] P. N. T. Unwin and R. Henderson, *J. Mol. Biol.* **94**, 425 (1975).
[3] R. Rambaud and M. Vessiere, *Bull. Soc. Chim. Fr.* p. 1567 (1961).
[4] L. Bartlett, W. Klyne, W. P. Mose, P. M. Scopes, G. Galskio, A. K. Mallams, B. C. L. Weedon, J. Szabolcs, and G. Toth., *J. Chem. Soc. C* p. 2527 (1969).
[5] E. J. Corey and A. Venkateswarlu, *J. Am. Chem. Soc.* **94**, 6190 (1972).

METHODS IN ENZYMOLOGY, VOL. 88

60% of the monosilyated product; subsequent oxidation with MnO_2 gave the aldehyde (9). Condensation of this aldehyde with Wittig salt (10)[6] gave the protected tetraenynol (11), which was then deprotected (Bu_4NF/THF)[5] and oxidized with MnO_2 to give the tetraenynal (12), 44% yield, which reacted with aqueous HBr in benzene to give all-*trans*-13-de-methyl-13-bromoretinal (2), 40% yield. It should be noted that since all intermediates after the deprotection were extremely unstable, subsequent reactions were carried out as quickly as possible until final product (2) was secured. The product (2) was purified by HPLC, Lichrosorb, 3% ether in hexane: λ_{max} (MeOH) 388 nm, $\epsilon = 23,800$; and λ_{max}(hexane) 380 nm, $\epsilon = 30,000$.

Because of the instability of the compounds they are stored at $-60°$ under argon and in the dark. All compounds were purified by HPLC just prior to incubation to ensure maximum regeneration of pigment.

Binding studies[7]

Analog bacteriorhodopsins (BR) were prepared by incubation of a 1:1 mixture of opsin and the all-*trans*-9-bromo- or 13-bromoretinal for 6 and 20 hr, respectively (rt. pH 7). When a suspension of 13-bromo-BR^{DA} (595 nm) was treated with all-*trans*-retinal in the dark (rt. pH 7) the 13-bromo chromophore was displaced; after 120 hr the λ_{max} was shifted to 568 nm λ_{max} of natural BR^{DA}. In contrast, the 9-bromo chromophore was

[6] O. Isler, *Helv. Chim. Acta* **39**, 463 (1956).
[7] M. G. Motto, M. Sheves, K. Tsujimoto, V. Balogh-Nair, and K. Nakanishi, *J. Am. Chem. Soc.* **102**, 7947 (1980).

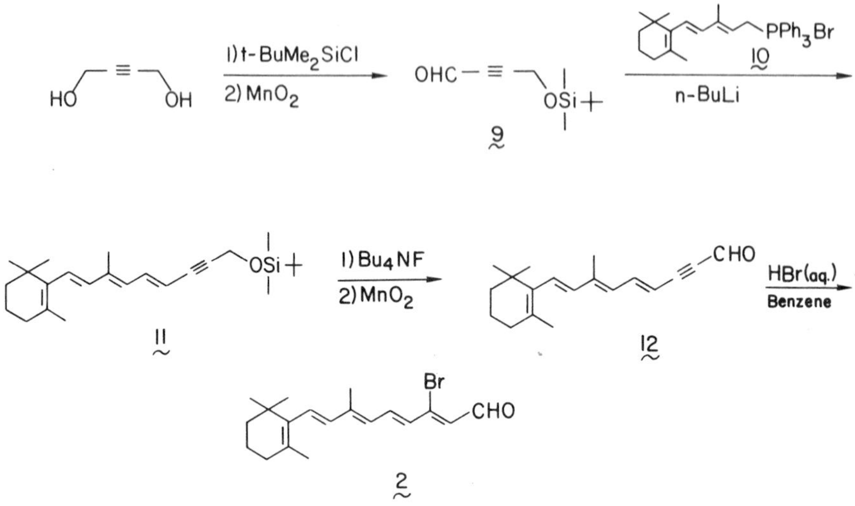

not displaced by the all-*trans*-retinal. Both the 9-bromo and the 13-bromo analogs underwent the following transformations between the light- and dark-adapted species:

13-demethyl-13-bromo BR

dark-adapted $\xrightleftharpoons[\text{30 hr}]{\text{30 min}}$ light-adapted

	dark-adapted	light-adapted
VIS.	585 nm	~15% reduction in intensity
CD.:	632 nm(−3.9)	~15% reduction in intensity
	560 nm (8.1)	

9-demethyl-9-bromo bR

dark-adapted $\xrightleftharpoons[\text{50 hr}]{\text{15 min}}$ light-adapted

	dark-adapted	light-adapted
VIS.	535 nm	545 nm
CD.:	585 nm(−1.6)	590 nm(−1.5)
	505 nm(+7.2)	512 nm(+7.8)

[22] Analysis of Photocycle and Orientation in Thin Layers

By RAFI KORENSTEIN AND BENNO HESS

The study of spectroscopic properties and photocycle kinetics of bacteriorhodopsin (BR) has been carried out initially with aqueous suspensions of purple membrane fragments. An alternative possibility is to study

METHODS IN ENZYMOLOGY, VOL. 88

the spectroscopy and the photochemistry of bacteriorhodopsin in hydrated solid thin layers of purple membrane. Such a preparation offers unique advantages, as described in the following sections.

Orientation. X-Ray studies[1-3] show that spreading of purple membrane on a slide, and consequent formation of thin layers after drying, yields a sample oriented with the plane of the fragments parallel to the surface of the support. An oriented sample makes it possible to perform differential absorption measurements with plane-polarized light. Such linear dichroic measurements can be carried out in the UV, visible, or the IR regions of the absorption spectrum, yielding information about the orientation of various groups in the protein. Moreover, complete immobilization of the purple membrane fragments in such a preparation makes it possible to determine very slow rotational movements of proteins in the membrane. The immobilization eliminates the decay contribution to the anisotropy factor in the photoselection studies, due to rotation of the whole purple fragments that takes place in aqueous solutions.

Hydration. Research on thin layers of purple membrane has demonstrated that it is possible to alter the hydration state of purple membrane by incubating the samples at different relative humidities.

Low-Temperature Studies. The technique of thin-layer photochemistry makes it possible to study the photocycle over a wide temperature region down to 4°K without using special water–organic solvent mixtures. In this way special solvent–purple membrane interactions are eliminated. Moreover, light scattering, which is pronounced in the UV absorption region in suspensions of purple membrane fragments, is minimized when employing thin layers of purple membrane, allowing also to record the kinetics of aromatic amino acid residues of bacteriorhodopsin.[4]

I. Experimental

Preparation. Thin layers of purple membrane are prepared by drying concentrated suspensions ($3.5 \times 10^{-4} M$) of the purple membrane in water (pH 7.2) on a glass slide. The drying is performed at atmospheric pressure, 45% relative humidity (r.h.), and 25°. The average thickness of the preparations is 0.5–3 μm as determined by scanning electron microscopy (see Fig. 1). Variable degrees of hydration of the purple membrane are obtained by equilibrating the samples with different relative air humi-

[1] A. E. Blaurock and W. Stoeckenius, *Nature (London), New Biol.* **233**, 149 (1971).
[2] A. E. Blaurock, *J. Mol. Biol.* **93**, 139 (1975).
[3] R. Henderson, *J. Mol. Biol.* **93**, 123 (1975).
[4] B. Hess and D. Kuschmitz, *FEBS Lett.* **100**, 334 (1979).

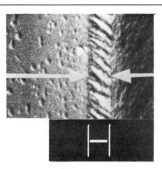

Fig. 1. Photograph of a thin layer of purple membrane on a glass slide. The picture is obtained by scanning electron microscopy (autoscan ETEC, Siemens), using a glass slide covered with the membrane and broken right across the thin layer. The broken slide is fixed on a holder and slightly spattered with gold in vacuum. The ridge of the slide and the purple membrane layer is analyzed with an angle of 45°. Identification of the layers was achieved by X-ray microanalysis defining the glass as well as the membrane surface. Arrows indicate the purple membrane layer. Magnification × 2000, calibration bar = 1 μm. Thus, the thickness of the preparation of approximately 1 μm accounts for approximately 200 monolayers of purple membrane with an average thickness of 50 Å.

dities produced by saturated salt solutions.[5] The glass slide with the preparations is inserted in a 1 × 1-cm cuvette and incubated in a desiccator at the required specific humidity for about 24 hr. Before measurement the cuvette is immediately sealed within the desiccator with a Teflon plus parafilm cover.

Kinetic Analysis of Relaxation Time Constants of the Photointermediates. The kinetics of the thermal decay of the various intermediates in the photocycle are analyzed in terms of a sum of exponentials. The initial values for the rate constant and their relative amplitudes are calculated by a computer program,[6] within the frame of a nonlinear approximation.[7,8] These values are then optimized by a least square program (Harwell Subroutine Library, VCO5A). The reliability of such a kinetic analysis is based on construction of a contour map of sum of errors for each multiexponential decay. An example of such an analysis is given in Figs. 2 a–c. The contours in Figs. 2b and c show the possible variation of the two rates (in a biexponential process), whenever performing an optimization of the rates by least squares. The variation of the rates will depend on its form and on how shallow or deep the minimum is. Each point in the contour curves (representing a constant value of sum of squares) is calculated

[5] A. Wexler and S. I. Hasegawa, *J. Res. Natl. Bur. Stand.* (U.S.) **53,** 19 (1954).
[6] K. H. Müller and T. Plesser, unpublished results.
[7] G. Meinardus and D. Schwedt, *Arch. Rat. Mech. Anal.* 17, 297 (1964).
[8] J. R. Rice, *Soc. Industr. Appl. Math.* **10,** 149 (1962).

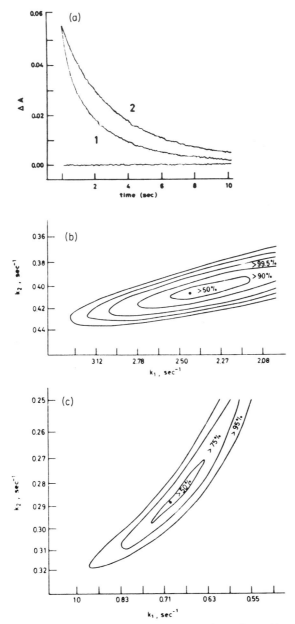

FIG. 2. (a) The decay kinetics of M_{412}. Curve 1: decay of M_{412} formed by a pulsed light (3 μ sec pulse width). Curve 2: decay of M_{412} formed by a continuous light (∞ pulse width). Both decay kinetics were measured at $-20°$. (b,c) Sum of squares contours as a function of the variation of the two rates found in the formation and decay kinetics of M_{412}. (b) Decay of M_{412} produced by a pulsed light. (c) Decay of M_{412} produced by a continuous light.[21]

by fixing the rates and fitting the amplitudes linearly by least squares. The percentages represent the probability of finding the minimum in the corresponding contour of sum of squares, found by statistical analysis using the F text.

II. Properties of Thin Layers

Hydration Effects on the Photocycle and on cis–trans Isomerization of Bacteriorhodopsin. Flash photolysis spectroscopy of thin-layer preparations of purple membrane at room temperature shows that bacteriorhodopsin undergoes a complete photocycle, involving the same intermediates characteristic of suspended purple membrane fragments in water. But the decay time of the "412" intermediate, as well as the formation of "660" and "610" intermediates, is determined by the specific hydration state of the membrane.[9] Preparations equilibrated with 94% relative humidity show the same kinetics and include the same transients as in purple membrane fragments suspended in water (Table I). The relaxation times for 412 decay are slowed down by lowering the hydration state of the preparation. The kinetics of 412 decay are analyzed in terms of the sum of two or three exponentials. The relaxation times and their relative amplitudes, at different hydration states, are given in Table II. The 660 and 610

TABLE I

RELAXATION TIME CONSTANTS (τ_i) OF THE 412, 660, AND 610 INTERMEDIATES[a,b]

Intermediate	Purple membrane in water[c] (τ, msec)	Thin purple membrane layers[d] (τ, msec)
412	1.7 ± 0.6 (0.21 ± 0.13)	1.8 ± 0.2 (0.3 ± 0.07)
	5.0 ± 0.8 (0.79 ± 0.13)	5.4 ± 0.9 (0.69 ± 0.07)
660	5.4 ± 0.4	5.6 ± 0.6
610	48 ± 2.1	46.0 ± 3.4

[a] Reprinted with permission from Korenstein and Hess, *Nature (London)* **270**, 184 (1977).
[b] The relaxation time constant is defined by $A(t) = A_i e^{-t/\tau_i}$, where the absorbance change is the sum of exponential terms. The relative amplitudes are given in brackets near the corresponding relaxation time constants. The experiments were done at room temperature ($22 \pm 1°$).
[c] Suspended purple membrane in water at pH 7.2.
[d] Thin purple membrane preparation equilibrated with 94% relative humidity.

[9] R. Korenstein and B. Hess, *Nature (London)* **270**, 184 (1977).

TABLE II

RELAXATION TIME CONSTANTS AND AMPLITUDES OF THE DECAY KINETICS OF 412 INTERMEDIATE IN THIN-LAYER
PREPARATIONS OF PURPLE MEMBRANE EQUILIBRATED IN VARIOUS RELATIVE AIR HUMIDITIES

Relative air humidity (%)	τ_1	τ_2	τ_3
90	3.5 ± 0.4 msec; (0.63 ± 0.05)[a]	13.1 ± 1.1 msec; (0.37 ± 0.03)[a]	—
83	10.2 ± 2.4 msec; (0.18 ± 0.03)	45 ± 3.3 msec; (0.52 ± 0.02)	252 ± 26 msec; (0.31 ± 0.05)[b]
75	30.5 ± 7.0 msec; (0.11 ± 0.04)	124 ± 33 msec; (0.56 ± 0.08)	503 ± 111 msec; (0.33 ± 0.08)
43	244 ± 100 msec; (0.30 ± 0.06)	1.25 ± 0.4 sec; (0.42 ± 0.04)	10.1 ± 3.0 sec; (0.28 ± 0.06)
10	0.9 ± 0.2 sec; (0.28 ± 0.11)	4.0 ± 1.8 sec; (0.35 ± 0.06)	28 ± 8 sec; (0.36 ± 0.09)
<1	2.3 ± 0.7 sec; (0.28 ± 0.03)	15.7 ± 4.5 sec; (0.32 ± 0.03)	154 ± 21 sec; (0.40 ± 0.4)

[a] Reprinted with permission from Korenstein and Hess, *Nature (London)* **270**, 184 (1977).
[b] Amplitudes are shown in parentheses.

are not observed at relative humidity lower than 90%. The results demonstrate that variation of adsorbed water on the purple membrane can change the relaxation times of M_{412} by four orders of magnitude.

Recently, on dehydration at room temperatures, an additional chromophore state with an absorption of 506 nm was described. Also, it was found that the 550-nm intermediate becomes undetectable with increasing dehydration.[10]

Using double pulse excitation methods at room temperature,[11] it has been shown that the back-photoreaction from M_{412} is initiated by the generation of a blue-shifted, photoproduct, M'_{390}. The latter undergoes an exponential thermal decay to BR_{570} with a half-life of 150 nsec (which is faster by a factor of 10^5 than the decay of M_{412}). Whereas M_{412} decay is modulated up to a factor of 10^4 by the hydration state of the membrane, the decay of M'_{390} is insensitive to the hydration parameter.[12]

The dark–light adaption process of bacteriorhodopsin, which corresponds to the 13-*cis*–-all-*trans* photoisomerization of the retinal moiety, is affected by the hydration state.[13] Fully hydrated samples (94% r.h.) undergo full conversion of 13-*cis* into all-*trans*. However, the 13-*cis*→all-*trans* photoisomerization is stopped at low hydration (10% r.h.). This suggests that hydration may change the efficiencies of competitive routes with the cis–trans isomerization process or may alter the $\phi_{cis-trans}/\phi_{trans-cis}$ quantum efficiency ratio. The effects of hydration changes of $BR^{all-trans}$ and BR^{13-cis} photocycles are summarized in Fig. 3 (see also Ref. 10).

A study of the infrared spectra in thin layers revealed that the state of hydration does not influence the parameters of amide bands with a full preservation of the shape of amide I bands and the half-width of the α-hel-

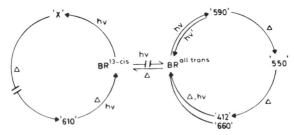

FIG. 3. Scheme of $BR^{all-trans}$ and BR^{13-cis} cycles and the *cis–trans* interconnecting route of the two cycles. The (⊢⤙) represent pathways that are blocked on reducing the hydration state of the purple membrane.[13]

[10] Y. A. Lazarev and E. L. Terpugov, *Biochim. Biophys. Acta* **590,** 324 (1980).

[11] O. Kalinsky, U. Lachish, and M. Ottolenghi, *Photochem. Photobiol.* **28,** 261 (1978).

[12] O. Kalinsky, M. Ottolenghi, B. Honig, and R. Korenstein, *Biochemistry* **20,** 649 (1981).

[13] R. Korenstein and B. Hess, *FEBS Lett.* **82,** 7 (1977).

ical component.[10] - For the influence of the state of hydration on the reaction of aromatic amino acid residues during the photocycle see Ref. 4, 9 and p. 190.

Branching Reactions in the Photocycle of Bacteriorhodopsin. Bacteriorhodopsin undergoes a reaction cycle involving several intermediates after light absorption. The photocycle can be described by the following scheme, showing the time sequence of intermediates of BR[14] (the numbers giving their approximate wavelength maxima[14-16]).

$$BR_{570} \rightarrow K_{590} \rightarrow L_{550} \rightarrow M_{412} \quad (O_{660})$$
$$\uparrow \text{all-}trans$$

Although the scheme describes the photocycle in terms of consecutive reactions, the possibility of branching reactions has already been raised in the early studies of the photocycle.[17-20] Indeed, an evidence for branching reactions emerges from stationary and flash low-temperature studies of fully hydrated thin purple membrane layers.

Continuous illumination of thin layers of purple membrane was carried out at low temperature, under those conditions where K_{550}, L_{550}, and M_{412} photointermediates are sufficiently long-lived, as to be detectable by conventional spectrometry. Irradiation of BR_{570} in fully hydrated thin layers, at $-180°$, yields photostationary state mixtures of BR_{570} and K_{590} (Fig. 4). The existence of an apparent isosbestic point at 578 nm suggests that at this temperature a photoequilibrium, mainly between two forms, is established. When the temperature of the irradiated systems is gradually raised in the dark, e.g., to $-120°$, a thermal relaxation takes place leading to the conversion of K_{590} into L_{550} (Fig. 5). In variance with the room temperature sequence, $L_{550} \rightarrow M_{412} \rightarrow BR_{570}$, the decay of L_{550} below $\sim -70°$ is not associated with the generation of the characteristic absorption maximum (~ 412 nm) due to the M_{412} intermediate. However, when the same BR_{570} is irradiated at $-60°$, the formation of M_{412} is observed.[12] It is thus evident that at low temperatures the thermal decay of L_{550} leads to the regeneration of BR_{570} in a process which bypasses the M_{412} state.

Furthermore, kinetic analysis of flash photolysis at $-20°$ shows that the M_{412} intermediate is formed and decays in a biexponential process. This suggests the existence of two forms of L_{550} that decay by parallel

[14] M. C. Kung, D. Devault, B. Hess, and D. Oesterhelt, *Biophys. J.* **15,** 907 (1975).
[15] H. Lozier, R. A. Bogomolni, and W. Stoeckenius, *Biophys. J.* **15,** 955 (1975).
[16] C. R. Goldschmidt, M. Ottolenghi, and R. Korenstein, *Biophys. J.* **16,** 839 (1976).
[17] M. A. Slifkin and S. R. Caplan, *Nature (London)* **253,** 56 (1975).
[18] M. Eisenbach, E. Bakker, R. Korenstein, and S. R. Caplan, *FEBS Lett.* **71,** 228 (1976).
[19] B. Hess and D. Kuschmitz, *FEBS Lett.* **74,** 20 (1977).
[20] J. B. Hurley, B. Becher, and T. G. Ebrey, *Nature (London)* **272,** 87 (1978).

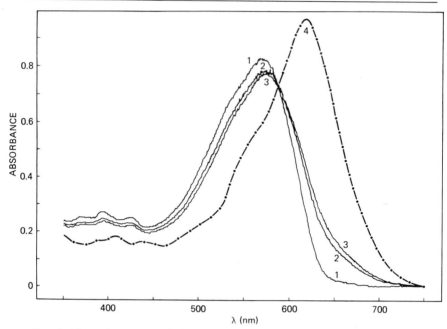

F<small>IG</small>. 4. Absorption spectra of photostationary state mixtures of BR_{570} and K_{590} in thin layers of purple membrane (equilibrated with 94% r.h.) at $-180°$. Curve 1, before irradiation; curve 2, photostationary state mixture of BR_{570} and K_{590} produced by irradiation at 578 nm; curve 3, photostationary state mixture of BR_{570} and K_{590} produced by irradiation at 546 nm; curve 4, the pure absorption spectrum of K_{590} obtained by extrapolation according to procedure introduced in Fischer.[20a]

paths yielding two conformations of M_{412}, which then decay to BR_{570}.[21] The existence of two conformers of M_{412} is supported by the observation of different difference spectra of $M_{412}-BR_{570}$ which are obtained by excitation with pulsed and continuous light. The dependence of the decay rates on the excitation pulse width suggests a possible equilibration of two conformers. Therefore, it is evident that parallel thermal reactions do occur in the photocycle. The various routes are summarized in Fig. 6.

Kinetic Analysis of the Aromatic Amino Acid Residues of Bacterio-rhodopsin. The interaction between the retinal chromophore of bacteriorhodopsin and aromatic amino acid residues of the opsin moiety was recognized when a reaction of the protein fluorescence following 570-nm light activation was discovered.[22] This observation was extended, suggesting a deprotonation and reprotonation of tyrosine and possibly of

[20a] E. Fischer, *J. Phys. Chem.* **71**, 3704 (1967).

[21] R. Korenstein, B. Hess, and D. Kuschmitz, *FEBS Lett.* **93**, 266 (1978).

[22] D. Oesterhelt and B. Hess, *Eur. J. Biochem.* **37**, 316 (1973).

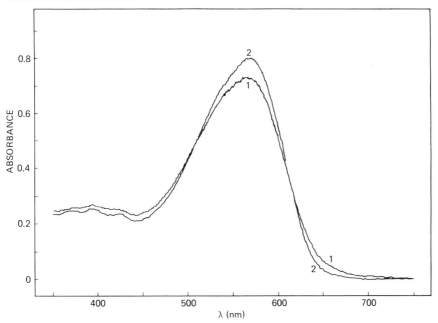

FIG. 5. Absorption spectra of a mixture of BR_{570} and L_{550} at $-120°$. Curve 1, the interconversion of K_{590} into L_{550} when raising the temperature of the sample in Fig. 4; curve 3, from $-180°$ to $-120°$; curve 2, absorption spectra of BR_{570} obtained after raising the temperature to $20°$ and cooling back down to $-120°$.

tryptophan during the photocycle.[4,23,24] These experiments indicate that photoactivation of the retinal chromophore leads to a disappearance of tyrosine and tryptophan analyzed at 275 nm with concomitant appearance of a band with a maximum of 295 nm, pointing indeed to a deprotonation of aromatic residues (obviously tyrosine), correlating with the kinetics of the photocycle of bacteriorhodopsin (see Chapter 32). The reactions are specific with respect to the absorption changes observed with vertically and horizontally polarized measuring light, illustrating the strong orientation of the residues involved as well as the change during the formation of the 295-nm component. The photochemical events involving aromatic amino acid residues of the protein occur between the decay of the L_{550} component in the formation of the M_{412} component of the retinal chromophore intermediates. The experiments also revealed photoinduced cross-reactions between the aromatic amino acid residues and intermediates of the photocycle. The kinetics of the aromatic amino

[23] R. A. Bogomolni, L. Stubbs, and J. K. Lanyi, *Biochemistry* **17**, 1037 (1978).
[24] B. Becher, F. Tokunaga, and Th.G. Ebrey, *Biochemistry* **17**, 2293 (1978).

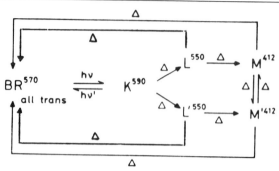

Fig. 6. Scheme of the photocycle of bacteriorhodopsin ($\triangle$ represents thermal reactions).

acid residues are strongly pH dependent. Furthermore, it is interesting to note that the relative humidity influences the rates of formation and decay of the 275-nm component analyzed at 94% and 85% humidity.[4]

Linear Dichroism of Bacteriorhodopsin. Linear dichroic measurements of bacteriorhodopsin[25] in thin purple membrane layers are performed at two planes, as shown in Fig. 7a). No dichroism in the absorption spectrum is observed when the polarized light axes are perpendicular to the plane of the membrane (90°, $\parallel$ and 90°, $\perp$). This would be expected if the chromophores are randomly oriented about the axis normal to the plane of the glass slide. However, a dichroism is observed (Fig. 7b) when the light axes form an angle of 45° with the plane of the glass slide (45°, $\parallel$ and 45°, $\perp$). The angle between the transition moment of the chromophore and the plane of the membrane can be derived from the measured dichroic ratio. The dichroic ratio for chromophores randomly oriented about the normal to the plane of the membrane is given by[26]

$$D = \sin^2 \alpha + 2 \tan^2 \varphi \cos^2 \alpha \tag{1}$$

where φ is the angle between the transition moment of the chromophore and the plane of the membrane, α is the angle of the polarized light axis with the plane of the membrane, and D is the dichroic ratio. Correcting for the index of refraction, n, using Snell's law, we obtain

$$D = 1 - \frac{\cos^2 \alpha}{n^2} + 2 \tan^2 \varphi \frac{\cos^2 \alpha}{n^2} \tag{2}$$

However, in order to account for the existence of multiple reflections in thin films a correction in the analysis of the dichroic data should be introduced. It can be shown that if the film thickness is larger compared to the

[25] R. Korenstein and B. Hess, *FEBS Lett.* **89**, 15 (1978).
[26] R. J. Cherry, H. Su Kwan, and D. Chapman, *Biochim. Biophys. Acta* **267**, 512 (1972).

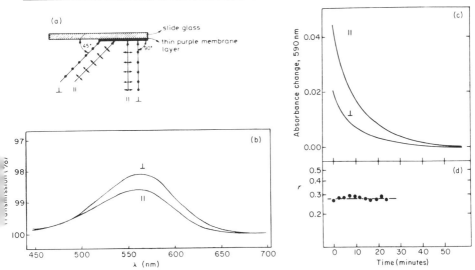

Fig. 7. (a) Scheme of the sample and the polarized light used in the linear dichroic absorption measurement. The glass slide was only half covered, the uncoated part serving as a reference. The purple membrane fragments were oriented with their membrane planes parallel to the glass slide plane. The parallel ($\parallel$) and the perpendicular ($\perp$) orientations of the polarized light were defined parallel to the long and the short axes of the rectangular slide. (b) Linear dichroic spectrum of purple membrane layer (equilibrated with a relative humidity of 45%) measured with polarized light axes forming an angle of 45° with the glass slide plane. (c) Transient dichroism in the decay kinetics of $BR^{LA} \to BR^{DA}$ measured in a thin layer of purple membrane (equilibrated with 94% relative humidity) at 40°. The axis of both the exciting and the analyzing light formed an angle of 90° with the glass slide plane. (d) Time dependence of the anisotropy parameter $r(t)$ as calculated from (c).[25]

wavelength of the analyzing light, the effect of multiple reflections can be neglected. However, when the thickness of the film is equal or smaller than the wavelength, the dichroic ratio will depend strongly on the film thickness.[27] Therefore, we measure the dichroic ratio in purple membrane samples of different thickness (d). The results are: $D = 0.89$ ($A = 0.6$, $d \sim 4 \mu m$); $D = 0.85$ ($A = 0.24$, $d \sim 1.3 \mu m$); $D = 0.78$ ($A = 0.015$, $d < 1 \mu m$). When substituting the various D values in Eq. (2), where $\alpha = 45°$ and $n = 1.5$,[28] we obtain values of 26.7°, 21.9°, and 4° for φ, correspondingly. Thus, it can be stated that $\varphi \leq 27°$. A value of 23.5° was measured in molecular films at an air–water interface,[29] whereas a value of 19°–20°

[27] D. den Engelsen, *J. Phys. Chem.* **76**, 3390 (1972).
[28] M. Brith-Lindner and K. Rosenheck, *FEBS Lett.* **76**, 41 (1977).
[29] R. A. Bogomolni, S. B. Hwang, Y. W. Treng, G. I. King, and W. Stoeckenius, *Biophys. J.* **17**, 98a (Abstr. W-POS-C13) (1977).

resulted from measurements in thin purple membrane layers.[30–32] The values might differ partly because of the disorder in the alignment of the purple membrane fragments (predominant in thick layers) or because of multiple reflection effects (predominant in thin layers).

Rotational Immobilization of Bacteriorhodopsin. Photoselection studies of the dichroic relaxation in bacteriorhodopsin are performed with thin purple membrane layers equilibrated with 94% relative humidity. Under such conditions bacteriorhodopsin undergoes a complete photocycle and has the same spectroscopic and kinetic characteristics found for purple membrane suspended in water.[13] In order to detect very slow rotations of bacteriorhodopsin within the purple matrix one has to use a phototransient of relaxation time slower than that made available by the 412 intermediate of the photocycle as well as immobilize the purple membrane fragments. Thus, we have chosen to measure the transient dichroism of the long $BR^{LA} \rightarrow BR^{DA}$ thermal transition (relaxation time constant of 14 min at 40°; LA = light-adapted, DA = dark adapted), which corresponds to an all-*trans*→13-*cis* isomerization of the retinal chromophore. Using thin purple membrane layers on glass slides the purple membrane fragments are completely immobilized to the glass surface.

A transient dichroism in the decay kinetics of $BR^{LA} \rightarrow BR^{DA}$ process is shown in Fig. 7c. The spectral changes at 590 nm are only due to one component (all-*trans*), since the photocycle spectral changes in the same spectral region take place in the millisecond time domain. Thus, the anisotropy factor* can be considered as independent of the $BR^{LA} \rightarrow BR^{DA}$ relaxation time.[30]

Analysis of the data shows that the anisotropy factor $r(t)$ remains constant over time range of 25 min (Figure 7d). The transient dichroic measurements are performed at 40°, after the purple membrane had undergone thermal transitions in the 23°–30° temperature range.[33] Thus, even under such conditions of reduced microviscosity no rotational freedom can be detected. A higher anisotropy factor ($r = 0.3$) observed when the exciting polarized light formed an angle of 45° with the glass slide plane as compared to the value obtained when the exciting light is perpendicular to

---

* The anisotropy factor $r(t)$ is defined as: $r(t) = \dfrac{A\|(t) - A\perp(t)}{A\|(t) + 2A\perp(t)}$, where $A\|$ and $A\perp$ are A_{590} values with the polarized light parallel and perpendicular to the exciting polarized light, respectively.

[30] M. P. Heyn, R. J. Cherry, and U. Müller, *J. Mol. Biol.* **117,** 607 (1977).
[31] A. U. Acuna and J. Gonzalez-Rodriguez, *Anal. Quim.* **75,** 630 (1979).
[32] D. D. Muccio and J. Y. Cassim, *J. Mol. Biol.* **35,** 595 (1979).
[33] R. Korenstein, W. V. Sherman and S. R. Caplan, *Biophys. Struct. Mechan.* **2,** 267 (1976).

the glass slide plane ($r = 0.27$), results from linear dichroic effects arising from the orientation of the purple membrane fragments with their planes parallel to the plane of the slide glass. Moreover, the value of 0.27 obtained for $r(t)$ is lower than the maximum theoretical value of 0.4, due to instrumental factors.

It can be concluded that bacteriorhodopsin is completely immobilized within the purple membrane matrix. The results eliminate the possibility for rotational freedom of bacteriorhodopsin monomer, trimer, or even a big cluster of trimers. The results also exclude the possibility for the chromophore to rotate about those axes which would lead to a dichroic decay, suggesting a strong retinal–opsin interaction. Using fluorescence depolarization, the microviscosity of the lipid domains in the purple membrane was found to be 5 poise.[33] Since the present study shows that bacteriorhodopsin is immobilized within the purple membrane matrix, it is clear that the microviscosity that bacteriorhodopsin experiences is mainly a result of protein–protein interactions and not lipid–protein interactions. These exceedingly strong protein–protein interactions within the purple membrane matrix play a key role in the cooperativity observed in the photocycle kinetics possibly modulating proton transfer processes.[34]

Acknowledgments

This assistance of Dr. K. Zierold and Mrs. M. Seiffert in the preparation of the electron micrographs is gratefully acknowledged.

[34] R. Korenstein, B. Hess, and M. Markus, *FEBS Lett.* **102**, 155 (1979).

[23] Cooperativity of Photocycle in Purple Membrane

By RAFI KORENSTEIN and BENNO HESS

The mechanism of coupling between the photocycle and the vectorial proton transfer process of bacteriorhodopsin (BR) of the native purple membrane of *Halobacterium halobium* implies an oriented and tight interaction between the chromophore and the protein conformation of bacteriorhodopsin within the structure of the purple membrane. Each intermediary state of the chromophore throughout the cycle is defined by a spectrum and corresponds to a well-definable conformation of the protein moiety. Bacteriorhodopsin molecules undergo such conformational changes while being arranged in a two-dimensional array, forming almost perfect crystal lattice of space group P_3 where three-protein molecules are in direct contact, forming a trimeric cluster.[1-5]

Such an organization of the molecules imposes strong protein–protein interactions, which occur through the large contact areas between helical domains. A consequence of these interactions is the complete immobilization of the protein within the purple membrane, allowing no rotational freedom to bacteriorhodopsin, even in the minute time range.[6] Such protein–protein interaction within the purple membrane matrix brings about an electronic coupling between the retinal chromophore, an interaction that can be viewed by the circular dichroic spectrum (CD) of the purple membrane. The visible circular dichroism of purple membrane consists of intense positive and negative bands with a crossover at the wavelength of the absorption maximum.[7,8] This CD spectrum was shown [7,9,10,11] to arise from two contributions: a positive band due to a retinal bound to the protein and positive and negative bands (bilobe) due to the exciton interaction between chromophores of neighboring proteins.

Further evidence for the interaction of bacteriorhodopsin molecules within the two-dimensional matrix of purple membrane is found in the sequential bleaching (in the presence of hydroxyl amine) and the subsequent regeneration of the purple membrane.[11] From these experiments it is evident that the bacteriorhodopsin molecules appear to bleach simultaneously within a trimer. Recent study on all-*trans*-retinal binding to the apomembrane[12] resulted in a Hill coefficient of 3.0. This Hill coefficient again suggests possible cooperativity in the binding of retinal within bacteriorhodopsin trimers. Furthermore, protein–protein modulation of the all-*trans*→13-*cis* thermal isomerization of the retinal is observed in the accelerated transformation of the light-adapted to the dark-adapted form of BR, when existing as monomers.[13]

Recently, a new two-dimensional crystal form of purple membrane has been obtained *in vitro*.[14] This new form of purple membrane is ortho-

[1] A. E. Blaurock and W. Stoeckenius, *Nature (London) New Biol.* **233**, 152 (1971).
[2] A. E. Blaurock, *J. Mol. Biol.* **93**, 139 (1975).
[3] R. Henderson, *J. Mol. Biol.* **93**, 123 (1975).
[4] P. N. T. Unwin and R. Henderson, *J. Mol. Biol.* **94**, 425 (1975).
[5] R. Henderson and P. N. T. Unwin, *Nature (London)* **257**, 28 (1975).
[6] R. Korenstein and B. Hess, *FEBS Lett.* **89**, 15 (1978).
[7] M. Heyn, P. Bauer, and N. Dencher, *Biochem. Biophys. Res. Commun.* **67**, 879 (1975).
[8] B. Becher and J. Cassim, *Biophys. J.* **15**, 66a (1975).
[9] P. Bauer, N. Dencher, and M. Heyn, *Biophys. Struct. Mech.* **2**, 79 (1976).
[10] B. Becher and T. Ebrey, *Biochem. Biophys. Res. Commun.* **69**, 1 (1976).
[11] B. Becher and J. Cassim, *Biophys. J.* **19**, 285 (1977).
[12] M. Rehorek and M. Heyn, *Biochemistry* **18**, 4977 (1979).
[13] T. G. Ebrey, B. Becher, B. Mao, P. Kilbridge, and B. Honig, *J. Mol. Biol.* **112**, 377 (1977).
[14] H. Michel, D. Oesterhelt, and R. Henderson, *Proc. Natl. Acad. Sci. U.S.A.* **77**, 338 (1980).

rhombic with space group $P22_12_1$. Thus at such a high level of protein organization, one would also expect to observe a kinetic coupling among bacteriorhodopsin molecules that constitute the various clusters in the hexagonal or the orthorhombic purple membrane. In photoselection studies, bacteriorhodopsin in the orthorhombic purple membrane was again shown[15] to be immobilized. Although the bacteriorhodopsin molecule has identical molecular structure in the native hexagonal and in the orthorhombic forms, protein–protein interactions occur through different helical domains in these two crystalline forms, thus allowing demonstration of cooperativity in both cases between different domains affecting the photocycle in the kinetic coupling between the interacting molecules.

Materials and Methods

In order to detect kinetic coupling among BR molecules both in the native hexagonal and in the orthorhombic purple membrane, it is necessary to correlate the kinetics of the thermal transitions of the photocycle with the population of a certain metastable state within a certain cluster. Since the M_{412} state is the longest-lived intermediate, it is possible by choosing the appropriate environmental conditions (temperature, pH, ionic strength) to obtain a high conversion yield of BR_{570} into M_{412} under photosteady light illumination, and thus by changing the light intensity to induce a variable number of BR molecules in the M_{412} state within a certain cluster. The dependence of the thermal decay of M_{412} on the occupancy of M_{412} (M_{412}/BR_{570}) is measured by turning the light off after reaching the various photostationary states of M_{412}.

Purple membrane was isolated[16] from *H. halobium* (mutant NRL R_1M_1). The kinetics are measured by the experimental setup shown in Fig. 1. The sample is excited by a 400-W W-I_2 lamp (LS_1), the light of which was filtered by a neutral density filter (F_1) and a cutoff Eppendorf filter 500 nm (F_2) (so that M_{412} was not excited by the light). The measuring light from a 400-W W-I_2 lamp (LS_2) passes a Bausch and Lomb monochromator (M), is focused onto the cuvette, and then filtered by a 412-nm interference filter (F_3) (8-nm band width) on the multiplier (PM) (EMI 9634 QR). The output of the multiplier is amplified and passed through a variable RC filter both to a slow-speed recorder and to a digital scope (Nicolet 1090). The temperature is measured by a Cu–constantan thermocouple connected to a Fluke 2100 Å digital thermometer ($\pm 0.1°$). The thermocouple is inserted into the sample solution at the illuminated area, in order to detect possible local heat during the photobleaching experi-

[15] R. Korenstein, B. Hess, H. Michel, and D. Oesterhelt, unpublished results.
[16] D. Oesterhelt, M. Muntzen, and L. Schuhmann, *Eur. J. Biochem.* **40**, 453 (1973).

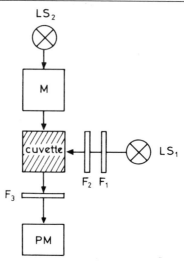

FIG. 1. Block diagram of experimental setup (for explanation, see text).[17]

ments. The light flux is measured with a Tektronix J16 digital photometer. Initial values for the rate constants and relative amplitudes are calculated according to Korenstein *et al.*[17] Optimizations are performed by a least square program from the Harwell Subroutine Library (VCO5A for the sum of exponentials and VAO5A for the cooperativity model).

Cooperativity in the Photocycle of the Native Hexagonal Purple Membrane

The measured decay of M_{412} from the different photostationary levels back to the equilibrium state ($BR_{570}^{all-trans}$) can be fitted by a sum of two exponentials, yielding the decay rate constants (k_a and k_b) shown in Table I. The total concentration of bacteriorhodopsin is denoted by $(BR)_0$. We see that the rates are strongly correlated with the degree of occupancy of M_{412}, namely, the $(M_{412})/(BR)_0$ ratio. It should be stressed that no significant changes in the formation rates of M_{412} were observed. The observation that M_{412} decays as a sum of two exponentials is explained by the existence of two conformers of M_{412}, which we denoted as M and M′.[17] These decays can be considered here as pseudo-first-order reactions, in spite of the protonation step involved, since the pH is kept constant owing to buffering. The trimeric structural organization of bacteriorhodopsin, together with the observed correlation between the overall decay rates and the degree of occupany of M_{412}, suggests that the corresponding in-

[17] R. Korenstein, B. Hess, and D. Kuschmitz, *FEBS Lett.* **93**, 266 (1978).

TABLE I

DECAY KINETICS[a] FROM DIFFERENT PHOTOSTATIONARY LEVELS OF M_{412} AT $-16°$ [20]

$[BR]_0^c$ (μM)	$[M_{412}]^d$ (μM)	$[M_{412}]/[BR]_0$ (%)	k_a (sec^{-1})	k_b (sec^{-1})
20	⎧ 9.0	45	2.78 ± .62 (0.25 ± .04)[b]	1.08 ± .01 (0.75 ± .05)[b]
	⎩ 1.9	9.5	1.54 ± .03 (0.61 ± .07)	0.64 ± .08 (0.39 ± .07)
5	2.45	49	2.59 ± .28 (0.32 ± .06)	0.95 ± .06 (0.68 ± .06)
30	2.55	8.5	1.06 ± .03 (0.63 ± .04)	0.42 ± .03 (0.37 ± .05)

[a] Purple membrane was suspended in aqueous solution of basal salts, 50 mM phosphate (pH 6.0).
[b] Amplitudes in parentheses.
[c] $\epsilon = 63,000$ (M^{-1} cm^{-1})
[d] Using a ratio of 0.5 between the difference extinction coefficient of M_{412} (at 412 nm) and the extinction coefficient of BR_{570}.

trinsic decay rates depend on the conformational state of the nearest neighboring molecules.

Since the bilobe in the CD spectrum of purple membrane arises from exciton interactions between the chromophores of protein molecules within or between trimers, its existence can serve as a measure of the degree of electronic coupling between adjacent chromophores. The exciton interaction can be decoupled by suspending the purple membrane in basal salts and then saturating the solution with diethyl ether.[18] In addition, the presence of ether induces rotation of the bacteriorhodopsin molecules within the membrane.[20] Therefore, a study of the photocycle under conditions where exciton coupling between the chromophores is abolished is useful to find out whether a correlation between kinetic and spectroscopic coupling exists. The experimental results, done in the same way as for the fully coupled system, are given in Table II. It is seen that the decay rates do not change essentially with the increase of M_{412} occupancy.

The findings indicate that the disappearance of exciton coupling between the chromophores is correlated with the disappearance of the kinetic coupling between the proteins.

Cooperativity in the Photocycle of the Orthorhombic Purple Membranes

Since strong interprotein interactions play a key role in the modulation of the photocycle kinetics of bacteriorhodopsin in the hexagonal purple

[18] T. Reed and B. Hess, *67th Annv. Meet. Am. Soc. Biol. Chem.* Abstr. 45 (1976).
[19] M. Heyn, P. Bauer, and N. Dencher, *FEBS-Symp.* **42,** 96 (1977).
[20] R. Korenstein, B. Hess, and M. Markus, *FEBS Lett.* **102,** 155 (1979).

TABLE II

Decay Kinetics[a] from Different Photostationary Levels of M_{412} at $20°$[20]

$[BR]_0$[c] (μM)	$[M_{412}]$[d] (μM)	$[M_{412}]/[BR]_0$ (%)	k_a (sec^{-1})	k_b (sec^{-1})
13.5 {	7.8	58	0.093 ± .010 (0.37 ± .05)[b]	0.027 ± .003 (0.62 ± .06)[b]
	3.8	28	0.087 ± .023 (0.27 ± .13)	0.027 ± .005 (0.73 ± .13)
3.4 {	1.9	14	0.118 ± .026 (0.13 ± .03)	0.027 ± .002 (0.87 ± .03)
	2.7	80	0.128 ± .011 (0.37 ± .04)	0.034 ± .002 (0.63 ± .04)
20.2	2.8	14	0.105 ± .033 (0.21 ± .14)	0.032 ± .004 (0.78 ± .12)

[a] Purple membrane was suspended in aqueous solution of basal salts, 50 mM phosphate buffer (pH 6.0) and saturated with diethyl ether.
[b,c,d] See legend to Table I.

membrane (see also Korenstein et al.[20]), the existence of similar kinetic coupling in the orthorhombic purple membrane is of interest. Analysis of the decay kinetics of M_{412} from the different photostationary concentration levels back to the equilibrium state (BR_{570}) yield again two apparent relaxation time constants. The relaxation time constants were dependent on the (M_{412})/(BR_{570}) ratio, as is given in Table III, indicating that the decay rates of bacteriorhodopsin in the orthorhombic crystalline state depend also on the conformational state of its nearest neighbors. Although the decay kinetics gave a fit by a sum of two exponentials, the fitting itself was only moderately good, as is demonstrated in the initial decay differ-

TABLE III

Decay Kinetics[a] from Different Photostationary[15] Levels of M_{412}

$[M_{412}]/[BR_{560}]$[c,d] (%)	τ_b (sec)	τ_a (sec)
5	13.0 ± 0.76 (0.54 ± .01)[b]	0.75 ± 0.02 (0.36 ± 0.1)
10	10.8 ± 0.51 (0.43 ± .01)	0.63 ± 0.03 (0.45 ± .01)
19	10.0 ± 0.20 (0.33 ± .01)	0.47 ± 0.07 (0.53 ± .01)
30	8.21 ± 0.23 (0.27 ± .01)	0.38 ± 0.08 (0.59 ± .01)
44	7.50 ± 0.20 (0.22 ± .01)	0.32 ± 0.01 (0.64 ± .02)

[a] Orthorhombic purple membrane (27 μM) in aqueous solution of 50 mM phosphate buffer, pH 7.0, 6°.
[b] Amplitudes are shown in parentheses.
[c] Using extinction coefficient of 59,000 (M^{-1}, cm^{-1}).
[d] Using a ratio of 0.6 between the difference extinction coefficient of M_{412} and extinction coefficient of BR_{560}.

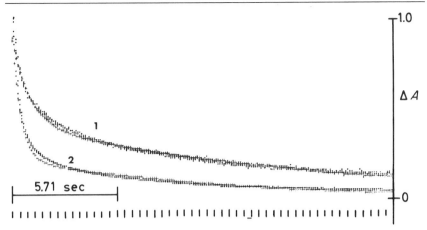

FIG. 2. Decay kinetics from different photostationary levels of M in aqueous suspensions of orthorhombic purple membrane (50 mM phosphate buffer, pH 7.0) at 6°. Curve 1, decay from photostationary state of $\triangle A_{412} = 0.420$; curve 2, decay from photostationary state of $\triangle A_{412} = 0.092$. Both initial amplitudes of the two decays are normalized. Dots (.) original data points. Cross (+) fitted data points for the experimental decay.

ence between original and fitted relaxations (Fig. 2). This would be expected if the two fitted apparent relaxation time constants were a combination of sequential and parallel intrinsic relaxation time constant, as in the cooperativity model of the photocycle for hexagonal purple membrane.[20] Thus these results suggest the existence of kinetic coupling between neighboring bacteriohodopsin molecules in the orthorhombic purple membrane, although the degree of coupling as compared to that existing in the hexagonal purple membrane cannot be determined quantitatively.

Analysis, Simulation, and Significance of Cooperative Kinetics

The search for cooperativity phenomena in membrane-bound systems is of general interest, since the high spatial and sterical ordering of membrane constituents within a two-dimensional array—whether fluid or immobilized—facilitate and favor protein–protein interactions. We here illustrate kinetic methods for the detection of cooperative functions, which in correlation with spectroscopic, structural, and chemical analysis help us to understand multiunit behavior in membrane systems, analog to the concept of cooperativity in soluble oligomeric enzymes. Indeed, the kinetics of the photocycle as well as the observation of two different conformations of the intermediary state, M and M'[17] allow us to derive a co-

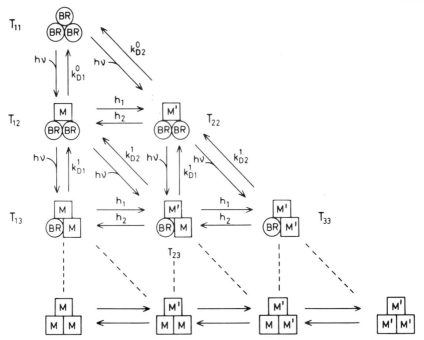

FIG. 3. Cooperativity model of the photocycle. Transitions between trimeric states in purple membrane assuming a two-state cycle. The broken lines indicate transitions to states that are neglected since the total conversion of BR into M was $\leq 66\%$ in the experiments.[20]

operativity model based on protein–protein interactions (Fig. 3), which can be analyzed by equations describing the time dependence of the concentrations of the trimeric states as a function of the total light power input per unit surface. A typical example of fitting is given in Fig. 4, demonstrating that the model favors the interpretation of the kinetic results.[20]

Direct evidence for interactions between the bacteriorhodopsin molecules in its native crystalline array is based on structural as well as functional properties of the purple membrane system. The correlation of all these properties under quasiphysiological conditions (although in open membrane sheets) is of greatest significance. Since kinetic cooperativity was observed in both the hexagonal and the orthorhombic purple membrane, it may be suggested that cooperativity effects are not confined to trimeric or dimeric clusters, but may extend through the whole two-dimensional lattice.

The observation of cooperativity in the purple membrane might be of importance for membrane functions in general (see Changeux and

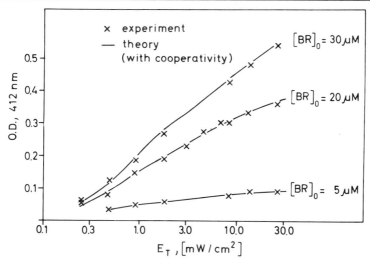

FIG. 4. Dependence of the photostationary absorbance of M_{412} on the total light power input per unit surface for three different concentrations of bacteriorhodopsin. The crosses are the experimental values. Curves correspond to the theory assuming cooperativity.[20]

Thiéry[21] and Emrich and Reich[22]) if physiological significance is understood in terms of the overall efficiency of coupled molecular functions in the oligomeric forms of the membrane organization. We visualize here cooperativity involved in transport and electrochemical gradient formation. In the case of the purple membrane all these considerations should apply. Here the cooperativity displayed by the increase of the rate constants with increasing occupancy of the M or M' state has a positive sign analogous to general autocatalysis, allowing the amplification of dynamic states upon suitable cooperative coupling. This positive cooperativity is relevant with respect to the relationship between the proton dissociation–associations functions coupled to the photocycle in the purple membrane. Recently, a high ratio of dissociated protons per M_{412} state in the range up to 3 at low light intensities and the decrease of this ratio to 1 at saturating light intensities were found.[23] This suggests an interaction between one molecule in the M_{412} state with its neighboring molecules, still in the BR_{570} state, inducing proton dissociation reactions in the latter molecules. Furthermore, it is significant that, as a function of the light intensities, the ratio of proton release per M_{412} state indicates negative cooperativity.

[21] J. P. Changeux and J. Thiéry, *BBA Libr.* **11**, 116 (1968).
[22] H. M. Emrich and R. Reich, *Z. Naturforsch., C: Biosci.* **29C**, 577 (1974).
[23] D. Kuschmitz and B. Hess, *Biochemistry* **21**, 5950 (1981).

[24] High-Performance Liquid Chromatography Method for Isolation of Membrane Proteins from Halobacterial Membrane

By TETSUYA KONISHI

Introduction

Purple membrane of *Halobacterium halobium* is a specially differentiated membrane region of the plasma membrane of *Halobacterium* which contains bacteriorhodopsin as a single protein component.[1] Bacteriorhodopsin molecules, which are light-driven proton pumps, are uniquely organized in the membrane, with trimers of bacteriorhodopsin arranged in the P3 two-dimensional protein lattice.[2] Thus the biogenesis of purple membrane is quite interesting for the study of integration and assembly of a functional protein in a membrane.

To follow membrane formation in purple membrane, a rapid and quantitative separation method for proteins is required. SDS–polyacrylamide gel electrophoresis (PAGE) has been used so far for this purpose; however the method is not highly quantitative, is time-consuming, and requires a skilled technique, so that it was hardly used to trace a successive change of certain protein species quantitatively. The recently developed technique of gel permeation high-performance liquid chromatography (HPLC) made it possible to separate and identify proteins of the molecular weight range 100,000–3000 within an hour.[3] By using a UV monitering device, quantitative determination of each protein is also possible.

In this chapter an application of gel permeation HPLC for the study of membrane protein biogenesis is described, where the successive change of the protein pattern of the plasma membrane of *Halobacterium halobium* S9 in the log growth phase was followed.

Cell Growth and Sample Preparation

The strain S9 of *H. halobium* was grown in a 1-liter culture flask containing 400 ml culture medium described by Oesterhelt and Stoeckenius[4]

---

[1] D. Oesterhelt and W. Stoeckenius, *Nature (London), New Biol.* **233,** 149 (1971).

[2] R. Henderson *J. Mol. Biol.* **93,** 123 (1975).

[3] N. Ui, *Anal. Biochem.* **97,** 65 (1979).

[4] D. Oesterhelt and W. Stoeckenius, this series, Vol. 31, p. 667.

METHODS IN ENZYMOLOGY, VOL. 88

in a shaking fermenter at 35° under illumination with a pair of 30-W fluorescent lamps.

At the several growth stages after the cell culture started, 5 ml of the culture medium was subjected to centrifugation at 10,000 g for 15 min to collect cells. The collected cell pellets were washed twice with basal salts solution, suspended in distilled water, then left overnight after adding ~300 units of freshly prepared DNase solution (Sigma, Type II). After DNase treatment, the cell lysate was centrifuged at 100,000 g for 60 min to recover membrane fraction as pellet. The membrane pellet was solubilized in 0.2 ml of 10% SDS with or without 1% mercaptoethanol at 60° for 20 min, then diluted with 0.4 ml of elution buffer. Before applying to the column, the sample solution was centrifuged at 3000 rpm using a desk-top centrifuge for 10 min to remove insoluble contaminants.

Gel Permeation HPLC by Using a G-3000 or G-4000 SW Column

A prepacked G-3000 or G-4000 SW column (Toyo Soda Co., Tokyo) equipped to a model 803A HPLC (Toyo Soda Co., Tokyo) was washed by letting 0.1% SDS in 0.2 M Tris-hydrochloride buffer (pH 8.0) flow through the column for 1 hr at a speed of 0.8 ml/min. (See Conclusion.) Buffer solutions was passed through a 0.4 μm Millipore filter and degassed *in vacuo* before use. Eluents from the column were recorded at 220 or 280 nm by UV monitor model UV 8 (Toyo Soda Co., Tokyo). Monitoring at 220 nm gave rise to much higher sensitivity than at 280 nm, but the protein specificity was lower in contrast to 280-nm monitoring. After the baseline stabilized, 2–10 μl of the solubilized membrane preparation (~0.01 A at 280 nm) was charged from the bottom of the column and eluted at a speed of 0.8 ml/min at room temperature. UV absorption was recorded on a strip chart recorder. Flow speed between 0.6 and 1.0 ml/min did not change the resolution.

Blue dextran solution was used to determine the void volume(V_0) of the column, and the total available volume(V_t) was estimated from the elution peak of mercaptoethanol or distilled water. The typical elution pattern of the membrane proteins of the cell at the stationary growth phase is shown in Fig. 1, where six major proteins were resolved. When the membrane was solubilized in the presence of mercaptoethanol, both the number and the elution volume of peaks were not essentially changed; however, the relative amount of each peak was changed slightly.

Molecular weight marker proteins were applied independently or as a mixture under the same elution condition to calibrate the molecular weight of proteins. Linear correlation between molecular weight and the elution volume was obtained in the range of 100,000–10,000 for a G-3000

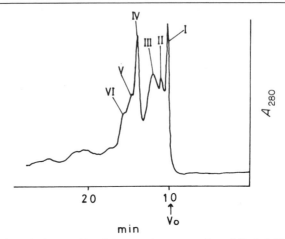

FIG. 1. Typical elution profile of the membrane proteins of *H. halobium* S9 from G-3000 SW column. Column: TSK G-3000 SW (7.5–600 mm), elution speed: 1.0 ml/min, eluents: 0.1% SDS in 0.2 *M* Tris–hydrochloride(pH 8.0), Record: at 280 nm.

SW column as shown in Fig. 2. Peak IV was determined as bacteriorhodopsin, strictly to say, bacterio-opsin, since the elution volume was identical to that of purified bacteriorhodopsin. However, the molecular weight indicated by the calibration curve in Fig. 2 was ~22,000, which was rather smaller than that previously reported.[4] Since the same molecular

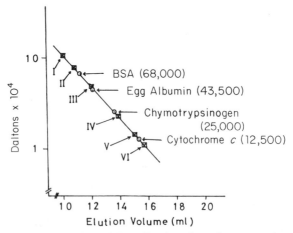

FIG. 2. Molecular-weight calibration of membrane proteins.

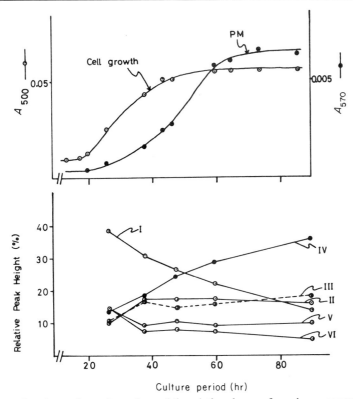

FIG. 3. Purple membrane formation and the relative change of membrane components in the course of cell growth.

weight range was occasionally obtained for bacteriorhodopsin by SDS PAGE, this may be due to the charge effect of this protein.

The change in the relative amount of membrane proteins in the course of purple membrane formation is summarized in Fig. 3, which shows the cell growth curve and purple membrane formation profile. It was clearly shown that bacterio-opsin formation coincided with the decrease of higher-molecular-weight protein peak, peak I. The elution volume of this peak was very close to V_0 in the case of the G-3000 SW column, so that the molecular weight was hardly obtained from the calibration curve. By using the G-4000 SW column instead, higher resolution in the higher-molecular-weight range was attained, as shown in Fig. 4, where peak I appeared in G-3000 SW was further resolved into two peaks, I and I'. The molecular weight of peak I derived from the calibration curve for the G-4000 SW column was ~220,000.

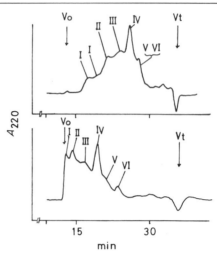

FIG. 4. Comparison of elution profile of membrane proteins from G-3000 SW and G-4000 SW columns. Record: at 220 nm. Elution speed: 0.8 ml/min.

Conclusion

By using a gel permeation column for HPLC, good resolution of membrane proteins of the plasma membrane of *Halobacterium halobium* was attained within half an hour and the method proved to be useful for studying the biosynthetic process of such membrane proteins as bacteriorhodopsin. Both G-3000 and G-4000 SW columns gave nice correlation between the molecular weight and the elution volume for molecular weight standard proteins under the condition used, but the molecular weights of membrane conponents derived from the calibration curve for G-4000 SW were slightly different from those obtained from the G-3000 SW column. The reason for this is unclear at present.

When a phosphate buffer system was used instead of Tris, V_t and the resolution were highly dependent on the salt concentration of the buffer.

Since a G-4000 SW column could cover the molecular-weight range of up to $\sim$500,000 in comparison with 100,000 of a G-3000 SW column, G-4000 SW seems to be useful for analyzing the initial stage of membrane biogenesis, where higher-molecular-weight precursor occasionally appears.

Although originally we used the elution solution described above, it was found that the column lifetime was shorter at higher pH above 8. So it is recommended to use lower pH than 7 for the elution buffer.

Using 0.2 M Tris–HCl (pH 7.0) containing 0.1 M NaCl and 0.1% SDS as an elution buffer, about 30 membrane components could be separated at a elution speed of 0.5 ml/min[5]. When the same sample was applied to SDS–PAGE, essentially the same number of components was identified. However, the present HPLC method was proved to be superior to SDS–PAGE with respect to the quantitative recovery of each peak, essentially of the low-molecular-weight components.

[5] Konishi, T. and Seki, M., paper presented at the "54th Japan Biochemistry Meeting (Sendai, 1981). Manuscript in preparation.

[25] Heterobifunctional Cross-Linking of Bacteriorhodopsin by Hydrophobic Azidoarylisothiocyanates

By Hans Sigrist and Peter Zahler

In the context of hydrophobic modification of membrane proteins, the described method centers on the nearest neighbor analysis of hydrophobically modified intrinsic membrane proteins through the use of site-specific heterobifunctional cross-linkers. Arylisothiocyanates (phenylisothiocyanate, naphthylisothiocyanate) are known to form covalent bonds with nucleophilic groups only in their nonprotonated form ($RS^- \gg RO^- > RNH_2$).[1] For modification of membrane proteins the pH-controlled reactivity of proteinaceous nucleophilic groups can therefore be used for selective labeling by arylisothiocyanates.[2] At neutral pH the ϵ-amino group of lysine residues exposed to the aqueous phase are protonated and accordingly not reactive with arylisothiocyanates. In contrast, the buried, bulk pH-independent amino functions may be in a reactive state (deprotonated), making modification with hydrophobic arylisothiocyanates feasible. Cysteine thiols, histidine imidazole NH, and α-amino groups exposed to the aqueous phase are partially deprotonated at neutral pH and are resultingly reactive with arylisothiocyanates. With regard to application to bacteriorhodopsin, it is known that the N-terminal amino acid is blocked (pyroglutamic acid) and that both cysteine and histidine residues are not present. Hydrophobically located lysine ϵ-amino groups are therefore favored for arylisothiocyanate modification.

The presence of a nonspecifically reactive heterofunction (azido) in

[1] L. Drobnika, P. Kristian, and J. Augustin, in "The Chemistry of Cyanates and Their Thio Derivatives" (S. Patai, ed.), Part 2, p. 1002. Wiley, New York, 1977.

[2] H. Sigrist and P. Zahler, in "Membranes and Transport: A critical review" (A. Martonosi, ed.), p. 173 Plenum Press, N.Y. in press.

[3] H. Sigrist and P. Zahler, FEBS Lett. 113, 307 (1980).

the arylisothiocyanate renders the probe a heterobifunctional cross-linker. The advantages of hydrophobic arylisothiocyanates for group-specific modification and the nonselective reactivity of photogenerated aryl-nitrenes[4] are thus combined. In the first step of cross-link formation the azidoarylisothiocyanates (*p*-azidophenylisothiocyanate, 5-isothiocyanato-1-naphthaleneazide) binds covalently to hydrophobically located nucleophiles. Light activation of the heterofunction (arylazide) then leads to the nonselective insertion into neighboring membrane constituents[6] (Fig. 1).

For bacteriorhodopsin the site of aryl(phenyl)isothiocyanate interaction has been identified as a hydrophobically located lysine ε-amino

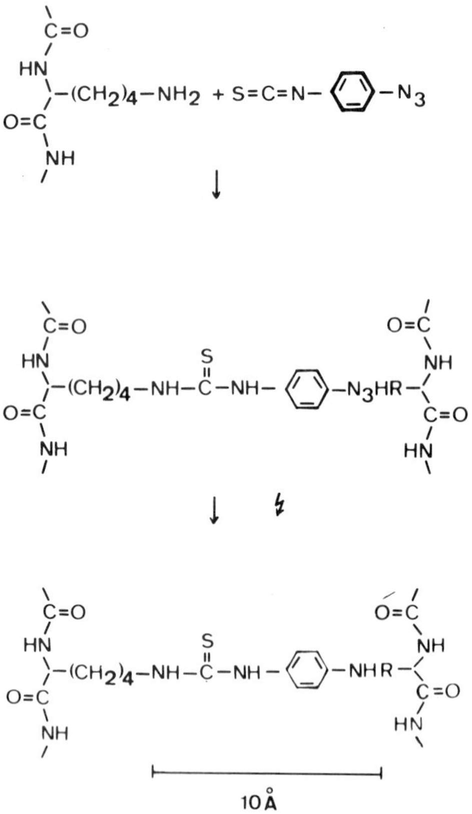

FIG. 1. Schematic representation of heterobifunctional cross-linking of membrane proteins by *p*-azidophenylisothiocyanate.

[4] W. G. Hanstein, this series, Vol. 56, p. 653.

group.[3,7] Both phenylisothiocyanate and naphthylisothiocyanate compete with phenyl[^{14}C]isothiocyanate for binding to this nucleophilic group.[3,8] The methodology reported here will focus on the description of intermolecular cross-linking of bacteriorhodopsin in purple membranes utilizing p-azidophenylisothiocyanate and 5-isothiocyanato-1-naphthaleneazide. Through the use of these probes cross-links can be formed that extend over the range of 10–11 Å. The equally interesting aspects of intramolecular and protein-lipid cross-linking will not be treated in this study.

Heterobifunctional Cross-Linking Reagents

The reagents p-azidophenylisothiocyanate and 5-isothiocyanato-1-naphthaleneazide were prepared as described by Sigrist et al.[8] Chemical characterization and purity of the probes were ascertained by thin-layer chromatography, melting point, ir and nmr spectroscopy, and elementary composition. The butylamine derivative of p-azidophenylisothiocyanate, N-butyl-N'-(p-azidophenyl)thiourea mimics the interaction product of the cross-linker with the side chain of lysine. The derivative, dissolved in ethanol, shows maximal absorption at 283 nm (molar extinction coefficient, 1 cm, ϵ = 16,000) with a significant shoulder at 320 nm. The butylamine derivative of 5-isothiocyanato-1-naphthaleneazide absorbs maximally at 308 nm (molar extinction coefficient, 1 cm, ϵ = 13,200). Both derivatives are photosensitive.[8]

Modification of Purple Membranes by Azidoarylisothiocyanates

Reagents and Solutions

Purple membranes[9] (0.2 mM bacteriorhodopsin[10]) suspended in 25 mM sodium phosphate buffer, pH 7.0
p-Azidophenylisothiocyanate in ethanol, 0.25 M
5-Isothiocyanato-1-naphthaleneazide
Phenyl[^{14}C]isothiocyanate, Amersham England, 7–12 Ci/mol, 10 mCi/ml ethanol

[5] H. Sigrist, P. R. Allegrini, R. J. Strasser, and P. Zahler, in "The Blue Light Syndrome" (H. Senger, ed.) p. 30 Springer Verlag, Berlin, 1980.
[6] C. Gitler and T. Bercovici, *Am. N.Y. Acad. Sci.* **346**, 199 (1980).
[7] H. Sigrist, P. R. Allegrini, N. Abdulaev, J. Schaller and P. Zahler, in preparation.
[8] H. Sigrist, C. Kempf, R. R. Allegrini, C. Schnippering, and P. Zahler, *Eur. J. Biochem.*, in press.
[9] D. Oesterhelt and W. Stoeckenius, this series, Vol. 31, p. 667.
[10] D. Oesterhelt and B. Hess, *Eur. J. Biochem.* **37**, 316 (1973).

The azidoarylisothiocyanate modification of purple membranes is performed under subdued light. The desired amount of cross-linker is added to a glass reaction vial, e.g., with respect to bacteriorhodopsin a 100(200)-fold molar excess of p-azidophenylisothiocyanate in ethanol (5-isothiocyanato-1-naphthaleneazide suspended in ethanol). The final concentration of ethanol in either incubation mixture should not exceed 8% (v/v). With other membrane systems lower amounts of ethanol may be advisable. On addition of purple membranes to the reagents, precipitation of the probes may occur because of their low solubility in aqueous media. It is therefore necessary to perform the modification under effective stirring at 37° for a minimum of 16 hr. Excess label is then partially removed by centrifugation (5 min, 1200 g) followed by exhaustive dialysis (72 hr) against double-distilled water, which is frequently changed. The modified sample is daily transferred into a new dialysis bag. Optionally, 10 mM cysteine, pH 7.5, may be present in the first changes of the dialysis medium. The removal of excess hydrophobic reagent is thus enhanced by reaction with the polar thiolate. Control samples, without the cross-linker present during the incubation, are identically treated. After dialysis the modified membranes are ready for photoactivation.

If a radioactive identification of cross-linked products is desired, the azidoarylisothiocyanate-labeled purple membranes are additionally modified by phenyl[^{14}C]isothiocyanate before photolysis. This second incubation (90 min, 37°) is performed as described earlier utilizing pheny[^{14}C]isothiocyanate (50–100 μCi/mg modified membrane protein) as monovalent radioactive marker. After dialysis (24 hr) against distilled water the double-labeled membranes are photolyzed.

Photolysis

Photolysis is carried out with consecutive flashes from an electronic flash unit (Optatron 350, Loewe Opta, West Germany) after removing the plastic window. As described by Kiehm and Ji,[11] the irradiation time of a single flash extends over 0.2 msec. The modified membranes are placed in a 1-cm quartz cuvette that is positioned 3 cm from the light source. The samples are thus exposed with the desired number of flashes. Photolysis can be performed as well with a mercury lamp (HBO 200 W, Osram). With this latter light source a 305-nm cutoff filter (Corning 7-60) was utilized.

[11] D. J. Kiehm and T. H. Ji, *J. Biol. Chem.* **252**, 8524 (1977).

Analysis of Intermolecular Cross-Linking

The cross-linked membranes are transferred to a glass centrifuge tube and combined with 5 vol ice-cold acetone/ammonia 5:1 (v/v). The precipitating protein is kept in ice for 5 min and sedimented at 4° (10 min, 3000 g). The sediment is resuspended in ice-cold acetone/ammonia by sonication (30 sec) using a bathtype sonicator (Laboratory Supplies Company; Hicksville, New York). The sedimentation procedure is repeated twice. The final pellet is solubilized in 2% SDS and analyzed for protein content,[12] radioactivity, and polymer formation by SDS gel electrophoresis[13] (10% acrylamide). Before application to the gels the solubilized samples are frozen and thawed once and then subjected to 2.5% mercaptoethanol for 10 min at 100°. This treatment is necessary for complete dissociation of non-cross-linked protein. After gel electrophoresis, Coomassie blue staining and destaining of the gels, the relative amount of polymers formed is calculated by comparing the areas of the densitometric traces of stained gels. Additionally, Coomassie blue-stained gels are cut into 1-mm slices and the individual slices are dissolved in 0.2 ml H_2O_2/ammonia (100:1, v/v) by overnight incubation at 80°. Scintillation fluid is then added and the relative amount of radioactivity present in the mono- and oligomers is determined. The results obtained for p-azidophenylisothiocyanate and 5-isothiocyanato-1-naphthaleneazide cross-linked purple membranes are presented in the accompanying table.

Discussion

The occupation of one hydrophobically located binding equivalent in bacteriorhodopsin has been reported for phenylisothiocyanate, a probe that interacts under labeling conditions as an emulsion with the purple membrane. On incubation (90 min, 37°) with a 1000-fold molar excess reagent, 0.85 mol phenylisothiocyanate is covalently bound to bacteriorhodopsin.[5] Similar to the monofunctional phenylisothiocyanate, the azidoarylisothiocyanates are sparingly soluble in aqueous media. While phenylisothiocyanate is present as an emulsion in the incubation medium, the bifunctional reagents are present as suspensions. Differences in the kinetics of the labeling process therefore reflect the different chemical parameters of the reagents. The extended incubation time (16 hr), combined with photolysis of the heterofunction results in protein polymerization.

[12] O. H. Lowry, N. S. Rosenbrough, A. L. Farr, and R. J. Randall, *J. Biol. Chem.* **193**, 265 (1951).
[13] G. Fairbanks, T. L. Steck, and D. H. F. Wallach, *Biochemistry* **10**, 2606 (1971).

TABLE I
HETEROBIFUNCTIONAL CROSS-LINKING OF BACTERIORHODOPSIN[a]

Cross-linking reagent	Monomer (%)	Dimer (%)	Trimer (%)	Polymeric aggregates (%)
N₃—⬡—NCS *p*-Azidophenylisothiocyanate	77 (61)	13 (13)	4 (5)	6 (19)
(naphthalene structure) 5-Isothiocyanato-1-naphthaleneazide	77 (66)	9 (15)	5 (8)	8 (10)
Control	91 (89)	2 (3)	0 (0)	7 (7)

[a] Purple membranes were modified with a 100-fold molar excess of *p*-azidophenylisothiocyanate and a 200-fold molar excess of 5-isothiocyanato-1-naphthaleneazide, respectively. The modified membranes including the control sample were photolyzed with 10 consecutive flashes of light (total exposure time 2 msec). The relative amount of recovered oligomers was determined by densitometric analysis of the Coomassie blue-stained gels. The percentage of radioactivity recovered in the electrophoretically separated protein bands is given in parentheses. The data should be considered in a qualitative rather than quantitative manner owing to possible differences in Coomassie blue staining efficiency.

Covalent reaction of both the arylisothiocyanate and corresponding arylnitrene into adjacent proteins is therewith demonstrated. As both azidoarylisothiocyanates compete for the hydrophobically located phenylisothiocyanate binding site,[3,8] it can be concluded that azidoarylisothiocyanates react covalently with the nucleophilic lysine ε-amino group, localized within the apolar domain of bacteriorhodopsin.[7,14] No selectivity is implied, however, for the nitrene insertion.

The presented method may illustrate the potential for the use of group-specific heterobifunctional cross-linking reagents in the study of membrane topography with regard to hydrophobic intermolecular relationships. Further structural information will be available upon applying the heterobifunctional reagents for intramolecular cross-linking of membrane proteins.

[14] Yu. A. Ovchinnikov, N. Abdulaev, M. Feigina, A. Kiselev, and N. Lobanov, *FEBS Lett.* **100**, 219 (1979).

Acknowledgments

Thanks are due to Dr. P. R. Alegrini, Dr. C. Kempf, and Mrs. Ch. Schnippering for collaboration in various stages of the work. The study was supported by the Swiss National Science Foundation (Grant 3.674-0.80) and by the Central Laboratories of the Swiss Blood Transfusion Service SRK, Bern, Switzerland.

[26] Measurement of Light-Regulated Phosphoproteins of *Halobacterium halobium*

By Elena Negri Spudich and John Lee Spudich

Halobacterium halobium cells contain a class of phosphoproteins that can be detected after *in vivo* labeling with [32 P]orthophosphate.[1] The extent of labeling of three of the phosphoproteins is regulated by light. Illumination of whole cells induces phosphate release from the three light-regulated ("LR") phosphoproteins. This effect is reversible; darkening the cells induces reincorporation of 32P label into LR proteins. The methods described below provide a quantitative assessment of the relative phosphate content of the LR phosphoproteins in cells exposed to various environmental conditions. Phosphate incorporated into proteins by either phosphorylation or nucleotidylylation reactions will be revealed by the methods and conditions described.

In addition to the LR phosphoproteins, the methods described reveal phosphate comigrating with bacterio-opsin during electrophoresis on SDS-polyacrylamide gels. This material copurifies with bacteriorhodopsin in the standard purple membrane isolation procedure described elsewhere in this series.[2] With the procedures described here that were developed for the LR phosphoproteins, the 32P incorporated into bacterio-opsin is relatively slight and will not necessarily be observed without isolation of the purple membrane.

Cell Culture and Preparation. To remove contaminating phosphate, all glassware used for medium and basal salt buffer preparation, cell culture, and cell preparation is soaked overnight in a chromic–sulfuric acid solution (Manostat Chromerge, American Scientific Products) and rinsed thoroughly in distilled water. Peptone medium is prepared as described by Lanyi and MacDonald elsewhere in this series.[3] After the medium is steri-

---

[1] J. L. Spudich and W. Stoeckenius, *J. Biol. Chem.* **255,** 5501 (1980).
[2] D. Oesterhelt and W. Stoeckenius, this series, Vol. 31, p. 667.
[3] J. K. Lanyi and R. E. MacDonald, this series, Vol. 56, Part G, p. 398.

METHODS IN ENZYMOLOGY, VOL. 88

lized by autoclaving and trace metal salts are added (as described in Lanyi and MacDonald[3]), we filter the complete medium through a sterile 0.45-μm-Nalgene filter unit.

A liquid stock culture is prepared by inoculating 7 ml medium in a glass culture tube (18 × 150 mm, Bellco) with a loop of cells from a peptone medium/1% Bacto-Agar (Difco) culture plate. This culture is incubated with vigorous shaking at 37° on a New Brunswick Model G-2 Gyrotory Shaker set at 240 rpm. Illumination during growth is from three cool-white fluorescent lamps mounted above the shaker that provide 5 × 10^3 ergs cm^{-2} sec^{-1} at the culture tube surface. After 7 days incubation this culture is refrigerated and used for 2 weeks as a liquid stock for inoculation of cultures for experimental use.

An experimental culture is prepared by diluting 0.05 ml from the refrigerated stock culture into 7 ml fresh peptone medium in a culture tube. These cells are used for labeling after 5 days incubation in the conditions described for the liquid stock culture, at which time the cell density is approximately 2.8 × 10^9 cells/ml. To prepare cells for labeling, 1 ml of the experimental culture is pelleted for 1 min in an Eppendorf Microfuge and resuspended in 1 ml of basal salt (peptone medium without peptone or added trace metal salts), containing 10 mM HEPES, pH 7.4, prewarmed to 37°. After a second such wash, the cells are resuspended to a final density of 10^9 cells in the basal salt–HEPES buffer in a Chromerge-washed tube.

^{32}P *In Vivo Labeling.* Immediately after preparation, 1 ml of the cell suspension in basal salt–HEPES buffer is added to a 10-mm light path rectangular disposable plastic cuvette (Evergreen Scientific), containing a cuvette magnetic stirrer (Fischer Scientific), and 25 μl of 0.02 N HCl containing 50 μCi of carrier-free [^{32}P]orthophosphate (prepared from New England Nuclear 5 mCi/ml in 0.02 N HCl). The suspension is incubated at 37° in the dark with vigorous stirring. After 15 min the cells are sufficiently radiolabeled for detection of the dark-adapted LR phosphoprotein banding pattern on autoradiograms prepared as described later. Considerably heavier labeling and a more detailed pattern are observed after 1 hr incubation of the cells. Cells incubated with radioactive phosphate for 30 min were used to prepare the data shown in Fig. 1.

Whole Cell Illumination. Light-induced phosphate release from the LR proteins can be observed after illuminating the cell suspension with orange or other bacteriorhodopsin-absorbing light, while maintaining the vigorous stirring with the cuvette in a thermostatable cuvette holder (e.g., the water-jacketed 10-mm cell holders sold by Varian Instruments for the SuperScan and Cary spectrophotometers are suitable). Illumination in the figure was with light from a 300-W tungsten halogen lamp (General Electric ENG) collimated with appropriate lenses to a 36-mm-diameter beam

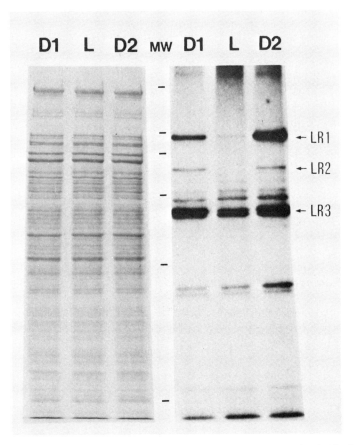

FIG. 1. *H. halobium* strain $R_1S_2^1$ was used in the procedure described. The leftmost three lanes show the Coomassie blue-stained cell extracts for cells dark-adapted (D1), illuminated for 5 min (L), then darkened for 10 min (D2). The rightmost three lanes show the autoradiogram of the same three gel lanes. Molecular-weight standards (MW) are from the top: myosin (200,000), β-galactosidase (116,000), phosphorylase a (94,000), bovine serum albumin (68,000), ovalbumin (43,000), and bacterioopsin (26,000).

filtered through 30 mm of water contained in a Bellco glass tissue culture flask, 5 mm of heat-absorbing glass (Edmund Scientific), a dichroic heat-reflecting mirror (Corion Corporation), and a Corning 3-69 long pass filter and finally focused on the cuvette at 2.0×10^6 ergs cm^{-2} sec^{-1}.

Protein Extraction and Electrophoresis. A 200-μl sample from the labeled cell suspension is transferred from the cuvette to a 15-ml Corex centrifuge tube containing 2 ml of ice-cold acetone. This procedure lyses the cells and precipitates cell proteins as well as the salts contained in the basal salt–HEPES buffer. The precipitated material is pelleted by centrifugation at 6000 g for 10 min at 4°. The supernatant is discarded into a radioactive waste bottle. To dissolve the salt in the pelleted acetone precipitate, 2 ml of ice-cold 50% acetone is added to the pellet. After the pellet is resuspended by vortexing for several minutes, the tube is recentrifuged. The supernatant is discarded and the pellet air dried for approximately 10 min to remove residual acetone. One hundred twenty microliters of the sodium lauryl sulfate (SDS) sample buffer described by Laemmli[4] is added to the dried tube, which is capped with aluminum foil, briefly vortexed, and placed in a boiling water bath for 3 min. Condensation droplets on the walls of the tube are forced to the bottom by a 30-sec centrifugation at 3000 g. This last step is necessary for reproducible protein concentrations.

Fifty microliters of the solubilized sample is loaded on a 10% SDS polyacrylamide gel prepared according to the method of Ames.[5] Electrophoresis is at 40 mA for 1 hr followed by 2 hr at 45 mA. The proteins are fixed, stained with Coomassie blue, and dried onto filter paper, as described by Ames.[5]

Radioautography and Scintillation Counting. The dried gel is processed by either (or both) of two procedures: radioautography, which gives a highly resolved labeling pattern, and scintillation counting of gel slices, which provides a quantitative assessment of incorporated label in specific regions of the gel.

An autoradiographic exposure of the gel to XRP-1 Kodak X-ray film at room temperature overnight (12–16 hr) is sufficient for a clear LR protein labeling pattern. The exposure time can be shortened to 2 hr by using a Dupont Cronex Quanta III intensifying screen and radioautography at −80°.

A gel slice to be examined by scintillation counting is placed in a glass vial and 0.5 ml of a 30% solution of H_2O_2, freshly prepared from refrigerated 50% H_2O_2 (Fisher H341), is added. After incubation of the tightly capped vial overnight at 60°, the vial is chilled for approximately 20 min at 4°. Ten milliliters of aqueous counting scintillant (ACS II, Amersham) is added and the vial is vigorously shaken. The vial may be counted immediately, but it is preferable to chill the vial in the dark at 4° approximately 12 h to reduce chemiluminescence.

Acknowledgments

This work was supported by Research Grant GM 27750 from the National Institute of General Medical Sciences.

[4] U. K. Laemmli, *Nature (London)* **227,** 680 (1970).
[5] G. F. Ames, *J. Biol. Chem.* **249,** 634 (1974).

[27] Emission Spectroscopy of Rhodopsin and Bacteriorhodopsin

By AARON LEWIS and G. J. PERREAULT

I. Introduction

The detection of emission from the excited state of retinylidene chromophores complexed to membrane proteins is an area which has only recently begun to yield reliable and important information. The retinylidene chromophores studied are found complexed to either rhodopsin, the primary light-absorbing entity in visual photoreceptors, or bacteriorhodopsin, the membrane-bound proton pump found in the bacterium, *Halobacterium halobium*. In both these systems the excited state of the retinylidene chromophore is involved both in the light absorption and in the initial molecular events that stimulate the cellular functions of these membrane proteins. Therefore, emission from this chromophore is an important and critical probe of the nature of the excited state in which light energy is converted into chemical energy and the chemical steps that lead to the transduction process.

There are two reasons why emission has been difficult to measure from these systems. First, the very efficient photochemistry rapidly (within <6 psec) depopulates the excited state causing the emission to be very weak.[1] Second, since the emission quantum yield is so low, even small amounts of an impurity could cause significant problems in identifying the origin of the emission. This latter problem plagued early workers in the field[2,3] and it is clear that the early reports[2,3] of bovine rhodopsin fluorescence probably arose from some other chemical species. This conclusion is based on the observation that *purified* rhodopsin, which behaves identically in terms of kinetics and other parameters to native rhodopsin, gave no evidence of the reported fluorescence.[1]

The first report of retinylidene chromophore emission, which clearly arose from the chromophore, rather than some extraneous impurity, was observed from the purple membrane by Lewis and co-workers.[4] This emission has been studied in detail in several subsequent investiga-

---

[1] G. E. Busch, M. L. Applebury, A. A. Lamola, and P. M. Rentzepis, *Proc. Natl. Acad. Sci. U.S.A.* **69**, 2802 (1972).

[2] A. V. Guzzo and G. L. Pool, *Science* **159**, 312 (1968).

[3] A. V. Guzzo and G. L. Pool, *Photochem. Photobiol.* **9**, 565 (1969).

[4] A. Lewis, J. P. Spoonhower, and G. J. Perreault, *Nature (London)* **260**, 675 (1976).

METHODS IN ENZYMOLOGY, VOL. 88

tions.[5-13] These studies have been aimed at critically assigning the emitting species, calculating the quantum yield of emission, elucidating the nature of the excited state responsible for the emission, studying the temperature dependence of the emission and interpreting all of these data to help determine the nature of the primary photochemistry in bacteriorhodopsin. More recently, convincing reports have begun to appear of fluorescence from photochemically generated invertebrate visual pigment intermediates[14,15] and there has even been a picosecond study of rhodopsin and isorhodopsin fluorescence.[16]

II. Bacteriorhodopsin

A. Low-Temperature Emission

(i) IS BACTERIORHODOPSIN (BR_{570}) RESPONSIBLE FOR THE EMISSION? Several experiments indicate that bacteriorhodopsin (BR_{570}) is the emitting species. In this section we will first describe these experiments and then attempt to integrate the results of Kriebel, Gilbro, and Wild[8-10] which have been interpreted[8-10] as indicating that a species distinct from BR_{570} is wholly or partially responsible for the emission.

The initial experiments suggesting that a single species was responsible for the emission were the low-temperature results of Lewis et al.[4] These experiments were performed at 77°K where only BR_{570} and the K species are thought to be present according to absorption measurements by several groups. The fluorescence at these temperatures in the original Lewis et al.[4] experiments exhibited three partially resolved peaks at approximately 680, 730, and 790 nm. The emission polarization was invariant throughout the entire spectrum which suggested that the emission

[5] R. R. Alfano, W. Yu, R. Govindjee, B. Becher, and T. G. Ebrey, *Biophys. J.* **16**, 541 (1976).

[6] R. Govindjee, B. Becher, and T. G. Ebrey, *Biophys. J.* **22**, 67 (1978).

[7] V. A. Sineshchekov and F. F. Litvin, *Biochim. Biophys. Acta* **462**, 450 (1977).

[8] T. Gillbro, A. N. Kriebel, and U. P. Wild, *FEBS Lett.* **78**, 57 (1977).

[9] T. Gillbro and A. N. Kriebel, *FEBS Lett.* **79**, 29 (1977).

[10] T. Gillbro, *Biochim. Biophys. Acta* **504**, 175 (1978).

[11] M. D. Hirsch, M. A. Marcus, A. Lewis, H. Mahr, and N. Frigo, *Biophys. J.* **16**, 1399 (1976).

[12] S. L. Shapiro, A. J. Campillo, A. Lewis, G. J. Perreault, J. P. Spoonhower, R. K. Clayton, and W. Stoeckenius, *Biophys. J.* **23**, 383 (1978).

[13] A. Lewis, *Biophys. J.* **25**, 79a (1979).

[14] N. Franceschini, K. Kirschfeld, and B. Minke, *Science* **213**, 1264 (1981).

[15] T. W. Cronin and T. H. Goldsmith, *Biophys. J.* **35**, 653 (1981).

[16] A. G. Doukas, P. Y. Lu, and R. R. Alfano, *Biophys. J.* **35**, 547 (1981).

arose from a single excited state of a single species. Picosecond experiments at 77°K also support this conclusion: the emission lifetime is 60 ± 15 psec at both 670 and 735 nm independent of whether single or multiple pulse excitation is used. In the single pulse experiment the excitation photon flux was sufficiently low that each molecule had only a probability of 0.008 of absorbing a photon, whereas in the multiple pulse experiments the probability was 0.8 by the end of the pulse train. The identity of the results in the single and multiple pulse experiments argues against the necessity of suggesting that a photoproduct is responsible for the emission.

The steady-state 77°K emission results of Govindjee et al.[6] also arrive at the conclusion that BR_{570} is the emitting species. These researchers[6] followed the time course of emission intensity at 680 nm excited at 590 nm after preillumination with 640 nm light. The intensity decreased from its initial value to a steady-state value of about 17% less. Since the emission intensity excited near the isosbestic point of BR_{570} and the primary photoproduct K is maximum when the concentration of BR_{570} is maximum, this result indicated that BR_{570} emits more strongly than K.[6] In another experiment,[6] the percentage decrease in emission as a function of wavelength of excitation was shown to follow the percentage of K produced at steady state by the excitation beam. The approximate equality further indicates that BR_{570} has an emission quantum yield much higher than that of K. Sineschekov and Litvin[7] have also measured the effects of manipulating the concentration of K at 77°K and arrive at the conclusion that BR_{570} is the principal emitting species. Although, they also report a "small" ($\sim 15\%$) *decrease* in the intensity of emission at 670 nm relative to 720 nm after extended preillumination.

(ii) Is there a pseudobacteriorhodopsin-emitting species? Gillbro et al.[8] have also reported preillumination experiments. However, they report a large ($\sim$ three-fold) *increase* in the intensity of the 670-nm band after preillumination with 514.5 nm laser light. They interpret their results in terms of a distinct new photochemical species which they call pseudobacteriorhodopsin. (P-BR) In support of this suggestion they report an excitation spectrum with a maximum at 597 nm which is significantly redshifted from the BR_{570} absorption maximum at 570 nm. However, it can be shown that the excitation spectrum of a BR/K mixture in photoequilibrium will be red-shifted because red light increases the percentage of BR_{570} in the sample which compensates for the decreasing absorption in the red. Thus, the position of the maximum will be at a point intermediate between the absorption maximum of BR and K (i.e., close to the isosbestic point). The predicted maximum for the excitation spectrum is 585 nm and 573 nm respectively depending on whether the absorption spectra of Ref.

17 or 18 are used. Therefore, the red-shift in the excitation spectrum does not necessarily prove that there is an emitting species distinct from BR_{570}.

Another important aspect of the experiments reported by Govindjee et al.,[6] Sineshchekov and Litvin,[7] and Gillbro et al.[8] is the invariance of the emission maxima as a function of either preillumination or excitation frequencies within the bulk of the absorption band of bacteriorhodopsin. Therefore, if emission is indeed being observed from a pseudobacteriorhodopsin (P-BR) species, then P-BR must have emission maxima and lifetime identical to those of BR and, based on our analysis of the excitation maximum, probably would have an absorption maximum identical to that of BR. This latter suggestion is in fact admitted in a discussion by Gillbro.[10] Thus, the only essential difference between the postulated P-BR species and BR is the relative intensity of emission.

(iii) SURVEY OF EMISSION DATA FROM VARIOUS GROUPS. Emission data for BR at low temperatures are summarized in Table I. All reported emission spectra have their most intense peak between 710 and 735 nm. The most extensive studies yet reported of the effect of environment on the emission maxima demonstrate that the most intense peak can vary in position over a range of ~ 10 nm (720–730 nm) depending on the method of preparation. Such an environmentally induced shift, together with small errors in correction of spectra for instrument response, probably accounts for most of the variation[4,7,8] in this peak position with the exception of the value of 710 nm reported by Govindjee et al.[6] Possible explanations for this outlying value could be the low resolution (10-nm interval) of the reported data and/or errors in the instrument response correction. Some investigators also report a band at ~ 790 nm[4,12] and Sineschekov and Litvin[7] have observed that this peak can be made to disappear or shift in frequency as a result of environmental factors. This peak is only apparent at ~ 790 nm, the value reported by Lewis et al.,[4] in a film at 77°K. The results are thus consistent since the samples used by Lewis et al.[4] were films on the tip of a glass rod immersed in liquid nitrogen. The 740-nm lowest energy peak of Govindjee et al.[6] has not been observed by any other workers.

The highest energy emission peak appears between 665–680 nm according to all workers.[4,6–8,12] Even though this variation is relatively large our data indicate that since this peak is a shoulder, the apparent maximum varies between ~ 665 and 680 nm depending on temperature (at 20°K the shoulder is well resolved and it occurs at 670 nm whereas at 100°K it is poorly resolved and appears at longer wavelength).[12] Preillumination experiments by Gillbro et al.[8] increased the intensity of this shoulder and

[17] B. Becher, F. Tokunaga, and T. G. Ebrey, Biochemistry 17, 2293 (1978).
[18] R. H. Lozier and W. Niederberger, Fed. Proc. Fed. Am. Soc. Exp. Biol. 36, 1805 (1977).

shifted it to shorter wavelengths. Thus, it is our considered opinion that the accurate position of this shoulder is 670 nm and lower temperatures, preillumination, and even possibly sample preparation cause a better definition of the shoulder and therefore a more accurate reading of the maximum. This above deduction is important because it suggests that before and after preillumination the species has a peak in its emission spectrum at 670 nm. This is also consistent with lifetime measurements discussed above which indicate that the species responsible for the 670-nm shoulder is also responsible for the principal emission between 720 and 730 nm.

(iv) SUMMARY A review of all the data certainly suggests that the initial species, BR_{570}, does indeed emit light. The steady-state preillumination experiments of Gillbro et al.[8-10] have been interpreted as suggesting that a new species, with the same band structure as the unilluminated molecules but with a more intense low-energy emission at 670 nm, is responsible for the emission. However, the Gillbro et al. conclusion[8-10] is not the only explanation for the effect of preillumination in creating this increase in 670-nm emission intensity. Alternate causes for this effect which agree better with the data discussed above are (1) laser annealing of the sample altering the nature of the sample preparation which has been shown to cause intensity alterations in the emission[7] or (2) conversion of a nonfluorescent or a species with weaker emission to a fluorescent species. (For example, suppose dark-adapted BR_{560} is a weaker emitter which is converted with laser light to the fluorescent BR_{570}.)

B. The Room Temperature Emission There is large variation in the room temperature emission maximum (Table II). Lewis et al.[4] reported it at 791 nm. However, Govindjee et al.[6] detect it at 700 nm. Sineshchekov and Litvin[7] report that the emission maximum at pH 5.7 is 660 nm. They also report a pH dependence: 645 nm at pH 10.2 and 720 nm at pH 2.7. Kriebel et al.[21] determined that a low (50 $\mu W/cm^2$) versus high (3 W/cm^2) incident intensity stationary sample showed a movement of the emission maximum from 714 to 735 nm. In fact, early data of Spoonhower,[22] obtained using xenon arc excitation, show a room temperature emission maximum at ~ 730 nm and dye laser excited emission spectrum with a broad maximum considerably to the red in agreement with Lewis et al.[4] However, recent experiments[23] with laser excitation and samples, which were flowed past the laser beam with a 10-μsec transit time to control

[19] G. J. Perreault and A. Lewis, Biophys. J. **21**, 171a (1978).

[20] G. J. Perreault and A. Lewis, Biophys. J. 25, 308a (1979).

[21] A. N. Kriebel, T. Gillbro, and U. P. Wild, Biochim. Biophys. Acta **546**, 106 (1979).

[22] J. Spoonhower, Ph.D. Dissertation, Cornell University, Ithaca, New York (1976).

[23] G. J. Perreault, unpublished results.

TABLE I

A SURVEY OF EMISSION DATA ON BACTERIORHODOPSIN AT LOW TEMPERATURE

Reference	Preparation		Temperature (°K)	Source/ intensity	Excitation wavelength (nm)		
					Range^a	For EMSP^b	Exc Max
Lewis et al.[4]	LA/CDK^g Deaerated: no effect	Suspension	77	Dye laser ~10^2 W/cm²	574–624	580	
Alfano et al.[5]	LA/CDK	Suspension	77	100-W tungsten low^h	500–640	580	
Govindjee et al.[6]	LA/CDK Preillumination 640	Suspension	77	100-W tungsten low^h	500–640	580	585
Sineshchekov and Litvin[7]	pH 7.0	Suspension	77	100-W tungsten low^h	400–700	≤530	585
	pH 7.0	Freeze-dried	77				
	pH 7.0	Film	77				635
	pH 1.4	Film	77				640
Gillbro et al.[8]	LA/CDK Preillumination Ar⁺ 515 nm 1 W/cm² 5 min		77	2.5 kW Xe			597 540 sh
	Further pre-illumination						
	Dye laser 623 nm 0.1 W/cm²						
Shapiro et al.[12]	LA/CDK 2x H_2O	Suspension	17	Ar⁺ laser ~10^2 W/cm²			515
Kriebel et al.[21]	H_2O		77	Xe 50 μW/cm²			570

^{a} Range of excitation wavelengths used.

^{b} Excitation wavelength used for emission spectra reported in this table.

^{c} Maximum of excitation spectra at given emission wavelength.

^{d} Ratio of quantum yield of emission from sample at low vs. room temperature. All samples at 7 except that of Shapiro et al.[12] where 17°K data are reported.

^{e} Polarization = $p = \dfrac{I_\parallel - I_\perp}{I_\parallel + I_\perp}$

    Anisotropy = $\dfrac{I_\perp}{I_\parallel}$

TABLE I
(continued)

EM wavelength[c] (nm)	Band positions nm (cm⁻¹)			Assigned emitting species	QY (LT)[d]/ QY(RT)	Polarization anisotropy[e]	COLL angle PMT[f]
	678	733	791			$p = 0.23-0.49$ $r = 0.63-0.34$	90° RCA C31034
	680	710	740	BR$_{570}$	15–20		
720	680	710–715	730–735	BR$_{570}$			
	665–670	720–730	Absent		10	$p = 0.45-0.47$	30°
	Absent	720	775				
	670 w[i]	720	795 w			$p = 0.45-0.47$	
	670	725					
0–740	678 w[i]	725			10		
	670 s[i]	720		P-BR	35		
	670	720			39		
	670	720	790	BR$_{570}$	73		180°
	670	720		P-BR			

[f] Photomultiplier.
[g] LA/CDK: light adapted at room temperature and cooled in darkness.
[h] Intensity sufficiently low that sample does not undergo significant photochemistry during time necessary to record spectrum.
[i] w = weak; s = strong

TABLE II
A SURVEY OF EMISSION DATA ON BACTERIORHODOPSIN AT ROOM TEMPERATURE

Reference	Preparation	Temperature (°K)	Source	Excitation wavelength (nm)		Quantum yield
				Range[a]	For EMSP[b]	
Lewis *et al.*[4]	Light adapted deaerated: no effect	RT	Ar^+ laser	515	515	—
Spoonhower[22]	Light adapted deaerated: no effect	RT	Xe lamp		570	(1.2–2.5 $\times 10^{-4}$
Govindjee *et al.*[6]	H_2O, light adapted	RT	W lamp	500–680	580	~5 × 10
Sineshchekov and Litvin[7]	H_2O, pH = 5.7	RT	W lamp	$\lambda \leq 530$	$\lambda \leq 530$	—
Gillbro *et al.*[8]		RT	Dye laser		573	<0.1
Kriebel *et al.*[12]	H_2O, light adapted	RT	Xe lamp	550–667	568	—
	H_2O, light adapted	RT	Dye laser	573	573	—

[a] Range of excitation wavelengths used.
[b] Excitation wavelength for emission spectra reported.
[c] $\frac{I_\perp}{I_\parallel}$
[d] $\frac{I_\parallel - I_\perp}{I_\parallel + I_\perp}$

photochemistry, have detected a room temperature emission but have not been able to reproduce the xenon excited emission maxima at short (700–730 nm) wavelengths. All of these data[23] with minimal photoalteration and laser excitation yield emission maxima at ~790 nm, in agreement with Lewis *et al.*[4] The independence of the laser excited emission maximum with photoalteration is certainly good evidence that BR_{570} does emit and that the room temperature emission cannot be solely due to a photoproduct. However, the difference between the low photoalteration laser-excited data and xenon arc-excited data requires further explanation.

One possible explanation for the lack of agreement between low intensity xenon-excited stationary and laser-excited flow experiments could be related to the different types of neighbor molecules under these two conditions. Namely, in the low photochemistry xenon-excited stationary case, most photoaltered molecules in the membrane lattice are in the intermediate M_{412}; whereas, in the flow case most of the photoaltered neigh-

TABLE II
(continued)

Em Anisotropy of polarization[c]	Em polarization[d]	Concentration (μM)	Positions of Bands nm (cm^{-1})			Assigned emitting species	Intensity
0.63	0.23				791	BR$_{570}$	1.4×10^2 –
0.34	0.49	10			(12,650)		1.4×10^3 W/cm^2
		10		730 (13,700)		BR$_{570}$	
—	—	20	700 (14,285)			—	
0.38–0.36	0.45–0.47	30	660 (15,150)			—	
		33		730 (13,700)		Unidentified	
—	—	30			714 (14,000)	BR	50 μW/cm^2
—	—	30			735 (13,600)	P-BR	3W/cm^2

boring molecules of the emitting bacteriorhodopsin are in either the K$_{610}$ or L$_{550}$ thermal intermediate states. This difference could perturb the exciton interaction that we know exists in the purple membrane[24] and/or could perturb the emitting molecule via long range conformational interactions with neighbors in the lattice. The one reported observation that the above suggestion does not account for is the 660-nm emission maximum of Sineshchekov and Litvin[7] which appears not to agree with any of the other reported room temperature data.

C. The Transition Responsible for Bacteriorhodopsin (BR$_{570}$) Emission Could Be Strongly Dipole Allowed It has been generally assumed[4,5,11] that bacteriorhodopsin emission emanates from a weakly allowed excited-to-ground-state transition from a state which lies lower in energy than the excited state responsible for the strongly allowed visible absorption. In this section, we demonstrate that these previous assumptions on the nature of the transition could be incorrect and that the emission could emanate from a strongly allowed transition.

Several observations supported the previous suggestion that the emission transition was only weakly allowed. For example, the original detec-

---

[24] M. P. Heyn, P. J. Bauer, and N. A. Dencher, Biochem. Biophys. Res. Commun. 67, 897 (1975).

tion of the bacteriorhodopsin emission demonstrated that there was little overlap between the allowed absorption and the emission.[4] This suggested that the transition may be arising from a low-lying forbidden excited state other than the allowed singlet into which light energy is absorbed. This idea was supported by the relatively low quantum efficiency (ϕ_f) of the room temperature emission calculated to be $1.8 (10^{-4})$.[12] In addition, measurement of the emission lifetime τ_m (15 psec at room temperature[11] and 40 psec at $90°K$[5]) appeared to strongly support these ideas of a low-lying dipole forbidden state. These measurements when coupled to emission quantum efficiency data as a function of temperature yield radiative lifetime (τ_m/ϕ_f) of 60–125 nsec. A comparison of these radiative lifetimes of the fluorescent state, (based on τ_m and ϕ_f measurements) with a value of 7 nsec calculated by integrating the absorption spectrum and using the Strickler–Berg relationships[25] demonstrated that there was a large discrepancy in the radiative lifetimes obtained by these two methods. This has been taken, in model polyene spectroscopy, as an indication of a low-lying forbidden singlet (Ag) state in polyenes such as diphenyloctatetraene.[26] Furthermore, a comparison of the data on BR_{570} and diphenyloctatetraene emission (such as nonoverlap of absorption and emission, low emission quantum efficiency, etc.)[26] suggested by analogy that the observed emission in these two molecules may be emanating from similar excited states.

However, a crucial element in all the above deductions on bacteriorhodopsin is the calculated room temperature emission quantum efficiency of $1.8(10^{-4})$.[12] This quantum efficiency calculation assumes that every photon absorbed does indeed populate the emitting state. However, such an assumption cannot be correct as recent data have indicated.

It is now known that after light absorption there are competing pathways for light emission and photochemistry. The above deduction is based on the observation that at room[27] and low temperatures ($77°K$)[28] the rate of K_{610} formation is close to 1.0 psec (measurements at $77°K$ were pulse width limited but showed the K_{610} forms in 6 psec)[28] whereas the observed emission lifetime changes drastically to a value of 40 psec at $90°K$[5] and 60 psec at $77°K$.[12] In addition to the above results suggesting split pathways of emission and photochemistry after vertical excitation,[29] there are also recent data on photochemical quantum yields of K_{610} formation and BR_{570} reformation which indicate that most of the molecules are channeled into photochemistry and not emission.[29] This is based on the

[25] S. J. Strickler and K. A. Berg, *J. Chem. Phys.* **37**, 814 (1962).

[26] B. Hudson and B. Kohler, *J. Chem. Phys.* **59**, 4984 (1973).

[27] E. P. Ippen, C. V. Shank, A. Lewis, and M. A. Marcus, *Science* **200**, 1279 (1978).

[28] M. L. Applebury, K. S. Peters, and P. M. Rentzepis, *Biophys. J.* **23**, 375 (1978).

[29] A. Lewis, *Proc. Natl. Acad. Sci. U.S.A.* **75**, 549 (1978).

fact that the sum of the forward ($\phi_{BR \to K}$) and reverse ($\phi_{K \to BR}$) photochemical quantum yields sum to one.[29-31] In photochemical reactions this has been taken to mean that all other pathways of decay are very small $< 10\%$ (the accuracy of the quantum yield measurements) and that BR_{570} and K_{610} share a common minimum in the excited state surface.[29,32] Therefore, as Lewis[33] has noted, $>90\%$ of the excited molecules never reach the emitting minimum and thus cannot be considered when calculating the emission quantum efficiency. In view of these results, it is possible that the room temperature emission quantum efficiency should be reevaluated to be $\geq 1.62(10^{-3})$ and the radiative lifetime to be ≤ 9.26 nsec. Similar corrections can be made to the previously calculated low-temperature radiative lifetimes, which bring these room temperature and low-temperature lifetimes in close agreement with the calculations based on the Strickler–Berg relationship.[25] Thus, it is clear that the above analysis of emission quantum efficiency and lifetime suggests that bacteriorhodopsin emission could emanate from a strongly dipole-allowed excited-to-ground-state transition.[13]

D. Temperature Dependence of the Emission There are two principal aspects to the temperature dependence of the emission. First, the lifetime and second, the relative quantum efficiency of the emission. In terms of the lifetime, the published measurements are at three temperatures 77°K,[12] 90°K,[5] and room temperature.[5,11,12] The 77°K and 90°K lifetimes, which are 60 ± 15 psec[12] and 40 ± 5 psec,[5] respectively, are perfectly consistent with the temperature dependence of the quantum efficiency in this region.[12] However, the measured room temperature lifetime[11] of 15 ± 3 psec is not in agreement with the 77°K or 90°K lifetimes based on the reported[11] relative quantum yields of 38/1 and 24/1 for 77°K and 90°K versus room temperature. Using these relative quantum yields, the predicted room temperature lifetime would be between 1 and 2 psec instead of the measured 15 ± 3 psec.[11] In another investigation, Alfano *et al.*[5] also report that they estimate the room temperature lifetime to be < 3 psec based on their low-temperature lifetime measurements. However, they indicate that an experiment performed by them at room temperature had a decay time for the envelope of 8 psec. Since this was approximately identical to the characteristics of their light gate and because of the discrepancy of this value with their low-temperature extrapolation, they did not feel justified to suggest the 8-psec value for the room temperature lifetime.

[30] B. Becher and T. G. Ebrey, *Biophys. J.* **17**, 185 (1977).

[31] C. R. Goldschmidt, O. Kalisky, T. Rosenfeld, and M. Ottolenghi, *Biophys. J.* **17**, 179 (1977).

[32] T. Rosenfeld, B. Honig, M. Ottolenghi, J. Hurley, and T. G. Ebrey, *Pure Appl. Chem.* **49**, 341 (1977).

In view of the extensive data of Shapiro *et al.*[12] eliminating the possibilities of alterations in the lifetime due to pulse intensity, preillumination, or presence of photoproduct, it appears likely that even though the experimental methods of Hirsch *et al.*[11] and Alfano *et al.*[12] were different, their combined data suggest that the room temperature lifetime is somewhere between 8 and 15 psec. Thus, based on our inability to interconnect the room and low-temperature lifetimes with the variable temperature quantum efficiency measurements, it is probable that the room and low-temperature emissions could be reflecting differing contributions from more than one region of the excited state surface.

Two generally accepted results from these variable temperature measurements are (1) the very large increase in emission intensity as the temperature is lowered (we estimate, based on 17°K measurements, that ~ 10% of the molecules are emitting at 4°K) and (2) the radically different temperature dependence of the emission lifetime when compared to the time scale for generating the photochemical product. The large increase in the quantum efficiency of emission as a function of lower temperatures is a characteristic of molecules undergoing twisting and isomerization in the excited state responsible for emission. Therefore, the data on bacteriorhodopsin could be signaling a large alteration in the retinal structure in the emitting state. It is also clear that the photochemistry precedes along an excited state coordinate distinct from that of emission. In view of the above suggestion that retinal structural alteration is integrally involved in the emission pathway it is thus certainly likely that the photochemical pathway has a large component of some other coordinate; for example, the excited state protein structural alteration suggested by Lewis.[29]

II. Rhodopsin Emission

Three papers have appeared with some very convincing results indicating detectable emission from visual photoreceptor pigments. The first two, Franceschini *et al.*[14] and Cronin and Goldsmith[15] are reports of emission from a stable photointermediate of invertebrate rhodopsin whereas Doukas *et al.*[16] have used picosecond spectroscopy to demonstrate vertebrate rhodopsin and isorhodopsin light emission.

Cronin and Goldsmith[15] have published a study on crayfish metarhodopsin emission from isolated rhabdoms, which are the photoreceptor organelles in invertebrates. They showed, using fluorescence microscopy, that initially the rhabdoms were essentially nonfluorescent. However, after illumination, an emission appeared whose time course and intensity were exactly related to the appearance of metarhodopsin. In addition, the excitation spectra of the emission also matched the metarhodopsin absorption and convincingly shifted as a function of pH (the absorption of crayfish metarhodopsin is ~ 525 nm at pH 7.5 and ~ 460 nm at pH 1.9).

The Stokes shift of the emission, the shape of the emission, and the location of the emission are all very similar to those reported for bacteriorhodopsin.[4,6] The quantum efficiency (1.6×10^{-3}) is somewhat higher than the value for bacteriorhodopsin of 1.8×10^{-4} reported by Shapiro et al.[12]

Franceschini et al.[14] have also reported emission from an invertebrate, a fly (*Musca domestica*). Photoreceptor cells are grouped in the fly in units of 8. Six of the cells numbered 1–6 exhibit a red emission with blue excitation that can be assigned to metarhodopsin; whereas, the emission from a subset of cells numbered 7 and cells numbered 8 are green and appear to come from a relatively photostable pigment. Excitation with ultraviolet alters the emission maximum and there are other subsets of rhabdoms numbered 7 and 8 that have different emission characteristics. It appears, however, from intracellular recordings, that the spectral sensitivity of the cell is correlated to its emission, suggesting that emission spectroscopy will become a color tag to map out various retinal types *in vivo*.[14]

Finally Doukas et al.[16] used picosecond spectroscopy to identify rhodopsin and isorhodopsin emission. They identified a rapid < 12-psec emission and, based on the criteria of repeatability, lack of emission in bleached samples, and consistency of emission lifetimes with photoproduct formation times, conclude that these vertebrate photopigments are indeed emitting detectable signals under their conditions. The relationship of these results to the observations in invertebrates[14,15] that the dark-adapted pigment does not fluoresce is still unknown and awaits clarification. A picosecond investigation on the invertebrate system should be most interesting.

III. Conclusion

In conclusion, until 1976 when Lewis et al.[4] reported emission from bacteriorhodopsin, there was little activity in fluorescence spectroscopy of rhodopsin and bacteriorhodopsin. This was mainly the result of the extreme difficulties encountered in detecting the weak emissions from these retinylidene membrane proteins. However, recent developments in lasers in terms of both tunability, stability, and pulse duration and the development of new streak camera detection systems coupled to fluorescence microscopes may well provide the basis for the application of emission spectroscopy to rhodopsin and bacteriorhodopsin.

Acknowledgements

The author acknowledges the support of the National Eye Institute, the Army, the National Aeronautics and Space Administration, the Naval Air Systems Command and the Solar Energy Research Institute at various times during the course of this investigation. A. L. was a John Simon Guggenheim Fellow.

[28] Preparation of Planar Membrane Monolayers for Spectroscopy and Electron Microscopy[1]

By KNUTE A. FISHER

Asymmetry is a well-established property of membranes in general and purple membrane in particular. Membrane sidedness is most often examined by methods that rely on chemical or enzymatic labeling techniques using whole cells, vesicles, reconstituted liposomes, or model bilayers. An alternative approach, derived from work with the human erythrocyte membrane,[2] is described here for purified nonvesicular fragments of bacterial purple membrane. This technique is based on the observation that most biomembranes bear a negative surface charge and will attach to a positively charged surface.[2,3] Since the surface of the membrane that is most negative will bind preferentially, one can produce highly oriented[4] membrane preparations given a planar, positive surface. The method described here involves three steps: (1) preparing the cationic surface, (2) applying the membrane to form a monolayer, and (3) processing the monolayers for spectroscopy and electron microscopy.

Materials and Reagents

Cover glass. Micro Cover Glasses, #1, 11×22 mm. Arthur H. Thomas Co., Philadelphia, Pa.; or other shape and thickness as desired.

Porcelain staining rack. Type B, 90° V-trough. Arthur H. Thomas Co.

Crystallizing dish. 70×50 mm (Corning No. 3140). Scientific Products, Evanston, Ill.

Detergent. Liqui-Nox (Alconox). Scientific Products.

Chromic–sulfuric acid. To a 4.08-kg bottle of concentrated sulfuric acid, add 25 ml Chromerge. Fisher Scientific Co., Fair Lawn, N.J.

[1] Portions of this work were supported by USPHS, NIH Grants GM 27057 and GM27049. This work was done during the tenure of an Established Investigatorship of the American Heart Association and with funds contributed in part by the AHA, California Affiliate.

[2] K. A. Fisher, *Science* **190**, 983 (1975).

[3] D. Mazia, G. Schatten, and W. Sale, *J. Cell Biol.* **66**, 198 (1975).

[4] Membranes are oriented as optically flat, planar surfaces, one membrane thick, with either their cytoplasmic or extracellular surfaces selectively attached; K. A. Fisher, *Electron Microsc., Proc. Int. Congr., 9th, 1978* p. 521 (1978); K. A. Fisher, K. Yanagimoto, and W. Stoeckenius, *J. Cell Biol.* **77**, 611 (1978).

Poly(L-*lysine*)*hydrobromide*. M.W. 1000 to 4000, Type II, Sigma Chem. Co., St. Louis, Mo.; or MW 1500 to 8000, Miles Laboratories, Inc., Elkhart, Ind.

Cationic Surface Preparation

Cover Glass Cleaning

(1) Notch or cut cover glass to desired shape with diamond marking pencil (Fig. 1).

(2) Transfer cover glass with forceps to V-racks in crystallizing dishes (Fig. 2), add hot tap water plus a few drops of detergent. Sonicate for 60 sec.[5]

(3) Lift rack out of detergent, pour off solution, replace rack, rinse for about 30 sec with running tap distilled water, lift rack, and pour off water.

(4) Cover racks of glasses with chromic–sulfuric acid and clean at 60 to 70° for 1–2 hr.

(5) Lift rack, pour off acid, briefly rinse rack of glasses in running ion-exchanged distilled water, then wash rack in a flow (about 500 ml min^{-1}) of glass distilled (preferred) or ion-exchanged distilled water for 10 min.[6]

(6) Remove cover glass from rack with *clean* forceps and quickly dry with compressed dry nitrogen gas.[7]

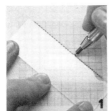

FIGS. 1–4. Cationic surface preparation.

[5] We use an 80-kHz, 80-W, bath-type sonicator, Power Supply Model G-80-80-1, Bath Model T-80-80-1RS, Laboratory Supplies Co., Hicksville, N.Y.

[6] After cleaning and rinsing, cover glass surface must be *totally hydrophilic*. Hydrophobic areas bind neither polylysine nor membrane. Contamination can be avoided by never touching glasses, racks, or forcep tips with fingers and by overflowing all washes and rinses. Forceps can be cleaned with organic solvents followed by 50% (v/v) nitric acid, glass distilled water, and nitrogen drying.

[7] The cleaned cover glass should be stored covered and used within 30 min; truly clean glass surfaces become partially hydrophobic quickly.

Poly(L-*Lysine*)*Application*

(1) Apply 25 μl 5 mM polylysine[8] from a hydrophobic pipette to one side of the cleaned, dry, 11 × 22-mm cover glass held horizontally— about 0.1 μl mm^{-2} surface area (Fig. 3). Dispense the solution quickly and evenly to the upper surface. The solution should flow quickly over the *entire* surface of the glass. For spectroscopy, coat both sides of an 8 × 22-mm cover glass.

(2) Gently gyrate the glass in the horizontal plane, coating it evenly for 30 sec at 20–22°. Wash away excess, unbound polylysine with glass distilled water for 30 sec either with a stream from a wash bottle or by immersion and gentle agitation in 150 ml. Some polylysine will also bind to the untreated side of the glass if rinsed by immersion.

(3) Dry with a burst of dry nitrogen gas from an air gun (Fig. 4).

(4) Use polylysine-treated cover glass (PL-glass) within 30 min.[9]

Purple Membrane Preparation and Application

(1) Isolate purple membrane (PM) fragments[10] and wash to remove sucrose, salt, and soluble proteins.[11] We typically transfer 1 ml of light-adapted PM[12] ($A_{568\,nm}$ = 1–2) to a 15-ml Corex centrifuge tube containing 14 ml of 10 mM buffer[13] and centrifuge it in an SS-34 rotor of a Sorvall RC2-B centrifuge at 18,000 rpm (about 40,000 g max.) for 40 min at 5°.

---

[8] The flaky white powder (if yellow do not use) is very hygroscopic. Store desiccated, below 0°. To use, warm to ambient temperature, open, and weigh quickly. Our solution is 5 mM in glass distilled water and is stable for several months when stored in stoppered 2-ml vials at 0–5° in the dark. The concentration of polylysine is well above surface saturation level for the volumes applied. More dilute solutions (to 0.5 mM) will also work. Higher-molecular-weight polycationic polymers may also be substituted. Polylysine will adsorb tenaciously to any surface bearing a net negative charge and therefore may be used to cationize hydrophilic plastics, plasma-etched carbon films, etc.

[9] Preparation and cleaning of 24 glasses takes about 2.5 hr, during which time the membranes can also be prepared. Shortcuts may be taken. Cleaning times and solutions may be adapted to the user's needs. Filtered compressed air may be used for drying. Strict adherence to the method as described, however, will produce surfaces capable of binding negatively charged cells, organelles, or membrane fragments as homogeneous planar monolayers.

[10] D. Oesterhelt and W. Stoeckenius, this series, Vol. 31, p. 667.

[11] Membranes in high salt will not bind to PL-glass, and soluble proteins will compete with membranes for binding sites on the cationic surface.

[12] R. H. Lozier, R. A. Bogomolni, and W. Stoeckenius, *Biophys. J.* **15**, 955 (1975).

[13] To attach the cytoplasmic side of PM to the cationic surface, wash and suspend the membrane in 10 mM phosphate buffer at pH 7; to attach the extracellular surface, suspend in 10 mM citrate buffer at pH 3.[4] PM turns blue and rapidly aggregates below pH 4; vortex vigorously and apply PM to PL-glass immediately.

(2) Resuspend PM pellet in 10 mM buffer to a final concentration of 3–4 mg bacteriorhodopsin ml^{-1} (molar extinction coefficient of BR[14] = 62,700 liters mol^{-1} cm^{-1}). Vortex vigorously to disperse aggregates.

(3) Apply 25 μl buffered PM at room temperature (20°–22°) to an 11 × 22-mm PL-glass held with *clean* forceps (Fig. 5) while gyrating the glass to coat its *entire* surface quickly (0.1 μl mm^{-2}; 0.3–0.4 μg BR mm^{-2}. For spectroscopy apply 20 μl to each side of the 8 × 22-mm glass. Continue to gyrate the sample for 30 sec.

(4) Wash glass vigorously with buffer or glass-distilled water from a wash bottle for 30 sec, 20–22°.

(5) Place glasses in V-racks for storage (Fig. 2).[15]

(6) Sonicate[5] PL-glass-bound PM for 15 sec; hold glass with forceps and plunge into 10 mM buffer in a 10 ml beaker in bath at 20–22° (Fig. 6).[16]

(7) Turn off sonicator with cover glass still submerged and overflow the beaker with water for about 30 sec.[17]

(8) Transfer sample to V-racks under water or in dilute buffer for storage. A true monolayer of PM on PL-glass is totally transparent. Nevertheless, PM attaches quite tenaciously and once attached remains attached. Glasses may be dried for shadowing electron microscopy (Fig. 7, left) or placed wet against thin copper (Fig. 7, right) and frozen for freeze-fracture electron microscopy. Replicas are harvested by floating onto hydrofluoric acid (Fig. 8).

FIGS. 5–8. PM monolayer preparation and replica harvesting.

[14] M. Rehorek and M. P. Heyn, *Biochemistry* **18**, 4977 (1979).

[15] At this point, attached membranes are overlapped and folded over, averaging two to three membrane layers in thickness (K. A. Fisher, unpublished observations).

[16] Sonication in different laboratories is the most variable step of the process. The conditions to remove overlapped or stacked membrane should be established for each sonication device.

[17] This step not only washes the sample but also removes amphiphiles and debris floating at the air–water interface. Glass surfaces are easily contaminated by passage through a dirty interface.

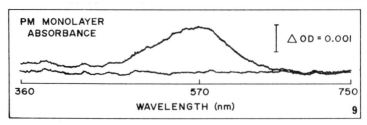

FIG. 9. PM monolayers: spectroscopy

Monolayer Spectroscopy and Electron Microscopy

Spectroscopy. An 8 × 22-mm glass, coated with PM on both sides to increase OD signal, is transferred to a buffer-filled cuvette. The glass is placed firmly against one side of the cuvette normal to the beam and scanned versus an identically oriented reference PL-glass that lacks bound membrane. A spectrophotometer with good signal-to-noise characteristics (e.g., 0.1–0.2 milli OD noise) such as the Aminco DW-2a can easily produce spectra of a PM monolayer that covers 50–70% of the total area of each side of the PL-glass (Fig. 9).[18]

Electron Microscopy. PL-glass oriented planar monolayers have been used for electron diffraction studies[19] and are especially suited to surface replication[4] and freeze-fracturing.[20]

REPLICA PREPARATIONS. Glasses are dried and shadowed with platinum-carbon *in vacuo*. The shadowed glass is cut into 2 × 2-mm strips and floated, replica side up, onto 1 part concentrated hydrofluoric acid plus one part distilled water in plastic petri dishes (Fig. 8). Replicas are transferred by loop (Fig. 8) through three distilled water washes and then picked up on grids. True monolayers do not require additional cleaning. Nitrogen-burst-dried or freeze-dried membranes are smooth (Fig. 10); slowly air-dried membranes have artifactual cracks on their extracellular surfaces (ES) and pits on their cytoplasmic surfaces (PS) (Fig. 11).[4]

FREEZE-FRACTURE PREPARATIONS. Hydrated glasses are placed against thin copper sheet (Fig. 7, right) and the sandwich rapidly frozen in Freon-22 at − 150° and stored in liquid nitrogen. The glass is pried from the copper under liquid nitrogen with a razor blade, placed fractured side

[18] K. A. Fisher, unpublished observations.
[19] S. B. Hayward, D. A. Grano, R. M. Glaeser, and K. A. Fisher, *Proc. Natl. Acad. Sci. U.S.A.* **75,** 4320 (1978).
[20] K. A. Fisher and W. Stoeckenius, *Science* **197,** 72 (1977).
[21] K. A. Fisher and D. Branton, this series, Vol. 32, p. 35.

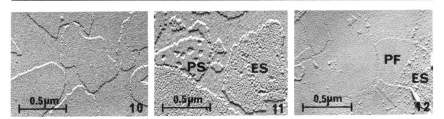

Figs. 10–12. PM monolayers: electron microscopy

up on a cold stage, and processed as in conventional freeze-fracturing, etching, and shadowing.[21] An example of a freeze-fractured PM monolayer is shown in Fig. 12. Cationic ferritin has been used to label the unfractured extracellular surfaces (ES) of the membrane (lower right) contiguous to the fractured, particulate, cytoplasmic fracture face (PF).

In summary, planar membrane monolayers can be produced easily and are well suited to spectroscopic and microscopic techniques. Because membranes can be selectively attached by either their cytoplasmic or extracellular surfaces, the oriented preparations are also useful for enzymatic or chemical modification studies and for surface analysis by physical techniques.

Acknowledgments

I thank Eleanor Crump for help in preparing the manuscript, Dr. Walther Stoeckenius for continued support, and Dr. Roberto Bogomolni for fruitful discussions and help with spectroscopic aspects of this study.

[29] Metal Decoration of the Purple Membrane

By Dorothea-Ch. Neugebauer

Principle

When a metal is evaporated onto a solid surface, no continuous film is formed at first, but thermal accommodation, nucleation, and crystallization result in individual metal grains 1–5 nm in diameter. A nonuniform distribution of nucleation sites leads to a corresponding distribution of metal grains. Their size and distribution depends on the surface properties of the solid and on the surface mobility of the metal. This phenomenon is called decoration.[1] Low melting metals such as gold, silver, and platinum have a high surface mobility and show a pronounced decoration effect. Since metal decoration is very sensitive toward differences in surface properties, it provides a simple method for visualizing such differences.

Metal decoration of the purple membrane has the following applications:

Distinction and Identification of the Two Faces of Individual Membranes. The cytoplasmic surface shows a hexagonal decoration pattern (Fig. 1), the other face a random distribution of metal grains (Fig. 2).[2] Freeze-fracturing[3] or shadowing with tantalum/tungsten after freeze-drying under carefully controlled conditions[4] or chemical labeling[5] could serve the same purpose, but none of these techniques is so simple and at the same time so fast and economical in the use of material as metal decoration.

Observation of the Crystalline Arrangement of Bacteriorhodopsin. The decoration pattern on the cytoplasmic surface reflects the crystalline arrangement of the bacteriorhodopsin (Fig. 1, 2). The contrast of the metal grains is large enough to make possible the investigation of the crystallinity of very small areas (4 × 4 unit cells) by optical diffraction (Fig. 3)[2] (cf. Neugebauer *et al.*[6]).

Detection of the Distribution of "Contaminants." Very thin layers of certain substances (e.g., glucose, saccharose) change the decoration pattern. Therefore, metal decoration has been used to investigate whether such substances are preferentially retained on one of the membrane surfaces.[2]

However, metal decoration cannot provide direct evidence on the chemical or physical nature of the investigated surface, since the factors governing nucleation are complex and not well understood. The different decoration patterns on the two faces of the purple membrane only show that the two faces differ in their distribution of nucleation sites for metal. For identifying the faces metal decoration had to be combined with more direct methods of surface labeling and with the results of freeze-fracturing.[7]

[1] G. A. Basset, J. W. Menter, and D. W. Pashley, *in* "Structure and Properties of Thin Films" (C. A. Neugebauer, J. B. Newkirk, and D. A. Vermileyea, eds.), p. 11. New York, 1959.

[2] D.-C. Neugebauer and H. P. Zingsheim, *J. Mol. Biol.* **123,** 235 (1978).

[3] D. Oesterhelt and W. Stoeckenius, *Nature (London) New Biol.* **233,** 152 (1971).

[4] D. Studer, H. Moor, and H. Gross, *J. Cell. Biol.* **90,** 153 (1980).

[5] R. Henderson, J. S. Jubb, and S. Whytock, *J. Mol. Biol.* **123,** 259 (1978).

[6] D. -C. Neugebauer, H. P. Zingsheim, and D. Oesterhelt, *J. Mol. Biol.* **123,** 127 (1978).

[7] H. P. Zingsheim, R. Henderson, and D. -C. Neugebauer, *J. Mol. Biol.* **123,** 275 (1978).

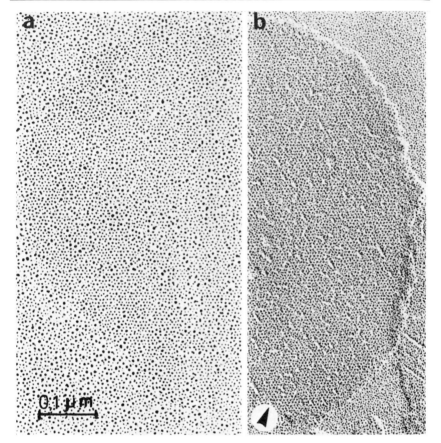

FIG. 1. Metal decoration reveals the hexagonal lattice of the bacteriorhodopsin on the cytoplasmic face of the purple membrane. (a) Decoration with a thin layer of silver (average thickness 1 nm). (b) Decoration as for (a), but only 0.6 nm of silver, followed by shadowing at 45° incidence with 0.6 nm of Ta/W. The arrow indicates the shadowing direction. The silver grains are much smaller than in (a) but are still clearly visible. The additional shadowing layer increases the contrast of the silver grains and reveals the surface relief of the membrane. Note that by decoration alone as in (a) it would be extremely difficult to detect the small fissures and cracks in the membrane surface. (From Neugebauer and Zingsheim[2] with permission of the publisher.)

Method

Specimen Preparation. Ideally, purple membrane for metal decoration is suspended in distilled water or a volatile buffer at a concentration of 1–2 mg protein/ml, although the presence of inorganic salts is compatible with metal decoration (e.g., 10–20 mM NaCl or KCl). Organic additives

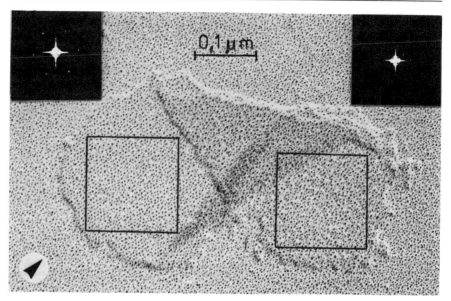

Fig. 2. The two faces of the purple membrane can be distinguished unequivocally by their decoration pattern. This membrane has folded over. The lattice periodicity of 6.2 nm is clearly revealed on the cytoplasmatic face. (See also the diffraction pattern of the boxed-in areas.) The statistical distribution of the silver particles on the other face is virtually identical to that on the carbon support film. After decoration with 0.6 nm of silver, the specimen was shadowed with 0.6 nm of Ta/W. The arrow indicates the shadowing direction. (From Neugebauer and Zingsheim[2] with permission of the publisher.)

to the solution, such as albumin or Tris buffer, must be washed out by centrifugation before the membranes are deposited on the support. Even extensive washing of the membranes after their deposition on the support does not remove albumin sufficiently.

The specimens may be prepared on any support that can be used in the electron microscope (e.g., carbon-, formvar-, polystyrene-film) or from which replicas can be prepared (e.g., mica, glass). The membrane suspension is brought into contact with the support for a few seconds. Subsequently the specimen is washed with distilled water either by rinsing with a few milliliters of water or by incubation on several droplets of water for a few minutes each time. The specimens may be air-dried or freeze-dried (cf. Studer et al.[4]).

Metal Decoration. Metal decoration can be carried out in any evaporation unit suited for metal shadowing of electron microscopic specimens. In contrast to shadowing, the metal is evaporated at a right angle to the specimen surface and much less metal is used for decoration than for

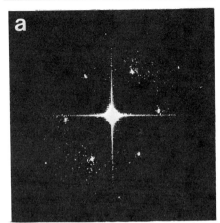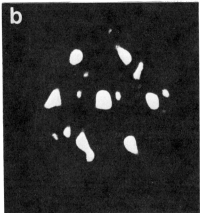

FIG. 3. Optical diffraction patterns from micrographs of decorated and shadowed purple membranes. (a) From an area comprising about 2000 unit cells. Loss of the third set of second-order diffraction spots is due to the directionallity of shadowing. (b) From 4 × 4-unit cell area. (From Neugebauer and Zingsheim[2] with permission of the publisher.)

shadowing. Gold and silver give equally good results on the purple membrane. The metal is evaporated by resistance heating from a molybdenum boat. A quartz crystal monitor is used to measure the average thickness of the metal film. The thickness of the metal film is critical. An average thickness of about 0.6 nm is best suited for the observation of the decorated purple membrane in the electron microscope (Fig. 4). This figure is measured with a Balzers Quartz Crystal Thin Film Monitor QSG 201. When using a different monitor it might be necessary to determine the optimal amount experimentally. It does not make any difference whether the metal is evaporated in one burst or in several steps. The decoration pattern of the purple membrane is independent of the specimen temperature between −80° and +30°.

Studer et al.[4] have also observed decoration with tantalum/tungsten on freeze-dried purple membranes. Due to the low surface mobility of these high-melting metals the decorating metal grains are so small that the effect can only be observed in the diffractograms of the microgaphs, which have very low contrast.

Comments

Contrast Enhancement. Optimal amounts of decorating metal (silver) give rather low contrast. Therefore, it is often useful to shadow the specimen after metal decoration at an angle of 45° with platinum/carbon or tan-

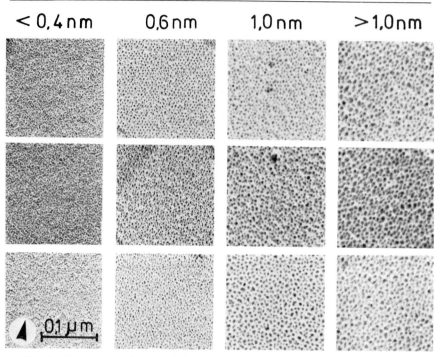

FIG. 4. The average thickness of the evaporated silver layer is critical for the visualization of the purple membrane lattice. Selected areas of folded-over membranes (similar to that shown in Fig. 2) decorated with silver and shadowed with 0.6 nm Ta/W. (Top row) The cytoplasmic face of the purple membrane. (Middle row) The extracellular face of the purple membrane. (Bottom row) Clean carbon film. With less than 0.4 nm of silver only the Ta/W shadowing layer is visible to the naked eye. The decoration pattern on the two membrane faces and the carbon film is indistinguishable. Optimal results for visual inspection require approximately 0.6 nm of silver. Higher amounts increase only the size of the silver grains. At more than 1 nm of silver the membrane faces and the carbon film become indistinguishable again. The arrow indicates the shadowing direction. (From Neugebauer and Zingsheim[2] with permission of the publisher.)

talum/tungsten (Fig. 1) (cf. Zingsheim et al.[8]). The shadowing enhances the contrast of the metal grains and, at the same time, allows us to visualize the membrane morphology. This is important for judging the decoration pattern, because the membranes may be more or less fragmented, depending on the properties of the support.[2] Obviously, fragmentation would interfere with the decoration pattern.

   Stability of the Specimens. The decorated specimens are not stable and should be looked at and photographed the same day. Although no

[8] H. P. Zingsheim, R. Abermann, and L. Bachmann, *J. Phys. E* **3**, 39 (1970).

changes can be observed during electron microscopic observation, the metal keeps moving. Already after 24 hr some metal grains have grown at the expense of others. After two days the specimens are almost useless for observing the hexagonal decoration pattern of the cytoplasmic face, and after a week most of the metal has collected around the edges of the membranes. Shadowing the specimen with tantalum/tungsten and coating it with a thin layer of carbon does not reduce this phenomenon markedly, nor does it make a difference whether the specimens are stored in a desiccator or under vacuum. However, on silver-decorated specimens, subsequently shadowed with tantalum/tungsten, the shadows of the silver grains are still at their original sites, although the silver has moved away.

[30] Neutron Scattering of Bacteriorhodopsin

By GLEN I. KING and BENNO P. SCHOENBORN

Neutron diffraction has proved to be a valuable technique for elucidating many aspects of the structure of biological membranes that are not easily determined by the other direct structural techniques, such as X-ray and electron diffraction. Basically, neutron scattering can be viewed as giving important complementary information to that obtained by the other techniques.[1] The large difference between the neutron-scattering factors of the two isotopes of hydrogen, ^1H and ^2H, is of particular importance. Thus this isotopic scattering difference can be used to locate particular membrane constituents by selectively deuterating them. The purple membrane of halobacteria is an ideal example showing the sort of unique structural information that can be obtained by this procedure.

Purple membrane has a number of unique features that make it an attractive system for neutron structural studies. First, it is a relatively simple system containing only a single protein, bacteriorhodopsin (BR) with known amino acid sequence.[2,3] Unlike most other membranes, purple membrane is isolated as a single planar patch rather than a vesicle. Most important, BR is arranged in the membrane plane in a highly organized hexagonal crystalline lattice. Finally, being a bacterial membrane, purple membrane can be grown in a fully deuterated environment[4] allowing for

[1] B. P. Schoenborn, *Biochim. Biophys. Acta* **475**, 41 (1976).

[2] Y. A. Ovchinnikov, N. G. Abdulaev, Y. M. Feigina, A. V. Kiselev, and N. A. Lobanov, *FEBS Lett.* **100**, 219 (1979).

[3] H. G. Khorana, G. E. Gerber, W. C. Herlihy, C. P. Gray, R. J. Anderegg, K. Nihei, and K. Biemann, *Proc. Natl. Acad. Sci. U.S.A.* **76**, 5046 (1979).

[4] H. L. Crespi and J. J. Katz, this series, Vol. 26, p. 627.

the possibility of obtaining any of its components in a perdeuterated form, and permitting functional reconstitutions that then can be used in determining the location of that component in the membrane. In particular, the location of the chromophore retinal can be determined relative to the known protein structure and sequence. Information about the protein environment of the chromophore is crucial for an understanding of its role in the proton pumping mechanism of the membrane.

Experimental Principles

Sample Preparation

To obtain the three-dimensional location of constituents of the purple membrane, the analysis is divided into the determination of orthogonal projections. To determine the average structure of the membrane as projected onto a line normal to the membrane plane (profile structure), it is necessary to obtain the continuous neutron-scattering data from the membrane sheets. This can be accomplished in two ways. The first involves orienting the membrane sheets parallel to each other (for example, with a magnetic field) with large random water spaces between them. This has the advantage of separating the continuous profile data from the in-plane scattering. An alternative is to collect data from unoriented dispersions of the membrane, where the in-plane and profile contributions are mixed together. In general, the first is the preferred method. However, even though purple membrane can be oriented in a magnetic field,[5] it is not possible to do so at a high enough membrane concentration to obtain an adequate signal. Fortunately, the in-plane scattering from purple membranes under the experimental conditions is negligible compared to the much stronger profile diffraction, so the latter can be obtained from random dispersions.

The data needed to determine the structure in the plane of the membrane (planar projection) can also be obtained. In this case it is necessary to form oriented samples of the membrane by the process of evaporation of excess water from a dispersion. This forms fairly well oriented stacks from which the planar crystalline reflections can be obtained by orienting the neutron beam normal to the membrane sheets. The resulting pattern from this geometry consists of a series of sharp concentric rings, which arise due to the random rotation of the different membranes in the plane of the stack, resulting in a two-dimensional powder pattern.

[5] D. L. Worcester, *Brookhaven Symp. Biol.* **27**, III-37 (1975).

Detector Systems

There are basically two different experimental arrangements that have been used to record neutron diffraction data from biological membrane systems. Since no suitable films exist for neutrons, they are of necessity electronic detection systems. The first type of system[6] used for membrane neutron diffraction was a step-scanning diffractometer with a single gas-filled proportional counter. This consists of a Soller collimator[7] between the exit of the beam pipe and the sample, with the diffracted neutrons being further collimated, and then monochromatized with a pyrolitic graphite, just before the detector. The disadvantage of this system for a pattern consisting of rings is that only a small portion of the data in each ring is being collected by the detector.

A more recent development in neutron detection systems is the two-dimensional position-sensitive detector.[8] This detector has the obvious advantage of collecting more of the pattern and is in general the system of choice. The two-dimensional detector proves to be ideal for measuring the diffuse continuous scatter from random dispersions. Here angular resolution of the pattern is not so important, but the ability to collect the whole pattern is. Data from dispersions of purple membrane that contain retinals in their protonated and perdeuterated state were collected on the detector system at the high flux beam reactor at Brookhaven National Laboratory.[9]

For the relatively weaker in-plane diffraction from purple membrane, both detector systems have been used for the data collection from the two isotopically substituted samples. Sufficiently well-resolved reflections could only be obtained on the two-dimensional detector system by narrowing down the collimator slits to such a degree that the counting times were unacceptably long. The Soller-slit system gave very good angular resolution, but unfortunately, also very long counting times. For stable systems like purple membrane,[10] good diffraction patterns have been obtained with the Soller slit system, since neutrons do not inflict any radiation damage and extended data collection time is possible (several days).

[6] B. P. Schoenborn, A. C. Nunes, and R. Nathans, *Ber. Bunsenges. Phys. Chem.* **74,** 1202 (1970).

[7] W. Soller, *Phys. Rev.* **24,** 158 (1924).

[8] J. Alberi, J. Fischer, V. Radeka, L. C. Rogers, and B. Schoenborn, *Nucl. Instrum. Methods* **127,** 507 (1975).

[9] G. I. King, W. Stoeckenius, H. L. Crespi, and B. P. Schoenborn, *J. Mol. Biol.* **130,** 395 (1979).

[10] G. I. King, P. C. Mowery, W. Stoeckenius, H. L. Crespi, and B. P. Schoenborn, *Proc. Natl. Acad. Sci. U.S.A.* **77,** 4726 (1980).

Data Reduction

The goal of the diffraction experiment is to obtain sets of properly scaled (*vide infra*) structure factors, from which the structure(s) can be obtained by Fourier transformation. The first step toward this goal is the determination of the absolute value of the structure factors from the observed intensity data. The absolute values are proportional to the square root of the observed intensities, with some of the proportionality factors being constants and some others being functions of the scattering angle. Since the analysis is generally done on a set of relative intensities, only the latter factors need be considered. The precise nature of these factors depends on the details of both the sample and experimental arrangements and are evaluated according to standard crystallographic theory. Once a set of corrected, integrated reflections (or the continuous modulus transform, in the case of dispersions) is obtained, the second phase of the structure analysis can begin.

Phase Determination

Since neither the profile nor the unit cell of the in-plane lattice of purple membrane have a center of symmetry, the phase problem arises in its most general form. Thus the phase function for the profile transform is a continuously varying function of the scattering angle and the in-plane structure factors can have *a priori* phase angles of any value. The solution to the phase problem for the two parts of the structure analysis (profile and in-plane) is quite different in both concept and methodology.

A general theoretical framework that has proven to be of great usefulness in one-dimensional diffraction theory is that based on what can be considered an information theory approach. That is, one asks the question of what the maximum amount of information contained in the diffraction pattern is. This can be given preciseness by considering the mathematical properties of Fourier transforms of spatially limited functions. In particular, the powerful theorems of analytic function theory can be used to attack the problem. It has been known for some time[11,12] that from the absolute value of the continuous Fourier transform of centrosymmetric structures, one can uniquely determine the structure, aside from inversions and translations. The former (replacement of a structure with its negative) must be ruled out by physical arguments, while the latter are generally (*vide infra*) irrelevant. For noncentrosymmetric structures it

[11] A. Calderon and R. Pepinsky, *in* "Computing Methods and the Phase Problem in X-ray Crystal Analysis" (R. Pepinsky, ed.), p. 000. State College, Pennsylvania, 1952.

[12] R. Hosemann and S. N. Bagchi, "Direct Analysis of Diffraction by Matter." North-Holland Publ., Amsterdam, 1962.

has been shown that this is not the case and that there are in general a number of different "structures" (i.e., mathematical functions) whose Fourier transforms have an absolute value agreeing with the observed transform modulus. Thus there is not sufficient information in the observed intensity function to uniquely determine the structure in this case. However, one of the authors[13] has shown how all the possible mathematical "structures" can be obtained from the data without showing which one is the real one. The best one can do in this case is to determine the common features, if any, of all the "structures," which are of necessity then features of the real structure. Fortunately, this method has been sufficient to determine the position of the chromophore retinal in the purple membrane profile via Fourier difference densities between the two isotopically substituted samples.[9]

With regard to difference densities, one has to be careful regarding the previously made statement about the irrelevance of translations. It must be determined that one of the structures giving rise to the difference density has not been translated relative to the other one, thus covering up a possible difference density feature at one edge of the structure. It is obvious that this can be accomplished by making sure there are no negative differences on the opposite side of the structure. Even though one might argue that a difference at the edge could still be hidden because of the sharp fall of the density there, the results from the difference profiles from purple membrane are unequivocal in showing that there is a significant feature in the center of the membrane that has a height of the right magnitude for perdeuterated retinal. This result, in combination with linear dichroism measurements[14,15] on purple membrane, leads to the general picture shown in Fig. 1.

The phasing of the in-plane data obviously cannot be handled in the same manner as the profile case. Fortunately, electron diffraction results[16] can be used to phase the Fourier difference density map between the protonated and perdeuterated retinal samples. That is, neutron phases can be calculated from the electron density map (they are quite close in value to the electron diffraction phases) and the difference map can be calculated using these phase angles with the observed modulus difference. The location of the retinal in-planar projection can thus be determined relative to the planar projection of the protein density map obtained from electron diffraction.[10] The spatial relationship of the retinal to

[13] G. I. King, *Acta Crystallogr., Sect. A* **A31**, 130 (1975).
[14] R. A. Bogomolni, S. B. Hwang, Y. W. Tseng, G. I. King, and W. Stoeckenius, *Biophys. J.* **17**, 98a (1977).
[15] M. P. Heyn, R. J. Cherry, and U. Mueller, *J. Mol. Biol.* **117**, 607 (1977).
[16] P. N. T. Unwin and R. Henderson, *J. Mol. Biol.* **94**, 425 (1975).

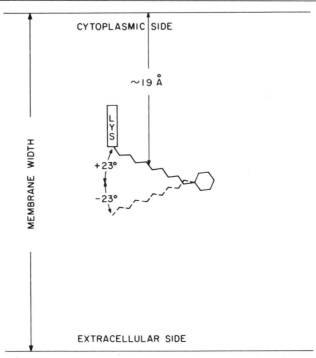

FIG. 1. Representation of the retinal location and possible orientations relative to the membrane profile. The indeterminate sign of the chromophore angle allows the two orientations shown. The correct orientation is not known. The hydrogen-dense β-ionone ring is situated in the center of the membrane. The diagram illustrates the minimum penetration of the polypeptide chain into the membrane when the lysine group is parallel to the helix axis.

the protein is shown in Fig. 2. Combining this picture with the profile results thus gives the three-dimensional orientation and location of the retinal relative to the protein alpha helices.

Model Building

The structure analysis is not complete merely because the proper Fourier maps have been calculated. The interpretation of the maps is seriously hindered by the termination artifacts present in them. Membrane Fourier maps are typically of quite low resolution, well below the atomic level resolutions obtainable in protein crystallography, for instance. Taking difference Fourier maps may help in the interpretation. For example, the difference profile maps of the two isotopically substituted retinal samples show a difference peak that is approximately Gaus-

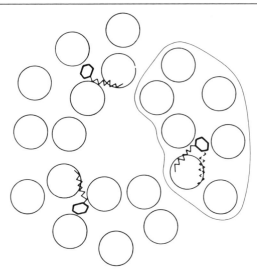

FIG. 2. Representation of the likely retinal orientation and position in the membrane plane relative to the protein's seven α helices.[16] A less likely alternative for the retinal chain position is indicated by the dashed lines. The crystallographic unit shown depicts a trimer, consistent with the Fourier map; the outline of one molecule is indicated.

sian in shape. A limited resolution Gaussian is merely a broadened Gaussian located at the same position, so the position of the retinal in the profile can be determined much more precisely than the limited resolution of the map might imply. However, for the two-dimensional planar-projection difference maps, the termination artifacts present a much more serious problem. There are numerous artifactual peaks present in the map that are difficult to distinguish from the real features. The use of model building is essential for sorting this all out.

In addition to being useful for resolving this second nonuniqueness problem (that of interpretation rather than phasing), model building also allows one to make sure that the structure factors used in the difference synthesis are on the same relative scale as well as making it possible to put the maps on an absolute density scale. The former is of obvious importance; the importance of the latter is perhaps not so obvious. To illustrate the point, the identification of the central peak in the difference profile as being due to the retinal relied on showing that its height on an absolute density scale had the value expected for perdeuterated retinal in the membrane.

Finally, with model studies it is possible to incorporate what is known about the membrane from other physical and chemical aproaches and thus begin to get a more detailed conception of the molecular arrangement in the membrane and its relationship to function. Knowledge of the amino acid sequence in bR makes it quite reasonable to consider construction of molecular models of the protein as an important step toward this goal. Knowing the location of the retinal molecule both in the sequence and spatially with respect to the protein tertiary structure places constraints on how the protein alpha helices must be connected to each other in this vital region of the protein. If enough groups in other parts of the membrane can similarly be located, this task can be completed. One approach to this is the correlation of Henderson's helical structure[16] with the amino acid sequence[2,3] using specifically labeled amino acids. Inspection of the sequence shows that certain amino acids, particularly valine and phenylalanine, are not uniformly distributed but are bunched. Engelman and Zaccai[17] are now attempting by neutron difference Fourier techniques to localize these specifically deuterated amino acids with the known features of the purple membrane structure.

Acknowledgments

The work described in this paper was supported by the NIH (HL06285), the NASA (NSG-07151), and the Department of Energy. G. K. is supported by a grant from the NIH (GM-26346).

[17] D. M. Engleman and G. Zaccae, *Proc. Natl. Acad. Sci. U.S.A.*, **77**, 5894 (1980).

[31] Transient Dichroism of Bacteriorhodopsin

By RICHARD J. CHERRY

Introduction

Optical excitation of bacteriorhodopsin (BR) produces a complex series of spectroscopic intermediates known as the photochemical cycle (for review, see Stoeckenius *et al.*[1]). When excitation is by linearly polarized light, only those chromophores whose transition dipole moments for absorption lie in or near the direction of excitation are excited. Thus an anisotropic distribution of excited BR molecules is photoselected from the initial isotropic distribution present in membrane suspensions. As a result, the absorbance changes arising from BR molecules entering the pho-

[1] W. Stoeckenius, R. H. Lozier, and R. A. Bogomolni, *Biochim. Biophys. Acta* **505**, 215 (1979).

tochemical cycle are dichroic. After excitation by a brief pulse of light, the dichroism decays because of rotational diffusion of the BR molecules. Transient dichroism can thus be used to measure the rotational motion of BR. These measurements provide a powerful method for studying self-association of BR. They have also proved particularly useful for testing theoretical treatments of rotational diffusion of membrane proteins.

Measurement of Transient Dichroism

A flash photolysis apparatus for measuring transient dichroism has been described in detail in a previous volume of this series.[2] Only a brief description of the essential features will therefore be given here. Excitation is by a flashlamp–pumped dye laser that with coumarin 6 (10^{-4} M in methanol) gives a vertically polarized pulse of duration 1–2 μsec and wavelength 540 nm (untuned). Absorption changes are measured by a continuous beam from a 100-W tungsten–halide lamp. After passing through the sample and monochromator, the initially unpolarized beam is split into horizontally and vertically polarized components by a polarizing beam splitter. The absorption changes for these two components are then measured using separate photomultipliers. The sample is contained in a 1-cm fluorimeter cell and ideally has an optical density of about 1 at 570 nm. Using laser intensities of a few millijoules per pulse, easily measurable absorption transients are obtained with a single shot. For accurate measurement of transient dichroism, however, signal averaging is desirable. The signals cannot be improved by increasing the laser intensity beyond a certain point, since the dichroism falls as saturation is approached.

Since the absorption spectra of the intermediates in the BR photochemical cycle are broad and occur at various wavelengths, transient absorption and dichroism are observable over most of the visible spectrum. The major signals for light-adapted BR, however, are the absorption increase at 412 nm resulting from formation of the M form and the loss of absorption at 570 nm resulting from BR molecules entering the cycle. For measurement of rotational diffusion, the 570-nm depletion signals have several advantages. First, they are larger and thus give better sensitivity and accuracy. Second, the transition dipole moments for excitation and measurement are parallel, leading to an important simplification in the interpretation of the data. (This may also be true of the 412-nm signals but remains to be demonstrated.) Finally, the relatively long wavelength of the measurement reduces problems arising from light scattering, which may occur in particular with reconstituted systems.

[2] R. J. Cherry, this series, Vol. 54, p. 47.

Analysis of Transient Dichroism Data

The signals obtained experimentally correspond to changes in the intensity of transmitted light following flash excitation. The signals are used to calculate the absorption anisotropy at time t after the flash, $(r(t))$, given by

$$r(t) = \frac{A_{\parallel}(t) - A_{\perp}(t)}{A_{\parallel}(t) + 2A_{\perp}(t)} \tag{1}$$

where $A_{\parallel}(t)$, $A_{\perp}^{-}(t)$ are, respectively, the absorbance changes for light polarized parallel and perpendicular with respect to the polarization of the exciting flash. $r(t)$ is independent of the signal lifetime and depends only on rotational motion when the absorption transient exhibits a single exponential decay.

The theoretical relationship between $r(t)$ and rotational diffusion has been treated by several authors (for review, see Rigler and Ehrenberg[3]). For the general case of anisotropic rotation, $r(t)$ is the sum of five exponentials. For integral membrane proteins the model shown in Fig. 1 has been used to interpret the experimental data. The model assumes that rotation can only occur around the membrane normal, i.e., tumbling about axes lying in the plane of the membrane does not occur. In this case it may be shown that[4]

$$r(t) = A_1 \exp(-D_{\parallel}t) + A_2 \exp(-4D_{\parallel}t) + A_3 \tag{2}$$

where $A_1 = (6/5) \sin^2 \theta \cos^2 \theta$; $A_2 = (3/10) \sin^4 \theta$; $A_3 = (1/10)(3 \cos^2 \theta - 1)$.[2] θ is the angle between the absorption transition dipole moment and

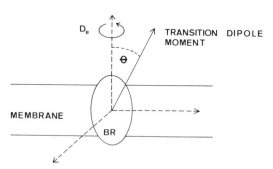

FIG. 1. Model used for interpretation of transient dichroism measurements with BR. It is assumed that rotation occurs only around the membrane normal. $D_{\parallel}$ is the rotational diffusion coefficient, while θ defines the direction of the transition dipole moment.

[3] R. Rigler and M. Ehrenberg, Q. Rev. Biophys. 6, 139 (1973).
[4] R. J. Cherry. Biochim. Biophys. Acta 559, 289 (1979).

the normal to the plane of the membrane and $D_\parallel$ is the diffusion coefficient for rotation about the membrane normal. In the preceding equation it is assumed that the transition dipole moments used for excitation and measurement are parallel.

The absolute values of the anisotropy are reduced in practice by instrumental factors and by excitation of a nonnegligible fraction of molecules. To a first approximation, these effects may be taken into account by rewriting Eq. (2) in the form

$$r(t) = \left[\frac{r_0}{A_1 + A_2 + A_3}\right][A_1 \exp(-D_\parallel t) + A_2 \exp(-4D_\parallel t) + A_3] \quad (3)$$

where r_0 is the experimental absorption anisotropy at time zero. The theoretical maximum of r_0 is 0.4, whereas, experimentally, r_0 is typically about 0.25 in rotational diffusion measurements. Higher values can be obtained by decreasing the laser intensity, but at the expense of decreasing the signal-to-noise ratio. In our laboratory, values of r_0 of up to 0.35 have been observed with purple membranes. The remaining difference between this value and the theoretical maximum is probably due to instrumental factors. Recently, Kouyama et al.[18] have reported a value of 0.395 ± 0.005. Thus the measured anisotropy is not decreased by rapid independent motion of the retinal chromophore, which appears to be held rigidly in the protein.

Transient Dichroism of Purple Membranes

In purple membranes no decay of the absorption anisotropy is detected other than a very slow component attributable to tumbling of the whole membrane fragments.[5-7] The conclusion that BR is immobilized is of course entirely consistent with the existence of the crystalline lattice.[8] However, the lattice is not present in the apo-brown membrane of cells grown in the presence of nicotine.[9] Rotational diffusion of BR is observed on reconstitution by the addition of retinal to the apoprotein, bacterioopsin, present in these membranes.[10]

[5] K. Razi Naqvi, J. Gonzalez Rodriguez, R. J. Cherry, and D. Chapman, Nature (London), New Biol. 245, 249 (1973).

[6] W. V. Sherman, M. A. Slifkin, and S. R. Caplan, Biochim. Biophys. Acta 423, 238 (1976).

[7] R. Korenstein and B. Hess, FEBS Lett. 89, 15 (1978).

[8] A. E. Blaurock and W. Stoeckenius, Nature (London) New Biol. 233, 152 (1971).

[9] M. Sumper, H. Reitmeier, and D. Oesterhelt, Angew. Chem., Int. Ed. Engl. 15, 187 (1976).

[10] R. J. Cherry, M. P. Heyn, and D. Oersterhelt, FEBS Lett. 78, 25 (1977).

Transient Dichroism of BR in Reconstituted Membranes

Transient dichroism of BR has been investigated in vesicles reconstituted from detergent-solubilized purple membrane and added phosphatidylcholine.[11-15] Typical $r(t)$ curves obtained with such vesicles are shown in Fig. 2. At temperatures well below the T_c (temperature of lipid gel to liquid–crystalline phase transition), only a very slow decay of $r(t)$ attributable to vesicle tumbling is observed. Immobilization of BR occurs as a result of recrystallization into a hexagonal lattice indistinguishable from that existing in the native purple membrane.[12] Above the T_c, the decay of $r(t)$ indicates that BR is mobile and hence that the proteins are disaggre-

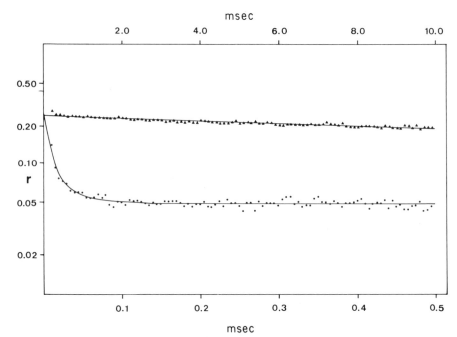

FIG. 2. Time dependence of absorption anisotropy calculated from 570-nm ground state depletion signals from BR–dimyristoyl phosphatidycholine vesicles above and below the lipid phase transition. Phospholipid–protein mole ratio = 110. ▲ 9° (upper time scale); ● 25° (lower time scale). The slight decay observed at 9° is due to vesicle rotation.

[11] R. J. Cherry, U. Müller, and G. Schneider, FEBS Lett. **80**, 465, (1977).
[12] R. J. Cherry, U. Müller, R. Henderson, and M. P. Heyn, J. Mol. Biol. **121**, 283 (1978).
[13] M. P. Heyn, R. J. Cherry, and U. Müller, J. Mol. Biol. **117**, 607 (1977).
[14] W. V. Sherman and S. R. Caplan, Biochim. Biophys. Acta **502**, 222 (1978).
[15] R. J. Cherry, U. Müller, C. Holenstein, and M. P. Heyn, Biochim. Biophys. Acta **596**, 145 (1980).

gated. [CD measurements[12] in fact demonstrate that BR is monomeric above the T_c at sufficiently high L/P (lipid:protein–mole ratio).] From the value of the residual constant anisotropy, the angle θ is calculated to be 78 ± 3° [see Eq. (3) and Heyn *et al.*[13]]. Since in all-*trans*-retinal the transition dipole moment lies along the long axis of the polyene chain, this angle defines the orientation of the chromophore. Similar though somewhat lower values of θ have been obtained from polarized absorption spectra of oriented purple membranes.[1,13,19]

In recent studies in our laboratory, the validity of Eq. (3) has been tested by curve fitting the experimental anisotropy decay curves.[20] With only three adjustable parameters, θ, $D_\parallel$ and r_0, it was found that Eq. (3) gives a good fit to the experimental data under conditions which favor completely monomeric BR, namely at L/P > ~100 and at temperatures sufficiently above the T_c. The value of θ obtained from the curve fitting was 78 ± 2° in good agreement with the earlier determination[13] but in disagreement with a value of about 30° which had also been obtained from transient dichroism data.[21] The rotational relaxation time $\phi_\parallel$ (defined as $1/D_\parallel$) was found to be 15 ± 5 μs over the temperature range 25°–37° and the L/P range 140–250. There is probably a weak dependence on temperature and L/P over these ranges.

At lower L/P and temperatures, deviations of the experimental $r(t)$ curves from Eq. (3) indicate the onset of aggregation. The midpoint of the transition from the monomer to the aggregated state of BR occurs several degrees below the T_c.[22] It has also been recently shown that the decay of $r(t)$ above the T_c is strongly reduced by the addition of cholesterol in excess of 20 mol% of the phospholipid.[15] This is due to BR aggregation accompanying a cholesterol-induced phase separation.

Transient Dichroism Induced by an Electric Field

Transient dichroism of purple membranes has also been observed by subjecting membrane suspensions to pulsed electric fields.[16,17] Dichroism results from orientation of the membrane fragments by the electric field.

[16] R. Shinar, S. Druckmann, M. Ottolenghi, and R. Korenstein, *Biophys. J.* **19**, 1 (1977).
[17] K. Tsuji and K. Rosenheck, *in* "Electro-optics and Dielectrics of Macromolecules and Colloids" (B.R. Jennings, ed.), p. 77. Plenum, New York, 1979.
[18] T. Kouyama, Y. Kimura, K. Kinosita, Jr., and A. Ikegami, *FEBS Lett.* **124**, 92 (1981).
[19] A. U. Acuna and J. Gonzalez-Rodriguez, *An. Quim.* **75**, 630 (1979).
[20] R. J. Cherry and R. E. Godfrey, *Biophys. J.* **36**, 257 (1981).
[21] W. Hoffmann, C. J. Restall, R. Hyla, and D. Chapman, *Biochim. Biophys. Acta* **602**, 531 (1980).
[22] M. P. Heyn, R. J. Cherry, and N. A. Dencher, *Biochemistry* **20**, 840 (1981).

The decay of dichroism after removal of the field is complex, exhibiting both fast and slow components.[16] The slow component (relaxation time ~100 msec) is simply due to rotation of the whole membrane fragments, while the interpretation of the faster component (relaxation time ~260 μsec) remains obscure. The maximum dichroism observed[17] is consistent with the measurements of chromophore orientation discussed earlier.

Acknowledgments

I wish to thank the Swiss National Science Foundation for financial support.

[32] Spectroscopic Methods for Protonation State Determination

By DIETRICH KUSCHMITZ and BENNO HESS

Spectroscopic methods for the determination of the protonation state of the purple membrane during the photocycle of the retinal chromophore are described for steady-state and transient-state conditions. The analysis of the aromatic spectral region indicates deprotonation of tyrosine and proton displacement from tryptophan. The simultaneous measurement of the protons released into the medium, using the pH indicator methylumbelliferone (MU), and of the state of the photocycle allows a quantification of the number of protons related to each photocycle.

The purple membrane of *Halobacterium halobium* acts as a light-driven proton pump, located at the bacteriorhodopsin molecule. The spectroscopic study of the photocycle as well as the cycle of deprotonation and protonation allows us to establish the ratio between the number of protons turned over and the turnover of the photocycle. The spectroscopic analysis of the deprotonation–protonation cycle requires intrinsic or extrinsic indicators that optically, by absorption or fluorescence, respond to proton dissociation reactions. Examples for both types of indicators are described: The use of absorption changes of intrinsic tyrosyl and tryptophyl residues of bacteriorhodopsin[1] and of fluorescence changes of externally added methylumbelliferone[2] as monitors for released and rebound protons, respectively, indicating changes of the protonation state of bacteriorhodopsin during the photocycle and for the determination of the proton to photocycle ratio.

[1] B. Hess and D. Kuschmitz, *FEBS Lett.* **100**, 334 (1979).
[2] D. Kuschmitz and B. Hess, *Biochemistry* **21**, 5950 (1981).

Intrinsic pH Indicators Tyrosine and Tryptophan

Tyrosyl and tryptophyl amino acid residues are intrinsic pH indicators in proteins.[3] The tyrosyl residue displays a dissociation reaction itself (in aqueous solution, $pK_a = 10.1$), the tryptophyl residues respond to charge perturbations in their neighborhood, operating through space.[4] Bacterio-rhodopsin contains 7 tryptophans and 11 tyrosines.[5]

Light–Dark Difference Spectra of Purple Membrane in the Absorption Range of Tyrosine and Tryptophan. Light–dark difference spectra in the UV absorption range can be obtained with purple membrane in suspensions and oriented in thin films (see Korenstein and Hess, Chapter 22, this volume). Examples of the spectral changes under continuous illumination of a purple membrane suspension and after applying a 1.5-μsec laser flash to a thin purple membrane preparation are given in Figs. 1a and 1b, respectively.[1] They show an absorption increase with maximum at 296 nm and an absorption decrease centered at 275 nm with more or less pronounced shoulders at 265–270 nm and 280 nm[6] and a further irregularity between 285 and 290 nm. The isosbestic points are at 290 nm and 255–260 nm. At lower wavelengths a second absorption increase with maximum around 240 nm is observed (not shown here, see Rafferty[7]). The spectrum can be further decomposed kinetically into fast- and slow-reacting components using laser flash excitation of 15-nsec pulse duration (Fig. 1c).[8]

Interpretation of the Difference Spectra. Difference spectra of the type shown in Fig. 1 are characteristic of three events[3]:

1. The dissociation of tyrosyl protons and the formation of tyrosyl-phenolate anions ($\lambda_{max} = 293, 240$ nm).
2. Charge perturbation of tryptophyl residues: displacement of protons from the neighborhood of the indole ring of tryptophan ($\lambda_{max} = 292$ nm, $\lambda_{min} = 263$ nm).
3. Solvent perturbation to a higher polarity environment of tyrosine ($\lambda_{min} = 276, 285$ nm) and tryptophan ($\lambda_{min} = 291, 283$ nm).

EXAMPLE: A difference spectrum of four tryptophans and one tyrosine (Fig. 2b) in ethanol at pH 6, simulating the protonated dark state and low

[3] J. W. Donovan, *in* "Physical Principles and Techniques of Protein Chemistry" (S. J. Leach, ed.), Part A, p. 101, Academic Press, New York, 1969.
[4] L. J. Andrews and L. S. Forster, *Biochemistry* **11**, 1875 (1972).
[5] Y. A. Ovchinnikov, N. G. Abdulaev, M. Y. Feigina, A. V. Kiselev, and N. A. Lobanov, *FEBS Lett.* **100**, 219 (1979).
[6] R. A. Bogomolni, L. Stubbs, and J. K. Lanyi, *Biochemistry* **17**, 1037 (1978).
[7] C. N. Rafferty, *Photochem. Photobiol.* **29**, 109 (1979).
[8] D. Kuschmitz and B. Hess, *FEBS Lett.* **138**, 137 (1982).

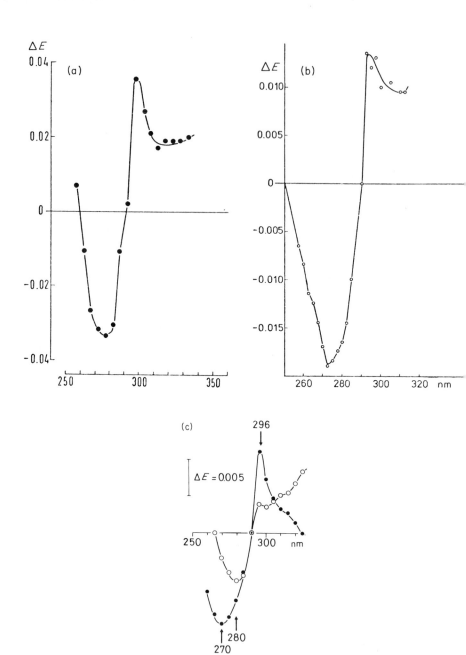

Fig. 1. (a) Steady state light–dark amplitude difference spectrum of a purple membrane suspension. 10 μM bacteriorhodopsin, pH 7.0, 4°. (b) Transient state difference spectrum of maximum amplitude change after a 1.5 μsec laser flash of 570 nm. Thin layers of purple membrane, OD (565) = 0.9, −2.5°. (c) Difference spectra of purple membrane suspension after 15 nsec 570 nm laser excitation. (●) Slow-reacting component, (○) fast-reacting component.

polarity, and in water at pH 12, simulating the deprotonated light state and high polarity, shows all characteristic light-induced absorption changes of bacteriorhodopsin.

Relation to the Photocycle. Variation of the 570-nm excitation light intensity and plotting the reciprocal absorption changes at 275, 296, and 412 nm versus the reciprocal light intensity yields straight lines.[1] This shows the direct relationship between the UV absorption changes and the photocyle and allows an extrapolation to light saturation for quantitative analysis. The direct relationship to the photocycle is furthermore indicated by cross-excitation experiments. Additional excitation of the M_{412} intermediate changes the UV absorption kinetics and vice versa.[1]

Differentiation from the Photocycle. The following methods were successfully used to differentiate between 275-nm, 296-nm absorption changes and M_{412} and also between 275- and 296-nm absorption changes.[1]

1. Kinetic measurements show that after flash excitation the UV absorption changes are faster than M_{412} formation in the order: $k_{on}(\Delta E_{275}) > k_{on}(\Delta E_{296}) > k_{on}(M_{412})$.

2. Measurements of the pH dependence show that at pH 11 the UV absorption changes are slower than M_{412} decay in the order: $k_{off}(M_{412}) > k_{off}(\Delta E_{296}) > k_{off}(\Delta E_{275})$.

3. Measurements of the linear dichroism (LD) show high LD at 275, 320, and 412 nm and a comparatively small LD at 296 nm.

Problems of Quantification. In comparison to the difference spectrum of a tyrosine–tryptophan model mixture (Fig. 2) the light–dark UV difference spectrum of bacteriorhodopsin (Fig. 1) is distorted, suggesting an overlapping absorption increase around 320 nm and an absorption decrease below 290 nm. This overlapping must be due to retinal chromophore interaction, trans-cis isomerization[8] as well as retinal–tryptophan interaction[9] as indicated by the linear dichroism[1] and circular dichroism bands[9-11] that are changed in the light state[11] and strongly changed in the apomembrane.[10] At 296 nm, however, both the LD and CD contributions are very small,[9-11] indicating only a minor interference with the retinal chromophore and allowing therefore a quantitative analysis. Using purple membrane suspensions in glycerol for minimizing light-scattering and polarity effects, an absorption change of 4000–5000 1 mol^{-1} per mole bacteriorhodopsin (extrapolated to light saturation) can be determined that

[9] D. D. Muccio and J. Y. Cassim, *J. Mol. Biol.* **135,** 595 (1979).
[10] D. D. Muccio and J. Y. Cassim, *Biophys. J.* **26,** 427 (1979).
[11] M. Yoshida, K. Ohno, Y. Takeuchi, and Y. Kagawa, *Biochem. Biophys. Res. Commun.* **75,** 1111 (1977).

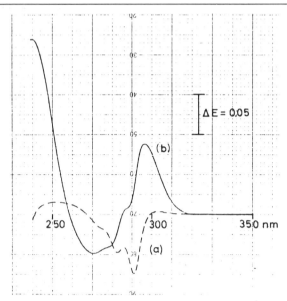

FIG. 2. (a) Difference spectrum between a solution of 80 μM tryptophan/20 μM tyrosine (pH 6) in ethanol (reference cuvette) and water (sample cuvette). (b) After adjustment of the water solution (sample cuvette) to pH 12.

accounts[3] for either two deprotonated tyrosines ($\Delta\epsilon = 2480$) or one deprotonated tyrosine and two tryptophans from which protons or positive charges are displaced ($\Delta\epsilon = 800-1100$). Because the 240-nm absorption increase is completely incompatible with two tyrosines, we quantify the 296-nm absorption change in terms of one deprotonated tyrosine and two tryptophans depleted from protons or positive charges per photocycling bacteriorhodopsin.

Determination of the Number of Protons Released from the Purple Membrane per Cycling Bacteriorhodopsin

The main problem in making spectrophotometric measurements of pH changes in the medium associated with the retinal photocycle using optically responding pH indicators is the optical interference between the indicator and the various intermediates of the retinal photocycle. Among various possible indicators the fluorescence change of methylumbelliferone was chosen, measured at 465 nm, which is near the isosbestic point for the BR_{570} dark state and the M_{412} spectrum. At pH 7.5 and high ionic strength only 10% optical interference takes place.

Properties of the pH Indicator Methylumbelliferone (MU). The criteria for the selection of a pH indicator as given by Chance and Scarpa[12] are high sensitivity, no binding to the material under investigation, and independence of the pK_a on the experimental conditions. The umbelliferone exhibits high fluorescence quantum yield (fluorescence maximum around 450 nm, excitation at 366 nm) in the dissociated state and a very low one in the undissociated state.[13] The absorption change at 420 nm between the dissociated and undissociated state is negligible. Binding experiments with respect to purple membrane showed that with unbuffered or buffered (20 mM potassium phosphate) purple membrane suspensions (pH 7.5) 95–100% of the indicator, added in equimolar concentrations (10 μM), could be recovered in the supernatant after centrifugation of purple membrane suspensions in water and solutions up to an ionic strength of 4 M sodium chloride. The remaining MU were recovered after only one additional washing of the purple membrane pellet with the same volume of the suspension medium. Only with 50 mM Tris-buffer (pH 7.5) the recovery in the first supernatant was 86 and 81% at an ionic strength of 2 M KCl and NaCl, respectively.

An interaction between the pH indicator and the purple membrane was furthermore excluded by the titration of 50 μM MU in the presence and absence of purple membrane (in 20 mM potassium phosphate at 2°). The presence of 12 μM bacteriorhodopsin quenched the MU fluorescence by about 20% but did not affect the pK_a of the titration curves significantly. The pK_a was 7.4 in the absence of purple membrane and 7.5 in its presence. The quenching is probably due to the absorption of bacteriorhodopsin at 366 nm, the excitation wavelength of the MU fluorescence. The pK_a of MU was not changed by the addition of 2 M NaCl. A true binding of the indicator would shift its pK_a by 3.4 units.[14] Thus, for the conditions used as given here, a binding of MU to purple membrane can be neglected.

Compensation of Intermolecular Filtering Effects. In order to compensate for intermolecular filtering effects due to the absorption of M_{412} at 366 nm developing on illumination, each experiment was calibrated by the addition of 10 nmol HCl per 2.05-ml cuvette volume. Experiments were then repeated in the presence of 50 mM Tris-buffer of ph 7.5 (Fig. 3). The buffering did not affect the light-induced absorption change at 420 nm but did decrease the fluorescence changes. At high salt concentrations about 10% of the fluorescence change observed in the absence of the buffer remained in the presence of the buffer (Fig. 3a). We interpret this

[12] B. Chance and A. Scarpa, this series, Vol. 24, p. 336.
[13] P. Fromherz, *Biochim. Biophys. Acta* **323**, 326 (1973).
[14] M. S. Fernandez, and P. Fromherz, *J. Phys. Chem.* **81**, 1755 (1977).

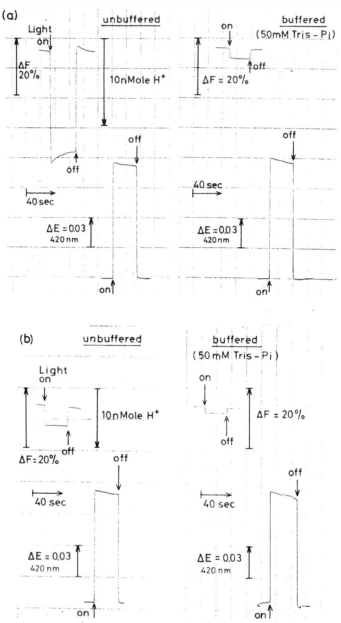

FIG. 3. Light-induced proton release and M_{412} formation of unbuffered (adjusted to pH 7.5 by fluorescence) and buffered (50 mM Tris-phosphate, pH 7.5) purple membrane suspension in (a) 2 M Nacl and (b) water. 11 μM bacteriorhodopsin, 10 μM MU, 0.5–1.0°, light intensity = 200 W/m².

as the unspecific change due to the intermolecular filtering effect. At low salt concentrations the unspecific fluorescence change was about 50% of the change seen without the buffer (Fig. 3b). The unspecific fluorescence change (ΔF_{buffer}) was linear with ΔE_{420} over the whole intensity range of actinic illumination (about 4–40% light saturation) for low as well as for high ionic strength. The fluorescence change, specific for an acidification, is defined as the difference between ΔF_{total} and ΔF_{buffer}.

   Light Effective Volume. Because the measuring cuvette was not totally illuminated by the actinic light, the light effective volume out of a total volume of 2.05 ml had to be determined, giving the nanomoles of M_{412} formed in the illuminated state. This was performed as indicated schematically in Fig. 5b by measuring the absorption difference of a purple membrane suspension in basal salt ether[15] on 578-nm illumination in the horizontal direction (ΔE_h) with the actinic light path (d_h) being 1 cm horizontally and simultaneously in a vertical direction (ΔE_v) where the actinic light path length is not known. The average vertical actinic light path length (d_v) is given by the ratio between the vertical and horizontal absorption change ($d_v = E_v d_h / E_h$). The average value in a $1 \times 1 = $ cm cuvette was 1.12 cm, giving an average light effective volume of 1.12 ml. The nanomoles of released protons in the illuminated state were determined in the total volume (2.05 ml) by calibration of the pH indicator response with known amounts of acid and base (see below).

   Estimations of H^+/M_{412} Ratios. Calculations were carried out with the molar extinction coefficient $\epsilon = 63,000 \text{ M}^{-1} \text{ cm}^{-1}$ for BR_{570}[15] and the differential molar absorption coefficient $\Delta\epsilon(420) = 23,000 \text{ M}^{-1} \text{ cm}^{-1}$ for the M_{412} intermediate.[16] Protonation and deprotonation reactions of the purple membrane (10–11 μM) were analyzed around 0° in unbuffered suspensions of purple membrane in water or NaCl or KCl solutions (up to 4 M) containing 10 μM MU. The pH indicator was titrated to half maximal fluorescence intensity corresponding to a bulk pH around 7.5 by addition of small amounts of either NaOH or KOH. Before or after the pH adjustment of the system the purple membrane was light adapted by continuous illumination with 578-nm light until the absorption at 570 nm remained constant. Fluorescence changes were measured with 366-nm excitation of MU; then absorption changes at 420 nm were measured without excitation of the indicator. After calibration the experiments were repeated in the presence of 50 mM Tris-buffer pH 7.5 for the determination of the unspecific fluorescence change. Examples for a purple membrane

[15] D. Oesterhelt and B. Hess, *Eur. J. Biochem.* **37**, 316 (1973).
[16] R. H. Lozier, W. Niederberger, R. A. Bogomolni, S. Hwang, and W. Stoeckenius, *Biochim. Biophys. Acta* **440**, 545 (1976).

suspension in 2 M NaCl and in water are shown in Fig. 3a and 3b, respectively. The experiments show that the 420-nm amplitude is only slightly affected by ionic strength, but that the amount of protons released in the illuminated state is considerably smaller at low compared to high ionic strength. The H^+/M_{412} ratios determined in the experiment of Fig. 3 are 1.75–1.95 (equilibrium and maximum value,respectively) and 0.4 in the 2 M NaCl and water suspension, respectively.

Remarks. The method can also be applied to H^+/M_{412} determinations under transient conditions.[17] Ratios are again calculated from the peak amplitudes of the 420-nm absorption change and from the specific fluorescence change of the indicator.[2]

Other Methods for H^+/M_{412} Determinations. Besides the glass electrode,[15,18–20] the following pH indicators (absorption changes) were used for the determination of pH changes in the medium on light activation of bacteriorhodopsin: umbelliferone,[16] brilliant yellow,[21] bromocresol green[22] and p-nitrophenol.[16,23]

Instrumentation

Spectrophotometry of Purple Membrane Thin Films. A single-beam spectrophotometer was used as given schematically in Fig. 4. The probes, thin purple membrane preparation on quartz glass slides adjusted to 94% humidity (Korenstein and Hess, Chapter 22, this volume), were placed into sealed cuvettes such that the axis of exciting and analyzing light formed an angle of 45° with the glass slide plane.

Spectrophotometry of Purple Membrane Suspensions. Measurements were performed with a "Dortmund" dual wavelength instrument,[15,24] as shown schematically in Fig. 5A. For the measurement of the M_{412} intermediate of the photocycle the instrument was equipped with two 400-W tungsten lamps, and the monochromators were adjusted to 420 nm (for M_{412} measurement) and 570 nm (for control of light adaptation). The photomultiplier was protected with guard filter (420 ± 30 nm, Schott). For the measurements in the UV spectral region a 200-W xenon–mercury

[17] B. Chance, M. Porte, B. Hess, and D. Oesterhelt, *Biophys. J.* **15**, 913 (1975).
[18] H. Garty, G. Klemperer, M. Eisenbach, and S. R. Caplan, *FEBS Lett.* **81**, 238 (1977).
[19] E. Bakker and S. R. Caplan, *Biochim. Biophys. Acta* **503**, 362 (1978).
[20] Y. Avi-Dor, R. Rott, and R. Schnaiderman, *Biochim. Biophys. Acta* **545**, 15 (1979).
[21] R. H. Lozier, R. A. Bogomolni, and W. Stoeckenius, *Biophys. J.* **15**, 955 (1975).
[22] N. Dencher and M. Wilms, *Biophys. Struct. Mech.* **1**, 259 (1975).
[23] R. Govindjee, T. G. Ebrey, and A. R. Crofts, *Biophys. J.* **30**, 231 (1980).
[24] B. Hess, H. Kleinhans, and H. Schlüter, *Hoppe-Seyler's Z. Physiol. Chem.* **351**, 515 (1970).

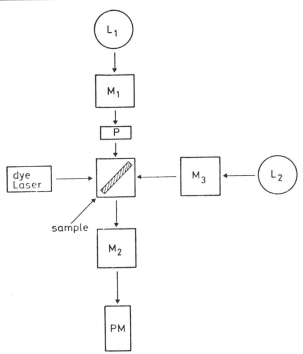

FIG. 4. Block diagram of the single-beam spectrophotometer. In the analyzing light path light from a 200-W xenon–mercury lamp (L_1), selected through two prism monochromators (M_1, M_2; Zeiss MQ4III) prior to and after the sample compartment, was used for UV and M_{412} measurements. For measurements of the linear dichroism a Glan–Taylor ultraviolet prism polarizer (P) was placed into the measuring light path prior to the sample. Light of a second 200-W xenon-mercury lamp (L_2), selected by a Bausch and Lomb monochromator (M_3), served for light adaptation of the sample and for cross-excitation at 412, 296, and 275 nm. The dye laser is a Rhodamine G6 dye laser from Zeiss or Quantel.

lamp and occasionally a prism monochromator (Zeiss MQ4) were used for the UV light path. The photomultiplier was protected with a broad-band filter (230–430 nm).

Fluorometry. The arrangement for the measurement of the fluorescence of the pH indicator methylumbelliferone (MU) is also given in Fig. 5a.

Illumination Technique. For continuous illumination and pulsed illumination down to 20 msec (performed with a commercial photoshutter of 2.4 msec opening and closing time) a 900-W xenon arc was used in the arrangement given schematically in Fig. 5a. Light intensity was measured with a Tektronix J16 Digitalphotometer. Flash illumination was performed with a Rhodamine G6 dye laser (Zeiss or Quantel).

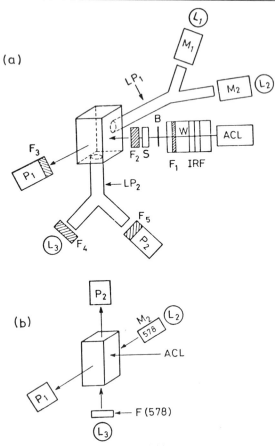

Fig. 5. (a) Block diagram of the dual wavelength instrument, fluorescence measurement and continuous illumination. Photometry: Light of two lamps (L_1, L_2) is selected by two monochromators (M_1, M_2; double monochromators of Bausch and Lomb) and chopped by rotating disks at 1 kHz (not shown), passed through light pipes (LP_1) onto the side of the cuvette toward a photomultiplier (P_1) equipped with a suitable filter to protect the photomultiplier against the actinic light (F_3). Fluorometry: Light of a 200-W xenon–mercury lamp (L_3), selected through a 366-nm Eppendorf filter (F_4) is guided through a light pipe (LP_2) onto the bottom of a 1 × 1-cm all quartz cuvette for fluorescence excitation. The emission is accepted through the same light pipe and passed through the second arm towards a photomultiplier (P_2) through an interference filter (F_5) of 465 nm (Schott). Actinic light: Actinic light is obtained by a 900-W xenon arc (ACL, Zeiss Ikon), passed through a water bath containing a 500-nm cutoff filter (F_1) and 3 infrared filters (IRF), an iris diaphragm (B), a collecting lens (omitted in the scheme), a photoshutter (S) and a 578-nm Eppendorf or 576-nm Schott interference filter (F_2) onto the side of the cuvette. (b) Experimental setup for the measurement of the light effective volume (for explanation see text).

Temperature Control. In all experiments carried out below room temperature, temperature was adjusted and controlled with equipment from L'Air Liquide (France).

Electronics. The electronics for the single-beam spectrophotometer and the dual wavelength instrument (time constants 80 nsec, 15 μsec and 3 msec, respectively) were made in Dortmund (modified according to Oesterhelt and Hess[15] and Hess *et al.*[24] They were used for absorption and fluorescence measurements under pulsed illumination. Output signals were recorded with transient recorders (Biomation 802, Nicolet 1090, 2090 and 1174). Fluorescence changes under continuous illumination were measured with a Keithley 153-μV ammeter, connected through a Rockland electronic filter (10-Hz) with a two-channel Siemens recorder, simultaneously connected with the output of the dual wavelength instrument measuring the M_{412} intermediate (420-nm light path).

[33] Monitoring of Protein Conformation Changes during Photocycle

By BRIAN BECHER

The absorption of light by the light-adapted purple membrane protein (BR) results in a photochemical reaction cycle through a series of intermediates.[1] Accompanying this cycle is the pumping of protons across the membrane.[2] In order to study this relationship in terms of possible protein conformation changes, the use of low-temperature ultraviolet (320–245 nm) absorption spectroscopy has proved valuable.[3]

At specific low temperatures, certain intermediates of the bacteriorhodopsin photocycle can be "trapped" or prevented from thermally converting to the next intermediate in the sequence. This method offers a number of distinct advantages in the study of protein changes in the photocycle intermediates. First, relatively large percentages of the pigment can be converted to the first three intermediates (K, L, and M). Samples can be prepared with 28% K, 65% L, or 100% M. In addition, the only other species found in significant amounts in these pigment mixtures is BR. This greatly simplifies analysis of the spectra of the intermediate. Furthermore, since the low-temperature technique traps the intermediates, the absorption measurements can be made carefully over rela-

---

[1] R. Lozier, R. A. Bogomolni, and W. Stoeckenius, *Biophys. J.* **15**, 955 (1975).

[2] D. Oesterhelt and W. Stoeckenius, *Proc. Natl. Acad. Sci. U.S.A.* **70**, 2853 (1973).

[3] B. Becher, F. Tokunaga, and T. G. Ebrey, *Biochemistry* **17**, 2293 (1978).

tively long periods of time. This is especially important in the measurement of proteins in the ultraviolet range where the signal-to-noise ratio is relatively low. Consequently, the low-temperature technique provides distinct advantages in terms of accuracy in ultraviolet absorption measurements of the bacteriorhodopsin protein.

A major problem with all ultraviolet absorption measurements of particulate (membrane) systems is artifacts resulting from nonselective and selective light scattering of the measuring beam by the sample.[4] This problem is compounded in the case of low-temperature measurements owing to freeze-cracking of the sample. Freeze-cracking of the sample can easily increase light scattering to the extent that reliable ultraviolet absorption measurements are impossible. Both of these complications can be greatly alleviated by use of 67% (v/v) glycerol in the membrane sample. This concentration of glycerol significantly decreases light scattering by more closely matching the index of refraction of the media with that of the membrane. In addition, glycerol addition prevents serious cracking of the sample at low temperatures if the sample is gradually cooled.

In order to monitor the protein changes during BR conversion to K and L, purple membrane is first suspended in a 2:1 glycerol–water (0.02 M phosphate buffer, pH 7.0) solution. For the study of the M conversion, purple membrane is first suspended in a 25% NaCl solution titrated to pH 10 with 0.1 M NaOH and then mixed with 2 parts of glycerol. Under these conditions the lifetime of the M intermediate greatly increases, allowing accurate spectral measurements.[5]

Low-temperature visible and ultraviolet absorption spectra of the K, L, and M intermediates are then recorded using a quartz Dewar flask. For the measurements of the conversion of BR to the K intermediate, purple membrane in the glycerol–buffer medium is placed in a stoppered 0.2-cm quartz cell, fully light adapted, and then, in the dark, lowered into an ethanol bath previously cooled to − 100° with liquid nitrogen. The cell is then removed from the ethanol bath and, after any ethanol remaining on the cell surface is rapidly blown off with a burst of nitrogen gas, lowered into the Dewar filled with liquid nitrogen. In this manner the absorption measurements of the sample are not significantly hampered by cracking of the sample on lowering its temperature to − 196°. Light-adapted purple membrane can be converted to the steady-state mixture of the K intermediate by 500–nm light (500-W slide projector light source and an interference filter). Measurement of the partial conversion of BR to the L intermediate at − 100° is done similarly except that the cell is placed in a Dewar flask

[4] P. Latimer and E. Rabinowitch, *Arch. Biochem. Biophys.* **84,** 428 (1959).
[5] B. Becher and T. G. Ebrey, *Biophys. J.* **17,** 185 (1977).

filled with ethanol cooled to $-100°$. A 640-nm interference filter is used in photoconverting the purple membrane to L, which remains stable at $-100°$. Complete conversion to M can be achieved by exposure of the NaCl–buffer–glycerol sample to 600-nm light at $-40°$.

The near ultraviolet absorption spectrum (320–245 nm) of light-adapted BR includes distinct maxima or shoulders at 290, 280, and 274 nm that are primarily attributed to $\pi-\pi^*$ transitions of the amino acids tryptophan and tyrosine. (Cystine is not found in bacteriorhodopsin.) In addition, light scattering and minor transitions of the retinal chromophore contribute to the 320–245-nm spectrum. The secondary, tertiary, or quarternary structure of the purple membrane protein may also alter the environment of the $\pi-\pi^*$ transitions of the aromatic amino acids relative to free amino acids from which extinction coefficients are estimated. In particular, side chains "buried" within the nonpolar environment of the protein are expected to have appreciably higher extinctions than those of free amino acids.[6]

In the case of BR conversion to K at $-196°$, no significant change in the near ultraviolet spectrum is found, indicating little protein conformation change at $-196°$ (as expected). However, the conversions of BR to L and BR to M result in large decreases in near ultraviolet extinction. A conversion to 65% L results in a decrease in 280-nm extinction of 6000 ± 1000 liters cm^{-1} mol^{-1}. Similarly, a 5000 ± 1000 liter cm^{-1} mol^{-1} decrease occurs on complete conversion to M. Most striking is the BR to L and BR to M *difference* spectra in the near ultraviolet that have shapes similar to the near ultraviolet absorption spectra of BR itself, including maxima or shoulders at 290, 280, 270 nm (Becher *et al.*[2]). These results strongly argue that a change in protein conformation occurs on conversion of BR to L or M. A conformation change in the protein could lead to the decreased absorbance at 280 nm by exposure of approximately 50% of the aromatic amino acids in the relatively nonpolar protein interior to the more polar (water) media. In addition, a conformational change could alter the relative orientation of originally interacting aromatic amino acid transitions, resulting in a loss of hyperchromism. Isomerization of the retinal chromophore is unlikely to contribute more than 2000 liters cm^{-1} mol^{-1} to the extinction decrease.

It is likely that protein conformation changes in the intermediates of the bacteriorhodopsin photocycle are directly involved in the proton pumping accompanying the cycle. Low-temperature ultraviolet spectroscopy is an accurate, relatively uncomplicated technique to measure these protein changes.

[6] J. Donovan, *in* "Physical Principles and Techniques of Protein Chemistry" (S. J. Leach, ed.), Part A, p. 101. Academic Press, New York, 1969.

[34] X-Ray Diffraction Studies of Purple Membranes Reconstituted from Brown Membrane

By Toshiaki Hamanaka, Kenji Hiraki, and Toshio Mitsui

Significant progress has been made in X-ray diffraction studies of bio-membranes. Readers may consult the review article[1] and the references cited in it for a general survey and the article by Akers[2] in this series for relevant X-ray techniques. The present article describes the specific techniques used in studies of the holo-brown membrane.

The brown membrane of *Halobacterium halobium* is considered a precursor of the purple membrane.[3] If the synthesis of retinal is inhibited by addition of nicotine to the growth medium, a differentiated domain similar to that of brown membrane is formed in the plasma membrane. This domain is called apo-brown membrane because of the absence of the retinal. By giving the exogenous retinal to the apo-brown membrane, reconstitution of bacteriorhodopsin (BR) is evidenced by the appearance of a 570-nm absorption band. The resultant membrane is referred to as holo-brown membrane. X-ray diffraction studies proved that there is no crystalline order in the apo-brown membrane, but that a two-dimensional hexagonal lattice appears in the holo-brown membrane.[4,5] Further, our research has attempted to ascertain the rate-limiting step in the formation of purple membrane patches in the holo-brown membrane: reconstitution of BR, cluster formation of BR or lattice formation.[6] The following discussion describes techniques used in these studies in some detail.

Sample Preparation

The method of isolation of the apo-brown membrane from *Halobacterium halobium* grown in the presence of 1 mM nicotine has been described by Sumper and Herrmann.[7]

Figure 1 illustrates the procedure for the preparation of X-ray specimens. The "unoriented specimen" on branch (1) in Fig. 1 means mem-

---

[1] T. Mitsui, *Adv. Biophys.* **10**, 97 (1978).
[2] C. K. Akers, this series, Vol. 32, p. 211.
[3] M. Sumper, H. Reitmeier, and D. Oesterhelt, *Angew. Chem., Int. Ed. Engl.* **15**, 187 (1976).
[4] K. Hiraki, T. Hamanaka, T. Mitsui, and Y. Kito, *Biochim. Biophys. Acta* **536**, 318 (1978).
[5] S.-B. Hwang, Y.-W. Tseng and W. Stoeckenius, *Photochem. Photobiol.* **33**, 429 (1981).
[6] K. Hiraki, T. Hamanaka, T. Mitsui, and Y. Kito, *Photochem. Photobiol.* **33**, 419 (1981).
[7] M. Sumper and G. Herrmann, *FEBS Lett.* **69**, 149 (1976).

METHODS IN ENZYMOLOGY, VOL. 88

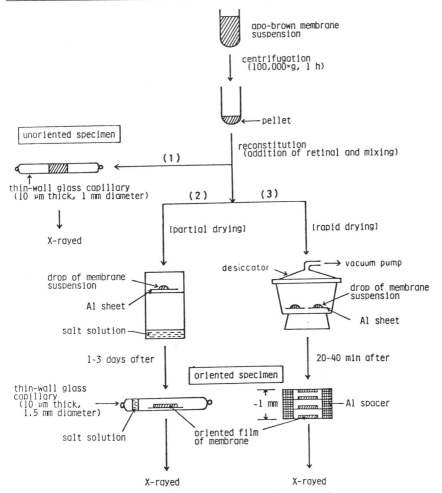

FIG. 1. Procedure for the preparation of X-ray specimens.

brane suspension in solution. In this case it is easy to control the environmental condition of membranes (i.e., pH, ionic strength, etc.). The membrane suspension is usually sealed in a thin-walled glass capillary (10-μm thickness and 1–1.5 mm diameter). As shown on branch (2) in Fig. 1, the oriented specimens were prepared by placing a membrane pellet on aluminum sheet under controlled relative humidity. The humidity was controlled by using the appropriate saturated salt solution.[8] Strips were

[8] R. C. Weast, ed., "Handbook of Chemistry and Physics," 57th ed., p. E-46. Chem. Rubber Publ. Co., Cleveland, Ohio, 1976–1977.

cut off from the partially dried material and sealed in a thin-walled glass capillary as shown on branch (2). The humidity in the capillary was kept constant by putting a small amount of the saturated salt solution in it. For the experiment on the time course of crystallization, oriented specimens were prepared by rapid vacuum drying as shown on branch (3). A larger amount of sample was desired for fast recording and four sheets of dried specimens were stacked with aluminum spacers, glued with cyanoacrylate adhesive. In this case the incident X-ray beam was perpendicular to the plane of the membrane.

X-Ray Measurement

The CuKα radiation (1.542 Å) was used in the experiment, which was filtered by 15-μm-thick Ni foil. The point focusing system is the most appropriate in studies of the X-ray diffraction from the two-dimensional lattice in the membrane. We used fine-focus rotating-anode X-ray generators made by Rigaku Denki Ltd.: RU-100 and FR-B units. They both give 100 × 100-μm^2 focal spot at the take-off angle of 6°. The RU-100 unit was operated at 40 kV, 30 mA (12kW/mm^2) and FR-B at 50 kV, 70 mA (35 kW/mm^2). They are the same types as Elliott GX6 (12 kW/mm^2) and GX13 (27 kW/mm^2), respectively.

We used an Elliott toroidal camera.[9] It gives a more intense beam than the two-mirror Franks camera. Its limiting resolution (about 200 Å in practice) is inferior to Franks camera (about 500 Å), but enough to measure the Bragg reflections by the purple membrane having the lattice constant of 63 Å. The X-ray path between the sample and the detector was evacuated to avoid air scattering.

Measurements of scattered X-ray intensities were made by the film method or a one-dimensional position-sensitive proportional counter. Fuji medical KX is a fine-grain film and gives photographs of high quality. Sakura cosmic ray film and Kodak medical film NS-54T are preferable to reduce exposure time. Their dynamic ranges are limited and therefore the multifilm method was used in all cases. The photographic density of the diffraction pattern was measured with the microdensitometer and converted to the X-ray intensity referring to a standard scale.

Advantages of the counter method are the large dynamic recording range and digital and fast recording. Disadvantages are the one-dimensional recording and relatively poor spatial resolution. We used a resistive-wire linear-position-sensitive detector[10,11] made by Rigaku Denki,

[9] A. Elliott, *J. Sci. Instrum.* **42**, 312 (1965).
[10] J. Borkowski and M. K. Kopp, *Rev. Sci. Instrum.* **39**, 1515 (1968).
[11] A. Gabriel and Y. Dupont, *Rev. Sci. Instrum.* **43**, 1600 (1972).

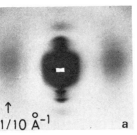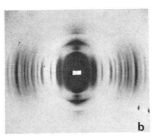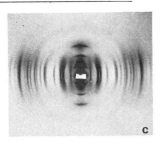

↑
$1/10 \, \overset{\circ}{A}^{-1}$ a   b   c

FIG. 2. X-Ray diffraction photographs obtained from oriented specimens: (a) the apo-brown membrane, (b) the holo-brown membrane, (c) the purple membrane. Incident X-ray beam was parallel to the plane of membranes, which corresponds to the horizontal direction on page. Specimen-to-film distance was 90 mm. The equator is defined as the horizontal line and the meridian as the vertical line, both passing through the center of each figure.

Ltd. Diffraction patterns by the purple membranes were recorded within 5 min. The spatial resolution was 200–400 μm along the wire of the counter. It was difficult to record the Bragg reflection peaks by the purple membrane completely separated at the specimen-to-counter distance of 90 mm with the toroidal optics.

Data

The suspension of membrane in a buffer solution gives the radially symmetric diffraction pattern that is a superposition of the reflections due to the in-plane structural order and the scattering related to the thickness profile of a membrane. The oriented specimen as depicted on branch (2) in Fig. 1 gives the in-plane reflections around the equator and the thickness profile scattering around the meridian, as shown in Fig. 2. Thus better signal-to-noise ratio can be obtained by using the oriented specimens.

Figure 2 shows example of X-ray diffraction photographs of oriented specimens of the apo-brown membrane (a), the holo-brown membrane (b), and the purple membrane (c). No sharp reflection lines are seen in Fig. 2a, proving that there is no crystalline order in protein arrangement in the apo-brown membrane. Many sharp reflections appear around the equator in Fig. 2b, indicating that BR molecules form a two-dimensional hexagonal lattice in the holo-brown membranes. Figure 2c shows the well-known diffraction pattern of oriented purple membranes. The lattice constants estimated from Figs. 2b, and c are the same (63 Å). The reflection lines along the meridian of each figure indicate that there is an order in stacking of membranes in the oriented specimens. For more detailed discussion of these patterns, readers are referred to other literature.[4,6]

[35] Light-Induced Volume Changes

By WILLIAM W. PARSON

Excitation of a suspension of purple membranes with a flash of light causes changes in the volume of the system. The volume changes have several origins. First, the release of heat from bacteriorhodopsin to the solvent can cause a thermal expansion or contraction. Measurements of the heat can provide information on the enthalpy changes that occur at various stages of bacteriorhodopsin's photocycle.[1] Second, the transfer of protons between bacteriorhodopsin and buffers in the solution can cause an expansion or contraction because of a change in the number of charged species in the system. The solvent surrounding a charged group is ordered so that its volume is less than it is in the absence of the charge. Measurements of volume changes thus can provide information on the kinetics[2] and stoichiometry[3] of proton release during the photocycle. In addition, there may be volume changes due to the movement of protons from one group to another on bacteriorhodopsin, or due to other internal transformations in the protein. Measuring these volume changes can give information on processes that may otherwise be very difficult to detect.[2]

Volume changes resulting from flash excitation can be measured with a capacitor microphone transducer mounted on a glass cell.[2,4,5] A current version of the device is shown in Fig. 1. One plate of the capacitor is a 35-mm-square, 0.3-mm-thick stainless-steel diaphragm (e); the other (stationary) plate is a 22-mm-diameter-round, 1.5-mm-thick piece of brass (b). The two plates are separated by a piece of 35-mm-square, 0.076-mm-thick polyimide ("Kapton") tape (d; Connecticut Hard Rubber Co.), which is stuck to the diaphragm. The stationary plate fits inside a cap (j) made by gluing together two pieces of 1.5-mm-thick epoxy board, one of which contains a hole 25 mm in diameter. The cap is fastened to the edges of the diaphragm by eight small (2-56) machine screws (i), holding the entire microphone together. (Nuts for the screws can be recessed into the top of the cap, or simply placed above the cap.) A sturdy wire (or a brass screw) soldered to the stationary plate is brought out through a hole in the cap (a). A fine wire to carry the signal from the microphone to a preamplifier is connected to this wire (k). A second fine wire for charging the capacitor is

[1] D. R. Ort and W. W. Parson, *Biophys. J.* 25, 355 (1979).
[2] D. R. Ort and W. W. Parson, *J. Biol. Chem.* 253, 6158 (1978).
[3] D. R. Ort and W. W. Parson, *Biophys. J.* 25, 341 (1979).
[4] J. B. Callis, M. Gouterman, and J. D. S. Danielson, *Rev. Sci. Instrum.* 40, 1599 (1969).
[5] J. B. Callis, W. W. Parson, and M. Gouterman, *Biochim. Biophys. Acta* 267, 348 (1972).

METHODS IN ENZYMOLOGY, VOL. 88

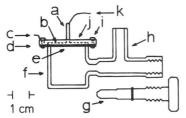

FIG. 1. Capacitor microphone apparatus. a, Heavy wire connected to stationary plate of capacitor; b, stationary plate (inside cap); c, connection for charging diaphragm; d, Kapton tape dielectric (stuck to upper face of diaphragm); e, steel diaphragm; f, pyrex cell; g, Teflon valve stem; h, threaded glass valve and solution reservoir; i, screws and nuts fastening cap to diaphragm; j, epoxy board cap; k, connection to preamplifier. See text for details.

connected to one corner of the diaphragm via one of the screws that pass through the cap (c).

The diaphragm of the microphone is glued to a cell made of 25-mm-square pyrex tubing (f). For filling the cell is equipped with an angle-design threaded glass needle valve with a 4-mm bore (h; Lab-Crest). Earlier versions of the apparatus[5] had a spring-loaded valve, but this is not needed. Epoxy cement is used to attach the microphone to the cell. (Cut the top of the cell with a glass saw and smooth the cut surface with sandpaper; leave the finish sufficiently coarse for good adhesion of the epoxy.) After the microphone has been mounted on the cell, the outer surface of the microphone, including the cap, mounting hardware, and the heavy wire, is coated with several layers of clear epoxy cement.

The assembled microphone typically has a capacitance of 130 to 170 pF. Its dc resistance should be $\geq 10^{13}$ ohms. Low resistance reveals electrical leakage that will make the volume measurements unacceptably noisy.

The diaphragm of the microphone is charged to $+300$ V dc with a battery. The stationary plate is connected through a 100 kohm resistor to a preamplifier with high input impedence and low noise (e.g., a Burr-Brown model 3420L FET-input operational amplifier in a noninverting configuration with a gain of 1). Leakage current from the microphone and the input bias current of the preamplifier are offset by an adjustable voltage source (e.g., a 10-turn potentiometer spanning ± 1.5 V), which is brought to the preamplifier input through a very large resistance (e.g., a 4.9×10^{10} ohms Eltec model 104 miniature resistor). The low-frequency cutoff of the microphone depends on the value of this resistance. With a 4.9×10^{10}-ohm resistor, a 150-pF capacitor has an RC relaxation time of 7.4 sec. Much longer relaxation times are impractical, because the apparatus becomes too sensitive to thermal drift and other sources of low-frequency noise.

The preamplifier is followed by an amplifier, which should have an input noise level of less than 25 μV over the frequency range 0–300 kHz (e.g., Tektronix AM502). To optimize the signal, it is useful for the amplifier to have selectable high- and low-frequency filtering and gain. The amplified and filtered signals can be sent to a digital signal averager for further enhancement of the signal/noise ratio. Beware that low-frequency filtering effectively introduces additional capacitance in series with the microphone. This can distort the signal by a double differentiation. The reciprocal of the low-frequency cutoff of the filter must be kept large compared to the time scale of the measurement.

For temperature control the microphone cell can be held in an aluminum or copper block that is heated or cooled by a thermoelectric module (EG & G Co.). This allows one to change the temperature more conveniently than is possible by simply circulating thermostated water through the block. The current through the thermoelectric module can be regulated by a proportional controller. The temperature sensor for the controller is best mounted in the block, rather than in the microphone cell itself. Mounting the sensor in the cell[5] gives overshoot in the control, because of the lag in thermal conduction between the block and the cell. A temperature sensor in the cell is not necessary if one routinely allows about 30 min for the cell to equilibrate with the block. However, the temperature in the cell will not be identical with that of the block if the block is much below or above room temperature; the temperature differential requires calibration.

The cell holder, preamplifier, and battery all should be housed in a sturdy metal box that is well shielded electrically and acoustically. Mount the box on a vibration-free platform, such as a balance-table slab on a pile of unfolded newspapers. The excitation light should enter the box through a glass or acrylic window that shields the microphone from acoustical noise and from dampness. A desiccant may be needed for keeping the interior dry, because any condensation of moisture on the surface of the microphone or preamplifier will make the signals noisy. Purging the box with dry air or N_2 helps. The mounting of the components in the box should be designed to minimize vibration of the offset-current resistor for the preamplifier and of the wires interconnecting the microphone, preamplifier, and battery; these vibrations can be another serious source of noise. Guard against vibrations introduced on cables and water lines connecting the box to external components.

The best excitation source for volume measurements is a flashlamp-pumped dye laser. Because the beam is well collimated, the laser can be placed several meters away from the microphone box. This helps to minimize acoustical and electrical noise associated with the flash. Rhodamine

6G (588 nm) or Green-9 (563 nm) are both useful dyes: Rhodamine 6G gives stronger lasing, but the wavelength of Green-9 is better matched to the absorption spectrum of bacteriorhodopsin.[3]

The volume cell must be filled completely with the suspension that is to be studied: Air bubbles will cause serious distortions of the signals. To remove dissolved air and small bubbles, overfill the cell into the side arm and then apply a mild vacuum with an aspirator pump. The expanded bubbles can be urged out of the cell by tapping a bottom corner of the cell gently against a hard surface. Another procedure is to warm the filled cell by placing it in warm water with the valve slightly open, then to close the valve and cool the cell by rinsing it with methanol. The contraction of the solution resulting from the cooling will cause air bubbles to expand. After the cell is freed of bubbles, it should be rinsed with water and then methanol (with the valve open), dried carefully, and placed in the temperature-controlled box. The valve should be left open until the cell temperature is close to that of the block. The valve then can be closed and the microphone connected to the battery and the preamplifier. To protect the amplifier, it is advisable to ground the input while one switches on the battery. The valve should be opened again during changes in temperature of the block.

One source of variability in the measurements is distortion of the capacitor microphone by the pressure that builds up when one closes the valve. Slow closing minimizes this variability. To check for this effect, measure the decay time of the signal after the imposition of a voltage step to the diaphragm. The capacitance, and hence the RC decay time, is inversely proportional to the gap between the plates of the microphone.

Procedures for calibrating the apparatus and additional details on the measurements are described in the literature.[1-5] By averaging the signals on 10–20 flashes, one can measure volume changes of less than 5 pl. If the solvent is pure H_2O at 20° (coefficient of thermal expansion 2×10^{-4} deg^{-1}), an expression of 5 pl will result from an enthalpy change of about 2.5×10^{-5} cal. A 588-nm excitation flash with a strength of 5×10^{-10} einsteins (3×10^{14} quanta) will introduce this amount of heat if it is absorbed by a solution of an inert absorber. The temperature change in the 10-ml solution is about $2.5° \times 10^{-6}$. Linearity of the response should be checked by using an inert absorber and varying the excitation intensity; the response is generally linear up to volume changes of at least 2 μl.

The response time of the apparatus is on the order of 100 μsec. However, this speed can be achieved only when the flash-induced volume changes are relatively small. A sudden, relatively large volume change causes oscillations with a period of about 150 μsec, which decay with a damping time of about 500 μsec. The oscillations probably result from pressure waves in the cell. They might be decreased by reducing the size of the cell, but this has not been explored extensively. Smaller cells are more difficult to build and use.

Acknowledgments

The apparatus described here was developed largely through the efforts of Drs. J. B. Callis and D. R. Ort. It is a pleasure to acknowledge their contributions, as well as additional helpful suggestions by Dr. H. Arata. Grants from the National Science Foundation supported the project.

[36] Application of Nanosecond X-Ray Diffraction Techniques to Bacteriorhodopsin

By ROBERT D. FRANKEL and JAMES M. FORSYTH

The bacteriorhodopsin (BR) in the purple membrane (PM) of the *Halobacterium halobium* has been clearly demonstrated to be a light-activated proton pump.[1] As with most biological macromolecules, atomic resolution structural studies involving diffraction techniques will be essential in discovering the molecular mechanisms underlying this function. The 7-Å resolution structure deduced by Henderson and Unwin from X-ray and electron diffraction data is a giant step in this direction.[2]

One particular issue under debate involves the magnitude of any conformational change that may take place during the BR photocycle. We hope to contribute to the resolution of this question by using the newly developed laser plasma X-ray source[3] to obtain subnanosecond X-ray diffraction patterns throughout the course of the 20-msec photocycle.

Generation of subnanosecond bursts of monochromatic X rays intense enough to obtain X-ray diffraction exposures of comparable intensity to those obtained in multihour exposures with rotating anode X-ray tubes requires use of plasmas produced by light from the most energetic and powerful lasers yet built. When such a high-powered laser pulse is focused on a solid, an electron avalanche is driven by the optical field, turning the solid into a highly ionized gas. If the laser pulse is of nanosecond duration or less, most of the light absorbed by the plasma cannot be transported away from the focal region during the pulse. As a result, very high particle temperatures are produced in a small region. With focused optical fields

[1] D. Oesterhelt and W. Stoeckenius, *Proc. Natl. Acad. Sci. U.S.A.* **70**, 2835 (1973).
[2] R. Henderson and P. N. T. Unwin, *Nature (London)* **257**, 28 (1975).
[3] R. D. Frankel and J. M. Forsyth, *Science* **204**, 622 (1979).

of greater than 10^{14} W/cm^2, plasma temperatures on the order of 1 keV are produced. Ions in such high-temperature plasmas are stripped of many electrons and radiate a substantial portion of the absorbed laser energy in the X-ray region of the spectrum. With the proper choice of target materials and laser intensity, a large part of the X-ray radiation may be channeled into a few narrow emission lines.[4] For X-ray diffraction studies, X-ray fluxes in the 4.5–1.5 Å region of the spectrum in a 1% wavelength interval are appropriate.

To generate the PM diffraction pattern discussed later, we used the single-beam Nd^{+3}: glass development laser at the University of Rochester's Laboratory for Laser Energetics (LLE).[5] This laser delivers single pulses of 1.054-μm light with full width at half maximum that can be varied from 50–700 psec. With newly added active mirror booster amplifiers[6] peak pulse powers of greater than 1 TW have been obtained with short pulses and a maximum energy of 250 J with long pulses. The laser output aperture was 15 cm. Our system repetition rate was 2 pulses/hr.

Laser pulses were brought to a focus of 100 μm in a 24-in. diameter vessel that was evacuated to 1×10^{-5} Torr. Plasmas must be produced in a vacuum to prevent air breakdown before the laser light reaches the surface of the target.

Attached to the target chamber was an X-ray camera (shown in Fig. 1)

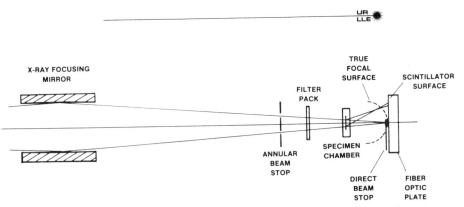

FIG. 1. Schematic diagram of X-ray diffraction camera. The true focal surface of the camera is a sphere tangent to the sample and also tangent to the flat phosphor plane.

[4] J. M. Forsyth, *Trans. Am. Crystallogr. Assoc.* **12**, 11 (1976).
[5] W. Seka, J. Soures, O. Lewis, J. Bunkenburg, D. Brown, S. Jacobs, G. Mourou, and J. Zimmerman, *Appl. Opt.* **19**, 409 (1980).
[6] J. A. Abate, L. Lund, D. Brown, S. Jacobs, S. Refermat, J. Kelly, M. Gavin, S. Waldbilling, and O. Lewis, *Appl. Opt.* **20**, 351 (1981).

consisting of a nickel-coated grazing reflection toroidal mirror[7] with a focal length of 62.5 cm and a mean angle incidence of 0.91°. The collection solid angle of the collector was 2.4×10^{-4} steradians. The camera also included an annular aperture in the reflected X-ray beam, a controlled environmental sample chamber, and a scintillator-image intensifier recording system. The 40-mm ZnS(Ag) scintillator was located 125 cm from the laser plasma source and was coupled to an Amperex model XX1360 25-mm channel plate intensifier through a fiberoptic faceplate. The intensifier had an adjustable luminosity gain of up to 7×10^4 and was housed in a mount permitting direct attachment of various 35-mm and Polaroid camera backs. A lead disk 2.5 mm in diameter was mounted just in front of the scintillator plane to block the direct beam from the X-ray mirror.

Our experiments so far have been performed with highly chlorinated targets; room temperature Saran ($C_2H_2Cl_2$) and pressed polycrystals of hexachloroethane (C_2Cl_6) held at 77 K (low temperature inhibits hexachloroethane sublimation in the vacuum chamber) have been used. In a Saran plasma produced by irradiation with an incident 175J, 700 psec laser pulse focused to $\sim 10^{15}$ W/cm^2, about $5 \times 10^{-2}\%$ of the incident laser energy (or $\sim 3 \times 10^{14}$ photons) is radiated by Cl^{+15} at 4.45 Å in a wavelength interval of $\Delta\lambda/\lambda$ of 8×10^{-3}. Since X-ray production by plasma ions is proportional to the ion density, the hexachloroethane produces about twice the X-ray flux of Saran. However, because Saran is simpler to use, it was the target used in most test shots.

A typical emission spectrum from a Saran plasma is shown in Fig. 2A. In addition to the intense line radiation at 4.45 Å we have X-ray lines at shorter wavelengths and a much weaker background continuum. Effective monochromatization of this emission is possible by means of thin Saran foils. Such filtered spectra are shown in Figs. 2B and 2C corresponding to 12.5-μm and 25-μm foil thicknesses, respectively. The 25-μm foil transmits 57% of the desired line radiation. Outside the spectral range shown in Fig. 2 we filter the long wavelength components with a 25-μm beryllium foil while the short wavelength radiation is not reflected by the toroidal mirror.

In test shots dried PM stacks 60–90 μm were placed in the converging X-ray beam. To prepare the stacks purified PM provided by Dr. Janos Lanyi was stored at 4° in 4 M NaCl until ready for use. The samples were washed free of NaCl by pelleting of the PM from solution in an ultracentrifuge at 50,000 g for 30 min, followed by decantation of the supernatant and resuspension of the PM pellet in double-distilled H_2O. This procedure

[7] B. L. Henke and J. W. M. DuMond, *J. Appl. Phys.* **26**, 903 (1955).

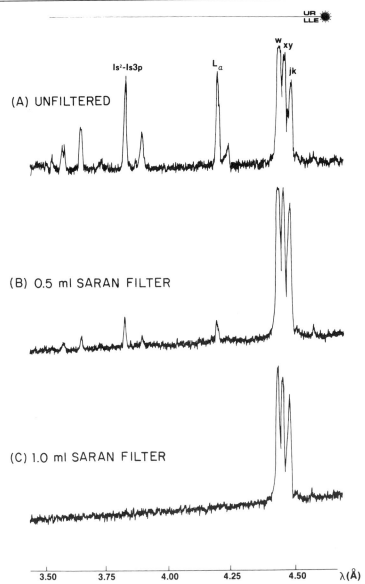

FIG. 2. Chlorine spectra from Saran targets on three different laser shots showing the effect of Saran filtration. All shots were approximately 170 J in 700 psec. The Cl^{+15} resonance line (w) and its associated satellites (x, y, j, k) are the components useful for X-ray diffraction. The $1s^2 - 1s3p$ is a higher energy Cl^{+15} emission line. The $L\alpha$ is radiated from Cl^{+16} ions. Filtration for (A) was 25 μm Be; for (B) 25 μm Be + 12.5 μm Saran; for (C) 25 μm Be + 25 μm Saran. In (C) the transmission of the resonance line and its satellites was 57%, while all other lines were almost completely absorbed.

was repeated three times. After the fourth centrifugation, the pellet was suspended in 0.5 ml of H_2O and a portion was extracted by a syringe and placed on a 25 μm Saran sheet to air-dry. The Saran also served to filter the beam.

Figure 3 shows an X-ray powder diffraction pattern from a PM stack recorded on 2475 high-speed recording film. It was obtained with a single 700-psec 220-J pulse focused on a hexachloroethane target; the camera collected $\sim$3–4 $\times$ 10^9 photons for diffraction. The identifiable reflections are noted. The diffraction rings are indexed on a two-dimensional P-3 hexagonal lattice.[8] Typical high-quality PM powder patterns have been obtained using rotating anode tube sources and require exposure times of 20–50 hr.[9,10] In these patterns strong rings out to 7 Å resolution are observed, while weaker rings out to 3.5 Å are also seen. In our pattern the lower-resolution orders are well resolved and quite intense. These extend

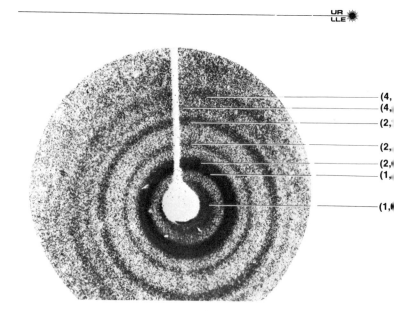

LASER PULSE ENERGY - 213 Joules   X-RAY SOURCE - Cl[+15] Laser Heated Plasma
LASER PULSE WIDTH - 700 psec   X-RAY WAVELENGTH - 4.45 Å

Fig. 3. Purple membrane diffraction pattern. The sample–phosphor separation was 2.7 cm. Identified are rings indexed on a hexagonal lattice. Higher-resolution rings are out of focus because of the spherical focal plane of the camera.

[8] A. E. Blaurock and W. Stoeckenius, *Nature (London) New Biol.* **233**, 152 (1971).
[9] R. Henderson, *J. Mol. Biol.* **93**, 123 (1975).
[10] A. E. Blaurock, *J. Mol. Biol.* **93**, 139 (1975).

out to a resolution of ~ 15 Å and are good enough for quantitative data reduction and structure determination. The high-resolution orders, although intrinsically as bright as the lower orders, are out of focus in Fig. 3. This is primarily because the scintillator used in this experiment was deposited on a flat fiber optic surface. The actual focal surface of the X-ray camera is a sphere tangent to the sample and to the mirror focus[7] as shown in Fig. 1. The diameter of the focal surface was 2.7 cm in the PM experiments. We will soon be incorporating a curved scintillator screen into the intensifier system that matches the focal sphere. This should yield intense high-resolution patterns out to 7 Å resolution with good statistics. In our upgraded system, patterns will be recorded with the intensifier coupled to a cooler SIT[11] tube detector to yield accurate, low-noise, quantitative data, while the laser will be upgraded to 400 J/pulse to yield about two times more X-ray flux.

In a time-resolved experiment, a master timing signal will generate an electrical pulse, and initiate a green laser stimulus pulse to a hydrated purple membrane stack situated in the humid sample chamber of the X-ray camera. Then, after a delay ranging from nanoseconds to tens of milliseconds, the main laser will fire, generating plasma-produced X-rays, which will yield an X-ray diffraction pattern. By varying the delay, a stroboscopic series of diffraction patterns will be obtained.

The information contained in these powder patterns can be used along with the phase information obtained from electron diffraction studies to yield a two-dimensional projection of the electron density of the PM perpendicular to the plane of the membrane. This view offers a great deal of information about BR because its seven helices are arrayed roughly perpendicular to the membrane plane, i.e., the projection looks down the helical chains. Hence, for example, in a dynamic experiment conformational changes involving movement of the helices may be readily apparent.

Acknowledgments

We are grateful to Dr. J. Lanyi, for the PM preparations, to F. Kirkpatrick and M. Nicholson for facilities for washing the PM, and to J. Abate, B. Flaherty, and T. Kessler for operating the laser.

This work is supported, in part, by the National Science Foundation, Grant PCM 79-04375, and by the National Institutes of Health, Grant 1R01 GM-27364-01.

This work was partially supported by the following sponsors: Exxon Research and Engineering Company, General Electric Company, Northeast Utilities Service Company, New York State Energy Research and Development Authority, The Standard Oil Company (Ohio), The University of Rochester, and Empire State Electric Energy Research Corporation. Such support does not imply endorsement of the content by any of the preceding parties.

[11] J. R. Milch, *IEEE Trans. Nucli. Sci.* **NS-26,** 338 (1979).

[37] X-Ray Diffraction Studies of Light Interactions with Bacteriorhodopsin

By J. Stamatoff, R. H. Lozier, and S. Gruner

The purple membrane of halobacteria offers exciting possibilities for structural investigation by X-ray diffraction methods. It is one of the few membranes that has an in-plane crystalline arrangement of proteins.[1] This arrangement has permitted diffraction investigation by crystallographic methods (producing a structural solution in the case of low-dose electron microscopy).[2] The structure is, of course, interesting in its own right. It reveals a trimer of bacteriorhodopsin molecules (the only protein present in the membrane) within one hexagonal unit cell ($a = 63$ Å). Each molecule consists of approximately seven helical segments that represent transmembrane. Three inner helices are orthogonal to the membrane and four outer helices are apparently tilted.

An obvious and exciting step in understanding the membrane's proton pumping mechanism is to establish structural changes that are associated with this function. Some suggested mechanisms would require small structural changes (e.g., the movement of one amino acid residue that would gate a proton hopping chain).[3] Other optical studies suggest that there are major structural rearrangements associated with the process.[4]

As a result of the retinal moiety bound to each bacteriorhodopsin molecule, purple membranes have one major absorption band in the visible. Changes in this band during the pumping process serve to identify spectral intermediates.[5] From extensive studies on retinal, it is probable that these spectral changes reflect very small structural changes in the atomic environment immediately surrounding the chromophore.[6] Thus these spectroscopic studies do not imply that major structural changes do take place in the pumping cycle. Rather, these spectroscopic states may be used to identify particular molecular states to be associated with diffraction patterns.

[1] A. E. Blaurock and W. Stoeckenius, *Nature (London) New Biol.* **233**, 152 (1971).

[2] R. Henderson and P. N. T. Unwin, *Nature (London)* **257**, 28 (1975).

[3] W. Stoeckenius, *in* "Membranes and Transduction" (R. A. Cone and J. E. Dowing, eds.), p. 39. Raven, New York, 1979; J. Nagle and H. J. Marowitz, *Proc. Natl. Acad. Sci. U.S.A.* **75**, 298 (1978).

[4] B. Becher, F. Tokunaga, and T. G. Ebrey, *Biochemistry* **17**, 2293 (1978).

[5] R. H. Lozier, R. A. Bogomolni, and W. Stoeckenius, *Biophys. J.* **15**, 955 (1975).

[6] B. Honig, U. Dinur, K. Nakanishi, V. Balogh-Nair, M. Gawinowicz, M. Arnaboldi, and M. G. Motto, *J. Am. Chem. Soc.* **101**, 7084 (1979).

METHODS IN ENZYMOLOGY, VOL. 88

Despite the rapid growth of interest in this proton translocating membrane, structural investigation of spectral intermediates by diffraction methods has been impeded for technical reasons. The intermediates cycle on a millisecond time scale,[5] requiring dramatic reduction in the time scale of a diffraction experiment. Further, the X-ray scattering cross section is much smaller than the optical absorption cross section. For example, a specimen of optimal thickness for diffraction studies has an optical density at 570 nm (the absorption maximum) that exceeds 100. Samples of these thicknesses are impossible to excite optically without producing heating or many photon process artifacts.

Over the past decade, rapid development of X-ray technology has made structural studies of intermediates feasible. The X-ray system at Bell Labs utilizes a large point-focusing monochromator[7] and stable position-sensitive detector. Figure 1a (solid line) shows a pattern collected using this system in 3 hr from dark-adapted ($OD_{570} \sim 8$) purple membranes. Figure 1a (dashed line) shows a $\frac{1}{2}$ hr pattern from the same speci-

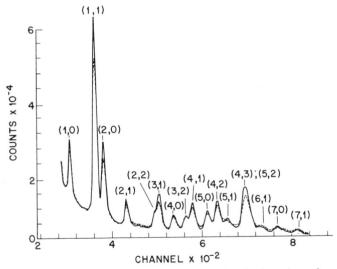

FIG. 1a. An X-ray diffraction pattern (solid line) of dark-adapted purple membranes ($OD_{570} \sim 8$) at 2°. The sample is layered so that the membrane planes are orthogonal to the beam. Indices for the two-dimensional hexagonal powder pattern are given as (h,k). The hexagonal unit cell dimension is 63 Å. The exposure time was 3.0 hr. Overlaid is a diffraction pattern (dashed line) of the same sample after a 15-min exposure to white light. The exposure time was 0.5 hr. The two patterns are scaled together such that the attenuated incident beams have the same intensity.

[7] D. W. Berreman, J. Stamatoff, and S. J. Kennedy, *Appl. Opt.* **16**, 2081 (1977).

men after 15 min irradiation with white light. The pattern is of equivalent quality as a film pattern from a thicker specimen of a 24–48-hr duration. Thus this system provides a reduction factor of 50–100 times conventional camera exposures. Half-hour exposure times permit a variety of experiments to be conducted. In this case, intensity changes appear that return to the original dark-adapted state diffraction pattern in a few hours.

These patterns may be analyzed in terms of structure by methods that require several major assumptions.[8] First, phase angles for the reflections are not known (nor likely to be known) for an X-ray experiment. These phase angles must be obtained from electron microscopy studies.[2] In principle, the proper phase angles for X-ray scattering may be derived from the EM phases by computation. However, these phases should be close to the electron-scattering values, and the phase angles used in this study are taken to be equal to the electron microscopy phases. Second, the specimens are layered so that the membrane plane is orthogonal to the incident X-ray beam. Nevertheless, individual sheets of purple membranes are disoriented about this axis. Thus the X-ray pattern is a powder pattern. Several reflections may be cylindrically averaged to produce one peak. A major assumption is that single-crystal electron microscopy patterns may be used to predict the ratios of overlapping reflections.

In-plane electron density maps calculated from the dark (solid) and light (dashed) patterns are shown in Fig. 1 b. Small rotation of the outer four helices relative to the inner helices is noted.

These experiments are not easily interpreted in terms of intermediates. The time scale for the changes ($\sim\frac{1}{2}$ hr decay constant) is too slow compared with decay rates of the slowest intermediate (M) lengthened by drying (~ 20 sec).[9] In addition, these light-induced changes are toward smaller maximum intensities for all reflections. Although there are changes in the relative intensities (resulting in positional shifts in the electron density map), the universal change toward small maximum intensities is indicative of smaller crystalline patches. This suggestion gains further support from the observation that the background between the peaks increases. Thus light in this experiment may cause intensity changes via dehydration by heating effects. Finally, such changes are not observed in wetter specimens.

Patterns may be recorded even more rapidly using area-sensitive X-ray detectors and synchrotron sources. Figure 2 is a pattern obtained at the Standford Synchrotron Radiation Laboratory using a silicon intensi-

[8] G. I. King, P. C. Mowery, W. Stoeckenius, H. L. Grespi, and B. P. Schoenborn, *Proc. Natl. Acad. Sci. U.S.A.* **77**, 4726 (1980).
[9] R. Korenstein and B. Hess, *Nature (London)* **270**, 184 (1977).

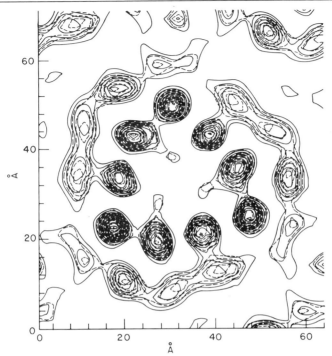

FIG. 1b. Electron density maps for the two patterns are shown. Intensities were obtained by removing a spline-fitted background curve. The same shape background curve was used for each pattern. The background stripped patterns were then corrected by a Lorentz factor of S^2. The resulting patterns were fit with an array of Gaussian functions. Integrated intensities were calculated from the Gaussian parameters. This procedure minimizes the problem of overlapping reflections. The maps were calculated using EM phases. Light-adapted (dashed contours) show small changes of the outer four helices relative to the inner three. These changes are probably heat-induced effects (see text).

fied target area detector.[10] The pattern was recorded in 200 sec from a sample of $OD_{570} \sim 2$. Thus there is a very significant enhancement due to the use of intense synchrotron radiation and an area detector (which permits integration over a large angular region of the powder rings).

Although these time scales (~ 200 sec) are still too long to record intermediate structure, use may be made of the cyclical nature of the pumping process. By pulsing light and recording at a defined interval following the pulse, an intermediate pattern may be integrated. Reduction of the total

[10] J. Milch, *IEEE Trans. Nucl. Sci.* **NS-26,** 338 (1979); S. Gruner, J. Milch, and G. Reynolds, *Rev. Sci. Inst.* in press.

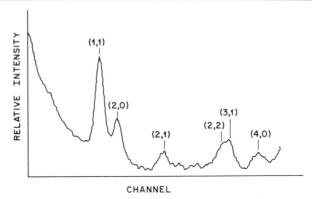

CHANNEL

Fig. 2. An X-ray diffraction pattern obtained using synchrotron radiation from a $OD_{570} \sim 2$ sample at room temperature. A SIT area detector was used to record the pattern which permits integration over a large angular range of the diffraction rings. X rays with an energy of 8 kev were used. The exposure time of 200 sec could be considerably reduced using lower-energy X rays.

integration time by the methods demonstrated earlier makes this strobo-scopic experiment feasible.

Detailed analysis of the X-ray patterns in terms of a unique structure is hampered by several assumptions (as described). However, changes in diffracted intensities may serve to determine whether or not there are major structural changes (if not to define them uniquely). Recent low-tem-perature results indicate the M intermediate is not associated with major changes.[11] However, small changes in the structural arrangements of large subunits have been reproducibly observed.[11]

In summary, rapid development of X-ray technology now permits studying spectral intermediates of bacteriorhodopsin. As demonstrated earlier, changes in the pattern can be directly interpreted in terms of small structural changes of large units. Thus this technique provides a sensitive and direct structural tool for investigating functional states of this mem-brane. Our results show that structural changes may be induced by heat associated with light but that changes directly due to light adaptation are not observed at this resolution. Similar studies on the intermediates[11] have proved useful in defining the mechanism of proton translocation.

Acknowledgments

We gratefully acknowledge valuable discussions with G. King and W. Stoeckenius. We thank T. Bilash and S. Davey for their superb technical assistance.

[11] J. Stamatoff, T. Bilash, T. Yamane, W. Stoeckenius, and R. Lozier, submitted for publi-cation in the *Biophys. J.*

[38] Orientation of Purple Membranes by Electric Field

By LAJOS KESZTHELYI

Introduction

Recently, studies have been made to orient* purple membranes (PM) in solution using electric fields.[1-4] In the experiments reported by Shinar *et al.*[1] Tsuji *et al.*[3] and Kimura *et al.*[4] the orienting process itself was studied, while Eisenbach *et al.*[2] applied the electric field to obtain oriented and immobilized layers of PMs that could be used as a device for light–electricity transduction.

The aim of the orientation experiments performed in our laboratory was to create a system of practically unperturbed PMs in solution to study the light-driven proton pumping function of the bacteriorhodopsin (BR) molecules embedded in the membranes.[5]

This section describes the procedures necessary to induce orientation of PMs and to check their orientation, and presents some typical data measured on this system.[6,7]

Application of the Electric Field

The simplest way to apply the electric field is to immerse Pt electrodes into the PM solution (Fig. 1). The solution is placed in a cuvette of 1 mm thickness; the electrodes are separated by a distance of $\simeq 8$ mm. Because of the effect of the electrodes the strength of the electric field E (V/cm) in

---

* *Orientation*, as a physical property, means the parallel ordering of the membrane fragments in such a way that the cytoplasmic side of all fragments faces in the same direction. Parallel ordering without sidedness is called *alignment*. Orientation cannot be achieved by sedimentation or magnetic fields.

[1] R. Shinar, S. Druckman, M. Ottolenghi, and R. Korenstein, *Biophys. J.* **19**, 1 (1977).
[2] M. Eisenbach, C. Weissmann, G. Tanny, and S. R. Caplan, *FEBS Lett.* **81**, 77 (1977).
[3] K. Tsuji and K. Rosenheck, *in* "Electrooptics and Dielectrics of Macromolecules and Colloids" (B. R. Jennings, ed.), p. 77. Plenum, New York, 1973.
[4] Y. Kimura, A. Ikegani, K. Ohno, S. Saigo, and Y. Takeuchi, *Photochem. Photobiol.* **33**, 435 (1981).
[5] L. Keszthelyi, *Biochim. Biophys. Acta* **598**, 429 (1980).
[6] L. Keszthelyi and P. Ormos, *FEBS Lett.* **109**, 189 (1980).
[7] P. Ormos, Zs. Dancsházy, and L. Keszthelyi, *Biophys. J.* **31**, 207 (1980).

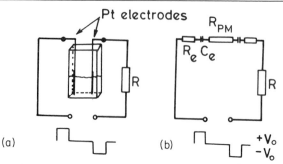

FIG. 1. Schematic of the circuit to apply the electric field. (a) The actual circuit through the PM solution (R the measuring resistance). (b) The equivalent circuit (R_e, C_e electrode resistance and capacitance, R_{PM} resistance of the PM solution.)

the solution is not V_o/D, where V_o is the applied voltage: the interface between the Pt electrodes and the solution represents an additional resistance and capacitance in the circuit (Fig. 1). The resistance R is used to measure the current in the circuit. Figure 2a shows voltage forms (V_R) obtained on R for different V_o values. The changing V_R means that the current through the PM solution changes with time and therefore implies a continuously changing electric field.

The electrode effects can be substantially decreased by using platinized Pt electrodes. The corresponding voltages are shown in Fig. 2B. The time course for E is much nearer to a rectangular form. Therefore, it is much more advantageous to always use platinized Pt electrodes to provide the orienting fields.

The resistance of the cell R_{PM} for PMs suspended in tridistilled water is usually $\sim 10^5$ Ω (electrodes separated by a distance of ~ 8 mm, the height of the column $\approx 4-5$ mm). If the pH of the solution is adjusted by adding NaOH or HCl, the range of R_{PM} is $2-10 \times 10^4$ ohms. These R_{PM} values are for PM solutions of rather high density: $A = 1.0-1.8$.

It is very important to apply the electric field symmetrically, i.e., to apply positive and negative fields of the same value and for the same time to avoid the unidirectional electrophoretic movement of the PMs. We used a positive voltage pulse for a time t, zero for a time 2 t, and then a negative pulse for another time t. The advantage of using this waveform, compared to a simpler t positive and t negative form, is that the orientation and the resulting decay can be observed twice. Using $t = 1-4$ sec and $E = 20-30$ V/cm, practically no deterioration of the sample could be observed for $20-30$ complete cycles.

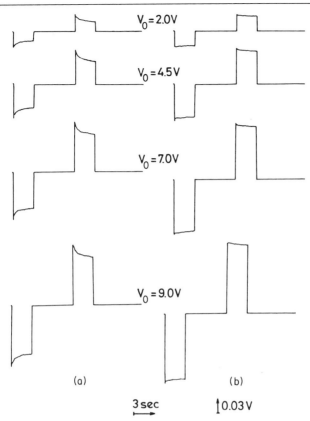

FIG. 2. Voltage form $V_R(t)$ on the resistance $R = 1$ kohm. The current through the cell is $i(t) = V_R(t)/R$, the electric field in the cell is $E(t) = R_{PM}/R \cdot V_R(t)/D$, where R_{PM} is the assumed ohmic resistance of the solution, and the distance of the electrodes, $D = 0.8$ cm. $V_R(t)$ was measured with (A) untreated and (B) platinized electrodes for different V_0. The electrode effects are clearly seen in (A) and are negligible for the platinized electrodes.

Measurement of the Orientation

Because the BR molecules in the purple membranes contain chromophores that are known to be oriented at $\theta \simeq 63-70°$ to the membrane normal,[8,9] linear dichroism can be used to measure the orientation.[10] The intensity of transmitted polarized light I changes when orientation

[8] M. P. Heyn, R. J. Cherry, and U. Müller, *J. Mol. Biol.* **117**, 607 (1977).

[9] R. Korenstein and B. Hess, *FEBS Lett.* **89**, 15 (1978).

[10] E. Fredericq and C. Houssier, "Electric Dichroism and Electric Birefringence." Oxford Univ. Press, (Clarendon), London and New York, 1973.

is established. The light intensity changes $\Delta I_\perp$ and $\Delta I_\parallel$, for polarizations perpendicular and parallel to the direction of the orienting field, respectively, are related to the reduced dichroisms as

$$\frac{\Delta A_\perp}{A} = -\frac{1}{A}\log\left(1 + \frac{\Delta I_\perp}{I}\right) = \phi(\beta)\left(\frac{3}{2}\sin^2\theta - 1\right)$$
$$\frac{\Delta A_\parallel}{A} = -\frac{1}{A}\log\left(1 + \frac{\Delta I_\parallel}{I}\right) = \phi(\beta)(3\cos^2\theta - 1)$$

(1)

where $\Delta A_\perp$ and $\Delta A_\parallel$ are the corresponding absorbance changes, I_0 is the incident light intensity, and $A = -\log(I/I_0)$. $\phi(\beta)$ is the so-called orientation function for permanent dipoles:

$$\phi(\beta) = 1 - \frac{3(\coth\beta - 1/\beta)}{\beta}$$

(2)

It has been shown[5] that at low electric field ($E < 20$–30 V/cm) the PMs are oriented mainly by their permanent dipole moment. Here $\beta = \mu BE/kT$, μ is the permanent dipole moment of the particle, B is a constant expressing that the internal field at the dipoles is different from the applied field E, k is the Boltzmann constant, and T is the temperature. From Eq. (1) it follows that the requirement

$$\frac{\Delta A_\perp}{A} = -\frac{1}{2}\frac{\Delta A_\parallel}{A}$$

(3)

must be satisfied independently of θ, which means that in the case of PMs where $\theta \simeq 63$–$70°$, $\Delta A_\perp/A > 0$ and $\Delta A_\parallel/A < 0$. Consequently, $\Delta I_\perp < 0$ and $\Delta I_\parallel > 0$ for any value of the orienting field [$\phi(\beta)$ is always positive]. In this argument it is assumed that the electric dipole moment and the normal of the purple membrane are parallel, which is very probably the case.

The results of such linear dichroism measurements are reproduced in Fig. 3. Figure 4 contains the evaluation of intensity changes in terms of Eq. (1). Equation (3) is satisfied to rather good accuracy.

There are several possible pitfalls that must be carefully avoided. The most important problem is that in addition to absorption the PMs also scatter light. The intensity of the scattered light from randomly distributed particles depends on the square of the size of the particles and its angular distribution for larger particles points more forward.[11] For oriented particles the intensity of the scattered light changes.

The excess scattering due to orientation does not depend on the polarization of the incoming light beam. Therefore it adds or subtracts equally[11] from $\Delta I_\perp$ and $\Delta I_\parallel$. This is illustrated in Fig. 5, where the light was col-

[11] S. P. Stoylov, *Adv. Colloid Interface Sci.* **3**, 45 (1971).

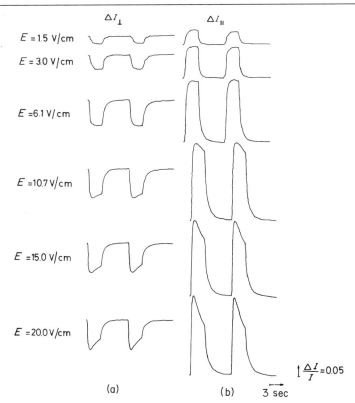

$\Delta I_\perp$ $\Delta I_\parallel$

$E = 1.5$ V/cm

$E = 3.0$ V/cm

$E = 6.1$ V/cm

$E = 10.7$ V/cm

$E = 15.0$ V/cm

$E = 20.0$ V/cm

$\uparrow \dfrac{\Delta I}{I} = 0.05$

(a)   (b) $\overrightarrow{3 \text{ sec}}$

FIG. 3. Time course of light intensity changes behind the absorbing PM solution for different field strengths. $\lambda = 570$ nm, $A = 0.6$ in 1-mm cuvette. (a) $\Delta I_\perp$, the measuring light was polarized perpendicular to the direction of electric field, and (b) $\Delta I_\parallel$, for parallel polarization. Light was collected from a cone of angle 22°.

lected behind the sample from a narrow cone of angle 5° (Fig. 6). The excess scattered light is equally missing from $\Delta I_\perp$ and $\Delta I_\parallel$. The results presented in Fig. 3 were obtained using a larger cone (22°) so that the major part of the scattered light was also collected (Fig. 6).

Additional problems arise if the sample contains large aggregates, as they will scatter a substantial amount of light. A gentle sonication in a water bath for 2–5 sec usually alleviates this problem.

The light scattering, while being a nuisance, also has advantages because the detection of the scattered light is an alternative method for measuring the orientation. It is applicable even for membrane fragments that do not contain a well-defined chromophore. Records of scattered light intensities for the same PMs as in Fig. 3 are shown in Fig. 7.

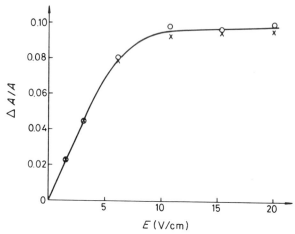

FIG. 4. Dependence of the reduced dichroism ($\Delta A_\perp/A$ . . . x and $-\frac{1}{2}\Delta A_\parallel/A$ . . . 0) on the electric field strength. Evaluation of the data in Fig. 3 according to Eq. (1).

Evaluation of the Data of Orientation Measurements

The data in Fig. 3 and calculations in Fig. 4 are rich in information. After switching off the electric field the decay of orientation is character-ized by the mean time of free relaxation τ, which, according to the the-ory,[12] is proportional to r^3, where r is the largest dimension of the asym-metric particle. The curve in Fig. 4 is described by the function $\phi(\beta)$. It is easy to see from Eq. (4) that the initial slope is proportional to μ, the di-pole moment of the PMs. The dipole moment of the PMs is the sum of individual dipole moments of the molecules in it, probably of the BR mol-ecules; consequently $\mu \sim r^2$.

Therefore, the average PM diameter $\bar{r}$ appears twice in the records of orientation as $\tau \sim \bar{r}^3$ and initial slope $S \sim \bar{r}^2$. If these rules are correct, then $R = \sqrt{S}/\sqrt[3]{\tau}$ must be constant. R has been checked experimentally

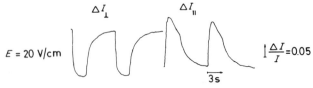

FIG. 5. Time course of light intensity changes behind the absorbing PM solution. The same as Fig. 3 at $E = 20$ V/cm, but the light was collected from a cone of angle 5°.

[12] F. Perrin, *J. Phys. Radium* [6] **7**, 390 (1926).

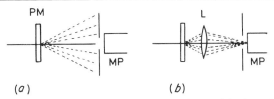

(a)   (b)

FIG. 6. Two geometrical arrangements to measure linear dichroism. (a) The scattered light is poorly collected. (b) A substantial part of the scattered light collected. (PM = purple membrane suspension, L = lens, MP = photomultiplier.)

by fragmenting PMs by sonication of variable duration. Table I contains data that show the approximate constancy of R, thereby illustrating that τ and S characterize the diameter of the particles. With $\tau \simeq 1$ sec, PMs may be oriented to near saturation by as small a field as $E = 6$–8 V/cm.

The linear dichroism records (Fig. 3) are rectangular at low electric fields, as expected, but decay in time at higher voltages. This phenomenon has been explained[5] as a consequence of the electrophoretic flow: The oriented fragments are tilted around axes in their plane when they are dragged. The charge state of the PMs plays an important role in the determination of the angle of the tilt, which can be calculated[5] to be $\simeq 30°$ for the highest field.

From the saturation value of $\Delta A_{\perp}/A$ or $\Delta A_{\parallel}/A$ the angle θ of the retinal

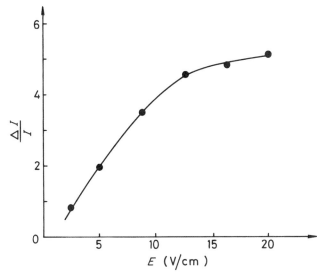

FIG. 7. Dependence of the scattered light intensity ($\Delta I/I$) on the electric field strength ($\lambda = 632$ nm). Light was collected from a cone of angle 22° and the direct beam shielded.

TABLE I
Relaxation Times (τ), Slopes (S), and the
Calculated Values of R (see text) for
Fragmented PMs

τ (sec)	S (arbitrary unit)	$R = \dfrac{\sqrt{S}}{\sqrt[3]{\tau}}$
1	2.5	1.6
0.02	0.25	1.8
0.005	0.05	1.2

transition moment, relative to the direction of the electric dipole moment, can be determined from Eq. (1). For the measurements shown in Fig. 3, $\theta \simeq 60°$. The highest value obtained in our laboratory for highly fragmented particles was $\theta = 63°$. The difference can be understood by assuming that the PMs, during their electrophoretic movement, deviate from rigid disk shape; consequently, the angle θ will be an average of slightly diffuse directions.

Experiments on Oriented PM Systems

The orientation of PMs in solution can be sustained for seconds by the electric field, and even after switching off the electric field the relaxation time of the orientation is long enough (it may be ~ 1 sec) that measurements taking a shorter time than ~ 1 sec can be performed. The photocycle of BR is ~ 20 msec. Consequently, the study of the photocycle, or more precisely, the unidirectional electric events of the proton pump, become possible.

For these studies a simple measuring system has been constructed (Fig. 8). To observe the electric signals associated with the photocycle of BR the circuit contains a resistance R in series with the PM cell. A voltage V_R appears on this resistance, as shown in Fig. 2b. For saturating fields $V_R \simeq 2-4$ V. The associated electric signals evolved by a laser pulse, however, are small; the order of magnitude of the long-living components turned out to be ~ 1 mV. The second PM system was used to compensate the large V_R with the help of a differential amplifier (Keithley 604). It had a built in baseline shift that was used for further adjustment of the small difference between V_{R_1} and V_{R_2}. (Note that the compensation should have been good at least for $10^4 : 1$). The system included a light beam to measure optically the photocycle at selected wavelengths and to check the orientation by observing the linear dichroism signal with polarized light.

The laser light evoked an electric signal from the differential amplifier

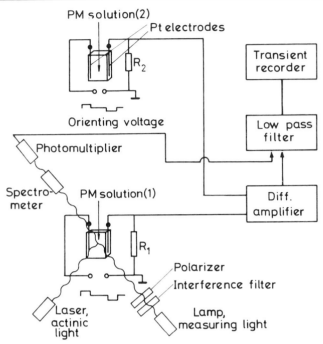

FIG. 8. Schematic of the measuring system for electric signals.

and the light signals went through a low pass filter to a transient recorder (product of the Central Research Institute of Physics, Hungary; smallest conversion time 0.1 μsec, conversion range 256 quanta/channel, 1024 channels) which was triggered directly by a small fraction of the laser light picked up by a photomultiplier. To avoid electric noise from the laser discharge it was located ~7 m from the electrically shielded PM cells and an optical coupler was used to connect the laser to the electronics. An Opton dye laser was used, λ = 580 nm, pulse length 1 μsec, energy 10 mJ. With these precautions no disturbing noise signal can be seen in the mV sensitivity range.

Representative electric signals measured with different ranges of time and amplitude and also measured in field on (a,b) and off (c,d) cases are reproduced in Fig. 9. The components I and II are practically the same; the difference between the components III (dashed line, b) must have its origin in the conductivity change caused by the transiently liberated protons. The field-independent components were assigned as displacement currents caused by the moving protons. The light signals were also measured on the same sample. In Table II we collect the time constants of the

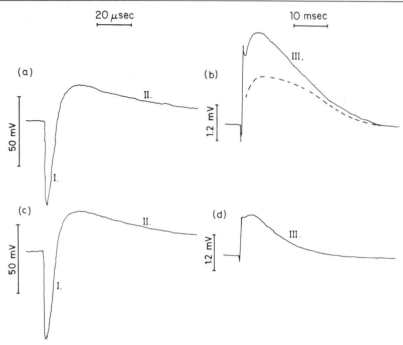

FIG. 9. Time dependence of the electric signals. (a,b) Signals measured at $t = 2.5$ sec (field-on case); (c,d) at $t = 3.3$ sec (field-off case). I–III designate the components of the signal. The dashed line in (b) is the difference between signals (b) and (d) i.e., the component originating from conductivity change. PM suspended in H_2O, $A = 1.8$, $T = 20°$, orienting voltage 8.5 V. Distance between electrodes $D = 0.8$ cm.

TABLE II

TIME CONSTANT DATA OF ELECTRIC AND LIGHT ABSORPTION SIGNALS OF
SUCCESSIVE TRANSITIONS IN THE PHOTOCHEMICAL CYCLE.[a]

Signal type	Transition: K → L τ_1 (μsec)	L → M τ_2 (μsec)	M → O τ_3 (msec)	O → BR τ_4 (msec)
Electric	4.4 (6.2)	81 (452)	8.0 (22.0)	2.5 (4.0)
A_{408}		81 (429)	7.9 (20.0)	
A_{522}	3.7 (5.1)	91 (445)	8.5 (20.0)	
A_{635}			9.2 (21.0)	2.2 (4.1)

[a] Data correspond to PM suspended in H_2O, in brackets in D_2O.

electric and light signals for PMs suspended in H_2O and D_2O. The correspondence of the time constants of the successive transitions was taken as important evidence that the proton pumping function of BR molecules is strictly correlated with the photocycle. Detailed analysis of the results and an experiment on the back-reaction from the M_{412} state can be found elsewhere.[6,7]

Conclusion

We have shown that the electric orientation of PMs is easily achieved by relatively small electric fields. Orientation in solution can be sustained for many seconds, which is more than enough to perform a variety of different experiments. The first steps have already been done but the detailed study of pH, salt, and temperature effects on natural and modified systems which would produce important new information still remains.

Acknowledgments

Contributions to the work treated herein from K. Barabás, J. Czégé, Zs. Dancsházy, A. Dér, M. Marden and P. Ormos are gratefully acknowledged. Thanks are due to D. Beece for help with the manuscript.

[39] Effect of Viscosity on the Photocycle of Bacteriorhodopsin

By LAURA EISENSTEIN

Proteins are dynamic systems. Transitions in a dynamic system should be influenced by damping, the damping factor being a function of the local viscosity in the region in which the transition is taking place. Protein reactions have been treated using transition state theory,[1] in which the effect of viscosity is not explicitly taken into account. Experimentalists have largely ignored the approach pioneered by Kramers[2-5] in which viscosity plays an essential role.

[1] S. Glasstone, K. J. Laidler, and H. Eyring, "The Theory of Rate Processes." McGraw-Hill, New York, 1941.
[2] H. A. Kramers, *Physica* (*The Hague*) **7**, 284 (1940).
[3] R. Landauer and J. A. Swanson, *Phys. Rev.* **121**, 1668 (1961).
[4] J. L. Skinner and P. G. Wolynes, *J. Chem. Phys.* **69**, 2143 (1978).
[5] S. Ishioka, *J. Phys. Soc. Jpn.* **48**, 367 (1980).

We have shown[6,7] that biomolecular reactions do depend on the solvent viscosity η and follow a modified Kramers law. The initial work was done for carbon monoxide and dioxygen binding to myoglobin but has recently been extended to the photocycle of bacteriorhodopsin.[8]

In transition state theory the rate of a reaction is written as

$$k = \nu\tau \exp(-G^{\neq}/k_B T) \tag{1}$$

Here $\nu = k_B T/h$ the attempt factor, τ is a transmission coefficient, k_B is the Boltzmann constant, $G^{\neq} = H^{\neq} - TS^{\neq}$ the activation Gibbs energy; $H^{\neq}$ and $S^{\neq}$ are activation enthalpy and entropy, and T is the temperature. For $H^{\neq}$ and $S^{\neq}$ independent of temperature and neglecting the temperature dependence of the attempt factor and transmission coefficient, Eq. (1) can be written as the Arrhenius law

$$k = A \exp(-H^{\neq}/k_B T) \tag{2}$$

where the preexponential $A = \nu\tau \exp(S^{\neq}/k_B)$. The derivation of Eq. (1) assumes that the system undergoing the transition is in thermal equilibrium with the surrounding heat bath; the coefficient τ is not explicitly given.

In transition state theory k is independent of viscosity. However, both the approach to equilibrium and the transmission coefficient depend on viscosity. At low viscosities the time of approach to equilibrium becomes larger than the average transition time and the reaction rate depends on the equilibration time, which is proportional to the viscosity. At higher viscosities the time required to reach equilibrium is shorter than the transition time; however the transmission coefficient is decreased owing to diffusion. Since diffusion is proportional to $1/\eta$, Kramers theory gives a rate that is inversely proportional to the viscosity at high viscosities. A simple model[6] shows that the high viscosity limit (overdamped case) should be applicable to biomolecular reactions even in aqueous solutions. The rate equation to be used for biomolecular reactions is therefore

$$k = \frac{A'}{\eta_{int}} \exp(-H^*/k_B T) \tag{3}$$

[6] D. Beece, L. Eisenstein, H. Frauenfelder, D. Good, M. C. Marden, L. Reinisch, A. H. Reynolds, L. B. Sorensen, and K. T. Yue, *Biochemistry* **19**, 5147 (1980).

[7] D. Beece, L. Eisenstein, H. Frauenfelder, D. Good, M. C. Marden, L. Reinisch, A. H. Reynolds, L. B. Sorensen, and K. T. Yue, *in* "Hemoglobin and Oxygen Binding" (C. Ho, ed.), p. 363. Elsevier, Amsterdam, 1982.

[8] D. Beece, S. F. Bowne, J. Czégé, L. Eisenstein, H. Frauenfelder, D. Good, M. C. Marden, J. Marque, P. Ormos, L. Reinisch, and K. T. Yue, *Photochem. Photobiol.* **33**, 517 (1981).

where η_{int} is the local viscosity that describes the coupling between the system undergoing the transition and the surrounding heat bath. Since the reaction sites are located in the interior of the protein, the pertinent coupling coefficient may be different from the external solvent viscosity. Experimental data suggest that the internal viscosity is related to the solvent viscosity η by[6]

$$\frac{1}{\eta_{int}} = \frac{1}{\eta^\kappa} + \frac{1}{\eta_0} \qquad (4)$$

The exponent κ indicates the extent to which the protein shields the external viscosity. Reactions with κ close to 1 are highly coupled to surface motions, while reactions with κ close to 0 are shielded by the protein structure. The second term in Eq. (4) allows reactions in the interior of the protein to take place even in the limit of extremely high solvent viscosities. This paper summarizes results on the viscosity dependence of the steps in the photocycle of bacteriorhodopsin and presents evidence for the validity of the modified Kramer's law [Eqs. (3) and (4)].

Materials and Methods

Measurements were carried out on suspensions of strain R_1S purple membrane obtained from R. Lozier and W. Stoeckenius. Viscosity was varied by changing the glycerol–water ratio of the suspending medium. The pH of the samples was stabilized with pH 6.5 phosphate buffer in water. The particular solutions used were 20, 40, 60, and 80% glycerol by volume. (Glycerol has no effect on the absorption spectrum of bacteriorhodopsin.) Figure 1 gives the temperature dependence of the viscosity for the solvents used. Extrapolation of the data of Douzou[9] on the pH of

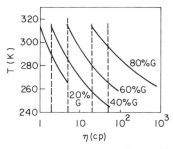

FIG. 1. Dependence of viscosity on temperature for the 20, 40, 60, and 80% glycerol-water solutions. Dotted lines are lines of constant viscosity. [From Beece *et al.*[8]]

[9] P. Douzou, "Cryobiochemistry." Academic Press, New York, 1977.

glycerol–water mixtures as a function of temperature shows that the change of pH in our entire experimental range is less than one unit. As a control we analyzed the kinetics of the cycle in pH 6.5 and 7.7 phosphate buffer solutions and found that the changes in the kinetics due to pH were less than 10% of the solvent-dependent changes.

The purple membrane suspension was placed in a cryostat and light-adapted with green light from a projector lamp. The optical density of the light-adapted samples was 0.55 at 570 nm in a 1-cm cell. The cycle was initiated with a 30-nsec 530-nm flash from a frequency-doubled Nd–glass laser. Sufficient laser energy was used to saturate the sample to obtain a maximum signal/noise ratio and to make the system insensitive to small fluctuations in laser energy. The subsequent changes in absorbance were monitored at each of four wavelengths: 400, 500, 580, and 660 nm and at temperatures between 240 and 315 K. For solvents that freeze at temperatures above 240 K, e.g., 20% glycerol, we performed measurements at temperatures above the samples freezing point. The data were recorded with a logarithmic time-base transient digitizer, permitting measurement of the signal continuously from 2 μsec to 1 ksec.[10]

Results and Discussion

Typical kinetic data are shown in Fig. 2. The measured intensity was transformed into a change in sample absorbance (ΔA) relative to bacteriorhodopsin (BR). The plots display log(ΔA) versus log(t/s). To accommodate both positive and negative changes in ΔA the infinite part between $+ 10^{-3}$ and $- 10^{-3}$ has been omitted.[11] The data represent averages of several flashes under identical conditions.

For any first-order reaction scheme the change in sample absorbance at time t and monitoring wavelength λ should be a sum of exponentials of the form:

$$\Delta A_\lambda(t) = \sum_i B_i(\lambda)e^{-\Lambda_i t} \tag{5}$$

where $B_i(\lambda)$ are wavelength-dependent amplitudes and the Λ_i are wavelength-independent macroscopic rates. We determine Λ_i by using a variable projection algorithm[12] that forces Λ_i to be the same at all wavelengths for a given temperature and solvent. The algorithm exploits the separabi-

[10] R. H. Austin, K. W. Beeson, L. Eisenstein, H. Frauenfelder, and I. C. Gunsalus, *Biochemistry* **14**, 5335 (1975).

[11] D. Beece, L. Eisenstein, H. Frauenfelder, D. Good, M. C. Marden, L. Reinisch, A. H. Reynolds, L. B. Sorensen, and K. T. Yue, *Biochemistry* **18**, 3421 (1979).

[12] F. Krogh, *Commun. ACM* **17**, 167 (1974).

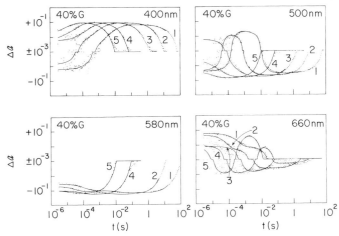

FIG. 2. Log(ΔA) versus log(t/s) for bacteriorhodopsin in 40% glycerol–water measured at 400, 500, 580, and 660 nm for temperatures between 245 and 315 K. (1) = 245 K, (2) = 260 K, (3) = 275 K, (4) = 295 K, and (5) = 315 K. To accomodate both positive and negative changes in ΔA the infinite part between $+10^{-3}$ and -10^{-3} has been omitted. The solid lines are the results of the fit as described in the text. [From Beece *et al.*[8]]

lity of the sum-of-exponentials; the linear parameters $B_i(\lambda)$ are uniquely determined from the data once the nonlinear parameters Λ_i are known. In almost all cases a satisfactory fit was obtained with four exponentials. For the higher temperature measurements, where our time resolution was not sufficient to record the fastest process, three exponentials were used. The observation of four processes implies that at least four intermediate states must occur in the reaction cycle. In the investigated temperature range we find that the cycle of bacteriorhodopsin can be approximated with the following linear scheme:[13]

$$BR \xrightarrow{h\nu} K \xrightarrow{k_1} L \xrightarrow{k_2} M \xrightarrow{k_3} O \xrightarrow{k_4} BR \qquad (6)$$

all thermal reactions being first order. The rates k_i are identical to the coefficients Λ_i determined from Eq. (5). The specific assignment requires some comments. If we take $\Lambda_1 > \Lambda_2 > \Lambda_3 > \Lambda_4$, we find unambiguously $k_1 = \Lambda_1$, $k_2 = \Lambda_2$. The assignment of k_3 and k_4 is ambiguous, and we have taken $k_3 = \Lambda_3$ and $k_4 = \Lambda_4$. The slowest rate thus is associated with the $O \rightarrow BR$ transition. This choice results in a spectrum for the O state with peaks near 400 and 660 nm. With increasing temperature the peak near 660 nm increases and the peak near 400 nm decreases, consistent with the O state being a mixture of two species in thermal equilibrium. This feature

[13] R. H. Lozier, R. A. Bogomolni, and W. Stoeckenius, *Biophys. J.* **15**, 955 (1975).

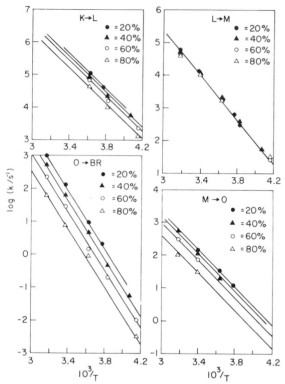

FIG. 3. Rates for the transitions K → L, L → M, M → O, and O → BR in 20, 40, 60, and 80% glycerol–water solutions versus $10^3/T$. Activation enthalpies and preexponentials deduced from these Arrhenius plots are given in Table I.

could result from a photocycle in which there are two consecutive M forms, the second being in (rapid) equilibrium with O,

$$
\begin{array}{c}
O \\
\updownarrow \\
K \to L \to M_1 \to M_2 \to BR
\end{array}
$$

Association of the slowest rate with the M → O transition ($k_3 = \Lambda_4$) results in a spectrum for O with a single peak near 660 nm that exhibits some temperature dependence. This temperature dependence is not eliminated by adding a branch M → BR. Our main conclusion, that large viscosity effects are observed in the latter half of the photocycle, are independent of the assignment of Λ_3 and Λ_4 to k_3 and k_4. The actual values of parameters corresponding to each transition does, of course, depend on the choice. If the slowest rate is associated with the M → O transition,

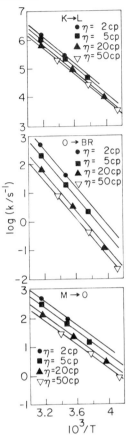

FIG. 4. Rates versus $10^3/T$ for the transitions K $\rightarrow$ L, M $\rightarrow$ O, O $\rightarrow$ BR at fixed viscosities η = 2 cP, 5 cP, 20 cP, and 50 cP. Each point on an isoviscosity curve represents a different glycerol–water solution.

then the parameters and rates presented here for M $\rightarrow$ O (O $\rightarrow$ BR) would actually correspond to O $\rightarrow$ BR (M $\rightarrow$ O).

The k_i obtained from the preceding analysis can be fit to the transition state theory [Eq. (2)]. The Arrhenius plots of Fig. 3 show that the solvent has a pronounced effect on the kinetics. In particular, the preexponential factors A_i decrease with increasing glycerol concentration. The magnitude of the effect is different for each reaction step; only the L $\rightarrow$ M transition is unaffected. These results are consistent with the findings of Dencher and Wilms,[14] in which the rate of M formation was found to be the same in water and 80% glycerol, and the rate of BR regeneration was

[14] N. Dencher and M. Wilms, *Biophys. Struct. Mech.* **1**, 259 (1975).

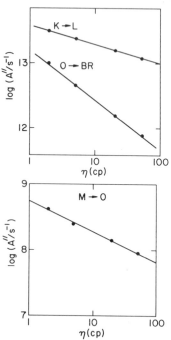

FIG. 5. Viscosity dependence of the preexponential factors $A'' = A'/\eta_{int}$ for the transitions K → L, M → O, and O → BR.

found to be a factor of 15 slower in 80% glycerol than in water. Table I summarizes our results for the different solvents.

The k_i for all solvents can be fit to a modified Kramer's law using Eqs. (3) and (4). Figure 4 gives the temperature dependence of the rates, for fixed solvent viscosities; Figure 5 gives the viscosity dependence of the preexponential factors A'/η_{int} in Eq. (3). Table II contains the activation enthalpies H^*, the factors A', and the exponents κ. For solvent viscosities smaller than 100 cP, $1/\eta_0$ [Eq. (2)] is found to be zero. The activation enthalpies found by this modified Kramers treatment are smaller than those found using transition state theory for all transitions other than L → M.[15] The value of κ_i is characteristic of the motion involved in each particular step. During the first half of the cycle, K → L and L → M, κ_i is small and thus points to processes "deep inside." During the later parts of the cycle, M → O and O → BR, κ_i is large. These transitions may involve major movements that are strongly influenced by the solvent.

[15] This effect can be easily understood. The data in Fig. 1 show that in a particular solvent the viscosity has Arrhenius temperature dependence $\eta \propto e^{\ H\eta/k_B T}$. Thus $H^{\neq} = H^* + \kappa H^\eta$. Our data are not good enough to resolve the solvent dependence of $H^{\neq}$, and we find that H^* is less than $H^{\neq}$ by an amount given by κ times the average value of H^η for the solvents studied.

TABLE I
ACTIVATION ENTHALPIES ($H^{\neq}$) AND PREEXPONENTIAL FACTORS (A) IN EQ. (2)
FOR THE TRANSITIONS $K \to L$, $L \to M$, $M \to O$, $O \to BR$ IN 20, 40, 60,
AND 80% GLYCEROL–WATER SOLUTIONS[a]

		$K \to L$	$L \to M$	$M \to O$	$O \to BR$
$H^{\neq}$(kJ mol⁻¹)		52	66	54	84
A(sec⁻¹)	20%	$8.3 \times 10^{+14}$	$1.7 \times 10^{+15}$	$4.5 \times 10^{+11}$	$2.1 \times 10^{+17}$
	40%	$6.5 \times 10^{+14}$	$1.7 \times 10^{+15}$	$3.3 \times 10^{+11}$	$1.0 \times 10^{+17}$
	60%	$4.7 \times 10^{+14}$	$1.7 \times 10^{+15}$	$2.0 \times 10^{+11}$	$4.5 \times 10^{+16}$
	80%	$2.9 \times 10^{+14}$	$1.7 \times 10^{+15}$	$9.1 \times 10^{+10}$	$1.6 \times 10^{+16}$

[a] The values presented for $H^{\neq}$ are averages of those found for the various solvents. The errors are about 10%.

TABLE II
PARAMETERS RESULTING FROM A FIT OF EQ. (3) $k_i = A_i'/\eta_{i,\text{int}} \, e^{-H_i^{\neq}/k_B T}$, AND

$$\text{EQ. (4)} \frac{1}{\eta_{i,\text{int}}} = \frac{1}{\eta \kappa_i} + \frac{1}{\eta_{i,0}} \text{ TO THE TRANSITION RATES}[a]$$

	$K \to L$	$L \to M$	$M \to O$	$O \to BR$
$H^{\neq}$(kJ mol⁻¹)	44	66	36	60
A'(sec⁻¹P^k)	9.2×10^{12}	1.7×10^{15}	6.3×10^{7}	4.5×10^{11}
κ	0.25 ± 0.1	0.0 ± 0.1	0.5 ± 0.2	0.8 ± 0.2

[a] Errors are about 20% except as noted. 1P = g cm⁻¹ sec⁻¹. $1/\eta_0$ is found to be zero for all transitions.

Recent theoretical work[16] has shown that the scaling of internal viscosities with solvent viscosity provides information on the dynamic behavior of proteins and yields the distribution of fast local protein modes that compete with the solvent coupled fluctuations and shield the influence of the solvent viscosity. To extract these protein modes requires data at high solvent viscosities for which the second term in Eq. (4) becomes important and transition rates become independent of solvent viscosity. Measurements at high solvent viscosities are now in progress.

Acknowledgments

The work described in this contribution was done in collaboration with D. Beece, J. Czégé, H. Frauenfelder, D. Good, M. C. Marden, J. Marque, P. Ormos, L. Reinisch, and K. T. Yue. I thank R. Lozier, R. LeVeque, and G. Golub for providing the source code for the variable projection algorithm. This work was supported in part by the U.S. Department of Health, Education, and Welfare under Grant GM 18051 and the National Science Foundation under Grants PCM79-05072 and INT78-27606.

[16] W. Doster, *Biophys. Chem.* (submitted for publication).

[40] Time-Resolved Protein Fluorescence Measurements of Intermediates in the Photocycle of Bacteriorhodopsin

By WILLIAM D. HOPEWELL and JOSEPH M. FUKUMOTO

Introduction

Bacteriorhodopsin (BR) is unique in its ability to use visible light to generate a transmembrane proton gradient. Hence BR functions as a light-driven proton pump.[1,2] The visible absorption of BR is a result of the presence of a retinal chromophore. On illumination, this chromophore is known to undergo cyclic spectral transformations that have been temporally correlated with the proton transport process.[3,4] However, it has become evident that chemical or photophysical transformations of the retinal alone cannot transport protons across the cell membrane. Assistance must be derived from functional groups of the surrounding protein moiety (bacterio-opsin), but the details of the interactions are not well understood. Protein–chromophore interactions are also the most likely cause of the photoinduced spectral variations in chromophore absorption, although to date their exact nature remains unspecified. For an overview discussion of recent time-resolved chromophore and protein studies see the article by M. A. El-Sayed in this volume, Chapter 73.

To explore the extent of aromatic amino acid participation in the BR photochemical cycle, a two laser photolysis-probe technique was used to monitor protein fluorescence intensity following laser photolysis of BR suspensions.[5] Signifiant decreases in fluorescence intensity were observed that correlated strongly to the L_{550} and M_{412} intermediates of the photocycle, while the remaining intermediates (K_{630}, N_{520}, and O_{640}) did not significantly alter the protein fluorescence intensity from that of the initial BR_{570} level.

The particular laser approach described here has taken advantage of the cyclic sequence of intermediate states in BR: however, the methodology should be applicable to the study of a number of similar phototransient phenomena with little modification. Linear, unidirectional photochemical sequences can be investigated, provided that care is taken to ensure the interrogation of an unphotolyzed, sample volume (i.e., with

[1] W. Stoeckenius, R. Lozier, and R. Bogomolni, *Biochim. Biophys. Acta* **505**, 215 (1979).
[2] R. Henderson, *Annu. Rev. Biophys. Bioeng.* **6**, 87 (1977).
[3] R. Lozier and W. Stoeckenius, *Biophys. J.* **15**, 955 (1975).
[4] B. Becher, F. Tokunaga, and T. Ebrey, *Biochemistry* **17**, 2293 (1973).
[5] J. Fukumoto, W. Hopewell, B. Karvaly, and M. El-Sayed, *Proc. Natl. Acad. Sci. U.S.A* **78**, 252 (1981).

METHODS IN ENZYMOLOGY, VOL. 88

appropriate sample flow techniques) with each individual pair of photolysis and probe pulses.

Materials

The method of Becher and Cassim[6] was used to isolate carotenoid-free BR that was used without sonification. Unbuffered samples at 22 μM concentration (ϵ_{570} = 63,000 liter mol^{-1} cm^{-1}) were prepared from doubly distilled deionized water. Samples were light adapted by a 2-hr exposure to room light and placed in a semimicro absorption cuvette with a 1 cm optical pathlength (Markson Ind., type-18). Polarization effects of the photolysis pulse were removed by photolyzing through the etched surface of the sample cuvette.

Methods

The photochemical cycle was initiated with a 6-nsec duration, 532-nm photolysis pulse from a frequency-doubled Q-switched Nd:YAG laser (Quanta Ray DCR). A diagram of the experimental layout is presented in Fig. 1a . This pulse was followed by a probe pulse, the frequency-doubled output of a N_2 pumped dye laser (Molectron UV1000 and DL200) at 288 nm, which excited the protein fluorescence at the reported ultraviolet absorption isosbestic.[7,8] Therefore, the monitored changes in flourescence intensity during the photocycle could not be attributed to variations in the value of the extinction coefficient at the fluorescent excitation wavelength. Time delays between the two pulses were adjustable over the

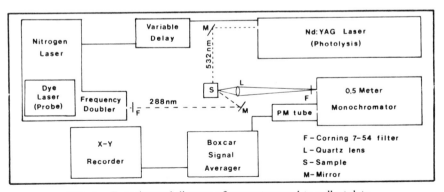

FIG. 1a. Experimental diagram of apparatus used to collect data.

[6] B. Becher and J. Cassim, *Prep. Biochem.* **5**, 161 (1975).
[7] C. Rafferty, *Photochem. Photobiol.* **29**, 109 (1979).
[8] B. Hess and D. Kuschmitz, *FEBS Lett.* **100**, 334 (1979).

range of 1 μsec to 20 msec and thus spanned the time scales of the K_{630}, L_{550}, M_{412}, N_{520}, and O_{640} phototransient states. The photolysis pulse intensities were sufficient to photolyze about 20% of the BR molecules present yet low enough to avoid multiple photon processes. Protein fluorescence from the front of the cell was collected with a quartz lens, filtered through a Corning 7-54 visible cutoff filter, and imaged onto the slit of a Jarrell Ash 0.5-m monochromator set to pass 320 nm $\pm$ 1 nm, the maximum in bacterio-opsin fluorescence. The fluorescence was detected using a high gain photomultiplier (EMI 9813GB). The resulting signal was fed to a boxcar signal averager (PAR 162) and the output displayed on a x-y recorder (Houston 2111). Pulsed ultraviolet excitation of all BR samples was limited to 20 min to reduce irreversible photochemistry. All measurements were carried out at a sample pH of approximately 6 and a temperature of 21 $\pm$ 2°.

Data were collected utilizing a particular sequence of excitation and probe pulses (Fig. 1b), which produced three levels of fluorescence intensity: (1) $I(t = 0)$, the protein fluorescence intensity obtained from unphotolyzed BR when excited with only 288-nm probe pulses; (2) $I(t = 200)$, the reference fluorescence level corresponding to the maximum in M_{412} population at room temperature, collected with a 200 μsec delay between the 532 nm photolysis and 288-nm probe pulses: (3) $I(t)$, the level of fluorescence recorded when t was the delay of interest between the 532-nm photolysis and 288-nm probe pulses. The delay (t) was varied from 1 μsec to 20 msec in order to record the protein fluorescence levels at probe pulse delays corresponding to the five photogenerated states (K_{630}, L_{550}, M_{412}, N_{520}, and O_{640}). A 15-min exposure of the sample to UV probe pulses and appropriate photolysis-probe pairs produced the intensity level sequence shown in Fig. 1b.

Information on the experimental parameters most critical to data quality was available from an examination of the constancy of $I(t = 0)$, $I(t = 200)$, and $I(t)$ levels. $I(t = 0)$ principally reported on irreversible UV pho-

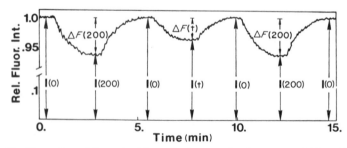

FIG. 1b. Illustration of the output from the boxcar signal averager. $\Delta F(t = 200)$ is the fluorescence intensity decrease when the photolysis and probe are separated by 200 μsec. $\Delta F(t)$ is the fluorescence intensity for the decay of interest.

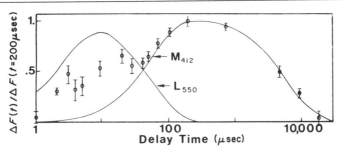

FIG. 2. Solid lines show the kinetic populations of the L and M intermediates, normalized to the maximum value of M at 200 μsec. The open circles with error bars are the observed changes in the fluorescence intensity, normalized to the maximum change, which also occurs at 200 μsec.

todegradation of the sample and long-term UV laser stability. $I(t = 200)$ monitored 532-nm photolysis pulse photodegradation, long-term photolysis pulse stability, photolysis-probe spatial overlap, and sample photocycling capacity. Repetition of an $I(t = 200)$ reference level once $I(t)$ was established ensured sample and experimental apparatus stability during the $I(t)$ recording interval (about 3 min).

Figure 2 shows the changes (open circles) in the protein fluorescence as a function of the delay between the photolysis and probe pulses. The values of the changes observed at variable probe delays, $(\Delta F(t)$, are normalized to the change recorded at a 200-μsec probe delay $(\Delta F(200))$, where the largest decrease was found to occur. Also shown in Fig. 2 are the time-dependent populations of L_{550} and M_{412} (solid lines) as calculated from the kinetic data of Henderson.[2] In using the Henderson data we assumed a linear unidirectional sequence of intermediate states evolving from the initial photolysis state. The elapsed time between successive photolysis-probe pulse pairs was always kept greater than 100 msec. This interval permitted an adequate repopulation of the BR_{570} light-adapted state (at room temperature) before the sample experienced the next photolysis-probe pulse pair. Under these conditions, the assumption of an effective linear, unidirectional sequence in BR does not appear unreasonable.

Discussion

From Fig. 2 it can be seen that photolysis-induced decreases in bacterio-opsin fluorescence occur exclusively in the L_{550} and M_{412} intermediate, with the M_{412} state exhibiting the lowest fluorescence quantum yield. Within the time resolution of this experiment and the accuracy of the published data[2] on the kinetics of the chromophore states, there is no detectable time shift between the formation of the L_{550} and M_{412} chromophore intermediates and the changes in protein fluorescence. The temporal inseparability of the variations in chromophore absorbance and opsin fluorescence may imply an intimate tryptophan and/or tyrosine-chromophore charge interactions during the formation and decay of the L_{550} and M_{412} intermediates. Alternately, deprotonation of 1–2 tyrosines or charge perturbation of 1–2 tryptophans could account for the magnitude of the reversable fluorescence decreases. These events could cause the formation of L_{550} and/or M_{412}, or result as a consequence of the formation of L_{550} and M_{412}. Unfortunately, additional evidence is needed to demonstrate which sequence of events is involved. Evidence from chemical modification studies demonstrates the necessity of intact aromatic residues for active proton transport on the L_{550} and M_{412} time scales.[9-11] The protein fluorescence decreases occur on precisely these time scales, and it seems reasonable that tryptophan and tyrosine residues may be involved in the active proton transport. Retinal electronic distributions comprising the K_{630}, N_{520}, and O_{640} states may be modulated by nonaromtic residues of the protein that do not affect the opsin fluorescence intensity.

Acknowledgments

The authors would like to thank Prof. M. A. El-Sayed for both interesting them in this problem and supporting this work. The financial support of the Dept of Energy (Office of Basic Energy Science, Grant DE-AS03-76SF00034) is greatly appreciated.

[9] T. Konishi and L. Packer, *FEBS Lett.* **79**, 369 (1977).
[10] T. Konishi and L. Packer. *FEBS Lett.* **92**, 1 (1978).
[11] T. Konishi, S. Tristram, and L. Packer, *Photochem. Photobiol.* **29**, 353 (1978).

[41] Recent Advances in the Study of Bacteriorhodopsin Dynamic Structure Using High-Field Solid-State Nuclear Magnetic Resonance Spectroscopy

By Eric Oldfield, Robert A. Kinsey, and Augustin Kintanar

There have been an impressive number of studies of the structure of proteins, both in the crystalline solid state[1] and in solution,[2,3] during the past decade. Direct information about protein structure most frequently

[1] T. L. Blundell and L. N. Johnson, "Protein Crystallography." Academic Press, New York, 1976.
[2] K. Wuthrich, "NMR in Biological Research: Peptides and Proteins." Am. Elsevier, New York, 1976.
[3] R. A. Dwek, "Nuclear Magnetic Resonance in Biochemistry; Applications to Enzyme Systems." Oxford Univ. Press (Clarendon), London and New York, 1973.

comes from X-ray crystallography of single crystals, while solution structural information has been obtained most directly by nuclear magnetic resonance (NMR) spectroscopy. Of the two methods, X-ray diffraction gives direct three-dimensional or *spatial* structural information, whereas NMR spectroscopy is more suited to determination of the *dynamical* aspects of protein structure.[4,5] Unfortunately, membrane proteins have not yet been crystallized in forms suitable for X-ray diffraction studies and do not give rise to high-resolution NMR spectra in solution, so that the application of these two powerful conventional structure determination techniques has yielded little information about the systems of interest to readers of this volume: bacteriorhodopsin and rhodopsin.

Fortunately, however, solid-state NMR methods are, in principle, capable of giving information about the structures of condensed phases,[6,7] and in this chapter we present a summary of our results on the dynamics of single types of aliphatic and aromatic amino acids in the purple membrane protein, bacteriorhodopsin, from the extreme halophile *Halobacterium halobium* R_1. This system has the desirable NMR characteristics of only one protein, bacteriorhodopsin, in the purple membrane,[8] its sequence is known,[9-11] and its three-dimensional structure is becoming available.[12,13] The system may also be enriched biosynthetically with a number of deuterated amino acids[13,14] without undue label "scrambling." Moreover, the system has been oriented using electric[15] or magnetic[16] fields or by drying down onto glass or mica surfaces,[13] and preliminary results on formation of microcrystals have been reported,[17] opening up the possibility of obtaining oriented samples for NMR spectroscopy, which permits in favorable cases determination of residue orientations.[18,19]

[4] F. R. N. Gurd and T. M. Rothgeb, *Adv. Protein Chem.* **33**, 73 (1979).

[5] G. Wagner, A. DeMarco, and K. Wüthrich, *Biophys. Struct. Mech.* **2**, 139 (1976).

[6] M. Mehring, *NMR: Basic Princ. Prog.* **11**, 1 (1976).

[7] H. W. Spiess, *NMR: Basic Princ. Prog.* **15**, 55 (1978).

[8] D. Oesterhelt and W. Stoeckenius, this series, Vol. 31, p. 667 (1974).

[9] G. E. Gerber, R. J. Anderegg, W. C. Herlihy, C. P. Gray, K. Biemann, and H. G. Khorana, *Proc. Natl. Acad. Sci. U.S.A.* **76**, 227 (1979).

[10] Yu. -A. Ovchinnikov, N. G. Abdulaev, M.-Yu. Feigina, A. V. Kiselev, and N. A. Lobanov, *FEBS Lett.* **100**, 219 (1979).

[11] J. E. Walker, A. F. Carne, and H. Schmitt, *Nature (London)* **278**, 653 (1979).

[12] R. Henderson and P. N. T. Unwin, *Nature (London)* **257**, 28 (1975).

[13] D. M. Engelman and G. Zaccai, *Proc. Natl. Acad. Sci. U.S.A.* **77**, 5894 (1980).

[14] R. A. Kinsey, A. Kintanar, M. -D. Tsai, R. L. Smith, N. Janes, and E. Oldfield, *J. Biol. Chem.* **256**, 4146 (1981).

[15] L. Keszthelyi, *Biochim. Biophys. Acta* **598**, 429 (1980).

[16] D-Ch. Neugebauer, A. E. Blaurock, and D. L. Worcester, *FEBS Lett.* **78**, 31 (1977).

[17] H. Michel and D. Oesterhelt, *Proc. Natl. Acad. Sci. U.S.A.* **77**, 1283 (1980).

[18] E. Oldfield and T. M. Rothgeb, *J. Am. Chem. Soc.* **102**, 3635 (1980).

[19] T. M. Rothgeb and E. Oldfield, *J. Biol. Chem.* **256**, 1432 (1981).

This Chapter demonstrates that with sufficiently sensitive NMR instrumentation it is now possible to study in some detail amino acid dynamics in this membrane protein. Such observations may eventually allow comparison of motions between proteins in membranes and in conventional three-dimensional crystals[20,21] and will, of course, facilitate *direct* observation of the effects of lipids and sterols on protein structure. Our results also directly complement the static structural information currently being obtained on *H. halobium* using neutron beam methods.[13]

Experimental Methods

Syntheses of ^2H-labeled amino acids. We have recently synthesized a wide range of ^2H-labeled amino acids, including the following species to be discussed: $[\delta_1, \delta_2, \epsilon_1, \epsilon_2, \zeta^{-2}H_5]$phenylalanine, $[\epsilon_1, \epsilon_2^{-2}H_2]$tyrosine, $[\delta_1, \epsilon_3, \zeta_2, \zeta_3, \eta_2^{-2}H_5]$tryptophan, and $[\alpha^{-2}H_1]$-, $[\beta^{-2}H_1]$-, and $[\gamma^{-2}H_6]$valine. The selectively labeled amino acids were synthesized using modifications of published procedures.[22,23] In addition, we obtained $[\alpha^{-2}H_2]$glycine and $[\beta^{-2}H_3]$alanine from Merck, Sharpe and Dohme (Montreal). The positions of ^2H-label incorporation are thus as shown in Fig. 1.

Production of Labeled Membranes. Halobacterium halobium strain R_1 was the kind gift of Professor T. Ebrey and was grown in a salt medium basically according to Onishi *et al.*[24] with the addition of 2% malate,[25] except that ^2H-labeled amino acids were substituted, one by one for the normal nonlabeled amino acids. For the Trp-labeled membrane system we incorporated $[^2H_5]$Trp at a level of 5.0 g/10 liters, since the growth medium does not normally contain tryptophan. Purple membranes were isolated according to Becher and Cassim[26] and were then exchanged with ^2H-depleted water (Aldrich Chemical Company, Milwaukee, Wisconsin) to remove some background HO^{2}H. Samples were generally exchanged twice, then finally concentrated by ultracentrifugation for 10 hr at 100,000 g prior to NMR spectroscopy.

Radiotracer Experiments. To determine the level of deuterated amino acid breakdown, and reincorporation into other amino-acids, 1-liter batches of cells were grown and harvested basically as for the ^2H-labeled

[20] H. Frauenfelder, G. Petsko, and D. Tsernoglou, *Nature (London)* **280**, 558 (1979).
[21] P. J. Artymiuk, C. C. F. Blake, D. E. P. Grace, S. J. Oatley, D. C. Phillips, and M. J. E. Sternberg, *Nature (London)* **280**, 563 (1979).
[22] H. R. Matthews, K. S. Matthews, and S. J. Opella, *Biochim. Biophys. Acta* **497**, 1 (1977).
[23] H. R. Snyder, J. F. Shekleton, and C. D. Lewis, *J. Am. Chem. Soc.* **67**, 310 (1945).
[24] H. Onishi, M. E. McCance, and N. E. Gibbons, *Can. J. Microbiol.* **11**, 365 (1965).
[25] A. Danon and W. Stoekenius, *Proc. Natl. Acad. Sci. U.S.A.* **71**, 1234 (1974).
[26] B. M. Becher and J. Y. Cassim, *Prep. Biochem.* **5**, 161 (1975).

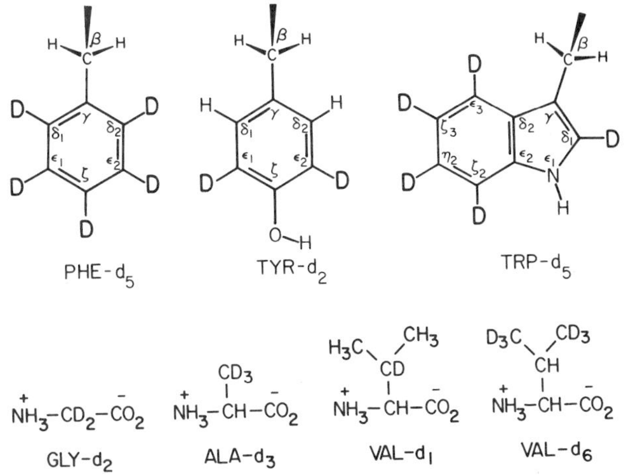

FIG. 1. Structures of the ²H-labeled amino acids discussed in this publication, showing position of ²H-label incorporation.

cells except that 50 μCi of either [¹⁴C]Tyr, [¹⁴C]Phe, [¹⁴C]Trp, [¹⁴C]Gly, [¹⁴C]Ala, or [¹⁴C]Val (New England Nuclear, Boston, Massachusetts) were added as radiotracers.

Spectroscopic Aspects. Nuclear magnetic resonance spectra were obtained using two "home-built" Fourier transform NMR spectrometers. The first one consists of an 8.5 Tesla 3.5-in. bore high-resolution superconducting solenoid (Oxford Instruments, Osney Mead, Oxford, U.K.), together with a variety of digital and radiofrequency electronics. We used a Nicolet 1180 computer, 293B pulse programmer, and Model NIC-2090 dual channel 50-nsec transient recorder (Nicolet Instrument Corporation, Madison, Wisconsin) for experiment control and rapid data acquisition, together with a dual Diablo Model 40 disk system for data storage (Diablo Systems, Inc., Haywood, California). In order to generate radio-frequency pulses of high enough power (~3 μsec 90° pulse widths) to cover the entire ²H NMR spectral width, we used an Amplifier Research (Amplifier Research, Souderton, Pennsylvania) Model 200L amplifier to drive a retuned Henry Radio (Henry Radio, Los Angeles) Model 2006 transmitter to a ~1000–1500 W output power level. The deuterium resonance frequency was 55.273 MHz. Deuterium NMR spectra were recorded on this instrument using an 800-μL sample volume and a quadrupole-echo[27,28] pulse sequence. The 90° pulse width varied between 2.0 and 3.5 μsec.

---

27 I. Solomon, *Phys. Rev.* **110**, 61 (1958).
28 J. H. Davis, K. R. Jeffrey, M. Bloom, M. I. Valic, and T. P. Higgs, *Chem. Phys. Lett.* **42**, 390 (1976).

The second instrument used was the medium-field spectrometer (5.2 Tesla) described previously,[29] except that it now uses a "homebuilt" 400 kHz data acquisition system based on a Digital Equipment Corporation LSI-11 microcomputer (Digital Equipment Corporation, Boston, Massachusetts) equipped with dual disks, together with a second Model 2006 transmitter. The 90° pulse widths (~ 2.0–3.5 μsec) and phase quadrature between the two radiofrequency pulses were established on both instruments by viewing quadrature free-induction decay signals of S-[*methyl-^{2}H$_3$*]methionine. Identical settings were used for ^{2}H-labeled membranes, and in essentially all cases no phase corrections were necessary after Fourier transformation. Sampling rates of 500 nsec per point were used at 8.5 Tesla, and 3 μsec per point at 5.2 Tesla. The zero frequencies of both instruments were established using a 1% D_2O reference, the zero frequency for the protein samples investigated being offset ~ 2 ppm downfield from this position for aromatic amino acids, or between 1 and 3 ppm upfield for aliphatics. Samples were run as solid high-speed pellets, probe temperature being regulated either by means of a liquid nitrogen boil-off system or by using a heated air flow.

Theoretical Background for NMR of Membrane Proteins. The allowed transitions for the spin I = 1 ^{2}H nucleus correspond to $+1 \leftrightarrow 0$ and $0 \leftrightarrow -1$ (Fig. 2A) and give rise to a "quadrupole splitting" ($\Delta\nu_Q$) of the absorption line, with separation between peak maxima of[29-31]

$$\Delta\nu_Q = \frac{3}{2} \frac{e^2 qQ}{h} \frac{3 \cos^2 \theta - 1}{2} \tag{1}$$

θ is the angle between the magnetic field H_0 and the principal axis of the electric field gradient tensor (frequently the C—D bond vector). All values of θ are possible for rigid polycrystalline solids and one therefore obtains a so-called powder pattern, Fig. 2B, having a peak separation corresponding to $\theta = 90°$, for which $\Delta\nu_Q = 3e^2 qQ/4h$, and a shoulder separation corresponding to $\theta = 0°$, for which $\Delta\nu_Q = 3e^2 qQ/2h$. A typical experimental example, [β-^{2}H$_1$]valine, is shown in Fig. 2C. For solid aliphatic compounds $\Delta\nu_Q$ values of 127 kHz ($\theta = 90°$) and 254 kHz ($\theta = 0°$), corresponding to an electric quadrupole coupling constant ($e^2 qQ/h$) of about 168 kHz[32,33] are expected, and observed (Fig. 2B,C). These results assume that there are no fast ($> 10^5 sec^{-1}$) large-amplitude motions of the pertinent C-^{2}H vectors in the solid amino acid. In the presence of such

[29] E. Oldfield, M. Meadows, D. Rice, and R. Jacobs, *Biochemistry* **17**, 2727 (1978).
[30] J. Seelig, *Q. Rev. Biophys.* **10**, 353 (1977).
[31] M. H. Cohen and F. Reif, *Solid State Phys.* **5**, 321 (1957).
[32] L. J. Burnett and B. H. Muller, *J. Chem. Phys.* **55**, 5829 (1971).
[33] W. Derbyshire, T. C. Gorvin, and D. Warner, *Mol. Phys.* **17**, 401 (1969).

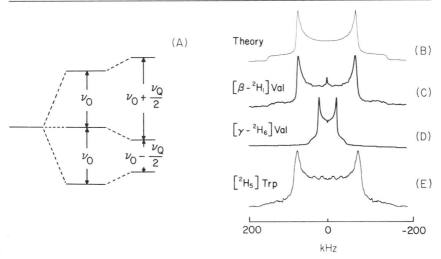

FIG. 2. Deuterium NMR energy level diagram and theoretical and experimental ^{2}H NMR spectra of polycrystalline amino acids. A, Energy level diagram showing Zeeman levels and presence of a first-order quadrupole perturbation. The transitions $-1 \leftrightarrow 0$ and $0 \leftrightarrow 1$ are shown, for an arbitrary crystal orientation. B, Theoretical ^{2}H NMR lineshape having $e^2qQ/h = 168$ kHz, $\eta = 0$, linewidth = 4000 Hz. C, Experimental spectrum of polycrystalline [β-^{2}H$_1$]valine, at 55.3 MHz, 23°. D, Experimental spectrum of polycrystalline [γ-^{2}H$_6$]valine, at 55.3 MHz, 23°. E, Experimental spectrum of polycrystalline [δ_1, ϵ_3, ζ_2, ζ_3, η_2-^{2}H$_5$]tryptophan, at 55.3 MHz, 23°.

fast large-amplitude motions it is necessary to take an average over the motions. Assuming that the asymmetry parameter (η) is zero, it can be shown that the motionally averaged splitting ($\Delta\nu$) is

$$\Delta\nu = \frac{3}{8} \frac{e^2qQ}{h} (3\cos^2\beta - 1)(3\cos^2\theta' - 1) \qquad (2)$$

where β is the angle between the principal axis of the electric field gradient tensor and the axis of motional averaging and θ' is the angle between the axis of motional averaging and H_0.

As an example of this motional averaging, consider the case of the C^{2}H$_3$ group in [γ-^{2}H$_6$]valine (Fig. 1, 2D). Methyl groups are expected to rotate rapidly about their C$_3$ axis at room temperature, in which case if we assume a tetrahedral geometry ($\beta = 109° 28'$), then a motionally averaged splitting ($\Delta\nu$) of about 42 kHz is predicted, in excellent agreement with the $\Delta\nu_Q \sim 39$ kHz observed experimentally in [γ-^{2}H$_6$]valine (Fig. 2D). Rotation about C$^\beta$—C$^\gamma$ is thus very rapid at room temperature ($\gg 10^5$ sec^{-1}). The unlikely possibility that fast rotation about C$^\alpha$—C$^\beta$ is the cause of the reduced splitting is eliminated by the observation of the full rigid lattice breadth in the ^{2}H NMR spectrum of [β-^{2}H$_1$]valine (Fig. 2C). The results of Fig. 2B–D therefore give us a benchmark with which to compare ^{2}H NMR spectra of aliphatic amino acid labeled membranes. In the case of aromatic residues quadrupole coupling constants are ~10% larger than for aliphatic species[34] and asymmetry parameters (η) for aromatic residues are nonzero; thus the full expression for the quadrupole splitting must be used[31]:

$$\Delta \nu_Q = \frac{3}{4} \frac{e^2 qQ}{h} [3 \cos^2 \theta - 1 - \eta \sin^2 \theta \cos 2\psi] \qquad (3)$$

where θ and ψ define the orientation of the principal axis of the electric field gradient tensor (usually the C—D bond vector) with respect to the laboratory coordinates. The computed quadrupole coupling constant of about 183 kHz for [^{2}H$_5$]Trp (Fig. 2E) is therefore considerably in excess of those found in aliphatic C—^{2}H systems using NMR methods.[34,35] In addition, computer simulation of the lineshape of Fig. 2E indicates an asymmetry parameter η = 0.05. These results are consistent with the increased $e^2 qQ/h$ values found in a variety of other aromatic compounds,[34–36] the observed trends for the electric field gradient (EFG) values for C—D bonds being sp > sp^2 > sp^3,[37] the average value for naphthalene and anthracene, perhaps the most reasonable published models for [^{2}H$_5$]Trp, being ~184 kHz.[38–40] In addition, it is well known that C—D bonds in aromatic systems may have nonzero asymmetry parameters,[41] η values of 0.053 ± 0.015 being typical.[34]

Motions in Some Aliphatic Side Chains. Our first goal with NMR studies of membrane protein structure is to obtain a rather broad overview of the *rates* and *types* of motions of amino acid side chains, then to develop methods for the resolution and assignment of individual atomic sites and to investigate the effects of various membrane constituents, such as cholesterol and phospholipids, on protein structure. The results we have presented give the reader some idea of the shapes and widths expected for ^{2}H NMR spectra of solid, polycrystalline ^{2}H-labeled amino acids. What, then, do membrane protein spectra look like, if we can see them?

[34] C. Brevard and J. P. Kintzinger, *in* "NMR and the Periodic Table" (R. K. Harris, and B. E. Mann, eds.) p. 119. Academic Press, New York, 1978.

[35] R. A. Kinsey, A. Kintanar, and E. Oldfield, *J. Biol. Chem.* **256**, 9028 (1981).

[36] R. G. Barnes, *Adv. Nucl. Quadrupole Reson.* **1**, 335 (1974).

[37] P. L. Olympia, Jr., I. Y. Wei, and B. M. Fung, *J. Chem. Phys.* **51**, 1610 (1969).

[38] M. Rinné, J. Depireux, and J. Duchesne, *J. Mol. Struct.* **1**, 178 (1967).

[39] D. M. Ellis and J. L. Bjorkstam, *J. Chem. Phys.* **46**, 4460 (1967).

[40] R. G. Barnes and J. W. Bloom, *J. Chem. Phys.* **57**, 3082 (1972).

[41] E. Oldfield, N. Janes, R. Kinsey, A. Kintanar, R. W. -K. Lee, T. M. Rothgeb, S. Schramm, R. Skarjune, R. Smith, and M. -D. Tsai, *Biochem. Soc. Trans.* **46**, 155 (1981).

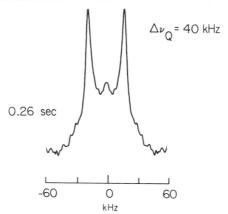

FIG. 3. 55.3 MHz ^{2}H NMR spectrum of [γ-^{2}H$_6$]valine labeled bacteriorhodopsin in the purple membrane of *H. halobium* R$_1$, − 100°, 260 msec data acquisition time.

We show in Fig. 3 the first requirement for the study of membrane proteins by NMR spectroscopy, the detection of signals in a reasonable period of time. The result of Fig. 3 was obtained in only 0.3 sec of data acquisition and is our fastest spectrum to date. Such rapid data acquisition was possible since the [γ-^{2}H$_6$]valine residues incorporated were highly deuterated; there are some 21 valines (42 methyl groups, 126 deuterons) in bacteriorhodopsin[9-11]; we used a low temperature (− 100°) to decrease the spin-lattice relaxation time (T_1), thereby permitting rapid pulsing; and the quadrupole splittings are relatively narrow. The result shown in Fig. 3 is nevertheless extraordinarily exciting, indicating the feasibility of observing any type of group in this membrane protein, although data acquisition still takes numerous hours for more dilute labels, especially when relaxation times are long. For example, the ^{2}H spin-lattice relaxation time of the [γ-^{2}H$_6$]valine label is only a few milliseconds near the T_1 minimum (∼ − 150°K, $\tau_c \approx \omega_0^{-1} \approx 10^{-9}$ sec), whereas T_1's of many seconds are obtained for rigidly bound residues ($\tau_c \gg \omega_0^{-1}$), as noted later, greatly increasing the periods of time required for data acquisition.

Nevertheless, we have successfully incorporated and obtained spectra of the following ^{2}H labeled aliphatic amino acids in the purple membrane: [α-^{2}H$_2$]glycine, [β-^{2}H$_3$]alanine, [α-^{2}H$_1$]valine, [β-^{2}H$_1$]valine, and [γ-^{2}H$_6$]-valine, and additional studies are underway on the labeling of *Escherichia coli* and *Acholeplasma laidlawii* B (PG9) with these amino acids, and several deuterated leucines and isoleucines. Details of such experiments are being[14,35,41] or will be reported in detail elsewhere.

In Fig. 4 we present a selection of NMR results on ^{2}H-labeled aliphatic amino acids in the purple membrane of *H. halobium* R$_1$, at 37° in deute-

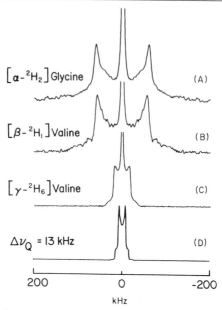

FIG. 4. 55.3 MHz ^{2}H NMR spectra of purple membranes of *H. halobium* R_1 at 37°, in ^{2}H-depleted water, together with a computer simulation of [γ-^{2}H$_6$]valine undergoing fast α–β and β–γ rotational diffusion. The labels were at the following sites: A, [α-^{2}H$_2$]glycine. B, [β-^{2}H$_1$]valine. C, [γ-^{2}H$_6$]valine. D, Computer simulation of [γ-^{2}H$_6$]valine undergoing fast rotational diffusion about C$^\alpha$—C$^\beta$ and C$^\beta$—C$^\gamma$. The quadrupole splitting $\Delta\nu_Q$ = 13 kHz.

rium-depleted water, which give some insight into the motions of these residues in bacteriorhodopsin. The spectrum of [α-^{2}H$_2$]glycine is essentially identical to that of the free amino acid[42] (and unpublished results of I. Baianu and E. O.), having e^2qQ/h = 168 kHz and η = 0.05 ± 0.02. The nonzero η is not unexpected, because of the adjacent CO and NH groupings. The large value of the quadrupole splitting rules out all but small-amplitude torsional or vibrational motions of the CD$_2$ group, consistent with the observed magnitude and temperature dependence of the ^{2}H spin-lattice relaxation time (unpublished results of A. K., R. A. K. and E. O). Glycine is thus "rigid" in the purple membrane.

Results with [β-^{2}H$_3$]alanine are complicated by the "scrambling" of this amino acid into other species [e.g., phenylalanine (unpublished results)], nevertheless it seems that the major ^{2}H resonance is attributable to the methyl deuterons of [β-^{2}H$_3$]alanine, which as in Figs. 2 and 3 are undergoing fast internal rotation resulting in a quadrupole splitting $\Delta\nu_Q$ = 39 ± 2 kHz. Many of the alanine side chains in bacteriorhodopsin are thus "rigid," with methyl rotation being the only fast large-amplitude mo-

[42] R. G. Barnes and J. W. Bloom, *Mol. Phys.* **25**, 493 (1973).

tion. A similar statement may be made for the valine side chains (Fig. 4B,C). The ^{2}H NMR spectrum of [β-^{2}H$_1$]valine labeled membranes (Fig. 4B) has $\Delta\nu_Q \sim$ 116kHz, about the same as that of the polycrystalline amino acid (Fig. 2C), and that of [γ-^{2}H$_6$]valine labeled purple membrane is also very similar, identical to that of the free amino acid (Fig. 4C, 2D).

Fast methyl group rotation averages the static quadrupole coupling constant by a value of $\sim\frac{1}{2}(3\cos^2 109.5° - 1)$, i.e., by a factor of about -0.3[14,32] and a motionally averaged spectrum having a breadth of $\sim$40 kHz is obtained. Observation of a motionally averaged quadrupole splitting of about 40 kHz for [γ-^{2}H$_6$]valine labeled sites strongly suggests that there is very fast motion only about C$^\beta$—C$^\gamma$, motion about C$^\alpha$—C$^\beta$ being very slow on our NMR timescale, such that no motional averaging of the quadrupole interaction occurs because of this motion. If motion about C$^\alpha$—C$^\beta$ were fast, then a methyl group splitting of about 13 kHz would be obtained, i.e., the quadrupole splitting would have to be reduced by another factor of $\sim$3 (Fig. 4D), as would the splitting of the [β-^{2}H$_1$]species. No such behavior is seen in the purple membrane of *H. halobium* in the temperature range $- 100°-53°$C.[14] The quadrupole splittings of [γ-^{2}H$_6$]valine-labeled bacteriorhodopsin in the purple membrane of *H. halobium* are in fact remarkably temperature independent,[14] ranging from the rigid lattice value $\Delta\nu_Q = 39 \pm 1$ kHz (due solely to fast Me rotation) below $\sim - 30°$, found also in the model system [γ-^{2}H$_6$]valine (Fig. 2D), to $\sim$33 kHz at 60°. Please note, however, that above $\sim$25° the spectra become considerably less "sharp" than those obtained at low temperature, either because of the onset of additional slow motions[43,44] or, perhaps more likely, because of the occurrence of a broader distribution of $\Delta\nu_Q$ values, resulting from the basic heterogeneity of the purple membrane. In any case, as viewed from the biosynthetically incorporated [γ-^{2}H$_6$]valine quadrupole splittings, the purple membrane remains a remarkably ordered structure over this wide range of temperature. We should point out that an "isotropic" component appears in all growth temperature spectra (Fig. 4A–C). We do not know the exact origin of this component, but attribute it in part to residual HO^{2}H. It disappears on sample freezing, or lyophilization, but could still arise from, e.g., mobile (surface) residues or from small membrane fragments.

The high signal-to-noise ratios obtained in Fig. 4 permit the direct study of amino acid side-chain dynamics via determination of NMR relaxation rates and of their temperature dependencies. For example, using a conventional inversion-recovery pulse sequence modified for solids, i.e. (180°-τ_3-90°-τ_1-90°$_{90°}$-τ_2-Echo-T), where T is the period of repetition of the

[43] R. F. Campbell, E. Meirovitch, and J. H. Freed, *J. Phys. Chem.* **83**, 525 (1979).
[44] E. Meirovitch and J. H. Freed, *Chem. Phys. Lett.* **64**, 311 (1979).

pulse sequence ($>5\ T_1$), τ_1 and τ_2 are fixed delays of $\sim 70\ \mu$sec, and τ_3 is a variable delay between the inverting (180°) and sampling (90°) pulses, we obtain for [γ-^{2}H$_6$]valine labeled purple membranes T_1 values of 7.4 msec at $-75°$, 35.7 msec at 0°, 72.7 msec at 55°.[14] These, and the results of additional experiments, when plotted in the form of an Arrhenius-type curve, yield an activation energy (ΔE) for the relaxation process—methyl group rotation, of $\sim 2.4\ +\ 0.2$ kcal mol^{-1}. This is in excellent agreement with the value $\Delta E = 2.6 \pm 0.2$ kcal mol^{-1} obtained previously by Anderson and Slichter,[45] who studied ^{1}H nuclear spin relaxation in solid n-alkanes. These workers also obtained a T_1 minimum (when $\omega_0\tau_c \sim 0.62$) at $-125°$, essentially identical to that we have obtained using our 5.2 Tesla instrument, at 34 MHz (unpublished results of E. Oldfield, A. Kintanar, M. Keniry, and B. Smith). These results suggest correlation times of $\approx 5 \times 10^{-11}$ sec for methyl group rotation, when analyzed using a simple relaxation model.[14]

Motions in Some Aromatic Side Chains. We have incorporated, and are incorporating, the following deuterated amino acids into bacteriorhodopsin in the purple membrane of *H. halobium* R$_1$, and a variety of other membrane systems: [α-^{2}H$_1$]phenylalanine, [δ_1, δ_2, ϵ_1, ϵ_2, ζ-^{2}H$_5$]phenylalanine, [ϵ_1, ϵ_2, ζ-^{2}H$_3$]phenylalanine, [ζ-^{2}H$_1$]phenylalanine, [α-^{2}H$_1$]tyrosine, [β-^{2}H$_2$]tyrosine, [ϵ_1, ϵ_2-^{2}H$_2$]tyrosine, [α-^{2}H$_1$]tryptophan and [δ_1, ϵ_3, ζ_2, ζ_3, η_2-^{2}H$_5$]tryptophan. Experiments with [ϵ-^{2}H$_1$]histidine are also under way with *E. coli*.

We show in Fig. 5 typical results obtained with [δ_1, ϵ_3, ζ_2, ζ_3, η_2-^{2}H$_5$]tryptophan, [δ_1, δ_2, ϵ_1, ϵ_2, ζ-^{2}H$_5$]phenylalanine and [ϵ_1, ϵ_2-^{2}H$_2$]tyrosine[35] at the growth temperature of the organism (37°). Note that the spectrum of [^{2}H$_5$]Trp is essentially identical to that of the free amino acid in the solid state (Fig. 2E), being simulated well by using $e^2qQ/h = 183.0$ kHz and $\eta = 0.05$.[35] Since the spectrum remains virtually the same between $-85°$ and $+85°$, it seems reasonable to characterize the Trp residues as "rigid," only undergoing torsional or librational motions of $\approx 5 - 10°$ amplitude.[35]

By contrast, the ^{2}H FT NMR spectra of the Phe and Tyr labeled purple membranes have rather unusual lineshapes. It is therefore of some interest to examine the likely motions of a Phe or Tyr residue in the solid state, as shown in Fig. 6.

We show in Fig. 6A,B the ^{2}H NMR spectrum of a "rigid" [^{2}H$_5$]phenylalanine residue together with its simulation, characterized by a deuteron quadrupole coupling constant $e^2qQ/h \sim 180$ kHz and $\eta = 0.05$. These re-

[45] J. E. Anderson and W. P. Slichter, *J. Phys. Chem.* **69**, 3099 (1965).

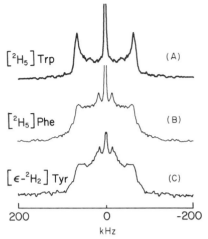

$[^2H_5]$Trp   (A)

$[^2H_5]$Phe   (B)

$[\epsilon\text{-}^2H_2]$ Tyr   (C)

200   0   -200
kHz

FIG. 5. Experimental ^2H NMR spectra, at 55.3 MHz, of aromatic amino acid labeled purple membranes in ^2H-depleted water, at 37°. The following labels were incorporated: A, $[\delta_1, \epsilon_3, \zeta_2, \zeta_3, \eta_2\text{-}^2H_5]$tryptophan. B, $[\delta_1, \delta_2, \epsilon_1, \epsilon_2, \zeta\text{-}^2H_5]$-phenylalanine. C, $[\epsilon_1, \epsilon_2\text{-}^2H_2]$tyrosine.

sults are in good agreement with the median values found for a series of monosubstituted benzenes of $e^2qQ/h \sim 181$ kHz and $\eta = 0.06$.[34,38,40,46] Let us now consider the effects of rapid ($\gg 10^5$ sec^{-1}) motions on the observed ^2H NMR spectrum. One possibility would be fast rotational diffusion about C^β—C^γ. In this case the phenyl ring would undergo rapid rotational diffusion about the C^γ—C^ζ axis, in which case the $C^{\delta1,\delta2,\epsilon1,\epsilon2}\text{-}^2H$ vectors would be at 60 ± 1° to the axis of motional averaging, and the C^ζ—^2H vector would be at 0°. It is a simple matter to calculate the observed spectrum, and the result is shown in Fig. 6C. It is quite dissimilar to that observed for $[^2H_5]$Phe in any native system investigated.

By contrast, a twofold "jump" model for phenylalanine motion, whereby the aromatic ring executes 180° reorientational "flips" about C^γ—C^ζ, and which has been detected previously in solution NMR studies of proteins by means of chemical shift data,[2,4,5,47] predicts a very different result (Fig. 6D). Assuming that motion is "fast" compared to the breadth of the rigid bond coupling, a motionally averaged tensor may be calculated by using the model of Soda and Chiba[48] for motional averaging of the deuterium quadrupole interaction by reorientation about a twofold axis.

[46] I. Y. Wei and B. M. Fung, *J. Chem. Phys.* **52**, 4917 (1970).
[47] C. M. Dobson, G. R. Moore, and R. J. P. Williams, *FEBS Lett.* **51**, 60 (1974).
[48] G. Soda and T. Chiba, *J. Chem. Phys.* **50**, 439 (1969).

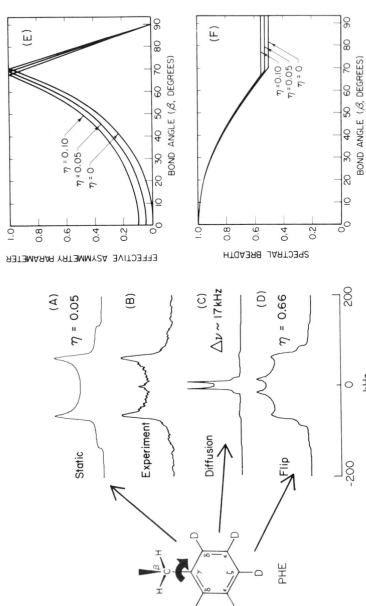

FIG. 6. ^{2}H HMR spectra and simulations of [δ_1, δ_2, ϵ_1, ϵ_2, ζ-^{2}H$_5$]phenylalanine, and theoretical plots of asymmetry parameter and spectral breadth as a function of flip-angle for a twofold flip. A, Computer simulation of rigid lattice lineshape using $e^2qQ/h = 180$ kHz, $\eta = 0.05$. B, Experimental 55.3 MHz ^{2}H NMR spectrum of [δ_1, δ_2, ϵ_1, ϵ_2, ζ-^{2}H$_5$]phenylalanine, at 23°C. C, Computer simulation of [δ_1, δ_2, ϵ_1, ϵ_2, ζ-^{2}H$_5$]phenylalanine undergoing fast ($\gg 2 \times 10^5$ sec^{-1}) rotational diffusion about C$^\gamma$—C$^\zeta$ C$_2$ axis. D, As in C but motion is a twofold "flip." E, Dependence of the effective asymmetry parameter (η_{eff}) on the interbond angle β for the case of motional averaging by twofold flipping for several values of the rigid-bond asymmetry parameter (η). F, Dependence of the reduced spectral breadth on the interbond angle β for the case of motional averaging by twofold flipping for several values of the rigid bond asymmetry.

The results for the averaged field gradient tensor components as a function of β, and the asymmetry parameter (η) of the static tensor are

$$V_{11} = -\tfrac{1}{2}q(1 + \eta)$$

$$V_{22} = +\tfrac{1}{4}q[(1 - 3 \cos 2\beta) + \eta(1 + \cos 2\beta)] \qquad (4)$$

$$V_{33} = +\tfrac{1}{4}q[(1 + 3 \cos 2\beta) + \eta(1 - \cos 2\beta)].$$

We show in Fig. 6E,F the new effective asymmetry parameters (η_{eff}) for the case of rapid 180° reorientations as a function of bond angle β together with the new maximum effective field gradients eq, i.e. the reduced spectral breadths $V_{33}-V_{11}$, from which it is a simple matter to calculate the new spectral lineshape for twofold flipping, as shown in Fig. 6D. The spectra now contain a new sharp narrow feature, corresponding to the separation between the singularities in the powder pattern, having $\Delta\nu_{Q1}$ ~ 30 kHz (Fig. 6D). This dominant feature is easily detected in some intact membrane spectra, as discussed later. Unfortunately, however, the mere presence of such a predicted peak does not guarantee its origin in the spectra of intact membranes. For example, in a system as heterogeneous as a biological membrane, it seems quite likely that there could be a variety of environments in which Phe and Tyr residues might undergo a variety of different motions. For example, a "rocking" motion about C^α—C^β together with torsional motions about the peptide bonds or "rigid body" motions of whole segments of surface-exposed residues might easily occur and give rise to a second, reduced quadrupole splitting.

Two pieces of evidence indicate that these explanations do not apply in the case of Phe and Tyr-labeled bacteriorhodopsin. First, spectra of [β-^{2}H$_2$]tyrosine exhibit the full rigid-lattice splitting of ~ 120 kHz and an asymmetry parameter $\eta = 0.00 \pm 0.02$, at the growth temperature of the organism. This rules out fast large-amplitude motions of the [β-^{2}H$_2$]tyrosine deuterons (unpublished results of A. Kintanar and E. Oldfield). Motional averaging of the tyrosine ring-deuteron spectrum must thus occur solely by motions about the C^γ—C^ζ axis. Similarly, ^{2}H NMR spectra of [ζ-^{2}H$_1$]phenylalanine, a *para*-substituted derivative, show essentially axially symmetric ($\Delta\nu_Q \sim 120$ kHz, $\eta = 0.05 \pm 0.02$) spectra at the temperature of growth, unlike the narrow, axially asymmetric spectra ($\eta \approx 0.65$) obtained for the [δ_1, δ_2, ϵ_1, ϵ_2, ζ-^{2}H$_5$]phenylalanine labeled samples (Fig. 5B and unpublished results of A. K. and E. O.). This arises from the fact that the *para*-deuteron's electric field gradient tensor principal axis lies directly along the axis of motional narrowing, i.e., $\beta = 0°$; consequently, there is essentially no averaging of the spectral linewidth, as shown in Fig. 6E, F.

Our results therefore indicate that tryptophan, phenylalanine, and tyrosine residues are rigid at low temperatures ($<$ $-30°$) while phenylalanine and tyrosine residues are each highly mobile at the temperature of growth of the *H. halobium* purple membrane (37°), undergoing fast ($> 10^5$ -10^6 sec^{-1}) twofold jumps about C^β—C^γ. Tryptophan residues do not undergo this type of motion even at 85°, immediately prior to protein denaturation. On denaturation at $\sim 90°$, "narrow-line" spectra (having linewidths ~ 5–10 kHz) are obtained for all three aromatic amino acids (unpublished results of R. A. K., A. K. and E. O.), suggesting fast large-amplitude motions.

The Picture to Date. Our results may be summarized as follows. Backbone (C^α) labeled amino acids in the purple membrane of *H. halobium* exhibit "rigid-lattice spectra," except for a possible small population of surface residues. There is no evidence for fast motion about C^α—C^β for any aliphatic or aromatic amino acid investigated to date, except for fast methyl rotations in alanine and a possible small population (~ 5–10%) of surface residues. In the case of valine-labeled purple membranes, motion about C^β—C^γ is fast ($> 10^6$ sec^{-1}) at all temperatures investigated (down to 120°K). The increased bulk of the benzenoid rings in Tyr and Phe greatly impede motion of these side chains. When they do begin to move (at about the 37° growth temperature of the organism) rotation is not diffusive but occurs by a twofold flipping process, as has been detected previously in solution NMR studies of soluble proteins. The additional bulk of the indole ring in tryptophan prevents even this motion and only small-angle librations are allowed, even at 85°. These results are supported by spin-lattice relaxation data (unpublished results of R. A. K., A. K., Becky Smith, Max Keniry, Herbert Gutowsky and E. O.) that show that all systems (except for the methyl labels) have spin-lattice relaxation times that decrease with increasing temperature, since correlation times (τ_c) are all $>> \omega_0^{-1}$.

A generalized picture of our observations is presented in Fig. 7. This picture of the dynamics of most aliphatic and aromatic amino acids in the purple membrane is one of a rather rigid protein; in most instances the ^2H NMR spectra of the protein are rather similar to those of the solid amino acid, at the same temperature, except for the Phe and Tyr residues that undergo twofold "flipping." Since it appears that Phe residues at least are located toward the center of the protein,[13] where H^+ translocation may occur, it seems possible that these motions could be of importance in the energy transduction process. Interestingly, as discussed earlier, the Trp residues do not undergo any such fast motions.

Since we have now obtained ^2H spectra of [γ-$^2H_6$]valine labeled membranes in as short a time as 300 msec, it is hoped that future studies, perhaps using electrically or magnetically ordered materials[15,16,18,19] as reso-

MOTION

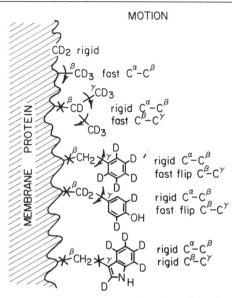

FIG. 7. Diagram showing the rates and types of motions of a variety of aliphatic and aromatic amino acid side chains of bacteriorhodopsin in the purple membrane of *H. halobium* R_1 at 37°, the temperature of growth.

lution-enhancing aids, may include time-resolved experiments aimed at determining the exact nature of conformational changes, both in the protein and in the retinal chromophore, during the photochemical cycle. At present, the major problem in carrying out such experiments is the availability of very high-field sperconducting magnets. Given fields of ~ 16–18 Tesla (~ 700–800 MHz ^{1}H frequency), within reach of current technology, time-resolved structural studies on this and other important energy-transduction systems should be quite feasible.

Acknowledgments

We thank Brenda Coles, Cathy Flynn, Peter Kolodziej, Ben Montez, Nathan Janes, Julie Nichols, Larry Pubentz, Tane Ray, Michael Rothgeb, Andrew Scheinman, Suzanne Schramm, Robert Skarjune, Rebecca Smith, Ming-Daw Tsai, Joe Vanderbranden, and Suzanne Volk for discussions and help with the experiments. This work was supported by the U.S. National Institutes of Health (Grant HL-19481), by the U.S. National Science Foundation (Grants PCM 78-23021, PCM 79-23170) and the Alfred P. Sloan Foundation; and in part by the University of Illinois National Science Foundation Regional Instrumentation Facility (Grant CHE 79-16100).

Eric Oldfield was an Alfred P. Sloan Research Fellow, 1978–1980; and is a USPHS Research Career Development Awardee, 1979–1984 (Grant CA-00595).

[42] Preparation of Oriented Multilamellar Arrays of Natural and Artificial Biological Membranes

By Noel A. Clark and Kenneth J. Rothschild

Introduction

For many probes of biological membrane structure and function, having the sample arranged into a stack of parallel planar membrane sheets enhances the information available. For example, techniques such as circular and linear dichroism and dispersion,[1] and nmr[2] and Raman[3] spectroscopy, when applied to planar multilamellar samples, can yield information on the orientational distribution and mean orientation of selected molecular units or subunits. In addition, well-ordered multilamellar samples possess quasi-long-range translational order along the direction normal to the layers, rendering them suitable for structure determination using X-ray[4-6] or neutron[7] diffraction techniques. This latter point is of particular interest, since most natural membrane systems lack any long-range order. A notable exception to this is purple membrane, which, by virtue of a high concentration of the single protein component, bacteriorhodopsin, occurs naturally as a two-dimensional crystal.[8] Exploitation of this feature in diffraction studies of planar single and multilamellar[9] samples has yielded the first glimpse of the internal structure of membrane proteins.[10] Furthermore, the recently reported three-dimensional crystallization of bacteriorhodopsin[11] promises that atom-resolved structures of isolated membrane proteins may soon be available. However, studies of structure and operation of the majority of functioning natural membrane

---

[1] A. D. Bangham and R. W. Horne, *J. Mol. Biol.* **8,** 660 (1964).

[2] R. G. Griffin, L. Powers, and P. S. Pershan, *Biochemistry* **17,** 2718 (1978).

[3] S. Jen, N. A. Clark, and P. S. Pershan, *J. Chem. Phys.* **66,** 4635 (1977).

[4] J. B. Stamatoff, S. Krimm, and N. R. Harvie, *Proc. Natl. Acad. Sci. U.S.A.* **72,** 531 (1975).

[5] L. Herbette, J. Marquardt, A. Scarpa, and J. K. Blasie, *Biophys. J.* **20,** 245 (1977).

[6] R. S. Khare, and C. R. Worthington, *Mol Cryst. Liq. Cryst.* **38,** 195 (1977).

[7] G. Zaccai, J. K. Blasie, and B. P. Schoenborn, *Proc. Natl. Acad. Sci. U.S.A.* **72,** 376 (1975).

[8] K. A. Fisher and W. Stoeckenius, *Science* **197,** 72 (1977).

[9] A. E. Blaurock and W. Stoeckenius, *Nature (London), New Biol.* **233,** 152 (1971); R. Henderson *J. Mol. Biol.* **93,** 123 (1975).

[10] R. Henderson and P. N. T. Unwin, *Nature (London)* **257,** 28 (1975).

[11] H. Michel and D. Oesterhelt, *Proc. Natl. Acad. Sci. U.S.A.* **77,** No. 3, 1283 (1980).

METHODS IN ENZYMOLOGY, VOL. 88

systems will require that the techniques noted earlier be applied to multi-lamellar noncrystalline arrays.

Membrane preparations to be oriented are typically available as suspensions of physically separate or loosely attached finite-sized membrane sheets. These sheets may be single or multilayer, aritificial or natural, and may be simply or multiply connected. Examples include cell plasma membrane fragments, purple membrane, whole cell envelopes such as erythrocyte ghosts or other closed natural membrane structures, rod outer segment disks, closed bilayer or multibilayer vesicles, and swollen multibilayer arrays. A variety of methods have been employed to produce oriented samples from such preparations. These include drying suspensions onto surfaces,[7,12] shearing between flat plates,[13] centrifugation,[7,14] preferential orientation at interfaces,[13] electric fields,[15] magnetic fields,[16] black lipid films,[17] multilayer deposition,[18] and annealing between flat plates.[19] Typically, these techniques represent some trade-off between desirable objectives: a large volume of material available in an ordered sample; high-quality orientational order; and minimum alteration caused by the orientation process.

Of these methods centrifugation and solvent evaporation (dehydration) have proved to be the most widely used. These are surface orientation methods, relying on the influence of a flat surface of an adjacent volume packed with anisotropically shaped membrane fragments.

We have recently considered in detail the physical processes affecting membrane orientation by centrifugation and drying.[20,21] This analysis is based on recent progress in understanding the elastic properties of liquid crystals, which are closely related to multilamellar membrane arrays. The essential results of the study may be stated as follows.

[12] Y. K. Levine, and M. H. F. Wilkins, *Nature (London), New Biol.* **230,** 69 (1971); W. E. Wright, P. K. Brown, and G. Wald, *J. Gen. Physiol.* **59,** 201 (1972).

[13] R. Bogomolni, S. B. Hwang, Y. W. Tseng, and W. Stoeckenius, *Biophys. J.* **21,** 183 (1978); J. I. Korenbrot and M. Pramik, *J. Membr. Biol.* **37,** 235 (1977); S. B. Hwang, U. I. Korenbrat, and W. Stoeckenius, *ibid.* **36,** 115 (1977).

[14] J. K. Blaise, M. M. Dewey, A. E. Blaurock, and C. R. Worthington, *J. Mol. Biol.* **14,** 143 (1965).

[15] K. Nagy, *Biochem. Biophys. Res. Commun.* **85,** 383 (1978).

[16] D. C. Neugebauer, A. E. Blaurock, and B. L. Worchester, *FEBS Lett.* **78,** 31 (1977).

[17] J. Yguerabide and L. Stryer, *Proc. Natl. Acad. Sci. U.S.A.* **68,** 1217 (1971).

[18] H. Kuhn, D. Mobius, and H. Bucher, *in* "Physical Methods of Chemistry" (A. Weissberger and B. W. Rossiter, eds.), Vol. I, Part 3B. Wiley, New York.

[19] L. Powers and N. A. Clark, *Proc. Natl. Acad. Sci. U.S.A.* **72,** 840 (1975).

[20] N. A. Clark, K. J. Rothchild, D. A. Luppold, and B. A. Simon, *Biophys. J.* **31,** 65 (1980).

[21] K. J. Rothschild, N. A. Clark, and K. Rosen, *Biophys. J.* **31,** 45 (1980).

(i) In a surface-ordered array of finite-sized membrane fragments there is an intrinsic limit, related to the mean fragment area, to the degree of orientational order achievable.

(ii) Centrifugation alone is not an optimum technique because the orienting stresses available, $\sigma \simeq 10^6$ dynes/cm^2, are not large enough to effectively eliminate spontaneous thermal membrane curvature fluctuation. Since flattening out thermally undulated sheets is accompanied by a decrease in system entropy, compression is required, and because such thermal fluctuations are ubiquitous, a fundamental limitation of centrifugation is implied.

(iii) In comparison to centrifugal compaction, solvent evaporation can exert via capillary suction much larger compacting pressures on a fragment array. For a gap, $r \simeq 10^{-6}$ cm, between two membrane sheets in aqueous suspension the capillary pressure at an air–water interface is $P = \gamma/r \simeq 10^8$ dyne/cm^2, sufficient to produce a tightly compacted array in which all the space between the sheets is eliminated. In fact, solvent evaporation is the *only* readily available technique by which multilamellar arrays can be effectively compressed. Evaporation, however, is accompanied by a host of problems, which complicate its use. Unlike centrifugation, which deposits sheets layer by layer, as fragments are concentrated by drying they are orientationally frozen into a texture that is only indirectly influenced by the surface. Furthermore, high solute concentrations that develop as an array are dried from a suspending medium containing small ions in solution, which constitutes a serious drawback, since many membrane preparations require or can be profitably studied with added salts. The drying process can also be complicated to some extent by the capillarity of the suspending medium, as generally a solution meniscus will be nonplanar, leading to thickness nonuniformity in the dried film.

Based on this analysis, we have recently proposed and demonstrated a new technique, *isopotential spin-dry (ISD) centrifugation,* for the preparation of well-oriented multilamellar arrays of natural and artificial membranes.[20] The method involves the use of specially designed inserts for the buckets of a standard vacuum ultracentrifuge. The membrane fragments to be oriented are sedimented from solution or suspension onto a substrate of a convenient material that forms a gravitational isopotential surface at high g. Sedimentation is accompanied by removal of the suspending medium at high g to produce oriented films with a selected degree of solvation. In addition, a method is described whereby small solute molecules can be maintained in constant concentration with the membrane

[22] K. J. Rothschild, S. Sanches, T. L. Hsiao, and N. A. Clark, *Biophys. J.* **31**, 53 (1980).

fragments during this process. Initial application of the method to the orientation of purple membrane fragments and bovine retinal rod outer segment disk membranes has demonstrated that exceptionally well-ordered multilamellar preparations are possible using the ISD method.[21-23]

Isopotential Spin-Dry Centrifugation

In isopotential spin-dry centrifugation the drying and centrifugation processes are combined in a simple extension of presently employed techniques, which permits layer-by-layer deposition as well as control of the surface configuration during evaporation. This procedure, in which drying occurs *during* centrifugation, is schematized in Fig. 1. In the simplest ISD process the suspending medium is salt free, and is slowly and completely evaporated while the centrifuge is spinning at high g. This can be done,

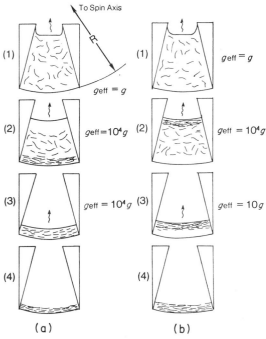

(a)   (b)

FIG. 1. (a) Isopotential spin-dry (ISD) centrifugation: Fragments are deposited layer by layer, interacting with each other only after the surface establishes orientational order. Surface deformation is suppressed by the large g. (b) Spin-dry process for fragments less dense than the suspending medium.

[23] P. D. Elworthy, *J. Chem. Soc.* p. 5385 (1961).

for example, by providing a sealed centrifuge cell containing the suspension with a small hole that allows slow solvent evaporation into an evacuated spin chamber. For this process to be effective, the surface on which the fragments are to be deposited must form a gravitation isopotential in the coordinate system of the spinning cell (i.e., be a cylinder coaxial with the spin axis). Drying while spinning takes advantage of the initial layer-by-layer deposition of centrifugation to form a loose, partially ordered array, followed by evaporative compression under controlled conditions where the suspension surface is precisely a gravitational isopotential. The suspension is effectively compressed between the isopotential substrate and isopotential suspension surface, which is very "hard" at high g (Fig. 1).

Our centrifugation studies have been carried out using a Beckman swinging bucket vacuum ultracentrifuge. The design of a centrifugation cell to produce oriented membrane fragment arrays from *salt-free* suspensions is shown in Fig. 2. The primary element is the isopotential centrifu-

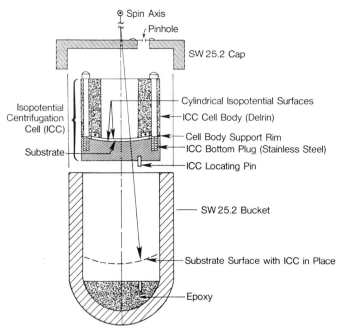

FIG. 2. Schematic of the isopotential centrifugation cell (ICC) used with the Beckman 100-ml SW 25.2 bucket, showing the epoxy filler, the ICC that fits snugly into the bucket, and the SW 25.2 cap, which seals the cell except for the leak through the pinhole to the centrifuge vacuum. Note that the delrin ICC body contacts the stainless steel bottom plug around its perimeter to minimize the compression of soft substrates.

gation cell (ICC), a single unit consisting of a bottom plug and cell body that forms the suspension cavity (SC) with the isopotential surface at the bottom. This unit is made to slide snugly but easily into the cylindrical part of the bucket. The bucket bottom is filled with an epoxy plug, machined accurately flat and normal to the bucket cylinder axis. The apparatus is made to fit directly into the bucket, since standard bucket inserts do not fit the bucket to close enough tolerance and mechanically relax at high g, leading to distortion of the ICC. The bottom ICC plug was machined to within 0.010 in. of being cylindrical and of the radius R_c equal to its distance from the spin axis when the buckets are fully swung out. The interchangeable substrate material on which the membrane fragment array is to be deposited is clamped between the cell body and bottom as they are fastened together, thereby bending the substrate to conform to the cylindrical bottom isopotential surface. This design for the ICC permits the use of a wide variety of substrate materials, including those commonly required in biophysical studies. We have successfully deposited oriented films on glass and quartz cover slips; plastic films from 6 μm up to 200 μm in thickness made from Mylar, Glad Wrap (polyethylene), and Teflon; aluminum foil; AgCl; and KRS-5.

For drying, a small pinhole (50 μm $\leqq$ diameter $\leqq$ 200 μ) was introduced into the cap of the bucket. Typical aqueous solvent volumes of 1–3 ml would evaporate in 4–10 h. This slow evaporation rate assured evaporation-limited compression of the partially compacted sedimented array, which forms relatively quickly ($\approx$ 10 min) after centrifugation commences.

The control of surface configuration while drying can be usefully exploited in other ways. Figure 3 shows how an ICC can be modified so that the salt or other solute concentration of the suspending fluid can be maintained constant during the sedimentation and compaction of a membrane fragment array. In this modified version (remote evaporation cell), evaporation takes place from a remote evaporation cavity (EC), which is connected to but distinct from the suspension cavity (SC) that holds the membrane fragment suspension. The ED and SC are connected by either a leak, a very narrow channel formed by the SC wall and the isopotential deposition surface (Fig. 3a), or a porus or semipermeable membrane (Fig. 3b). Evaporation from the SC is suppressed by capping it, although a small channel (CH) remains in the cap to allow pressure equalization as evaporation proceeds. In operation (Fig. 3c–e) both cavities are filled, the SC with the suspension and the EC with the suspending fluid with the level in the EC slightly higher. When the centrifuge is activated, hydrostatic pressure will induce flow of solvent from the EC to the SC until fluid surfaces in the EC and SC lie on a common gravitational isopotential.

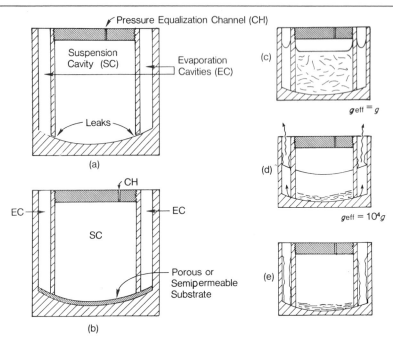

FIG. 3. Remote evaporation isopotential centrifugation cells for sedimenting and compacting a membrane array while maintaining a constant salt concentration. Evaporation takes place in cavities (ECs) connected to the sedimentation cavity (SC) by (a) leaks or a porous membrane. (b) Evaporation from the SC is suppressed by a cap with a small channel (CH) for pressure equalization. (c)–(e) Schematic of the operating cycle for these ICCs.

Centrifugation also sediments the membrane fragments to form a non-compacted array on the bottom (isopotential) surface of the SC. Controlled evaporation will occur at the solvent surface in the ED, and as it does, continued centrifugation will force solvent and solute from the SC to the EC, maintaining their surfaces on nearly a common isopotential that approaches the bottom isopotential.

It is also possible to control the degree of hydration in the final compacted array. The solvent evaporation rate will be determined by the partial pressures of the evaporating components which are maintained over the solution surface. These may be varied over the course of sedimentation and compaction. The ultimate solvent content of the membrane array will be determined by the equality of the chemical potential of the remaining solvent with that of the vapor in equilibrium and thus by the partial pressures of the solvent components in the vapor. For example, the control of the degree of hydration of aqueous membrane arrays by manipulation of the chemical potential of water has been demonstrated.[24] Similar techniques are readily applicable in connection with the methods described here.

Results and Conclusion

Our experience with ISD preparations to date indicates that, in the trade-off of desirable features mentioned in the Introduction, ISD samples can be optimized with regard to sample volume and orientational ordering. With regard to modification of membrane structure and function during the orientation, the key process is reduction of solvent chemical potential, which can have a significant, but at this point not very well understood, effect on equilibrium membrane structure. Our studies of high-stability systems such as purple membrane indicate that all three objectives can be achieved.[20] Multilamellar films of bovine retinal rod outer segment (ROS) disk membranes prepared by ISD from suspensions in distilled water exhibit excellent orientation and volume sufficient for a variety of spectroscopic probes.[21,22] X-ray diffraction and freeze-fracture studies of these ROS disk films are currently in progress. Preliminary results indicate first an exceptional degree of one-dimensional translational ordering, with up to 15 observable Bragg reflections from the bilamellar repeat, (S. Gruner, K. J. Rothschild, and N. A. Clark, Biophys. J., to appear), and second, possible evidence for phase separation of the protein and lipidic components. This latter question of modification of the membrane structure and function during the array formation process will clearly require extensive future study.

Acknowledgments

This research was supported in part by grants from the National Institutes of Health, National Eye Institute, National Science Foundation, Division of Materials Research, and Army Research Office, Grant DAAG-29-79-C-0174.

[43] Light-Induced pH Gradients Measured with Spin-Labeled Amine and Carboxylic Acid Probes: Application to *Halobacterium halobium* Cell Envelope Vesicles

By R. J. MEHLHORN and I. PROBST

Introduction

Several techniques have been used to measure transmembrane pH gradients. The probes and methods commonly employed to detect proton movement and to quantitize pH gradients, however, have inherent limitations. Valid data on stationary-state pH gradients are provided by the flow dialysis technique,[1,2] but the time course of the initial stages of the energization process is generally inadequately resolved. A fluorimetric technique, which rapidly monitors the kinetics of membrane energization, uses fluorescent amines such as acridine to determine pH gradients.[3] These dyes, however, interact with membrane surfaces and the observed fluorescence changes may not necessarily be the result of ΔpH-driven amine uptake into the intravesicular space.[4]

The introduction of spin-labeled pH probes[5,6] has proved advantageous because EPR spectroscopy monitors probe partitioning between the aqueous and the membrane phases, which result in different signals. It is possible, by means of paramagnetic broadening agents in the extravesicular phase, to visualize specifically the intravesicular spin probe population, whose change in concentration is easily and rapidly recorded.

The light-induced formation of a proton gradient across *Halobacterium halobium* vesicle membranes results from the activity of bacterio-

---

Abbreviations: TA, 4-amino-2,2,6,6-tetramethylpiperidine-*N*-oxyl; CA, 2,2,5,5-tetramethyl-3-pyrrolin-1-oxyl-3-carboxylic acid; Tempone, 2,2,6,6-tetramethyl-4-oxopiperidine-*N*-oxyl; CAT$_1$, 2,2,6,6-tetramethyl-4-(trimethylammonium)piperidine-*N*-oxyl bromide; MES, 2-(*n*-morpholino)ethanesulfonic acid; HEPES, *N*-2-hydroxyethylpiperazine-*N'*-2-ethanesulfonic; TAPS: 2-hydroxyl-1-bishydroxymethylamino-1-propanesulfonic acid.

[1] S. Ramos, S. Schuldiner, and H. R. Kaback, *Proc. Natl. Acad. Sci. U.S.A.* **73**, 1892 (1976).
[2] J. K. Lanyi, S. L. Helgerson, and M. P. Silverman, *Arch. Biochem. Biophys.* **193**, 329 (1978).
[3] H. Rottenberg and C. P. Lee, *Biochemistry* **14**, 2675 (1975).
[4] G. F. W. Searle, J. Barber, and J. D. Mills, *Biochim. Biophys. Acta* **461**, 413 (1977).
[5] R. D. Kornberg, M. G. McNamee and H. M. McConnell, *Proc. Natl. Acad. Sci. U.S.A.* **69**, 1508 (1972).
[6] A. T. Quintanilha and R. J. Mehlhorn, *FEBS Lett.* **91**, 104 (1978).

rhodopsin, a light-driven proton pump.[7] The electrochemical gradient of H^+ thus established provides energy that may be used to form secondary ion gradients such as for sodium and potassium.[8,9] Light-induced transmembrane proton movement will, therefore, monitor cation transport, provided that the cation movement is linked to proton transport.

We report here the application of two nitroxide spin probes, a carboxylic acid (CA) and a primary amine (TA), to determine pH gradients across *H. halobium* vesicle membranes. It is found that the transmembrane equilibration of spin probes rapidly and precisely monitors light-induced proton extrusion from the vesicles. The intravesicular volume was also determined by spin probe methods. These two measurements permit quantitation of light-induced pH gradients, which reach up to 2 pH units over the external pH range covered (pH 5–9).

Methods

H. halobium (strain R_1) cell envelope vesicles were prepared as in Lanyi and MacDonald[10] and stored in 3.6 *M* KCl at 4°. The extravesicular pH of vesicle preparations was adjusted by addition of 50 μl pretitrated 1 *M* buffer solutions dissolved in 3 *M* KCl to 450 μl of the vesicle suspension. To permit equilibration of pH in the dark, samples were incubated for 30 min at 4°, and the pH was then measured with a glass pH electrode (Sensorex S 900C). The external medium was buffered with MES below pH 6.8, with HEPES for the pH range from pH 7.0–8.2, and with TAPS for pH values higher than pH 8.2. None of the subsequent additions altered the external pH.

The spin-labeled probes, CA (Eastman Chemical Co.) and TA (Aldrich Chemical Co.) were purchased, CAT_1 was synthesized as in Mehlhorn and Packer,[11] and Tempone was a gift from A. D. Keith. TA and CA

[7] Oesterhelt D. and W. Stoeckenius, *Proc. Natl. Acad. Sci. U.S.A.* **70**, 2835 (1973).

[8] J. K. Lanyi and R. E. MacDonald, *Biochemisty* **15**, 4608 (1976).

[9] G. Wagner, R. Hartmann and D. Oesterhelt, *Eur. J. Biochem.* **89**, 169 (1978).

[10] J. K. Lanyi and R. E. MacDonald, this series, Vol. 56, p. 398.

[11] R. J. Mehlhorn and L. Packer, *Biochim. Biophys. Acta* **423**, 382 (1976).

were dissolved in 3 M KCl (pH 7.0). Cell envelope vesicles at a concentration of 40 mg protein/ml were mixed with 0.5–2 μl of the spin probe stock solution and 5–20 μl of a 1 M potassium ferricyanide solution. The final KCl concentration varied between 2.2 and 3.3 M. Additions were made under dim light and the mixture was kept in the dark for at least 2 min before use. In 100 μl microcapillaries 40-μl samples were illuminated in the cavity of a Varian E-109 E spectrometer with a Quartzline lamp at 12.5 mW/cm².

Room temperature EPR measurements were made at a microwave power of 10 mW, modulation amplitude 1 G (0.4 G for Tempone measurements), time constant 0.128 sec, and scan rate 0.4 G/sec. A PDP 11/34 computer interfaced with the EPR instrument was used for subtraction and double integration of spectra and for drawing the difference spectra on the EPR recorder.

Intravesicular pH values were calculated as in Rottenberg[12] from values of the internal and external TA and CA concentrations; pK_a values, determined by titration, were found to be 9.37 (TA) and 4.4 (CA) in 3 M KCl.

Results

Typical experiments showing TA and CA EPR spectra in the presence of vesicles are shown in Fig. 1. Ferricyanide was employed to quench the EPR signal arising from the extravesicular spin probe population.[6,13] Even in the presence of 100 mM ferricyanide, spin probe signals were unaffected for several hours, indicating that the vesicles are highly impermeable to ferricyanide. Both spin probes diffuse rapidly into the vesicles and their intravesicular EPR signal is clearly superimposed on the broadened extravesicular component.

The EPR signal of TA and CA are influenced by increasing ionic strength; KCl concentrations higher than 500 mM led to a decrease in the EPR line heights (7% at 1 M KCl, 23% at 3 M KCl). Therefore, measurements of line heights and integrated signal intensities were conducted at fixed ionic strength.

Membrane vesicles did not reduce the nitroxide radical, either in the dark or when illuminated, but starved cell preparations still contained sufficient oxidizable substrate to rapidly reduce either of the nitroxide spin probes.

To test the possibility that the membrane is permeable to the charged

---

[12] H. Rottenberg, *J. Bioenerg.* **7**, 61 (1975).
[13] P. D. Morse, *Biochem. Biophys. Res. Commun.* **77**, 1486 (1977).

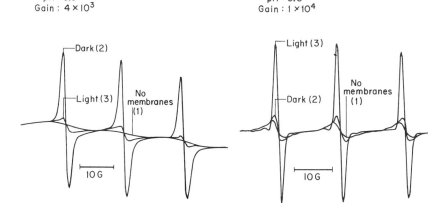

FIG. 1. Spin probe distribution across *H. halobium* vesicle membranes in the dark and light. Conditions as indicated on the figure with 14 mg protein/ml, 3 *M* KCl and 50 m*M* MES buffer,

form of the amine, a permanently charged analog of TA, CAT_1, was added to the vesicles but no free internal signal was observed either in the light or in the dark over the time intervals used in this study. It is concluded that protonated TA is not permeable. Another paramagnetic broadening agent, $NiCl_2$, which very effectively quenches the EPR spectrum of CA at a low concentration (30 m*M*), was unsuitable for quantitative determinations as it caused a continuous decrease of the internal EPR signal, probably because of a slow permeation of the bivalent cation into the vesicles.

Light-induced H^+ release from *H. halobium* vesicles results in the accumulation of CA, a weak acid, inside the membrane-enclosed aqueous space, and in the extrusion of the weak base, TA, into the medium. These changes in internal spin probe concentration are reflected in the EPR spectra (Fig. 1 light spectra). The uptake of CA into and the release of TA from the vesicles during illumination can be observed directly as an increase or decrease in the low field line height (Fig. 2). The effect is fully reversed in the dark. At an external pH of 7.0, the full extent of spin probe uptake on release was reached after about 20 sec of illumination. This compares quite well with independent measurements of H^+ release using a glass pH electrode under similar light intensity. The total amount of spin probe that traverses the membrane upon illumination was found to be linearly proportional to the vesicle concentration between 2 and 20 mg pro-

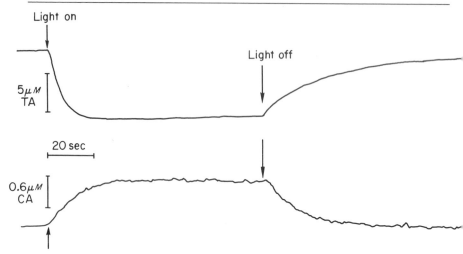

FIG. 2. Kinetics of light-induced spin probe movement across *H. halobium* vesicle membranes. Conditions as in Fig. 1, vesicles (5.2 mg protein/ml) were externally buffered at pH 7.0.

tein/ml. Therefore, light attentuation did not limit the extent of probe movement at any of the vesicle concentrations used in these experiments. Since addition of proton-conducting ionophores, such as 25 nmol/mg protein carbonyl cyanide trifluoromethoxyphenylhydrazone together with 10 nmol/mg protein nigericin completely abolished light-induced probe movement, it can be concluded that only transmembrane pH gradients are detected.

Spin probe uptake or release was independent of the total concentration of probe up to 20 μM CA and 1 mM TA (Fig. 3). It was found that the vesicles could withstand high ferricyanide concentrations, as the light-induced pH gradient was unaffected by as much as 400 mM ferricyanide.

The operational vesicle volume was determined as the ferricyanide-impermeable space[6] using either TA or Tempone as spin probes. Tempone, which is an uncharged molecule, does not show a pH-dependent distribution between external and internal space. Tempone was chosen as a general volume probe because its narrow line width is more effectively quenched by paramagnetic ions than the other probes and hence it can be used at lower concentrations of quenching agents if necessary.

A residual EPR signal of the extravesicular spin probe population remains with ferricyanide concentrations of 200 mM or less (Fig. 1, curve 2). To correct for this, the ferricyanide-quenched signal obtained in the absence of vesicles was computer-subtracted from the spectra exhibited by spin-labeled vesicles in the dark (Fig. 1, spectrum 2 minus spectrum 1). A difference spectrum of this kind is presented in Fig. 5, curve A. The low field line height of the difference spectrum was then expressed as a percentage of the total spin probe present and directly taken as the ferricyanide-inaccessible intravesicular volume (internal volume as a percentage of the total sample volume). The low field line height of the unquenched spin probe at comparable ionic strength served as the 100% value. Identical results were obtained when either TA or Tempone were used, thereby excluding the existence of measurable pH gradients in the dark. The absence of a dark pH gradient was also confirmed by the finding that unpouplers had no effect on the dark TA spectrum. Increasing ferricyanide concentrations lead to a decrease in the operational vesicle volume (corrected for osmotic effects), perhaps because of effects on the structural integrity of the membranes during ferricyanide addition to the sample. Also, the addition of acid or alkaline buffer solution (or dilute HCl and KOH) resulted in a rapid and irreversible decrease in the apparent internal volume (Fig. 4). This finding emphasizes the importance of determining both ΔpH and internal volume under identical assay conditions in each sample to obtain accurate quantitative data for pH gradients. The volume of unbuffered vesicles under most conditions was calculated to be 3 μl/mg protein, which agrees with values obtained by other methods.[10,14] No light-induced volume changes could be detected using Tempone as a volume probe and 20 μM acetate or 1 mM NH$_4$Cl as EPR-silent pH probe analog.

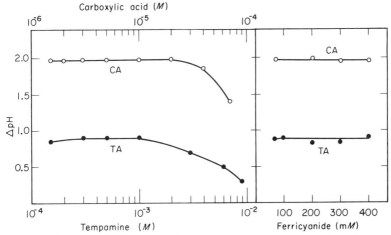

FIG. 3. Light-induced pH gradients measured as a function of spin probe and potassium ferricyanide concentration. Vesicles (16 mg protein/ml) were externally buffered at pH 5.0.

[14] M. Eisenback, S. Cooper, H. Garty, R. M. Johnstone, H. Rottenberg and S. R. Caplan, in "Bioenergetics of Membranes" (L. Packer, G. C. Papageorgiou, and A. Trebst, eds.), p. 101. Elsevier, North-Holland, Amsterdam, 1977.

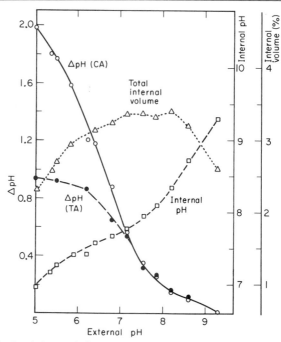

Fig. 4. Calculated changes in internal pH and ΔpH as a function of external pH. Vesicles (13.3 mg protein/ml), CA: 20 μM, TA: 1 mM, potassium ferricyanide: 100 mM.

The amount of spin probe that traverses the membrane in equilibration with the pH gradient was obtained by computer-subtraction of the light and dark EPR spectra (Fig. 1, spectrum 3 minus spectrum 2) or by direct measurement of the light-induced change in the line height of the internal signal. Again, the low field line height was taken as a measure of the percentage of spin probe that traversed the membrane. Results obtained with both methods were identical. In a typical example, at a vesicle concentration of 9 mg/ml protein and a ΔpH of 0.6 units, 7.5% of the total CA was taken up, compared to 1.9% of the total TA released from the vesicles. In another experiment, conducted at a bulk pH of 5.0 and at an operational vesicle volume of 62.1%, 2.17% of the total CA was taken up into the vesicles. This increased the intravesicular probe concentration from an initial value of 20 μM in the dark to 571 μM in the light and correspondingly decreased the extravesicular concentration from 20 to 7.73 μM.

At an external pH of 5.0, the uncharged population of CA must be considered. The formula that can be used for this purpose[12] is as follows:

$$\frac{(CA)_{in}}{(CA)_{out}} = \frac{1/Ka + 1/(H^+)_{in}}{1/Ka + 1/(H^+)_{out}}$$

With the experimental values given above and in the Methods section, this becomes:

$$73.9 = \frac{10^{4.4} + 1/(H^+)_{in}}{10^{4.4} + 10^{5.0}}$$

This yields $1/(H^+)_{in} = 9.24 \times 10^6$ and hence $\Delta pH = 1.97$.

Light-induced pH gradients across KCl-loaded vesicles decreased sharply with increasing external pH reaching zero above pH 9 (Fig. 4). These data confirm values reported previously for whole cells[15,16] and vesicles.[16,17] The calculated steady state intravesicular pH increased continually over the range of external pH examined. At internal pH values lower than pH 7.0, the calculated pH derived from TA measurements deviated from that obtained with the CA probe. pH gradients larger than one unit could not be monitored accurately by the apparent distribution of TA, as discussed below.

Figure 5A represents the dark difference spectrum obtained by subtracting the ferricyanide-quenched TA signal in the absence of vesicles from the spectrum in the presence of vesicles, yielding the signal exclusive of the broadened probe in bulk water. The spectrum is essentially identical to a spectrum of TA obtained in water, indicating that most of the probe is tumbling freely in the intravesicular aqueous compartment. When these envelopes were sonicated in the presence of ferricyanide, the spectrum shown in Fig. 5C resulted. A reference spectrum of the probe with ferricyanide in the absence of vesicles is shown in Fig. 5B. A computer subtraction of these two spectra is presented in Fig. 5D. This spectrum has the asymmetrically broadened lines characteristic of an intermediate state of immobilization of the probe as might be expected if it were intercalated among lipid headgroups (similar spectra of TA are observed in asolectin lipid vesicles). Figure 5E shows a computer-subtracted spectrum of CA that was obtained by procedures analogous to those used in producing Fig. 5D, except that nickel chloride was substituted for ferricyanide. It is suggested by the very asymmetrical nature of this spectrum that the bound CA is extremely immobilized in this instance. A careful analysis of line heights of the probes for sonicated versus unsonicated vesicles in the presence of quenching agents (representative spectra are shown in Fig. 5) has revealed that 8% of the TA line height is due to bound probe while less than 2% of CA is due to a bound component. Another approach toward estimating binding of the probe is to note that the bound

[15] E. P. Bakker, H. Rottenberg and S. R. Caplan, *Biochim. Biophys. Acta* **440**, (1976).
[16] H. Michel and D. Oesterhelt, *FEBS Lett.* **65**, 175 (1976).
[17] B. I. Kanner and E. Racker, *Biochem. Biophys. Res. Commun.* **64**, 1054 (1975).

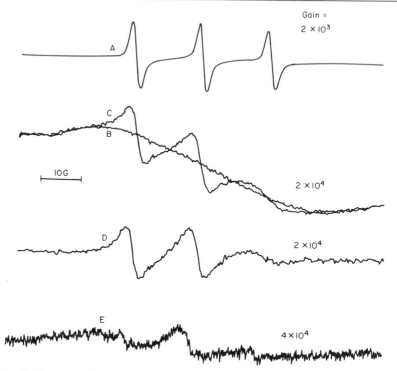

FIG. 5. Epr spectra demonstrating the extent of membrane binding by spin probes. Vesicles: 18 mg protein/ml in 3 M KCl in the dark with 1 mM TA or CA and 300 mM $K_3Fe(CN)_6$ or $NiCl_2$ as indicated. (A) Intravesicular TA signal, obtained by computer subtraction of a ferricyanide-quenched spectrum in the absence of membranes (shown in curve B but taken at a lower gain) from a ferricyanide quenched signal in the presence of intact envelope vesicles. (B) Signal of TA in the presence of ferricyanide. (C) The spectrum obtained when envelope vesicles and TA are sonicated in the presence of ferricyanide. (D) The membrane-bound spectrum of TA obtained by subtraction of B from C. (E) The membrane-bound spectrum of CA obtained by sonicating envelope vesicles in the presence of $NiCl_2$ and subtracting a spectrum in the absence of membranes.

component is sufficiently immobilized to exhibit a substantial reduction of the third line height relative to the low field line height. This characteristic of binding allows the contribution of the bound component to the line height to be estimated from the spectra of the probe during illumination by assuming that the line heights of free and bound probe are additive. Using this assumption, it was demonstrated that in the light less than 2% of the line height represents bound CA.

The experiments reported here do not discriminate between spin probes bound to the extravesicular and intravesicular membrane surfaces. Hence it is not possible to infer accurately the correct probe concentrations inside and outside the vesicles. However, by using the computer to perform double integrations of bound and free spectra, line heights of the different spectra were related to probe concentrations in the different environments and the maximum possible error in the pH calculation, which would arise if all the CA were bound to the cytoplasmic membrane interface, was estimated as 0.08 pH unit. Unlike CA, TA is extruded from the vesicles on illumination and under these conditions the residual signal observed in the light could include contributions from a substantial admixture of membrane-bound TA. Hence, TA may not be useful for estimating large pH gradients in systems where the probe is extruded from the intravesicular compartment. Thus CA is the probe of choice for proton-extruding vesicular systems that bind TA and produce large pH gradients.

Sodium Transport Across H. halobium Vesicles

On illumination, NaCl-loaded vesicles will extrude Na^+ via a $\Delta\mu H$-driven sodium/proton antiport system.[8] Figure 6 shows that the activity of the sodium/proton antiport influences the light-induced TA extrusion from *H. halobium* vesicles. The initial light-induced TA (H^+) extrusion is followed by a TA (H^+) influx. This continues until the onset of a final phase of TA extrusion that is the result of Na^+ depletion from within the vesicles and the establishment of steady state ΔpH. With subsequent illuminations, the Na^+-depleted vesicles show only the acidification pattern, which is also observed with KCl-loaded vesicles.

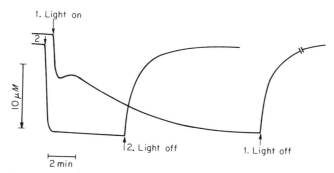

FIG. 6. Demonstrating of sodium/proton antiport activity by following light-induced TA release. Vesicles (9 mg protein/ml) were incubated in 2.9 M KCl/0.1 M NaCl for 48 hr. TA: 0.5 mM, sodium ferricyanide: 100 mM.

Summary

It has been shown that: (a) the EPR-detectable distribution of a spin-labeled weak acid provides an accurate method to determine the intravesicular pH in *H. halobium,* whereas a spin-labeled amine is of more limited utility for such measurements because spectra obtained may need to be corrected by computer subtraction methods for amounts of the probe that are membrane bound; (b) precise quantitation of the pH gradient necessitates the simultaneous determination of the intravesicular volume, which is readily measured as the ferricyanide-inaccessible space using the spin probe Tempone; (c) spin probe movement rapidly monitors light-induced cation transport, provided this movement is linked to or causes proton movement. The quantitative data obtained with this technique reveal a maximum light-induced pH gradient of two units across *H. halobium* membranes at acidic bulk pH values.

Acknowledgments

This work was supported by NIH (GM-24273), by the Energy and Environment Division of Lawrence Berkeley Laboratory, U.S. Department of Energy, under Contract No. W-7405-ENG-48, and a fellowship from the Deutsche Forschungsgemeinschaft to I.P. We thank L. Packer and J. Lanyi for useful discussions.

[44] Secondary Ion Movements in *Halobacterium halobium*

By Gottfried Wagner

Conversion of light energy into chemical energy in *Halobacterium halobium* is based on the action of the electrogenic proton pump bacteriorhodopsin.[1,2] The proton ejection from the cell creates an increase in the membrane electrical potential ($\Delta\Psi$) and in the transmembrane proton gradient (ΔpH).[3-5] Both forms of energy are taken to drive cellular ATP synthesis, while a change in the membrane electrical potential also induces ionic shifts of Na^+, K^+, and other ions.[6] Because halobacteria live in

[1] D. Oesterhelt and W. Stoeckenius, *Nature (London) New Biol.* **233,** 149 (1971).
[2] D. Oesterhelt and W. Stoeckenius, *Proc. Natl. Acad. Sci. U.S.A.* **70,** 2853 (1973).
[3] G. Wagner and A. B. Hope, *Aust. J. Plant Physiol.* **3,** 665 (1976).
[4] H. Michel and D. Oesterhelt, *FEBS Lett.* **65,** 175 (1976).
[5] E. P. Bakker, H. Rottenberg, and S. R. Caplan, *Biochim. Biophys. Acta* **440,** 557 (1976).
[6] J. K. Lanyi, *Microbiol. Rev.* **42,** 682 (1978).

concentrated salt solutions, these ionic shifts can reach the molar range and are of relevance when energy or signal transducing processes are studied in detail.

Cell Culture and Standardization

Halobacteria with a constant yield of bacteriorhodopsin[7] may be cultured according to Oesterhelt and Stoeckenius.[8] For ion transport measurements, a shake culture of halobacteria is centrifuged for 10 min at 13,500 g and 4° to form a tight pellet. The supernatant is replaced by a certain amount of basal salt medium,[7] and the cells are gently resuspended to a defined density (e.g., 5×10^9 cells/ml). Cell density is best determined by cell counting techniques, e.g., by use of a Coulter counter[9,10] or on the basis of total cell volume (hematocrit, radioactive isotope, and dielectric dispersion techniques; see later). For routine measurements, determination of the calibrated protein content or photometric turbidity may suffice.[7]

Ion-Selective Electrode Measurements

Recording of concentration changes of protons, potassium, and other ions in the suspending basal salt medium, e.g., due to cellular ion pumping in light, can be carried out in a thermostated glass vessel; one of a variety of commercially available pH meters connected to an appropriate recorder may be used, but recordings from the pH electrode should be tested against possible light artifacts and salt errors.

For potassium measurements, the highly K^+-selective electrode IS 561-K^+ (Philips) is often used plugged into a PHM 64 (now 84) research pH meter (Radiometer, Copenhagen). The electrode must be shielded against light, and noise and drift problems are greatly reduced if the reference electrode (e.g., calomel) is placed externally in basal salt solution and connected via a basal salt-agar bridge to the reaction vessel. The reaction mixture should be stirred, e.g., magnetically, to reduce unstirred layer effects and nonuniform irradiation.

Total change in H^+ and K^+ concentration during the experiment may be calibrated by use of reference solutions or internal standardization be-

[7] R. Hartmann and D. Oesterhelt, *Eur. J. Biochem.* **77**, 325 (1977).
[8] D. Oesterhelt and W. Stoeckenius, this series, Vol. 31, p. 667.
[9] K. Shibata, this series, Vol. 24, p. 171.
[10] U. Zimmermann, J. Schultz, and G. Pilwat, *Biophys. J.* **13**, 1005 (1973).

fore and after the experiment; as follows from the Nernst equation, the signal heights are a nonlinear function with change in concentration.[11,12]

Atomic Absorption Spectroscopy and Flame Photometry

Intracellular ion concentrations in collected samples may be determined by atomic absorption spectroscopy or flame photometry. Employing a centrifugation technique for sample collection,[13] defined amounts of a cell suspension are placed on top of a thin layer of silicone oil with adjusted density (>1.15 g/ml) in transparent polypropyl microcentrifuge tubes.[4] The tubes are placed horizontally in a microcentrifuge with a transparent cover and thus may be irradiated from the top. After a 5-min run at 13,500 g, the cells are spun down to form a tight pellet, and the tubes are cut across the silicon oil layer to prevent cross contamination between pellet and supernatant.[4] The parts containing the cell pellet are shaken in standard amounts of diethylether for removal of silicon oil.[12] The cells are lysed and homogeneously dissolved in double-distilled water. The signals from the atomic absorption spectrometer or flame photometer are calibrated by reference solutions or internal standardization; blanks are subtracted. The intracellular concentration of ions ($[j]$) is given by[11]

$$[j]_{\text{intracellular}} = \frac{[j]_{\text{pellet}} - \left(\begin{array}{c}\text{fraction} \\ \text{extracellular} \times [j]_{\text{suspending}} \\ \text{space} \quad \text{medium}\end{array}\right)}{1 - \text{fraction extracellular space}} \tag{1}$$

The amount of medium dragged through the silicon layer during centrifugation compared to the total pellet volume (fraction extracellular space) has to be determined separately, e.g., by radioactive isotope techniques.

Radioactive Isotope Techniques

Radioactive isotopes may be used to determine both ion transport processes and the fraction of extracellular space to total pellet volume. For determination of sample volumes, ^{3}H labeled H_2O as a membrane-permeant agent and a ^{14}C labeled macromolecular sugar like dextran as a

[11] J. K. Lanyi and M. P. Silverman, *Can. J. Microbiol.* **18**, 993 (1972).
[12] G. Wagner, R. Hartmann, and D. Oesterhelt, *Eur. J. Biochem.* **89**, 169 (1978).
[13] M. Klingenberg and E. Pfaff, this series, Vol. 10, p. 680.

membrane-impermeant agent are often used in a double-label experiment.[3-5] Cells are either centrifuged as described earlier or filtrated[14] and thus separated from the suspending medium. Using a liquid scintillation counter with energy discriminating counting channels, the radioactivities in the suspending medium compared to those in the cell sample are determined taking care of quench corrections and spillover rates of ^{14}C counts into the ^3H channel. From the calibrated count rates per volume fraction of suspending medium and the calibrated count rates in the cell sample, the ^3H_2O-permeable and the ^{14}C-dextran-permeable volume of the cell sample may be figured out, the difference of which by definition is the intracellular volume (osmotic volume). Distributions of radioactively detectable ions like $^{22}Na^+$, $^{42}K^+$ (resp. $^{86}Rb^+$), $^{36}Cl^-$, or $^{45}Ca^{2+}$ and thus ion transport rates per membrane area, or concentrations per intracellular volume, may be determined accordingly.

Dielectric Dispersion Measurements

Electrical cell properties, needed to understand membrane transport processes in detail, may be determined by the elegant, independent technique. The method is based on the phenomenon that dielectric constant and electrical conductance of spheres,[15] spheroids,[16] and halobacterial cells in suspension change from one constant level to another as the frequency of an alternating field across the suspension is varied (dielectric dispersion).

At low frequencies, lines of current flow around the suspended cells because of the high resistance of the cell membrane (Fig. 1A). This allows

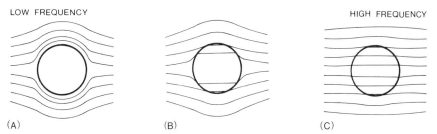

LOW FREQUENCY   HIGH FREQUENCY

(A)   (B)   (C)

FIG. 1. Frequency dependence of path of current flow around and through a spherical cell in suspension. (Modified after Pauly[15].)

[14] R. A. Dilley, this series, Vol. 24, p. 68.
[15] H. Pauly, *IRE Trans. Med. Electron.* **9**, 93 (1962).
[16] J. Bernhardt and H. Pauly, *Biophysik* **10**, 89 (1973).

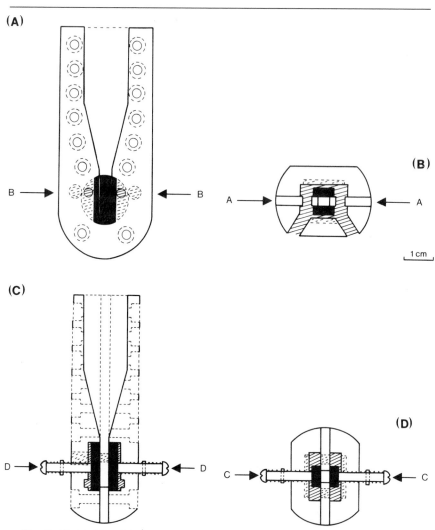

(A)

(B)

(C)

(D)

1 cm

FIG. 2. Measuring cell for dielectric dispersion of halobacterial samples in four sectional views as indicated by the lettered arrows. The outer shape of the measuring cell, made out of three sheets of Perspex and held together by steel bolts, is rounded up to fit into a centrifuge tube of a swing-out rotor (A and B). Halobacterial cell suspensions may be spun down to form a tight pellet between the electrodes (black inlets, A–D). The stainless steel electrodes and the entrapped cell pellet may be thermostated by a circular flow of distilled water (hatched channels). The measuring cell is connected to the Schering bridge by two steel bolts and nuts functioning as wires (C and D).

us to calculate volume fractions of cell suspensions,[17] e.g., in order to prove results from hematocrit, tracer, or Coulter counter techniques (see earlier).

With increasing frequency, more and more lines of current pass the cell membrane (Fig. 1B), and finally, the cell membrane becomes short-circuited (Fig. 1C). At this extreme, the effective cell conductivity may be determined that then mainly gives the intracellular conductance.

At intermediate frequencies, the effective cell conductivity is also a function of membrane conductance. Based on electrochemical considerations of earlier workers, procedures to determine intracellular conductance, membrane conductance, and membrane capacitance in different systems, including bacteria and rod outer segments, were reviewed recently.[18] A suitable measuring cell, used for dielectric dispersion measurements in *H. halobium* in the frequency range of 80–250 MHz, is shown in Fig. 2. The cell is to be connected directly to a Hewlett–Packard 250 B RX meter (Schering bridge) in order to measure the parallel capacitance and resistance of the sample. Internal conductivity and membrane conductance of intact cells of *H. halobium* may be determined from these data, e.g., as a function of energy or signal-transducing ion transport processes in light.

Acknowledgment

This work was supported by the Deutsche Forschungsgemeinschaft.

[17] M. Ginzburg, B. Lepkipfer, A. Porath, and B. Z. Ginzburg, *Biophys. Struct. Mechanism* **4**, 237 (1978).
[18] O. F. Schanne and E. R. P. -Ceretti, "Impedance Measurements in Biological Cells." Wiley (Interscience), New York, 1978.

[45] Photophosphorylation and Reconstitution of Photophosphorylation in Halobacterial Cells

By Dieter Oesterhelt

Introduction

In halobacteria all possible types of phosphorylation occur: electron transport phosphorylation (oxidative phosphorylation), bacteriorhodop-sin-mediated phosphorylation (photophosphorylation), and substrate-level phosphorylation induced by arginine degradation.[1-6] All phosphorylation reactions can be demonstrated in one and the same cell suspension, as shown in Fig. 1. Oxidative phosphorylation occurs in the dark on aeration of the cell suspension and photophosphorylation under anaerobic conditions on illumination. Under anaerobic conditions in the dark, arginine mediates substrate-level phosphorylation. Other experimental conditions leading to ATP synthesis in halobacteria are sudden acidification of the medium or changes in the potassium ion gradient across the cell membrane.[4,7] As expected from the well-established schemes of phosphorylation, only oxidative phosphorylation is inhibited by cyanide or other inhibitors of electron transport while oxidative as well as photophos-

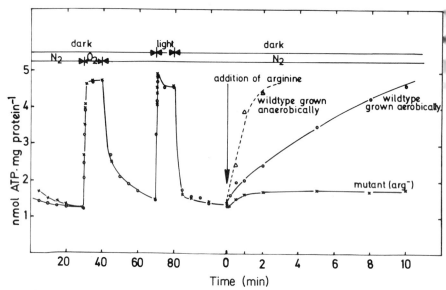

FIG. 1. Oxidative phosphorylation, photophosphorylation, and substrate level phosphorylation in *H. halobium* (taken from Hartmann *et al.*[6]).

[1] D. Oesterhelt, *in* "Membrane Proteins in Transport and Phosphorylation" (G. S. Azzone, M. E. Klingenberg, E. Quagliariello, and N. Siliprandi, eds.), p. 79. Elsevier/North-Holland, Amsterdam, 1974.
[2] A. Danon and W. Stoeckenius, *Proc. Natl. Acad. Sci. U.S.A.* **71,** 1234 (1974).
[3] D. Oesterhelt, *Ciba Found. Symp.* [*N.S.*] **31,** 147 (1975).
[4] R. Hartmann and D. Oesterhelt, *FEBS Lett.* **78,** 25 (1977).
[5] A. Danon and S. R. Caplan, *Biochim. Biophys. Acta* **423,** 133 (1976).
[6] R. Hartmann, H.-D. Sickinger, and D. Oesterhelt, *Proc. Natl. Acad. Sci. U.S.A.* **77,** 3821 (1980).
[7] G. Wagner, R. Hartmann, and D. Oesterhelt, *Eur. J. Biochem.* **89,** 169 (1978).

phorylation are inhibited by uncouplers such as carbonyl cyanide 3-chlorophenylhydrazone (CCCP) or carbonyl cyanide 3-fluorophenylhydrazone (FCCP) and inhibitors of the ATP synthase like dicyclohexylcarbodiimide (DCCD).

Reconstitution of photophosphorylation in halobacteria can be demonstrated in two different ways. If cells are grown in the presence of the alkaloid nicotine, they are unable to synthesize retinal, because the conversion of lycopene to β-carotene, the immediate precursor of retinal, is blocked.[8,9] Such cells, although they grow normally, will not contain bacteriorhodopsin and therefore do not photophosphorylate. However, the cells still synthesize bacterio-opsin to a limited extent and on addition of retinal or retinal analog compounds to a cell suspension bacteriorhodopsins can be formed and thus photophosphorylation is reconstituted. A second way that avoids the use of the alkaloid involves mutant cells unable to synthesize retinal. Photophosphorylation is not observed in these cells but again can be reconstituted by the addition of retinal or appropriate retinal analogue compounds.

This article describes the experimental methods for measurement of ATP levels in halobacterial cells influenced by the previously mentioned processes.

Strains and Growth Conditions

Halobacterium halobium R_1M_1[4] or strain S9[10] was used as cell types producing large amounts of bacteriorhodopsin. Mutant strain A0 151,[11] strain JW5,[12] or strain W 296[13] were used as retinal-deficient strains. Cells with high bacteriorhodopsin content are obtained by growth on complete medium[14] under limited aeration, which is achieved by growing a 700-ml culture (5% inoculum) in 2-liter conical flasks (alternatively, 35 ml in a 100-ml flask) at 39° on a rotary shaker at 100 rpm for 6 days (final O.D. at 578 of 0.9–1.0). Low levels of bacteriorhodopsin are obtained at high rates of aeration during growth which is achieved by growth at 180 rpm for 30 hr (O.D. 578 of 0.8–0.9) under otherwise identical conditions. Bacteriorhodopsin-deficient cells are grown under limited aeration as described previously in a medium containing 1 mM nicotine (strains R_1M_1

[8] M. Sumper, H. Reitmeier, and D. Oesterhelt, *Angew. Chem.* **88**, 203 (1976).
[9] D. Oesterhelt and V. Christoffel, *Biochem. Soc. Trans.* **4**, 556 (1976).
[10] Obtained from L. Jan.
[11] U. Bergner, Diplomarbeit, Universität Würzburg, (1978).
[12] Obtained from J. Weber.
[13] Obtained from G. Weidinger.
[14] D. Oesterhelt and W. Stoeckenius, this series, Vol. 31, p. 667.

and S9) or without nicotine for the retinal-deficient strains. To obtain cells with varying bacteriorhodopsin content, 500 ml cultures are shaken in 1-liter conical flasks at 100 rpm and the cells are harvested after 1, 2, 3, 4, or 5 days.

The medium contains in 1000 ml: NaCl, 250.0 g; $MgSO_4 \times 7 H_2O$, 20.0 g; trisodium citrate $2H_2O$, 3.0 g; KCl, 2.0 g; Oxoid Bacteriological peptone L37 (Oxoid, Postfach 1127, 4230 Wesel) 10 g. The medium is autoclaved for 20 min at 120°. Basal salt does not contain the peptone and citrate.

Turbidity of cell suspensions is measured with a photometer Eppendorf model 1101 M. Under all growth conditions described the average dimensions of a halobacterial cell are 5 μm in length and 0.6 μm in diameter (1.4×10^{-12} ml). The cell volume can directly be estimated from cell count and total cell volume of a given suspension. One-ml cell suspension of an O.D. 578 (Eppendorf photometer) of 5.5 has a cell volume of 7.5 μl and a protein content of 2.75 mg protein.[15,16]

Determination of Bacteriorhodopsin

The bacteriorhodopsin content of the cells is determined by difference spectroscopy of a sample of lysed cells against a standard sample where bacteriorodopsin is first bleached with hydroxylamine in light.[17] The standard is prepared by illumination of a cell suspension in the presence of 0.2 M hydroxylamine until no further decrease in absorbance at 570 nm is observed. Cell densities of the sample and the standard are adjusted to 5.5 mg protein/ml. Cells are lysed by spinning down the cells at 13,000 g for 5 min and resuspension of the pellet in water under addition of a trace of DNase to reduce viscosity. Then a difference spectrum is recorded on a spectrophotometer with light-scattering attachment. For calculation of bacteriorhodopsin concentration an absorption coefficient of 63,000 cm^{-1} mol^{-1} is used. The standard can be used for several weeks.

Incubation of Cells and Sampling

Cells are spun down at 13,000 g for 5 min and resuspended carefully in basal salt containing 250 g NaCl, 20 g $MgSO_4 \times 7 H_2O$ and 2 g KCl/liter. Careful suspension of the cell is necessary to avoid breakage of the mechanically very labile cells. The cell suspension is then adjusted to a tur-

[15] R. Hartmann, H. -D. Sickinger, and D. Oesterhelt, FEBS Lett. 82, (1977).
[16] H. Michel and D. Oesterhelt, FEBS Lett. 65, 175 (1976).
[17] D. Oesterhelt and L. Schuhmann, FEBS Lett. 44, 262 (1974).

bidity of 4 or 5.5 at 578 nm corresponding to a protein concentration of 2 or 2.75 mg/ml. The cell suspensions can be buffered with Tris–HCl or Tris–maleate up to a concentration of 12.5 mM without an influence on the rate and extent of photophosphorylation in contrast to other buffering substances like citrate and imidazole. For the determination of ATP 10 ml of such a cell suspension in basal salt is stirred in a cylindric glass chamber, 2 cm in diameter with a water jacket at 25°. For anerobic conditions nitrogen is passed over the surface of the suspension and the glass chamber is closed with parafilm (American Can Corp., Nina, Wisconsin) except for a small hole for sampling and, if necessary, another one for a pH electrode to monitor pH changes during illumination. For maintaining dark conditions it is sufficient to use dimmed room lights less than 0.1 mW/cm², no fluorescent tubes. Illumination is provided by a 150-W slide projector (Rollei) with an OG 515 cutoff filter (Schott) at a light intensity of 50 mW/cm² at the position of the sample. If higher intensities are necessary, a 250-W projector (Norris Trumpf Halogen 250) at a light intensity of 100 mW/cm² can be used. The light intensity is measured with a bolometer (Kipp und Zonen) connected to a Knick microvoltmeter. Lower light intensities are provided with neutral density glass filters (Schott). The turbidity of the cell suspension is determined at 578 nm before and after the experiments in order to check for cell lysis. Only absorbance changes smaller than 10% can be tolerated. For determination of action spectra a 900-W xenon lamp is used. The light beam passes through a 20-cm water bath, a 400-nm cutoff filter and is then focused through a light guide onto the glass chamber with the cell suspension. The desired wavelength is selected by use of interference filters with a bandwidth of 5 nm (Anders, Diendorf, FRG). Absolute light intensity is determined by a radiometer detector head (model 580-25 A) equipped with a narrow beam attachment (model 580-00-11). The same instrument is used to determine light absorption of cell suspensions in an integrating sphere (50-cm diameter; cf. Oesterhelt and Krippahl[18]). The pH of the cell suspension can be monitored with a glass electrode attached to a standard pH-meter and a pen recorder (e.g., Servogor S, Metrawatt). For the determination of ATP 0.1-ml samples of the cell suspension are taken at various times with an automatic pipette preferentially by a pipette Beckton Dickinson and quickly pipetted into 5 ml ice-cold sodium phosphate buffer (10 mM, pH 7.4 containing 0.1 mM EDTA). This treatment lyses halobacterial cells immediately and the ATP in the lysate is stable for a day at 0°. It is important to push down the piston of the syringe before inserting it into the chamber. This avoids injection of air, which would result in a raise of ATP level.

---

[18] D. Oesterhelt and G. Krippahl, *FEBS Lett.* **36,** (1973).

ATP Assay

ATP is determined by the luciferin–luciferase assay. The lysate (0.5 ml) is pipetted into a cuvette, which is then placed into a bioluminescence device (see later). The reaction is started by injecting 0.1 ml enzyme (see later) and 0.4 ml buffer (0.1 M Tris–HCl, pH 7.4, with 10 mM MgSO$_4$) with an automatic syringe into the cuvette. The obtained light signal is recorded by a strip-chart recorder or by an automatic integrating device and the amount of ATP is determined by use of a calibration curve. For this purpose increasing amounts of a 2 μM ATP solution in phosphate buffer 10 mM pH 7.4 are pipetted to a constant amount of the same bacterial lysate of low ATP content and ATP measured as described earlier. The range from 0 to 50 μl ATP solution correspond to a range of 0–100 pmol ATP. For ATP measurement any commercial available bioluminescence device can be used that has a sensitivity to detect the range of 10–100 pmol ATP. The actual amount of ATP in a sample obtained as a readout in mV (see Fig. 2) is calculated with help of the calibration curve and the relationship (see above) between protein content, internal cell volume, and turbidity.[6,15,16] Typical values of the maximal ATP levels in illuminated cells are 6 nmol ATP/mg protein corresponding to 2.2 mmol

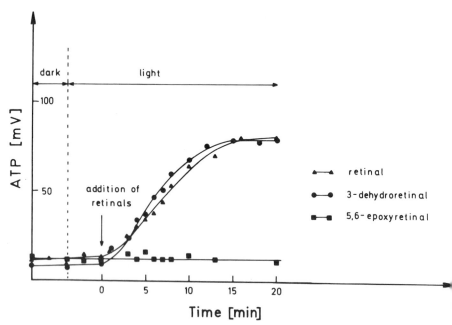

FIG. 2. Reconstitution of photophosphorylation (taken from Oesterhelt and Christoffel[9]).

ATP/liter of cell volume or 3.5 mmol/kg cell water. Under anaerobic conditions in the dark the cellular ATP level drops to 25–30% of that obtained in the light or by aeration. If the cells are kept anaerobically in the dark overnight at room temperature, the ATP level decreases further to 5–10%.[4]

Reconstitution of Bacteriorhodopsin in Retinal-Deficient Cells

Cells are grown and harvested as described earlier, then resuspended in 10 ml either of basal salt or of the same salt solution containing 150 g NaCl/liter instead of 250 g/liter. Reconstitution of bacteriorhodopsin is slow at the high sodium chloride concentration, but the reduced amount of salt in the suspending medium allows faster reconstitution. Therefore the ATP level of the cells that does not change in the dark and on illumination is seen to raise within 10–20 min after addition of the retinal solution (10 μl of a 10 mM solution). Figure 2 shows a typical experiment taken from Oesterhelt and Christoffel.[9] While retinal and 3-dehydroretinal induce photophosphorylation equally well, 5,6-epoxyretinal turns out to be an inactive pigment.

Alternatively, the cells are resuspended in basal salt and one sample is used as a control to demonstrate the absence of photophosphorylation, while to the second sample retinal is added and photophosphorylation measured after complete reconstitution within 1–2 hr. Reconstitution in intact cells can also be followed by difference spectroscopy.[8]

Preparation of Luceferin-Luciferase

One g desiccated firefly lanterns (Sigma FET) is ground in a precooled mortar to a fine powder. One hundred ml ice-cold sodium phosphate buffer (0.1 M, pH 7.4 containing 1 mM EDTA) is slowly added under stirring. The suspension is stirred an additional 15 min in the cold and then centrifuged for 15 min at 30,000 g. The sediment is discarded, 1 ml 1 M MgSO$_4$ is added to the supernatant (crude luciferase), and the enzyme is kept overnight at 4–8° to consume endogenous ATP and stored in 10–15 ml portions at $-20°$. A white-yellowish sediment appearing on rethawing is removed by centrifugation.

[46] Simultaneous Measurements of Proton Movement and Membrane Potential Changes in the Wild-Type and Mutant *Halobacterium halobium* Vesicles

By Naoki Kamo, Toni Racanelli, and Lester Packer

We will describe here a method of simultaneous measurement of H^+ movement and membrane potential changes in *H. halobium* envelope vesicles upon illumination. The membrane potential is estimated by measuring the distribution of tetraphenylphosphonium cation (abbreviated as TPP^+) in the medium and inside the vesicles. A change in TPP^+ concentration in the medium is monitored with a TPP^+-selective electrode. The TPP^+-selective electrode method allows the simultaneous use of electrodes sensitive to other cations such as H^+, K^+, and Na^+.

Principle of Determination of Membrane Potentials Using TPP^+ Electrode

Once TPP^+ has permeated the membrane,[1-5] the distribution of TPP^+ across the membrane at equilibrium is governed by the Nernst equation[6]:

$$\Delta\Psi = (RT/F) \ln C_{out}/C_{in} \qquad (1)$$

where $\Delta\Psi$ stands for the membrane potential with respect to the medium, and C_{in} and C_{out} stand for the TPP^+ concentration in the medium and inside the vesicles, respectively. In Eq. (1), R, T, and F have their usual thermodynamic significance. The relationship between the electrode potential E and a TPP^+ concentration in a medium C_{out} is expressed as

$$E = (\text{const}) + (RT/F) \ln C_{out} \qquad (2)$$

The mass conservation of TPP^+ can be written as

$$VC_0 = VC_{out} + vC_{in} \qquad (3)$$

[1] V. P. Skulachev, *MTP Int. Rev. Sci.: Biochem., Ser. One* **3**, 31 (1975).
[2] N. Kamo, M. Muratsugu, R. Hongo, and Y. Kobatake, *J. Membr. Biol.* **49**, 1015 (1979).
[3] M. Muratsugu, N. Kamo, Y. Kobatake, and K. Kimura, *Bioelectrochem. Bioenerg.* **6**, 477 (1979).
[4] D. Lichtshtein, H. R. Kaback, and A. J. Blume, *Proc. Natl. Acad. Sci. U.S.A.* **76**, 650 (1979).
[5] E. Komor and W. Tanner, *Eur. J. Biochem.* **70**, 197 (1976).
[6] H. Rottenberg, this series, Vol. 55, p. 547.

METHODS IN ENZYMOLOGY, VOL. 88

where V and v are the volume of the medium and the vesicles, respectively. In this equation VC_0 is the amount of TPP$^+$ added. From Eqs. (1), (2), and (3), we obtain[7,8]

$$\Delta\Psi = (RT/F)\ln(v/V) - (RT/F)\ln[\exp F(E - E_0)/RT - 1] \quad (4)$$

where E_0 is the electrode potential when the TPP$^+$ concentration is C_0.
As described later, the sensor membrane of the TPP$^+$ electrode is composed of poly (vinyl chloride) (PVC) film containing lipophilic anion, tetraphenylboron (TPB$^-$) as an ion exchanger. Since PVC is hydrophobic and the anion is embedded, the electrode is sensitive only to lipophilic cations such as TPP$^+$ and dibenzyldimethylammonium ion.[7]

Construction of TPP$^+$ Selective Electrode[2]

After addition of 3 ml of 10^{-2} M sodium tetraphenylboron (NaTPB) in to 10 ml of THF solution containing 0.5 g PVC (purchased from Aldrich Chem., Milwaukee, high molecular weight) and 1.5 ml of dioctylphthalate acting as plasticizer, the solvent was air-evaporated overnight in a flat petri dish (60 cm^2 in area). The membrane thus obtained was transparent and 0.15–0.2 mm thick. A piece of the membrane was glued to a PVC tube with THF. A 10^{-2} M TPP solution (purchased from Dojindo, Kumamoto, Japan) was placed inside the tube as an internal reference solution, into which an Ag-AgCl electrode was inserted. Before use, the electrode was soaked overnight in a concentrated TPP$^+$ solution ($\sim 10^2$ M) for conditioning. During this conditioning process, it is supposed that exchange occurs between TPP$^+$ and Na$^+$ in the sensor membrane (a counterion of TBP$^-$ as an ion exchanger) and the insoluble salt of TPP$^+$-TPB$^-$ forms. The electromotive force (emf) between the TPP$^+$ electrode and a calomel reference electrode in a TPP$^+$ solution of varying concentrations is measured by an electrometer (Keithley, 610C). The response of the electrode follows Eq. (2) until the concentration decreases to $\sim 10^{-6}$ M (even in 4 M NaCl) or 5×10^{-7} if the electrode is good. The electrode can respond in 2–7 sec to a doubling of the TPP$^+$ concentration. When stored in distilled water, the electrode is accurate for at least one year.

[7] T. Shinbo, N. Kamo, K. Kurihara, and Y. Kobatake, *Arch. Biochem. Biophys.* **187,** 414 (1978).

[8] M. Muratsugu, N. Kamo, K. Kurihara, and Y. Kobatake, *Biochim. Biophys. Acta* **464,** 613 (1977).

Simultaneous Measurements of Proton Movement and Membrane
Potential of Envelope Vesicles of *H. halobium*

The proton movement can be monitored easily with a glass pH electrode. Since the reference electrode of the pH meter is usually not
grounded, we used another grounded reference electrode for the TPP$^+$
electrode. For experiments with *H. halobium* vesicles, media are well defined and limited, e.g., 4 *M* NaCl, 3 *M* NaCl, or 3 *M* KCl. Then a reference electrode for a TPP$^+$ electrode was constructed as follows: Agar
(1.5–2%) was added to the salt medium and heated until dissolved. A
glass Pasteur pipette was half filled with the hot salt–agar solution. When
this had cooled, the salt medium was layered on top of the salt–agar. An
Ag–AgCl electrode was then inserted into the salt medium.

Two strains of *H. halobium*, S$_9$ and R$_1$mR^9 (a gift from Dr. Y. Mukohata, Osaka University), were used in the preparation of envelope vesicles as in Lanyi and MacDonald.[10]

The membrane potential of S$_9$, measured under varying vesicle concentrations, indicates that $\Delta\Psi$ is independent of protein concentration
(see accompanying table). In NaCl media a large membrane potential and
small acidification in the medium were observed upon illumination of S$_9$

MEMBRANE POTENTIAL OF S$_9$ ENVELOPE VESICLE
UPON ILLUMINATION UNDER VARYING
PROTEIN CONCENTRATIONS[a]

Protein concentration (mg/ml)	Estimated membrane potential (mV)
0.2	150
0.4	150
0.6	154
0.8	155
1.0	155

[a] Medium was 4.0 *M* NaCl (pH 7.0). Illumination was by
a GE 300-W tungsten projector lamp through a heat
filter; 7-cm water layer and yellow (430 nm) filter. TPP$^+$
concentration was 5 μM. Intravesicular volume was assumed to be 3 μl/mg protein.[11] Protein was assayed
with Lowry method using bovine serum albumin as a
standard.

[9] A. Matsuno-Yagi and Y. Mukohata, *Biochem. Biophys. Res. Commun.* **78**, 237 (1977);
Arch. Biochem. Biophys. **199**, 293 (1980).
[10] J. K. Lanyi and R. E. MacDonald, *Biochemistry* **15**, 4608 (1976); this series, Vol. 56, p.
398.

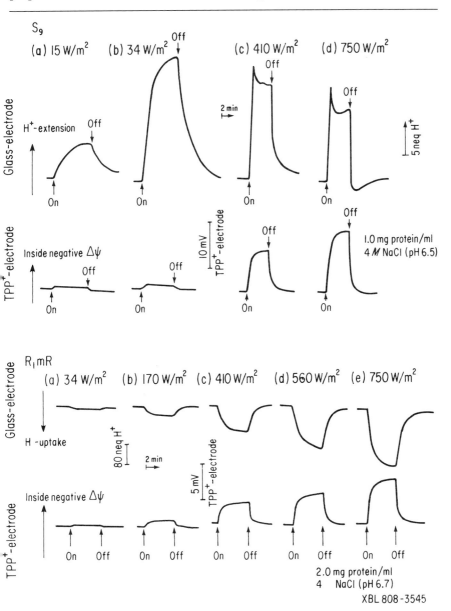

FIG. 1. Simultaneous measurements of H⁺ movement and TPP⁺ uptake under varying light intensity. (Top) S_9 vesicles suspended in 4 M NaCl (pH 6.5). Protein concentration was 1.0 mg/ml. (Bottom) R_1mR vesicles suspended in 4 M NaCl (pH 6.7) and protein concentration was 2.0 mg/ml. The initial concentration of TPP⁺ was 25 μM.[17,18]

vesicles, when pH of the medium was smaller than 6.5–6.7. In KCl media, a small membrane potential and large acidification were observed. These are consistent with the results obtained by Renthal and Lanyi.[12] In NaCl media whose pH were alkaline, alkalinization in the medium (H^+ uptake) was observed due to electrogenic Na/H exchange.[13] R_1mR suspended in NaCl medium consistently showed H^+ uptake upon illumination, because of the proposed light-driven Na pump.[14–16]

Traces of simultaneous measurements of H^+ movement and TPP^+ uptake by S_9 and R_1mR vesicles under varying light intensity are shown in Fig. 1. S_9 vesicles suspended in 4 M NaCl showed an increase in H^+ extrusion as light intensity was increased: at higher light intensities (above ~ 410 W/m^2) a biphasic response was observed. At about the same light intensity, proton uptake was observed in R_1mR and the extent of ΔpH in NaCl medium was reduced (appearance of biphasic ΔpH response) in S_9.[17,18] In both strains, $\Delta\Psi$, negative interior after illumination, was found. This method is also applicable to intact whole cells.[16,17]

[11] R. E. MacDonald and J. K. Lanyi, Biochemistry 14, 2882 (1975).
[12] R. Renthal and J. K. Lanyi, Biochemistry 15, 2136 (1976).
[13] J. K. Lanyi, R. Renthal, and R. E. MacDonald, Biochemistry 15, 1603 (1976).
[14] R. V. Greene and J. K. Lanyi, J. Biol. Chem. 254, 10986 (1979).
[15] E. V. Lindley and R. E. MacDonald, Biochem. Biophys. Res. Commun. 88, 297 (1979).
[16] Y. Mukohata and Y. Kaji, Arch. Biochem. Biophys. (in press).
[17] N. Kamo, T. Racanelli, and L. Packer, Proc. Int. Photosynth. Congr., 5th, 561, Ed. G. Akoyunoglou, Balaban Internat. Sci. Services, Philadelphia (1980).
[18] N. Kamo, T. Racanelli, and L. Packer, Membrane Biochem. 4, 175 (1982).

[47] The Classification of Halobacteria

By Terje Torsvik and Ian Dundas

Obligately halophilic bacteria, requiring extremely high salt concentrations for growth, are currently placed in the family Halobacteriaceae.[1] This family contains only two genera, the genus *Halobacterium* with two species, *H. salinarium* and *H. halobium*, and the genus *Halococcus* with only one species, *H. morrhuae*. Three other species of Halobacteriaceae are designated as *incerta sedis*. Early isolates of halobacteria were sometimes only perfunctorily described and are difficult to place taxonomically. Until recently the halobacteria seemed to form a very homogeneous bacterial group, but this may have been largely due to uniformly utilized

[1] N. E. Gibbons, in "Bergey's Manual of Determinative Bacteriology" (R. E. Buchanan and N. E. Gibbons, eds.), 8th ed., p. 269. Williams & Wilkins, Baltimore, Maryland, 1974.

enrichment and isolation procedures.[2] The recent isolation of many new and different strains of halobacteria[3-7] and the difficulties found in attempting to integrate these, and also some old strains,[8,9] in existing taxonomic schemes,[1] indicate that the currently accepted definitions of the genera *Halobacterium* and *Halococcus* may be too restrictive and could be fruitfully redefined. The recognition that the organisms in the family Halobacteriaceae belong to the group of bacteria designated as Archaebacteria[10,11] allows one to redefine the family criteria. Although the assignment of a generic name seldom presents difficulties, a major problem in the classification of halobacteria is the proper assignment of species names. This problem has led to a proliferation of new species names and to the dubious assignment of some new isolates to established species. In addition some strains have been given new species designations owing to mislabeling or to subjective decisions by the laboratories carrying the strain. Until accumulating taxonomic data permits a more definite species differentiation, one may be justified in regarding most species as just strain epithets. Accordingly no attempt will be made at this stage to redefine species criteria, and descriptions of designated halobacteria strains are given for recognition purposes only (table).

It is extremely important to verify the provenience and prior history of any strain of halobacteria to be used in future research.

Family Definition of Halobacteriaceae

The following two criteria are necessary and sufficient for including a microbial strain in the Halobacteriaceae:

1. *It must belong to the "Archaebacteria" group of microorganisms,* having characteristic ribosomes, in particular the 16S rRNA component.[11]

2. *It must have extremely halophilic procaryotic cells,* requiring salt concentrations higher than 2 *M* for growth, with Na^+ or Mg^{2+} as the domi-

---

[2] K. Eimhjellen, *Zentralbl. Bakteriol., Parasitenkd., Infektionskr. Hyg., Abt. 1, Suppl.* **1**, 126 (1965).

[3] M. F. Mullakhanbhai and H. Larsen, *Arch. Microbiol.* **104**, 207 (1975).

[4] G. A. Tomlinson and L. I. Hochstein, *Can. J. Microbiol.* **22**, 587 (1976).

[5] C. Gonzales, C. Gutierrez, and C. Ramirez, *Can. J. Microbiol.* **24**, 710 (1978).

[6] B. J. Tindall, A. A. Mills, and W. D. Grant, *J. Gen. Microbiol.* **116**, 257 (1980).

[7] F. Rodriguez-Valera, F. Ruiz-Berraquero, and A. Ramos-Cormenzana, *J. Gen. Microbiol.* **119**, 535 (1980).

[8] H. F. M. Petter, *Proc. K. Ned. Akad. Wet.* **34**, 90 (1931).

[9] B. Elazari-Volcani, *in* "Bergey's Manual of Determinative Bacteriology" (R. S. Breed, E. G. D. Murray, and N. R. Smith, eds.), 7th ed., p. 207 Bailliere, London, 1957.

[10] C. R. Woese, L. J. Magrum, and G. E. Fox, *J. Mol. Evol.* **11**, 245 (1978).

[11] L. J. Magrum, K. R. Luehrsen, and C. R. Woese, *J. Mol. Evol.* **11**, 1 (1978).

SPECIES CRITERIA OF HALOBACTERIA

Halobacteria species	Morphology	Mol % GC[a]	NaCl, M optimum	Mg^{2+}, M optimum	Growth on glucose	Gas from NO$_3$	Indole	Polymyxine sensitivity	Bacitracine sensitivity	Retinal	Refs.
Halobacterium											
H. salinarium	Rod	67 + 59	3.4–4.3	0.1–0.5	–	–	+	+	+	+	1,16–18,22
H. cutirubrum	Rod	67 + 59	3.4–4.3	0.1–0.5	–	–	+	+	+	+	1,16–18,22
H. halobium	Rod	66 + 57	3.4–4.3	0.1–0.5	–	–	+	+	+	+	1,16–18,22
H. trapanicum	Rod				–	+	–			–	1,8,9,22
H. morrhuae (Amoebobacter)	Irregular	67 + 59			–					+	1,22
Alkalophilic halobacteria	Rod	59	4.0	0.002	–	–		–			6
H. saccharovorum	Rod		4.3	0.1	+[c]	–		+	–	–	4,5
H. vallismortis	Pleomorph		4.3	0.1	+[c]	+	+	–	+		5
H. volcanii	Disk	63.4	1.7–2.5	0.2–1	+[d,e]	–	–			–	3,34
H. marismortui	Rod				+[d]	+	–				1,5,9
H. R-4	Disk	60.4[b]	2.7		+[d]	–	+				7
Halococcus											
H. morrhuae	Coccus	67 + 59	3.4–4.3	0.1–0.5	–	–	+	±		±	1,22

[a] Major + minor component.
[b] F. Rodriquez-Valera, personal communication.
[c] With 0.5% casamino acids.
[d] As sole carbon and energy source.
[e] H. Larsen, personal communication.

nating extracellular cation and with an intracellular salt concentration of similar strength to that of the environment but with K^+ as the dominating intracellular cation.[12,13]

Ancillary properties to these two basic criteria are as follows: Ribosomes are insensitive to a number of antibiotics known to inhibit eubacterial 70 S ribosomes,[14] cell walls devoid of peptidoglycan,[15] cells insensitive to penicillin, ampicillin, and other antibiotics inhibiting peptidoglycan synthesis.[16] The cells contain characteristic lipids with dihydrophytol type (C-20) side chains bound to the glycerol by ether linkages.[17] Enzymes are dependent on high salt concentrations for activity and stability, with proteins containing a marked excess of acidic amino acids.[15]

The GC mol % of DNA from the Halobacteriaceae investigated to date varies between 59 and 68 (see accompanying table). Many Halobacteriaceae have a major and a minor DNA component with the minor component representing 10 to 30% of the total. The GC mol % of these components vary within rather narrow limits, being 66–68 for the major and 57–60 for the minor component.[18]

All strains contain squalene and menaquinone (vitamin MK-8) and most strains contain dihydrogenated menaquinone (vitamin MK-8 H_2).[19]

All Halobacteriaceae investigated to date are proteolytic and have a typically heterotrophic aerobic respiratory metabolism. Some strains are able to respire carbohydrate aerobically, as sole source of carbon and energy,[7] or in conjunction with amino acids.[4,5] Anaerobically some strains are able to respire carbohydrates with nitrate as the electron acceptor[5] and some strains are able to obtain energy by the anaerobic degradation of arginine or citrulline to ornithine.[20] Some halobacteria are very versatile, possessing several of these different mechanisms for energy metabolism. Halobacteriaceae as isolated from nature usually contain characteristic C-50 carotenoid pigments, bacterioruberins.[17] Colorless mutants occur spontaneously with high frequency in pure cultures and colorless strains may exist in natural environments. Retinal, and by inference bacteriorho-

[12] J. H. B. Christian and J. A. Waltho, *Biochim. Biophys. Acta* **65**, 506 (1962).
[13] S. T. Bayley and R. A. Morton, *in* "Strategies of Microbial Life in Extreme Environments" (M. Shilo, ed.), p. 109. Dahlem Konferenzen, Berlin, 1979.
[14] T. Pecher and A. Böck, *FEMS Microbiol. Lett.* **10**, 295 (1981).
[15] H. Larsen, *Adv. Microb. Physiol.* **1**, 97 (1967).
[16] M. F. Mesher and J. L. Strominger, *J. Gen. Microbiol.* **89**, 357 (1975).
[17] M. Kates, *Prog. Chem. Fats Other Lipids* **15**, 301 (1978).
[18] R. L. Moore and B. J. McCarthy, *J. Bacteriol.* **99**, 248 (1969).
[19] M. D. Collins and D. Jones, *Microbiol. Rev.* **45**, 316 (1981).
[20] R. Hartmann, H.-D. Sickinger, and D. Oesterhelt, *Proc. Natl. Acad. Sci. U.S.A.* **77**, 3821 (1980).

dopsin, has been found in members of both genera of Halobacteriaceae,[21,22] but both the presence and the concentration of bacteriorhodopsin in any one strain is greatly dependent on growth conditions. Accordingly neither carotenoid pigmentation nor the presence of bacteriorhodopsin are very useful taxonomic criteria.

Morphology of halobacteria is greatly dependent on growth conditions and physiological age of the cells, and morphology is accordingly a taxonomic criterion of dubious value.

Genus definition of Halococcus

The following criterion is necessary and sufficient for including a strain of Halobacteriaceae in this genus:

1. The wall consists of a rigid layer of sulfated heteropolysaccharide.[23-25]

Ancillary properties to this criterion are the ability of the cells to withstand being suspended in salt-free media without structural disintegration[15] and the ability of the cells to maintain viability in seawater.[26]

Members of this genus never exhibit a regular rod shape; they may vary from regular spheres organized as diplococci or tetrad packages to roughly globular cells arranged in irregular clusters.

Genus Definition of Halobacterium

The following criterion is necessary and sufficient for including a strain of Halobacteriaceae in this genus:

1. The cell wall consists of glycoprotein.[27-29]

Ancillary properties to this criterion are the characteristic hexagonal pattern exhibited by the cell surface[3,15] and the extreme mechanical fragil-

[21] W. Stoeckenius, R. H. Lozier, and R. A. Bogomolni, *Biochim. Biophys. Acta* **505**, 215 (1979).
[22] S. C. Kushwaha, M. B. Gochnauer, D. J. Kushner, and M. Kates, *Can. J. Microbiol.* **20**, 241 (1974).
[23] A. D. Brown and K. Y. Cho, *J. Gen. Microbiol.* **62**, 267 (1970).
[24] R. Reistad, *Arch. Microbiol.* **102**, 71 (1975).
[25] J. Steber and K. H. Schleifer, *Arch. Microbiol.* **105**, 173 (1975).
[26] F. Rodriguez-Valera, F. Ruiz-Berraquero, and A. Ramos-Cormenzana, *Appl. Environ. Microbiol.* **38**, 164 (1979).
[27] M. Koncewicz, *Biochem. J.* **130**, 40 (1972).
[28] M. F. Mesher, J. L. Strominger, and S. W. Watson, *J. Bacteriol.* **120**, 945 (1974).
[29] M. F. Mesher and J. L. Strominger, in "Energetics and Structure of Halophilic Microorganisms" (S. R. Caplan and M. Ginzburg, eds.), p. 503. Elsevier/North-Holland Biomedical Press, Amsterdam, 1978.

ity of the cells. Suspension of the cells in solutions with about 2 M NaCl causes the complete loss of cell wall rigidity, the cells assuming a spherical shape. Lowering the NaCl concentration below 1 M causes the disintegration of the cell wall and the lysis of the cells. Subjecting bacterial colonies to mild mechanical stress also results in massive lysis of cells.

The classical shape of a *Halobacterium* is a slender rod. Some members of this genus do not exhibit a regular rod morphology, growing normally as more or less regular round, oblong, or square plates. All *halobacteria* form spheres when the salt in the suspending medium is lowered to the point where the cell wall looses its rigidity. When the organism has not been isolated in pure culture, such as is the case for the strikingly square *Halobacterium* described by Walsby,[30] morphology remains the only useful criterion for identification. Although many *halobacteria* are motile by means of polar flagella, motility may be influenced by cultural conditions and is accordingly of limited taxonomic value. The presence of characteristic gas vacuoles in some *halobacteria* has been shown to be a plasmid-linked trait[31,32] and is as such also of limited value for species differentiation.

The requirements of *Halobacterium* strains as to the ionic composition, salt concentration, and pH of their growth media often reflect the conditions in the natural environment from which they were isolated. Strains isolated from highly alkaline hypersaline desert lakes may require high pH, low magnesium and high sodium chloride in their media, while strains isolated from solar salterns require neutral pH, moderate magnesium and high sodium chloride, and strains isolated from the Dead Sea may have moderate requirements for sodium chloride and high requirements for magnesium in their growth media.[3,6] Such environmentally determined traits are of major taxonomic value in distinguishing *Halobacterium* strains.

Key for the Classification of *Halobacterium* Strains

Group 1. Strains unable to utilize carbohydrates as carbon and energy sources.
  A. Grow optimally at pH 6–8, no growth at pH 9.5: *H. salinarium, H. cutirubrum, H. halobium, H. trapanicum, H. morrhuae* "Amoebobacter morrhuae" (Penso).
  B. Grow optimally at pH 9.5, no growth below pH 8.5: Alkalophilic halobacterium SP-1.

[30] A. E. Walsby, *Nature (London)* **283**, 69 (1980).
[31] G. Weidinger, G. Klotz, and W. Goebel, *Plasmid* **2**, 377 (1979).
[32] R. D. Simon, *Nature (London)* **273**, 314 (1978).

Group 2.   Strains able to utilize carbohydrates as carbon and energy
           sources.
    A. Complex growth requirements: *H. saccharovorum,*
       H. vallismortis.
    B. Simple growth requirements: *H. volcanii, H. maris-*
       mortui, H. R-4.

Description of the Strains

Group 1A. These are the classical rod-shaped halobacteria; they do
not utilize carbohydrates and do not form acid from glucose. Most strains
(except *H. trapanicum*[8]) have complex growth requirements. They grow
optimally in media with 20–25% w/v NaCl (3.4–4.3 M) and 0.1–0.5 M
Mg^{2+} at pH 6–8.

The two species *H. salinarium* and *H. cutirubrum* have been considered to be identical[1] but both species names are currently in use. The ability to reduce nitrate to nitrite and differences in DNA homologies[1,33] have
been used to differentiate between *H. salinarium* and *H. halobium.* The
formation of nitrite may be influenced by the salt concentration, and in
view of the large content of extrachromosomal DNA[18] and the diversity of
plasmids[32,34] found in these bacteria, both nitrite formation and the demonstrated differences in DNA homology are taxonomic criteria of doubt-
ful value for these strains. Differences in the structure of the cell envelope[35,36] and in the membrane lipid composition[17] have been reported, but
may be influenced by growth conditions. *H. salinarium, H. cutirubrum,*
and *H. halobium* are undoubtedly distinctly different strains, but the
available data on their differences do not seem to warrant species differentiation, and at the present stage all three strains may be regarded as
belonging to a single species. This view is also supported by data from
phage typing. Strains designated *H. salinarium, H. cutirubrum* and *H. ha-*
lobium may all be infected by the same phage.[37]

Two additional strains, *H. trapanicum* and *H. morrhuae* "Amoebobacter morrhuae" (Penso), probably belong to this group, they are however poorly described and are difficult to place taxonomically. *H. trapani-*
cum was originally described as being able to grow in a mineral medium
with asparagine as the sole carbon and nitrogen source, and as able to reduce nitrate to nitrite.[8] "Amoebobacter" contains the typical *Halobac-*

[33] R. L. Moore and B. J. McCarthy, *J. Bacteriol.* **99,** 255 (1969).
[34] F. Pfeifer, G. Weidinger, and W. Goebel, *J. Bacteriol.* **145,** 369 (1981).
[35] W. Stoeckenius and R. Rowen, *J. Cell Biol.* **34,** 365 (1967).
[36] H. Steensland and H. Larsen, *J. Gen. Microbiol.* **55,** 325 (1969).
[37] T. Torsvik, unpublished observations.

terium cell wall,[36] but is characterized by never assuming a regular rod shape.[1]

Group 1B. The alkalophilic halobacteria are not only distinguished by their ability to grow at high pH; they also require unusually low magnesium concentrations for growth (less than 0.5 mM) and grow optimally in a medium with 4 M NaCl and 2 mM Mg^{2+} at pH 9.5. They are also sensitive to the vibriostatic agent 0/129, a rare property among *Halobacteria*.[6] Their special requirements as to salt composition and pH seem to justify placing the alkalophilic bacteria in a separate group.

Group 2A. Bacteria in this group have similar salt requirements to those in group 1A, but are distinguished by being saccharolytic. The complex growth requirements of *H. saccharovorum* and *H. vallismortis* distinguishes these strains from the bacteria in group 2B. These two organisms can be separated by their ability to form gas from nitrate, by indole formation, and by their sensitivity toward polymyxine and bacitracine (see the table).

Group 2B. This group contains the saccharolytic bacteria with simple growth requirements. *H. volcanii, H. marismortui* and *H. R-4* can all grow with glucose as the sole source of carbon and energy. *H. volcanii* and *H. marismortui* have both been isolated from the Dead Sea and can be separated by the ability to form gas from nitrate and by their morphology (see table). *H. volcanii* grows well with 1.5 M MgCl and 1.7 M NaCl in the media,[3] a salt composition similar to that of its natural environment. *H. marismortui* grows in media made up with Dead Sea water.[9] Both *H. volcanii* and *H. marismortui* are thus able to grow at NaCl concentrations well below those needed for growth of most other halobacteria. The reported inability of *H. marismortui* to grow in media with less than 15% NaCl[9] is probably due to magnesium deficiency in the employed media. The original *H. marismortui*, isolated by Volcani,[9] was later reported lost.[1] A strain with similar properties has been isolated from the Dead Sea and has been referred to both as *Halobacterium* of the Dead Sea[38] and as *H. marismortui*.[39] *H. R-4* is characterized by its unusually rapid growth, its nutritional versatility, and its ability to hydrolyze gelatine, caseine, and starch.[7]

The square bacterium described by Walsby[30] has not been grown in axenic culture. It was found in a hypersaline pool at sea level by Nabq in the Sinai peninsula, an extreme environment typical for halobacteria. Its waferlike morphology suggests absence of turgor, indicating similar extra-

[38] M. M. Werber and M. Mevarech, *Arch. Biochem. Biophys.* **186**, 60 (1978).

[39] M. Ginzburg, *in* "Energetics and Structure of Halophilic Microorganisms" (S. R. Caplan and M. Ginzburg, eds.), p. 561. Elsevier/North-Holland Biomedical Press, Amsterdam, 1978.

cellular and intracellular osmotic tension. The cell wall exhibits the characteristic hexagonal pattern[15] typical for many *Halobacterium* cells. The cells contain gas vacuoles and cell division occurs in two planes, alternating at right angles.

Practical Considerations in the Classification of Halobacteriaceae

Disintegration and lysis of cells on suspension in salt-free media remains the easiest test for distinguishing between *Halobacterium* and *Halococcus* strains. This test is not sufficient to separate *Halobacterium* strains from the many halophilic, extremely halotolerant bacteria with similar morphology, which may be isolated from hypersaline environments but do not belong to the Halobacteriaceae. Satisfying the two main criteria for Halobacteriaceae, archaebacterial affiliation and high intracellular salt concentrations, by characterization of the ribosomal 16 S rRNA and direct determination of intracellular salt is not easy. Antibiotic screening and determination of the salt relationship of cytoplasmic enzymes are easy, if indirect methods for satisfying the classification criteria. Resistance to Ampicillin and Penicillin (100 μg/ml)[16] correlates with the absence of peptidoglycan for *Halobacterium* and sensitivity to Anisomycin (2 μg/ml)[14] is related to the characteristic ribosomes of these bacteria. A large number of cytoplasmic enzymes of halobacteria are unstable at low salt concentrations, maintaining stability and high activities at high salt concentrations[15] as opposed to similar enzymes from halophilic organisms not belonging to the Halobacteriaceae. The salt dependency of cytoplasmic enzymes is thus correlated to the intracellular salt concentrations. Determination of the glycoprotein nature of the cell wall[27] and of the presence of lipids with ether linkages in the cytoplasmic membranes[17] are also practical procedures for establishing the archaebacterial affiliation of bacterial strains.

The importance of environmentally determined traits for the classification of Halobacteria has been clearly demonstrated by Edgerton and Brimblecombe.[40] The growth niches as described by chemical parameters clearly reflect the marine origin of *H. salinarium* and *H. halobium* strains while indicating a more heterogeneous natural environment for the strains of *H. volcanii*.

---

[40] M. E. Edgerton and P. Brimblecombe, *Can. J. Microbiol.* **27**, 899 (1981).

[48] The *Halobacterium* Group: Microbiological Methods

By HANS JURGEN WEBER, SYAM SARMA, and TERRANCE LEIGHTON

Recently there has been considerable interest in utilizing halobacterial strains for biophysical, physiological, biochemical, and genetic studies of light–energy transduction, membrane structure and function, mechanisms regulating adaptation to extreme environments, and those determinants that are uniquely associated with the archaebacterial line of evolution. These investigations require well-defined microbiological methods for the isolation, cultivation, and maintenance of *Halobacterium* strains. We describe here techniques we have found to be useful for these purposes.

The Halobacterium Group

Classical taxonomy and systematics define the genus *Halobacterium* by two distinct species: *halobium* and *salinarium* (*cutirubrum*).[1] These species have been distinguished on the basis of their physiological phenotypes. Recently Fox *et al.*[2] have characterized the *Halobacterium* group by 16 S ribosomal RNA sequence analysis. This technique has great value in taxonomic studies as it is capable of establishing phylogenetic relationships at the genotypic rather than phenotypic level. Ribosomal 16 S RNA sequence catalogues have been obtained from several representative strains of *H. halobium, H. cutirubrum, H. sp., H. volcanii, Amebobacter morrhuae,* and *Halococcus morrhuae.*[2] The *H. halobium, H. salinarium, H. cutirubrum,* and *Amebobacter morrhuae* ribosomal 16 S RNA sequence catalogues are sufficiently similar to suggest that these strains are all members of a single species. The *Halococcus morrhuae* and *H. volcanii* 16 S ribosomal RNA sequence divergence is sufficient to place these strains as separate species within the *Halobacterium* group. Hence this cluster of organisms includes rod and coccus forms of moderate and extreme halophiles which are related phylogenetically to another archaebacterial group—the *Methanobacteriales.*

[1] N. E. Gibbons, *in* "Bergey's Manual of Determinative Bacteriology" (R. E. Buchanan and N. E. Gibbons, eds.), 8th ed., p. 269. Williams & Wilkins, Baltimore, Maryland, 1974.

[2] G. E. Fox, E. Stackebrandt, R. B. Hespell, J. Gibson, J. Maniloff, T. A. Dyer, R. S. Wolfe, W. E. Balch, R. S. Tanner, L. J. Magrum, L. B. Zablen, R. Blakemore, R. Gupta, L. Bonen, B. J. Lewis, D. A. Stahl, K. R. Luehrsen, K. N. Chen, and C. R. Woese, *Science* **209**, 457 (1980).

METHODS IN ENZYMOLOGY, VOL. 88

In concordance with these studies we have observed spontaneous variation of phenotypic characters in *H. halobium* (Weber and Bogomolni, this volume Ch. 50; Weber *et al.*[2a]). A segregation pedigree has been established for these characters which includes spontaneously occurring variants of *H. halobium* wild-type strain NRL identical in phenotype to reference strains *H. salinarium* and *H. cutirubrum*. Goebel and co-workers[3,4] have demonstrated that many of these phenotypic instabilities are correlated with complex alterations in Halobacterial plasmid DNA sequences. These observations suggest that rigorous criteria should be established to monitor constantly each wild-type halobacterial strain for genotypic and phenotypic purity (see following sections). Maintenance or cultivation of halobacterial strains under inappropriate conditions may allow the inadvertent selection of spontaneously occurring pseudo-wild-type strains that are uniquely adapted to a particular laboratory environment.

We should emphasize that there has not been an exhaustive microbiological survey of *Halobacterium* strains present in the great variety of extant moderate and extreme saline environments. Hence at present we have a very limited appreciation for the diversity and ecology of this important microbial group.

Experimental Methods

Peptone Medium

The preparation and composition of a peptone–basal salts medium (RM medium) is described elsewhere in this volume (Weber and Bogomolni Ch. 50). The commercial source of the peptone is critical. Oxoid or Inolex brands are the only peptone preparations we have found to give consistently satisfactory results. With some lots of peptone small amounts of precipitates will form during the preparation of the medium. This material can be removed by centrifugation at 6000 g for 30 min (20°) prior to heat or filter sterilization of the medium. RM solid medium is prepared with Difco or Inolex Noble Agar (1.5% v/v) as described (Weber and Bogomolni, this volume Ch. 50).

RM liquid and solid media are very good general-purpose, undefined media for the cultivation and maintenance of *Halobacterium* strains. For growth of moderate halophiles, such as *H. volcanii,* the NaCl concentration should be reduced to approximately 120 g/liter.

[2a] H. J. Weber, S. Sarma, and T. Leighton, in preparation.
[3] F. Pfeifer, G. Weidinger, and W. Goebel, *J. Bacteriol.* **145,** 369 (1981).
[4] F. Pfeifer, G. Weidinger, and W. Goebel, *J. Bacteriol.* **145,** 375 (1981).

Halobacterium strains may be isolated and cultured from natural environments by a variety of techniques. Liquid medium may be prepared by centrifuging the "natural medium" at 6000 g for 30 min (20°) followed by heat or membrane sterilization. Solid medium can be prepared using 1.5% w/v Noble Agar as described. More rapid growth in these "natural media" is obtained by supplementation with 0.5% (w/v) peptone.

Minimal Medium

Grey and Fitt[5] have described a synthetic growth medium for *H. cutirubrum*. We have modified this medium to allow for a more extensive search for auxotrophic mutants. Quantities listed are grams required for 1 liter of medium.

Amino Acid Solution I (1) (grams)

L-Arginine, 0.4
L-Cysteine-HCl monohydrate, 0.05
L-Isoleucine, 0.44
L-Leucine, 0.80
L-Lysine, 0.85
L-Methionine, 0.37
L-Threonine, 0.50
L-Valine, 1.00
L-Serine, 0.61

Dissolve the amino acids *sequentially and in the order indicated* in 120 ml of double distilled water, adjust to pH 7.0 with 10 N NaOH, and bring the volume up to 150 ml.

Amino Acid Solution II (2) (grams)

L-Glutamic acid, 1.30
L-Tyrosine, 0.20

Dissolve the amino acids *sequentially and in the order indicated* in 30 ml of double distilled water, adjust to pH 10.5 with 10 N NaOH, and bring the volume up to 40 ml.

Basal Salts Solution (3) (grams)

KCl, 2.00
KH$_2$PO$_4$, 0.15
K$_2$HPO$_4$·3H$_2$O, 0.15

5 V. L. Grey and P. S. Fitt, *Can. J. Microbiol.* **22,** 440 (1976).

KNO$_3$, 0.10
MgSO$_4$·7H$_2$O, 20.000
NaCl, 210.00
NH$_4$Cl, 5.00
Sodium citrate·2H$_2$O, 0.50

Dissolve the components *sequentially and in the order indicated* in 700 ml of double distilled water, adjust to pH 7.0 with 10 *N* NaOH, and bring the volume up to 800 ml.

Trace Metals (4) (grams)

CaCl$_2$·2H$_2$O, 0.07
CuSO$_4$·5H$_2$O, 0.005
FeCl$_2$·4H$_2$O, 0.230
MnSO$_4$·H$_2$O, 0.030
ZnSO$_4$·7H$_2$O, 0.044

Dissolve the components *sequentially and in the order indicated* in 100 ml of double distilled water.

Vitamins (5) (grams)

Thiamine, 0.100
Folic acid, 0.100
Biotin, 0.010

Dissolve the components in 10 ml of double distilled water. Distribute into 1 ml aliquots and store at −20°.

Glycerol (6)
AR reagents or spectroscopic grade

To assemble MM liquid medium combine solutions 1, 2, and 3, add 1 ml of trace metal solution (4), 0.1 ml of vitamin solution (5), and 0.8 ml of glycerol. Bring the volume of the medium up to 1 liter and adjust to pH 7.0. Sterilize the medium by filtration through a 0.45-μm membrane filter. To prepare 1 liter of MM solid medium add 17.5 g of Noble Agar to 800 ml of basal salt solution (3) and let stand for 2 hr at 20°. Add amino acid solutions (1) and (2), and 1 ml of trace metal solution (4), and adjust to pH 7.0 with 10 *N* NaOH. Autoclave the medium for 15 min at 15 lb/in^2. When the medium has cooled to 65°, add 0.1 ml of vitamin solution (5) and 0.8 ml of glycerol. For growth of moderate halophiles the NaCl concentration should be reduced to approximately 120 g/liter. We have not explored systematically whether all the MM components are obligately required by all moderate and extreme halophiles. Similarly there may be

Halobacterium strains that have nutritional requirements in addition to these provided by MM medium.

Strain Maintenance

Halobacterium strains can be maintained on RM agar plates (4°) for 6 months to 1 year. The plates should be sealed tightly with parafilm to prevent desiccation. RM agar slants tightly sealed with parafilm and stored at 4° will preserve *Halobacterium* strains for 1–2 yrs. *Halobacterium* log phase liquid cultures (RM or MM media) can be stored at −10°C for longer times.

Phenotypic and Genotypic Characterization

The inherent genetic instability of *Halobacterium* strains requires that new isolates from the natural environment or strains received from culture collections be monitored carefully and continuously to ensure that genetic drift of laboratory strains does not occur. The most precise and unfortunately the most tedious methods for strain characterization are techniques that determine halobacterial nucleic acid sequence. Methods available for genotypic characterization of strains include restriction enzyme analysis of plasmid DNA fractions,[3,4] or the analysis of ribosomal 16 S RNA oligonucleotide fingerprints.[2] Methods available for the phenotypic characterization of strains include cell shape and size; motility; colony morphology; colony pigmentation; cell growth rates in defined and undefined media; minimal, optimal and maximal temperatures of growth; and individual strain requirements for amino acids and vitamins. The quantitative distribution of a variety of pigments, including carotenoids, bacteriorhodopsin and pigment P_{588}, can be easily assayed in colonies by microspectrophotometric techniques (see Weber and Bogomolni Article [50], and Bogomolni and Weber Article [58], this volume). At the minimum, a number of these phenotypic criteria should be employed regularly to verify the properties of laboratory halobacterial strains. *Halobacterium halobium* strain NRL can adapt readily to growth at 49° (RM medium), anaerobic growth in the dark (RM medium containing 0.1% L-arginine,[6] and growth in RM medium saturated with NaCl. In all of these cases selection for spontaneously occurring variants of the wild-type strain occurs (Weber, unpublished data). Hence whenever *Halobacterium* strains are exposed to a new environmental condition, one must establish whether genetic alteration of the wild-type strain has occurred and been selected for by an environmental pertubation.

[6] R. Hartmann, H. D. Sickinger, and D. Oesterhelt, *Proc. Natl. Acad. Sci. U.S.A.* **77**, 3821 (1980).

[49] Plasmids in Halobacteria: Restriction Maps

By Gottfried Weidinger, Felicitas Pfeifer, and Werner Goebel

Halobacterium halobium and other species of the genus *Halobacterium* have been shown to carry extrachromosomal DNA elements of various sizes.[1-3] In *H. halobium* this DNA was shown to be indistinguishable from the satellite DNA previously identified in this and other Halobacteria.[4,5] Although there is no convincing evidence up to now that the extrachromosomal DNA found in halobacteria is capable of regulated autonomous replication, it will be designated in the following as plasmid. The involvement of plasmids in the genetic determination of gas vacuoles has been claimed in *H. salinarium*[1] and *H. halobium*[2] on different experimental grounds. Experiments performed in this laboratory have further demonstrated specific alterations in the plasmid pHH1 (see below) in opsin-negative as well as in ruberin-negative mutants of *H. halobium*.[6] In the following the isolation procedure and the physical mapping of the major plasmid species of *H. halobium* NRC817 and the comparison of pHH1 with the plasmids of two other *H. halobium* strains 670 and 671 from the Deutsche Stammsammlung (DSM), and one *H. cutirubrum* strain NRC34001, are described.

Methods

Growth of Bacteria. All strains used are grown in a medium containing 4 M NaCl, 0.12 M MgSO$_4$, 30 m$\tilde{M}$ KCl, 10 mM Na$_3$-citrate, and 10 g per liter bactopeptone, pH 7.2. Cells are grown with shaking and illumination for 7 days.

Purification of Plasmids

A 1-liter culture of *H. halobium* grown to stationary phase in the above-described medium is harvested by centrifugation. Cells are resus-

---

[1] R. D. Simon, *Nature (London)* **273**, 314 (1978).
[2] G. Weidinger and W. Goebel, *Plasmid* **2**, 377 (1979).
[3] F. Pfeifer, G. Weidinger, and W. Goebel, *J. Bacteriol.*, **145**, 369 (1981).
[4] R. L. Moore and B. J. McCarthy, *J. Bacteriol.* **99**, 249 (1969).
[5] R. L. Moore and B. J. McCarthy, *J. Bacteriol.* **99**, 255 (1969).
[6] F. Pfeifer, G. Weidinger, and W. Goebel, *J. Bacteriol.*, **145**, 375 (1981).

pended in 40 ml buffer containing 4 M NaCl, 0.12 M MgSO$_4$, 30 mM KCl, 10 mM Na$_3$-citrate, pH 7.2, and lysed by the addition of 7 mM Na-deoxycholate (30 min at 0°). Centrifugation of the lysed cells at 10,000 g for 30 min removes most of the chromosomal DNA. The cleared lysate is diluted with 2 vol of H$_2$O and the DNA then concentrated by adding polyethylene glycol (PEG-6000) to a final concentration of 10% (4°, 5 hr). The precipitate is dissolved in 15 ml TEN buffer (40 mM Tris, pH 8.0, 5 mM EDTA, and 50 mM NaCl, pH 8.0) and subjected to a two-step CsCl–ethidium bromide centrifugation, using first a TV850 rotor and then a TI50 rotor. After fractionation of the final gradient the fractions containing covalently closed circular (CCC) DNA are pooled, ethidium bromide is removed by isopropanol, the pooled fractions are dialyzed against 10 mM Tris, pH 7.5, 1 mM EDTA, and the DNA is concentrated by ethanol precipitation. The final yield is about 150 μg of CCC-DNA.

Enzyme Reactions

Reactions with *Hind*III are carried out in 50 mM Tris, 10 mM NaCl, 10 mM MgCl$_2$, pH 7.5, for 1 hr at 37° and with *Pst*I in 10 mM Tris, 10 mM Tris, 10 M MgCl$_2$, pH 7.4.

Digestions with *Hind*III and *Pst*I are performed sequentially after heat-inactivating the first enzyme (10 min, 68°) and adjusting the buffer. For ligation with T4 ligase (0.1 units/μg DNA) the reaction mixture is adjusted to 66 mM Tris, pH 7.5, 33 mM NaCl, 7 mM MgCl$_2$, 50 μM ATP and 1 mg/ml dithiothreitol. Cohesive end ligations between *Pst*I or *Hind*III fragments of pHH1 and *Pst*I- or *Hind*III-cleaved pBR322 or pHC79 are run for 16–24 hr at 10°. Labeling of the 5' termini of *Hind*III fragments of pHH1 is performed in 50 mM Tris, pH 7.4, 0.1 mM EDTA, 10 mM MgCl$_2$ containing 50 μCi [γ-^{32}P]ATP (specific activity 20 mCi/μmol), 2–5 μg pHH1 DNA and T4 polynucleotide kinase (0.1 units/μg DNA). Incubation is continued for 30 min at 37°. Reaction is stopped with 20 mM EDTA and the ^{32}P-labelled *Hind*III fragments are precipitated twice with ethanol before being separated by agarose gel electrophoresis using 1% gels and 40 mM Tris, pH 8, 20 mM NaH$_2$PO$_4$, 18 mM NaCl, and 2 mM EDTA as electrophoresis buffer. Labeled cRNA from cloned *Pst*I fragments of pHH1 is prepared in a mixture containing 40 mM Tris, pH 7.9, 150 mM KCl, 10 mM MgCl$_2$, 0.5 mM EDTA, 0.5 mg/ml BSA, 5 × 10^{-4} of ATP, GTP, and CTP, 50 μCi [α-^{32}P]UTP (specific activity 400 Ci/mmol), and 2 μg plasmid DNA. Reaction is carried out for 90 min at 37° and stopped by the addition of 20 mM EDTA and 0.5% SDS. The labeled RNA is purified on a Sephadex G-50 column.

Determination of the Physical Map of Plasmid pHH1

a. Plasmid pHH1 DNA is cleaved by *Hind*III into 11 fragments (H1 to H11) and by *Pst*I into 22 fragments (P1 to P22) (Fig. 1). Digestion of pHH1 by *Eco*RI yields over 40-fragments. After cleavage of pHH1 with both *Hind*III and *Eco*RI, two *Hind*III fragments (H3 and H10) are not further cut by *Pst*I; 14 *Pst*I fragments are cut out of the other 9 *Hind*III fragments; the other *Pst*I fragments are cleaved by *Hind*III giving *Pst*I/*Hind*III subfragments, which can be used for determining the linked *Hind*III fragments.

b. *Hind*III-cleaved pHH1 DNA is ^{32}P-labeled with polynucleotide kinase, and the fragments are separated on agarose gels, individually isolated from the gel by the hydroxyapatite method,[7] and subfragmented by *Pst*I. The results of this method are as follows: H3 and H10 are not further cut by *Pst*I, P15 is contained in H9; P4, P16 and P21 in H4. Two *Pst*I/*Hind*III subfragments of H4 and H7 yield P6, two from H6 and H4 yield P5, and two from H8 and H9 yield P7, indicating the linkage of these *Hind*III fragments.

c. *Pst*I fragments of pHH1 are cloned in pBR322.[8] ^{32}P-labeled cRNA is prepared and hybridized to the filter-bound *Hind*III fragments of pHH1. Hybridization of cRNA of P8 is obtained with H6 and H1, of P7 with H8 and H9, of P14 with H2 and H9, and of P12 with H7 and H11. It is further demonstrated that P2 contains H10 as an internal part and the two *Pst*I/*Hind*III subfragments of H1 and H8. Labeled cRNA prepared from both H3 and H5 hybridize with P1. Their molecular weights together with one *Hind*III/*Pst*I subfragment of H2 add up to that of P1. The arrangement of the *Pst*I fragments within the large H1 fragment is determined in a cosmoid clone that contained in pHC79[9] *Hind*III fragments H1 together with H5, by partial digestion with *Pst*I. The arrangement of the *Pst*I fragments in H2 is based on the hybridization of cRNA from *Pst*I fragments cloned in pBR322 with *Eco*RI-cleaved pHH1. The physical map of pHH1 resulting from these data is given in Fig. 2.

Molecular Weights of Plasmids from H. halobium and H. cutirubrum

H. halobium NRC817 contains one major plasmid pHH1, with a molecular weight of 100×10^6, determined by measurements of the contour lengths of open circular DNA and the molecular weights of restriction fragments of pHH1. Besides pHH1, minor ccc components of larger and

[7] W. Oertel, R. Kollek, E. Beck, and W. Goebel, *Mol. Gen. Genet.* **171**, 277 (1979).
[8] R. Bolivar and R. L. Rodriguez, *Gene* **2**, 95 (1977).
[9] J. Collins and B. Hohn, *Proc. Natl. Acad. Sci. U.S.A.* **75**, 4242 (1978).

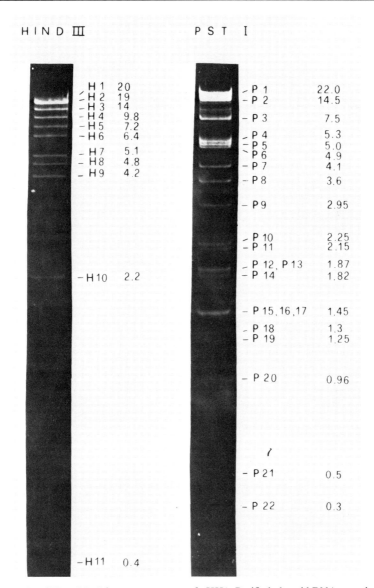

FIG. 1. *Hind*III and *Pst*I fragment patterns of pHH1. Purified plasmid DNA was cleaved with the restriction enzymes under the conditions described and separated on 1% agarose gels. The numbers indicate the molecular weights of the restriction fragments in megadalton.

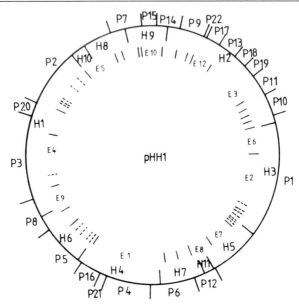

FIG. 2. Circular map of plasmid pHH1. The broken lines indicate regions of pHH1 where the arrangement of the EcoRI fragments is not yet quite sure.

lower molecular weights than pHH1 are found in this strain.[2] The DSM strain *H. halobium* 670 carries as a major component a plasmid pHH2 slightly larger than pHH1, whereas another DSM strain 671, which lacks gas vacuoles, harbors a considerably smaller plasmid pHH3. *H. cutirubrum* NRC34001 carries a plasmid, pHC1, similar in size to pHH1.

Comparison of the Plasmids from H. halobium and H. cutirubrum

The plasmids from the DSM strains of *H. halobium* and *H. cutirubrum* yield similar patterns of restriction fragments as pHH1 when they are cleaved with *Eco*RI, *Hind*III or *Pst*I (Fig. 3). Hybridization of nick-translated pHC1 DNA with *Pst*I-cleaved pHH1 shows homology with all *Pst*I fragments of pHH1 except P8, which seems to be missing in pHC1. Since only P2, 7, 15, 14, 9, 22, 13, 17, 11, 10, 19 of pHH1 are also indistinguishable in size to pHC1, it follows that the arrangement of the other parts of pHC1 despite homologous sequences is different.

Plasmid pHH2 of *H. halobium* DSM 670, slightly larger than pHH1, is indistinguishable from the latter one in the region indicated by the solid line. The other parts, although highly homologous to pHH1, have different arrangements. Plasmid pHH3 has lost a substantial part of the se-

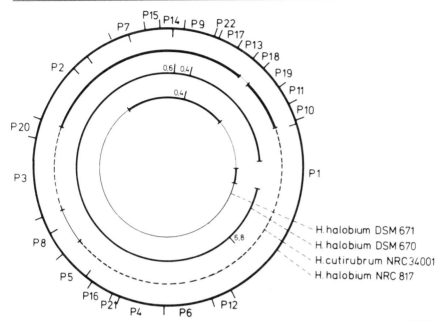

FIG. 3. Comparison of the circular maps of plasmids pHH1 of *H. halobium* NRC817, pHC1 of *H. cutirubrum* NRC34001, pHH2 of *H. halobium* DSM 670, and pHH3 of *H. halobium* DSM 671. Solid lines in the latter three plasmids represent regions of complete homology, yielding restriction fragments indistinguishable from pHH1. The dashed regions of pHC1 show extended homology with pHH1 but the restriction fragments from this regions are different from those of pHH1. The thin lines represent regions of pHH1 that are obviously deleted in the other plasmids as judged from the lack of hybridization of these plasmids with the corresponding restriction fragments of pHH1.

quences present in the other three plasmids, but hybridization data indicate that the remaining part of pHH3 is homologous with the indicated segment of pHH1. Plasmids with differently arranged segments, with insertions and deletions in pHH1, have been also obtained from various mutants of *H. halobium* NRL817.[6]

[50] The Isolation of *Halobacterium* Mutant Strains with Defects in Pigment Synthesis

By HANS JURGEN WEBER and ROBERTO A. BOGOMOLNI

The microorganism *Halobacterium halobium* produces a variety of carotenoid pigments that make up its bright pink color. Two of them, BR and P_{588} (also called "halorhodopsin," this volume) are believed to be retinal-containing proteins.[1-4] They seem to act as light-driven ion pumps[1,5] and may convert light energy into ATP.[1,6] Pigment mutants were isolated that allowed a separate photochemical characterization of P_{588} in its native environment.[3,4] It is reasonable to assume that such mutants will also be helpful in defining the specific contribution of either pigment to the energy balance of *Halobacterium* cells. Another line of investigation that could greatly benefit from an application of genetic techniques is the elucidication of BR as well as P_{588} biosynthesis and their regulation. Optimal production of both pigments requires synchronized synthesis of their respective chromophores and apoproteins. Feedback regulation processes are likely to occur[7,8]; however, their exact nature remains to be defined (see Fig. 1).

The isolation and characterization of *Halobacterium* mutant strains provide a first step in the development of genetic techniques for these organisms. We present here a protocol that outlines how pigment mutants can be obtained and their properties be described by spectroscopic as well as functional criteria.

Materials

Basal salt solution (BS) consists of 3×10^{-2} M KCl, 10^{-2} M sodium citrate, 3.6 M NaCl, 8×10^{-2} M MgSO$_4$ in distilled water. Liquid-rich medium (RM) is BS with 15 g/liter of peptone at pH 7.0. Routinely, RM is heat sterilized in 100-ml quantities for 15 min and then stored in the cold. For specific experiments, previously heated RM can be filtered through sterile filter units (Nalgene, pore size 0.45 μm) to remove dark sediments that originate upon dissolving and heating peptone in BS.

RM-agar plates were prepared from RM with 1.5% agar. The agar was allowed to swell in nonsterile medium for about 2 hr before heat sterilization.

Abbreviations: ves, gas vesicles; bra, bacterio-opsin; brb, P_{588}-opsin (or "halo-opsin"); ret, retinal; rub, bacterioruberins (negative subscripts combined with these abbreviations indicate the absence, no signs, the presence of these characters in bacterial cells); WT, wild-type strain NRL (*National Research Laboratory*, Ottawa, Canada). BR, bacteriorhodopsin; P_{588}, a new retinal-containing pigment that absorbs maximally at λ = 588 nm.[3,4]

[1] W. Stoeckenius, R. H. Lozier, and R. A. Bogomolni, *Biochim. Biophys. Acta* **505**, 215 (1979).
[2] A. Matsuno-Yagi and Y. Mukohata, *Biochem. Biophys. Res. Commun.* **78**, 237 (1977).
[3] J. K. Lanyi and H. J. Weber, *J. Biol. Chem.* **255**, 243 (1980).
[4] H. J. Weber and R. A. Bogomolni, *Photochem. Photobiol.* **33**, 601 (1981).
[5] E. V. Lindley and R. E. MacDonald, *Biochem. Biophys. Res. Commun.* **88**, 491 (1979).
[6] A. Matsuno-Yagi and Y. Mukohata, *Arch. Biochem. Biophys.* **199**, 297 (1980).
[7] M. Sumper and G. Hermann, *FEBS Lett.* **69**, 149 (1976).
[8] M. Sumper and G. Hermann, *FEBS Lett.* **71**, 333 (1976).

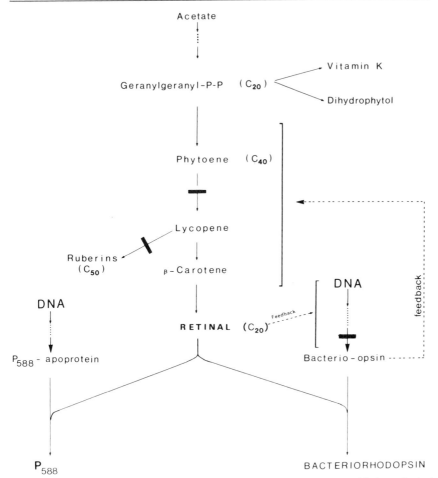

FIG. 1. Biosynthetic paths of *Halobacterium halobium* pigments. In this hypothetical biosynthetic scheme the numbers in parentheses indicate the carbon backbone of the respective carotenoid. The heavy bars across arrows indicate biosynthetic blocks that prohibit formation of certain pigments. Possible feedback regulation is indicated; its precise targets, however, are unknown.

Solutions of *p*-trifluoromethoxycarbonylcyanidephenylhydrazone (FCCP, Sigma Chem. Co.) were prepared in ethanol at 20 m*M*. A few microliters were added to bacterial suspensions to make the final concentration 50 μM. The amount of ethanol in the mixture never exceeded 0.7%.

Peptone was from Inolex Corp., Noble Agar from Difco, and all standard chemicals from Mallinckrodt.

Bacterial Growth in Liquid Cultures for Comparative Analysis

Our procedure for growing *Halobacterium* strains is based on the consideration that cells should have the physiological capacity for developing the set of pigments they are genetically capable of synthesizing. We therefore chose to grow them under nutritionally rich conditions. Pigment production starts toward the end of logarithmic phase and maximizes when the oxygen supply is reduced. It is therefore necessary to reproduce certain experimental parameters such as duration of logarithmic growth phase, duration of incubation in stationary phase, and availability of oxygen. Finally, pigment synthesis should occur synchronously in parallel cultures, which implies that the bacteria exhibit roughly the same doubling times. We satisfied these needs by consecutively diluting and incubating bacterial cultures, thus diluting out slowly growing cells. The resulting bacterial populations exhibit maximum growth rates, are homogeneous with respect to size and shape, give optimal yields (total protein content at time of harvest), and produce easily detectable quantities of pigments.

Experimental Methods. Cells from agar plates are resuspended in 2 ml of RM and grown at 37° until they reach a density of $1-5 \times 10^8$/ml. They are then diluted 10,000-fold into fresh RM (10 ml in 125-ml Erlenmeyer flasks) and grown to late log phase ($\sim 4 \times 10^8$ cells/ml), which usually takes 72 hr (37°, rotary shaker, New Brunswick, shaking rate 150 rpm). This operation is repeated once and then a desired volume of RM inoculated at a density of 10^7 cells/ml. Care has to be taken that surface-to-volume ratios and shaker speeds allow for optimal aeration. We routinely fill culture flasks one-tenth to one-fifth of their total volume. At cell densities of about 5×10^7/ml, 37°, and shaking rates of 150 rpm the cell mass doubles every 4.5–5 hr. After growth has proceeded for 96 hr, cultures are further incubated for 48 hr without shaking in order to reduce oxygen supply. Light is not absolutely necessary for bacterial growth or pigment formation. It appears, however, that light intensities of 10^5-10^6 erg sec^{-1} cm^{-2} maximize pigment production. Cells are harvested by centrifugation at 5000 g for 20 min, resuspended in the same amount of BS, and again centrifuged. Finally, the bacteria are gently resuspended in BS (one-seventh of the culture volume) for spectroscopic analysis.

Isolation of Pigment Mutants

Halobacteria cells form well-defined colonies on RM-agar. Agar plates can therefore be seeded with large numbers of cells and the resulting colonies viewed through a binocular microscope at $4 \times -25 \times$ magnification. It is very easy to recognize mutant colonies by their differences in color, especially when they are compared with a majority of parent colonies under reflected lighting conditions.

Experimental Methods. Bacteria are grown in RM as described before. The final culture is incubated for 96 hr, the cell count determined microscopically, and the cells diluted into BS to 10^4 cells/ml. Spreading of 0.05–0.1 ml on RM-agar results in about 500–1000 colonies per plate indicating that the plating efficiency of these cells is 100%. It is necessary to prevent the agar from drying; therefore plates are kept in sealed plastic bags throughout the incubation period at 37°. We recommend a first inspection of the colonies after an incubation of 15 days followed by a second viewing after a total of 25 days. Essentially, color mutants of all shades can be isolated, depending on the starting strain. Under reflecting lighting conditions and with a black background, ves⁻ mutants appear as dark-orange colonies, WT as bright pink, and rub⁻ mutants as purplish white colonies. The latter two types of colonies are not transparent. The presence of gas vesicles influences the color significantly since they act as an internal light screen, thereby reflecting incoming light. Gas vesicles can be easily detected by viewing single cells with a phase contrast microscope. The highly refractile vesicles appear bright against the dark background of the cell.

Mutant strains are first purified by repeated streaking on RM-agar. They can be kept on RM-agar for about 6 months provided the plates are stored in sealed plastic bags at 4°. Mutant phenotypes should be recorded and checked visually after restreaking.

Absorption Spectroscopy of Cell Populations

Spectroscopy provides a quick, reliable, and nondestructive method to characterize differences in pigment composition. Because bacterial suspensions scatter light beams, an "integrating sphere" is needed for the measurement of absorption spectra.[9] This device prevents the loss of measuring light except for absorption by the sample. The dominant features of a WT difference spectrum (see Fig. 3A) are the regions of 570 nm (BR); 544 nm, 506 nm, 477 nm (bacterioruberin); 342 nm (lycopene). We analyzed some of our mutant strains by standard biochemical techniques and were able to confirm previous data that had suggested spectroscopic discrimination of lycopene, β-carotene, bacterioruberin, and BR in halobacteria cells[10,11]

[9] R. A. Bogomolni, R. A. Baker, R. H. Lozier, and W. Stoeckenius, *Biochemistry* **19**, 2152 (1980).
[10] S. C. Kushwaha, M. Kates, and H. J. Weber, *Can. J. Microbiol.* **26**, 1011 (1980).
[11] H. J. Weber *et al.*, unpublished.

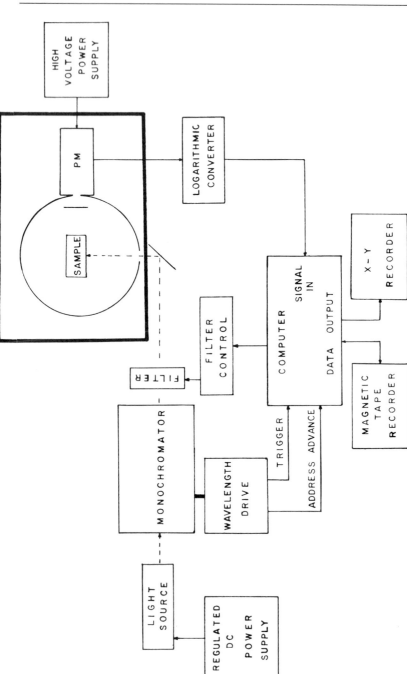

Fig. 2. (A) Single-beam integrating sphere spectrophotometer. The integrating sphere is depicted graphically in the rectangle. It consists of a hollow sphere that is uniformly coated inside with a nonabsorbing scattering material (e.g., MgSO₄). Input optics, diaphragm sample holder, and photomultiplier are permanently attached to the sphere. The whole unit can be easily mounted onto the sample compartments of commercially available instruments (e.g., Cary 14, Aminco DW-2A). (B) Microspectrometer attachment. The sphere and detector enclosed in the

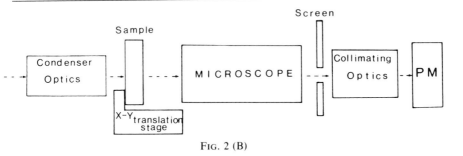

FIG. 2 (B)

Experimental Methods. The instrument we used in our experiments is schematized in Fig 2A. It is equipped with a compact filament tungsten halogen lamp powered by a highly stable dc power supply (Lambda LE 103 FM). The image of the filament is focused onto the entrance slit of a monochromator which is driven by a stepper motor wavelength-scanning device (Jarrel Ash). A filter (Corning 3-69) driven by a rotary solenoid is used above 600 nm to reflect second-order stray light. The output beam from the monochromator is deflected upward by a mirror into the integrating sphere. This modification allowed us to investigate light-absorbing material which may settle at the bottom of the cuvette during repeated scans. Cell suspensions in BS (4 ml; 10^{10} cells/ml) are filled into a cylindrical cuvette and positioned inside the sphere so that front surface reflections from the cuvette exit through the diaphragm. The photomultiplier is protected from direct light reflections by an opal–quartz mirror. Its output is fed into a logarithmic current-to-voltage converter (Analog Devices 755P), then digitized and stored in a computer (Nicolet 1180). A trigger signal from the scanning mechanism synchronizes the computer with the monochromator wavelength. The order sorting filter position is also controlled by the computer. The spectrophotometer is operated in single-beam mode from 750 nm to 340 nm with BS as a reference. Spectra are recorded, scaled, and subtracted by computer. Difference spectra for each analyzed mutant strain are stored on magnetic disks and can be easily recalled for detailed analysis (e.g., to superimpose various spectra for comparison or to obtain derivative spectra for the detection of poorly resolved pigments). Figure 3A shows representative difference spectra. Note the reproducibility between samples from different experiments (top trace, Fig. 3A), which is a result of rigorously controlled growth conditions (see earlier).

The presence of bacteriorhodopsin can be assessed independently by "light–dark" difference spectroscopy (see Fig. 3B). An absorption spectrum of a dark-adapted culture is first recorded and stored in the computer. Without changing the position of the sample it is then illuminated

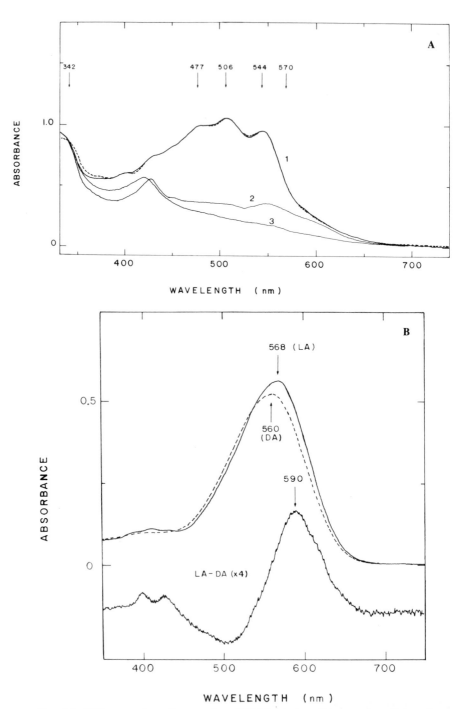

Fig. 3A. Difference absorption spectra of *Halobacteria* cell suspensions (integrating sphere setup). Trace 1: The solid and dotted curves indicate two WT spectra taken from two independent experiments. Trace 2: Spectrum of mutant cells, phenotype (ves, bra, brb, ret, rub⁻). Trace 3: Spectrum of mutant cells, phenotype (ves⁻, bra⁻, brb, ret, rub⁻). (B) Differ-

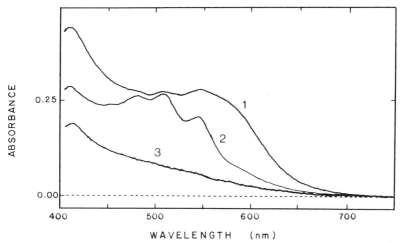

Fig. 3C: Microspectrophotometry of bacterial colonies on agar plates. Trace 1: Spectrum of mutant clone, phenotype (ves⁻, bra, brb, ret, rub). Trace 2: Spectrum of mutant clone, phenotype (ves⁻, bra, brb, ret, rub). Trace 3: Spectrum of mutant clone phenotype (ves⁻, bra⁻, brb, ret, rub⁻). The first two traces represent the characteristics of two different mutant clones, both of which show reduced amounts of bacterioruberin but one (trace 1) is an overproducer of BR.

for about 10 min using a standard microscope lamp with a 3-69 Corning cutoff filter. Then a second spectrum is taken. Calculation of the light–dark difference spectrum indicates the presence and quantity of bacteriorhodopsin.[9]

Spectroscopic Measurements on Single Colonies

Bacterial colonies consist of densely packed cells that display little light scattering if they lack gas vesicles. They can be analyzed between 800 and 400 nm through agar and plate bottom, using a refined spectrophotometer that we have recently developed. As reference spectrum RM-agar and plate are measured in close proximity to the colony. The spectrophotometer consists of a combination of lenses to focus the measuring beam onto a bacterial colony and an x-y stage for mounting and precise positioning of agar plates (see Fig. 2B). The image of the colony is provided by a set of lenses assembled in a standard microscope tube and

ence absorption spectra of aqueous purple membrane suspensions (integrating sphere set up). DA: Spectrum for dark-adapted purple membrane. LA: The same for the light-adapted form. LA-DA: Difference spectrum of light-adapted–dark-adapted BR. From the magnitude of the peak at λ = 590 nm the amount of BR can be calculated.[9]

projected onto a photomultiplier (Hamamatsu R928). The difference spectra measured with this instrument (Fig. 3C) show comparable signal-to-noise ratios with respect to spectra obtained from cell suspensions. Both methods are nondestructive and the physiology of bacteria cells can be further investigated.

Discrimination between Mutants that Lack Retinal Synthesis (ret⁻, bra, brp) and Mutants with Defective Pigment Apoprotein (ret, bra⁺, brb or ret, bra, brb⁻)

Currently, the two most intensely studied pigments in *Halobacteria* are BR and P_{588}. They both employ retinal as the chromophore. Consequently, the inability of cells to synthesize retinal is manifested in pleiotropic effects and leaves mutant cells without functional, light-mediated ion translocation. Halobacteria cells, however, can take up externally added retinal. After growth for 48 hr in RM, all-*trans*-retinal (50 mM in isopropanol) is added to the culture to a concentration of 25 μM. It is advisable to repeat this addition after 24 hr because retinal may deteriorate at 37° in the light. After harvest, the cells can be analyzed in various ways.

The presence of BR can be determined as already described but most cells contain too little P_{588} to be observed directly by difference spectroscopy. The detection of P_{588} requires flash spectroscopy techniques (see Article [58], this volume) or measurement of light-induced pH changes of cell suspensions (see below). If pigment production can be detected after addition of retinal but not in its absence, the mutants are classified as ret⁻, bra, brb. The lack of pigment production even in the presence of added retinal suggests phenotypes such as ret⁻, bra⁻, brb⁻ or ret, bra⁻, brb⁻.

Discrimination between BR and P_{558} in Cell Suspensions by Measuring Light-Induced pH Changes

Halobacterium strains that contain BR acidify the medium in the light (see Fig. 4A). This reflects the dominance of outwardly directed proton pumping by BR.[12] Mutant cells that do not synthesize BR but still contain P_{588} show light-induced alkalinization of the medium (see Fig. 4C). This finding was explained as a consequence of electrogenic, light-driven Na⁺ extrusion mediated by P_{588}.[5,13] A membrane potential (inside negative) is thus generated that serves as the driving force for the inwardly directed proton movement.[5,13] In accordance with this interpretation, proton conductors (e.g., FCCP) enhance light-connected H⁺ inflow (see Fig. 4D).

[12] R. A. Bogomolni, R. A. Baker, R. H. Lozier, and W. Stoeckenius, *Biochim. Biophys. Acta* **440**, 68 (1976).
[13] R. V. Greene and J. K. Lanyi, *J. Biol. Chem.* **254**, 10986 (1979).

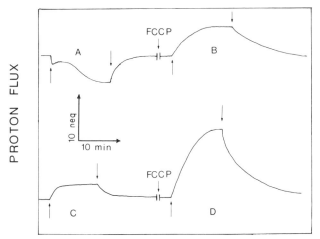

FIG. 4. Light-induced pH changes of *Halobacterium* cell suspensions. (A) Mutant strain with phenotype (ves⁻, bra, brb, ret, rub), pH 6.0. (B) The same in the presence of the proton-ionophore FCCP (50 μM). (C) Mutant strain with phenotype (ves⁻, bra⁻, brb, ret, rub⁻), pH 6.0. (D) Same as (C) but in the presence of FCCP (50 μM). Arrows indicate light on (↑) and light off (↓).

When FCCP is added to WT cells (bra, brb), steady state levels of extruded protons (due to BR) decrease while levels of uptaken protons (due to P_{588}) increase, now dominating net H^+ movement in the light (see Fig. 4B). We adopted an assay for the measurement of BR and/or P_{588} that is based on these time-resolved pH changes of bacterial suspensions in the light and in the presence or absence of proton ionophores. Note that ret⁻ mutants can also be investigated after growing them in the presence of retinal.

Experimental Methods. We routinely use a thermostatted (25°) stainless-steel chamber in which a pH electrode is inserted. The chamber can be illuminated through a glass window and is mounted onto an optical bench together with a 300-W quartz halogen lamp and appropriate lenses. Heat filters are a 3-cm circulating water filter and three infrared reflecting glasses (each 3 mm thick). After passing an additional 3-69 Corning cutoff filter, light energies should be around 10^6 erg sec⁻¹ cm⁻² as measured at the front window of the chamber (YSI Kettering Radiometer, Model 65). The glass electrode (Beckmann Instruments No. 39003) is connected to a pH meter (Corning Model 12) and the pH recorded on a strip chart recorder as a function of time. Cell suspensions (5 × 10^9 cells/ml in BS) are filled into the chamber, continuously purged with N_2, and equilibrated for about 30 min until the base line is stable. Buffer capacities are determined by adding small amounts of 20 mM acid or base. They are around 0.6–1.8 × 10^{-6} Eq H^+/pH at pH 6.0 under our experimental conditions. Subsequent exposure to light lasts for 20–30 min during which time ΔpH is recorded.

Strategy for the Isolation of Specific bra⁻ Mutants

Starting with WT strain NRL, we isolated mutants in which pigment P_{588} can be spectroscopically studied. It requires the loss of bacterioruberin, BR, and for convenience, the gas vesicles. Since spontaneous mutations from rub → rub⁻ and bra → bra⁻ occur with a frequency of about 10^{-4} each, the isolation has to proceed in several steps. It is advisable to screen first for rub⁻ mutants, because bacterioruberin is responsible for the dominating orange color of WT cells. Other carotenoid mutants can then be easily isolated during further screenings.

Many mutants harbor defects in retinal synthesis but continue to synthesize bacterio-opsin (members of class M2A, see below). They can be conveniently used to study the regeneration of BR with retinal or retinal analogs. Since this is possible in whole cells as well as in envelope vesicles, the capability of regenerated BR to translocate protons can be studied.

Other mutants are deficient in bacterio-opsin production and show only the action of pigment P_{588} (members of class M2B, see below). Ret⁻ strains are then isolated from them and used for regeneration studies of pigment P_{588} (see this volume).

Experimental Methods. Start with WT colonies (ves, bra, brb, ret, rub) on RM-agar and screen for purplish white colonies. Analysis shows that they constitute class M1, phenotype (ves, bra, brb, ret, rub⁻). Plates containing colonies of one representative M1 strain are now visually screened for white clones. They were found to represent class M2, phenotype (ves, bra$^\pm$, brb, ret$^\pm$, rub⁻). It is indicated by superscripts that they can have defects in either bacterio-opsin or retinal synthesis. They are distinguished by growing them in liquid RM with and without retinal (parallel cultures) followed by spectroscopic analysis. This allows a division into classes M2A, phenotype (ves, bra, brb, ret⁻, rub⁻) and class M2B, phenotype (ves, bra⁻, brb, ret, rub⁻). Plating on RM-agar is repeated with representatives of class M2B and transparent colonies are isolated. They belong to class M3B, phenotype (ves⁻, bra⁻, brb, ret, rub⁻) and are convenient material for the investigation of P_{88} by spectroscopic techniques.[3,4]

Acknowledgments

This work was supported by Grants GM-28767 and GM-27057 from the National Institutes of Health, General Medical Sciences.

[51] The Brown Membrane of *Halobacterium halobium:* The Biosynthetic Precursor of the Purple Membrane

By MANFRED SUMPER

The purple membrane *in vivo* exists as a differentiated insular region of the cell membrane[1-3] Under growth conditions inducing purple membrane biogenesis, these regions may cover more than 50% of the cell membrane area. Thus bacterio-opsin is by far the most abundant membrane protein of halobacteria. This makes the purple membrane an appealing model system for studying the biogenesis of a differentiated membrane region.

Purple membrane biogenesis is induced by a low oxygen supply in the growth medium.[3] These conditions turn on the synthesis of both the membrane components bacterio-opsin and retinal in a strictly coordinated manner.[4,5] Another domain of the halobacterial plasma membrane termed the "brown membrane" was demonstrated to be involved in the biogenesis of the purple membrane.[4,6] This membrane fraction has a lower buoyant density (1.14) than purple membrane (1.18) and contains bacteriorhodopsin and a cytochrome *b*-type protein besides other membrane protein species. Newly synthesized bacterio-opsin is initially found in the brown membrane fraction and can only "crystallize" to form the purple membrane patches after reaction with retinal.

This article deals with the preparation of the brown membrane and its *in vivo* conversion into the purple membrane.

Culture Conditions

In order to minimize interference in the spectroscopic analysis of bacteriorhodopsin, mutants of *H. halobium* deficient in bacterioruberin synthesis (e.g., R_1M_1) are preferred for the isolation of the brown membrane.

Halobacteria are grown as described by Oesterhelt and Stoeckenius.[7] Bacteriorhodopsin synthesis is only initiated under growth conditions of

[1] D. Oesterhelt and W. Stoeckenius, *Nature (London), New Biol.* **233,** 149 (1971).
[2] A. E. Blaurock and W. Stoeckenius, *Nature (London), New Biol.* **233,** 152 (1971).
[3] D. Oesterhelt and W. Stoeckenius, *Proc. Natl. Acad. Sci. U.S.A.* **70,** 2853 (1973).
[4] M. Sumper, H. Reitmeier, and D. Oesterhelt, *Angew. Chem., Int. Ed. Engl.* **15,** 187 (1976).
[5] M. Sumper and G. Herrmann, *FEBS Lett.* **71,** 333 (1976).
[6] M. Sumper and G. Herrmann, *FEBS Lett.* **69,** 149 (1976).
[7] D. Oesterhelt and W. Stoeckenius, this series, Vol. 31, p. 667.

METHODS IN ENZYMOLOGY, VOL. 88

limited aeration. This requirement is met as follows:

Small scale: 220 ml complex medium (containing in 1000 ml: NaCl, 250 g; $MgSO_4 \cdot 7H_2O$, 20 g; KCl, 2 g; and Oxoid Bacteriological Peptone L 37, 10 g) adjusted to pH 6.5–7.0 in a 500-ml Erlenmeyer flask is inoculated with 4.0 ml of a stationary growth phase culture and shaken at 100 rpm at 39°.

Large scale: 10-liter medium; growth conditions as described by Oesterhelt and Stoeckenius.[7]

Under these conditions bacteriorhodopsin synthesis is initiated after a growth period of 2 days and continues for at least 3 days more.

Fragmentation of the Cell Membrane

Cells from a 220-ml culture grown for 4 days are harvested by centrifugation for 15 min at 15,000 g. Cells are lysed in 10–20 ml 50 mM Tris-HCl, pH 7.5, containing 10 μg/ml DNase. The lysate is centrifuged at 150,000 g for 2 hr. The sediment containing the crude membrane fraction is resuspended in 1–2 ml of 50 mM Tris–HCl, pH 7.5, and layered over a linear 20–45% sucrose density gradient containing the above buffer in Beckman SW41 tubes. A centrifugation time of at least 6 hr is required at 39,000 rpm. Figure 1 shows a typical gradient tube with the membrane fragments A, B, C, and D. Fraction A, containing all the lycopene of the original plasma membrane, is termed red membrane RM 340, B is a yel-

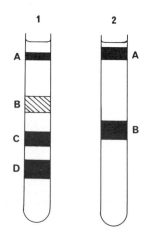

FIG. 1. Separation of halobacterial cell membrane fragments by sucrose density gradient centrifugation. (1) Membrane fragments from normal cells. (2) Membrane fragments from nicotine-treated cells.

low-colored membrane fraction, C is the brown membrane fraction, and D is the purple membrane fraction.

Centrifugation of the same cell lysate in sucrose density gradients containing 0.5 M NaCl results in a single-colored membrane fraction with absorption maxima at 340 nm (lycopene), 419 nm (cytochrome), and 570 nm (bacteriorhodopsin). Therefore, fragmentation of the cell membrane only takes place under low ionic strength conditions.

Preparation of Apo-Brown Membrane from Retinal-free Cells

An increased yield of brown membrane is observed when cells are grown in the presence of the alkaloid nicotine. Nicotine selectively inhibits synthesis of retinal by blocking the cyclization of lycopene to β-carotene[8] without affecting the growth rate of the cells. Cells grown in the presence of 0.5 mM nicotine contain only about 2% retinal compared with control cells. One mM nicotine depresses the retinal content to less than 1%.[4] Synthesis of bacterio-opsin, however, remains inducible by limited aeration.

Growth conditions for cells in the presence of 1 mM nicotine ("nicotine cells") are exactly as described above for normal cells. After a growth period of 5–6 days, cells are harvested by centrifugation and the membrane fractions are prepared in exactly the same way. Figure 1 shows a typical gradient tube with the membrane fragments A and B from nicotine cells. A is again the membrane fraction RM 340, whereas the yellowish fraction B represents the apo-brown membrane (lacking the chromophore of bacteriorhodopsin) containing the total amount of bacterio-opsin synthesized in nicotine cells. A (retinal-free) apo-purple membrane fraction is not detectable. Thus bacterio-opsin accumulating in nicotine cells does not form the crystalline protein lattice of the purple membrane.

Properties of the Brown Membrane

The brown membrane fraction exhibits two characteristic absorption maxima. One maximum at 419 nm must be assigned to a cytochrome b type protein because a heme group can be extracted from this fraction that, in the form of its pyridinium complex, shows characteristic absorption bands at 418, 524, and 554 nm in the difference spectrum (reduced/oxidized).[4] The absorption at 570 nm reflects the bacteriorhodopsin content of the brown membrane. The absorption spectrum of the apo-brown membrane lacks the 570-nm maximum. Addition of retinal (so-

[8] C. D. Howes and P. P. Batra, *Biochim. Biophys. Acta* **222**, 174 (1970).

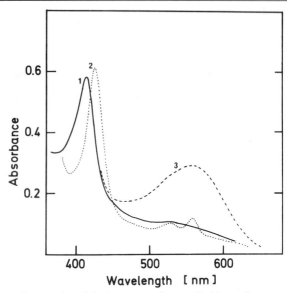

FIG. 2. Spectral properties of the brown membrane. (1) Absorption spectrum of the apo-brown membrane isolated from "nicotine cells." (2) Dithionite-reduced form of the apo-brown membrane. (3) Absorption spectrum of the apo-brown membrane after the addition of retinal.

lution in isopropanol) within a few minutes converts the bacterio-opsin of the apo-brown membrane to bacteriorhodopsin, thereby forming the 570-nm chromophore (Fig. 2).

In Vivo Conversion of the Bacterio-opsin to Bacteriorhodopsin and the Reconstitution of Purple Membrane Formation in Nicotine Cells

Retinal (or retinal analogs) added to a nicotine cell suspension readily diffuses into the cell membrane and reconstitutes the bacteriorhodopsin chromophore: Nicotine cells grown for 5 days are harvested by centrifugation, resuspended in basal salt solution (per 1000 ml: NaCl, 250 g; $MgSO_4\cdot7H_2O$, 20 g; KCl, 2 g) containing 0.5% L-alanine and 1 mM nicotine (final cell concentration: O.D.$_{578}$ = 1.5). Two-hundred-milliliter aliquots are shaken in 500-ml Erlenmeyer flasks at 39° and 100 rpm. Retinal (100 μl of a 10^{-2} M solution in isopropanol) is added, and after different times the cells are lysed. The membrane fractions are analyzed by sucrose density gradient centrifugation as described earlier. Within 1–2 hr most of the bacterio-opsin is converted to bacteriorhodopsin. "Crystallization" of bacteriorhodopsin to form purple membrane patches is initiated after about 2 hr. In the absence of L-alanine in the medium or in the presence of the uncoupler carbonyl cyanide *m*-chlorophenylhydrazone ($2 \times 10^{-5} M$), reconstitution of bacteriorhodopsin in the brown membrane proceeds as normal; however, no purple membrane formation can be detected by sucrose density gradient analysis. In contrast, the bacterio-opsin of the isolated apo-brown membrane was shown by X-ray analysis spontaneously to form a crystalline lattice after reconstitution with retinal.[9] This discrepancy to our *in vivo* experimental results might be explained by a "concentration effect": Nicotine cells contain a reduced amount of bacterio-opsin[6] that after conversion to bacteriorhodopsin is unlikely to form more than crystalline micropatches within the brown membrane. Such micropatches would not be detectable by sucrose density gradient analysis. A continued synthesis of bacterio-opsin appears necessary to increase the area of the crystalline patches, thus allowing fragmentation of the membrane and subsequent detection on sucrose density gradients. In addition, this interpretation would explain the energy dependence (i.e., protein synthesis dependence) found for the *in vivo* crystallization.

Pulse Labeling of Bacterio-opsin with Radioactive Amino Acids

Halobacteria will readily take up amino acids. Chase can be performed by dilution with unlabeled material. It is possible to follow by this technique the initial incorporation of newly synthesized bacterio-opsin into the brown membrane and its subsequent transfer to the purple membrane patches.

Efficient incorporation of radioactively labeled amino acids into bacterio-opsin is achieved by the following procedure: Cells (220 ml of growth medium) grown for 3–4 days are collected by centrifugation (10 min, 15,000 *g*) and resuspended in 200 ml of basal salt solution (see above), containing 0.5% L-alanine. This suspension is shaken at 40° for 3 hr. Ten ml of the cell suspension is centrifuged at 20,000 *g* for 5 min and the cells are resuspended in 400 μl basal salt solution, containing 0.5% L-alanine. The 25-fold concentrated cell suspension is gently stirred in a plastic test tube at 40° and illuminated by a 150-W slide projector with an OG 515 nm cutoff filter (Schott). After 5 min preincubation, a labeled amino acid (e.g., 10–40 μCi [^{35}S]methionine, 500 Ci/mmol) is added. Incorporation into bacterio-opsin is detectable after a pulse length as short as 60 sec (detection by SDS–polyacrylamide gel electrophoresis of cell lysates or of isolated membrane fractions).

---

[9] K. Hiraki, T. Hamanaka, T. Mitsui, and Y. Kito, *Biochim. Biophys. Acta* **536**, 318 (1978).

[52] Lipids of *Thermoplasma*[1]

By Thomas A. Langworthy

Thermoplasma acidophilum, an obligately thermoacidophilic archaebacterium,[2] grows optimally at pH 2 and 59°. It is capable of multiplying between pH 0.5–4 and 40–62°, although growth is very slow at the extremes. *Thermoplasma* is also a mycoplasma,[3] lacking a cell wall, thereby exposing its membrane directly to its hot acid environment. It is an aerobic heterotroph originally isolated by Darland *et al.*[4] from acidic regions of self-heating coal refuse piles, still its only known natural habitat. Details of the ecology, physiology, and phyletic relationship of *Thermoplasma* to *Sulfolobus acidocaldarius* (another extreme thermoacidophile), halophilic, and methanogenic bacteria may be found in recent reviews.[2,5–8]

Thermoplasma has several interesting features, some of which are analogous to the extremely halophilic bacteria. (1) *Thermoplasma* requires an extreme ionic environment, specifically H⁺ ions, not only for growth and reproduction, but also for maintenance of cellular integrity.[9,10] *Thermoplasma* is lysed by neutrality. Cells begin to leak cytoplasmic constituents at pH 5.5 and cells or membranes disintegrate unpon adjustment to pH 7–8. Other monovalent or divalent cations will not substitute for either growth or cell stabilization. (2) The intracellular pH of *Thermo-*

[1] Portions of the work described herein were supported by Grant PCM-7809351 from the National Science Foundation.

[2] G. E. Fox, E. Stackebrandt, R. B. Hespell, J. Gibson, J. Maniloff, T. A. Dyer, R. S. Wolfe, W. E. Balch, R. S. Tanner, L. J. Magrum, L. B. Zablen, R. Blakemore, R. Gupta, L. Bonen, B. J. Lewis, D. A. Stahl, K. R. Luehrsen, K. N. Chen, and C. R. Woese, *Science* **209**, 457 (1980).
[3] R. E. Buchanan and N. E. Gibbons, eds., "Bergey's Manual of Determinative Bacteriology," 8th ed., p. 952. Williams & Wilkins, Baltimore, Maryland, 1974.
[4] G. Darland, T. D. Brock, W. Samsonoff, and S. F. Conti, *Science* **170**, 1416 (1970).
[5] R. T. Belly, B. B. Bohlool, and T. D. Brock, *Ann. N.Y. Acad. Sci.* **225**, 94 (1973).
[6] T. A. Langworthy, *in* "Biochemistry of Thermophily" (S. M. Friedman, ed.), p. 11. Academic Press, New York, 1978.
[7] T. D. Brock, "Thermophilic Microorganisms and Life at High Temperatures," p. 92. Springer-Verlag, Berlin and New York, 1978.
[8] T. A. Langworthy, *in* "The Mycoplasmas" (M. F. Barile and S. Razin, eds.), Vol. 1, p. 495. Academic Press, New York, 1979.
[9] R. T. Belly and T. D. Brock, *J. Gen. Microbiol.* **73**, 465 (1972).
[10] P. F. Smith, T. A. Langworthy, W. R. Mayberry, and A. E. Hougland, *J. Bacteriol.* **116**, 1019 (1973).

plasma is near neutrality (pH 5.5–6.5).[11,12] Thus the internal and external H^+-ion concentrations are not in equilibrium, but H^+ ions are excluded from the cell interior, creating a pH gradient of approximately four units across the membrane. (3) Fatty acids are not synthesized. The apolar residues of the glycolipids and acidic lipids, and thereby the hydrophobic membrane interior, are comprised of diglyceryl tetraethers,[13–15] which provide *Thermoplasma* with a lipid "monolayer" membrane assembly rather than a normal lipid bilayer.[8,16,17]

Although di-*O*-phytanyl glycerol ethers[18] are present in small quantities (< 10%) in *Thermoplasma* lipids,[16,17] diglyceryl tetraethers account for about 3% of the cell dry weight and 15% of the membrane. The tetraethers are made up of two *sn*-2,3-glycerol molecules bridged through ether linkages by two identical C_{40} isopranoid-branched, fully saturated, terminal diols (Fig. 1). The C_{40} diols are made up of two C_{20} perhydrophytol units that are joined at the geminal 16,16′ position.[14–19] Thus the tetraethers themselves are the structural equivalent of bisphytanyl glycerol in which two molecules of di-*O*-phytanyl glycerol have been covalently condensed between the *gem*-dimethyl terminal ends of the phytanyl chains. In *Thermoplasma* the biphytanyl chains may also contain up to two cyclopentane rings (Fig. 1). The C_{40} hydrocarbon chains are present as identical pairs that may be either the acyclic biphytanyl ($C_{40}H_{82}$), monocyclic ($C_{40}H_{80}$), or bicyclic ($C_{40}H_{78}$), giving rise to three diglyceryl tetraether species in *Thermoplasma*: $C_{86}H_{172}O_6$, $C_{86}H_{168}O_6$, and $C_{86}H_{164}O_6$ having molecular weights of 1300, 1296, and 1292, respectively. The acyclic tetraether predominates (65%) with lesser amounts of the monocyclic (33%) and bicyclic (2%) species, although the distribution is influenced by the temperature of growth.

Described herein are basic procedures for the cultivation, isolation, and characterization of the diglyceryl tetraethers of *Thermoplasma*.

[11] J. C. Hsung and A. Haug, *Biochim. Biophys. Acta* **389**, 477 (1975).
[12] D. G. Searcy, *Biochim. Biophys. Acta* **451**, 278 (1976).
[13] T. A. Langworthy, *Biochim. Biophys Acta* **487**, 37 (1977).
[14] M. de Rosa, S. de Rosa, A. Gambacorta, L. Minale, and J. D. Bu'Lock, *Phytochemistry* **16**, 1961 (1977).
[15] L. L. Yang and A. Haug, *Biochim. Biophys. Acta* **573**, 308 (1979).
[16] T. A. Langworthy, *Life Sci. Res. Rep.* **13**, 417 (1979).
[17] T. A. Langworthy, *in* "Microbial Membrane Lipids" (S. Razin and S. Rottem, eds.), p. 45. "Current Topics in Membranes and Transport," Vol. 17. Academic Press, New York (1982).
[18] Kates *et al.*, Article [13], this volume.
[19] M. de Rosa, A. Gambacorta, B. Nicolaus, S. Sodano, and J. D. Bu'Lock, *Phytochemistry* **19**, 833 (1980).

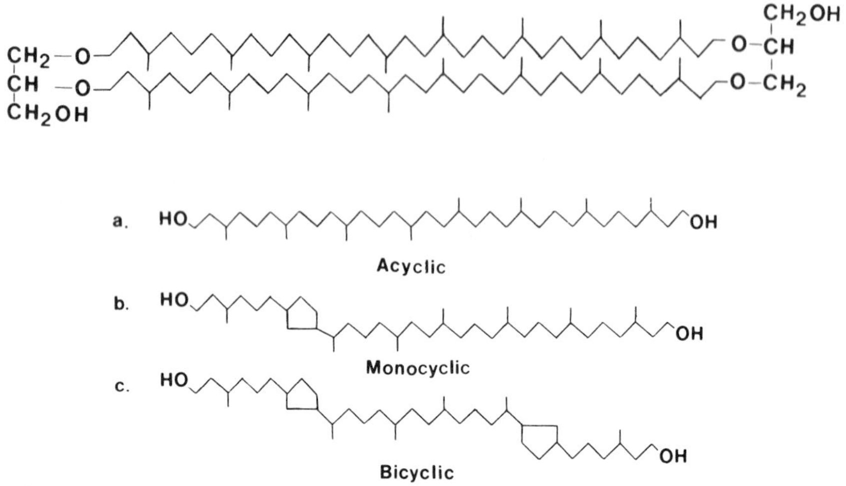

FIG. 1. Structure of the diglyceryl tetraethers of *Thermoplasma*. The identical pairs of C_{40} diols comprising the ether linkages may be either (a) $C_{40}H_{82}O_2$ (acyclic), (b) $C_{40}H_{80}O_2$ (monocyclic), or (c) $C_{40}H_{78}O_2$ (bicyclic). Shown here is the bisphytanyl glycerol tetraether species that possesses the acyclic hydrocarbon chains.

Culture Conditions

Thermoplasma acidophilum, isolate 122-1B2 (originally obtained from the American Type Culture Collection, ATCC 25905), is grown in a medium[5] that contains per 1000 ml: KH_2PO_4, 3.0 g; $MgSO_4$, 0.5 g; $CaCl_2 \cdot 2H_2O$, 0.25 g; $(NH_4)_2SO_4$, 0.2 g; yeast extract, 1.0 g followed by adjustment to pH 2 with 10 N H_2SO_4. After autoclaving, 10 g of glucose, contained in 25 ml, which has been sterilized separately, is added to give a final concentration of 1.0%.[20]

A 10% volume of a 24–48-hr liquid culture is used as the inoculum and cultures are incubated at 59°. Lesser inoculum sizes result in poor growth and reduced cell yields.[10] Cultures may be incubated statically, but growth and yields are maximal when cultures are shaken (vigorously forced aeration reduces growth). Cultures are conveniently incubated in a PsychroTherm incubator shaker (New Brunswick Scientific, Edison,

---

[20] *Thermoplasma* has an absolute requirement for yeast extract for growth. No other compounds have been found to substitute. In addition, growth and yields are variable, depending on the lot number of yeast extract employed. Several lot numbers of either Difco or Scott Laboratories yeast extract should be tested and those supporting the best growth reserved for preparation of *Thermoplasma* medium. The active component(s) in yeast extract required for growth appears to be polypeptide in nature. See P. F. Smith, T. A. Langworthy, and M. R. Smith, *J. Bacteriol.* **124**, 884 (1975).

N.J.) and stock culture tubes can be incubated in an electric block heater (Hallikainen Instruments, Richmond, Calif.). Incubation in covered water baths, is less convenient for routine work because of evaporation at 59°. Although the large-scale preparation of cells has been accomplished in heated 60-liter polyethylene bottles equipped with reflux condensers[21] and up to 115 liters grown in garbage cans,[22] for most purposes sufficient quantities of cells can be conveniently grown in 2-liter volumes of medium, contained in 3-liter flasks, which are shaken at 120 rpm in an incubator shaker. After 40–48 hr incubation, at which point early stationary phase is reached (O.D.$_{540}$ = 0.3–0.4), cultures are brought to room temperature. After the desired quantity of cells has been accumulated, cells are concentrated in a Sharples centrifuge, sedimented, and washed two times with deionized water at 10,000 g for 10 min in a Sorvall RC2-B centrifuge. The wet cells can be used directly or lyophilized. Under these conditions the generation time for *Thermoplasma* is about 5 hr and the yield of cells approximates 50 mg dry weight per liter of medium.

Preservation of stock cultures in the laboratory presents some problems. Liquid cultures lose viability in 2–3 weeks at room temperature and much more rapidly when refrigerated. Viable cells can rarely be recovered after lyophylization and neutralization is precluded by cell lysis. Cells can sometimes be recovered from the frozen state. In addition, growth on agar is difficult, because of both hydrolysis and dehydration.[5] The most reliable procedure is continuous passage of 24–48 hr statically grown cultures every 2–3 days, although cultures may more conveniently be held at room temperature for up to 5 days between transfers.

Isolation of Diglyceryl Tetraethers

General Procedure. Diglyceryl tetraethers are recovered from the nonsaponifiable fraction obtained by acid methanolysis of either whole cells, total lipids, or individual complex lipid species.[13,23] Acid methanolysis is carried out in either 2.5% anhydrous methanolic-HCl prepared by bubbling 2.5 g of HCl gas into 100 ml of redistilled methanol or in 1 N HCl freshly prepared from concentrated aqueous HCl. Larger samples, such as whole cells or more than 50 mg of lipid, are heated under reflux for 18 hr. Smaller samples are heated at 100° for 18 hr in Teflon-lined screw cap tubes. After cooling, a 10% volume of water is added and the methanolysate extracted three times with an equal volume of n-hexane or petro-

[21] S. Sturm, U. Schönefeld, W. Zillig, D. Janekovic, and K. O. Stetter, *Zentralbl. Bakteriol., Abt. 1: Orig. [Reihe] C* **1**, 12 (1980).

[22] D. G. Searcy, *Biochim. Biophys. Acta* **395**, 535 (1975).

[23] T. A. Langworthy, P. F. Smith, and W. R. Mayberry, *J. Bacteriol.* **112**, 1193 (1972).

leum ether. The combined upper n-hexane phase is washed with 10% Na_2CO_3, and the solvent is removed on a rotary evaporator or, for smaller volumes, under a stream of N_2. The residue is taken up in a small volume of n-hexane or chloroform and the tetraethers are purified by thin-layer chromatography.

Thin-layer chromatography (TLC) is carried out on 0.25-mm layers of silica gel H activated by heating at 105° for 1 hr. Samples (1–15 mg) containing tetraethers are applied 2 cm from the edge of a 20 × 20-cm plate as a series of closely spaced spots. Plates are developed in n-hexane–diethyl ether–acetic acid (80:20:1, v/v). Tetraethers migrate with an R_f 0.16–0.25. In this solvent di-O-phytanyl glycerol migrates at R_f 0.6, whereas fatty acid esters and neutral products migrate near the solvent front. Plates may also be developed in chloroform–diethyl ether (9:1, v/v) with tetraethers, di-O-phytanyl glycerol and fatty acid esters having R_f values of 0.45–0.55, 0.90, and 0.95, respectively. Plates are developed until the solvent front has reached 18–19 cm, which requires about 30 min. Plates are removed and allowed to air-dry for several hours in a hood or until solvent has evaporated. Tetraethers are located by placing the plate in a tank containing iodine crystals. Tetraethers may also be detected by spraying with a fine mist of water with the tetraethers appearing as a white band on an opalescent background. The band is marked and the plate is placed in a hood until the iodine vapors or water have evaporated from the plate. The band is scraped from the plate and the tetraethers are eluted from the silica gel with 15 ml of chloroform–methanol (2:1, v/v) on a sintered glass filter. The tetraethers are recovered after evaporation of the solvent as a clear viscous oil. If tetraethers are not to be recovered from the plates, components may be detected by spraying with 50% methanolic-H_2SO_4 followed by charing at 105°.

Individual tetraether species are only slightly resolved in the above solvent systems on silica gel H that give rise to the slight extension of the tetraether spot. However, de Rosa et al.[14,19] have reported resolution of tetraether species by careful TLC on Kieselgel 60-F254 (Merck) developed in chloroform–diethyl ether (9:1, v/v). The acyclic (R_f 0.55), monocyclic (R_f 0.49), and bicyclic (R_f 0.43) tetraethers may still overlap but may be more fully resolved by multiple elution and rechromatography.

Tetraether mixtures have been also reported[19] to be resolved on 2.5 × 40-cm glass columns packed in chloroform with 80 g of Kieselgel (70-230 mesh) previously activated by heating at 150°. Sample (0.3 g), which was preabsorbed on 5 g of Kieselgel, was applied to the column and tetraethers eluted with 3 liters of a linear gradient of 0–10% diethyl ether in chloroform. The acyclic, monocyclic, and bicyclic tetraethers were reported to elute at chloroform–diethyl ether ratios of 98:2, 97:3, and 95:5, respectively.

Isolation from Whole Cells

Either wet or dry cells (0.5–1.0 g) are refluxed in 1 *N* methanolic HCl (50 ml) for 18 hr in a 250-ml round-bottomed flask. After cooling, an equal volume of *n*-hexane is added and the mixture is transferred to a separatory funnel. Water (10 ml) is added and the tetraethers are recovered as described earlier.

For rapid detection, tetraethers can be recovered directly from the wet cell pellets of 50- to 100-ml *Thermoplasma* cultures.[23] Radiolabeled cultures may also be employed by adding to the culture either 2–5 μCi of [2-^{14}C]acetate, yielding fully labeled tetraethers, or [2-^{14}C]mevalonic acid dibenzylethylenediamine (DBED) salt, which specifically labels the C_{40}-isopranoid chains. After washing, cell pellets are taken up in 0.5 ml of water and transferred to a 15- × 120-mm screw cap tube. Methanolic HCl (5 ml) is added, tubes capped, and heated at 100° for 18 hr. After cooling, 1.0 ml of water is added followed by *n*-hexane extraction (5 ml). After workup the entire sample is applied as a single spot on the TLC plate. Enough material is present such that, following development, tetraethers can be seen directly on the plate or detected as described earlier. Radiolabeled ethers are detected by exposure to X-ray film (Kodak, X-Omat R film, XR-5). When using [2-^{14}C]mevalonate, this procedure may serve as rapid *presumptive* evidence for the presence of di-*O*-phytanyl glycerol or diglyceryl tetraethers in an organism.

Isolation from Extracted Lipids

Tetraethers may be isolated by acid methanolysis of the total extractable lipids,[24] fractionated glycolipid or phospholipid classes, or individual complex lipids. The procedure given here is for 1–5 g dry weight of cells.

Total lipids are extracted from freeze-dried cells by stirring at room temperature with 30 vol/cell weight of chloroform–methanol (2:1, v/v). The extract is filtered through sintered glass and the cell residue is reextracted. Combined extracts are taken to dryness on a rotary evaporator.

Nonlipid contaminants are removed by passage of the lipid, dissolved in chloroform–methanol–water (60:30:4.5, v/v), through a 1.0 g column of Sephadex G-25 packed in the same solvent in a 50-ml burette plugged with glass wool.[25] Prior to column preparation, the Sephadex is soaked in water to remove fines, filtered, and dried with acetone. After elution of

[24] Total lipids account for about 3% of the cell dry weight and are comprised of about 17% neutral lipids, 25% glycolipids, and 57% phospholipids. Individual complex lipids have not been well characterized.[23] The phospholipids, however, exist as phosphoglycolipids containing glycerolphosphate and carbohydrate asymmetrically substituted on opposite sides of the tetraether molecule (T. A. Langworthy, unpublished).

[25] M. A. Wells and J. C. Dittmer, *Biochemistry* **2**, 1259 (1963).

the first 15 ml of solvent at a flow rate of about 0.1 ml/min, 5 ml of chloroform–methanol (2:1, v/v) is passed through the column and combined with the first.

The bulk Sephadex eluate is added directly to a column (2.5 × 22 cm) of diethylaminoethyl (DEAE)-cellulose in the acetate form, prepared with chloroform–methanol (7:3, v/v), according to Rouser et al.[26] The column is eluted with 500 ml of chloroform–methanol (7:3, v/v) for combined neutral lipids and glycolipids; with 250 ml chloroform–acetic acid (3:1, v/v) for removal of pigments; with 250 ml methanol for removal of acetic acid from the previous solvent; and finally with 500 ml of chloroform–methanol (7:3, v/v) containing 10 ml ammonium hydroxide and 2 g of ammonium acetate for phospholipids. To remove salts from the phospholipid fraction, the eluate is dried, taken up in 100 ml of the lower phase of chloroform–water (2:1:0.3, v/v) and partitioned two to three times against 30 ml of the upper phase of the same solvent.

Neutral lipids and glycolipids are separated on a silicic acid column (2 × 8-cm column of Unisil, Clarkson Chemical Co., Williamsport, PA) by elution with 250 ml of chloroform for neutral lipids and 250 ml chloroform–methanol, 2:1 (v/v) for glycolipids.

The fractionation into lipid classes employs DEAE-cellulose, since the phospholipids of *Thermoplasma* may be significantly contaminated by glycolipids when fractionation is carried out solely on a silicic acid column using chloroform (neutral lipids), acetone (glycolipids), and methanol (phospholipids) as eluting solvents. An alternative to the previously described procedure is fractionation, first on silicic acid with chloroform (neutral lipids) and then methanol (combined glycolipids and phospholipids), followed by fractionation of the methanol eluate on DEAE-cellulose to separate the glycolipids and phospholipids.

Fractionated lipid classes are taken up in chloroform-methanol (2:1, v/v) and separated into individual components by TLC on silica gel H developed in the following solvent systems. Neutral lipids are separated by development in the two-step solvent system, isopropyl ether–acetic acid (96:4, v/v), followed by n-hexane–diethyl ether–acetic acid (90:10:1, v/v) in the same direction. Glycolipids are separated using chloroform–methanol (9:1, v/v) and phospholipids by development in chloroform–methanol–water (65:25:4, v/v) or by chloroform–methanol–acetic acid–water (100:20:12:5, v/v). Components are detected by iodine vapors, scraped, and eluted from the silica gel using chloroform–methanol (2:1, v/v).

The same procedures are employed for extraction of isotopically la-

[26] G. Rouser, G. Kritchevsky, A. Yamamoto, G. Simon, C. Galli, and A. J. Bauman, this series, Vol. 14, p. 272.

beled lipids from small cultures (50–100 ml), with two exceptions. Lipids are extracted from wet cell pellets with 25 ml of chloroform-methanol (2:1, v/v) and the lipids are fractionated on 1 × 2-cm DEAE-cellulose and silicic acid columns using 25 ml of the eluting solvents.

Degradation and Preparation of C_{40} Hydrocarbon Chain Derivatives

Tetraethers are degraded to the C_{40} alkyl dichlorides and glycerol by boron trichloride or degraded by hydriodic acid to the C_{40} alkyl diiodides and iodinated glycerol. Alkyl iodides are either reduced to the C_{40} alkanes or acetylated to form C_{40} diacetates, which in turn may be saponified to yield the C_{40} diols. Except where noted, the following procedures are modifications of those described by Kates et al.[27] and employ 1–10 mg of tetraether or derivatives using 13 × 100-mm Teflon-lined, screw cap tubes.

Alkyl Diiodides. Tetraether is dried in the bottom of a screw cap tube and 1.0 ml of 57% hydriodic acid is added. Tubes are flushed with N_2, capped and heated at 100° for 18 hr. The diiodides are extracted twice with 3 ml of n-hexane. The n-hexane phase is washed in succession with water, 10% Na_2CO_3 and 50% $Na_2S_2O_3$, and then dried under N_2.

Alkanes. To the alkyl diiodides obtained as above and dried in the bottom of a screw cap tube are added 0.1–0.2 g of zinc dust followed by 1.0 ml of acetic acid.[28] The loosely capped tube is heated at 100° for 18 hr. After cooling, 10% Na_2CO_3 is added carefully until the acetic acid is neutralized followed by 1 ml of water. The alkanes are extracted twice with 3 ml of n-hexane, which is then washed with 1.0 ml of 10% Na_2CO_3, and the solvent is removed under N_2.

Alkyl Diacetates.
To the alkyl diiodides, dried in the bottom of a 100-ml round-bottomed flask is, added 200 mg of silver acetate and 15 ml of acetic acid. The mixture is stirred and heated under reflux for 24 hr. The mixture is diluted with 50 ml of diethyl ether followed by centrifugation to sediment silver iodide. The supernatant fluid is transferred to a separatory funnel and washed successively with 50 ml water, saturated $NaHCO_3$, and 50% $Na_2S_2O_3$, and the solvent is finally removed in vacuo.

Alkyl Diols. Alkyl diols are generated from the alkyl diacetates by saponification in 2.0 ml of 0.5 M KOH in methanol at 100° for 3 hr. The alkyl diols are recovered by extracting twice with 3.0 ml of n-hexane and solvent removed by drying under N_2.

[27] M. Kates, L. S. Yengoyan, and P. S. Sastry, Biochim. Biophys. Acta 98, 252 (1965).
[28] R. V. Panganamola, C. F. Sievert, and D. G. Cornwell, Chem. Phys. Lipids 7, 336 (1971).

Alkyl Dichlorides. To the tetraether dried in a 13 × 100-mm screw cap tube is added 0.5 ml of chloroform and 0.5 ml of liquified boron trichloride. Liquified boron trichloride is prepared in small quantities by blowing boron trichloride gas into the bottom of a test tube deeply seated in a dry ice–acetone bath ($-70°$). After addition of the boron trichloride, the tube is capped tightly and the contents are mixed and allowed to stand at room temperature overnight. Reagent is evaporated under a stream of N_2. Methanol (0.5 ml) is added and again evaporated. Water (1.0 ml) is added to the residue and the alkyl dichlorides are extracted with 3 ml of *n*-hexane.

The alkyl chain derivatives may be further purified if desired by TLC on silica gel H developed in *n*-hexane for alkyl diiodides and dichlorides or chloroform–diethyl ether (9 : 1, v/v) for alkyl diols and diacetates.

Analysis of C_{40} Hydrocarbon Chains

Hydrocarbon chains and derivatives may be analyzed by gas-liquid chromatography isothermally at 320° on 183-cm × 6-mm columns packed with either 5.5% SE-30 or 3.0% OV-11 on 80–100 mesh Gas Chrom Q (Applied Science Laboratories, State College, PA). Alkanes are the most convenient derivatives for identification. Shown in the accompanying table are the relative retention times (*n*-hexatriacontane, $C_{36}H_{74}$ = 1.00; elution time 7.11 and 7.45 min on the above columns, respectively) and the equivalent chain lengths for the three C_{40} hydrocarbons in *Thermoplasma* tetraethers. The equivalent chain length values are obtained by reference to logarithmic plots of the retention time versus chain length for an *n*-C_{20}–C_{44} hydrocarbon mixture. Retention times increase as the number of cyclopentyl rings increases. Relative percentages of the alkanes may be computed from the area under the peaks. Since each tetraether

EQUIVALENT CHAIN LENGTH (ECL) AND RELATIVE RETENTION TIME (*n*-HEXATRIACONTANE, $C_{36}H_{74}$ = 1.00) OF THE $C_{40}H_{82-78}$ ALKANES FROM THE DIGLYCERYL TETRAETHERS OF *Thermoplasma*[a]

	3.0% OV-11		5.5% SE-30	
Hydrocarbon	ECL	$t_{R'36}$	ECL	$t_{R'36}$
$C_{36}H_{74}$	36	1.00	36	1.00
$C_{40}H_{82}$ (acyclic)	34.8	0.79	35.1	0.87
$C_{40}H_{80}$ (monocyclic)	36.4	1.09	36.5	1.10
$C_{40}H_{78}$ (bicyclic)	38.1	1.53	27.9	1.44

[a] Gas-liquid chromatography at 320° on 3.0% OV-11 or 5.5% SE-30.

species has an identical pair of alkyl chains, the relative proportion of alkane derivatives is equivalent to the distribution of tetraether species in the original tetraether mixture. In addition to serving as a marker for relative retention times, n-hexatriacontane may be gainfully employed as an internal standard. Columns packed with 5% OV-101 or 5% SP2100 give separations similar to that shown for 5.5% SE-30.

Similarly, the alkyl diacetates may be analyzed directly, but the alkyl diols must first be converted to either the trimethylsilyl or trifluoroacetate derivatives. Trimethylsilyl derivatives are prepared by incubation for 30 min in a mixture (0.1–0.3 ml) of pyridine-hexamethyldisilazane-N,O-bis-(trimethylsilyl)trifluoroacetamide–trimethylchlorosilane (2:2:1:1, v/v) and the reagent is removed under N_2. Trifluoroacetates are prepared in the same way by incubation in trifluoroacetic anhydride. All of the diol derivatives elute with the same pattern as the alkanes but with longer retention times.

Alkyl diiodides and dichlorides are not as useful for gas chromatographic analysis because of loss of the halides at the high temperatures employed and unstable double bond formation. This leads to the formation of broad as well as multiple peaks.

Quantitation of Diglyceryl Tetraethers

Tetraethers may be quantitated gravimetrically after drying to a constant weight *in vacuo*. Based on the average molecular weight distribution of the three tetraether species, 1.298 mg equals 1.0 μmol of the tetraethers from *Thermoplasma*.

Tetraethers can be estimated, although with variable accuracy, by peracetylation of the tetraether followed by assay of the acetate ester equivalents. Tetraethers (1–2 mg) are dried in a screw cap tube and 1.0 ml of dry pyridine-redistilled acetic anhydride (3:1, v/v) is added. The tube is heated at 100° for 1 hr, the solvent is evaporated under N_2, and the contents are dried *in vacuo*. The acetate ester equivalents are determined by the ferric hydroxamate procedure fully detailed by Kates.[29] If a single ester equivalent standard is used, such as methyl stearate, the tetraether concentration is equal to only half of the colorimetric value obtained because of the presence of two ester equivalents in the tetraether molecule. If a diester, such as dipalmitin, is employed as a standard, the color values obtained are proportional.

A quantitative estimate may also be made by analysis of the glycerol released following degradation of 2–3 mg of tetraether in liquified boron trichloride. After evaporation of the reaction solvent under a stream of

[29] M. Kates, *Lab. Tech. Biochem. Mol. Biol.* **3**, 269 (1972).

N_2, 1.0 ml of water is added. The glycerol released is assayed enzymatically by measuring the reduction of NAD^+ by α-glycerol-3-phosphate dehydrogenase (EC 1.1.1.8) after phosphorylation by glycerol kinase (EC 2.7.1.30), as fully described by Wieland.[30] The diglyceryl tetraether concentration is equivalent to half the amount of glycerol detected.

Comments

Diglyceryl tetraethers are found also in the extreme thermoacidophile, *Sulfolobus acidocaldarius*.[14,19,31] These tetraethers are more specialized than in *Thermoplasma*. The hydrocarbon chains are more highly cyclized, having higher proportions of monocyclic and bicyclic hydrocarbons, and they contain tricyclic and tetracyclic chains as well. Additionally, a second tetraether type is present in which a branched nonitol replaces one of the glycerol molecules in the tetraether structure.[32] Whereas the extremely halophilic bacteria, as well as the methanogenic bacteria, possess di-*O*-phytanyl glycerol ethers, certain species of methanogens, notably *Methanobacterium* and *Methanospirillum,* contain, in addition, up to 62% diglycerol tetraethers.[33,34] The only tetraether type so far detected in methanogens is the acyclic bisphytanyl tetraether. The assay and identification of phytanyl ethers and derived tetraethers therefore provides a valuable distinguishing chemical marker for detecting members of this unusual group of organisms.

[30] O. Wieland, *in* "Methods of Enzymatic Analysis" (H. U. Bergmeyer, ed.), p. 211. Academic Press New York, 1963.

[31] T. A. Langworthy, W. R. Mayberry, and P. F. Smith. *J. Bacteriol.* **119,** 106 (1974).

[32] M. de Rosa, S. de Rosa, A. Gambacorta, and J. D. Bu'Lock, *Phytochemistry* **19,** 249 (1980).

[33] T. G. Tornabene and T. A. Langworthy, *Science* **203,** 51 (1979).

[34] S. C. Kushwaha, M. Kates, G. D. Sprott, and I. C. P. Smith. *Biochim. Biophys. Acta* **664,** 156 (1981).

[53] Isolation of the White Membrane of Crystalline Bacterio-opsin from *Halobacterium halobium* R_1mW Lacking Carotenoid

By YASUO MUKOHATA and YASUO SUGIYAMA

Introduction

Under low tension of oxygen, some strains of *Halobacterium halobium* such as R_1 and R_1M_1 form purple membrane patches[1,2] in their cytoplasmic membranes. The purple membrane is composed of three-quarters parts of bacteriorhodopsin by weight and one-quarter lipid.[1,3] Bacteriorhodopsin involves retinal as the chromophore.[1] A spontaneous mutant R_1mW[4,5] isolated from R_1M_1 lacks pathways for carotenoid synthesis and thus shows yellowish white color because of respiratory enzymes. R_1mW does not respond to light, whereas R_1 and R_1M_1, which involve both bacteriorhodopsin and halorhodopsin,[5] and R_1mR,[4,6] which involves only the latter, all respond to light by changing the membrane potential, cellular ATP level, and the external pH of cell suspensions. After all-*trans*-retinal is added to an R_1mW suspension, however, the cells become purple and develop the activities to respond to light.[7] The light-induced pH response thus developed by retinal reveals that R_1mW involves haloopsin (as indicated by the pH increase due to halorhodopsin formed[4-6]) and bacterioopsin (pH decrease due to bacteriorhodopsin formed[8]). The former is known to be in very small quantity, whereas the latter is found by electronmicroscopy as white membrane patches[7,9] in freeze-fractured cytoplasmic membranes. The white membrane is composed of bacterio-opsin[7] in a hexagonal crystalline array almost identical to that of bacteriorhodopsin in the purple membrane. The white membrane can be isolated and shown[7] to be a material suitable for the study of molecular dynamics of bacteriorhodopsin proton pump.

[1] D. Oesterhelt and W. Stoeckenius, *Nature (London), New Biol.* **233**, 149 (1971).

[2] A. E. Blaurock and W. Stoeckenius, *Nature (London), New Biol.* **233**, 152 (1971).

[3] S. C. Kushwaha and M. Kates, *Biochim. Biophys. Acta* **316**, 235 (1973).

[4] A. Matsuno-Yagi and Y. Mukohata, *Arch. Biochem. Biophys.* **199**, 297 (1980).

[5] Y. Mukohata, A. Matsuno-Yagi, and Y. Kaji, *in* "Saline Environment" (M. Masui *et al.*, eds.), p. 27. Univ. of Tokyo Press Tokyo, 1980.

[6] A. Matsuno-Yagi and Y. Mukohata, *Biochem. Biophys. Res. Commun.* **78**, 237 (1977).

[7] Y. Mukohata, Y. Sugiyama, Y. Kaji, J. Usukura, and E. Yamada, *Photochem. Photobiol.* **33**, 593 (1981).

[8] D. Oesterhelt and W. Stoeckenius, *Proc. Natl. Acad. Sci. U.S.A.* **70**, 2853 (1973).

[9] J. Usukura, E. Yamada, and Y. Mukohata, *Photochem. Photobiol.* **33**, 475 (1981).

Culture of R_1mW Cells

A single transparent pale yellow colony of R_1mW on an agar plate is transferred in a test tube containing 5 ml of the culture medium (below) and incubated for 1 week at 37°. This is again transferred in 150 ml of culture medium in a 300-ml Erlenmeyer flask with an aluminum cap, cultured for 1 week at 37° with moderate stirring, then used as an inoculum for an 8-liter culture. The culture medium contains in 8 liters, 2 kg NaCl, 16 g KCl, 80 g $MgSO_4\cdot7H_2O$, 1.6 g $CaCl_2\cdot2H_2O$, 24 g trisodium citrate·$2H_2O$ and 26.4 g peptone (Daigo Eiyo Co., Osaka) at pH 7.4 adjusted by a glass electrode pH meter with NaOH. The medium in a 10-liter glass bottle is autoclaved and then placed on a large magnetic stirrer to give a moderate flow in the medium. Temperature is controlled at 35–40°C by a dip-in temperature controller and a heater (200 W) for a tropical fish aquarium. Aeration is provided by bubbling filtered air from an air pump at a rate less than 60 liters/hr. Illumination is not required but still convenient to watch the growth.

After inoculation, cells start to grow with a doubling time of about 13 hr for the initial 30 hr; then the growth slows down (Fig. 1). The bacterio-opsin content in namomoles per milligram protein of cell lysate increases after the cell growth reaches the end of the logarithmic phase and remains almost constant after 50 hr of culturing. Cells at the stationary phase are then harvested by centrifugation at 5000 g for 30 min and the pellet is suspended in the basal salt solution (culture medium minus citrate and peptone, pH 7.4) to be about 100 ml, which gives a concentration of 20–30 mg of lysate protein/ml.

Isolation of the White Membrane

The following procedures are to be run at about 5° except for centrifugation at 10°. To 100 ml of a stock suspension of R_1mW from an 8-liter culture, 2 mg of DNase I (Sigma, DN-CL) is added and the mixture is dialyzed overnight in a cellophane tube against 1 liter solution of 0.1 M NaCl and 10 mM Tris–HCl (pH 7.6) which is once replaced.

Clear lysate in faint brown is centifuged at 60,000 g for 30 min and the supernatant is discarded. The pellet is suspended in 360 ml of 0.1 M NaCl and centrifuged at 60,000 g for 1 hr. This washing is repeated to take off cell membrane (corresponding to red membranes in R_1, R_1M_1, and R_1mR) debris separable from the white membrane. The purity of the white membrane largely depends on this washing process. The pellet is suspended in distilled water and centrifuged at 60,000 g for 3 hr. The crude white membrane pellet (in clear faint brown) is then suspended in 12 ml of distilled

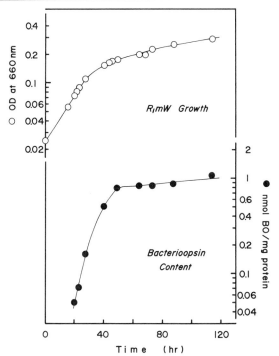

FIG. 1. Growth of *H. halobium* R₁mW and its production of bacterio-opsin. The growth (○) was followed by the optical density of the culture at 660 nm. The bacterio-opsin content (●) was titrated with all-*trans*-retinal by the absorbance increase at 560 nm (see text for details).

water and 3 ml each of the suspension is placed on top of the sucrose density gradient tube (Spinco SW27), where from 25 to 55% (5% step) sucrose solutions are layered. The sucrose density gradient centrifugation is run at 25,000 rpm for 20 hr. The white membrane is collected at a buoyant density of a little less than 1.18 g/ml.

In Fig. 2 electrophoretic patterns on SDS–polyacrylamide gels are shown of the white membrane preparations before and after the density gradient centrifugation. The R_f value of the main peak (bacterio-opsin) is 0.64 on 10% gels in the presence of 0.1% SDS. In Fig. 3 the absorption spectra of the white membrane and the purple membrane formed after addition of all-*trans*-retinal are shown. The latter spectrum corresponds to that of the dark-adapted bacteriorhodopsin. Contamination of 410-nm components (mainly cytochrome *b*) is inevitable. The contaminants are estimated spectrophotometrically to be in an order of several mol % of bacterio-opsin.

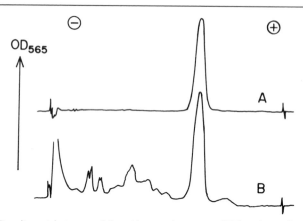

FIG. 2. Densitometric traces of the white membranes on SDS–polyacrylamide gels after electrophoresis. The white membrane specimens were sampled before (B) and after (A) sucrose density gradient centrifugation, heated at 100° for 5 min in the presence of 2% SDS and 2% mercaptoethanol, and then electrophoresed on polyacrylamide gels (10%) in the presence of SDS (0.1%). Proteins were stained with Coomassie blue and traced densitometrically at 565 nm.

Comments

For preservation of the strain, either slants or plates can be used. Stock suspensions in 20–30 mg protein/ml are diluted about 10^8 times by the basal salt solution and then 0.1 ml of this diluted suspension is poured on an agar plate (1.8% agar in the culture medium) 10 cm in diameter. After the plate is incubated for 10 days at 37°, 100–1000 colonies/plate will be found. The plate is then stored at 4° in the dark.

Cells also grow on arginine fermentation.[10] By an addition of 0.5% arginine hydrochloride to the culture medium, the initial growth rate becomes much higher than the control, but bacterio-opsin content (per mg lysate protein) decreases one-third of the control. Not only sufficient aeration but also fermentative energy supply seem to depress bacterio-opsin formation in R_1mW as much as bacteriorhodopsin in R_1L_3.[8]

Bacterio-opsin content is determined spectrophotometrically after addition of sufficient amount of retinal to cell lysate. Cells are sampled from the culture, centrifuged at 5000 g for 30 min, resuspended in a small volume of the basal salt solution, and then lysed by dialysis (in cellophane tube) against distilled water for 20 hr. One part of the lysate is used to assay for proteins by the Lowry procedure and all-*trans*-retinal (about 0.5 mM in ethanol) is added to the other. The amount of added retinal

[10] R. Hartmann, H.-D. Sickinger, and D. Oesterhelt, *Proc. Natl. Acad. Sci. U.S.A.* **77**, 3821 (1980).

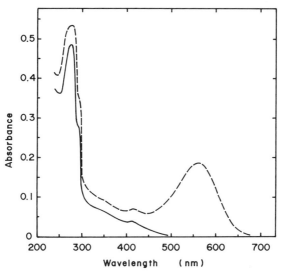

FIG. 3. Absorption spectra of the white and purple membranes. The white membrane (solid line) was examined in 1 M NaCl and 10 mM phosphate buffer at pH 7.0 and then converted to the purple membrane (broken line) by incubation with 3.5μM all-*trans*-retinal at 25° for 30 min in the dark.

should be sufficient so as to give an absorption peak at 381 nm for free retinal. The final retinal concentration may be 2 nmol/mg protein. Bacterio-opsin is estimated from the absorbance increase at 560 nm for newly formed dark-adapted bacteriorhodopsin using the molar extinction coefficient of 53,000 M^{-1} cm^{-1}.[7]

Some lysates give a white membrane band at a buoyant density of nearly 1.17 g/ml. After all-*trans*-retinal was added to the lysate and incubated, two purple bands appear at 1.17 and 1.18 g/ml. If retinal is added to cells prior to lysis, incubated, and then lysed, only one purple band is found at 1.18 g/ml. All of these bands, however, only show a single peak for bacterio(rhod)opsin on SDS–polyacrylamide gels by electrophoresis. This would suggest that the rigidity of the white membrane (or packing of bacterio-opsin and lipid in the isolated white membrane) somewhat differs from that of the isolated purple membrane. Factors affecting the rigidity may involve pH, temperature, and ionic strength of preparation media and have not been pointed out.

[54] Identification and Detection of Electron Transfer Components in Methanogens

By Jack R. Lancaster, Jr.

Introduction

The methanogenic bacteria are members of a class of organisms called the Archaebacteria. Along with the halobacteria and the thermoacidophiles *Sulfolobus* and *Thermoplasma*, this group has recently been postulated to represent a unique line of phylogenetic descent, the members of which are only distantly related to the other prokaryotes.[1] One common feature of these organisms is that their respective habitats are not unlike those that existed in Archaen times.

The methanogenic bacteria are characterized by the ability to derive all cellular energy requirements from the reduction of CO_2 by H_2 to form CH_4.[2] Most species can grow autotrophically, i.e., possess the additional ability to derive all cellular carbon from the fixation of CO_2.

Recent evidence[3-7] suggests a role for transmembrane ion gradients in methanogenesis, implicating the involvement of a membrane-bound electron transfer chain in energy coupling to methane production. Such a possibility has in the past been difficult to rationalize in light of the fact that methanogens contain neither cytochromes (however, see Ref. 8 for a report of a *b*-type cytochrome in *Methanosarcina barkeri*) nor quinones, which are present in all other organisms that carry out electron transport phosphorylation.[9]

This article describes recent results in the detection of such electron transport components in methanogens, with particular emphasis on the

[1] G. E. Fox, E. Stackebrandt, R. B. Hespell, J. Givson, J. Maniloff, T. A. Dyer, R. S. Wolfe, W. E. Balch, R. S. Tanner, L. J. Magrum, L. B. Zablen, R. Blakemore, R. Gupta, L. Bonen, B. J. Lewis, D. A. Stahl, K. R. Luehrsen, K. N. Chen, and C. R. Woese, *Science* **209**, 4576 (1980).

[2] R. S. Wolfe and I. S. Higgins, *Int. Rev. Biochem.* **21**, 267 (1979).

[3] H. J. Doddema, T. J. Hutten, C. van der Drift, and G. D. Vogels, *J. Bacteriol.* **136**, 19 (1978).

[4] D. O. Mountfort, *Biochem. Biophys. Res. Commun.* **85**, 1346 (1978).

[5] F. D. Sauer, J. D. Erfle, and S. Mahadevan, *Biochem. J.* **178**, 165 (1978).

[6] H. J. Doddema, C. van der Drift, G. D. Vogels, and M. Veenhuis, *J. Bacteriol.* **140**, 1081 (1979).

[7] F. D. Sauer, S. Mahadevan, and J. D. Erfle, *Biochem. Biophys. Res. Commun.* **95**, 715 (1980).

[8] W. Kühn, K. Fiebig, R. Walther, and G. Gottschalk, *FEBS Lett.* **105**, 271 (1979).

[9] R. K. Thauer, K. Jungermann, and K. Decker, *Bacteriol. Rev.* **41**, 100 (1977).

METHODS IN ENZYMOLOGY, VOL. 88

use of the technique of liquid helium electron paramagnetic resonance spectroscopy. Other soluble electron transfer components, including the unique cofactors F_{420} and coenzyme M as well as soluble oxidoreductase enzymes, are not covered here; for this the reader is referred to the recent literature.[2,9,10]

EPR Detection of Components in *M. bryantii*

Low-temperature EPR spectroscopy is undoubtedly the method of choice for studying electron transfer components in methanogens. Since methanogens contain no highly absorbant species such as hemes, optical spectroscopy is of only limited use. EPR has proved extremely useful in establishing structure–function relationships in other electron transfer systems, providing information that can be obtained by virtually no other technique.[14,15]

Table I describes the EPR-observable components found to date in the species *Methanobacterium bryantii,* which are described in more detail later.

Iron–Sulfur Centers. The soluble fraction exhibits a low-temperature signal characteristic of a "HiPip"-type iron sulfur center (paramagnetic in the oxidized state), being a rather isotropic signal with a principal upward feature at $g = 2.02$. This species is precipitated with 65% saturation in ammonium sulfate, is nondialyzable, and is reducible by dithionite. Very similar signals are exhibited by hydrogenases from a variety of bacteria.[16,17] Hydrogenase activity has been detected in methanogens.[17]

The addition of dithionite to the membrane fraction results in the appearance of several low-temperature features near $g = 2.0$ attributable to at least two "$g = 1.94$"-type iron–sulfur centers. It is impossible to tell at this point from the EPR spectra how many centers there are, since there

---

[10] W. E. Balch, G. E. Fox, L. J. Magram, C. R. Woese, and R. S. Wolfe, *Micr. Rev.* **43,** 260 (1979).

[11] J. R. Lancaster, *FEBS Lett.* **115,** 285 (1980).

[12] T. W. Kirby, J. R. Lancaster, Jr., and I. Fridovich, *Arch. Biochem. Biophys.* (in press).

[13] J. R. Lancaster, Jr., *Biochem. Biophys. Res. Commun.* (in press).

[14] H. Beinert, this series, Vol. 54 Part E, Article [11].

[15] T. Ohnishi, *in* "Membrane Proteins in Energy Transduction" (R. A. Capaldi, ed.), p. 1. Dekker, New York, 1979.

[16] R. Cammack, W. L. Maharajh, and K. Schneider, *in* "Interaction Between Iron and Proteins in Oxygen and Electron Transport" (C. Ho, ed.). Elesevier/North-Holland, Amsterdam (in press).

[17] M. W. W. Adams, L. E. Mortenson, and J.-S. Chen, *Biochim. Biophys. Acta* **594,** 105 (1981).

EPR-DETECTABLE ELECTRON TRANSFER COMPONENTS IN *M. bryantii* OBSERVED TO DATE

	HiPiP-type Fe/S center	Iron-containing superoxide dismutase	Ni(III) species[a]	Fe/S centers	Radical (FAD?)
Paramagnetic state Subcellular location[b] Shape	Oxidized Soluble Fairly isotropic, with slight distortions	Oxidized Soluble Rhombic	Oxidized Membrane Rhombic	Reduced Membrane Rhombic	Dithionite addition Membrane Isotropic
Temperature range	$<20°K$	Visible at $20°K$	Visible and still sharp at $77°K$	$<20°K$	$8–20°K$
Redox properties	Reducible by dithionite	—	Reducible by dithionite	Reducible by dithionite	Produced by dithionite addition
Other properties	—	$D = 2.4 \text{ cm}^{-1}$ $E = 0.6 \text{ cm}^{-1}$ Shows characteristic increase in rhombicity upon F^- or N_3^- addition	Octahedral symmetry; very specific environment; possibly F_{430}	At least two species	Relatively fast-relaxing
References	11	12	11	11	11,13

[a] Tentative identification; see text and reference for details.
[b] Based on a 150,000 g, 45 min ultracentrifugation.

are usually only slight variations in g-value between the features of these species. It also is not known what type(s) of centers are present ($Fe_2S_2^*$ or $Fe_4S_4^*$).

Superoxide Dismutase. The soluble fraction also contains an iron-containing superoxide dismutase, which has recently been purified and characterized.[12] The enzyme has EPR and activity characteristics virtually identical to the corresponding enzyme from *E. coli,* including the characteristic effect on the EPR signal of the addition of the inhibitors fluoride and azide [an increase in the rhombicity (E/D) to very close to its maximal value of $\frac{1}{3}$].

Membrane-Bound Radical Species. Dithionite addition to the membrane fraction results in the additional appearance of an isotropic $g = 2.0$ radical-type signal. The most likely candidate for this species is the semiquinone form of flavin adenine dinucleotide, which has recently been identified in the membrane fraction of *M. bryantii.*[13] The fast-relaxing nature of this signal (being visible at 8°K) is consistent with this interpretation.

Nickel(III) Species. The oxidized membrane fraction contains a unique and well-resolved rhombic signal with all three g values significantly above 2.0. The features are very sharp, even at liquid nitrogen temperatures, and the signal is easily saturated at 20°K. These observations, coupled with the relatively wide spread of g values, argue against an iron-sulfur cluster and suggest instead a transition metal complex involving a system of d electrons.

In transition metals the splitting of resonances away from the free electron value ($g_e = 2.0023$) is a result primarily of vector contributions by orbital angular momentum, which is a function of the ground state distribution of the unpaired electron(s) among the d orbitals. This distribution will, of course, be a function of the number electrons and thus the oxidation state, but also the energetic relationships between the orbitals themselves. In an isolated atom these orbitals are degenerate, i.e., of equivalent energy. Transition metals, however, are able to bind and organize ligands in specific geometries around them, and it is the interactions between these ligands and the d orbitals (specifically oriented in space with respect to the arrangement of the ligands) that stabilizes some orbitals energetically relative to others and thus results in nondegeneracy. EPR thus is a very powerful tool for both the qualitative identification of metals and the determination of the nature of ligand complexation.

With regard to the use of EPR for the determination of possible candidates for the signal described above among the common transition elements, the absence of nuclear hyperfine structure in this sharp signal immediately eliminates vanadium, manganese, cobalt, copper, and molybdenum as candidates. Titanium(III) as well as chromium(V) are d^1 systems and usually give features at or below $g = 2$, depending on the geometry of the species. Chromium(III) (d^3) gives similar EPR spectra. Appearance of features *below* $g = 2$ for these systems is in most cases primarily a result of the sign of the spin-orbit coupling constant λ, which is positive for ions with less than a half-filled shell and negative for ions with more than a half-filled shell.[18] Thus, with commonly encountered geometries, the signal in methanogen membranes might be expected to arise from a configuration of d^5, d^7, or d^9. Reasonable candidates thus would be Mn(II) and Fe(III) (for d^5), Fe(I), Co(II), and Ni(III) (for d^7); and Cu(II) and Ni(I) (for d^9). As described earlier the lack of nuclear hyperfine leaves only the Fe(III), Fe(I) and Ni(III), Ni(I) species. The fact that the signal is from an *oxidized* species (i.e., the signal disappears reduction by dithionite) eliminates Fe(I) and Ni(I) from consideration.

In octahedral coordination both low-spin Fe(III) and low-spin Ni(III) can give values for g_1 very similar to the two low-field features of the signal in *M. bryantii*. It is in fact the value for g_{11} that clearly discriminates between these two final possibilities. The unpaired electron in low-spin Fe(III) in an octahedral field resides within the t_{2g} set of d orbitals (d_{xz}, d_{yz}, and d_{xy}). Substantial quantum mixing of these orbitals results in departures of the observed g values away from g_e, in a theoretically predictable manner.[19] In particular, the value for g_{zz} (or g_{11}) will always be observed at values *less* than $g = 2.00$. Assuming the formalism developed by Griffith[19] and widely applied to low-spin hemes,[20] one can in fact predict the value for g_{11} given the other two g values; such a calculation results in $g_{11} = 1.94$, a value clearly in disagreement with the observed value of $g_{11} = 2.02$.

The unpaired electron in low-spin Ni(III) in octahedral symmetry with tetragonal distortion is localized primarily in the d_{z^2} orbital. In this case, no quantum mixing of atomic d orbitals will occur along the z axis and the value for g_{zz} is 2.0. In fact, deviations in the *positive* direction are commonly observed because of vibronic mixing of the d_{z^2} and $d_{x^2-y^2}$, ground states induced by axial binding.[21] Work with model complexes has shown that very similar EPR spectra to the one reported earlier are exhibited by

[18] J. S. Griffith, "The Theory of Transition-Metal Ions," p. 111. Cambridge Univ. Press, London and New York, 1964.

[19] J. S. Griffith, *Nature (London)* **180**, 30 (1957).

[20] W. E. Blumberg and J. Peisach, *in* "Probes of Structure and Function of Macromolecules and Membranes" (B. Chance, T. Yonetani, and A. S. Mildvan, eds.), Vol. 2, p. 215. Academic Press, New York, 1971.

[21] H. Kon and N. Kataoka, *in* "Electron Spin Resonance of Metal Complexes" (T. F. Yen, ed.), p. 59. Plenum, New York, 1969.

Ni(III) systems[22-24] and that the redox potential for Ni(II)–Ni(III) couple can be shifted to quite negative values, making Ni(III) a stable oxidation state.[25] It should be pointed out that the oxidation state and ligand field configuration in this Ni(III) species is the same as that for Co(II) in COB(II) alamin. The properties of this component would be expected to be similar to those in B_{12} and thus is an excellent candidate for group transfer and/or ligand oxidation-reduction. In general, complexation of a ligand to a transition metal results in (among other effects[26]) an increase in the susceptibility of the ligand to nucleophilic attack due to a polarization of the electron density of the ligand toward the metal ion. In addition, this metal ion itself might function as an electron carrier.

Finally, a chromophoric factor of unknown function has been isolated from this bacterium (factor F_{430}), which has recently been shown to contain nickel, which is probably present in a tetrapyrrole structure.[27-29] This is consistent with the postulated geometry of the nickel species described earlier.[30]

[22] F. V. Lovecchio, E. S. Gore, and D. H. Busch, *J. Am. Chem. Soc.* **96**, 3109 (1974).
[23] A. G. Lappin, C. K. Murray, and D. W. Margerum, *Inorg. Chem.* **17**, 1630 (1978).
[24] Y. Sugiura and Y. Mino, *Inorg. Chem.* **18**, 1336 (1979).
[25] D. H. Busch, *Acc. Chem. Res.* **11**, 392 (1978).
[26] M. M. Jones and J. R. Hix, Jr., *Inorg. Biochem.* **1**, 361 (1973).
[27] G. Diekert, B. Klee, and R. K. Thauer, *Arch. Microbiol.* **124**, 103 (1980).
[28] W. B. Whitman and R. S. Wolfe, *Biochem. Biophys. Res. Commun.* **92**, 1196 (1980).
[29] G. Diekert, R. Jaenchen, and R. K. Thauer, *FEBS Lett.* **119**, 118 (1980).
[30] Supported in part by Grant DE-ACO2-81ER10875 from the Department of Energy.

[55] Anaerobic Growth of Halobacteria

By DIETER OESTERHELT

Introduction

Structural and functional studies on bacteriorhodopsin are limited so far to biophysical and biochemical methods. Genetic methods would provide a powerful addition. The development of proper selection techniques for bacteriorhodopsin mutants is a necessary step for the application of classic genetic methods. Light energy conversion by bacteriorhodopsin is a bioenergetic pathway alternative to respiration, and therefore aerobic growth of halobacterial cells will not depend on bacteriorhodopsin. On the other hand, strictly anaerobic conditions will prevent synthesis of bacteriorhodopsin because retinal biosynthesis involves oxidative cleavage of β-carotene. Furthermore, oxygen might be essential for some other metabolic pathways beside respiration and could also prevent anaerobic growth. Therefore conditions for bacteriorhodopsin-dependent growth have to be worked out.

A new energy-transducing pathway in halobacteria has been described[1] involving arginine-mediated substrate level phosphorylation. This reaction allows the cells to grow anaerobically, indicating that oxygen is not necessary for metabolic reactions other than respiration. Anaerobic growth without arginine is observed if enough bacteriorhodopsin is transferred to an anaerobic culture with the inoculum or if retinal is added to the culture. This bacteriorhodopsin-dependent growth allows the differentiation between wild-type cells and bacteriorhodopsin mutants. The experimental conditions for anaerobic growth on agar plates or in suspension are described in this article.

Strains, Growth Medium, and Standard Growth Conditions

In Article [45] of this volume the strains, growth medium, and growth conditions for cells with low and high levels of bacteriorhodopsin and the analytical methods are described.

Arginine-Dependent Anaerobic Growth

For anaerobic growth a gas Pak anaerobic system (Becton Dickinson, Heidelberg/Wieblingen) is used that contains up to 1.5 liters of complete medium with 0.5% arginine at pH 7. The inoculum of 5% freshly grown aerobic cells is added and the vessels are allowed to become anaerobic by reduction of the oxygen with hydrogen at a platinum catalyst. The cell suspension is stirred magnetically at 40° and a septum in the fermenting vessel allows sampling during growth. Under these conditions the cells duplicate in about 40 hr and reach about half of the density of aerobic cell suspensions.

Cell colonies are grown on agar plates prepared from arginine (0.5%) containing medium with 1.5% agar. The petri dishes are incubated in the gas Pak system or the anaerobic chamber described later (see Fig. 1).

Light-Dependent Anaerobic Growth

When cells are grown anaerobically under illumination, a 100-ml Erlenmeyer flask containing 35 ml of medium is placed inside a gas Pak vessel and the cell suspensions are stirred. Light from two 150-W projectors

[1] R. Hartmann, H.-D. Sickinger, and D. Oesterhelt, *Proc. Natl. Acad. Sci. U.S.A.* **77**, 3821 (1980).

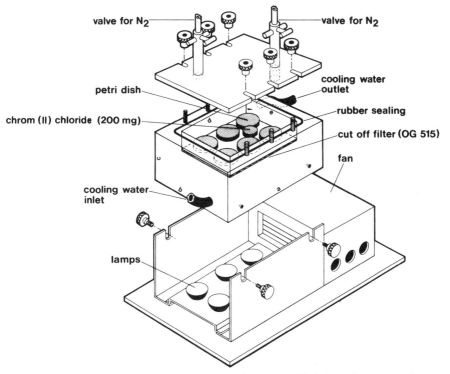

FIG. 1. Chamber for light-dependent anaerobic growth of halobacteria on agar plates. The dimensions (in mm) are $a = 450$, $b = 330$, $c = 115$, $d = 300$, $e = 200$, $f = 25$.

is filtered through orange glass filters (OG 515, Schott, Mainz) and directed into the cell suspension. If a 10% inoculum of bacteriorhodopsin-rich cells is used, the culture will reach the original density within about 150 hr with no growth being observed in the dark control. However, the cellular bacteriorhodopsin content, which is proportional to the cell count, indicates that traces of oxygen are present in the anaerobic system that allow the β-carotenase* reaction to produce the small amounts of retinal necessary for bacteriorhodopsin formation. Retinal synthesis can be blocked by 1 mM nicotine in the medium, or it (retinal synthesis) can be blocked if the anaerobic system is placed inside a vacuum desiccator filled with oxygen-free nitrogen. Under these conditions *de novo* synthesis of bacteriorhodopsin is no longer observed and the cell growth depends on the total amount of bacteriorhodopsin added with the inoculum.

Wild-type cells (S9) or retinal-deficient mutants (W 296) will also grow from very low inoculating cell counts (down to 1 cell/ml) if supplemented

* This is the enzyme that carries the oxidative cleavage of β-carotene.

with 30–300 μM retinal added in isopropanol (not more than 0.5% of culture volume). Instead of using a gas Pak anaerobic system for growth, cell cultures may simply be grown in Erlenmeyer flasks sealed under anaerobic conditions (Anaerobic environmental enclosure model 3650, National Appliance Company; Portland, Oregon). The flasks have a side arm for measurement of cell density in a Klett photometer and are incubated in a lab shaker (model LSR, Kuehner AG, Basel, Switzerland) under illumination from a 1200-W light source (OG 515, Schott, and heat filter). The irradiance at the place of the flasks is about 100 mW/cm².

Light-dependent cell growth on agar plates is observed in the apparatus shown in Fig. 1. A chamber containing up to six petri dishes is surrounded by a cooling jacket and its bottom is covered with an orange glass cutoff filter (OG 515, Schott). The chamber is made anaerobic by flushing with a stream of oxygen-free nitrogen and residual traces of oxygen are removed by chromium(II) chloride (200 mg) dissolved in water immediately before closing the chamber. The petri dishes are placed without covers bottom up into the chamber and are illuminated by six lamps (HaloStar 6443 BF, 12V, 50 W, Osram) cooled with a fan. Growth of colonies takes about 10 days if the cells are replicated from a master plate to distinguish wild-type cells from bacteriorhodopsin mutants. Dark controls show very small growth, presumably because of the arginine content of the complex medium used for preparation of the agar plates. Although no growth from single cells is observed under these conditions, the method is useful for the selection of bacteriorhodopsin mutants with the replication technique.

[56] Photobehavior of Halobacterium halobium

By Norbert A. Dencher and Eilo Hildebrand

Halobacterium halobium is a rodlike cell about 4–10 μm long and with a diameter of about 0.7 μm. It carries five to eight flagella on both poles and swims with a speed of about 2.5 μm sec⁻¹ equally well in both directions parallel to its long axis.[1] From time to time the bacterium spontaneously reverses its swimming direction. According to other flagellated bacteria, whose motor behavior has been studied in more detail, the reversal is the result of a change in the rotational sense of the flagella. Spontaneous reversal normally occurs every 15–50 sec and in strains selected for high motility approximately every 4 sec.[2] The frequency of reversals

---

[1] E. Hildebrand and N. Dencher, Nature (London) 257, 46 (1975).

[2] J. L. Spudich and W. Stoeckenius, Photobiochem. Photobiophys. 1, 43 (1979).

can be modulated by changing the light conditions. Sudden increase of blue or UV light evokes a reversal response within one to a few seconds, depending on stimulus amplitude and duration and on the background light.[1] Removal of blue light causes suppression of spontaneous reversals for a period of some 10 sec in high-motility strains.[2] Blue light can thus be considered as a repellent, and indeed halobacteria avoid blue and UV light. Moreover, decrease of visible (yellow-green) light also elicits a reversal response,[1] and increasing the intensity of that light leads to transient suppression of spontaneous reversals.[2] Visible light acts as an attractant and halobacteria accumulate in the light of appropriate wavelength.[3,4]

According to their different action spectra the two photosystems are called PS 370 and PS 565. They differ also with respect to absolute sensitivity, adaptation, time of appearance, and other features.[1,3,5]

There is now strong evidence that both photosystems contain retinal as the chromophore.[6,7] Although the action spectrum of PS 565 corresponds fairly well with the absorption spectrum of bacteriorhodopsin (BR) in the visible range the nature of the photopigment is not known with certainty. Also the pigment of PS 370 has not been isolated or spectroscopically detected so far. Carotenoids act as accessory pigments in this photosystem.[6]

Photokinesis (i.e., light-dependent changes of swimming rate) has not been found up to now in *Halobacterium*.[3] However, orientation of the bacteria with respect to the direction of incident light has been recently reported.[8]

Because of the small size of the cell the molecular basis of photobehavior (sensory transduction, signal transmission, and effector control) can be presently studied only by behavioral analysis. However, genetic dissection and biochemical investigation of mutants that are defective in light-dependent motor responses seem to be a promising tool.

Culture Methods and Conditions of Behavioral Studies

Halobacterium halobium is grown in 20-ml sterile culture medium in 100-ml Erlenmeyer flasks under semiaerobic conditions (shaking fre-

[3] N. A. Dencher, *in* "Energetics and Structure of Halophilic Microorganisms" (S. R. Caplan and M. Ginzburg, eds.), p. 67. Elsevier/North-Holland, Amsterdam, 1978 (rev.).

[4] W. Nultsch and M. Häder, *Ber. Dtsch. Bot. Ges.* **91**, 441 (1978).

[5] E. Hildebrand, *in* "Taxis and Behavior" (G. Hazelbauer, ed.), p. 35. Chapman & Hall, London, 1978 (rev.).

[6] N. A. Dencher and E. Hildebrand, *Z. Naturforsch. C: Biosci.* **34c**, 841 (1979).

[7] W. Sperling and A. Schimz, *Biophys. Struct. Mech.* **6**, 165 (1980).

[8] K. W. Foster and H. C. Berg, *Biophys. J.* **25**, 313a (1979) (abstr.).

quency 100 rpm) at 37–40° under illumination with white fluorescent light of approximately 100 μW cm^{-2} at the Erlenmeyer. The growth medium contains 250 g NaCl, 20 g MgSO$_4$·7H$_2$O, 3 g trisodium citrate·2H$_2$O, 2 g KCl, 10 g Oxoid bacteriological peptone (Code L 37) in 1000 ml. The pH should be adjusted to 6.8.

The culture medium can be inoculated with cells from the stationary phase of a liquid culture or from agar plates (1.5% agar). After 5 days at the late exponential growth phase, bacteria can be used for experiments. Both photosystems are developed at this stage and remain active during the stationary phase. High-motility strains can be selected after the method described by Spudich and Stoeckenius.[2]

If the photobehavior should be studied in solutions different from the culture medium, cells can be harvested by 1.5-min centrifugation at 10,000 g and resuspended in the experimental saline. Bacteria are motile and show photoresponses at pH 5.5–9.0 and in the temperature range between 10°–45°. Optimum conditions are pH 7.0–8.0 and 30–40°C.

Individual Cell Method

Direct Observation. The simplest method to analyze the photobehavior of *H. halobium* is direct observation of individual cells through the microscope. Phase-contrast optics as well as differential interference contrast (DIC), or dark field illumination are suitable to visualize the bacteria. DIC, however, is not always recommendable because of the polarized background light.

The culture, which has been grown up to a density of 10^8–10^9 cells/ml, should be diluted with culture medium 1:5 before the suspension is placed in the experimental chamber or simply between a microscopic slide and a cover slip. Slides with a 0.2-mm-deep hole or a corresponding chamber have proved useful. For measurements in the UV light below 340 nm the chamber should be made from quartz. It is recommended to provide the microscopic stage with a thermostat that allows us to work at temperatures up to 40°. Observation should be made at red background light ($\lambda > 650$ nm) if the blue/UV light response (PS 370) is under study. Bright white background light seems most appropriate if the visible light system (PS 565) will be examined. We found a "Neofluar" 63× phase contrast objective (C. Zeiss) in connection with 10× eyepieces most suitable for observation. In the case of dark field illumination planacromatic lenses 40× to 70× are recommended. Quartz optics (e.g., "Ultrafluar" 100× phase contrast (C. Zeiss)) are required for measurements in the UV below 340 nm and care should be taken that no UV-absorbing material is inserted in the stimulating light beam.

Stimulating light should be monochromatic. It may be applied either by means of a beam splitter through the condenser of the microscope (indispensable in the case of dark field technique[2]) or by an incident light illuminator through the objective.[1] For both photosystems a 200-W mercury lamp connected either with an UV/vis monochromator or interference filters of appropriate wavelength transmission was found most suitable. However, also a 150-W xenon lamp or 100–500-W halogen lamps can be used. The bandwidth of monochromatic light depends on the experimental question. It should be, of course, as narrow as possible if high spectral resolution is required. The intensity of the stimulating light can be varied by neutral density filters. The duration of stimuli is controlled by electronic shutters. Either light pulses of 5 msec to 2 sec duration, in the case of PS 370, and transient decrease of light intensity of 100 msec to 2 sec duration in the case of PS 565, or steps of increasing or decreasing light, respectively, can be applied to induce the motor response. When a bacterium experiences such a stimulus, it stops swimming after a latent period of about 1 sec, following the onset of the stimulus, and, after a short phase of tumbling, starts swimming again, but nearly in the opposite direction. Sometimes weaker responses can be seen that consist only of a stop reaction and some tumbling; thereafter the bacterium resumes its original movement. Suppression of reversal can be produced either by a sudden decrease of blue/UV light or by an increase of yellow-green light. As the measured parameter either the latent period can be obtained by a stopwatch or the stimulus intensity that is needed to result in a given change of the reversal frequency (e.g., induction of reversal after a latent period of 2 sec) can be measured.

The intensity of the stimulating light should be measured in the plane of the preparation with a calibrated light detector (e.g., U.D.T. Model $40 \times$ Opto-meter, connected with pin 10 or pin 10 UV, respectively). The minimum light intensity at 370 nm required to elicit a reversal response was found to be about 10 nW mm^{-2}. An intensity decrease of at least 80 nW mm^{-2} at 565 nm is necessary.[1]

Video and Photographic Recording. To facilitate individual observation a video recording system is useful for different reasons. (1) If an IR-sensitive video camera (e.g., National WV 1350) is used in connection with the microscope, recordings can be made at dim red background light ($\lambda \geq 700$ nm), which does not interfere with one of the photosystems. (2) Swimming tracks of several individuals can be simultaneously stored on video tape. (3) If the tape recorder allows single frame display (e.g., National NV 8030), photoresponses can be analyzed with a higher temporal resolution.

A less expensive method for two-dimensional recording is long time exposure photography in connection with dark field microscopy. The movement of the cells appears as continuous black traces on the film. Stroboscopic dark field illumination allows us to calculate the swimming rate of the cells at every time. However, this modification has not been applied to *H. halobium* so far, and it seems questionable whether the swimming rate is high enough to guarantee spatial resolution on the photographs.

Three-Dimensional Automatic Tracking. The most elegant but rather expensive method to record bacterial movement is to use the three-dimensional automatic tracking microscope originally designed for *Escherichia coli.*[9] The principle of this sophisticated apparatus is to keep one individual cell automatically in focus by means of a photometrically controlled movable microscopic stage and to compute the readout of the three compensatory operating servo systems. This method requires an approximately spherical shape of the organism, and therefore strains of *H. halobium* having extremely short cells are most suited for this apparatus.

Population Method

As already mentioned, halobacteria avoid UV light but accumulate in the yellow-green. This feature can be used to study photobehavior by means of a mass method that allows us to obtain results of a high degree of statistical significance. The method is based on the photometer principle. The cell density in an irradiated area of a suspension of bacteria is continuously monitored by a photocell or photomultiplier or simply by counting the cells in the irradiated part of the microscopic field.[3,4] With this method it is impossible to distinguish whether accumulation or dispersal is the result of reversal responses, photokinesis, or phototaxis (i.e., orientation with respect to the direction of incident light). However, once the basic response is known and provided all samples are equal with respect to photobehavior, the method can be applied to study the effect of certain parameters on the photobehavior.

Applications

Identification of Photopigments. With the help of direct observation of individual cells[1] and the population method,[4] action spectra of both photosystems have been determined. PS 370 shows two pronounced sensitivity maxima in the UV region at 370 and 280 nm and some secondary peaks between 400 and 530 nm. The action spectrum shows some similarities

---

[9] H. C. Berg, *Rev. Sci. Instrum.* **42**, 868 (1971).

with the absorption spectrum of a retinylidene protein that can be obtained *in vitro* from bacteriorhodopsin (BR) after certain treatment.[1] The action spectrum of PS 565 resembles the absorption band in the visible of BR, but no reversal responses were observed when the light intensity at wavelengths below 450 nm was reduced.[1] It was argued that the lack of sensitivity in this part of the spectrum may be due to the counteraction of PS 370, which suppresses the frequency of reversals when blue or UV light is removed.[2]

The next step to identify the photopigments was biochemical blocking and modification of the photosystems in living cells. For that purpose halobacteria, strain R_1L_3, were grown in the presence of 1 mM L-nicotine, which inhibits retinal synthesis. After repeated (six to seven times) transfer of cells from the stationary phase to culture medium containing nicotine the sensitivity to all kinds of photostimuli disappeared.[6] Both photosystems could be reconstituted within some 10 min after adding *trans*-retinal in ethanolic solution to a final concentration of about 10^{-6} M.[3,6] The concentration of ethanol in the reconstitution medium was kept below 1%.

The final evidence that retinal acts as the chromophore in both photosystems came from experiments in which the reconstitution was carried out with retinal$_2$ (3,4-dehydroretinal) instead of retinal$_1$. In this case the main maxima of both photosystems (370 and 565 nm, respectively) were shifted 15–20 nm toward longer wavelengths.[7]

Retinal added to nicotine cells of R_1L_3 reconstitutes only the two main sensitivity bands at 280 and 370 nm in PS 370. However, if carotenoids extracted from strain R_1 are added together with retinal, PS 370 is sensitive also at wavelengths between 450 and 570 nm.[3,6] Carotenoids by themselves cannot trigger the response. They are therefore considered as accessory pigments to the retinal protein pigment of PS 370.

Photosensory Transduction

The mechanism by which the light stimulus is transduced and the nature of the signal that transmits information to the bacterial flagella are presently obscure. Some results point toward the membrane potential to be involved.[4,5,10] It would be especially interesting to study the influence of certain ions and of specific ionophores on the photobehavior of halobacteria. Simultaneous stimulation of PS 370 and PS 565 shows that signals from both systems are integrated.[2,5]

[10] G. Wagner, G. Geissler, R. Linhardt, A. Mollwo, and A. Vonhof, *in* "Plant Membrane Transport: Current Conceptual Issues" (R. M. Spanswick, W. J. Lucas, and J. Dainty, eds.), p. 641. Elsevier/North-Holland, Amsterdam, 1980.

Halobacterium is also sensitive to certain chemicals, most of them acting as attractants,[11] and it was recently shown that signals from photosensory and chemosensory systems are integrated by the cell.[2]

Genetic Dissection

Several strains of *H. halobium* with an altered content of carotenoids are available. Also mutants that are defective in certain steps of the photosensory pathway have been recently isolated and are presently under study.[12] Up to now no mutation was found that affects only one of the photosystems leaving the other intact. Such a mutant would be very helpful to unravel the biochemical relationship of both photosystems. In general, genetic dissection of the photosensory pathway and the examination of behavioral mutants by means of the methods mentioned earlier seem to be a promising tool to analyze the mechanisms of photobehavior in *Halobacterium*.

[11] A. Schimz and E. Hildebrand, *J. Bacteriol.* **140**, 749 (1979).
[12] A. Schimz and E. Hildebrand, *Hoppe-Seyler's Z. Physiol. Chem.* **360**, 1190 (1979) (abstr.).

[57] Demonstration of Primary Sodium Transport Activity in *Halobacterium halobium* Envelope Vesicles

By EDWARD V. LINDLEY, PHILLIP N. HOWLES, and RUSSELL E. MACDONALD

Introduction

Two classes of membrane envelope vesicles of *Halobacterium halobium* may be prepared; each catalyzes distinctly different light-driven ion translocation phenomena. The first class of vesicles, L vesicles, contains an active light-driven proton pump, bacteriorhodopsin (BR).[1] The method of preparation of these L vesicles, and their characteristic activities, such as light-driven proton efflux, H^+/Na^+-antiporter activity, and amino acid uptake during actinic illumination at 570 nm have been reviewed elsewhere.[2,3]

[1] W. Stoeckenius, R. H. Lozier, and R. A. Bogomolni, *Biochim. Biophys. Acta* **505**, 215 (1978).
[2] J. K. Lanyi and R. E. MacDonald, this series, Vol. 56, Article [35].
[3] J. K. Lanyi, *Microbiol. Rev.* **42**, 682 (1978).

The second class of vesicles, M vesicles, were first described by Kanner and Racker.[4] They sediment slowly and contain an active ion pump called NaP_{590}[5] or halorhodopsin (HR)[6] that extrudes sodium on illumination with maximal activity at 590 nm.[7] The efflux of Na^+ is the primary light-driven event[10,11] and is manifested in several ways: (1) formation of a chemical gradient of Na^+ $(\Delta\mu_{Na^+})$; (2) formation of an electrical potential difference across the vesicle membrane $(\Delta\psi)$; (3) influx of protons from the external medium in response to the $\Delta\psi$. Each of these phenomena provides several convenient methods for assay of the M vesicles during their preparation.

Preparation of M and L Vesicles

Halobacterium halobium R_1 cells are grown as described by Lanyi and MacDonald in this series[2] using the chemically defined medium and are illuminated at low O_2 tension as described. The cells are harvested by centrifugation at 8000 g for 30 min and washed twice by resuspending the pellet in 4 M NaCl, 25 mM Tris, pH 6.8, and centrifuging as above.

The washed pellet is suspended in 4 M NaCl, 25 mM Tris, pH 6.8 buffer to A_{450} = 30–50. Aliquots (30 ml) are sonically disrupted using a series of 30-sec treatments at 150 W (Bronwill Biosonic IV, large probe) at 5–20°. Sonication is repeated until < 1% intact cells remain as monitored by phase contrast microscopy. Disruption of the cells with the French press (2000 lb in^{-2}) or with other sonic disruptors (e.g., Bronson or Raytheon) will produce M vesicle preparations of equally good quality (conditions such as treatment time and cell concentration should be slightly modified). The M and L vesicles are separated by differential centrifugation, accomplished by underlaying the sonicated suspension with

[4] B. I. Kanner and E. Racker, *Biochem. Biophys. Res. Commun.* **64**, 1054 (1975).

[5] R. V. Greene, R. E. MacDonald, and G. J. Perreault, *J. Biol. Chem.* **255**, 3245 (1980).

[6] Y. Mukohata, A. Matsuno-Yagi, and Y. Kaji, *in* ''Saline Environment'' (H. Morishita and M. Masui, eds.), p. 31. Japan Science Society, Tokyo, Japan (1980).

[7] Using a tunable dye laser the maximum activity for halorhodopsin-mediated proton influx was at 590 nm for M vesicles.[5] Using flash photolysis and integrating sphere spectrophotometry[8,9] a maximum absorbance at 588 nm was obtained for the halorhodopsin pigment of mutant cells that lacked bacteriorhodopsin or carotenoid pigment. This discrepancy is only apparent since the activity spectrum was not corrected for scattering or screening effects due to absorption by other pigments present in the M vesicles.

[8] J. K. Lanyi and H. J. Weber, *J. Biol. Chem.* **255**, 243 (1980).

[9] H. J. Weber and R. A. Bogmolni, *Photochem. Photobiol.* **33**, 601 (1981).

[10] R. E. MacDonald, R. V. Greene, R. D. Clark, and E. V. Lindley, *J. Biol. Chem* **254**, 11831 (1979).

[11] R. V. Greene and J. K. Lanyi, *J. Biol. Chem.* **254**, 10986 (1979).

4 M NaCl, 17% sodium tartrate, and centrifuging at 27,000 g for 40 min to obtain two intermediate fractions and a pellet of unbroken cells and large debris. The upper, slower-sedimenting fraction is mainly M vesicles (containing predominantly HR), and the fraction that accumulates at the tartrate interface is the L vesicles (containing predominantly BR). These vesicle fractions are individually collected by aspiration and concentrated by procedures that differ only in the centrifugal force used to pellet them: L vesicles, 27,000 g for 40 min; M vesicles, 70,000 g for 60 min. Typically, the vesicles in the tartrate fractions are diluted about twofold with 4 M NaCl, 1 mM HEPES, pH 6.1, and centrifuged as above; the pellets are resuspended in the HEPES salt buffer and washed twice. The pellet from the last wash is finally resuspended to $\simeq 25$ mg vesicle protein/ml in 4 M NaCl, HEPES, pH 6.1. The preparations retain light-driven ion translocation activity for several months at 4°; storage at room temperature for several days has little effect on activity, but prolonged exposure to room light appears to be harmful.[12] (See Fig. 1 for details.)

Preparation of Vesicles from H. Halobium Mutant Strains that Have Halorhodopsin Activity

Two mutant strains of *H. halobium*, R_1mR[13] and ET-15,[14] have been described that have halorhodopsin as their only photoactivated ion pump. Membrane vesicles can be prepared from these organisms following growth and light induction conditions as described for the *H. halobium* R_1 strain. Vesicles are prepared by sonicating the harvested cells in 4 M NaCl, 25 mM Tris, pH 6.8, buffer and are collected by centrifugation onto a 17% sodium tartrate cushion as described earlier. The vesicles contained in both the slowly and rapidly sedimenting fractions obtained from these mutants are functionally identical in their halorhodopsin content, and both lack detectable bacteriorhodopsin activity. Vesicles prepared from the R_1mR and ET-15 may be stored in the same manner as described for the M and L vesicles without appreciable loss of photoactivated ion pumping activity for several months.

[12] This technique is not uniformly reproducible; i.e., contamination of the M vesicle fraction with L vesicles is variable. Attempts to eliminate this using linear or exponential sucrose or Ficoll (Pharmacia Fine Chemicals) gradients ocasionally gained improved separation. Variation in illumination time, cell concentration, and growth medium composition did not improve the yield of M vesicles (usually 20–50% of the total vesicle material).

[13] E. V. Lindley and R. E. MacDonald, *Biochem. Biophys. Res. Commun.* **88**, 491 (1979).

[14] B. F. Luisi, J. K. Lanyi, and H. J. Weber, *FEBS Lett.* **117**, 354 (1980).

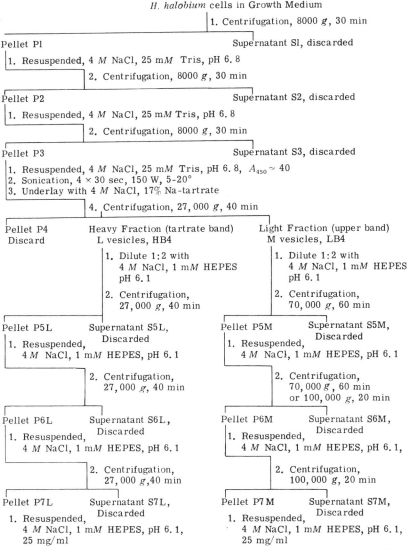

FIG. 1. Preparation procedure for M and L vesicles derived from *H. halobium*. All centrifugation steps are done at 15°. Membrane vesicles from mutant strains R_1mR and ET-15 that contain halorhodopsin are prepared using the same scheme except that only the heavy band on the Na-tartrate cushion is used (step 4).

Properties of the M Vesicles

The M vesicles are similar to the more completely characterized L vesicles in that they contain purple membrane patches that may be isolated and reconstituted into artificial liposome vesicles.[15] These liposomes exhibit similar properties to those made with bacteriorhodopsin isolated from L vesicles; i.e., the external medium becomes more alkaline on actinic illumination with yellow light. However, while this bacteriorhodopsin is in the M vesicles it remains largely inactive, even when conditions that inhibit the halorhodopsin (see below) but not the bacteriorhodopsin are imposed. The BR of M vesicles exhibits a distinct absorption peak at 570 nm in complete agreement with the L vesicle absorption spectrum. Reasons for the inactivity of the bacteriorhodopsin while in the M vesicles are unclear; however, the BR can be activated *in situ* by heating at 70° for 5 min or by acidifying the vesicles to pH 3 and returning them to pH 6.

In collaboration with E. Racker we have made preliminary attempts to isolate and reconstitute the halorhodopsin of M vesicles into artifical lipid vesicles. Methods such as distilled water extraction, extraction into detergents, and salt composition manipulation (hypotonic shock) have each been unsuccessful in these reconstitution attempts, although the "inactive" BR present in these vesicles is readily reconstituted and actively pumps protons on illumination.

Assay of Halorhodopsin Activity

The assays for M vesicles may be done with the vesicles suspended in 4 M NaCl or with vesicles that are osmotically lysed and resealed by 1 : 100 dilution in 3 M salt mixtures of the desired composition and are collected by centrifugation (100,000 g, 15 min). The M vesicles are then resuspended in the 3 M salt ("popping" buffer) to a protein concentration of 10–30 mg ml^{-1}. Optimum conditions for this loading procedure have not been determined, but we have used the method described above to load vesicles with di- and trinucleotides, radioactive amino acids, and various ions. The simplest assumption that the vesicle preparations are uniformly loaded appears to be satisfactory, although the rate and extent of the lysing and resealing process can be decreased under some conditions (e.g., in the presence of divalent cations). For best results the vesicles should equilibrate in the "popping" buffer at least 24 hr before their collection by centrifugation, since resealing may not be uniform.

The assumption that the intravesicular water volume of the M vesicles

---

[15] S. Hwang and W. Stoekenius, *J. Memb. Biol.* **33**, 325 (1977).

is 3 μl mg^{-1} is consistent with the results obtained when the concentrations of the various compounds or ions loaded into the vesicles were measured and calculated using the directly determined water volume as reported for L vesicles.[16,17]

Assay of Light-Dependent Proton Gradient. Direct measurement of the external pH changes can be followed simply with an electronically shielded glass electrode (Sargent S-30070-10) attached to a suitable pH meter (Corning Model 12) that is connected to a strip chart recorder (1–20 mV span). The reaction mixture, typically 1.5 ml of the 3 M salt buffer containing vesicles at 0.1–0.3 mg ml^{-1}, is placed in a thermostatted (30°) glass reaction vessel (Bolab, Inc.) that is lined with reflective foil. Adjustments of the pH are made by addition of 0.1 N HCl using a syringe; buffering capacity of the reaction mixture is determined by adding an aliquot (e.g., 5 nEq) of standard HCl. When illuminated (2 $\times$ 10^6 erg sec^{-1} cm^{-2}) with light from a Sylvania ELH 300-W bulb passed through a Corning 3-69 cutoff filter and a heat-absorbing filter, the pH of the reaction mixture becomes more alkaline, indicating uptake of protons by the vesicles. This conclusion is supported by the observation that the internal pH of the vesicles becomes more acid.[10] When the light is turned off, the pH usually remains at the illuminated level. This suggests that the dark exit of the protons is somewhat restricted, i.e., the vesicles are "tight" to protons. The degree of tightness is variable among the various preparations but conditions affecting this property have not been characterized. Protons may be released from these tight vesicles on the addition of 1–10 μM uncoupler (1799, SF 6847, FCCP, CCCP are each effective). Vesicles treated with uncoupler always display enhanced proton uptake on illumination. Typical proton uptake activity for M vesicles in the presence of 10 μM 1799 are 30–50 nEq H$^+$ mg^{-1} vesicle protein; in the absence of uncoupler proton uptake is variable but usually about 25 nEq H$^+$ mg^{-1}.[10] As noted, crude preparations of M vesicles are frequently contaminated with L vesicles. These preparations usually *extrude* protons on illumination. High concentration (20 μM 1799) of uncoupler severely inhibit the proton extrusion in L vesicles, whereas M vesicles retain their high level of proton uptake. Proton extrusion in L vesicles is not particularly sensitive to low (0.5 μM) levels of 1799, whereas M vesicles demonstrate a marked increase in uptake of protons in the presence of even low concentrations of 1799. Thus L vesicle contamination can be estimated in an M vesicle preparation by noting the increase in light-driven proton uptake, i.e., an

[16] R. E. MacDonald and J. K. Lanyi, *Biochemistry* **14**, 2882 (1979).
[17] M. Eisenbach, S. Cooper, H. Garty, R. M. Johnstone, H. Rottenberg, and S. R. Caplan, *Biochim. Biophys. Acta* **465**, 599 (1977).

increase in external pH, when the 1799 concentration is increased from 0.5 to 20 μM. Proton uptake by M vesicles can be abolished by the addition of 2 μg ml^{-1} gramicidin or by 5 μM valinomycin (if K$^+$ is present), whereas in L vesicles, proton extrusion is only slightly inhibited, or enhanced, respectively. Thus L vesicle contamination can be estimated using these ionophores.

Assay by Measurement of Membrane Potential Difference. Two methods of monitoring the formation of membrane potential difference ($\Delta\psi$) on illumination of M vesicles have been used. These involve accumulation of membrane permeant cations: isotopically labeled [^{3}H]TPMP$^+$ (New England Nuclear) or a fluorescent dye, di-O-C$_5$-(3). The isotope accumulation is monitored by membrane filtration and counting techniques as done for a typical transport experiment.[2,16] Vesicles at 0.1 mg ml^{-1} in 3 M NaCl, 1 mM Hepes, pH 6.3, are treated with the [^{3}H]TPMP$^+$ (0.5 μCi ml^{-1}, 0.55 μM) to initiate the experiment. The vesicles are then illuminated at 2.5 × 10^6 ergs sec^{-1} cm^{-2} for 1–10 min followed by a dark period. Aliquots of 100 μl are taken, placed on a 3-mm-diameter spot on Gelman GA-6 or GA-8 filters (presoaked in 3 M NaCl, 1 mM TPMP$^+$); suction is applied to dry the spot and then the spot is washed four times with 0.3 ml of 3 M NaCl. Radioactivity retained by the filters is determined using conventional isotope detection techniques and the membrane potential difference is quantitatively determined from the distribution of labeled [^{3}H]TPMP$^+$ during illumination by applying the Nernst potential equation given by Rottenberg[18] using the vesicular volume of 3 μl mg^{-1} vesicle protein.

A second method of monitoring the formation of a $\Delta\psi$ involves the partitioning of the fluorescent cyanine dye di-O-C$_5$-(3). Waggoner[19] has discussed the characteristics of such dyes when applied as probes of membrane potential difference. Typically, M vesicles at 0.3 mg ml^{-1} are loaded with and suspended in 2.7 M NaCl, 0.3 M KCl, 1 mM Hepes, pH 6.3. Dye is added to the vesicle suspension at 0.5 μM; it is essential that the dye be added to the vesicle suspension to minimize adsorption of the dye to the glass of the cuvette walls. The actinic beam enters the sample cuvette antiparallel to the excitation beam (483 nm) obtained from the spectrofluorimeter (Aminco SPF-500). Scattered light effects may be reduced by placing a vertical polarizer between the sample and the emission monochromator (501 nm). Fluorescence quenching of the dye is calibrated with valinomycin-induced K$^+$-diffusion potentials of known magni-

[18] H. Rottenberg, this series, Vol. 55, Article [65].
[19] A. Waggoner, this series, Vol. 55, Article [75].

tude as described by Renthal and Lanyi.[20] The membrane potential difference formed after 2 min illumination at 2.5×10^6 ergs sec^{-1} cm^{-2} is about -80 mV. L vesicles contain BR, which on illumination causes the extrusion of protons and creates a membrane potential difference. This $\Delta\psi$ may be abolished by the addition of a proton uncoupler such as 1799. In contrast, in M vesicles, the membrane potential is created by the extrusion of Na^+ and cannot be completely eliminated by a protonophore, although it is reduced to an extent (typically $\simeq 40$ mV) that is dependent on the pH of the assay mixture.[10]

Membrane potential differences may also be estimated using the more sensitive technique of flow dialysis,[10] which is described by Kaback and co-workers in this series.[21]

Assay of Na^+ Efflux by Isotope Exit. M vesicles may be loaded with $^{22}Na^+$ by incubating concentrated suspensions (10 mg ml^{-1}) with 12.5 μCi $mmol^{-1}$ of $^{22}Na^+$ for 48–72 hr. The $^{22}Na^+$-loaded vesicles are then diluted 33-fold into label-free3 M NaCl, 1 mM Hepes, pH 6.3. Efflux of the label during illumination is monitored by the standard membrane filtration methods as described earlier for [^{3}H]TPMP$^+$ using Gelman GA-6 filters. The efflux of $^{22}Na^+$ in the dark under these conditions is appreciable and must be corrected for in order to estimate the light-driven efflux.

Assay of Na^+ Efflux by Atomic Absorption. Internal Na^+ may be measured directly with atomic absorption by collecting an aliquot of vesicles, washing them free of external Na^+ and extracting the internal Na^+ into water as described by Lanyi and Silverman.[22] Aliquots (100 μl) of M vesicles (0.1–0.3 mg ml^{-1} in 3 M salt buffer) are taken as described above and placed on Millipore EG celotate filters presoaked in 3.3 M KCl, which is isosmotic with 3 M NaCl. The vesicles on the filter are then washed four times with 0.3 ml of 3.3 M KCl, and the internal Na^+ is extracted by placing the damp-dry filter in a plastic tube with 2 ml distilled water, 40 μg ml^{-1} Triton X-100. After at least 4 hr of incubation at room temperature with occasional mixing, the ion concentration of the extract can be determined directly using an atomic absorption spectrophotometer and the appropriate Na^+ calibration curve over the range from 0.1 to 4.0 μg ml^{-1}.

[20] R. Renthal and J. K. Lanyi, *Biochemistry* **15**, 2136 (1976).
[21] S. Ramos, S. Schuldiner, and H. R. Kaback, this series, Vol. 55, Article [74].
[22] J. K. Lanyi and M. P. Silverman, *J. Biol. Chem.* **254**, 4750 (1979).

[58] Assay of Pigment P_{588} and Its Discrimination from Bacteriorhodopsin by Flash Spectroscopy Techniques

By ROBERTO A. BOGOMOLNI and HANS JURGEN WEBER

Introduction

Recently, a new photochemically active pigment was spectrophotometrically characterized in a *Halobacterium halobium* strain.[1,2] It is called pigment P_{588}^s (s = salt) (or halorhodopsin, see this volume) because of its absorption maximum at 588 nm. This pigment contains most likely retinal as a chromophore and is differentiated from bacteriorhodopsin (BR) by a number of criteria[1-4]:

1. *Halobacterium halobium* mutant strains can be isolated that lack BR but contain P_{588}, thereby suggesting that the two pigments are encoded for by different genes.

2. When such mutant membranes in high salt media (2 M NaCl and above) are probed with pulsed actinic light, photochemical activity different from that of BR is observed.

3. This photochemical activity is dependent on the presence of NaCl in the medium. On exposure to buffers lacking NaCl the pigment converts into a different photoactive form, P_{565}^w, (w = water) with only one time-resolved intermediate.

4. Heating of membranes that contain P_{588} leads to pigment inactivation at temperatures where purple membranes are stable.

There are reasons to believe that P_{588} is identical with or part of the light-driven sodium pump postulated earlier by MacDonald and collaborators.[5] The evidence is based on identical action spectra for P_{588} photocycling,[2] for light-mediated ion translocation in BR missing mutants strains,[2,6] and for the light-dependent generation of membrane potential in certain envelope vesicle preparations.[7] In addition, loss of absorbance at 588 nm and of photocycling activity during chemical bleaching correlates quantitatively with loss of light-driven ion translocation, whereas addition of all-*trans*-retinal to bleached samples regenerates all three functions.[1,2]

[1] J. K. Lanyi and H. J. Weber, *J. Biol. Chem.* **255**, (1980).
[2] H. J. Weber and R. A. Bogomolni, *Photochem. Photobiol.* **33**, 601 (1981).
[3] H. J. Weber, M. E. Taylor, J. Belliveau, and R. A. Bogomolni, in preparation.
[4] A. Matsuno-Yagi and Y. Mukohata, *Arch. Biochem. Biophys.* **199**, 297 (1980).
[5] E. V. Lindley and R. E. MacDonald, *Biochem. Biophys. Res. Commun.* **88**, 491 (1979).
[6] R. V. Green, R. E. MacDonald, and G. J. Perreault, *J. Biol. Chem.* **255**, 3245 (1980).
[7] R. V. Green and J. K. Lanyi, *J. Biol. Chem.* **254**, 10986 (1979).

METHODS IN ENZYMOLOGY, VOL. 88

Further insight into the mechanism of light-driven Na^+ translocation will depend on the separation of P_{588} from other membrane proteins. An important requirement for such an undertaking is the availability of a functional assay with which the purification of P_{588} can be monitored. We have measured P_{588}-activity by flash spectroscopy techniques, which are specific, quick, and nondestructive.

Measurement of P_{588}^s in Envelope Vesicles and of P_{565}^w in Media of Lacking NaCl.

In concentrated salt solutions (above 2 M NaCl) P_{588}^s cycles via several photointermediates after excitation.[2] A transient bleaching at λ_{max} = 600 nm is followed by absorption increases at λ_{max} = 500 nm and λ_{max} = 380 nm (Fig. 1) representing the two photointermediates P_{500}^s and P_{380}^s. Although absorbance changes at each λ_{max} are indicative of photochemical activity of P_{588}^s, we recommend measuring the transient depletion at λ_{max} = 600 nm, which has the highest differential molar extinction coefficient for the bleaching reaction (Fig. 1). Suspensions of envelope vesicles scatter visible light, and the energy density of the actinic light beam decreases with the amount of envelope vesicles in the light path. For each geometrical configuration a standard curve has therefore to be determined in order to correlate the amount of P_{588} in membranes with the magnitude of ΔA_{600}. If the percentage of P_{588}^s photocycling is constant under these conditions (e.g., quantum yield independent of light intensity), ΔA_{600} per milligram total protein can be defined as the specific activity of P_{588}^s. Measurements of this kind were used for detection of mutant strains with high P_{588}^s content as well as to monitor vesicle separation on ficoll density gradients. Excellent linear correlation is obtained up to a vesicle concentration of 4.7 mg/ml protein (regression coefficient 0.999). Above that, light scattering largely distorts the effective path length for the actinic beam. If the absence of BR in this material is not established, control measurements at λ_{max} = 420 nm should be performed. According to Fig. 1, P_{588}^s shows minimal light-induced absorbance changes at this wavelength, whereas the M intermediate of BR absorbs close to maximum.[8] If the presence of BR is indicated by significant absorbance changes at 420 nm (Fig. 2), flash spectroscopy is not suitable to assay for P_{588}^s.

P_{588}^s can be transformed into a different photoactive pigment P_{565}^w by exposing membranes to media which lack NaCl.[3] This transition is reversible and characterized by the occurrence of a single time-resolved ($t_{1/2}$ < 10 μsec) photointermediate P_{640}^w (Fig. 1, trace B). This unexpected result allowed us to monitor P_{565}^w by measuring light-induced absorbance

---

[8] R. H. Lozier, R. A. Bogomolni, and W. Stoeckenius, *Biophys. J.* **15**, 955 (1975).

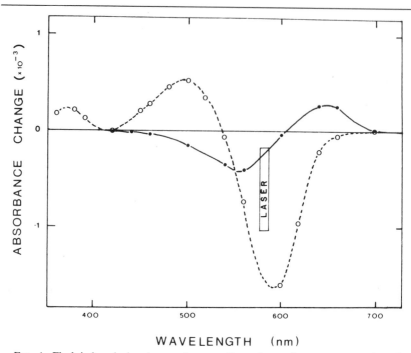

WAVELENGTH   (nm)

FIG. 1. Flash-induced absorbance changes of membrane fragment suspensions from strain JW-1 (formerly ET-15). ●——● 0.05 *M* Tris buffer, pH 8. ○——○ 4 M NaCl 0.05 *M* Tris buffer, pH 8. Protein concentration for both samples, 2.8 mg/ml; temperature 21°. The data points are the average of 1024 flashes, repetition rate 4 Hz. Laser actinic wavelength 579 nm.

changes in low salt at 640 nm or after addition of NaCl at 600 nm. The latter method may be preferred at low pigment concentrations, since P^s_{588} has a higher molecular extinction coefficient than P^w_{565} (see Fig. 1). We relied on measurements of ΔA_{640} during membrane fractionation studies on sucrose density gradients. It should be noted here that this procedure is conveniently performed in media of low ionic strength, as are other separation techniques, such as column chromatography, solubilization of P^w_{565} in detergents, etc.

Experimental Methods

We used a nitrogen-laser-pumped dye laser (Molectron) as actinic light source. The actinic wavelength was adjusted to 580 nm. Light output was 0.5–0.8 mJ per flash of 7 nsec duration. The laser beam, incident at 90° with respect to the measuring beam, was focused onto a rectangular cu-

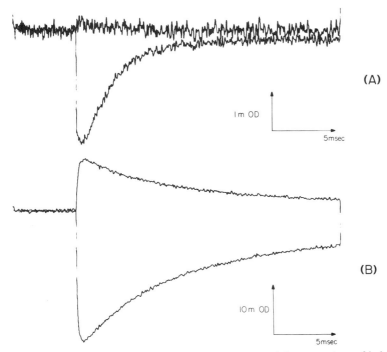

FIG. 2. Flash-induced absorbance changes in envelope vesicle suspensions of halobacteria strains in 4 M NaCl, pH 6.5, temperature 22°. Laser actinic wavelength 579 nm (7 nsec). Repetition rate 4 Hz. (A) Absorbance changes of P$_{588}^s$-containing strain. The upper trace is the absorbance change at 420 nm (average of 1024 flashes). The lower trace is the absorbance change at 600 nm (average of 512 flashes). (B) Absorbance changes of a vesicle suspension of BR-containing strain R$_1$. The upper and lower traces are the average of 64 flashes and correspond, as above, to 420 and 600 nm, respectively.

vette through a cylindrical lens that produced an image that covered most of the narrow measuring beam within the cuvette. The measuring beam (dimensions: 0.2–0.3 cm diameter at the focal plane) was passed through two interference filters (10-nm, bandwidth) to reject actinic scattered light and focused onto the photomultiplier. For further instrumental details see R. Lozier in this volume.

Measurement of P$_{588}^s$ in Bacterial Colonies

Recently we developed an instrument that allows a quick characterization of bacterial clones by spectroscopic techniques. This method can be applied to all translucent mutants (e.g., strains that lack gas vesicles) and

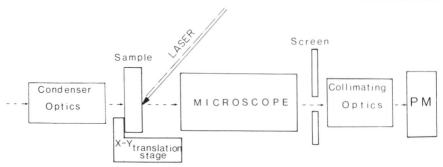

Fig. 3. Flash microspectrophotometer. The sample located on the *x-y* translation stage is an agar plate with the cover removed such that the signal colony faces the incident measuring beam, which after passing through the colony and agar base is focused by the optical system on the photomultiplier. The laser is incident at about 45° from the back of the plate, through the agar base. This angle was chosen to minimize collection by the objective of the unwanted front surface reflection of the laser beam. For other instrumental details see Weber and Bogomolni, this volume.

is fully described in another contribution (H. J. Weber and R. A. Bogomolni, Article [50], this volume). A slight modification of this setup allows the measurement of light-induced absorption changes within one colony, thereby indicating the presence or absence of photobiologically active pigment. For this purpose, pulsed actinic light (λ_{act} = 580 nm) is focused onto a colony through the back of the agar plate (see Fig. 3). The measuring beam (λ = 600 nm) forms an angle of about 45° with the actinic beam and passes several interference filters to avoid flash artifacts. It is important to adjust the two beams such that they are superimposed and focused onto the colony, with the measuring beam being the smaller one.

We were able to detect the presence of photochemically active pigments above a noise level of ΔA_{600} = 1 × 10^{-4}. Typical colonies, such as mutant strain JW-1 (formerly ET15), display signals of ΔA_{600} = − 1 × 10^{-3}. We used this spectrophotometer for the detection of P_{588}^s overproducers as indicated by transient absorbance changes at 600 nm. Transient absorbance decreases at 600 nm are generally indicative of the presence of either BR, P_{588}^s, or both. As mentioned earlier, an additional measurement at λ_{max} = 420 nm can be used as a clue for the presence or absence of BR (Fig. 2). Therefore, if only ΔA_{600} is used for the assay, BR has to be absent. In BR-producing strains, however, a diagnostic wavelength for the presence of P_{588}^s is 500 nm, where a transient absorbance increase is indicative of intermediate P_{500}^s (Fig. 1). At this wavelength BR shows absorbance changes opposite in sign and different in kinetics from those shown by P_{588}^s[8] (Fig. 1). The possibility of measuring photobiologically active pigments with high sensitivity *in vivo* allows an immediate classification of mutant strains with respect to their pigment content and improves current technology significantly.

Acknowledgments

This work was supported in part by GM-28767, GM-27057, and NASA NSG-7151.

[59] Spectrophotometric Determination of Halorhodopsin in *Halobacterium halobium* Membranes

By Janos K. Lanyi

Observations of unexpected kinds of cation transport in *H. halobium* cell envelope preparations had prompted MacDonald and co-workers[1-3] to propose the existence of a light-activated electrogenic transport system for sodium ions. Functional identification of such a transport system in envelope preparations was based on (a) uncoupler-insensitive, light-dependent sodium ion extrusion,[1,3] (b) uncoupler-insensitive, sodium-ion-dependent, and light-dependent creation of negative inside membrane electrical potential,[1,3] or (c) uncoupler-enhanced, light-dependent passive proton uptake.[2-4] In bacteriorhodopsin-deficient *H. halobium* strains these observations uniquely identify a primary electrogenic transport system for an ion other than proton. When bacteriorhodopsin is present in the membranes the observations are more complicated but can be interpreted[3,4] as reflecting the existence of two primary pumps: one for protons (bacteriorhodopsin) and one proposed for sodium ions (halorhodopsin). The hypothesis of a second light-driven pump readily explains earlier results of anomalous effects of uncouplers in cell envelope vesicles from *H. halobium*.[5] It provides interpretation also to results by Mukohata and co-workers,[6,7] who separated light-driven proton extrusion and uptake in intact *H. halobium* cells of different strains and attributed proton uptake and light-dependent ATP synthesis to a pigment different from bacterio-

[1] E. V. Lindley and R. E. MacDonald, *Biochem. Biophys. Res. Commun.* **88**, 491 (1979).

[2] R. E. MacDonald, J. K. Lanyi, and R. V. Greene, *Biophys. J.* **25**, 205a (1979).

[3] R. E. MacDonald, R. V. Greene, R. D. Clark, and E. V. Lindley, *J. Biol. Chem.* **254**, 11831 (1979).

[4] R. V. Greene and J. K. Lanyi, *J. Biol Chem.* **254**, 10986 (1979).

[5] B. I. Kanner and E. Racker, *Biochem. Biophys. Res. Commun.* **64**, 1054 (1975).

[6] A. Matsuno-Yagi and Y. Mukohata, *Biochem. Biophys. Res. Commun.* **78**, 237 (1977).

[7] A. Matsuno-Yagi and Y. Mukohata, *Arch. Biochem. Biophys.* **199**, 297 (1980).

METHODS IN ENZYMOLOGY, VOL. 88

rhodopsin, whereas proton extrusion could be linked to bacteriorhodopsin.

The preceding results had implied the existence of a pigment in *H. halobium* membranes with functional, and therefore molecular and spectroscopic, properties different from those of bacteriorhodopsin. Action spectra for membrane potential and pH difference, obtained with 10-nm-bandwidth interference filters had indicated[4] a band red-shifted 15–20 nm from the main band of bacteriorhodopsin. More recently, action spectra with laser illumination (bandwidth 0.05 nm) for proton uptake also yielded a red-shifted band, at 590 nm.[8] Spectrophotometric demonstration of such a pigment was impossible, however, until *H. halobium* strains were isolated that lacked not only bacteriorhodopsin but also the red carotenoids normally found in the membranes. These strains include ET-15,[9] R_1mW,[7] as well as L-33 and L-7 recently isolated in our laboratory. *Halobacterium halobium* ET-15 and L-33 produce retinal, and envelope vesicles[10] prepared from them contain about 0.13 and 0.50 nmol/mg protein of halorhodopsin, respectively. *H. halobium* R_1mW and L-7 lack retinal, and the latter is defective in bacterio-opsin as well.

The high turbidity of the cell envelope preparations and their high cytochrome and low halorhodopsin content preclude a direct spectrophotometric demonstration of the pigment. Exposing the envelope vesicles to distilled water decreases turbidity as the membranes disintegrate,[11,12] but absorption from the respiratory pigments still predominates. On the other hand, removal of the retinal from halorhodopsin by bleaching in the presence of hydroxylamine provides a convenient reference for difference spectroscopy.[13] The membranes from *H. halobium* ET-15 or L-33 are suspended, to 1–4 mg/ml protein, in 4 *M* NaCl containing 0.1 *M* freshly prepared hydroxylamine, adjusted to pH 7.0. Bleaching by illumination is similar to the procedure described for bacteriorhodopsin[14] but must be carried out for 3–5 times longer periods of time. At light intensities of 1–2 × 10[7] ergs cm^{-2} sec^{-1}, bleaching takes 8–12 hr. More concentrated envelope suspensions take longer to bleach. Irreversible destruction of the pigment during the illumination is minimized by cooling to 12° and the use of cutoff filters with transmission above 530 nm, but such destruction can be as high as 50%. After bleaching, the membranes are washed with 4 or

[8] R. V. Greene, R. E. MacDonald, and G. J. Perrault, *J. Biol. Chem.* **255**, 3245 (1980).
[9] This strain was isolated by H. J. Weber and will be described in detail elsewhere.
[10] J. K. Lanyi and R. E. MacDonald, this series, Vol. 56, p. 398.
[11] J. K. Lanyi, *Bacteriol. Rev.* **38**, 272 (1974).
[12] D. J. Kushner, *in* "Microbial Life in Extreme Environments" (D. J. Kushner, ed.), p. 317. Academic Press, New York, 1978.
[13] J. K. Lanyi and H. J. Weber, *J. Biol. Chem.* **255**, 243 (1980).
[14] D. Oesterhelt, L. Schuhmann, and H. Gruber, *FEBS Lett.* **44**, 257 (1974).

5 M NaCl repeatedly to remove excess hydroxylamine, as described elsewhere,[13] and can be stored in saline at 5° for several weeks.

As with bacteriorhodopsin, the absorption spectrum of halorhodopsin is reconstituted when retinal is added to bleached membranes. The vesicle preparation in 4 M NaCl (at 2–10 mg/ml protein, as the sensitivity and signal to noise ratio of the spectrophotometer used allows) is divided into two portions and placed into 1-cm path length cuvettes. *trans*-Retinal, dissolved in methanol,[15] is added to the sample cuvette, and the same volume of methanol is added to the reference cuvette. Retinal reconstitution in such preparations takes 4–6 hr at room temperature, and overnight incubation with the retinal may be required. The samples should be kept in the dark during this process. The slow rate of reconstitution is specific for retinal, as retinal analogs that reconstitute rapidly with bacterio-opsin (e.g., vitamin A_2 aldehyde) reconstitute with the halorhodopsin apoprotein within an hour.

Difference spectra obtained in this way show the appearance of an absorption band at 588 nm and the diminution of the absorbance of free retinal at 340–385 nm.[13] Up to about 30 min of reconstitution, the presence of an isosbestic point at 430 nm indicates that the reaction can be described with two absorbing species, i.e., free retinal and the product, the 588-nm pigment. At later times an isosbestic point no longer exists, as apparently additional reactions take place. It was possible to calculate an extinction coefficient for the 588-nm pigment, and a value of 48,000 M^{-1} cm^{-1} was obtained assuming that the chromophore contains one retinal per molecule.[13]

Retinal reconstitution of halorhodopsin is simpler and more direct with cell envelope preparations from retinal-deficient *H. halobium* strains. It appears that several such strains produce sufficient amounts of apoprotein to yield detectable quantities of the 588-nm pigment on retinal addition. In the R_1mW strain the 588-nm band is observed fleetingly after retinal addition, since the large excess of bacterio-opsin, which reconstitutes somewhat more slowly and gives a 568-nm band, later obscures halorhodopsin.[7] In contrast, the strain L-7 lacks bacterio-opsin, and after retinal addition to envelope vesicles prepared from these cells only the 588-nm pigment is observed. Figure 1A shows difference spectra between samples incubated 24 hr with various amounts of retinal and a reference that had received methanol only. The retinal-dependent appearance of the 588-nm band is easily observed, as is unreacted free retinal when it is

---

[15] Retinal is usually stored in isopropanol in order to avoid acetal formation, but this solvent is poorly miscible with the saline solution used in the reconstitution experiments. Since the acetal formation does not interfere with reconstitution of either bacteriorhodopsin or halorhodopsin, the reaction is best carried out with retinal added in methanol.

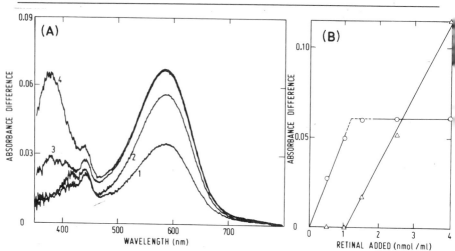

Fig. 1. Retinal reconstitution of halorhodopsin pigment in *H. halobium* L-7 envelope vesicles. The envelopes were diluted to 8.5 mg/ml protein in 4 *M* NaCl and divided into several portions. Different amounts of *trans*-retinal were added (final methanol concentration 0.04%) to some of these, and one received methanol only. After 24 hr incubation at room temperature in the dark, difference spectra were measured between the retinal and methanol samples in an Aminco DW-2a spectrophotometer. (A) Traces 1–4 from samples that received 0.5, 1.0, 1.5, and 2.5 nmol/ml retinal, respectively. (B) Absorption changes at 588 nm (○) and at 380 nm (△) are plotted versus the amount of retinal added.

added in excess amounts. Figure 1B shows a titration curve from such reconstitution, with the following features: (a) the amount of retinal that produces maximal amount of 588-nm pigment is about 0.14 nmol/mg protein; above this amount excess retinal remains after the reconstitution reaction; (b) extinction coefficients of 52,000 and 39,000 M^{-1} cm^{-1} can be calculated for the 588-nm pigment and free retinal, respectively. These values correspond approximately to the earlier estimate[13] of 48,000 M^{-1} cm^{-1} for the former, and a published value[16] of 42,000 M^{-1} cm^{-1} for the latter in ethanol. Since these samples are highly turbid, the values obtained must be regarded still as approximate. It seems safe to conclude, nevertheless, that the extinction coefficient of halorhodopsin is about 50,000 M^{-1} cm^{-1} and that L-7 membranes contain about 0.14 nmol/mg protein of the apoprotein.

A rapid method for detecting halorhodopsin and estimating its quantity in envelope vesicles is based on the pH dependence of its absorption spectrum. Unlike bacteriorhodopsin in the envelope preparations, the absorption band of halorhodopsin shows a blue shift of 40 nm above pH 10

[16] C. D. Robeson, W. P. Blum, J. M. Dieterle, J. D. Cawley, and J. G. Baxter, *J. Am. Chem. Soc.* **77**, 4120 (1955).

and an amplitude decrease.[13] This effect is reversible on neutralization. The changes cause a net decrease in the absorbance of the vesicles near 600 nm, where very little light absorption by other components is seen. Thus, titrating the contents of the reference cuvette with measured amounts of 1 N NaOH, while the sample cuvette remains at neutral pH, will result in the appearance of a difference band near 600 nm. The exact magnitude of this band is difficult to estimate, since under these conditions the light scattering of the membrane suspension also changes somewhat, giving an altered baseline, but a reasonable estimate of the amount of 588-nm absorption is obtained by multiplying the absorption change at 600 nm by 1.5.

Reconstitution with [³H]retinal, followed by reduction of the Schiff's base formed, allows the identification of the apoprotein of halorhodopsin; it appears to be a polypeptide somewhat smaller than bacterio-opsin.[17]

---

[17] J. K. Lanyi and D. Oesterhelt, *J. Biol. Chem.* **257,** 2674 (1982).

Section II

General Methods for Retinal Proteins

A. Bacteriorhodopsin and Rhodopsin Molecular Structure
Articles 60 through 62

B. Model Chromophores
Articles 63 through 71

C. Physical and Chemical Methods
Articles 72 through 91

[60] Comparison of Bacteriorhodopsin and Rhodopsin Molecular Structure

By B. A. WALLACE

Bacteriorhodopsin and rhodopsin are integral membrane proteins that play a role in light energy transduction processes, bacteriorhodopsin as a light-driven proton pump in the purple membrane of *Halobacterium halobium* and rhodopsin in visual transmission in photoreceptor cells. The abundance of these pigments in specialized regions of the membrane and their relative ease of purification has rendered them amenable to structural studies. A large amount of information has emerged concerning their molecular structures, including disposition in the membrane (from labeling and digestion studies), sequences, and secondary structure and shape (from spectroscopy and diffraction). A more detailed picture of the bacteriorhodopsin molecule has been obtained as a result of its propensity to organize into two-dimensional crystals, which has permitted low-resolution diffraction and image analyses that complement information derived by other techniques. Rhodopsin has yet to be induced to form regular crystalline arrays, so information concerning its structure is much less direct; however, a wide variety of chemical and physical studies have yielded substantial information concerning its structure.

In this review the available structural information is summarized and possible models for bacteriorhodopsin and rhodopsin are discussed. In addition, the two protein conformations are compared and contrasted and then related to other integral membrane protein structures in order to indicate general motifs for folding of proteins in the hydrophobic environment of the lipid bilayer.

The Structure of Bacteriorhodopsin

Bacteriorhodopsin is the sole protein component of the purple membrane of *Halobacterium halobium* and is produced by the cell in response to oxygen starvation.[1] It functions as a light-driven proton pump, which creates an electrochemical gradient across the plasma membrane that can be used by the cells for ATP synthesis.[2]

Bacteriorhodopsin consists of a single polypeptide chain of molecular

---

[1] D. Oesterhelt and W. Stoeckenius, *Nature (London), New Biol.* **233**, 149 (1971).

[2] D. Oesterhelt and W. Stoeckenius, *Proc. Natl. Acad. Sci. U.S.A.* **70**, 2853 (1973).

METHODS IN ENZYMOLOGY, VOL. 88

weight 26,500[3] and a chromophore, retinal.[1] Not only are the primary and secondary structures of the molecule known, but because of its unique property of forming two-dimensional ordered hexagonal arrays in the plane of the membrane,[4] it has also been possible to characterize its three-dimensional structure at low resolution.

The complete amino acid sequence of bacteriorhodopsin has been determined independently by two laboratories.[3,5] The sequence (Fig. 1) is largely hydrophobic but does contain a number of charged amino acids. No His or Cys residues are present and the N-terminus is a blocked pyroglutamic acid. Proteolytic digestion experiments[6-8] have established it to be a transmembrane protein with the N-terminus exposed at the extracellular surface and the C-terminus exposed on the cytoplasmic surface. The susceptibility (or lack of susceptibility) of various portions of the molecule to cleavage by proteolytic enzymes indicates the polypeptide segments that are either exposed at the membrane surfaces or buried. It appears that at least seven N-terminal amino acids[7] and 21 C-terminal amino acids[9] are accessible to attack and thus must be exposed to the aqueous medium. However, the bulk of the protein remains membrane-bound and fully active[10] after digestion. Chemical labeling studies with membrane-impermeable probes have also suggested which protein segments traverse the membrane and may indicate certain residues as essential to protein function.[11-13]

One retinal molecule is bound stoichiometrically to each protein through a Schiff base linkage to a lysine-ε-amino group.[1] The location of the retinal, both in the sequence and in three-dimensional structure, has

[3] Y. A. Ovchinnikov, N. G. Abdulaev, M. Y. Feigina, A. V. Kiselev, and N. A. Lobanov, FEBS Lett. **100**, 219 (1979).

[4] A. Blaurock and W. Stoeckenius, Nature (London), New Biol. **233**, 152 (1971).

[5] H. G. Khorana, G. E. Gerber, W. C. Herlihy, C. P. Gray, R. J. Anderegg, K. Nihei, and K. Bieman, Proc. Natl. Acad. Sci. U.S.A. **76**, 5046 (1979).

[6] G. E. Gerber, C. P. Gray, D. Wildenauer, and H. G. Khorana, Proc. Natl. Acad. Sci. U.S.A. **74**, 5426 (1977).

[7] J. E. Walker, A. F. Carne, and H. W. Schmitt, Nature (London), **278**, 653 (1979).

[8] Y. A. Ovchinnikov, N. G. Abdulaev, M. Y. Feigina, A. V. Kiselev, and N. A. Lobanov, FEBS Lett. **84**, 1 (1977).

[9] B. A. Wallace and R. Henderson, in "Proceedings in Life Sciences: Electron Microscopy and Molecular Dimensions" (W. Baumeister, ed.), p. 57. Springer-Verlag, Berlin and New York, 1980.

[10] N. G. Abdulaev, M. Y. Feigina, A. V. Kiselev, Y. A. Ovchinikov, L. A. Drachev, A. D. Kaulen, L. V. Khitrina, and V. P. Skulachev, FEBS Lett. **90**, 190 (1978).

[11] L. Packer, S. Tristram, J. M. Herz, C. Russel, and C. L. Borders, FEBS Lett. **108**, 243 (1979).

[12] R. Renthal, G. Harris, and R. Parrish, Biochim. Biophys. Acta **547**, 258 (1979).

[13] M. Campos-Cavieres, T. A. Moore, and R. N. Perham, Biochem. J. **179**, 233 (1979).

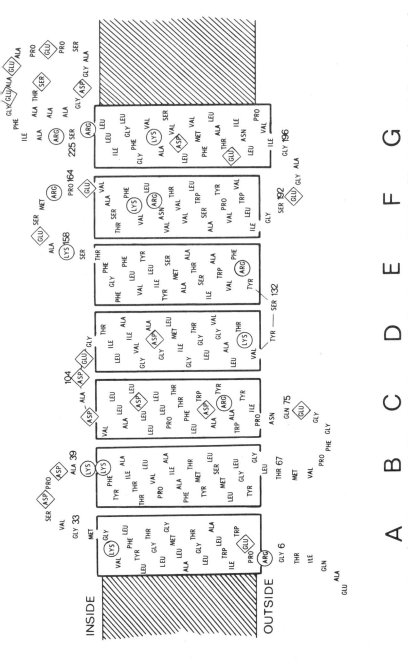

FIG. 1. The amino acid sequence[3,5] of bacteriorhodopsin, indicating the proposed[30] folding of the chain into 7 helices, designated A through G. The hatched marks indicate the approximate location of the hydrocarbon region. ○ and ◇ indicate positive and negative charges respectively.

been investigated. An early report indicated that retinal was bound to Lys-41;[14] however, more recent studies have demonstrated that the correct binding site is Lys-215.[15] The location of the chromophore perpendicular to the membrane has been controversial. Neutron diffraction experiments[16] placed the retinal near the center of the bilayer, whereas fluorescence energy transfer measurements[17] suggested it was within 10 Å of a negatively charged surface (presumably the cytoplasmic). There is more agreement on the location of the retinal in the plane of the membrane: both neutron diffraction studies of a deuterated retinal[18] and electron diffraction studies of a heavy-atom-labeled retinal[19] locate the chromophore in the internal portion of the molecule, nearer one end. (The approximate location is indicated in Fig. 2.)

Spectroscopic and X-ray diffraction studies have indicated not only the proportions of secondary structure present in the molecule, but also the orientation of these features relative to the membrane surface. X-Ray diffraction of oriented hydrates pellets[20,21] first suggested the presence of close-packed helices (with parameters slightly altered from those of typical α-helices) and indicated that these features were oriented perpendicular to the plane of the bilayer. Infrared dichroism studies[22] of oriented multilamellar films also detected somewhat distorted helices as the major structural motif and estimated their average tilt to be less than 26° from the membrane normal. Circular dichroism studies[23] determined that the amount of helix present was 70–80%, with the remainder of the structure being essentially random coil.

Elegant image reconstruction studies by Henderson and Unwin[24] have provided a three-dimensional view of the bacteriorhodopsin molecule at ~7–10 Å resolution (Fig. 3). The molecule is composed of seven closely packed helical rods, oriented nearly perpendicular to the membrane surface, that account for ~80% of the total molecule. The remainder of the

[14] J. Bridgen and I. D. Walker, Biochemistry 15, 792 (1976).
[15] J. H. Bayley, K. S. Huang, R. Radhakrishnan, A. H. Ross, V. Takagaki, and H. G. Khorana, Proc. Natl. Acad. Sci. 78, 2225 (1981); N. V. Katre, P. K. Wolber, W. Stoeckenius, and R. M. Stroud. Proc. Natl. Acad. Sci. 78, 4068 (1981).
[16] G. I. King, W. Stoeckenius, H. Crespi, and B. P. Schoenborn, J. Mol. Biol. 130, 395 (1979).
[17] D. D. Thomas and L. Stryer, Fed. Proc., Fed. Am. Soc. Exp. Biol. 39, 1847 (1980).
[18] G. I. King, P. C. Mowery, W. Stoeckenius, H. Crespi, and B. P. Schoenborn, Proc. Natl. Acad. Sci. 77, 4726 (1980).
[19] R. Henderson, T. Tsujimoto, and K. Nakanishi, personal communication.
[20] R. Henderson, J. Mol. Biol. 93, 123 (1975).
[21] A. Blaurock, J. Mol. Biol. 93, 139 (1975).
[22] K. Rothschild and N. A. Clark, Biophys. J. 25, 473 (1979).
[23] M. M. Long, D. W. Urry, and W. Stoeckenius, Biochim Biophys. Acta 75, 725 (1977).
[24] R. Henderson and P. N. T. Unwin, Nature (London) 257, 30 (1975).

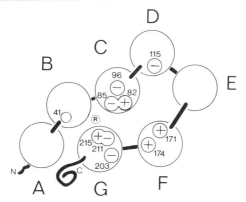

FIG. 2. Proposed model[30] of bacteriorhodopsin viewed from inside the cell. The thick connecting lines between helices indicate the path of the polypeptide. The buried charges face toward the interior of the protein, and the suggested location of the retinal Ⓡ is indicated.

structure (i.e., the portions of the polypeptide linking two helices together) is not represented in the model. Variation of the solvent density in X-ray diffraction experiments[20] demonstrated that there are no large protrusions of the polypeptide into the aqueous medium and thus most of the molecule must be located within the confines of the bilayer. The absence of the linker regions that are exposed to the aqueous medium in the map results both because some of these portions of the molecule are disordered (i.e. the C-terminus)[25] and because experimentally some of the data (the "missing core") would not have been detected. Attempts to manipulate the data further[26] may indicate that some linker regions are ordered and can be visualized.

In the native purple membrane, bacteriorhodopsin is arranged in a hexagonal lattice with three molecules per unit cell.[20] The trimer appears to be relatively stable, although the monomer apparently is the functional unit. Definition of the molecular boundaries in the plane of the membrane was made possible by studies of another two-dimensional crystal form (orthorhombic),[27] in which there are different contacts between molecules. The orientation of the map with respect to the membrane surface has been determined by correlations of electron diffraction with freeze fracture results.[28,29] In Fig. 3 the top of the model corresponds to the cyto-

[25] B. A. Wallace and R. Henderson, *Biophys. J.* (in press).

[26] D. A. Agard and R. M. Stroud, *Biophys. J.* (in press).

[27] H. Michel, D. Oesterhelt, and R. Henderson, *Proc. Natl. Acad. Sci. U.S.A.* **77**, 338 (1980).

[28] R. Henderson, J. S. Jubb, and S. Whytock, *J. Mol. Biol.* **123**, 259 (1978).

[29] S. B. Hayward, D. A. Grano, R. M. Glaeser, and K. A. Fisher, *Proc. Natl. Acad. Sci. U.S.A.* **75**, 4320 (1978).

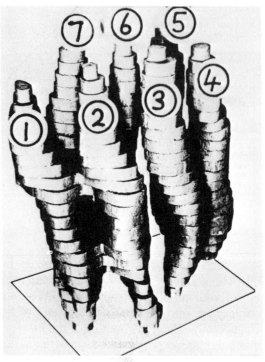

FIG. 3. Three-dimensional map of bacteriorhodopsin,[24] indicating the convention used for numbering the seven rods of helical density.[30]

plasmic surface; thus the relative orientations of the sequence and the physical model are known.

Although these studies have provided a view of the molecular shape and secondary structure, it is not yet of sufficient resolution to permit fitting of the sequence to the electron density map. To examine the topology of the molecule, model building[30] has been done based on electron diffraction, sequence, and chemical data. It is obvious from the primary structure that seven regions of the sequence could form helices (Fig. 1) that correspond to the seven rods of helical density in the three-dimensional map (Fig. 3).

There are 7! (5040) possible ways of correlating the seven helices in the sequence with the seven helices in the map. (There would have been 10,080 if their relative orientation had been unknown). Based on criteria of connectivity, charge neutralization, and scattering density, one of

[30] D. M. Engelman, R. Henderson, A. D. McLachlan, and B. A. Wallace, *Proc. Natl. Acad. Sci. U.S.A.* **77**, 2023 (1980).

the 5040 models arises as most probable (Fig. 2). This model incorporates a number of features desirable for a membrane-embedded proton pump, including a hydrophobic surface in contact with lipid, a hydrophilic interior that could form a channel, and a globular structure in which all covalently connected helices would be adjacent directly after synthesis. Of course, these arguments are not proof and only indicate the most probable of a number of possible means of folding the polypeptide. The model has been tested experimentally by difference Fourier analyses of electron, X-ray, and neutron diffraction patterns of a number of specifically modified purple membranes. Various derivatives (including ones at the retinal binding[18,19] site, C-terminus,[25] one particular lysine,[31] and of valine and phenylalanine residues[32]) have been located both in the sequence and in the plane of the membrane (projection map). All these results are consistent with the model proposed. The locations of other heavy atom labels[33] and the sites of the linker regions[26] are also currently under investigation and may aid in defining the topology of the molecule.

The ultimate goal of structural studies is a detailed molecular model of the protein. In the future, it is expected that a higher resolution model of bacteriorhodopsin may be obtained by extending the electron diffraction analysis using larger, reconstituted sheets of purple membrane,[34] or by X-ray crystallography (which will require production of three-dimensional crystals). To date, small, disordered crystals of purple membranes have been prepared[35,36] that are, as yet, unsuitable for single-crystal studies.

In summary, the combination of a variety of chemical and physical techniques has provided a low-resolution view of the bacteriorhodopsin molecular structure that presents the most detailed picture of any membrane protein known to date.

The Structure of Rhodopsin

Rhodopsin is the photoreceptor pigment of retinal rod cells that is responsible for black and white vision and is the most abundant protein component of vertebrate rod outer segment disks.[37,38] Its function is to absorb light and, by a yet poorly understood mechanism, to convert the en-

[31] B. A. Wallace and R. Henderson, *Biophys. J.* **33**, 261a (1981).
[32] D. M. Engelman and G. Zaccai, *Fed. Proc., Fed. Am. Soc. Exp. Biol.* **39**, 1885 (1980).
[33] M. E. Dumont, J. W. Wiggins, and S. B. Hayward, *Proc. Natl. Acad. Sci.* **78**, 2947 (1981).
[34] R. J. Cherry, U. Müller, R. Henderson, and M. P. Heyn, *J. Mol. Biol.* **121**, 283 (1978).
[35] R. Henderson and D. Shotton, *J. Mol. Biol.* **139**, 99 (1980).
[36] H. Michel and D. Oesterhelt, *Proc. Natl. Acad. Sci. U.S.A.* **77**, 1283 (1980).
[37] H. Heitzman, *Nature (London), New Biol.* **235**, 114 (1972).
[38] H. G. Smith, G. W. Stubbs, and B. J. Litman, *Exp. Eye Res.* **20**, 211 (1975).

ergy of the absorbed photons into a change in membrane potential that then modulates synaptic activity.[39] It has been proposed[40] that rhodopsin may act by regulating the release of calcium ions from the internal compartment of the disk, and the calcium may then cause hyperpolarization of the plasma membrane. Other studies have indicated that cGMP[41] is the transmitter. The primary event in the process is a *cis-trans* isomerization of the stoichiometrically bound chromophore retinal.[42] Details of the photocycle have been reviewed recently.[39]

To date, most structural studies have concentrated on bovine and frog rhodopsins, which exhibit similar features, including apparent molecular weight (35,000–40,000),[43] susceptibility to proteolytic cleavage, shape, and secondary structure.

That rhodopsin spans the disk membrane and exposes part of its chain at both the extradisk (cytoplasmic) and intradisk surfaces has been demonstrated by labeling and proteolytic cleavage of native and reconstituted membranes.[44] Ferritin–lectin staining of both frog and bovine disks indicates that the carbohydrate moieties are located only on the intradisk surface.[45] Chemical labeling with membrane-impermeable probes[46] has indicated an abundance of accessible sites on the extradisk space. Few sites are accessible to these probes in the intradisk space and much of the polypeptide chain is embedded within the bilayer. The C-terminus is located in the cytoplasm.[47]

Digestion experiments utilizing several proteolytic enzymes have provided information on the topology of folding. Subtilisin[48] cleaves the frog protein at several sites (Fig. 4), yielding three membrane-bound fragments (of apparent molecular weight 20,200, 11,000, and 16,200, designated S1, S2, and S3, respectively) that represent ~80% of the total amino acids found in the intact protein. These three fragments are not overlapping partial cleavage products. The apparent discrepancy in the sum of the molecular weights of the fragments versus the intact molecule (which is observed for cleavage of rhodopsin by nearly all proteolytic enzymes) is

[39] B. Honig, *Annu. Rev. Phys. Chem.* **29**, 31 (1978).
[40] W. A. Hagins, *Annu. Rev. Biophys. Bioeng.* **1**, 131 (1972).
[41] D. Hood and T. G. Ebrey, *Vision Res.* **14**, 436 (1974).
[42] G. Wald, *Science* **162**, 230 (1968).
[43] J. E. Gaw, E. A. Dratz, C. A. Vandenberg, S. E. Staughter, E. A. Kelton, M. J. Lipschultz, and S. Schwartz, *Biophys. J.* **17**, 80a (1977).
[44] B. K. Fung and W. L. Hubbell, *Biochemistry* **17**, 4404 (1978).
[45] P. Rohlich, *Nature (London)* **263**, 789 (1976).
[46] P. P. Nemes, G. P. Miljanich, D. L. White, and E. A. Dratz, *Biochemistry* **19**, 2067 (1980).
[47] P. A. Hargrave and S. L. Fong, *J. Supramol. Struct.* **6**, 99 (1977).
[48] E. A. Dratz, G. P. Miljanich, P. P. Nemes, J. Gaw, and S. Schwartz, *Photochem. Photobiol.* **29**, 661 (1979).

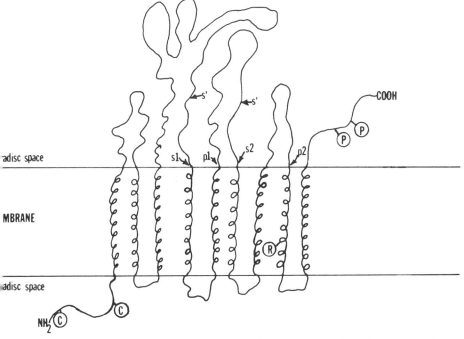

FIG. 4. Schematic drawing of rhodopsin, indicating approximate sites of cleavage with subtilisin (s1, s2, s') and papain (p1, p2) and the location of carbohydrate Ⓒ, retinal Ⓡ, and phosphorylated Ⓟ residues.

explained by the aberrant running of membrane proteins on SDS gels, which renders estimates of their molecular weights inaccurate.[49] Additional cleavage sites (labeled S' in Fig. 4) remove small, aqueous-soluble pieces from the major fragments. All cleavage sites are on the cytoplasmic surface, so the three major fragments must span the membrane. The S1 fragment contains the N-terminus and all the carbohydrate binding sites, S3 is the C-terminal peptide and contains the retinal binding site. The sequential order of the peptides is S1, S2, S3. Similar results were obtained for bovine rhodopsin, although all fragments were slightly smaller. Thermolysin is reported[50,51] to cleave the protein into two fragments of apparent molecular weights around 30,000 and 18,000 (T1 and T2), which remain associated in the membrane and in nondenaturing detergents. Pober and Stryer[50] suggested that the dissociation of the fragments in the

[49] S. P. Grefraths and J. A. Reynolds, *Proc. Natl. Acad. Sci. U.S.A.* **71,** 3913 (1974).
[50] J. S. Pober and L. Stryer, *J. Mol. Biol.* **95,** 477 (1975).
[51] B. K. Fung and W. L. Hubell, *Biochemistry* **17,** 4396 (1978).

presence of light might indicate that a significant conformation change occurred in visual excitation; Fung and Hubell[51] reported that the dissociation was not light-linked but occurred only on denaturation. The T1 fragment contains the N-terminus and carbohydrate sites (and thus must overlap S1), and the T2 polypeptide has the retinal binding site (as does the S3 fragment). Several studies[51-54] have examined the cleavage patterns of bovine rhodopsin with papain. Although the same cleavage sites have not been found by all laboratories during this work, general patterns arise that are similar to the thermolysin cleavage results. In most cases[52,53] three fragments of approximate molecular weights 23,000, 15,500, and 6,000 (P1, P2, P3) are generated (Fig. 4). That the fragments remain associated after cleavage without much structural rearrangement is demonstrated by their unchanged absorption spectrum, nearly identical CD spectrum, and appearance of the membranes in freeze fracture electron microscopy.[52] The 23,000 dalton fragment also contains the N-terminus and carbohydrate sites, whereas the lowest-molecular-weight fragment likely is the C-terminus and contains several sites of phosphorylation. Since the 15,500-dalton fragment contains the retinal binding site and the linear order of fragments is P1, P2, P3, comparison with the subtilisin results indicates the retinal must be located between ∼ 16,000 daltons and 6000 daltons from the C-terminus. Thus, digestion and labeling experiments have demonstrated that rhodopsin spans the membrane several times, with the C-terminus exposed on the cytoplasmic side and the N-terminus located in the intradisk space.

Much of the primary structure has been determined to date.[55] The sequenced segments include the hydrophilic peptides exposed in the aqueous compartments of the rod outer segments, the N-terminus and C-terminal F2 fragment and sites of carbohydrate binding (Asn^2 and Asn^{15}) and phosphorylation. The C terminal amino acids in the F2 fragment could easily form two hydrophobic helices of length similar to those found in bacteriorhodopsin. The retinal binding site is found on a lysine 53 residues from the C-terminus, in what could be the penultimate helix.[56] The two carbohydrate moities are smaller than the sugars of any known glycoprotein, their major structure being[57]:

[52] B. J. Litman, *Photochem. Photobiol.* **29,** 671 (1979).

[53] G. J. Sale, P. Towner, and M. Akhtar, *Biochemistry* **16,** 5641 (1977).

[54] P. Trayhurn, P. Mandel, and N. Virmaux, *FEBS Lett.* **38,** 351 (1974).

[55] P. A. Hargrave, J. H. McDowell, E. C. Siemiatkowski-Juszczak, S-L. Fong, H. Kühn, J. K. Wang, D. R. Curtis, J. K. Mohana Rao, P. Argos, and R. J. Feldman, *Vision Res.* (in press).

[56] J. K. Wang, J. H. McDowell, and P. A. Hargrave, *Biochemistry* **19,** 5111 (1980).

[57] M. N. Fukuda, D. S. Papermaster, and P. A. Hargrave, *J. Biol. Chem.* **254,** 8201 (1979).

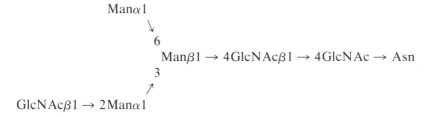

Manα1
　　$\searrow$
　　6
　　Manβ1 $\rightarrow$ 4GlcNAcβ1 $\rightarrow$ 4GlcNAc $\rightarrow$ Asn
　　3
　　$\nearrow$
GlcNAcβ1 $\rightarrow$ 2Manα1

Details of the secondary structure have been obtained from infrared and circular dichroism spectroscopy of oriented specimens.[58,59] The molecule is composed of $\sim 50\%$ α-helix, with the remainder being almost entirely random coil. Linear dichroism studies using either disks that were aligned in a magnetic field[58] or rhodopsin incorporated into oriented multilamellar films,[59] demonstrated that the helices are aligned nearly normal to the membrane surface with an average tilt of less than 40°. The presence of close-packed helices is also indicated by the appearance of X-ray reflections at 5 Å and 10 Å.[60] Peptide bond deuterium exchange studied by IR dichroism[58] indicates that the nonhelical regions are freely accessible to solvent and suggests that the helices may be the part of the molecule that is embedded in the bilayer. From these data the number of transmembrane segments can be estimated: Since $\geq 50\%$ of the mass of the molecule is helical, approximately 200 amino acids must be located in these helices. To span the hydrophobic portion of a membrane ~ 50 Å thick[61] would require 23–30 amino acids, so it is likely that there are either seven or nine transmembrane helices (since the protein must cross the bilayer an odd number of times). These oriented helices are the source of the diamagnetic anisotropy of the molecule,[62] which enables the disk to align in magnetic fields.

The shape of the rhodopsin molecule and its location within the membrane have been investigated by X-ray[60,63] and neutron[61,64] diffraction. The molecule appears to be elongated perpendicular to the bilayer with an axial ratio of $\sim 2:1$. Approximately one-half of its mass is located in the

[58] M. Michel-Villaz, H. Saibil, and M. Chabre, *Proc. Natl. Acad. Sci. U.S.A.* **76**, 4405 (1979).

[59] K. J. Rothchild, R. Sanches, T. C. Hsao, and N. A. Clark, *Biophys. J.* **31**, 53 (1980).

[60] E. A. Dratz, G. P. Miljanich, P. P. Nemes, J. E. Gaw, and S. Schwartz, *Photochem. Photobiol.* **29**, 661 (1979).

[61] M. Yeager, B. P. Schoenborn, D. M. Engelman, P. Moore, and L. Stryer, *J. Mol. Biol.* **137**, 315 (1980).

[62] M. Chabre, *Proc. Natl. Acad. Sci. U.S.A.* **75**, 5471 (1978); D. L. Worcester, *Proc. Natl. Acad. Sci. U.S.A.* **75**, 5475 (1978);

[63] M. Chabre, *Biochim. Biophys. Acta* **382**, 322 (1975).

[64] H. R. Saibil, M. Chabre, and D. L. Worcester, *Nature (London)* **262**, 266 (1976).

hydrophobic center of the bilayer[64]; it spans the bilayer with a relatively uniform cross section.[60] The hydrophilic portion is asymmetrically disposed with respect to the bilayer, with almost all the external mass located in the extradiskal space. This result is consistent with freeze fracture experiments[65] that find the intramembrane particles (presumably the protein) associated with the cytoplasmic leaflet of the bilayer and with labeling results using membrane-impermeable reagents[46] that suggest 35–55% of the amino groups are accessible on the extradisk surface. Contrast-matching experiments[61] by neutron diffraction suggest that the molecule does not contain bulk aqueous channels, although such a channel had been proposed to explain earlier hydrogen exchange data.[67] Finally, fluorescence energy transfer[66] experiments have indicated the retinal moiety is found closer to the inside of the disk than to the outside and is located ~20 Å from the aqueous surface.

In summary, the physical measurement data can be correalated with labeling, digestion, and spectroscopic results to form a tentative model for the rhodopsin molecule (Fig. 4).

A Comparison of Bacteriorhodopsin and Rhodopsin

Bacteriorhodopsin and rhodopsin are both integral membrane proteins whose function is light-activated: Bacteriorhodopsin is a proton pump, whereas rhodopsin functions in visual excitation. These two proteins exhibit a number of common features and display some structural similarities with a variety of other membrane proteins as well.

Bacteriorhodopsin and rhodopsin are by far the must abundant proteins in the specialized regions of the membranes in which they are found: Purple membranes contain only a single protein component,[1] whereas ~90% of the protein in disk membrane is rhodopsin.[30] These high concentrations result in increased probabilities of pigment-pigment interactions and in bacteriorhodopsin where there is a paucity of lipids, direct protein–protein contacts and chromophore exciton interactions.[68]

The major structural motif encountered in these proteins seems to be the presence of hydrophobic, membrane-embedded α-helices whose axes run perpendicular to the plane of the lipid bilayer. Bacteriorhodopsin and rhodopsin are oriented in the same manner with respect to the cell sur-

[65] Y. S. Chen and W. L. Hubbell, *Exp. Eye Res.* **17**, 517 (1973).
[66] D. D. Thomas, W. F. Carlsen, and L. Stryer, *Biophys. J.* **25**, 75a (1979).
[67] N. Downer and S. W. Englander, *J. Biol. Chem.* **252**, 8092 (1977).
[68] B. Becher and J. Cassim, *Biophys. J.* **15**, 65 (1975).

face: Their C-termini are located on the cytoplasmic surface and their N-terminal are exposed at the opposite side of the membrane, a feature that may be a consequence of their biosynthetic insertion mechanisms. As a result, each spans the bilayer an odd integral number of times. The hydrophobic transmembrane segments are organized almost entirely into close-packed helices. Both proteins have approximately the same amount of polypeptide embedded in the membrane ($\sim$ 20,000 daltons). For bacteriorhodopsin, this amounts to 80–90% of the total protein mass, the remainder being small N- and C-termini and linker regions, which are mostly in random coil configurations.[69] For rhodopsin, the termini and linker regions are also mostly random coil, but in the case of this molecule, they account for up to 50% of the mass of the protein, resulting in an asymmetric structure. Both these molecules are relatively compact and, being composed of secondary structures that are thermodynamically favorable, are very stable. After proteolytic cleavage of the linker regions, the resulting fragments do not dissociate or change conformation in the membrane and, even after denaturation in detergents and organic solvents, bacteriorhodopsin is capable of refolding into an active native conformation.[70]

Both proteins may contain or form channels through which they transmit ions or molecules from one surface of the membrane to the other. Proton movement probably occurs along a channel of amino acid side chains in bacteriorhodopsin, since this molecule does not appear to contain an aqueous channel.[71] Light-induced permeability changes in rhodopsin-containing systems have indicated its capacity either to form channels[72] or otherwise regulate the release of transmitter substances.[40,41] There are indications that rhodopsin also is unlikely to contain any regions of bulk water in a channel.

The most notable feature the proteins have in common is their chromophore, retinal, which is essential to their photoactivated processes. The discovery that the bacterial pigment also contained retinal prompted its designation as "bacteriorhodopsin" even though its function is unlike the vertebrate photoreceptor pigment.[1] In both proteins the retinal is bound stoichiometrically in a Schiff base linkage to a lysine of the polypeptide chain. Despite the chromophore being at least partially embedded within the hydrophobic environment of the bilayer in both proteins, the two pigments have very different spectral characteristics: Specific struc-

[69] B. A. Wallace, unpublished observations.
[70] K. S. Huang, H. Bayley, M.-J. Liao, E. London, and H. G. Khorana, *J. Biol. Chem.* **256**, 3802 (1981).
[71] G. Zaccai and D. Gilmore, *J. Mol. Biol.* **132**, 181 (1979).
[72] M. Montal, *Biochem. Soc. Trans.* **4**, 560 (1976).

tural details of protein binding sites give rise to different absorption max-ima[73] (495 nm for rhodopsin, 570 nm for bacteriorhodopsin) and different intermediates in their photocycles. Furthermore, the vertebrate protein utilizes the 11-*cis* form of the chromophore, whereas the 13-*cis* isomer is found in the bacterial protein. As a result, the two pigments.exhibit different capacities for using modified retinals.[39]

Thus these two proteins have a number of structural features in common. Certain features, such as the chromophore and concentration in specialized regions of the membrane, may be associated with their specific light-energy transducing functions. However, many of the similarities result more because they are members of the class of integral membrane proteins than because of their specific functional properties. Certain features may be favored in proteins that traverse the membrane, simply because of requirements for biosynthetic insertion and thermodynamic stability.

Membrane Protein Folding

Membrane proteins generally function in one of three ways: to transport molecules and ions (i.e., channels), to transfer information (i.e., receptors), or to anchor and concentrate proteins or carbohydrates that act primarily in the surrounding aqueous phase. Clearly, the first type must span the membrane; the latter two types may or may not, but all their structures must be influenced by surrounding lipid, since they are embedded in a hydrophobic environment that cannot solvate peptide amino and carbonyl groups, as does the aqueous solvent for cytoplasmic proteins. This restriction, the additional energy required for burying unneutralized charges in the lipid hydrocarbon region, and the steric constraints and potential hydrophobic interactions of lipid fatty acid chains with hydrophobic amino acid side chains result in folding motifs for membrane proteins that differ substantially from those in soluble proteins. Not only are the energetics of folding different, but the functional properties of the membranes, such as asymmetry and directional orientation of activity, impose certain structural requirements.

That proteins fold in different manners in membranes and aqueous solutions may be clearly illustrated in the case of bacteriorhodopsin: Using an empirical scheme for predicting secondary structure from amino acid sequence, which relies on a data base of cytoplasmic proteins, quite accu-

[73] K. Nakanishi, V. Balogh-Nair, M. Arnaboldi, K. Tsujimoto, and B. Honig, *J. Am. Chem. Soc.* **102**, 7945 (1980); B. Honig, U. Dinur, K. Nakanishi, V. Balogh-Nair, M. A. Gawinowics, M. Arnaboldi, and M. G. Motto, *J. Am. Chem. Soc.* **101**, 7084 (1979).

rate predictions are obtained for a large number of aqueous-soluble proteins.[74] However, when applied to bacteriorhodopsin, it predicts mostly β-sheet, which is clearly at odds with the experimental results. Similar discrepancies are found for other membrane proteins.[75,76]

The hydrocarbon chains of phospholipid molecules present a very different structural matrix from water for the solvation of proteins and favor dispositions of protein that place hydrophobic amino acids within the confines of the bilayer and residues with a net charge at the surface. The primary structures of a number of membrane proteins (i.e., bacteriorhodopsin, glycophorin,[77] cytochrome b_5,[76] the H^+-ATPase proteolipid,[78] and bacteriophage fd coat protein[79]) exhibit distinctly hydrophobic regions that could span the membrane and hydrophilic regions that may be in contact with the surrounding aqueous medium.

Since the lipid "solvent" cannot form hydrogen bonds with peptide backbone, secondary structures that satisfy the hydrogen-bonding potentials of the amino and carbonyl groups are even more favorable in the hydrophobic environment than in aqueous solutions. It is expected, therefore, that α-helices and β-sheets, which permit formation of hydrogen bonds between all backbone groups should yield stable structures of low free energy. Amino acids that are not normally found with high frequency in these types of secondary structures in aqueous solutions may well exist in them in membranes. β-Turns and random coil structures in which some amino and carbonyl groups are unpaired are most likely to be found in the portions of the molecule external to the bilayer. Indeed the dominant structural motif found for the bilayer-embedded portions of membrane proteins appears to be the α-helix (possible with a slightly distorted geometry). The transmembrane segments of bacteriorhodopsin, rhodopsin, glycophorin,[80] Na^+, K^+-ATPase,[81] and H^+-ATPase proteolipid,[82] among others, are essentially all α-helix. β-Sheet-like structures should also be favorable; however, they seem to be found with much lower frequency (although fd coat protein is a possible example[83]).

[74] P. Chou and G. Fasman, *Biochemistry* **13**, 211 (1974).
[75] N. Green and M. Flanagan, *Biochem. J.* **153**, 729 (1976).
[76] H. A. Dailey and P. Strittmatter, *J. Biol. Chem.* **253**, 8203 (1978).
[77] M. Tomita and V. T. Marchesi, *Proc. Natl. Acad. Sci. U.S.A.* **72**, 2964 (1976).
[78] W. Sebald and E. Wachter, *in* "Energy Conversion in Biollogical Membranes" (G. Schafer and M. Klingenberg, eds.), p. 228. Springer-Verlag, Berlin and New York, 1978.
[79] V. Nakashima and W. Koningsberg, *J. Mol. Biol.* **88**, 598 (1974).
[80] T. H. Schulte and V. T. Marchesi, *Biochemistry* **18**, 275 (1979).
[81] B. A. Wallace, A. Spencer, and I. S. Edelman, unpublished observations.
[82] D. Mao, E. Wachter, and B. A. Wallace, *Biophys. J.* **37**, 325a (1982).
[83] Y. Nozaki, B. K. Chamberlain, R. E. Webster, and C. Tanford, *Biochim. Biophys. Acta* **510**, 18 (1978).

The tertiary structures of membrane proteins are also influenced by lipid molecules. Since some charged amino acid side chains are located within the confines of the bilayer (for instance, in bacteriorhodopsin, see Fig. 1), they may be neutralized by forming ion pairs either within the same helix or sheet or with residues in adjacent helices or sheets, producing structures in which the portion of the molecule in the bilayer is compact and globular. Such a molecule would then have a hydrophilic interior (possibly a channel) and hydrophobic surfaces in contact with the lipid, as may be the case for bacteriorhodopsin.[30,32] Alternatively, in multimeric membrane proteins, intermolecular ion pairs could be formed resulting in channels composed of several polypeptide chains. Finally, folding within the bilayer need not constrain the aqueous-soluble positions of the chain, so extended molecules in the direction perpendicular to the bilayer are possible.

In summary, bacteriorhodopsin and rhodopsin exhibit general structural features that should be common to many integral membrane proteins, regardless of their mode of action. In order to elucidate the specific features associated with their functional properties, more details of their three-dimensional structures must be known. However, even the low-resolution information currently available has been sufficient to provide tentative models of these two important pigments and suggest likely patterns for folding of other integral membrane proteins.

Acknowledgments

The work was supported by GM27292 from the National Institutes of Health and by a grant from the Jane Coffin Childs Memorial Fund for Medical Research. The author is a recipient of an Irma T. Hirschl Career Scientist Award.

[61] Protein–Chromophore Interactions as Spectroscopic and Photochemical Determinants

By BARRY HONIG and THOMAS G. EBREY

Introduction

The problem of accounting for the absorption maxima of visual pigments has been a subject of great interest for many years.[1] Pigments have absorption maxima ranging from 430 to 580 nm, a variation that must be

[1] For a recent review, see B. Honig, *Annu. Rev. Phys. Chem.* **29**, 31 (1978).

understood in terms of the opsin, since the chromophore is always 11-*cis*-retinal. Retinal itself absorbs light in the ultraviolet ($\sim$380 nm), so that the various opsins must be capable of both shifting its absorption to the visible and regulating absorption within the visible region of the spectrum. It is known from Raman studies[2] that the retinal is bound to the protein via a protonated Schiff base linkage to the ε-amino group of a lysine. This accounts for a part of the wavelength shift, since protonated Schiff bases of retinal absorb at about 450 nm when isolated in solution.[3] However, it does not explain an absorption maxima near 500 nm of most rod pigments, and maxima ranging out to 580 nm for the reddest cone pigments.

An identical problem arises in trying to account for the absorption maximum of the purple membrane protein of *H. halobium*, which is frequently referred to as bacteriorhodopsin. Bacteriorhodopsin also uses retinal as a chromophore, it is also bound to the protein via a protonated Schiff base linkage to the ε-amino group of a lysine,[4,5] and it is red-shifted to 570 nm (in the range of the reddest visual pigments). It seems likely that similar mechanisms for shifting absorption maxima are operating in both the visual pigments and bacteriorhodopsin. Understanding the mechanisms involved is important in its own right but in addition allows us to interpret the light-driven spectroscopic changes that these pigments undergo.

A second problem we will address concerns possible energy storage mechanisms. In bacteriorhodopsin part of the photon's energy must be stored in order to drive the proton pump, whereas in visual pigments, although the need is less obvious, considerable photochemical energy storage is also achieved.[6–8] Since the primary photoproduct in both visual pigments and bacteriorhodopsin is a high-energy, red-shifted species, it is necessary to consider models capable of explaining both types of observations.

[2] A. Oseroff and R. Callender, *Biochemistry* **13**, 4243 (1974).

[3] See, e.g., P. Blatz and J. Mohler, *Biochemistry* **14**, 2304 (1975).

[4] A. Lewis, J. Spoonhower, R. A. Bogomolni, R. Lozier, and W. Stoeckenius, *Proc. Natl. Acad. Sci. U.S.A.* **71**, 4462 (1974).

[5] B. Aton, A. Doukas, R. Callender, B. Becher, and T. Ebrey, *Biochemistry* **16**, 2995 (1977).

[6] T. Rosenfeld, B. Honig, M. Ottolenghi, J. Hurley, and T. Ebrey, *Pure Appl. Chem.* **49**, 341 (1977).

[7] B. Honig, T. Ebrey, R. Callender, U. Dinur, and M. Ottolenghi, *Proc. Natl. Acad. Sci. U.S.A.* **76**, 2503 (1979).

[8] A. Cooper, *Nature (London)* **282**, 531 (1979).

Polyene Spectra

The factors that determine the spectroscopic properties of linear polyenes, such as retinal, are well understood.[9] It appears to be a general principle that long-wavelength absorption is correlated with increased electron delocalization and decreased single-double bond alternation. This behavior can be seen when comparing cyanine dyes, which have an odd number of atoms, to polyenes, which have an even number of atoms. Cyanines, whose π electrons are largely delocalized, have a long wavelength absorption maximum equal to about 1000 Å times the number of double bonds and have approximately equal bond lengths; that is, single and double bonds are equivalent. Polyenes tend to absorb at much shorter wavelengths than cyanines of the same length and exhibit considerable bond alternation. The spectroscopic properties of conjugated hydrocarbon chains are related to the extent that their π electrons are delocalized.[10] Any mechanism that increases delocalization and reduces bond alternation, even in a polyene with an even number of atoms, should induce bathochromic shifts in the absorption maxima.

When the Schiff base of retinal is protonated, a positive charge is partially delocalized throughout the π system. Resonance structures[11] such as those of Fig. 1b will contribute and hence increase π electron delocalization and decrease bond alternation. The latter effect can be measured from the frequency down shift of the C=C stretching vibration on protonation.[12] The bathochromic shift that occurs on protonation can also be qualitatively understood in terms of increased π electron delocalization. However, the magnitude of the shift is difficult to determine experimentally, since the protonated Schiff base cation does not exist in an isolated form and can only be measured in solution where significant environmental effects are to be expected.

Theoretical calculations[13,14] confirm the original prediction of Blatz[3] that the isolated protonated Schiff base of retinal should absorb near 600 nm. However, solutions of protonated Schiff bases absorb at ~450 nm, much to the blue of this theoretical value. This is due in large part to the state of association of the counterion. In nonpolar solvents, where the salt is not dissociated, the effect of a counterion is to pull the

[9] See, e.g., L. Salem, "Molecular Orbital Theory of Conjugated Systems." Benjamin, New York, 1966.
[10] H. Labhart, *J. Chem. Phys.* **27**, 957 (1957).
[11] A. Kropf and R. Hubbard, *Ann. N.Y. Acad. Sci.* **74**, 266 (1958).
[12] See discussion in R. Callender and B. Honig, *Annu. Rev. Biophys. Bioeng.* **6**, 33 (1977).
[13] H. Suzuki, T. Komatsu, and H. Kitajima, *J. Phys. Soc. Jpn.* **37**, 177 (1974).
[14] B. Honig, A. Greenberg, U. Dinur, and T. Ebrey, *Biochemistry* **15**, 4593 (1976).

positive charge to the nitrogen, increase the contribution of structure 1a, increase bond alternation, and thus induce a blue shift.[1,3,14]

Pigment Color

The discussion of the previous section suggests that it is possible to view pigment spectra as being blue shifted from the ~600 nm expected for the isolated protonated Schiff base, or red shifted from the corresponding solution value ~450 nm which includes the effect of the counterion. The solution value seems more appropriate for the stable form of a particular pigment, since the positively charged Schiff base is almost certainly balanced in the different proteins by a negatively charged amino acid.[14] (The generalization that buried net charges are not found in proteins should be particularly valid in the low dielectric environment provided by membranes.) However, this constraint need not be applied to the unstable transient photoproducts of both visual pigments and bacteriorhodopsin.

It is possible to use the resonance structures of Fig. 1 to consider the types of effects that can influence pigment spectra. The effect of a counterion has already been rationalized in these terms. If we assume a fixed counterion, variability in absorption maxima can arise from the positioning of additional charged groups around the retinal chromophore.[14] For example, a negative charge located near the β-ionone ring will increase the contribution of structure 1b and thus induce a red shift whereas a positive charge in the same location will induce a blue shift. It is not always possible, however, to rely completely on resonance diagrams to predict the effect of a particular environmental perturbation. In general, external charges will interact with many positions along the polyene chain, thus stabilizing some resonance structures while destabilizing others. In order to predict the effect of a charge at a particular location, it is, in general,

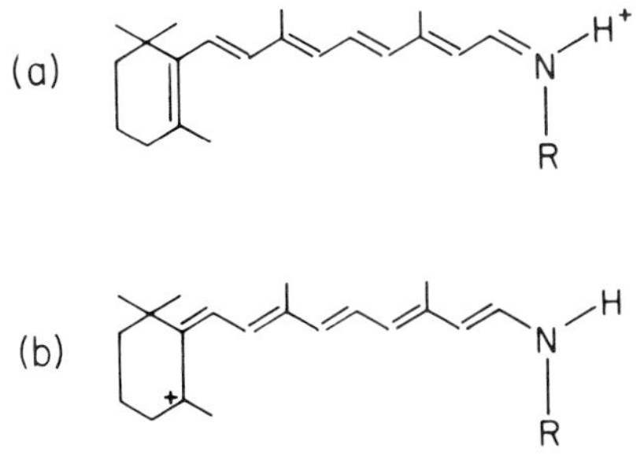

FIG. 1. Resonance structures of the protonated Schiff base of retinal: (a) the "classical" polyene like structure (b) a structure that contributes to enhanced π electron delocalization.

necessary to carry out a quantum mechanical calculation of transition energies. Semiempirical schemes (and corresponding computer programs) for this purpose are widely available, but must be used with extreme caution.[14,15]

Studies using artificial pigments based on dihydro retinals have succeeded in locating the position of the wavelength determining external charges in both visual pigments[16] and bacteriorhodopsin.[17] The relevant models are shown in Fig. 2. Based on this limited sample, it appears that the general mechanism of color regulation involves the placement of charged amino acids in different positions around the retinal chromophore. Future studies will reveal where these charges are located in other pigments. However, at this stage it is possible to use the crude models of the binding sites in Fig. 2 to interpret the spectral changes and energy storage mechanisms that result from the primary photochemistry.

Energy Storage Mechanisms

The unstable photoproducts produced as a result of the photochemical transformations of the parent pigments are generally characterized by

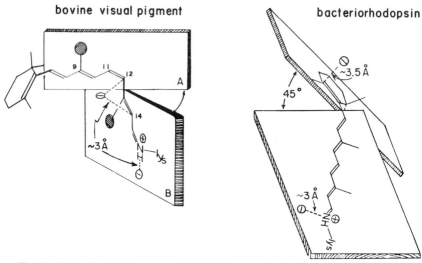

FIG. 2. External point charge models for (a) bovine rhodopsin and (b) bacteriorhodopsin.

[15] U. Dinur, B. Honig, and K. Schulten, Chem. Phys. Lett. 72, 493 (1980).
[16] B. Honig, U. Dinur, K. Nakanishi, V. Balogh-Nair, M. Gawinowicz, M. Arnaboldi, and M. Motto, J. Am. Chem. Soc. 101, 7084 (1979).
[17] K. Nakanishi, V. Balogh-Nair, M. Arnaboldi, K. Tsujimoto, and B. Honig, J. Am. Chem. Soc. 102, 7945 (1980).

their absorption maxima. The primary photoproduct at 77°K is always a red-shifted species (bathorhodopsin in visual pigments, K in the purple membrane) that decays through a series of intermediates. Although at present there is insufficient data on which detailed models of the various photoproducts can be based, the qualitative arguments made in the previous sections as well as the models of Fig. 2 can be used to evaluate different possibilities. For example, deprotonation of the Schiff base will always result in a spectral blue shift. Indeed, it is unlikely that unprotonated Schiff bases can be made to absorb above 430 nm. A second useful generalization is that charges near either end of the polyene chain are more effective in shifting absorption maxima than charges in the middle. This can be seen in Fig. 2, where the effect of the external charge in bacteriorhodopsin is much greater than that in bovine rhodopsin.

These considerations allow a plausible model for the primary photochemical event to be constructed for both pigments. In each case the primary photoproduct is a species which is red-shifted with respect to the parent pigment. Since there is a great deal of evidence that an isomerization event is an essential component of the photochemistry of both rhodopsin[3,18,19] and bacteriorhodopsin[3,20–23] (11-*cis*→all-*trans* in rhodopsin; all-*trans*→13-*cis* in bacteriorhodopsin), it is necessary to find a model in which an additional red shift is produced by an isomerization.

A second, crucial requirement is that the model account for the significant fraction of the photon's energy that is stored in the primary photoproduct. In rhodopsin we estimated[7] a lower limit for this value of > 20 kcal/mol (see Fig. 3) and a recent photocalorimetric measurement[8] has pushed this figure up to the somewhat surprising magnitude of ~35 kcal/mol. Although the corresponding number has not yet been measured in bacteriorhodopsin, its function as a proton pump requires considerable photochemical energy storage.

It is useful to consider the factors that could produce such a high-energy species. One possibility is that a new chemical entity is produced, but there is no evidence for any photochemical change in the chromophore other than isomerization. A possible model for energy storage is that steric constraints force bathorhodopsin into a conformation that is

[18] R. Hubbard and A. Kropf, *Proc. Natl. Acad. Sci. U.S.A.* **44**, 130 (1958).
[19] T. Yoshizawa and G. Wald, *Nature (London)* **197**, 1279 (1963).
[20] M. Pettei, A. Yudd, K. Nakanishi, R. Henselman, and W. Stoeckenius, *Biochemistry* **16**, 1955 (1977).
[21] J. Hurley, T. Ebrey, B. Honig, and M. Ottolenghi, *Nature (London)* **270**, 540 (1977).
[22] J. Hurley, B. Becher, and T. Ebrey, *Nature (London)* **272**, 87 (1978).
[23] M. Tsuda, *Fed. Proc., Fed. Am. Soc. Exp. Biol.* **39**, 1846 (1980).

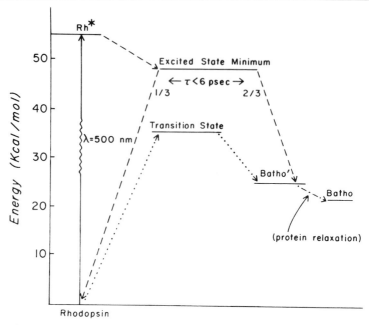

FIG. 3. Energy level diagram for primary processes in vision. Batho' is the transient observed by Peters *et al.*[24] Bathorhodopsin is the longer-lived species. The chromophore of rhodopsin has an 11-*cis* conformation. Batho' and Batho have "transoid" chromophores; they differ by a small conformational change in the protein, perhaps involving the movement of a proton(s). The excited-state minimum and the transition state are configurations twisted by approximately 90° about the 11═12 double bond: (---) pathway of radiationless deexcitation involving torsional motion about the 11═12 double bond; (.....) pathway of thermal population of bathorhodopsin through the transition state. The activation energy is determined from the thermal noise in r o d s: (-·-) relaxation of the protein (proton transfer) after *cis-trans* isomerization.

twisted about one or more double bonds.[6,14,25] Since the intrinsic barrier to isomerization about double bonds is ~25 kcal/mol,[26] such a mechanism could conceivably lead to considerable destabilization. It would also result in the required spectroscopic effect, since twisting about double bonds is known to produce red shifts.[14,25] The difficulty with this model is that, given all the single bonds in the chromophore, it is difficult to conceive of a stable conformer that is significantly twisted about double bonds, even at low temperature. However, strain energy could certainly contribute some fraction of the total energy stored in the primary photoproduct.

[24] K. Peters, M. Applebury, and P. Rentzepis, *Proc. Natl. Acad. Sci. U.S.A.* **74**, 3119 (1977).

[25] T. Kakitani and H. Kakitani, *J. Phys. Soc. Jpn.* **38**, 1455 (1975).

[26] R. Hubbard, *J. Biol. Chem.* **241**, 1814 (1960).

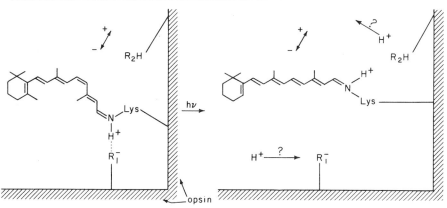

FIG. 4. Model for the early events in visual excitation. The 11-*cis* chromophore of rhodopsin is depicted with its Schiff base forming a salt bridge with a negative counterion. The additional charge pair near the 11═12 double bond represents the group that regulates the absorption maxima of different pigments. The photochemical event is an isomerization about the 11—12 double bond in rhodopsin (probably about the 13—14 bond in bacteriorhodopsin), but any isomerization in any direction will produce charge separation as shown in the first step in the figure. The pK values of the Schiff base and those of other groups on the protein, such as R_1 and R_2, are strongly affected by photoisomerization because a salt bridge is broken, a positive charge has moved near R_2 and R_1 is now a bare negative charge. Possible proton transfer steps resulting from charge separation are depicted. For bacteriorhodopsin, the isomerization is *trans-cis* rather than *cis-trans*, but all other events are assumed to be equivalent.

Another possibility that arises from a photoisomerization is based on the observation that charge separation is a natural consequence of a photoisomerization. Assuming that most of the physical motion involves the lighter, Schiff base end of the chromophore (which is bound to the highly flexible lysine), isomerization will necessarily remove the positively charged Schiff base from its negative counterion[7] (see Fig. 4). A spectral red shift, as is evident from the preceding discussion, is a natural consequence of such an event. Furthermore, charge separation in a low dielectric medium will result in a significant increase in free energy. Although the effective dielectric constant (ϵ) is not known, separating charge by a few angstroms can easily lead to values of ~ 20 kcal/mol (for $\epsilon = 3$), and considerably higher values are possible if ϵ is smaller or has a distance dependence.[7] Thus it seems likely that electrostatic factors make the major contribution but that "conformational strain energy" is also significant.

Concluding Remarks

Studies on pigment spectra have demonstrated that appropriately positioned charged or polar amino acids are responsible for producing the colors of visual pigments. Electrostatic interactions also appear to play an important role in the light transduction mechanism. A related problem, not considered in this work, concerns the nature of the photoisomerization in the excited state of the chromophore. The quantum yield for this process is 0.67 in the pigments, an order of magnitude greater than the solution value.[6] It is not unlikely that the same charges responsible for wavelength regulation and energy storage play a catalytic role in the photochemistry. Electrostatic interactions are believed to play a central role in many enzyme mechanisms,[5] and it is to be expected that their relative contribution will be even greater in the low dielectric medium provided by the interior of membranes. Thus our emphasis on these factors in membrane proteins such as rhodopsin and bacteriorhodopsin is not unwarranted.

[62] Molecular Aspects of the Photocycles of Rhodopsin and Bacteriorhodopsin: A Comparative Overview

By Michael Ottolenghi

I. Introduction

In spite of their different biological roles, quantum detection in visual rhodopsins and energy conversion (photosynthesis) in bacteriorhodopsin, both pigments share the same basic chromophore system: a retinyl polyene, bound to the opsin via a protonated Schiff base linkage with a lysine ε-amino group. It is thus not too surprising that both kinds of rhodopsins exhibit photocycles showing a variety of common features. It is the purpose of this overview to discuss, in a comparative way, molecular aspects of the photocycles of visual rhodopsins and of bacteriorhodopsin. We shall emphasize the nature and photophysical aspects of the primary event. Several features associated with later stages of both photoreactions, especially those related to the biological function of the pigments, will also be discussed.

The Primary Photochemical Events

Bathorhodopsins and Hypsorhodopsins: Kinetic Aspects

The primary event in visual rhodopsins is associated with the generation of the red-shifted phototransient bathorhodopsin (or prelumirhodopsin), first identified by Yoshizawa and co-workers as a thermally stable photoproduct (λ_{max} = 548 nm) in the photolysis of cattle rhodopsin (λ_{max} = 498 nm) at 77°K.[1] Analogous batho-species characterize all visual pigments,[1] as well as a variety of synthetic analogs.[2] Bathorhodopsin (BATHO) is the only precursor of the subsequent thermal intermediates in the bleaching sequence (see Fig. 1) and may be photochemically converted to either rhodopsin or isorhodopsin, characterized by 11-*cis*- and 9-*cis*-retinyl chromophores, respectively. Such observations, first made at 77°K,[1] were shown to be applicable at physiological temperatures.[3]

Batho-species denoted as K_{610} and K^c are also the first phototransients that can be stabilized at low temperatures in the photocycles of all-*trans* (BR^t_{570})[4] and 13-*cis* (BR^c_{550})[5] bacteriorhodopsin, respectively. The former is the only retinal isomer extracted from the light-adapted form of the pigment (BR^{LA}_{570}), while the latter is the 50% component (along with 50%

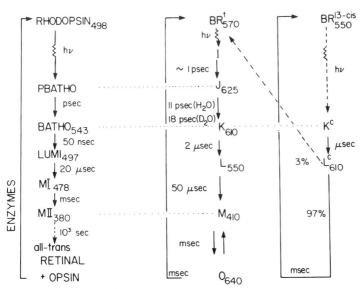

FIG. 1. The photocycles of rhodopsin and bacteriorhodopsins (all-*trans* and 13-*cis*). Indexes refer to wavelengths of maximum absorption. Time notations are approximate room temperature values. Horizontal dotted lines indicate analogous intermediates.

[1] T. Yoshizawa, in "Handbook of Sensory Physiology—Photochemistry of Vision" (H. J. A. Dartnall, ed.), p. 146. Springer-Verlag, Berlin and New York, 1972.
[2] A. Kropf, *Nature (London)* **264**, 92 (1976).
[3] C. R. Goldschmidt, T. Rosenfeld, and M. Ottolenghi, *Nature (London)* **263**, 169 (1976).
[4] W. Stoeckenius and R. H. Lozier, *J. Supramol. Struct.* **2**, 269 (1974).
[5] F. Tokunaga, T. Iwasa, and T. Yoshizawa, *FEBS Lett.* **102**, 155 (1979).

BR_{570}^t) of the dark-adapted modification, BR_{560}^{DA}.[6] The photocycles of bacteriorhodopsin are shown in Fig. 1.

Picosecond Absorption Kinetics. A basic feature, common to bovine rhodopsin and to BR_{570}^t, is the kinetics associated with the generation of the respective batho-intermediates, BATHO and K_{610}. In their picosecond experiments with bovine rhodopsin performed in the range $4-30°K$,[7] Rentzepis and co-workers observed an ultrafast absorbance decay at 570 nm ($\tau = 36$ psec at $4°K$) characterized by a non-Arrhenius temperature dependence and by a marked isotope effect. The reaction was attributed to a primary photogenerated red-absorbing species (denoted here as prebathorhodopsin or PBATHO), decaying into bathorhodopsin. Very similar phenomena were observed by Applebury *et al.*[8] for the generation of K_{610} from a red-shifted precursor that we denote as J_{625}. The latter process is slower relative to the generation of BATHO in the bovine visual pigment, being resolvable even at room temperature ($\tau = 11$ psec). It also exhibits a smaller isotope effect ($k_H/k_D = 2.4$, relative to 7.3 in the bovine system). Nevertheless, it appears that the two processes, i.e., PBATHO $\rightarrow$ BATHO and $J_{625} \rightarrow K_{610}$, represent the same type of transformation.

By applying a 615-nm subpicosecond excitation and detection technique at room temperature, Ippen *et al.*[9] have time resolved the 1.0 ± 0.5 psec growing-in of a red-shifted photoproduct of BR_{570}^t, claimed to be stable for more than 50 psec. This species was identified by them as K_{610}, in evident inconsistency with the deuteration-sensitive 11-psec generation of K_{610} from J_{625}, as claimed by Applebury *et al.*[8] The apparent discrepancy between the two observations was resolved[10] by recalling that the 615-nm monitoring wavelength of Ippen *et al.*[9] is close to the isosbestic point between the spectra of K_{610} and J_{625}, as measured by Applebury *et al.*[8] It was therefore suggested[10] that the ~ 1 psec growing-in process should be attributed to the formation of J_{625}, rather than to that of K_{610}, according to the scheme shown in Fig. 1. A "dark reaction" analogous to the I $\rightarrow J_{625}$ step (i.e., one preceding the deuteration sensitive formation of BATHO) has not yet been observed in visual rhodopsins.

Fluorescence Kinetics. Being unstable even at $4°K$, any attempt to

[6] M. J. Pattei, A. P. Yudd, K. Nakanishi, K. Henselman, and W. Stoeckenius, *Biochemistry* **16**, 1955 (1977).

[7] K. Peters, M. L. Applebury, and P. M. Rentzepis, *Proc. Natl. Acad. Sci. U.S.A.* **74**, 3119 (1977).

[8] M. L. Applebury, K. S. Peters, and P. M. Rentzepis, *Biophys. J.* **23**, 375 (1978).

[9] E. P. Ippen, A. Shank, A. Lewis, and M. A. Marcus, *Science* **200**, 1279 (1978).

[10] U. Dinur, B. Honig, and M. Ottolenghi, *in* "Developments in Biophysical Research." Ed. A. Borsellino, P. Omodeo, R. Strom, A. Vecli, and E. Wanke, pp. 209–222 (1980). Plenum, New York, and London.

identify the short-lived precursors of bathorhodopsins (PBATHO in visual pigments and I and J_{625} in BR_{570}^t) should consider their possible assignment to excited states, thus calling for the investigation of fluorescence phenomena. Although no emission following the excitation of visual pigments in their main visible absorption bands has been observed, a weak fluorescence from BR_{570}^t samples[11–13] has induced considerable interest as a potential tool in elucidating the primary photochemical event in bacteriorhodopsin. The question as to whether the emission (which is markedly enhanced at low temperatures) is due to the excitation of BR_{570}^t,[11–14] or to that of a photoproduct different from any of the commonly known intermediates in the photocycle of BR_{570}^t,[15,16] is still controversial. Nevertheless, it has been argued[10] that even if the former possibility applies, the fluorescent state observed at low temperatures cannot be identified either with J_{625} or with I. This conclusion is based on the following observations:

1. The rates of the $J_{625} \rightarrow K_{610}$ process, as measured at low temperatures,[8] are lower than those of the fluorescence decay.[12,14] Moreover, the marked temperature dependence of the emission yield, ϕ_f,[14] cannot be accounted for by identifying J_{625} with the fluorescent state.

2. The identification of I as the low-temperature fluorescent state[14] is readily ruled out by the observation that J_{625} is formed in less than 6 psec over the entire range 1.8–298°K,[8] while the fluorescence lifetimes are substantially longer, e.g., ~40 psec at 77°K.[14]

It is therefore evident that the observed emission in light-adapted bacteriorhodopsin (if due to BR_{570}^t and not to a photoproduct) originates from an excited state produced in a photochemically unimportant side path.

Visual Hypsorhodopsins. Additional features distinguishing the primary photoprocesses in visual rhodopsins from those in bacteriorhodopsin are associated with the blue-shifted ($\lambda_{max} \cong 440$ nm) intermediate hypsorhodopsin (HYPSO). This species was detected by Yoshizawa and co-workers as a stable product of the photolysis of the rhodopsins (and in some cases isorhodopsins), from cattle,[1] squid,[17] and frog[18] at 4.2°K.

[11] A. Lewis, J. Spoonhower, and G. Perrault, *Nature (London)* **260**, 675 (1976).

[12] R. R. Alfano, W. Yu, R. Govindjee, B. Becher, and T. G. Ebrey, *Biophys. J.* **16**, 541 (1976).

[13] R. Govindjee, B. Becher, and T. G. Ebrey, *Biophys. J.* **22**, 67 (1978).

[14] S. L. Shapiro, A. J. Campillo, A. Lewis, J. Perreault, J. P. Spoonhower, R. K. Klayton, and W. Stoeckenius, *Biophys. J.* **23**, 383 (1978).

[15] T. Gillbro, A. N. Kriebel, and U. P. Wild, *FEBS Lett.* **78**, 57 (1977).

[16] A. N. Kreibel, T. Gillbro, and U. P. Wild, *Biochim. Biophys. Acta* **546**, 106 (1979).

[17] Y. Schichida, F. Tokunaga, and T. Yoshizawa, *Photochem. Photobiol.* **29**, 343 (1979).

[18] S. Horiuchi, F. Tokunaga, and T. Yoshizawa, *Biochim. Biophys. Acta* **503**, 402 (1978).

Above ~ 22°K the cattle HYPSO thermally decays to BATHO. However, the question as to whether the BATHO observed at 4.2°K is a (primary) photoproduct of rhodopsin (or isorhodopsin), or a (secondary) species formed from the excitation of rhodopsin (or isorhodopsin) remains unanswered. Experiments with picosecond time resolution performed by Schichida et al.[19,20] with squid rhodopsin at room temperature revealed a growing in of the BATHO absorbance exactly matching the 45-psec decay of HYPSO. At 77°K it was also possible to resolve the growing in of hypsorhodopsin itself (τ = 70 psec), suggesting the existence of an earlier precursor (PHYPSO) whose absorbance at 540 nm is identical to that of rhodopsin.

The observations concerning the HYPSO intermediate in the bovine system are still controversial. Such a species has not been observed by Rentzepis and co-workers, neither at room temperature[21] nor at low temperatures.[7] Recently, working at room temperature under experimental conditions (i.e., sample preparation and excitation mode) identical to those of Sundstrom et al.,[21] Kobayashi[22] has also observed the HYPSO intermediate with bovine rhodopsin. The proposed sequence (PHYPSO) $\xrightarrow{\sim 15 \text{ psec}}$ HYPSO $\xrightarrow{50 \text{ psec}}$ (BATHO) is similar to that reported for the squid system.[19,20] As suggested by Kobayashi,[22] it is possible that the failure to observe HYPSO at room temperature[21] is due to the single monitoring wavelength employed (570 nm), which is almost isosbestic for the three species PHYPSO, HYPSO, and BATHO. However, this would not explain the fact that HYPSO was not detected under picosecond laser excitation at lower temperatures, where a broad range of monitoring wavelengths was employed.[7] As long as new experimental evidence is unavailable, it appears that Scheme I might account for the primary events in visual rhodopsins. The dependence of the branching ratio (from PBATHO?) to HYPSO and BATHO on sample, temperature, and excitation conditions might account for the preceding differences in the experimental observations.

Quantum Yields. Basic features, common to bovine rhodopsin and to bacteriorhodopsin, are associated with the values of the quantum yields

---

[19] Y. Schichida, T. Yoshizawa, T. Kobayashi, H. Ohtani, and S. Nagakura, *FEBS Lett.* **80**, 214 (1977).

[20] Y. Schichida, T. Kobayashi, H. Ohtani, T. Yoshizawa, and S. Nagakura, *Photochem. Photobiol.* **27**, 335 (1978).

[21] V. Sundstrom, P. M. Rentzepis, K. Peters, and M. L. Applebury, *Nature (London)* **267**, 645 (1977).

[22] T. Kobayashi, *FEBS Lett.* **106**, 313 (1979).

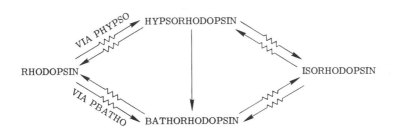

SCHEME I

of the primary photoprocesses:

$$\text{Rhodopsin} \underset{h\nu}{\overset{h\nu}{\rightleftharpoons}} \text{BATHO}$$

and

$$\text{BR}'_{570} \underset{h\nu}{\overset{h\nu}{\rightleftharpoons}} \text{K}_{610}$$

which we denote as $\phi_{R\to B}$ (or $\phi_{BR\to K}$) and $\phi_{B\to R}$ (or $\phi_{K\to BR}$) for the forward and the back photoreactions, respectively. The main observations are[23,24]:

1. Although differing in their absolute values, the yields of the forward and the back photoprocesses in both systems sum to unity, i.e., $\phi_{BR\to K} + \phi_{K\to BR} \cong 0.25 + 0.75 \cong 1$ and $\phi_{R\to B} + \phi_{B\to R} = 0.67 + 0.30 = 0.97$. [In the case of the visual rhodopsin, excitation of BATHO also yields isorhodopsin (I) in a minor side path $\phi_{B\to I} = 0.1$].

2. The values of $\phi_{R\to B}$, $\phi_{B\to R}$, $\phi_{B\to I}$, $\phi_{BR\to K}$, $\phi_{K\to BR}$ are independent of excitation wavelength and temperature over a range of 200°K.

These remarkable features, characterizing the photochemical behavior of both rhodopsin and bacteriorhodopsin, are in clear contrast with those of model compounds such as free retinal or protonated Schiff bases in solution.[23-25] For the latter there are no simple relationships between

[23] T. Rosenfeld, B. Honig, M. Ottolenghi, J. Hurley, and T. G. Ebrey, *Pure Appl. Chem.* **49**, 341 (1977).

[24] J. Hurley, T. G. Ebrey, B. Honig, and M. Ottolenghi, *Nature (London)* **270**, 540 (1977).

[25] M. Ottolenghi, *Adv. Photochem.* **12**, 97 (1980).

the yields of the forward and back reactions. Moreover, the yields exhibit complicated temperature and wavelength effects. As discussed later, all this not only points out a common primary event in visual and in halo-bacterial rhodopsins, but also constitutes the basis for the formulation of a joint model for the primary event in which the opsin plays the role of a "photochemical enzyme."

Molecular Mechanisms for the Primary Events

Models for the Structure of I, J_{625}, PBATHO, BATHO, and HYPSO. The observations described in the previous sections have led to the suggestion of essentially identical models for the primary photochemical events and for the structure of the batho-intermediates in both rhodopsin and bacteriorhodopsin. The three principal classes of suggested mechanisms are summarized in the accompanying table. Although all of them incorporate a proton translocation step in order to account for the observations of Applebury et al.[8] and Peters et al.,[7] they differ in defining the exact nature, as well as the sequence, of the primary events. While disagreeing on the nature of the groups and the (excited) potential surface associated with proton translocation, both approaches A and B consider the latter reaction as an essential primary step. On the other hand, in the isomerization model (C) proton transfer is a secondary step, taking place in the protein after (primary) isomerization to a thermalized ground-state photoproduct. This approach views proton transfer in the opsin as a relaxation of the protein environment around the isomerized chromophore, in some respects analogous to a resolvation process.

Extensive evidence, mainly based on the photoequilibria[1] attainable between rhodopsin (11-*cis*), bathorhodopsin and isorhodopsins (9-*cis* as well as 7-*cis*), supports an all-transoid chromophore structure in BATHO.[23,25,33] Similar evidence is not available for K_{610}. Nevertheless, on the basis of the analogies between BATHO and K_{610} in terms of the

[26] M. R. Fransen, W. C. M. Luyten, J. Can Thuijl, J. Lugtenburg, P. A. A. Jansen, P. J. G. Van Breugel, and F. J. M. Daemen, *Nature (London)* **260**, 726 (1976).

[27] K. Van der Meer, J. J. C. Mulden, and J. Lugtenburg, *Photochem. Photobiol.* **24**, 363 (1976).

[28] J. Fevrot, J. M. Leclerg, R. Roberge, C. Sandorfy, and D. Vocelle, *Photochem. Photobiol.* **29**, 99 (1979).

[29] A. Warshel, *Proc. Natl. Acad. Sci. U.S.A.* **75**, 2558 (1978).

[30] A. Lewis, *Proc. Natl. Acad. Sci. U.S.A.* **75**, 549 (1978).

[31] J. Hurley, T. G. Ebrey, B. Honig, and M. Ottolenghi, *Nature (London)* 540 (1977).

[32] B. Honig, T. Ebrey, R. Callender, U. Dinur, and M. Ottolenghi, *Proc. Natl. Acad. Sci. U.S.A.* **76**, 2503 (1979).

[33] B. Aton, R. Callender, and B. Honig, *Nature (London)* **273**, 784 (1978).

basic red shift, the quantum yields, and the picosecond phenomena, it was claimed that isomerization is an essential feature in the formation of K_{610} as well.[31] Since the later M_{410} intermediate in the BR_{570}^t photocycle (Fig. 1) contains an isomerized (13-*cis*) chromophore[6] (as does the analogous (all-*trans*) MII_{380} intermediate in the visual photocycle) it was suggested[31] that the primary event in light-adapted bacterorhodopsin is BR_{570}^t (all-*trans*) $\xrightarrow{h\nu}$ K_{610} (13-*cis*). The isomerization model[32] accounts for the red shifts in both BATHO and K_{610} in terms of charge separation between the protonated Schiff base and its original counterion A_1^- (see the table). This explains the fact that a red shift occurs independently of whether the primary event is [11-*cis*] → [all-*trans*] (rhodopsin) or [all-*trans*] → [13-*cis*] (bacteriorhodopsin). Charge separation was also suggested as the main mechanism of energy storage (~30 kcal) in BATHO.[32] As an additional reason for the high energy content in BATHO, chromophore distortions from planarity (induced by conformational changes in the lysine residue) have recently been suggested.[34]

Charge separation in the primary event also accounts for the observation that the 13-*cis* component (BR_{550}^{13-cis}) of dark-adapted bacteriorhodopsin[35,36] is characterized by its own batho-photoproduct K^c. However, adopting the *cis-trans* isomerization model also for BR_{550}^{13-cis}, the existence of two independent photocycles for the two forms of bacteriorhodopsin leads to a mechanistic complexity that is absent in the case of visual rhodopsins. As suggested by Lozier *et al.*,[37] this implies isomerization about at least two bonds, e.g., BR_{570}^t (13-*trans*, X-*trans*) $\xrightarrow{h\nu}$ K_{610} (13-*cis*, X-*trans*) and BR_{550}^{13-cis} (13-*cis*, X-*cis*) $\xrightarrow{h\nu}$ K^c (13-*trans*, X-*cis*), where X denotes any other polyene single bond or the C=N bond. A conceptually similar approach has been suggested by Schulten[38]: BR_{570}^t (13-*trans*, 14s-*trans*) $\xrightarrow{h\nu}$ K_{610} (13-*cis*, 14s-*trans*) and BR_{550}^{13-cis} (13-*cis*, 14s-*trans*, 15-*cis*) $\xrightarrow{h\nu}$ K^c (13-*trans*, 14s-*cis*, 15-*cis*).

In addition to their incapability to account for *cis-trans* isomerization as an essential feature of the batho-products, models A and B encounter some specific difficulties that have been extensively discussed.[23,25,39] For example, model A is inconsistent with the accumulated evidence from

[34] R. R. Birge and L. M. Hubbard, *J. Am. Chem. Soc.* **102**, 2195 (1980).

[35] O. Kalisky, C. R. Goldschmidt, and M. Ottolenghi, *Biophys. J.* **19**, 185 (1977).

[36] W. Sperling, P. Karl, C. N. Rafferty, and N. A. Dencher, *Biophys. Struct. Mech.* **3**, 79 (1977).

[37] R. H. Lozier, W. Niederberger, M. Ottolenghi, G. Sivorinovsky and W. Stoeckenius, *in* "Energetics and Structure of Halophilic Microorganisms" (S. R. Caplan and M. Ginzburg, eds.), p. 123. Elsevier North-Holland, New York, 1978.

[38] K. Schulten, *in* "Energetics and Structure of Holophilic Microorganisms" (S. R. Caplan and M. Ginzburg, eds.), p. 331. Elsevier North-Holland, New York, 1978.

[39] B. Honig, *Annu. Rev. Phys. Chem.* **29**, 31 (1978).

Assignments for Chromophore Structure, Photochemical Intermediates,

Model and reference	Basic Chromophore structure[a,b]	Sequence of primary photoreactions[c,d]	J_{625} and PBATHO
A. Proton translocation to Schiff base[7,26–28]	SB	Proton translocation in in S_1, from protein to SB	SB in thermalized S_1
B. Proton translocation in the opsin			
Warshel[29]	PSB	1. Partial (90°) polyene isomerization in S_1 2. Intraprotein proton translocation in nonthermalized S_0	Partially isomerized PSB in nonthermalized S_0
Lewis[30]	PSB	1. Intraprotein proton translocation in nonthermalized S_1 and S_0 2. Undefined change in polyene conformation	S_1, or non-relaxed S_0, but PSB not isomerized
C. Polyene isomerization[23,31,32]	PSB	1. Ultrafast relaxation to "common minimum" in S_1 (I?) 2. Thermalization to isomerized S_0 3. Protein relaxation (intraprotein proton transfer) in thermalized S_0	Isomerized PSB in thermalized S_0 with nonrelaxed opsin

[a] SB ≡ unprotonated Schiff base
[b] PSB ≡ protonated Schiff base
[c] S_1 denotes the lowest excited $^1B_u^+$ state
[d] S_0 denotes the $^1A_g^-$ ground state

AND PRIMARY EVENTS IN RHODOPSIN AND BACTERIORHODOPSIN

K_{610} and BATHO
(structures refer to BATHO only)

PSB in thermalized S_0 (polyene not isomerized)	
PSB in thermalized S_0 with proton translocated in protein, change in polyene conformation nonessential	
PSB in thermalized S_0, with proton translocated in protein and distortion about C_9–C_{10} and C_{11}–C_{12} bonds	
Isomerized PSB in thermalized S_0 with relaxed opsin (intraprotein proton transfer)	

resonance Raman spectroscopy showing that rhodopsin and BR_{570}^{t} are protonated Schiff bases and that the state of protonation is unchanged upon the formation of BATHO (see Refs.[40–42] for reviews on the application of resonance Raman spectroscopy to rhodopsins). Model B(a) yields an unreasonably low ground-state barrier for the transition from rhodopsin or bacteriorhodopsin to their batho-products. (Both products are viewed in this model as "charge-stabilized" rather than "isomerization-stabilized"[32] intermediates.)

Criteria for testing the feasibility of the various models were also derived from an analysis[10,43] of the picosecond absorption and emission kinetics, as described in the previous sections. The major conclusion is that J_{625} cannot be identified with an excited state or with an unthermalized ground state, retaining the basic (all-*trans*) conformation of BR_{570}^{t}, as suggested by models A and B. The same applies to BATHO in the rhodopsin system. The main arguments in this respect are as follows: (a) A theoretical analysis[43] has shown that the only feasible excited state is the allowed S_1, $^1B_u^+$, state. Such a state should be highly fluorescent, in variance with the experimental observations (see preceding discussion). (b) The theoretically predicted $S_i \leftarrow S_1$ absorption spectrum[10] is substantially different from that of J_{625}. Moreover, it can be readily verified that the spectrum of J_{625} differs from that expected for a vibrationally excited S_0 species, identified with J_{625} in models B(a,b).

To conclude, in models A and B, the proton translocation process originates either in an excited state (S_1) or in a "hot" ground state (S_0^*), after complete thermalization to an extremely shallow (< 1 kcal) potential minimum, from which a proton is transferred by activated barrier crossing or (at low temperatures) by tunneling. This implies the unlikely assumption that optical excitation, which results in a very large excess of vibrational energy, will lead to selective thermalization in the preceding trap. On the other hand, according to model C proton transfer occurs between protein groups that are vibrationally detached from the chromophore. The reaction, which in a way is similar to a "solvent relaxation" process, takes place around a fully thermalized ground state. It accounts for the deuteration kinetic effects, after complete dissipation of the excess of all vibronic energy.

The table shows a plausible way of proton transfer associated with the PBATHO $\rightarrow$ BATHO, or $J_{625} \rightarrow K_{610}$, processes according to model C.

[40] R. H. Callender and B. Honig, *Annu. Rev. Biophys. Bioeng.* **6**, 33 (1977).
[41] R. Mathies in "Chemical and Biological Applications of Lasers" (C. B. Moore, ed.), Vol. 4. Academic Press, New York, 1979.
[42] A. Warshel, *Annu. Rev. Biophys. Bioeng.* **6**, 273 (1977).
[43] U. Dinur, B. Honig, and M. Ottolenghi, *Photochem. Photobiol.* **33**, 523 (1981).

After separation from the original counterion (A_1^-), the Schiff base approaches a second acid group A_2H (PBATHO or J_{625} formation). Subsequent deprotonation of A_2H forms BATHO (or K_{610}). The generation of the blue-absorbing HYPSO was tentatively attributed[32] to a species in which the Schiff base itself, rather than A_2H, loses a proton at the stage of PBATHO. Deuteration effects on the picosecond generation decay of HYPSO may be used to test this hypothesis.

The *"Common Excited State"* *Hypothesis.* Based on the previously discussed absolute quantum yield values ($\phi_{R\to B}$, $\phi_{B\to R}$, $\phi_{BR\to K}$, $\phi_{K\to BR}$), as well as on their independence of temperature and excitation wavelength, a model was suggested accounting for the primary event in both rhodopsin and bacteriorhodopsin.[23,31] It was argued that such an event is controlled by the quantitative population of a common-excited-state minimum along the coordinate between the pigment and its bathophotoproduct (see Fig. 2). From such a minimum partition takes place to the two corresponding ground states. In the case of BR_{570}^t the partition ratio is 0.75 back to BR_{570}^t and 0.25 to K_{610}. In the bovine rhodopsin system the fraction decaying to BATHO is 0.7 while that repopulating the original ground state is 0.3. The existence of a "common minimum" along the $C_{11}-C_{12}$ coordinate in the lowest excited $^1B_u^+$ (S_1) state of rhodopsin has been recently confirmed by extensive PPP/CI[44] and INDO-CISD[34] theoretical calculations. Weiss and Warshel[45] have recently argued that surface crossing from such a minimum to the ground state is too fast for actual trapping in the common excited state. However, the semiempirical molecular dynamics procedure of Birge and Hubbard[34] predicts that the excited-state species is trapped during isomerization in an activated complex ("common excited state") for about 0.5 psec. In the case of rhodopsin the complex oscillates between two components, that preferentially decay (on a 1–2-psec time scale) to form a distorted all-*trans* product (BATHO) or the original 11-*cis* chromophore, both in their thermalized ground states.

Very recently, Suzuki and Callender[46] have accurately redetermined the parameter $\phi_{B\to R}$ for bovine rhodopsin, obtaining for the sum $\phi_{R\to B} + \phi_{B\to R}$ a value that is somewhat larger than 1, ($\phi_{R\to B} + \phi_{B\to R} = 0.67 + 0.5 = 1.17$). Within the framework of the "isomerization–common-excited-state" model this observation may be interpreted in two ways: (a) We have seen that rhodopsin and BATHO have different protein configurations and that protein relaxation (represented by the process PBATHO $\to$ BATHO) is slower than internal conversion to the ground

[44] U. Dinur, Ph.D. Thesis, Jerusalem (1979).
[45] R. M. Weiss and A. Warshel, *J. Am. Chem. Soc.* **101**, 6131 (1979).
[46] P. Suzuki and R. H. Callender, *Biophys. J.* **34**, 261 (1981).

state. Thus the "common excited state" produced by the excitation of rhodopsin differs in its protein environment from that produced from bathorhodopsin. This may lead to deviations of the yield sum, $\phi_{R \to B} + \phi_{B \to R}$, from unity. (b) The model of Rosenfeld et al.[23] assumes complete thermalization in the "common-excited-state" minimum. It thus implies that a large number of oscillations occur along the (11-cis)–(all-trans) coordinate before crossing to the ground state and trapping in either cis (back reaction) or trans (isomerization) configuration. As suggested by Suzuki and Callender, a deviation of $\phi_{R \to B} + \phi_{B \to R}$ from unity may imply that only a few oscillations occur in the excited state as the result of a relatively high probability of crossing to the ground state. This situation, which essentially implies incomplete thermalization in the common excited state, has been predicted theoretically by Birge and Hubbard.[34]

In both rhodopsin and bacteriorhodopsin, Rentzepis and co-workers[7,8] as well as Monger et al.[47] failed to resolve the ultrafast repopulation of the original ground-state pigments from the respective excited states. Given the significant quantum yields for ground-state repopulation, this implies that such a process is concluded within times that are shorter than the ~6-psec resolution of the preceding measurements. The "common-excited-state" hypothesis then requires that the partitions leading to the formation of J_{625} and BATHO also occur in less than ~6 psec. However, since the deuteration sensitive growing in of both batho-species at low temperatures is considerably slower, it is evident that neither J_{625} nor PBATHO can be identified with the photochemically important excited state. Thus the above considerations, as well as the very close similarity between the spectra of K_{610} and J_{625} (or of BATHO and PBATHO), strongly support the identification of PBATHO and J_{625} as perturbed ground-state chromophores (see Fig. 2).

The question arises as to the identification of the primary species I. Its assignment to a higher electronic state, S_2 ($^1A_g^-$ or $^1A_g^+$), was ruled out on the basis of theoretical considerations.[10,43] Feasible assignments are to vibrationally unrelaxed S_1 or S_0 states. Alternatively it is also reasonable to identify I with the "common excited state." This obviously requires that: (a) the original ground state of BR_{570}^t is repopulated in ~1 psec (the observed rise time of J_{625}); (b) that a ~1-psec fluorescence component (different from the observed long-lived low-temperature emission) exists for BR_{570}^t. Both predictions can, in principle, be experimentally verified. [It should be noted that the first prediction is consistent with the trajectory calculations of Birge and Hubbard.[34] As to the second, fluorescence

[47] T. R. Monger, R. R. Alfano, and R. H. Callender, Biophys. J. 27, 105 (1979).

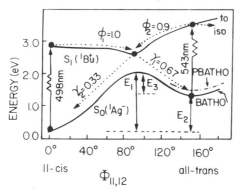

FIG. 2. Potential energy surfaces describing the primary processes in visual rhodopsin as predicted by the isomerization model.[23,24] The curves presented are those calculated by Birge and Hubbard.[34] The values of E_1, E_2, and E_3 empirically estimated by Honig et al.[32] are $E_1 \cong 37$ kcal/mol, $E_2 \cong 30$ kcal/mol, and $E_3 \cong 7$ kcal/mol. The theoretically calculated values (including the contributions of both counterion and lysine conformational distortion) are $E_1 = 26$ kcal/mol, $E_2 = 42$ kcal/mol, $E_3 = 16$ kcal/mol.[34] Note that according to the isomerization model $\gamma_1 + \gamma_2 = 1$, $\phi_{R \to B} = \gamma_1 \phi_1$, $\phi_{B \to R} = \gamma_2 \phi_2$, $\phi_1 = 1$, and $\phi_2 + \phi_3 = 1$ [ϕ_3 is the small (0.1) yield to isorhodopsin]. A similar diagram was postulated for the photochemical interconversion between BR^t_{570} and K_{610}, probably involving the C_{13}–C_{14} torsional coordinate.[24]

yields of $\phi_f \cong 2 \times 10^{-5}$ and $\phi_f \cong 2 \times 10^{-4}$ have been reported at 300°K by Shapiro et al.[14] and by Alfano et al.,[12] respectively. Using the expected radiative lifetime for an allowed S_1, $^1B_u^+$, state ($\tau_r = 6$ nsec), these numbers yield $\tau_f \sim 1$ psec and $\tau_f \sim 0.1$ psec for the fluorescence lifetime ($\tau_f = \phi_f \tau_r$) respectively. If the lower quantum yield estimate applies, then the calculated value of τ_f is consistent with the rise time of J_{625}.]

Very recently, using 30-psec laser pulses, Kryukov et al.[48] investigated the reverse photoreaction $K_{610} \xrightarrow{h\nu} BR^t_{570}$ in deuterated bacteriorhodopsin at 13°K. They found the process to be faster (below 30 psec) than the forward ~ 80-psec generation of (deuterated) K_{610}.[8,48] Assigning both precursors, that of K_{610} (i.e., J_{625}) and that of BR_{570} (yet undetected) in the back photoreaction, to excited states, they interpreted the different lifetimes of such states as being inconsistent with the common-excited-state hypothesis.[23,24] (The latter requires identical rates for the excited-state decay in the forward and in the back photoprocesses.) Consequently, they suggested a double-well potential surface in S_1, with a considerable barrier separating the two minima corresponding to BR^t_{570} and K_{610}. Obviously, such conclusions are inconsistent with the assignment of

[48] P. G. Kryukov, Yu. A. Lazarev, Yu. M. Matveetz, A. V. Sharkov, and E. L. Terpugov, submitted for publication.

J_{625} (as well as of the possible precursor of BR_{570}^t in the back reaction) to isomerized ground-state polyenes with an unrelaxed protein environment. Since the actual excited-state branching reaction (as, for example, the decay of I) was not resolved by Kryukov *et al.*, their data are not discriminative in respect to the shape of the excited-state potential energy surface.

Comparative Aspects of Later Stages in the Photocycles

Apart from the analogies previously discussed for the primary events, additional common features exist between later stages in the photocycles of visual rhodopsins and light-adapted bacteriorhodopsin. We shall briefly discuss the main features that bear directly on the function of the pigments.

Photoreversibility

A basic characteristic of the visual photocycle is that all intermediates, up to the stage of metarhodopsin II (MII_{380}), convert to rhodopsin (and partially to 9-*cis*-isorhodopsin) in an essentially identical back photoreaction.[1,49,50] Chemical extractions,[51] as well as resonance Raman spectroscopy,[52,53] unambiguously indicate that MII_{380} contains an (unprotonated) all-*trans* Schiff base (while the Raman spectrum of its precursor MI_{478} resembles that of an all-*trans* protonated Schiff base[53]). These characteristics were taken as additional evidence for an essentially all-transoid chromophore even at the early stage of BATHO.[23,25] Note, however, that (in variance with MI_{478}) the resonance Raman spectrum of BATHO is different from that of a model protonated all-*trans* Schiff base.[41] This is attributed to a strained all-*trans* conformation[54] accounting for the high energy content of BATHO as postulated by model C.[23,31,32]

Analogous back photoreactions yielding the corresponding original pigment, have also been observed for essentially all transients in the photocycles of BR_{570}^t and BR_{550}^{13-cis} (for a review, see Ref. 25). In one case, that of the M_{410} intermediate in the photocycle of BR_{570}^t, the back photoreac-

[49] B. N. Baker and T. P. Williams, *Vision Res.* **11**, 449 (1971).
[50] T. P. Williams, *Acc. Chem. Res.* **8**, 107 (1975).
[51] R. Hubbard and R. St. George, *J. Gen. Physiol.* **41**, 501 (1957).
[52] A. G. Doukas, B. Aton, R. H. Callender, and T. G. Ebrey, *Biochemistry* **17**, 2430 (1978).
[53] M. Sulkes, A. Lewis, A. Lemley, and R. Cookingham, *Proc. Natl. Acad. Sci. U.S.A.* **73**, 4266 (1976).
[54] T. Yoshizawa and G. Wald, *Nature (London)* **197**, 1279 (1963).

tion was time resolved, showing the following sequence of events[55]:

$$M_{410} \xrightarrow{h\nu} M'_{390} \xrightarrow{150 \text{ nsec}} BR'_{570} \longrightarrow BR^t_{570}$$

The first (unresolved) step, yielding the species M'_{390}, which is very close in absorption to M_{410}, was attributed to a back (13-*cis* → *trans*) isomerization. Proton transfer to the Schiff base (which is unprotonated in M_{410}[56,57]) occurs subsequently, yielding a species (BR'_{570}) that is spectroscopically undistinguishable from BR^t_{570}. The photoreaction of M_{410} has been used as an important mechanistic tool in relation to the following points[58]:

First, the proton pump mechanism in BR^t_{570} is associated with the ejection of protons to the outside of the membrane at a stage immediately following the formation of M_{412} (see Ref. 25 for a review). It was shown[37,59] that the back photoreaction from M_{410} inhibits pumping by inducing a reverse transfer in which protons are returned to the membrane from the aqueous side. Second, reprotonation of the Schiff base in the back photoreaction (i.e., formation of BR^t_{570}) involves an internal proton donor, most probably a tyrosine residue that makes part of a proton wire connecting the retinyl polyene and the outside medium.[58] 3) The back photoreaction of M_{410} was also used as a criterion for discriminating between various generalized mechanisms for pumping.[60] Mechanisms based on a (primary) light-induced accessibility change of the chromophore (class I models)[60] were compared with a model (class II) based primarily on pK changes.[58] According to the latter mechanism, the Schiff base in BR^t_{570} is accessible to the outside surface of the membrane. The primary photoisomerization of the chromophore reduces its pK without changing its accessibility to the outside. Accessibility changes to the inside membrane surface are required for completion of the pumping cycle but occur only at a later stage. An analysis of the back photoreaction from M_{410} tends to favor such models over the previous (class I) approach.

In the case of the photocycle of BR^{13-cis}_{550}[35,36] the back photoreaction from the intermediate L^C_{610} was investigated as a tool for elucidating the mechanism of light adaptation ($BR^{13-cis}_{550} \xrightarrow{h\nu} BR^t_{570}$) in bacteriorhodopsin.[37] By using double-pulse laser excitation, it was shown that branching takes

[55] O. Kalisky, U. Lachish, and M. Ottolenghi, *Photochem. Photobiol.* **28**, 261 (1978).

[56] A. Lewis, J. Spoonhower, R. A. Bogomolni, R. Lozier, and W. Stoeckenius, *Proc. Natl. Acad. Sci. U.S.A.* **71**, 4462 (1974).

[57] B. Aton, A. G. Doukas, R. H. Callender, B. Becher, and T. G. Ebrey, *Biochemistry* **16**, 2995 (1977).

[58] O. Kalisky, M. Ottolenghi, B. Honig, and R. Korenstein, *Biochemistry* **20**, 649 (1981).

[59] B. Karvaly and Z. Dancshazy, *FEBS Lett.* **76**, 36 (1977).

[60] B. Honig, in "Energetics and Structure of Halophilic Microorganisms" (S. R. Caplan and M. Ginzburg, eds.), p. 109. Elsevier North-Holland, New York, 1978.

place at the stage of L_{610}^C yielding in parallel processes both BR_{550}^{13-cis} (~97%) and BR_{570}^t) (~3%) (see Fig. 1).

Branching

Although being photoreversible, reversible thermal back reactions to the original parent pigments from intermediates in the visual or bacteriorhodopsin photocycles do not occur under physiological conditions. However, the occurrence of such processes at low temperatures has led to important mechanistic implications.

In the case of chicken iodopsin, the corresponding batho-intermediate reverts at low temperatures to the original pigment, efficiently competing with the forward (cycling) process.[61] The activation energy for this reaction has been used for estimating the energy (~30 kcal/mol) stored in bathorhodopsin.[32] According to the isomerization model (C) as illustrated in Fig. 2, the very fact that the back process, BATHO → Rhodopsin, (a double-bond, *trans* → 11-*cis* isomerization) takes place with an activation energy of only ~7 kcal/mol is clear evidence for energy storage in BATHO.

Very recently, a complicated branching scheme (Scheme II) was suggested for the photocycle of BR_{570}^t in order to account for a variety of environmental (mainly temperature and pH) factors[73]:

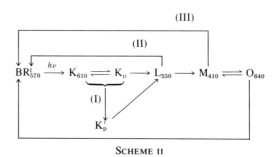

SCHEME II

Under normal conditions around room temperature a simple consecutive sequence takes place, as shown in Fig. 1. The branching associated with reaction (III) occurs at low temperatures or high pH, leading in both cases to a biphasic regeneration of BR_{570}^t. The branching associated with process (I) leads to complex kinetic patterns and to the two red-absorbing intermediates K_p and K_p' at temperatures around $-70°C$.[58] However, in

[61] R. Hubbard, D. Bounds, and T. Yoshizawa, *Cold Spring Harbor Symp. Quant. Biol.* **30**, 301 (1965).

variance with chicken iodopsin, it does not lead to a "short cut" in the photocycle, i.e., to the regeneration of BR_{570}^t.

Of special interest is the branching reaction (II), which, under neutral pH conditions, predominates below $\sim -60°$.[58] It was shown, however, that the forward $L_{550} \rightarrow M_{410}$ process is enhanced at high (> 10) pH so as to be able to compete with process (II) even at low temperatures. The effect, reflected as a titration-like dependence of the yield of M_{410}, was attributed to the catalytic action of a protein group of $pK \cong 10$.[58] This observation, along with evidence[62,63] indicating that a tyrosine moiety deprotonates during the $L_{550} \rightarrow M_{412}$ transition, suggested that the formation of a tyrosine ion is a prerequisite for deprotonation of the Schiff base during the photocycle.[58]

MII_{380} and M_{412}

It is important to summarize the main analogies between MII_{380} and M_{412}—first, from the limited aspect of the polyene chromophore:

1. Both species are deprotonated Schiff bases, isomerized in respect to their (protonated) parent pigments. In each case there is an equilibrium between the deprotonated, blue-absorbing chromophore and a corresponding protonated, red-absorbing modification, i.e.,

$$MI_{478} \overset{H^+}{\rightleftharpoons} MII_{380}$$

$$O_{640} \rightleftharpoons M_{410}$$

(Note, however, that in the BR_{570}^t photocycle M_{410} precedes O_{640} while in visual rhodopsins MI_{478} precedes MII_{380}.) The first equilibrium[64,65] is pH dependent ($pK = 6.4$). The transition from MI_{478} to MII_{380}, although being associated with deprotonation of the Schiff base, involves a net uptake of a proton from the external medium. Although the amount of O_{640} decreases at high pH,[66] it is unclear whether this effect is due to a pH dependence of the $O_{640} \rightleftharpoons M_{410}$ equilibrium.[37]

2. Both MII_{380} and M_{412} undergo a back photoreaction yielding a primary photoproduct, close in absorbance to the corresponding M species (isomerization), which subsequently decays to the parent pigment in a thermal reaction assigned to protonation of the Schiff base (see preceding paragraph 1).

Apart from the similarities between the chromophore structure and

[62] R. A. Bogomolni, L. Stubbs, and J. K. Lanyi, *Biochemistry* **17**, 1037 (1978).
[63] B. Hess and D. Kuschmitz, *FEBS Lett.* **100**, 334 (1979).
[64] R. G. Matthews, R. Hubbard, P. K. Brown, and J. Wald, *J. Gen. Physiol.* **47**, 215 (1963).
[65] S. Ostroy, F. Erhardt, and E. W. Abrahamson, *Biochim. Biophys. Acta* **112**, 265 (1966).
[66] R. H. Lozier and W. Niederberger, *Fed. Proc., Fed. Am. Soc. Exp. Biol.* **36**, 1805 (1977).

the photochemical behavior of MII_{380} and M_{410}, accumulated evidence is now available indicating that both intermediates play a key role in determining the biological function of the respective pigments. This conclusion deserves special attention in view of the different roles played by the two systems: the release of a diffusible transmitter in the visual process and the pumping of protons in bacteriorhodopsin. In the following sections we shall briefly review the relation between each of the M intermediates and the corresponding pigment function.

Vision. In spite of the lack of a detailed correlation between the photochemical events as monitored spectroscopically and the electrical events in the retina,[67,68] it is generally believed that the transmitter release and the corresponding "late receptor potential (LRP)"[67,68] are initiated by protein conformational changes occurring as a result of the generation of MII_{380}.[64,69] The evidence in this respect is indirect. It is mainly based on the observation that the activation parameters for the preceding thermal steps (those associated (see Fig. 1) with $BATHO_{543}$, $LUMI_{497}$ and MI_{478}) do not suggest large protein conformational changes.[70] Such changes do, however, occur at the stage of MII_{380} (see Ref. 25 for a brief recent review). It appears that a rearrangement in the tertiary structure of the opsin has taken place during the $MI_{478} \rightarrow MII_{380}$ transition and that such a change, although relatively moderate and reversible, is sufficient for inducing the transmitter release, thus triggering the transduction process. The detailed relationships between the change in the chromophore structure, the protein conformational change, and the transmitter release, are at present unknown.

Bacteriorhodopsin. There is evidence indicating that in the case of bacteriorhodopsin the retinyl moiety is not only responsible for light absorption and for energy storage, but may be directly involved as a vectorial gate in the proton pump mechanism. Moreover, the initial proton release is correlated with the generation of M_{410}. The main observations in this respect are as follows: (a) The generation of M_{410} involves deprotonation of the Schiff base.[56,57] (b) Evidence demonstrating additional proton translocation during the $L_{550} \rightarrow M_{410}$ transition is provided by light-induced absorbance changes in the UV, indicating that a tyrosine residue is deprotonated at this state.[62,63] (c) The ejection of protons to the outside of the membrane occurs on a time scale ($\sim 10^{-4}$ sec) comparable to that of the rate of M_{410} formation.[71,72] No faster (down to $\sim 10^{-6}$ sec) proton re-

[67] W. A. Hagins, *Annu. Rev. Biophys. Bioen.* **1**, 131 (1972).
[68] W. A. Hagins and S. Yoshikani, *Biophys. J.* **17**, 196a (1977).
[69] E. W. Abrahamson and S. E. Ostroy, *Prog. Biophys. Mol. Biol.* **17**, 179 (1967).
[70] E. W. Abrahamson, *Acc. Chem. Res.* **8**, 101 (1975).
[71] R. H. Lozier, W. Niederberger, R. A. Bogomolni, S.-B. Hwang, and W. Stoeckenius, *Biochim. Biophys. Acta* **440**, 545 (1976).
[72] D. R. Ort and W. W. Parson, *J. Biol. Chem.* **253**, 6158 (1978); *Biophys. J.* **25**, 341 (1979).

lease or uptake phenomena have been observed (see discussions in Refs. 25, 37). (d) Evidence has been obtained[37] showing that no proton pumping activity is associated with the photocycle of BR_{13-cis},[35,36] which lacks a deprotonated species comparable to M_{410} (Fig. 1).

Based on an analysis of the preceding observations and also (see earlier relevant discussions) on the titration-like pH dependence of the branching at the stage of L_{550} and on the back photoreactions from M_{410}, a scheme (Fig. 3) was suggested[58] to describe the role of the retinyl moiety in the pump mechanism. The primary photoisomerization event in the polyene induces a change in the proton affinity of a neighboring tyrosine, which as a result transfers its proton to an acceptor group A_2^-. The negatively charged tyrosinate subsequently catalyzes the transfer of the Schiff base proton to another acceptor group (A_1^-), yielding the M_{410} species. The existence of the internal proton traps, A_1^- and A_2^-, is implied by the time delay observed between the appearance of protons in the outside medium and the (preceding) deprotonations of the tyrosine and of the Schiff base. It is also consistent with the generation of M_{410} even in the absence of the external aqueous medium.[74] On the basis of the time scale ($\sim 10^{-4}$ sec) as-

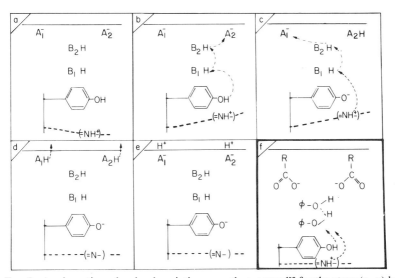

FIG. 3. A schematic molecular description recently proposed[57] for the steps (a–e) leading to the photoejection of up to two protons to the outside of the purple membrane. The upper horizontal line represents the external aqueous surface of the cell. f shows plausible identifications for groups A_1^-, A_2^-, B_1H, and B_2H.

[73] O. Kalisky and M. Ottolenghi, *Photochem. Photobiol.* **35**, 109 (1982).
[74] R. Korenstein and B. Hess, *Nature (London)* **270**, 184 (1977).

sociated with proton release, the pK values of A_1^- and A_2^- were estimated[58] as lower than ~ 6. This implies that the pK of the tyrosine moiety and of the Schiff base are reduced during the photocycle from ~ 10 and > 12, respectively, to below ~ 5. The mechanism implies the translocation of up to two protons per photon and is thus consistent with the observations of a maximum of two pumped protons per photocycle.[72,75,76] The groups B_1H and B_2H are introduced as internal proton donors to the Schiff base (and possibly to the tyrosinate), accounting for the generation of M'_{390} in the back photoreaction (see preceding discussion). Since B_1H must connect the Schiff base with the outside (rather than the inside) of the membrane[58] it was suggested that such groups play the additional role of mediating the initial proton transfer to A_1^- and A_2^-. Consequently, it was postulated that they constitute a short segment of a proton wire of the type suggested by Nagle and Morowitz[77] (Fig. 3).

Summarizing Remarks

The photocycles of visual rhodopsins and of bacteriorhodopsin are both initiated by the generation of a red-shifted intermediate (BATHO and K_{610}, respectively) that stores a large fraction of the photon energy. Accumulated evidence indicates that in BATHO and K_{610} the retinyl moiety has undergone a double-bond isomerization process, about $C_{11}-C_{12}$ (rhodopsin) and $C_{13}-C_{14}$ (BR^t_{570}). The formation of the batho-intermediates occurs via a deuteration-sensitive thermal process, from precursors (PBATHO or J_{625}) that are identified as thermalized ground-state molecules. The latter process involves proton translocation in the opsin with no changes in the retinyl structure. Thus the primary ground-state products, PBATHO and J_{625}, are already isomerized in respect to their parent pigments.

In both rhodopsin and bacteriorhodopsin the excited-state surface responsible for the photochemical act is barrierless, with a common minimum along the coordinate separating the pigment and its batho-product. This accounts for the high yields and for the specificity of the two photoreactions. It is possible that in the case of BR^t_{570} the "common excited state" can be identified with an ultrafast (~ 1 psec) precursor of J_{625}. In both systems the role of the protein is to catalyze isomerization, thus exhibiting a "photochemical enzyme" activity.

[75] B. Hess and D. Kuschmitz, in "Frontiers in Biological Energetics" Vol. 1, p. 257. Academic Press, New York, 1978.
[76] R. A. Bogomolni, R. H. Lozier, G. Sivorinovsky, and W. Stoeckenius, Biophys. J. **25** (Abstr. W-AM-Po101) (1974).
[77] J. F. Nagle and H. J. Morowitz, Proc. Natl. Acad. Sci. U.S.A. **75,** 298 (1978).

The controversies associated with the detection of the early interme-diate, hypsorhodopsin, in visual pigments are discussed. It is possible that HYPSO is a species formed from PBATHO by reversible deprotonation of the Schiff base. An additional difference between visual and halobac-terial pigments is the occurrence of two distinct photocycles for the two isomers BR_{570}^t and BR_{550}^{13-cis}. This phenomenon is accounted for by photo-isomerizations about at least two polyene bonds.

In spite of their different biological roles, the functions of rhodopsin and all-*trans*-bacteriorhodopsin are both closely related to the later inter-mediates, MII_{380} and M_{410}, in which the Schiff base nitrogen has under-gone a deprotonation process. In visual rhodopsins the MII_{380} formation is associated with a substantial conformational change in the opsin. How-ever, the exact relationship between such changes and the release of the diffusible transmitter is unknown. In bacteriorhodopsin the relation be-tween M_{410} and the pump mechanism appears to be more direct. It is plau-sible that the Schiff base functions as a molecular switch in the proton translocation process. A plausible mechanism, involving the participation of several opsin acid residues, is discussed.

Acknowledgments

The author is indebted to the Israeli Commission for Basic Research and to the U.S.–Is-rael Binational Science Foundation for supporting this work.

[63] Use of High-Performance Liquid Chromatography to Separate and Identify Retinals

By RICHARD A. BLATCHLY and KOJI NAKANISHI

Introduction

High-performance liquid chromatography (HPLC) is ideally suited to the task of separating various isomers of retinal or retinal analogs for the following reasons:

1. Very high resolution. A typical analytical column (4 × 250 mm) has 10,000 theoretical plates. Baseline separation of two or three components can be effected on a mixture that is one spot on TLC.

2. High sensitivity. High ϵ-values of most retinals and analogs allows detection (differential UV) of about 1 ng.

3. Mild conditions. The sample is applied to a solvent-wetted column,

and is completely shielded from light, thus minimizing thermal, chemical, or photochemical isomerization.

4. Convenience. A typical separation is effected in less than 20 min, and UV detection provides a chromatogram that can be integrated to give absolute quantities of the compounds present.

In the last two years several review articles dealing with HPLC of retinals and retinal analogs (in particular, vitamin A and vitamin A esters) have appeared.[1-4] Thus this article will deal with analytical HPLC as it relates to less common retinal analogs. It will also deal with separation of larger quantities of retinals and analogs using techniques of preparative and semipreparative LC. The latter is vital to anyone initiating biochemical studies in which a relatively large sample is needed, i.e., from 1 mg for visual pigments to several grams for toxicity tests.

Instrumentation and Precautions

Our work with retinals uses Waters LC pumps and Schoeffels variable-wavelength detectors equipped with a scanning motor enabling us to obtain UV spectra on eluants before collection, thus minimizing the chance of sample decomposition. The analytical work involves micro-Porasil, micro-CN, and micro-Bondapak columns, all by Waters; semi-prep columns are Whatman ODS-10 reverse-phase (10 × 500 mm) and Altex LiChrosorb packed columns (10 × 250 mm). All analytical and semi-prep columns were protected by using an appropriate Whatman pre-column. These pre-columns can be easily repacked if sediment or decomposed compound cause them to deteriorate. Prep LC uses the Waters 500 LC system and columns. The internal R.I. detector of the prep LC is used for > 50 mg; for < 50 mg a modified (to decrease sensitivity) Uvidec fixed-wavelength detector is used.

Analytical Section

The high resolution and speed of HPLC make it an ideal technique for analyzing complex mixtures of retinal isomers. In most cases, careful choice of conditions enables one to achieve baseline separation of all the components. Generally, the separation is carried out using a silica gel column (5–10 μm particle size) and ether–hexane solvent (~2–12% ether in

[1] M. A. Adams and K. Nakanishi, *J. Liq. Chromatogr.* **2**, 1097 (1979).
[2] G. W. T. Groenendijk, this series, Vol. 67, p. 203.
[3] A. M. McCormick, J. L. Napoli, and H. F. DeLuca, this series, Vol. 67, 220.
[4] R. F. Taylor and M. Ikawa, this series, Vol. 67, p. 233.

hexane). Elution time and order vary slightly with conditions, but the mono-*cis* compounds normally are eluted in the order: 13-*cis*, 11-*cis*, 9-*cis*, 7-*cis*, followed by all-*trans*.[5] However, 9-bromo-9-demethylretinal, when irradiated in acetonitrile by fluorescent light, gives a mixture that elutes (10% ether–hexane) in the order: 13-*cis*, 7-*cis*, unknown, all-*trans*, 9-*cis*.[6] As can be expected, there is no standard elution sequence among the various isomers, particularly in the case of retinal analogs, and hence it is imperative that all isomers be fully characterized by ^{1}H-NMR measurements.

Studies of retinal photoisomerization have used HPLC analysis extensively. Early studies[7] using TLC or UV methods were imprecise or inaccurate; the low resolution of each method made it impossible to detect the large number of isomers that we now know result from photochemistry of retinal. Use of HPLC analysis not only enables one to determine accurately the relative quantities of isomers, but also requires only extremely small amounts of material. Quantum yields can be determined with an accuracy of ± 15% from photolysis of 5 nmol of retinals, with 10% conversion.[8]

When making artificial rhodopsins, particularly from sensitive retinal analogs, it is necessary to ensure that the retinal is not altered on binding. Originally, the rhodopsin was thermally denatured, and the retinal was extracted and analyzed by TLC or HPLC.[5,9,10]

This procedure, however, leads to artifacts due to thermal isomerization or decomposition of the chromophore. Thus a nonthermal and nonisomerizing process was developed, involving methylene chloride denaturation of the rhodopsin, with concomitant extraction of the chromophore.[11] Along with the chromophore, however, some detergent and lipids are also extracted; they would stick to silica gel columns, destroying them. However, use of a bonded-phase column, compatible with either normal or reverse-phase conditions, allows this mixture to be injected directly for analysis (1% ether–hexane). The lipid and detergent are subsequently washed off the column with chloroform, methanol, water, followed by methanol, chloroform, and ether–hexane for reequilibration.

[5] A. E. Asato and R. S. H. Liu, *J. Am. Chem. Soc.* **97**, 4128 (1975).
[6] M. G. Motto *et al.*, unpublished.
[7] For example, see A. Kropf and R. Hubbard, *Photochem. Photobiol.*, **12**, 249 (1970).
[8] W. A. Waddell, R. K. Crouch, K. Nakanishi, and N. J. Turro, *J. Am. Chem. Soc.* **98**, 4189 (1976).
[9] W. K. Chan, K. Nakanishi, T. G. Ebrey, and B. Honig, *J. Am. Chem. Soc.* **96**, 3642 (1974).
[10] J. P. Rotmans and A. Kropf, *Vision Res.* **15**, 1301 (1975).
[11] F. G. Pilkiewicz, M. J. Pettei, A. P. Yudd, and K. Nakanishi, *Exp. Eye Res.* **24**, 421 (1977).

A modification of this method, applicable if the rhodopsin is a suspension rather than in detergent solution, involves the same methylene chloride extraction, followed by drying over $MgSO_4$ and direct analysis on HPLC (silical gel, ether–hexane solvents).[4] This is particularly helpful if, as in the case of some artificial chromophores, the separation is not satisfactory on the micro-CN column. Yoshizawa and co-workers used this technique to great advantage in isolating 7-*cis*-retinal from rhodopsin irradiated at low temperatures.[12]

Preparative Section

Preparation of pure sample of a retinal or retinal analog is one of the most important aspects of retinal synthesis. A minor component present in a retinal sample may produce a spurious pigment, or worse, may completely inhibit pigment formation. There are three options, depending on scale.

1. Preparative liquid chromatography (prep LC) is capable of performing separations on the largest quantity of material. Use of large-diameter (5-cm) columns and solvent pumps capable of delivering solvent at

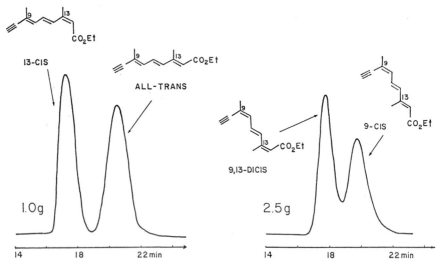

Fig. 1. Separation of acetylenic ester side-chain analogs. Waters Prep LC-500, two silica gel columns, 1% ether in hexane, 250 ml/min, R.I. detector.

[12] A. Maeda, T. Ogurusu, Y. Shichida, R. Tokunaga, and T. Yoshizawa, *FEBS Lett.* **92**, 77 (1978).

0.5 liter/min enables one to separate samples of greater than 50 g. Although it has no theoretical lower limit on sample size, the large solvent volume (flow rate is typically 250 ml/min) gives a practical limit of ~ 10–25 mg. A mixture of 13-*cis*- and all-*trans*-retinal (32 mg) were separated with quantitative recovery of 100% pure isomers.[13]

A typical separation is exemplified in Fig. 1, that of the acetylenic side-chain precursor used in a recently published synthesis of retinal.[14] Up to 3 g of a mixture of all-*trans* and 13-*cis* was easily separated in less than 25 min, using 1% ether in hexane.

An example of a large-scale separation on the prep LC is given in Fig. 2. Reduction of the acetylenic diol II[15] with LAH gave a mixture of four components. Injection of 37 g of the reaction mixture yielded four pure fractions in less than 25 min. A comparable separation using flash chromatography would have taken about 20 hr.

2. While flash chromotography[16] may not be useful for large separations, it is extremely useful for routine, low-resolution separations of samples for the 5 mg to 2 g range; it is also an excellent technique for pre-purification of retinal samples before HPLC and for purification of synthetic intermediates. This technique involves the use of fine mesh silica gel

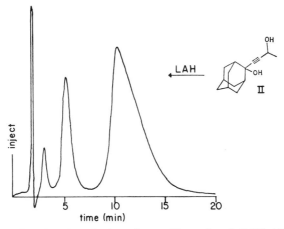

FIG. 2. Separation of crude reduction mixture, Waters Prep LC-500, 10% ether in hexane, 250 ml/min, two silica gel columns, R.I. detector.

[13] M. J. Pettei, F. G. Pilkiewicz, and K. Nakanishi, *Tetrahedron Lett.* p. 2083 (1977).
[14] F. Derguini, V. Balogh-Nair, and K. Nakanishi, *Tetrahedron Lett.* p. 4899 (1979).
[15] R. A. Blatchly, J. Carriker, V. Balogh-Nair, and K. Nakanishi, *J. Am. Chem. Soc.* **102,** 2495 (1980).
[16] W. C. Still, M. Kahn, and A. Mitra, *J. Org. Chem.* **43,** 2923 (1978).

(TLC grade), thick-walled glass columns, and solvent mixtures slightly less polar than those used for HPLC. A moderately low air pressure gives a flow rate fast enough to complete the separation in less than 20 min.

3. Semiprep LC entails use of a normal LC pump and detector with a larger diameter (e.g., 10 mm) column. This enables one to load significantly more sample (1–5 mg) without overloading the column. However, fraction size is typically less than 10 ml, as opposed to about 1 liter, as with prep LC. Resolution is nearly identical to analytical HPLC, and is much better than that of prep LC. These features make semiprep LC ideal for separations of up to 50 mg of a mixture.

[64] Synthetic Analogs of Retinal, Bacteriorhodopsin, and Bovine Rhodopsin

By VALERIA BALOGH-NAIR and KOJI NAKANISHI

The synthetic routes leading to analogs of retinal in most cases follow methodology developed for the synthesis of vitamin A and carotenoids. The classical reference book, entitled "Carotenoids," edited by Otto Isler,[1] gives an excellent summary of the state of arts with references up to the middle of 1970. The total syntheses described there make use of the Wittig and Horner condensation of carbonyl compounds with triphenylphosphonium halides or dialkyl phosphonates, Grignard or Nef reaction of carbonyl compounds with metal acetylides, aldol condensations, Reformatsky reaction of carbonyl compounds with α- or γ-haloesters and nitriles, Knovenagel–Doebner condensation of aldehydes with compounds possessing activated hydrogens, Wurtz-type reactions, etc., to mention only the few most commonly utilized reaction types. Recent reviews on the enol ether synthesis of polyenes,[2] technical syntheses of vitamin A and carotenoids,[3] and synthesis of carotenoids and polyenes[4] appeared in 1976. Other recent and useful synthetic procedures are outlined in the following paragraphs.

The alkylation of allylic sulfones and their subsequent 1,2-elimination[5] lead to a number of synthetic procedures for the preparation of vitamin A (Scheme 1).

[1] H. Mayer and O. Isler, in "Carotenoids" (O. Isler, ed.), p. 325. Birkhäuser, Basel, 1971.
[2] S. M. Makin, Pure Appl. Chem. **47,** 173 (1976).
[3] F. Kienzle, Pure Appl. Chem. **47,** 183 (1976).
[4] B. C. L. Weedon, Pure Appl. Chem. **47,** 161 (1976).
[5] M. Julia and D. Arnould, Bull. Soc. Chim. Fr. p. 746 (1973); S. Torii, K. Uneyama, and M. Isihara, Chem. Lett. p. 479 (1975); K. Uneyama and S. Torii, Tetrahedron Lett. p. 443

METHODS IN ENZYMOLOGY, VOL. 88

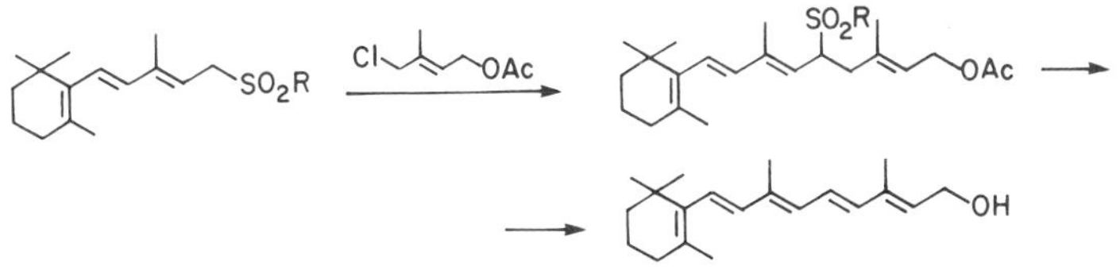

SCHEME I

A stereospecific synthesis of vitamin A from 2,2,6-trimethylcyclohex-
anone[6] used as a key reaction a novel vanadium(V)-catalyzed rearrangement of an ethynyl-substituted 2,2,6-trimethylcyclohexenyl derivative to
obtain an 8-oxo compound, which then could be transformed into a 7,8
double bond (Scheme 2).

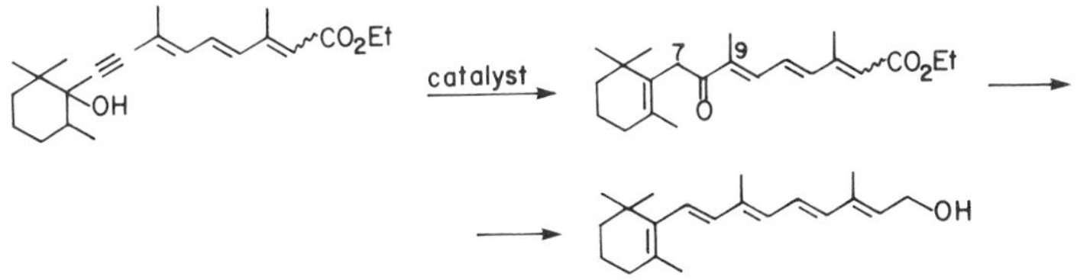

SCHEME II

A versatile synthesis of retinal and its analogs containing modified rings
was accomplished by condensation of key intermediates 3,7-dimethyl-2,4,
6-nonatrien-8-ynoates, i.e., the entire side chain of retinal, to cyclic ke-
tones[7] (Scheme 3).

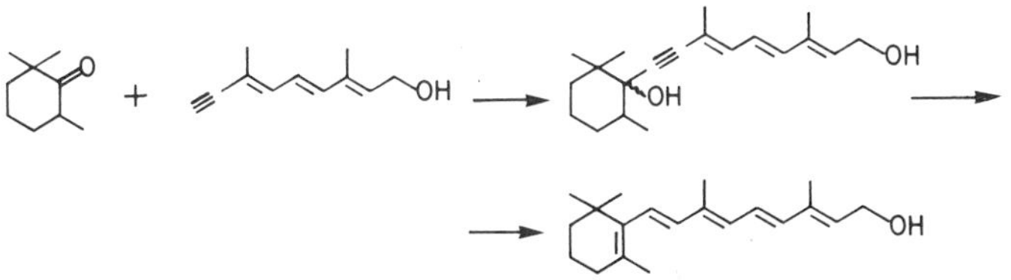

SCHEME III

(1976); G. L. Olson, H.-C. Cheung, K. D. Morgan, C. Neukom, and G. Saucy, *J. Org.
Chem.* **41**, 3287 (1976); P. S. Manchand, M. Rosenberger, G. Saucy, P. A. Wehrli, H.
Wong, L. Chambers, M. P. Ferro, and W. Jackson, *Helv. Chim. Acta* **59**, 387 (1976); M.
Julia and D. Uguen, *Bull. Soc. Chim. Fr.* p. 513 (1976); A. Fischly, H. Mayer, W. Simon,
and H.-J. Stoller, *Helv. Chim. Acta* 59, 397 (1976); P. Chabardes, J. P. Decor, and J.
Varagnat, *Tetrahedron* 33, 2799 (1977); P. S. Manchand, H. S. Wong, and J. F. Blount, *J.
Org. Chem.* **43**, 4769 (1978).
[6] G. L. Olson, H.-C. Cheung, K. D. Morgan, R. Borer, and G. Saucy, *Helv. Chim. Acta*
59, 567 (1976).
[7] F. Derguini, V. Balogh-Nair, and K. Nakanishi, *Tetrahedron Lett.* p. 4899 (1979).

Directed aldol condensation coupled with separation of isomers by preparative liquid chromatography (PLC) was found to be a rapid and efficient procedure for the preparation of pure *cis* and *trans* isomers of the methylated analogs of retinal[8] (Scheme 4).

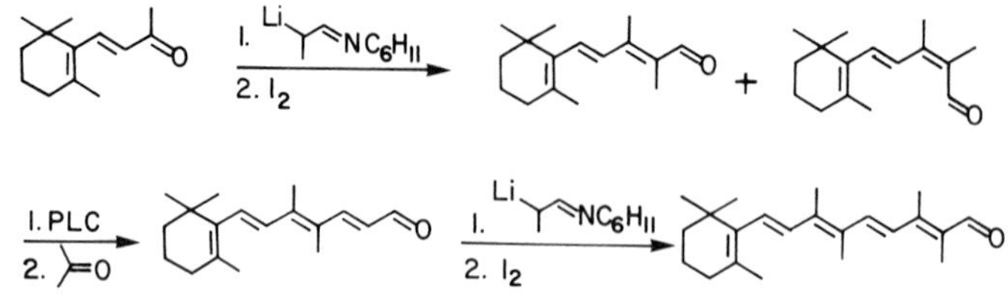

SCHEME IV

Condensation of the α-anion of sodium 3-methyl-2-butenoate[9] or the dianion obtained from the metallation of 3-methyl-but-3-en-1-ol,[10] as isoprene units, with β-ionylidene acetaldehyde gave key intermediates used in the synthesis of vitamin A (Scheme 5).

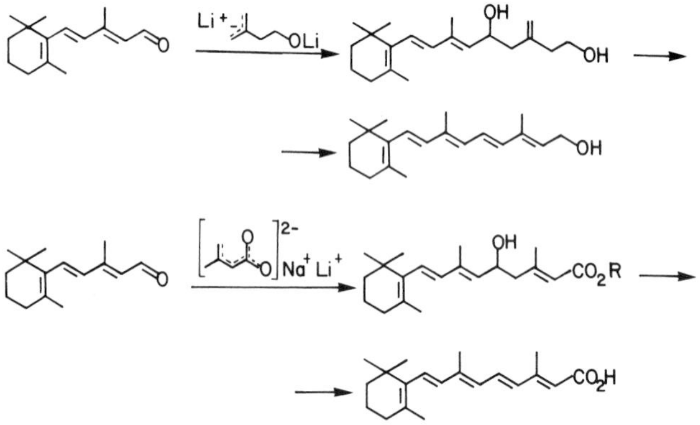

SCHEME V

The reaction of acetals with enol ethers provides a valuable supplement to the aldol condensation. Thus the reaction between α,β-unsaturated aldehyde acetals and vinyl ethers or alkoxydienes has been used with considerable success in polyene synthesis.[2] In Mukaiyama's proce-

[8] S. P. Tanis, R. H. Brown, and K. Nakanishi, *Tetrahedron Lett.* p. 869 (1978).
[9] G. Cainelli, G. Cardillo, M. Contento, P. Graselli, and A. U. Ronchi, *Gazz. Chim. Ital.* **103**, 117 (1973).
[10] G. Cardillo, M. Contento, S. Sandri, and M. Panunzio, *J. Chem. Soc., Perkin Trans.* p. 1729 (1979).

dure[11] for the synthesis of vitamin A, dienoxysilanes in presence of TiCl$_4$/Ti(i-Pr)$_4$ catalysts are used; this selectively leads to cross-aldol-type addition products in high yields (Scheme 6).

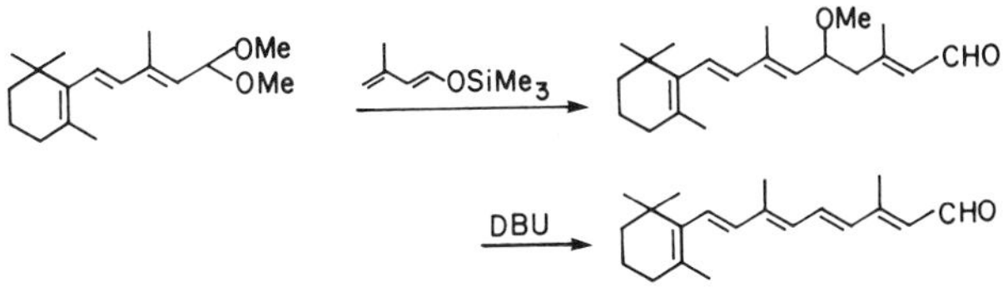

SCHEME VI

Russian workers reported the use of trimethyl and triphenysilyl protecting groups[12] in vitamin A synthesis. Stereoselective synthesis of the hitherto unknown 7-*cis* isomers of retinal was achieved by Liu's group via a photochemical reaction[13] (Scheme 7).

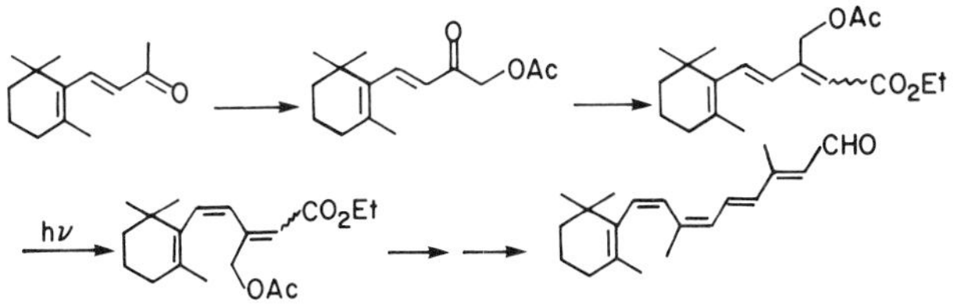

SCHEME VII

Cis isomers of retinal and modified retinals suitable for formation of visual pigment analogs can be prepared by irradiation of the corresponding *trans* isomers. Irradiation conditions, such as the polarity of the solvent, determine which *cis* isomer(s) is the major product; e.g., photoisomerization of the all-*trans*-retinal in nonpolar solvents gives 13-*cis* and 9-*cis* retinals as major products, whereas in polar solvents the 11-*cis* isomer is predominant along with the 9- and 13-*cis* isomers.

Recent reviews in Japanese on the binding of the synthetic analogs of

[11] T. Mukaiyama and A. Ishida, *Chem. Lett.* p. 1201 (1975).

[12] B. I. Mitsner, I. A. Vasilenko, N. A. Sokolova, E. N. Zvonkova, G. A. Serebrennikova and R. P. Evstigneeva, *Biol. Akt. Soedin. Elem. IV B Gruppy* p. 36 (1977); N. A. Sokolova, B. I. Mitsner, N. Yu. Gorina, R. P. Evstigneeva, S. S. Schukolyukov, E. P. Chizhevich, and V. P. Korchagin, *Bioorg. Khim.* **3**, 1234 (1977).

[13] V. Ramamurthy and R. S. H. Liu, *Tetrahedron* **31**, 201 (1975); A. E. Asato and R. S. H. Liu, *J. Am. Chem. Soc.* **97**, 4128 (1975); V. Ramamurthy, G. Tustin, C. Yau, and R. S. H. Liu, *Tetrahedron 31,* 193 (1975); A. Kini, H. Matsumoto, and R. S. H. Liu, *J. Am. Chem. Soc.* **101**, 5078 (1979).

retinal to visual opsins were published by Yoshizawa *et al.*,[14] and a partial list of the known bacteriorhodopsin analogs is described by Stoeckenius *et al.*[15] Tables I–IV list references for the synthesis and binding studies carried out to obtain bacterio- and bovine rhodopsin analogs. Table I lists the absorption maxima of artificial bacteriorhodopsin prepared from the all-*trans* isomers of various model retinals, whereas Table II, lists the analogs of retinal that do not yield bacteriorhodopsins. Similarly, Table III shows the structures of retinals and absorption maxima of visual pigment analogs formed either from the 11-*cis* or 9-*cis* isomer (indicated by *); Table IV lists synthetic derivatives not capable of forming rhodopsins.

Precautions

Since many of the reactions referred to in this chapter require the use of anhydrous conditions and inert atmosphere, some experience on the part of the operator is essential. In addition, the preparation of highly modified retinals requires good experience in organic synthesis and also in the handling of light-, air-, acid-sensitive, and often thermally labile compounds. In order to avoid photoisomerization and/or photodecomposition of synthetic intermediates and retinals, synthesis and purification have to be carried out in a dark room (dim red light—e.g., red fluorescent tubes with red filters from Illumination Tech. Inc., Fairfield, N.J., are convenient to use). Glassware should not be washed with "Chromerge" or similar cleaning solutions containing chromium ions, and use of solvent containing traces of acid should be avoided, unless the reaction specifically calls for the use of acid as reagent or catalyst. Oxidative degradation of intermediates and retinals should be minimized by handling and storage under inert atmosphere such as argon gas or pure nitrogen. Many of the synthetic analogs are thermally more labile than retinal; therefore they should be stored under inert atmosphere at low temperatures (− 70°), either in pure state or in hexane solution, depending on case.

Chromatographic techniques, such as thin-layer, flash-column,[16] and preparative liquid chromatography are useful for the purification of synthetic intermediates and retinals. Although these techniques permit the separation of geometrical isomers, thin-layer chromatography is not sufficient to ascertain the isomeric purity of synthetic retinals; high-pressure liquid chromatography (HPLC) should be used for this purpose.

[14] T. Yoshizawa and Y. Shichida, *Kagaku no Ryoiki* **32**, 159 (1978); T. Yoshizawa and O. Muto, *Taisha* (Metabolism), Tokyo, **18**, 49 (1981).
[15] W. Stoeckenius, R. H. Lozier, and R. A. Bogomolni, *Biochim. Biophys. Acta* **505**, 215 (1978).
[16] W. C. Still, M. Kahn, and A. Mitra, *J. Org. Chem.* **43**, 2923 (1978).

TABLE I
Retinal Analogs that Bind with Bacterioopsin, and the
Absorption Maxima of Pigments Formed

Analogs	Absorption maxima	Analogs	Absorption maxima
	$530^{a,b}$		$593^{c,d,e}$
	560^{f}		a
	476^{g}		400^{g}
	325^{g}		$510^{a,b}$
	$540^{a,h,i}$		$520^{e,h,j}$
	$480^{i,k}$		$535^{i,k}$
	$480^{a,b}$		$535^{i,k}$
	525^{i}		$595^{i,k}$

[a] D. Oesterhelt and V. Christoffel, *Biochem. Soc. Trans.* **4**, 556 (1976).

[b] M. A. Markus, A. Lewis, E. Racker, and H. Crespi, *Biochem. Biophys. Res. Commun.* **78**, 669 (1977).

[c] W. Stoeckenius and R. Rowan, *J. Cell Biol.* **34**, 305 (1967); W. Stoeckenius and W. H. Kunau, *ibid.* **38**, 337 (1968).

[d] W. Stoeckenius, R. H. Lozier, and R. A. Bogomolni, *Biochim. Biophys. Acta* **505**, 215 (1978).

[e] A. E. Blaurock and W. Stoeckenius, *Nature (London), New Biol.* **233**, 152 (1971); F. Tokunaga and T. Ebrey, *Biochemistry* **17**, 1915 (1978).

[f] F. Tokunaga, R. Govindjee, T. G. Ebrey, and R. Crouch, *Biophys. J.* **19**, 191 (1977).

[g] K. Nakanishi, V. Balogh-Nair, M. Arnaboldi, K. Tsujimoto, and B. Honig, *J. Am. Chem. Soc.* **102**, 7945 (1980).

[h] N. A. Sokolova, B. I. Mitsner, and V. I. Zakis, *Bioorg. Khim.* **5**, 1053 (1979).

[i] V. Balogh-Nair, J. D. Carriker, B. Honig, V. Kamat, M. G. Motto, K. Nakanishi, R. Sen, M. Sheves, M. A. Tanis, and K. Tsujimoto, *Photochem. Photobiol.* **33**, 483 (1981).

[j] M. Sumper and G. Herrman, *FEBS Lett.* **71**, 333 (1976).

[k] M. G. Motto, M. Sheves, K. Tsujimoto, V. Balogh-Nair, and K. Nakanishi, *J. Am. Chem. Soc.* **102**, 7947 (1980).

TABLE II
RETINAL ANALOGS THAT DO NOT AFFORD BACTERIORHODOPSIN ANALOGS

Analog	Ref.	Analog	Ref.
(structure)	a	*(structure)*	b
(structure, labeled 4)	b	*(structure, labeled 3)*	b
(structure, Br)	c	*(structure)*	d
9-cis retinal	e	11-cis retinal	e
9,13-dicis retinal	e	Retinol	a

[a] W. Stoeckenius, R. H. Lozier, and R. A. Bogomolni, *Biochim. Biophys. Acta* **505**, 215 (1978).

[b] M. A. Markus, A. Lewis, E. Racker, and H. Crespi, *Biochem. Biophys. Res. Commun.* **78**, 669 (1977).

[c] M. G. Motto, M. Sheves, K. Tsujimoto, V. Balogh-Nair, and K. Nakanishi, unpublished results.

[d] M. G. Motto, M. Sheves, K. Tsujimoto, V. Balogh-Nair, and K. Nakanishi, *J. Am. Chem. Soc.* **102**, 7947 (1980).

[e] W. Stoeckenius and R. Rowan, *J. Cell Biol.* **34**, 305 (1967); W. Stoeckenius and W. H. Kunau, *ibid.* **38**, 337 (1968).

TABLE III
RETINAL ANALOGS THAT BIND TO BOVINE OPSIN AND THE ABSORPTION MAXIMA OF THE VISUAL PIGMENT ANALOGS FORMED

Analog	Absorption maxima	Analog	Absorption maxima
(structure)	467^a	*(structure)*	465^{h-j}
(structure)	420^b*	*(structure)*	$495^{h, i, k-m}$
(structure)	497^c	*(structure)*	483^{h-j}
(structure)	485^d*	*(structure)*	492^e*
(structure)	502^e	*(structure)*	$465^{i, n-p}*$

TABLE III (continued)

Analog	Absorption maxima	Analog	Absorption maxima
(structure)	500[f,*]	(structure)	420[p,q,*]
(structure)	485[g]	(structure)	345[p,q]
(structure)	315[p,r]	(structure)	470[w]
(structure)	315[p,r]	(structure)	467[x,y]
(structure)	517[s,t]	(structure)	465[z]
(structure)	500[u]	(structure)	411[x]
(structure)	527[u]	(structure)	490[aa]
(structure)	465[v,*]	(structure)	485[j,*]
(structure)	520[v]	(structure)	485[v,*]
(structure)	460[bb,*]	(structure)	489[dd]
(structure)	410[cc,*]	(structure)	488[dd]
(structure)	490[dd]	(structure)	483[dd]
9-cis retinal	485[ee]	7,9-dicis retinal	460[hh]
9,13-dicis retinal	481[ff]	7,9,13-tricis retinal	455[gg]
7-cis retinal	450[gg]		

(continued)

TABLE III (*continued*)

[a] S. E. Houghton, D. R. Lewin, and G. A. J. Pitt, unpublished work.
[b] M. Ito, K. Hirata, A. Kodama, K. Tsukida, H. Matsumoto, K. Hirouchi, and T. Yoshizawa, *Chem. Pharm. Bull.* **26,** 925 (1978).
[c] A. Kropf, *Abstr. Annu. Meet. Biophys. Soc. J.* p. 281 (1975).
[d] K. Nakanishi *et al.,* unpublished results.
[e] T. Ebrey, R. Govindjee, B. Honig, E. Pollock, W. Chan, A. Yudd, and K. Nakanishi, *Biochemistry* **14,** 3933 (1975).
[f] S. P. Tanis, R. H. Brown, and K. Nakanishi, *Tetrahedron Lett.* p. 869 (1978); K. Nakanishi *et al.,* unpublished work.
[g] A. Kropf, *Nature (London)* **264,** 92 (1976).
[h] P. J. Van den Tempel and H. O. Huisman, *Tetrahedron* **22,** 293 (1966).
[i] P. E. Blatz, M. Lin, P. Balasubramaniyan, V. Balasubramaniyan, and P. B. Dewhurst, *J. Am. Chem. Soc.* **91,** 5930 (1969).
[j] A. Kropf, B. P. Wittenberger, S. P. Goff, and A. S. Waggoner, *Exp. Eye Res.* **17,** 591 (1973).
[k] W. Gärtner, H. Hopf, W. E. Hull, D. Oesterhelt, D. Scheutzov, and P. Towner, *Tetrahedron Lett.* p. 347 (1980).
[l] R. Nelson, J. K. deRiel, and A. Kropf, *Proc. Natl. Acad. Sci. U.S.A.* **66,** 531 (1970).
[m] W. H. Waddel, M. Uemura, and J. L. West, *Tetrahedron Lett.* p. 3223 (1978).
[n] P. E. Blatz, P. B. Dewhurst, P. Balasubramaniyan, V. Balasubramaniyan, and M. Lin, *Photochem. Photobiol.* **11,** 1 (1970).
[o] P. E. Blatz, P. Balasubramaniyan, and V. Balasubramaniyan, *J. Am. Chem. Soc.* **90,** 3282 (1968); P. E. Blatz, P. B. Dewhurst, P. Balasubramaniyan, and V. Balasubramaniyan, *Nature (London)* **219,** 169 (1968).
[p] M. Arnaboldi, M. G. Motto, K. Tsujimoto, V. Balogh-Nair, and K. Nakanishi, *J. Am. Chem. Soc.* **101,** 7082 (1979).
[q] K. Nakanishi, V. Balogh-Nair, M. A. Gawinowicz, M. Arnaboldi, M. G. Motto, and B. Honig, *Photochem. Photobiol.* **29,** 657 (1979); B. Honig, U. Dinur, K. Nakanishi, V. Balogh-Nair, M. A. Gawinowicz, M. Arnaboldi, and M. G. Motto, *J. Am. Chem. Soc.* **101,** 7084 (1979).
[r] M. A. Gawinowicz, V. Balogh-Nair, J. S. Sabol, and K. Nakanishi, *J. Am. Chem. Soc.* **99,** 7720 (1977).
[s] G. Wald, *Fed. Proc., Fed. Am. Soc. Exp. Biol.* **12,** 606 (1953).
[t] R. S. H. Liu, A. E. Asato, and M. Denny, *J. Am. Chem. Soc.* **99,** 8095 (1977); H. Matsumoto, A. E. Asato, and R. S. H. Liu, *Photochem. Photobiol.* **29,** 695 (1979).
[u] A. E. Asato, H. Matsumoto, M. Denny, and R. S. H. Liu, *J. Am. Chem. Soc.* **100,** 5957 (1978).
[v] M. G. Motto, M. Sheves, K. Tsujimoto V. Balogh-Nair, and K. Nakanishi, *J. Am. Chem. Soc.* **102,** 7947 (1980).
[w] M. J. Curtis, G. A. J. Pitt, J. McC. Howell, *in* "Recent Progress in Photobiology" (E. J. Bowen, ed.), p. 119. Blackwell, Oxford, 1965; N. A. Sokolova, B. I. Mitsner, and V. I. Zakis, *Bioorg. Khim.* **5,** 1053 (1979).
[x] D. R. Lewin and N. J. Thomson, *Biochem. J.* **103,** 36P (1967).
[y] M. Ito, A. Kodama, M. Murata, M. Kobayashi, K. Tsukida, Y. Shichida, and T. Yoshizawa, *J. Nutr. Sci. Vitaminol.* **25,** 343 (1979).
[z] M. Azuma, K. Azuma, and Y. Kito, *Biochim. Biophys. Acta* **295,** 520 (1973).
[aa] A. Kropf, personal communication.
[bb] K. Nakanishi, A. P. Yudd, R. Crouch, G. L. Olson, H.-C. Cheung, R. Govindjee, T. G. Ebrey, and D. J. Patel, *J. Am. Chem. Soc.* **98,** 236 (1976).

TABLE III (*continued*)

cc R. A. Blatchly, J. D. Carriker, V. Balogh-Nair, and K. Nakanishi, *J. Am. Chem. Soc.* **102**, 2495 (1980).

dd H. Akita, S. P. Tanis, M. Adams, V. Balogh-Nair, and K. Nakanishi, *J. Am. Chem. Soc.* **102**, 6370 (1980).

ee R. Hubbard and G. Wald, *J. Gen. Phys.* **36**, 269 (1952).

ff R. Crouch, V. Purvin, K. Nakanishi, and T. Ebrey, *Proc. Natl. Acad. Sci. U.S.A.* **72**, 1538 (1975).

gg W. J. DeGrip, R. S. H. Liu, A. E. Asato, and V. Ramamurthy, *Nature (London)* **262**, 416 (1976).

hh A. Kini, H. Matsumoto, and R. S. H. Liu, *J. Am. Chem. Soc.* **101**, 5078 (1979).

TABLE IV

RETINAL ANALOGS WHICH DO NOT FORM VISUAL PIGMENT ANALOGS
WITH BOVINE OPSIN

Analog	Ref.	Analog	Ref.
	a		*c**
	a		*c**
	b		*c*
	b		*d*
	a		*a*

a P. E. Blatz, M. Lin, P. Balasubramaniyan, V. Balasubramaniyan, and P. B. Dewhurst, *J. Am. Chem. Soc.* **91**, 5930 (1969).

b A. Kropf, B. P. Wittenberger, S. P. Goff, and A. S. Waggoner, *Exp. Eye Res.* **17**, 591 (1973).

c M. G. Motto, M. Sheves, K. Tsujimoto, V. Balogh-Nair, and K. Nakanishi, unpublished results.

d T. Ebrey, R. Govindjee, B. Honig, E. Pollock, W. Chan, A. Yudd, and K. Nakanishi, *Biochemistry* **14**, 3933 (1975).

Before each recombination experiment with the protein, the purity of the retinal analog should be ascertained by HPLC. (Only use of single geometrical isomers containing less than 2–3% contamination is recommended.) This is a necessary precaution, as some of the retinals have been found to be very unstable even though they were stored at low temperature under inert atmosphere.

The structure of retinals used for reconstitution experiments should be established in an unambiguous manner by spectroscopic measurements, such as MS, UV, IR, and NMR spectroscopy. (The order of elution on HPLC by itself is not sufficient to determine the geometry of retinal isomers.) In preparing solutions for NMR experiments the solvents should be free from traces of acid. Use of commercial $CDCl_3$ as such may result in total decomposition of some of the retinal analogs during the measurements. The samples also should be well protected from light during measurements.

[65] Synthesis and Photochemistry of Stereoisomers of Retinal

By ROBERT S. H. LIU and ALFRED E. ASATO

Introduction

Six (all-*trans*, 9-*cis*, 11-*cis*, 13-*cis*, 9,13-di*cis*, and 11,13-di*cis*) of the 16 possible geometric isomers of retinal have been known for some time. The first four were isolated from irradiation mixtures of all-*trans*-retinal, and all six were subsequently synthesized by several independent routes. Their preparations have been adequately reviewed[2] and details described elsewhere[3]; therefore, no further discussion will be made here.

However, in the recent several years because of the discovery of a general chemical route to several precursors to those retinal isomers containing the hindered 7-*cis* geometry[4] and because of the availability of better separation methods such as high-pressure liquid chromatography, many more geometric isomers are now known. Among those containing the 7-*cis* geometry, five (7-*cis*, 7,9-di*cis*, 7,9,13-tri*cis*,[5,6] 7,13-di*cis*,[5,7] and 7,11-di*cis*[8]) have been isolated and their geometry unambiguously characterized. The two remaining 7-*trans* isomers (9,11-di*cis*[9] and 9,11,13-tri*cis*[10]) have also been prepared. Additionally, the 7,11,13-tri*cis*,[8] 7,9,11-

[1] The work was supported by grants from the U.S. Public Health Services (Grant No. AM-17806 and CP-75933).

[2] H. Mayer and O. Isler, *in* "Carotenoids" (O. Isler, ed.), p. 325. Birkhaeuser, Basel, 1971.
[3] R. Hubbard, P. K. Brown, and D. Bownds, this series, Vol. 18, p. 615.
[4] V. Ramamurthy, G. Tustin, C. C. Yau, and R. S. H. Liu, *Tetrahedron* **31**, 193 (1975).
[5] V. Ramamurthy and R. S. H. Liu, *Tetrahedron* **31**, 201 (1975).
[6] A. E. Asato and R. S. H. Liu, *J. Am. Chem. Soc.* **97**, 4128 (1975).
[7] A. Kini and R. S. H. Liu, unpublished results.
[8] A. Kini, H. Matsumoto, and R. S. H. Liu, *J. Am. Chem. Soc.* **101**, 5078 (1979).
[9] A. Kini, H. Matsumoto, and R. S. H. Liu, *Bioorg. Chem.* **9**, 406 (1980).
[10] C. G. Knudson, S. C. Carey, and W. H. Okamura, *J. Am. Chem. Soc.* **102**, 6355 (1980).

tri*cis*, and the all-*cis* isomers[7] of retinonitrile have been synthesized. The aldehyde form of the first was found to be unstable at room temperature,[8] and thus far no attempt has been made to prepare the aldehyde form of the last two.

Nearly all of these new isomers of retinal yield pigment analogues when incubated with rod outer segment extract of cattle retinae. The absorption properties of these rhodopsin isomers are summarized elsewhere.[11]

Below the most direct routes to some of the hindered isomers of retinal are described followed by a brief discussion on the methods of characterization and separation of isomers of retinal. But first there will be a few words on the nomenclature of the polyene geometry.

Nomenclature

Listed in Fig. 1 are the structures of all 16 geometric isomers, together with the common and the IUPAC names for the polyene geometry. We

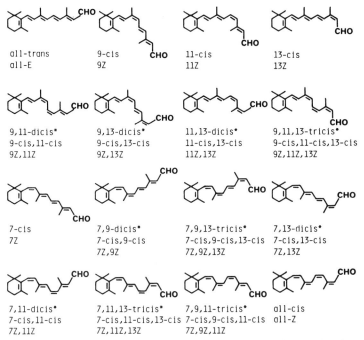

FIG. 1. Sixteen possible geometric isomers of retinal. The prefixes designating the polyene geometry are listed below each structure. Those labeled with an asterisk should be considered as common names.

[11] R. S. H. Liu and H. Matsumoto, this series, Vol. 81, p. 694 (1982).

wish to point out that the popular usage of the prefixes *di-* or *tri-* to indicate the presence of a multi-*cis* linkage in the side chain of a retinal isomer does not evolve from IUPAC rules on naming of stereoisomers. The corresponding names should therefore be considered as common names.

Synthesis of 7-*cis*-, 7,9-di*cis*-, 7,13-di*cis*-, and 7,9,13-tri*cis*-Retinal

Reaction Scheme

*Photosensitized Isomerization of β-Ionylideneacetonitrile: The preparation of 7-*cis*- and 7,9-di*cis*-β-Ionylideneacetonitrile, I.* In a typical preparative-scale photoisomerization experiment a chloroform (0.5 liter) solution of β-ionylideneacetonitrile,[12] I (10.0 g, 46.5 mmol, all-*trans*/9-*cis* ≈ 6), and 0.25 g of benzanthrone (E_t = 47 kcal/mol, Aldrich Chemical Co., recrystallized from alcohol) was irradiated for 24 h at ambient temperature in an immersion photochemical reactor vessel[13] under a blanket of nitrogen or argon with a Hanovia 200-W medium-pressure Hg lamp and uranium glass insert filter.[14] After the irradiation was determined to be complete by NMR or HPLC analysis, the irradiated mixture was transferred to a 1-liter round-bottomed flask and concentrated *in vacuo* on a rotary evaporator at room temperature. The resultant yellow mixture was triturated with hexane (0.1 liter), refrigerated for 1 hr, and filtered from precipitated benzanthrone (hexane wash). The process was twice re-

---

[12] Prepared by reaction of β-ionone (Givaudan Corp.) with cyanoacetic acid (Aldrich) according to the procedure of H. O. Huisman, A. Smit, S. Vromen, and L. G. M. Fischer, *Recl. Trav. Chim. Pays-Bas* **71**, 913 (1952).

[13] A suitable photochemical reactor vessel is available from Ace Glass, Inc. For a description of the use of the reactor see R. Srinivasan, *Photochem. Synth.* **1**, 14 (1971); V. Ramamurthy and R. S. H. Liu, *ibid.* **2**, 70 (1976).

[14] UV cut off = 350 nm. Available from Corning Glass Company.

peated to remove the bulk of the sensitizer. The desired 7-*cis* isomers of I were obtained in quantitative yield in a ratio of 7-*cis*/7,9-di*cis* = 2.

The Preparation of 7-cis- *and* 7,9-di*cis*-β-*Ionylideneacetaldehyde, II.* A 200-ml, three-necked, round-bottomed flask was equipped with a gas inlet tube, gas exit tube, and septum-capped tube. After being thoroughly flushed with argon (30 min), the flask was charged with a solution of 1.12 g (5.21 mmol) of a mixture of 7-*cis*- and 7,9-di*cis*-I in hexane (50 ml) and cooled to −78° (acetone–dry ice bath). To the magnetically stirred solution was then added a large excess of diisobutylaluminum hydride (DIBAH, 10 ml, 1.0 M in hexane, 10.0 mmol, Aldrich) by syringe over 5 min. Immediately after the addition of DIBAH the cooling bath was removed and the clear yellow solution was stirred to room temperature over 2 hr. At this time the solution was cooled to 0° and a slurry of silica gel 60 (12 g) deactivated with water (1.5 ml) in hexane (20 ml) was added *in small portions* over 10 min.[15] After stirring to room temperature for 30 min the orange mixture was transferred to a column packed with silica gel 60 (50 g) and eluted with ~0.3 liter of 40% ether–hexane. After concentrating the resultant eluate *in vacuo* on a rotary evaporator there was obtained in high purity 0.93 g (82%) of a mixture of 7-*cis*- and 7,9-di*cis*-II as an orange oil: 7-*cis*/7,9-di*cis* = 2; TLC (silica gel 60 F-254, chloroform, R_f = 0.41); NMR (100 MHz, CDCl₃), 7-*cis:* δ2.16, s (CH₃-9), 6.05, d, J = 12 Hz (H-8); 7,9-di*cis:* δ 1.93, s (CH₃-9), 6.86, J = 12 Hz (H-8).[4]

Separation of 7-*cis*-II from 7,9-di*cis*-II was effected by preparative HPLC (Waters LC, System 500, two cycles, 9% ether–hexane, 0.2 liter/min).

The Preparation of Methyl 7,9-Di*cis*- *and* 7,9,13-Tri*cis*-*Retinoate, III.* A 100-ml, three-necked, round-bottomed flask equipped with a pressure-equalized addition funnel, gas inlet tube, gas exit tube, and thermometer was purged with argon (30 min) and charged with sodium hydride (0.32 g, 13.6 mmol, 0.651 g of 50% sodium hydride dispersion in mineral oil, Ventron Chemical Co.). The mineral oil was removed by three times triturating with hexane (10 ml) and careful removal of the hexane solution with a syringe. A solution of 3.63 g (14.5 mmol) of methyl 4-diethylphosphono-senecioate (E/Z = 1.25)[16] in THF (50 ml) was added, with magnetic stirring for 5 min, to the sodium hydride suspended in dry THF (10 ml) and

---

[15] Work-up procedure according to A. Kini.[8] Initially, the addition of wet silica gel to the reaction mixture which contains an excess of DIBAH results in a vigorous reaction with gas evolution. Thereafter, the hydrolysis proceeds smoothly. Larger-scale preparation (30 mmol and above) are best worked up by adding the reaction mixture with stirring to a mixture of 0.5 liter ice–H₂O containing 0.1–0.15 liter 20% H₂SO₄.

[16] G. Pattenden and B. C. L. Weedon, *J. Chem. Soc. C.* p. 1984 (1968); J. B. Davis, L. M. Jackman, P. T. Siddons, and B. C. L. Weedon, *ibid.* p. 2154 (1966).

cooled to 0°. Anion formation commenced on warming the reaction mixture of ~5–7° and proceeded smoothly to completion (30–45 min) at this temperature.[17] The resultant clear brown solution was cooled to −40° (carbon tetrachloride–dry ice) and a solution of 2.11 g (9.68 mmol) of 7,9-di*cis*-II in THF (10 ml) was added dropwise over 5 min. After addition the red-brown reaction mixture was stirred to 0° for 3 hr. At this time analysis of the crude products by TLC (sil gel 60 F-254, chloroform) showed the presence of 7,9-di*cis*-III (R_f = 0.58) and 7,9,13-tri*cis*-III (R_f = 0.63) and the absence of starting, 7,9-di*cis*-II (R_f = 0.41).

The reaction mixture was subsequently worked up by dilution with water (50 ml) and thorough extraction with 50% ether–hexane (four 20-ml portions). After backwashing the combined organic extracts with water (20 ml) followed by brine solution (20 ml), the orange solution was dried (MgSO$_4$), filtered, and concentrated *in vacuo* on a rotary evaporator to give 5.0 g of crude product as a red oil. Column chromatography of the crude material on silica gel 60 (160 g) using 0.7 liter of 7% ether–hexane afforded 2.03 g (66.8%) of a 1 : 1 mixture of 7,9-di*cis*-III and 7,9,13-tri*cis*-III. Further purification by preparative HPLC (Waters LC/System 500, 2 columns, 2 cycles, 2.5% ether-hexane, 0.2 liter/min) afforded pure samples of 7,9-di*cis*-III and 7,9,13-tri*cis*-III as crystalline, lemon yellow solids, m.p. 102–103°, UV 337 nm (42,860), and 119.5–121°, UV (41,400), respectively (recrystallized from methanol).

7,9-cis- and 7,9,13-tricis-Retinol. A 100-ml round-bottomed flask equipped with a gas inlet tube, gas exit tube, and septum-capped tube was flushed with argon (30 min) and charged with lithium aluminum hydride (0.48 g, 6.64 mmol, 52.6% dispersion in mineral oil, Ventron). After cooling to −78° (acetone–dry ice bath), anhydrous ether (5 ml) was added by syringe, followed by a solution of 1.0 g (3.18 mmol) of a mixture of 7,9-di*cis*- and 7,9,13-tri*cis*-III in ether (10 ml). The reaction mixture was magnetically stirred for 45 min at this temperature followed by 1.5 hr at 0°. At this time the reaction mixture was worked up, starting at −78° by the cautious consecutive addition of (a) water (0.2 ml), (b) 15% sodium hydroxide solution (0.2 ml), and (c) water (0.7 ml).[18] The resultant yellow heterogeneous mixture was stirred to 0°for 30 min, whereupon a granular precipitate of aluminum salts was formed. The mixture was filtered through a 1-cm bed of MgSO$_4$, the colorless filter cake thoroughly washed with ether (~0.1 liter), and the filtrate concentrated *in vacuo* on a rotary evaporator to give 1.0 g of crude isomeric retinols as a viscous, pale yellow oil. The

---

[17] Alternatively, generation of the lithio derivative of methyl 4-diethylphosphonosenecioate using lithium diisopropylamide in THF-hexane is convenient and practicable up to ~100 mmol.

[18] V. M. Mićović and M. L. J. Mihailović, *J. Org. Chem.* **18,** 1170 (1953).

TABLE I

¹H-NMR Data of Isomers of Retinal, Obtained in CDCl₃ Unless Specified Otherwise

Isomer	CH₃-1	CH₃-5	CH₃-9	CH₃-13	H₇	H₈	H₁₀	H₁₁	H₁₂	H₁₄	H₁₅	$J_{7,8}$	$J_{10,11}$	$J_{11,12}$	$J_{14,15}$
all-t[a]	1.04	1.72	2.03	2.33	6.36	6.18	6.20	7.15	6.37	5.98	10.12	16.5	12.0	15.4	8.0
7-c[b]	1.05	1.52	1.93	2.31	5.99	6.11	6.25	7.08	6.33	5.96	10.10	11.3	11.5	15.0	8.2
9-c[a]	1.05	1.75	2.00	2.30	6.31	6.64	6.06	7.20	6.27	5.94	10.07	15.9	11.8	15.4	8.2
11-c[a]	1.02	1.71	1.99	2.36	6.32	6.14	6.54	6.69	5.92	6.07	10.10	16.0	13.0	11.5	8.0
13-c[a]	1.04	1.72	2.02	2.14	6.35	6.18	6.23	7.05	7.28	5.85	10.20	16.0	11.0	15.0	8.0
7,9-c[c]	1.06	1.47	1.89	2.31	6.11	6.58	6.02	7.16	6.26	5.96	10.10	11.8	11.8	15.6	8.0
7,11-c[d]	0.98	1.46	1.80	2.17	5.82	6.00	?	?	?	5.80	9.93	12.3	?	?	7.8
7,13-c[d]	1.07	1.54	1.94	2.10	5.98	6.11	6.36	6.92	7.22	5.80	10.20	11.5	11.2	15.5	8.0
9,11-c[e]	1.04	1.74	2.04	2.36	6.26	6.64	6.40	6.72	5.86	5.98	10.06	16.0	11.5	11.5	7.6
9,13-c[a]	1.05	1.77	2.05	2.15	6.36	6.68	6.16	7.16	7.25	5.87	10.27	16.0	10.5	15.0	8.0
11,13-c[a]	1.01	1.68	1.96	2.07	6.28	6.08	6.20	6.77	6.11	5.98	9.71	16.0	12.5	11.8	8.1
7,9,13-c[c]	1.04	1.47	1.88	2.12	5.97	6.52	6.13	6.99	7.12	5.83	10.20	12.1	10.5	14.5	8.0
7,11,13-c[d] (aldehyde form, unstable)															
9,11,13-c[f]	1.04	1.71	1.91	2.03	6.28	6.62	6.07	6.85	5.99	5.96	9.68	15.4	11.7	11.7	7.9

[a] all-t, all-*trans*; c, *cis*. Data of D. Patel, *Nature (London)*, **221**, 825 (1969).
[b] W. J. DeGrip, R. S. H. Liu, A. E. Asato, and V. Ramamurthy, *Nature (London)* **262**, 416 (1976).
[c] A. E. Asato and R. S. J. Liu, *J. Am. Chem. Soc.* **97**, 4128 (1975).
[d] A. Kini, H. Matsumoto, and R. S. H. Liu, *J. Am. Chem. Soc.* **101**, 5078 (1979). Solvent: CCl₄ with 5% dioxane-d₈.
[e] A. Kini, H. Matsumoto, and R. S. H. Liu, *Bioorg. Chem.* **9**, 406 (1980). Solvent: CCl₄ with 5% dioxane-d₈.
[f] C. G. Knudsen, S. C. Carey, and W. H. Okamura, *J. Am. Chem. Soc.* **102**, 6355 (1980).

TABLE II
THE CARBON CHEMICAL SHIFT DATA OF RETINAL ISOMERS (ppm)

Isomer	C-16,17	C-18	C-19	C-20	C-1	C-2	C-3	C-4	C-5	C-6
All-t	29.0	21.7	13.0	13.0	34.1	39.6	19.3	33.2	130.3	137.6
	27.3	20.0	11.0	11.0	32.9	38.4	17.9	31.6	128.6	136.5
7-c	28.7	21.6	13.1	14.8	34.6	39.2	19.2	32.1	129.4	136.3
9-c	29.0	21.8	20.9	13.2	34.1	39.7	19.3	33.2	130.4	138.1
	27.3	20.0	19.0	11.1	32.8	38.3	17.9	31.6	128.7	136.6
11-c	29.0	21.6	11.9	17.4	34.6	39.6	19.5	33.3	130.0	138.1
	27.3	19.9	10.4	15.9	32.9	38.4	17.9	31.6	128.5	136.5
13-c	29.0	21.7	13.0	21.1	34.3	39.7	19.3	33.2	130.3	137.6
	27.3	19.9	11.0	19.0	32.9	38.4	17.9	31.6	128.5	136.4
7,9-c	28.9	21.7	22.0	13.1	34.3	39.2	19.1	32.3	130.2	136.5
7,9,13-c	28.9	21.7	22.0	20.9	34.3	39.2	19.1	32.3	130.2	136.5

[a] G. Englert, *Helv. Chim. Acta* **58**, 2367 (1975). Solvent: $CDCl_3$.
[b] R. Rowan, III and B. D. Sykes, *J. Am. Chem. Soc.* **96**, 7000 (1974).
[c] R. S. H. Liu and A. E. Asato, unpublished results.
[d] R. S. Becker, S. Berger, D. K. Dalling, D. M. Grant, and R. J. Pugmire, *J. Am. Chem. Soc.* **96**, 7008 (1974).

crude reaction product was used immediately in the following preparation without further purification.

7,9-dicis- and 7,9,13-tricis-Retinal, IV. To a magnetically stirred suspension of active MnO_2[19] (12.0 g) in dichloromethane (70 ml) was added at one time a solution of crude retinols (see earlier) in dichloromethane (10 ml). After vigorous stirring at room temperature (argon atmosphere) for 1.5 hr the reaction mixture was vacuum filtered through a 2-cm filter cake of Celite and washed with ether (0.2 liter) until the filtrate was colorless. Concentrations of the filtrate *in vacuo* afforded 0.6 g of essentially pure 7,9-di*cis*- and, 7,9,13-tri*cis*-IV. Separation of the isomers could be effected by column chromatography on silica gel using 25% $CHCl_3$–hexane[6] or 10% ether–hexane.

7-cis- and 7,13-dicis-Retinal. The procedure for preparation of these two isomers is identical to that described for the 7,9-di*cis*- and 7,9,13-tri*cis* isomers except that the starting triene aldehyde is 7-*cis*-II. Separation of retinal isomers was again effected by column chromatography on silica gel using 25% $CHCl_3$–hexane.

7-cis-Retinal from Direct Irradiation of the All-*trans* Isomer

The procedure for preparation of the 9-*cis*, 11-*cis*, 13-*cis* isomers of retinal by direct irradiation of the all-*trans* was described in detail in the

[19] J. Attenburrow, A. F. B. Cameron, J. H. Chapman, R. M. Evans, B. A. Hems, A. B. A. Jansen, and T. Walker, *J. Chem. Soc.*, 1094 (1952).

C-7	C-8	C-9	C-10	C-11	C-12	C-13	C-14	C-15	Solvent
129.6	137.1	141.1	129.4	132.4	134.5	154.5	128.9	190.7	CDCl$_3$^{a}
127.8	136.3	139.4	128.7	131.1	133.8	153.0	127.8	189.1	Acetone^b
128.9	134.5	142.1	130.8	132.3	134.2	154.5	130.8	190.8	CDCl$_3$^{c}
131.1	129.4	140.0	127.9	131.2	133.8	154.3	128.9	190.6	CDCl$_3$^{a}
129.3	128.5	138.1	127.2	129.9	133.2	153.2	127.8	189.2	Acetone^b
129.3	138.3	141.1	126.6	130.9	131.2	154.6	130.5	190.8	Dioxane^d
127.9	136.5	139.7	125.0	129.6	129.6	153.9	128.7	189.2	Acetone^b
129.6	137.0	141.3	129.4	133.4	126.5	154.2	127.7	189.6	CDCl$_3$^{a}
127.7	136.2	140.5	128.8	132.0	125.8	152.6	126.6	187.9	Acetone^b
127.1	128.9	142.0	129.4	131.6	131.6	154.6	131.7	190.7	CDCl$_3$^{c}
127.2	131.7	142.2	129.4	132.4	125.1	154.0	127.4	186.5	CDCl$_3$^{c}

classical paper by Brown and Wald.[20] A slight modification of the procedure involving the use of acetonitrile instead of ethanol as a solvent not only led to an increased yield of the 11-*cis* isomer but also made possible the isolation of the 7-*cis* isomer from the mixture.[21] This procedure represents the most direct route to small amounts of the hindered isomer.

Irradiation Procedure. A solution of 35 mg of all-*trans*-retinal (Sigma Chemical) dissolved in acetonitrile (35 ml) in a Pyrex vessel (or vessels) was placed directly behind a Corning 3-74 filter plate and irradiated externally with a 450-W Hanovia medium-pressure mercury lamp. The progress of the reaction was followed by periodic HPLC analyses ($\frac{1}{2}$ in × 25 cm 10-μm Lichrosorb column, 5% ether–hexane). After 10 hr, the mixture contained mainly five isomers in relative amounts of 13-*cis* : 11-*cis* : 9-*cis* : 7-*cis* : all-*trans* = 3.7 : 7.0 : 2.6 : 1 : 4.3. The irradiated mixture was concentrated by evaporation on a rotary evaporator. The isomers were separated by preparative HPLC using conditions identical to those for analysis. Isolated yield of 7-*cis*-retinal: 0.78 mg.

Characterization of Isomers of Retinal

As in any organic structure determination, nuclear magnetic resonance spectroscopy has become the most reliable method for characterizing the stereochemistry of the polyene chain of the retinal isomers. The popularity is partially due to the nondestructive nature of the method and the increased sensitivity of the newer spectrometers.

Usually a medium-field spectrometer (80–100 MHz) is suffcent to

[20] P. K. Brown and G. Wald, *J. Biol. Chem.* **222**, 865 (1965).
[21] M. Denny and R. S. H. Liu, *J. Am. Chem. Soc.* **99**, 4865 (1977).

TABLE III
UV-vis ABSORPTION DATA OF ISOMERS OF RETINAL

Isomer	Hexane		Ethanol	
	λmax	ε	λmax	ε
All-*trans*[a]	368	48,000	383	42,884
7-*cis*	359[b]	44,100	377[g]	38,000
9-*cis*[a]	363	37,660	373	36,010
11-*cis*[a]	365	26,360	379.5	24,940
13-*cis*[a]	363	38,770	375	35,500
7,9-di*cis*	351[b]	42,500		
7,11-di*cis*	355[c]	18,800	374[g]	16,000
7,13-di*cis*[b,d]	357	—		
9,11-di*cis*[e]	352	30,600	368[g]	27,000
9,13-di*cis*[a]	359	34,170	368	32,380
11,13-di*cis*[a]	356	22,010	373	19,880
7,9,13-tri*cis*	346[b]	36,600		
9,11,13-tri*cis*[f]	302	15,500	302	14,300

[a] From R. Hubbard, P. K. Brown, and D. Bownds, this series, Vol. 18, p. 628.
[b] W. DeGrip, R. S. H. Liu, A. E. Asato, and V. Ramamurthy, *Nature* (*London*), **262**, 416 (1976).
[c] A. Kini, H. Matsumoto, and R. S. H. Liu, *J. Am. Chem. Soc.* **101**, 5078 (1979).
[d] A. Kini, H. Matsumoto, and R. S. H. Liu, unpublished results.
[e] A. Kini, H. Matsumoto, and R. S. H. Liu, *Bioorg. Chem.* **9**, 406 (1980).
[f] C. G. Knudsen, S. C. Carey, and W. H. Okamura, *J. Am. Chem. Soc.* **102**, 6355 (1980).
[g] M. Denny and R. S. H. Liu, unpublished results.

spread out all vinyl hydrogen signals for complete stereochemical assignments. Occasional difficulties can be overcome by the use of different solvents,[8] addition of shift reagents,[5] and use of a higher field spectrometer.[22] In Table I is a complete listing of the key ¹H-NMR parameters associated with the known isomers of retinal. In Table II is a listing of the ¹³C chemical shifts of the retinal isomers from their noise decoupled spectra. The list is not as complete as those for ¹H-NMR.

A word of caution in the use of the common deuterated chloroform (CDCl₃) as an NMR solvent is appropriate. Because of the acidic nature of the solvent, we found occasional isomerization of the hindered isomers of retinal when allowed to remain in the solvent for a prolonged period. Therefore the storage of the retinal isomers in CDCl₃ should be avoided. The use of other solvents, such as deuterated acetone and carbon tetrachloride (dioxane-d₈ (5%v/v) may be added if an internal lock is required for a given spectrometer), should be considered.

Further confirmation of retinal stereochemistry was provided by ultra-

[22] D. Patel, *Nature* (*London*) **221**, 825 (1969).

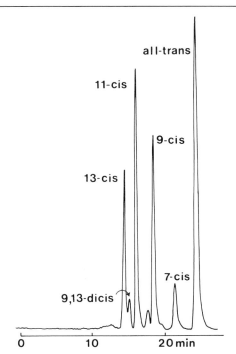

FIG. 2. A high-pressure liquid chromatogram of a mixture of retinal isomers obtained during direct irradiation of all-*trans*-retinal in acetonitrile. Analytical conditions: 10 × 250 nm 5-μm Lichrosorb Si column, 5% ether in hexane and detecting beam, 360 nm. The peak before that of 9-*cis* has been identified as 9,11-di*cis*-retinal by its UV absorption and interaction with cattle opsin.[27] The retention time of the shoulder immediately after the 9-*cis* peak is identical to that of 7,9-di*cis*-retinal. Additionally, the retention times of 7,13-di*cis*- and 7,9,13-tri*cis*-retinal are the same as that of 13-*cis*, whereas that of 7,11-di*cis*-retinal is slightly longer than that of 11-*cis*.

violet-visible spectroscopy. In Table III are listed pertinent data associated with the absorption properties of the retinal isomers.

Comparison of HPLC retention time is another fairly reliable and sensitive method for providing some information on the nature of an unknown isomer. While the subject is dealt with in detail in a separate chapter, we show in Fig. 2 a chromatogram of a mixture containing eight of the 13 known isomers.

3-Dehydroretinal (Vitamin A$_2$)

Six (all-*trans*, 9-*cis*, 11-*cis*, 13-*cis*, 9,13-di*cis* and 11,13-di*cis*) isomers of 3-dehydroretinal have been previously prepared.[23] Some of their spec-

[23] U. Schwieter, G. Saucy, M. Montavon, C. Planta, R. Rüegg, and O. Isler, *Helv. Chim. Acta* **45**, 517, 528, 541 (1962).

tral properties were summarized before.[3] Since then, only one new isomer has been isolated: the 7-cis[24] isomer along with the 9-cis[25] were found to be the major photoisomers of all-trans-3-dehydroretinal when irradiated in acetonitrile. Their ^{1}H-NMR data, the HPLC conditions for their separation,[24] and their interaction with cattle opsin are in the literature.[25,26]

[24] R. S. H. Liu, A. E. Asato, and M. Denny, J. Am. Chem. Soc. **99**, 8095 (1977).
[25] M. Azuma, K. Azuma, and Y. Kito, Biochim. Biophys. Acta **295**, 520 (1973).
[26] H. Matsumoto, A. E. Asato, and R. S. H. Liu, Photochem. Photobiol. **29**, 695 (1979).
[27] M. Denny, M. Chun, and R. S. H. Liu, Photochem. Photobiol. **33**, 267 (1981).

[66] Synthetic Pigments of Rhodopsin and Bacteriorhodopsin

By THOMAS G. EBREY

For rhodopsin and bacteriorhodopsin, two types of nonnatural pigments can be envisioned—those with a modified chromophore or those having a modified apoprotein. With regard to the latter, so far only one naturally occurring type of bacterio-opsin has been isolated, but it may be possible to isolate mutants of the pigment from *Halobacterium halobium*. A number of different rhodopsins have been found. Studies on artificial pigments from two native components have been made by combining the physiological chromophore, retinal$_2$, with a native visual pigment apoprotein such as cattle opsin, which does not contain retinal$_2$ *in vivo*.[1,2] One exciting future possibility is the cloning of rhodopsin and bacteriorhodopsin DNA and the subsequent creation of mutant DNAs that can be used to generate mutant apoproteins. Here we will restrict our discussion to the creation of artifical pigments using modified chromophores.

Chromophores

All synthetic chromophores that have been used in attempts to reconstitute artificial pigments are chemical modifications, to some degree or another, of retinal. The changes involved can be minor, such as the addition or deletion of a double bond, or major, such as the removal of all the methyl groups or of the β-ionone ring. A fairly recent listing of the synthetic chromophores for which reconstitution with opsin has been tried is

[1] G. Wald, Fed. Proc. Fed. Am. Soc. Exp. Biol. **12**, 606 (1953).
[2] G. Wald, P. Brown, and P. Smith, Science **118**, 505 (1953).

given in Ref. 3. A similar list is found in Ref. 4 for bacterio-opsin. The chromophores that have been studied have been prepared for a variety of reasons. A few have been rationally designed to test specific hypotheses, and others have been synthesized for reasons unrelated to the role of retinal as a chromophore. For example, some modified retinals have been made as anticancer drugs. Later investigators have used these for artificial pigment studies. Finally, some chromophores have been made and tested because they were relatively easy to synthesize or the investigator had an intuitive notion that they might be worth investigating. Indeed, almost all artifical pigments formed from synthetic chromophores and opsin or bacterio-opsin have turned out to have quite interesting properties.

Most syntheses produce the all-*trans* isomer. Thus, bacteriorhodopsin analogs, which utilize the all-*trans* isomer, are usually more stable and easier to make in large quantities than rhodopsin analogs, which require the 7-*cis*, 9-*cis*, 9,13-di*cis*, or 11-*cis* isomers of the synthetic chromophore.

Generally it is difficult to obtain more than a few milligrams of a given analog. The greater availability of the all-*trans* isomers, together with the observation that all bacteriorhodopsin analogs examined so far do not bleach irreversibly, make detailed studies of bacteriorhodopsin analogs more convenient than similar studies of rhodopsin analogs.

Even for bacteriorhodopsin analog studies, the 13-*cis* isomer is sometimes needed. Photoisomerization of the all-*trans* isomer usually will produce reasonable quantities of the 13-*cis* isomer (best yields in a nonpolar solvent).[5,6] Often tests for pigment formation with opsin are made by adding a mixture of isomers, presumably including the 11-*cis* and 9-*cis* isomers. To prepare optimum amounts of the physiological 11-*cis* isomer, irradiation in a polar solvent such as ethanol is usually best.[6] Although early studies used thin-layer chromatography to separate isomers from a mixture, high-performance liquid chromatography (HPLC) is now the method of choice (see Article [63], this volume[7]).

Preparation of the Apoproteins

Opsin. The first step in the preparation of opsin is the purification of the photopigment-containing membrane. Almost all work has been done

[3] A. Knowles and H. Dartnall, *in* "The Eye" (H. Dawson, ed.), Vol. 2B p. 102. Academic Press, New York, 1977.
[4] W. Stoeckenius, R. H. Lozier, and R. A. Bogomolni, *Biochim. Biophys. Acta* **505**, 215 (1979).
[5] W. Waddell, R. Crouch, K. Nakanishi, and N. Turro, *J. Am. Chem. Soc.* **98**, 4189 (1976).
[6] W. Waddell and J. West, *J. Phys. Chem.* **84**, 134 (1980).
[7] R. A. Blatchly and K. Nakanishi, Article [63], this volume.

with bovine material, but other vertebrate pigment preparations have occasionally been used. The reader is referred to specific articles on the preparation of photoreceptor membranes elsewhere in this volume.[8] To bleach rhodopsin, aliquots of a hydroxylamine (NH_2OH) solution are added to the photoreceptor membrane preparation to make the final concentration $\sim 0.1\ M$. Since NH_2OH is unstable near neutral pH, the NH_2OH solution should have its pH adjusted to ~ 6 just before mixing with the membrane. The sample is then exposed to yellow or orange light, which quickly bleaches the rhodopsin to opsin liberating retinal as its oxime. After the bleaching process is complete, the excess hydroxylamine must be removed by washing the membranes several times with buffer and/or by dialysis. If the bleaching is done with strong white light rather than yellow or orange light, the retinyl oxime is destroyed, which in some experiments may be undesirable. An alternative way to remove retinyl oxime is to freeze-dry the bleached membrane preparation followed by several washings with hexane. In addition to removing much of the retinyl oxime, some of the lipids are removed. In general, regeneration of pigments with opsin is more efficient if the apoprotein has not been solubilized with a detergent; in fact, regeneration is essentially impossible with many detergents.

It has recently been reported that the chromophore can be removed from at least one invertebrate pigment, squid, by bleaching in the presence of hydroxylamine. Regeneration of a pigment from added 11-cis-retinal has also been shown.[9]

Bacterio-opsin. It is much more difficult to prepare the apoprotein of bacteriorhodopsin by irradiation in the presence of NH_2OH,[10,11] because light does not normally cause the chromophore to become dissociated from bacteriorhodopsin. Rather, the pigment undergoes a photocyclic reaction. Hydroxylamine can attack the Schiff base linkage in the presence of light and retinyl oxime will be produced, but even with bright light (400-W slide projectors are convenient) this process can take several hours to bleach >85% of the pigment. High concentrations ($0.5 \rightarrow 2\ M$) of NH_2OH are usually used. The NH_2OH seems to attack the Schiff base during the time when the M intermediate is present and techniques that increase the lifetime of M seem to shorten the time needed to bleach. The easy way to lengthen the lifetime of M is to increase the pH, but this also accelerates the destruction of the NH_2OH. A reasonable compromise seems to be at about pH = 7, adjusting the pH of the NH_2OH solu-

[8] This volume [15].
[9] T. Seki, R. Hara, and T. Hara, *Vision Res.* **20**, 79 (1980).
[10] D. Oesterhelt, L. Schuhmann, and H. Gruber, *FEBS Lett.* **44**, 257 (1974).
[11] T. Ebrey, B. Becher, B. Mao, P. Kilbride, and B. Honig, *J. Mol. Biol.* **112**, 377 (1977).

tion just before mixing with the membrane.[11] More NH_2OH can be added after 1 hr. After bleaching, the free hydroxylamine should be removed by multiple washing with buffer and/or by dialysis. Most of the retinyl oxime can be removed by freeze-drying the preparation and washing several times with hexane.[12] Extraction of the retinyl oxime is enhanced if the bleached membrane in hexane is sonicated briefly. If the retinyl oxime is not extracted, it will eventually hydrolyze to yield free retinal, which will spontaneously recombine with the bacterio-opsin to regenerate the native pigment. A further difficulty with obtaining bacteriorhodopsin is that, unlike rhodopsin, it is extremely difficult to bleach 100% of the pigment. Even with repeated irradiation and extraction cycles a few percent of the pigment remains unbleached. For some experiments this is of little consequence; in many others it serves as an extremely aggravating contaminant.

One possible solution to this problem is the use of *H. halobium* mutants that are blocked at early stages in their carotenoid synthesis so they do not produce retinal. These mutants do produce bacterio-opsin, but because there is no retinal, a pigment is not formed.[13] Using these mutants, the apomembrane can be purified and used to regenerate pigments that will be free from possible interference of the native pigment.

Pigment Regeneration

Opsin-Based Pigments. Opsin concentrations are most conveniently quantitated in terms of either the milliliters of rhodopsin of a given absorbance that are bleached to make the opsin or the milliliters of rhodopsin of a given absorbance that can be regenerated using excess 11-*cis*-retinal. Although neither is ideal, they provide a fairly accurate measure of opsin concentration and allow one to predict roughly how much pigment can be regenerated with an artificial chromophore.

Since retinal and its dervatives are not soluble in water, their chromophores are dissolved in an organic solvent, usually ethanol, or in a mixture of ethanol with a small amount of detergent to allow mixing with an aqueous solution of opsin buffered to about pH = 7 (e.g., 100 mM phosphate buffer). The ethanolic solution should be very concentrated so that only a small aliquot of the solution, less than 100 μl, need be added to 1-ml solution of opsin. The final ethanol concentration should be small, less than 5%. To observe pigment regeneration it is often convenient to measure the absorbance against an identical opsin sample. After adding the chromophore, the absorbance due to pigment formation can be mea-

---

[12] F. Tokunaga and T. Ebrey, *Biochemistry* **17**, 1915 (1978).
[13] H. J. Weber and R. Bogomolni, this volume, [50].

sured with the distortion of light-scattering, etc. reduced. Regeneration is usually done at about room temperature and higher temperatures can slowly lead to opsin denaturation, depending on the chromophore and the temperature, regeneration will require 15 min to many hours. It is important to ascertain if chromophore binds strongly to the apoprotein. While it is difficult to determine K_m, the appearance of significant free chromophore in the presence of excess opsin indicates a low K_m; of course care must be taken to avoid the addition of an isomer of the free chromophore that cannot recombine with the opsin. Finally, it is a good practice to ensure that the pigment formed has bound the desired chromophore and not a degradation product. This can be checked by reextracting the chromophore from the pigment with methylene chloride (1 ml/ml of pigment solution—emulsify with a syringe) drying, redissolving, in hexane, and running an HPLC to compare with that of the original chromophore.[14,15] Detergents can often lead to artifactual isomerization during the reextraction process.[14]

Bacterio-opsin-Based Pigments. Basically, analogs of bacteriorhodopsin are regenerated in much the same way as described earlier for analogs of rhodopsin. A small aliquot of an ethanolic solution of the chromophore is added to a buffered (∼7.0) solution of the apomembrane and allowed to sit from 1 hr to several days at room temperatures.[12,16] Often a regeneration intermediate will precede the formation of the final pigment, and occasionally the transformation to the pigment is extremely slow. In this case, irradiation of the intermediate seems greatly to accelerate formation of the final pigment. Other procedures mentioned earlier, such as reextraction of chromophore after regeneration, are also useful to follow. Sonication of the methylene chloride emulsion used for reextraction can increase the yield of extraction without isomerizing the chromophore.[15]

Tests of Analog Pigment Functionality

Rhodopsin Analogs. Given the ability of a chromophore to couple with opsin to form a pigment, there are then two tests to determine if the pigment is functional. One test, which is limited in its significance but is experimentally straightforward, is to examine the photochemical activity of the pigment. This test is to try to bleach the pigment by irradiating with light that is absorbed by it and see if the absorption spectrum shifts to shorter wavelengths, indicating that the chromophore has become de-

---

[14] R. Crouch, V. Parvin, K. Nakanishi, and T. Ebrey, *Proc. Natl. Acad. Sci. U.S.A.* **72,** 1538 (1975).

[15] M. Tsuda and T. Ebrey, *Biophys. J.* **30,** 149 (1980).

[16] F. Tokunaga, R. Govindjee, T. Ebrey, and R. Crouch, *Biophys. J.* **19,** 191 (1977).

tached. Hydroxylamine, which is often used to increase the rate of decay of the later bleaching intermediates of rhodopsin, must be used with caution, since many of the artificial pigments are denatured by hydroxylamine in the dark. If no bleaching is seen with this simple test, then a more elaborate search for photochemical activity can be made. Such techniques as low-temperature, flash, and picosecond spectroscopy can be used. (See the relevant sections in this volume.)

However, photochemical activity does not necessarily imply that the pigment could substitute for rhodopsin's functional role in vision. Unfortunately, the exact role of rhodopsin in visual excitation and adaptation is not known. The ability of rhodopsin after light absorption to control a set of enzymes that regulate cGMP levels in photoreceptors strongly suggests that this enzyme activation is at least one of the functional roles of rhodopsin *in vivo*.[17] Irradiated artificial pigments regenerated in bleached bovine rod outer segment membranes can be tested for their ability to activate these enzymes, in particular the rod outer segment phosphodiesterase.[18]

Bacteriorhodopsin Analogs. As with rhodopsin, there are two reasonably straightforward tests to determine if the artificial bacteriorhodopsin pigment is functional. The first is to look for photochemical activity. This is more difficult than for rhodopsin because light usually cycles rather than dissociates the chromophore in bacteriorhodopsin analogs. Hence, low temperature and flash photolysis methods must be used. (See the relevant sections of this volume.) The second test for functionality of bacteriorhodopsin analogs is to determine if light can cause either a release of protons from the membrane or, even better, the transport of protons across a membrane.[16] These techniques as well as the methods of incorporating bacteriorhodopsin into lipid bilayer vesicles are discussed elsewhere in this volume. A source of possible confusion with either technique may occur if, as for the native membrane, only the all-*trans* chromophore has an extensive photochemical cycle and is able to pump protons. Then little photochemical activity might be seen if the dark transformation of the artificial pigment from the all-*trans* pigment to the 13-*cis* pigment is very rapid[19] and if there is very little all-*trans* pigment in the equilibrium dark-adapted state.

[17] J. Pober and M. Bitensky, *Adv. Cyclic Nucleotide Res.* **11**, 265 (1979).
[18] T. Ebrey, G. Sassenrath, M. Tsuda, J. West, and W. Waddell *FEBS Lett.* **116**, 217 (1980).
[19] B. Mao, R. Govindjee, T. Ebrey, M. Arnaboldi, V. Balogh-Nair, K. Nakanishi, and R. Crouch, *Biochemistry* **20**, 428 (1981).

[67] Energy Levels of the Visual Chromophores by Two-Photon Spectroscopy

By ROBERT R. BIRGE

Introduction

The level ordering of the low-lying excited singlet states of the visual chromophores has been the subject of extensive spectroscopic study and some controversy.[1] The controversy is characterized by numerous conflicting assignments in the literature and is associated with the inherent difficulty of spectroscopically assigning the excited states of molecules, like the visual chromophores, which are subject to severe inhomogeneous broadening in their electronic spectra. The excited singlet state manifold of retinal provides a particularly challenging system to study because there are three low-lying states [$"^1B_u^{*+}"$ $(\pi\pi^*)$; $"^1A_g^{*-}"$ $(\pi\pi^*)$, and $^1n\pi^*$] that are very close in energy.[1] Furthermore, the relative level ordering appears to be solvent dependent.[2]

The location of the low-lying "forbidden" $^1A_g^{*-}$-like $\pi\pi^*$ state in the visual chromophores is believed to be of significant importance in defining the photochemical properties of these molecules.[1,3,4] Recent calculations indicate that the interaction of this state with a lowest-lying $^1B_u^{*+}$ state may be directly responsible for producing a barrierless excited state potential energy surface for cis-trans isomerization of the chromophore in rhodopsin.[4] The extremely rapid formation of bathorhodopsin may therefore be a consequence of the photochemical lability of the $"^1A_g^{*-}"$ state.[4]

The pioneering spectroscopic investigations of Hudson, Kohler, Christensen, and co-workers indicate that the forbidden $^1A_g^{*-}$ state is the lowest-lying excited singlet in long-chain linear polyenes.[5-7] It is not possible, however, to generalize this level ordering to the visual chromo-

[1] R. R. Birge, Annu. Rev. Biophys. Bioeng. 10, 315 (1981).
[2] T. Takemura, P. K. Das, G. Hug, and R. S. Becker, J. Am. Chem. Soc. 100, 2626 (1978).
[3] R. R. Birge, K. Schulten, and M. Karplus, Chem. Phys. Lett. 31, 451 (1975).
[4] R. R. Birge and L. M. Hubbard, J. Am. Chem. Soc. 102, 2195 (1980); Biophys. J. 34, 517 (1981).
[5] B. S. Hudson and B. E. Kohler, J. Chem. Phys. 59, 4984 (1973); Annu. Rev. Phys. Chem. 25, 437 (1974).
[6] R. L. Christensen and B. E. Kohler, Photochem. Photobiol. 18, 293 (1973); J. Phys. Chem. 80, 2197 (1976).
[7] M. F. Granville, G. R. Holtom, B. E. Kohler, R. L. Christensen, and K. D'Amico, J. Chem. Phys. 70, 593 (1979).

phores because this covalent state is predicted to be highly sensitive to conformation and polarity.[3]

Two-photon spectroscopy has proved to be one of the most versatile methods of studying forbidden $^1A_g^{*-}$ and $^1A_g^{*-}$-like states.[7-13] This technique has verified the lowest energy position of this excited state in the linear polyenes diphenylbutadiene,[8] diphenylhexatriene,[9,10] diphenyloctatetraene,[11] and octatetraene[7] and in the visual chromophores all-*trans*-retinol[12] and all-*trans*-retinal.[13] Two-photon spectroscopy is particularly useful for studying the visual chromophores[12,13] because inhomogeneous broadening prevents the use of the high-resolution, low-temperature matrix techniques that have successfully been used to study the $^1A_g^{*-}$ state in the linear polyenes.[5,6]

The purpose of this chapter is to introduce the reader to the technique of two-photon spectroscopy and demonstrate its application to the determination of level ordering of the visual chromophores. In keeping with the objectives of this volume, it is intended primarily for individuals unfamiliar with nonlinear spectroscopy. Those readers interested in a more detailed discussion of the experimental and theoretical aspects of two-photon spectroscopy should read the excellent reviews by McClain[14-15] as well as the material presented in Refs. 16-18.

Two-Photon Selection Rules

The two-photon absorption process is schematically shown in Fig. 1. This diagram is drawn arbitrarily, assuming that the final state accessed

[8] R. L. Swofford and W. M. McClain, *J. Chem. Phys.* **59**, 5740 (1973); J. A. Bennett and R. R. Birge, *J. Chem. Phys.* **73**, 4234 (1980).

[9] G. R. Holtom and W. M. McClain, *Chem. Phys. Lett.* **44**, 436 (1976); G. Holtom, R. J. M. Anderson, and W. W. McClain, 32nd Symposium on Molecular Spectroscopy, Ohio State University, Columbus, OH (1977).

[10] H. L. -B. Fang, R. J. Thrash, and G. E. Leroi, *Chem. Phys. Lett.* **57**, 59 (1978).

[11] H. L. -B. Fang, R. J. Thrash, and G. E. Leroi, *J. Chem. Phys.* **67**, 3389 (1977).

[12] R. R. Birge, J. A. Bennett, B. M. Pierce, and T. M. Thomas, *J. Am. Chem. Soc.* **100**, 1533 (1978).

[13a] R. R. Birge, J. A. Bennett, H. L. -B. Fang, and G. E. Leroi, *Springer Ser. Chem. Phys.* **3**, 347 (1978).

[13b] R. R. Birge, J. A. Bennett, L. M. Hubbard, H. L. Fang, B. M. Pierce, D. S. Kliger, and G. E. Leroi, *J. Am. Chem. Soc.* (in press).

[14] W. M. McClain *Acc. Chem. Res.* **7**, 129 (1974).

[14a] W. M. McClain, *J. Chem. Phys.* **55**, 2789 (1971).

[15] W. M. McClain and R. A. Harris, *Excited States*, **3**, 1 (1977).

[16] R. R. Birge, *in* "Ultrasensitive Spectroscopic Techniques," (D. S. Kliger, ed.) Academic Press, New York (in press).

[17] D. S. Kliger, *Acc. Chem. Res.* **13**, 129 (1980).

[18] R. R. Birge and B. M. Pierce, *J. Chem. Phys.* **70**, 165 (1979).

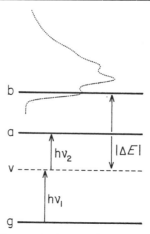

FIG. 1. The energy level diagram of a hypothetical molecule showing a two-photon transition to a lowest-lying forbidden excited state (a). The properties of the virtual state (v) are determined by the properties of a nearby one-photon allowed excited state (b).

by the two-photon excitation is the lowest-lying excited singlet. This is not a necessary condition and two-photon spectroscopy can be successfully used to probe any two-photon allowed excited state as long as the laser excitation does not intercept allowed electronic states via one-photon (linear) processes.

It is conceptually useful to analyze the two-photon absorption process as a combination of two, essentially simultaneous, one-photon absorption events. The first photon is "absorbed" by an extremely short-lived virtual state that has a lifetime of approximately 10^{-16} sec. (The virtual state does not actually absorb the photon. The process is a special case of Dirac's dispersion theory and represents a type of scattering phenomenon.[19]) A second photon must "arrive" before the virtual state scatters the first photon. If the sum of the photon energies corresponds to an excited state vibronic level of the correct symmetry, both photons are absorbed. The extremely short lifetime of the virtual state requires the use of an intense radiation field to induce experimentally a two-photon absorption. The virtual state also plays an important role under lower light intensities, where its interaction with the radiation field is responsible for Raman scattering.

A simplified analysis of the two-photon excitation process in terms of two allowed one-photon absorption processes is useful for assigning selection rules. One proceeds by recognizing that the virtual state has definite symmetry properties and must be "prepared" by an allowed one-

[19] M. Göppert-Mayer, *Ann. Phys. (Leipzig)* [5] **9**, 273 (1931).

photon transition from the ground state. Linear polyenes are represented by the C_{2h} point group, and the $\pi\pi^*$ states of these molecules can be classified under one of four possible symmetries: $^1A_g^{*-}$, $^1A_g^{*+}$, $^1B_u^{*-}$, or $^1B_u^{*+}$. (The superscript "$+$" and "$-$" labels derive from orbital pairing relationships as discussed in Refs. 5 and 18.) The ground state has $^1A_g^{*-}$ symmetry and couples via one-photon selection rules only with $^1B_u^{*+}$ states. Accordingly, the virtual state must have $^1B_u^{*+}$ symmetry. The second photon must access the final state by a one-photon allowed process, and therefore the final state must have $^1A_g^{*-}$ symmetry. Accordingly, only $^1A_g^{*-}$ states are two-photon allowed in linear polyenes.[18] Retinyl polyenes belong to the C_1 point group. The fact that all electronic states have the same symmetry (A) means that all the excited singlets are formally allowed in both one-photon and two-photon spectroscopy. Nevertheless, the excited singlet states of the visual chromophores maintain many of the characteristics of linear polyene excited states and it is useful to describe these states by reference to the C_{2h} point group (e.g., "$^1A_g^{*-}$", "$^1A_g^{*+}$", "$^1B_u^{*+}$", "$^1B_u^{*-}$"). The approximate symmetry classifications are given in quotation marks and are derived by correlating the properties of a given electronic state with those of the analogous state in a linear polyene of C_{2h} symmetry. In other words, "$^1A_g^{*-}$" should be interpreted as $^1A_g^{*-}$-like. As shown in Fig. 2, the lowest-lying "$^1A_g^{*-}$" states of both all-*trans*-retinol and all-*trans*-retinal are calculated to be sufficiently more two-photon allowed than the nearby "$^1B_u^{*+}$" states to make two-photon spectroscopy well suited for studying these compounds (see below).

Two-Photon Cross Sections

The allowedness of a two-photon transition is quantified in terms of the two-photon absorptivity (or "cross section"). This variable has units of cm^4 sec molecule^{-1} photon^{-1} and can be viewed from a semiclassical perspective as the product of the one-photon cross section for excitation into the virtual state, $\sigma_{\lambda_1}^{(v\leftarrow o)}$, the lifetime of the virtual state, Δt_v, and the one-photon cross section for excitation from the virtual state into the final state $\sigma_{\lambda_2}^{(f\leftarrow v)}$. The one-photon cross section (units of cm^2 molecule^{-1}) is related to the one-photon molar absorptivity ϵ_λ by the equation

$$\sigma_\lambda = 1000\ \ell n\ 10\epsilon_\lambda/N_A = 3.82 \times 10^{-21}\ \epsilon_\lambda \tag{1}$$

where N_A is Avogadro's number. The lifetime of the virtual state is determined by Heisenberg's uncertainty principle,

$$\Delta t_v(sec) = h/(4\pi\ \Delta\bar{\nu}) = 2.65 \times 10^{-12}/\Delta\bar{\nu}^{(i,v)}, \tag{2}$$

where $\Delta\bar{\nu}^{(i,v)}$ is the separation in wavenumbers between the virtual state

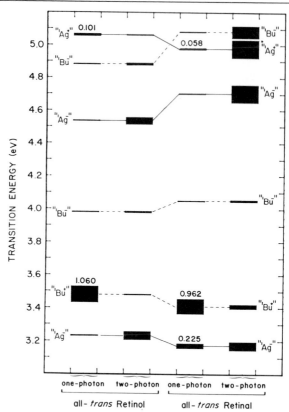

FIG. 2. A comparison of the calculated one-photon oscillator strengths and the two-photon absorptivities for the low-lying $\pi\pi^*$ electronic states of all-*trans*-retinol and all-*trans*-retinal. The electronic states are represented by horizontal bars where the vertical position indicates the transition energy of the state and the vertical width is proportional to either the one-photon oscillator strength or the two-photon absorptivity. The one-photon oscillator strengths are indicated above the allowed electronic states. The two-photon absorptivities are calculated for two linearly polarized photons of equal energy ($E_\lambda = E_\mu = \Delta E/2$) (from Ref. 18).

and the nearest strongly one-photon allowed (intermediate) excited singlet. [Equation (2) assumes there is only one nearby allowed state.] An approximate estimate of the absorptivity maximum of a two-photon transition can therefore be calculated using the following relationship,

$$\delta_{max}^{(f\leftarrow o)} = \sigma_{max}^{(f\leftarrow v)}\sigma_{max}^{(v\leftarrow o)}\,\Delta t_v \tag{3a}$$

$$= (3.87 \times 10^{-53})\,\epsilon_{max}^{(f\leftarrow v)}\,\epsilon_{max}^{(v\leftarrow o)}/\Delta\bar{\nu}\ ^{(i,v)} \tag{3b}$$

Application of Eq. (3) proceeds by recognizing that the optical properties of the virtual state are determined by the one-photon allowed intermediate state. We illustrate its application using all-*trans*-retinol. The low-lying strongly allowed "$^1B_u^{*+}$" state has an absorption maximum at $\sim 30,000$ cm^{-1} with a molar absorptivity of $\sim 35,000$ M^{-1} cm^{-1} (EPA, 77°K).[12] The lowest-lying "$^1A_g^{*-}$" state has an excitation maximum at $\sim 28,400$ cm^{-1} (EPA, 77°K).[12] If this state is accessed using 14,200 cm^{-1} laser (two-photon) excitation, the virtual state is 15,800 cm^{-1} below the "$^1B_u^{*+}$" state. Accordingly, $\epsilon_{max}^{(v \leftarrow 0)} \cong 35,500$ M^{-1} cm^{-1} and $\Delta\bar{\nu} = 15,800$ cm^{-1}. The unknown quantity is $\epsilon_{max}^{(f \leftarrow v)}$, which we will arbitrarily assume is equal to $\epsilon_{max}^{(v \leftarrow 0)}$, since the ground state and the "$^1A_g^{*-}$" excited state have identical symmetry. Equation (3b) predicts a maximum two-photon absorptivity of

$$\delta_{max}^{(f \leftarrow 0)} = (3.87 \times 10^{-53})(35,500)^2/15,800 \qquad (4a)$$

$$= 3 \times 10^{-48} \text{ cm}^4 \text{ sec molecule}^{-1} \text{ photon}^{-1} \qquad (4b)$$

The observed value is 2×10^{-49} cm^4 sec molecule^{-1} photon^{-1}.[12] The difference is due in large part to the semiclassical simplifications inherent in Eq. (3), but the preceding treatment serves to illustrate two important aspects of two-photon spectroscopy. First, the two-photon absorptivity of a given electronic state is a function of the properties of both the final state and the one-photon allowed intermediate state(s) that determine(s) the properties of the virtual state. The presence of a low-lying strongly one-photon allowed "$^1B_u^{*+}$" state is partially responsible for the strongly two-photon allowed character of the lowest-lying "$^1A_g^{*-}$" state in all-*trans*-retinol[12] and all-*trans*-retinal.[13,18] Second, as the wavelength of the laser radiation approaches the wavelength of an allowed electronic transition of the molecule, $\Delta\bar{\nu}$ approaches zero and the two-photon absorptivity increases dramatically. This condition leads to what is called resonance enhancement of the two-photon absorptivity[18] and can be exploited by using two lasers. One laser is tuned slightly to the red of an allowed electronic transition ($\Delta\bar{\nu} \sim 100$–5000 cm^{-1}) and the second laser scans the two-photon excitation spectrum. This enhancement technique is the nonlinear analogy of resonance Raman spectroscopy.

Two-Photon Polarization Effects

The semiclassical treatment of two-photon spectroscopy introduced earlier provides a useful introduction to the technique. It should be emphasized, however, that more exact quantum mechanical treatments are available and are necessary to describe this phenomenon properly.[15,18] For example, a significant amount of information concerning the electronic properties of a two-photon allowed state can be obtained by observing the effect of photon polarization on the two-photon absorptivity of randomly oriented molecules.[14–15,18] Accordingly, oriented samples are usually unnecessary to assign the symmetry properties of an excited state, and this is one of the unique features of two-photon spectroscopy. The calculation of photon polarization and propagation effects requires the use of orientationally averaged perturbation treatments,[15] which are beyond the scope of this discussion. McClain has presented complete assignment rules for the major point groups,[14a] and Birge and Pierce have described the application of two-photon polarization effects to the analysis of the visual chromophores.[18]

Experimental Techniques in Two-Photon Spectroscopy

The common experimental techniques for observing two-photon spectra include the excitation method, the thermal lens method, the multiphoton ionization technique, the optical-acoustical technique, the double-resonance method, and the direct absorption method.[14–17] The excitation method represents one of the most sensitive techniques, but it requires that the solute under investigation have a reasonable quantum yield of fluorescence ($\phi_f > 0.01$). Most, but not all, of the visual chromophores have quantum yields in low-temperature polar solvent environments (EPA, 77°K) that satisfy this criterion. A notable exception is 11-cis-retinal. The thermal lens method, although somewhat less sensitive than the excitation method, can be used on any molecule and is particularly useful for studying ambient temperature solvent effects on two-photon absorption maxima.[17] The double-resonance technique is currently being used in our laboratory to study the two-photon absorption spectra of the chromophores in rhodopsin and bacteriorhodopsin.

We will limit our discussion to an analysis of the two-photon excitation method because this is the most common experimental technique and was the method used to obtain the two-photon spectra of the visual chromophores.[12,13] A typical experimental arrangement for observing two-photon excitation spectra is shown in Fig. 3. The absorption of two photons is observed by monitoring the fluorescence (or in some cases, phosphorescence) of the solute molecule. The laser excitation is directed into the sample perpendicular to the photomultiplier opening to minimize the amount of scattered laser light reaching the detector. The principal experimental difficulty in two-photon excitation spectroscopy is to observe the relatively weak solute emission in the presence of the scattered laser excitation. The following calculation illustrates the problem.

The number of photons emitted as fluorescence by the solute (per laser pulse), N_{em}, is given by

$$N_{em} \cong \tfrac{1}{2}\phi_{f\lambda}^{(2)} N_0 \langle \delta_\lambda \rangle CI_0 L, \qquad (5)$$

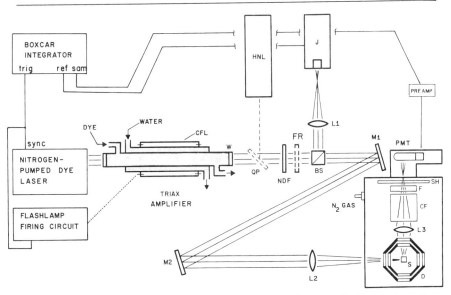

FIG. 3. A schematic diagram of a tunable dye laser two-photon excitation spectrometer. Symbols are as follows: TRIAX = coaxial arrangement of dye flow tube, water cooling jacket, and flashlamp; CFL = coaxial xenon flashlamp; W = antireflection coated window; HNL = He–Ne laser used for optical alignment; QP = quartz plate, removed during operation of dye laser; FR = Fresnel rhomb or $\frac{1}{4}$ wave plate (for polarization studies); NDF = neutral density filter(s), used if necessary to attenuate beam to avoid sample cracking; BS = cube beamsplitter; L1 = condensing lens (200 mm f.l.) used to direct a portion of pulsed laser beam onto the pyroelectric detector of the joulemeter; J = joulemeter (pyroelectric detector/FET preamplifier combination, rf-shielded); M1 = M2 = dielectric mirrors; L2 = condensing lens (6-in f.l.) for focusing laser pulse into sample cell; D = optical dewar; S = sample cell; L3 = light-gathering quartz lens (2-in. f.l.) to direct fluorescence from sample onto photomultiplier tube; CF = chemical filter usually containing aqueous $CuSO_4$ solution; F = broadband or interference filter; SH = shutter; PMT = photomultiplier tube; N_2GAS = dry gas flow to prevent fogging of windows of dewar. The flashlamp-pumped amplifier is used only for laser excitation wavelengths above 750 nm to increase peak power. Although the amplifier is shown in a single-pass configuration, it is easily modified for double- or triple-pass configurations. [Reproduced from the Ph.D. thesis of J. A. Bennett, University of California, Riverside, Calif. (1980).]

where $\phi_{f\lambda}^{(2)}$ is the quantum yield of fluorescence for the absorption of two photons of wavelength λ and the other terms are defined later. (For convenience we assume a single laser excitation wavelength λ and that fluorescence is the only radiative mode.) Note that we have defined $\phi_{f\lambda}^{(2)}$ such that this term equals unity when two absorbed photons always produce one photon of emission. This definition is in keeping with the concept of a quantum yield representing a conversion efficiency. The sum of all the quantum yields representing processes depopulating a given excited state should then equal unity. In practice, it is usually assumed that the quantum yield of fluorescence for two-photon excitation $\phi_{f\lambda}^{(2)}$ is identical to that for one-photon excitation $\phi_{f\lambda}^{(1)}$, provided the total energy of excitation is the same (i.e., $\lambda = 2\lambda'$). All-*trans*-retinal may be an exception to this "rule," as discussed later. In order to calculate the number of solute photons emitted, we will assume the following solute properties and typical experimental conditions:

$\langle \delta_\lambda \rangle = 10^{-50}$ cm^4 sec molecule^{-1} photon^{-1}

$\phi_{f\lambda}^{(2)} = 0.4$

N_0 (incident photons per pulse) = 10^{15} photons pulse^{-1}

I_0 (incident intensity) = 10^{-26} photons cm^{-2} sec^{-1}

L (path length of focused beam waist) = 0.2 cm

C (solute number density) = 10^{17} molecules cm^{-3}

Equation (5) predicts that the solute will emit 4×10^6 photons per pulse. These photons are radiated over a solid angle of 4π steradians and must be discriminated against a background of 10^{15} laser photons. Because optically imperfect low-temperature solvent glasses produce a significant amount of scattered light, very effective filtering is necessary. We have found that a combination of a chemical filter (CF) and an interference filter provides an optimum degree of discrimination against scattered laser light without undue absorption of the solute emission.[16] We typically use a 5-cm-long quartz cell containing 0.4 M CuSO$_4$/H$_2$O as the chemical filter. The preceding filtering techniques as well as the efficiency of the optics typically yield a total collection efficiency of $\sim 10^{-4}$. That is, ~ 400 photons/pulse arrive at the detector. Given the fact that these photons arrive during a very short period of time (determined by the laser pulse width and the fluorescence lifetime of the solute), a relatively strong signal is obtained. However, if the quantum yield for emission is only 0.01, only 10 photons will arrive at the detector. Efficient boxcar integration or gated photon counting techniques[16] are then usually necessary to extract the signal from the ambient photomultiplier dark noise.

All-*trans*-retinol

The first two-photon spectrum of a visual chromophore was reported in 1978.[12] This investigation of all-*trans*-retinol used the two-photon excitation technique to demonstrate that this chromophore has a lowest-lying, strongly two-photon allowed ($\delta = 2 \times 10^{-49}$ cm^4 sec molecule^{-1} photon^{-1}) "$^1A_g^{*-}$" excited state. The two-photon excitation maximum was observed ~ 1600 cm^{-1} to the red of the one-photon ("$^1B_u^{*+}$") absorption maximum in EPA at 77°K (Fig. 4). A comparison of maxima cannot be used as a reliable indicator of system origin energy separations because Frank–Condon factors associated with the "$^1A_g^{*-}$" state are expected to shift the

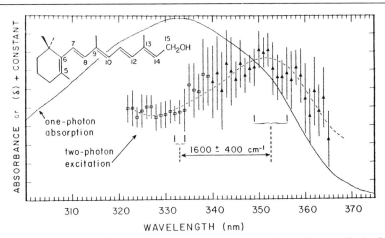

FIG. 4. Comparison of the two-photon excitation spectrum (dashed line, vertical axis linear in two-photon absorptivity) and one-photon absorption spectrum (solid line, vertical axis linear in molar absorptivity) of all-*trans*-retinol in EPA at 77°k. The laser excitation wavelength is twice the wavelength indicated. [adapted from Ref. 12].

energy vibronically of the inhomogeneously broadened two-photon excitation maximum to higher energy than that characteristic of the "$^1B_u^{*+}$" state contour.[12] Accordingly, the system origin of the "$^1A_g^{*-}$" state in all-*trans*-retinol is probably 2000–2500 cm^{-1} below the "$^1B_u^{*+}$" state system origin in EPA at 77°K.

All-*trans*-retinal

The two-photon excitation spectrum of all-*trans*-retinal[13] is shown in Fig. 5. The interpretation of the excitation spectrum is complicated by the observation that all-*trans*-retinal displays a wavelength dependence in its (one-photon induced) fluorescence quantum yield.[2,13] Becker and co-workers have recently suggested that the wavelength dependence of ϕ_f in retinal is associated with the simultaneous presence of two retinal species in solution, a hydrogen-bonded species that fluoresces and a nonhydrogen-bonded species that does not fluoresce.[2] These authors conclude that an inner-filter effect associated with the nonhydrogen-bonded species is responsible for producing the "apparent" wavelength dependence.[2] This inner-filter effect would not be observed in two-photon spectroscopy because a negligible amount of the laser irradiation is absorbed during a two-photon excitation experiment (see discussion in Ref. 13b). Accordingly, if the inner-filter effect hypothesis is correct, no quantum yield correction should be made to the observed two-photon excitation spectrum. The spectra shown in Fig. 5 indicate that all-*trans*-retinal has a "$^1A_g^{*-}$" $\pi\pi^*$

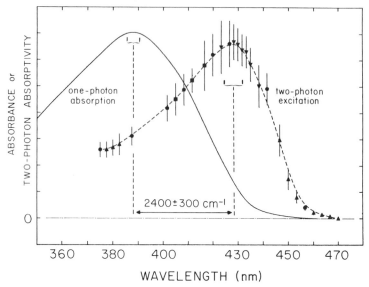

FIG. 5. Comparison of the two-photon excitation spectrum (dashed line) and one-photon absorption spectrum of all-*trans*-retinal in EPA at 77°K. [Reproduced from the Ph.D. thesis of J. A. Bennett, University of California, Riverside (1980); see also Ref 13b.]

excitation maximum roughly 2400 cm^{-1} below the "$^1B_u^{*+}$" $\pi\pi^*$ absorption maximum in EPA at 77°K. Molecular orbital calculations indicate that the two-photon absorptivity of the low-lying $n\pi^*$ state is extremely small.[13b] Accordingly, the $n\pi^*$ state is definitely not associated with the two-photon excitation spectrum shown in Fig. 5.[13b] The solvent effect data reported by Becker and co-workers, however, suggested that the $n\pi^*$ state is nearly degenerate with the "$^1A_g^{*-}$" state and that its level ordering relative to the "$^1A_g^{*-}$" state is solvent dependent.[2] The experimental results for all-*trans*-retinal can be summarized as follows:

$$\text{``}^1B_u^{*+}\text{''} > \text{``}^1A_g^{*-}\text{''} > n\pi^* \qquad \text{(nonpolar solvents)}$$

or
$$\left.\begin{array}{c} \text{``}^1B_u^{*+}\text{''} > n\pi^* > \text{``}^1A_g^{*-}\text{''} \\[6pt] n\pi^* > \text{``}^1B_u^{*+}\text{''} > \text{``}^1A_g^{*-}\text{''} \end{array}\right\} \quad \text{(hydrogen bonding solvents)}$$

Acknowledgments

The author gratefully acknowledges many helpful discussions with Professors Bruce Hudson, David Kliger, Bryan Kohler, George Leroi, Martin McClain, and Dr. James Bennett on two-photon spectroscopy. The work originating from the author's laboratory was supported in part by the National Institutes of Health; the Albert G. and Herman B. Memorial Fund of Fight for Sight, Inc., New York City; and Research Corporation.

[68] Photoisomerization Kinetics of Retinal Analogs

By DAVID S. KLIGER

When rhodopsin is irradiated at liquid nitrogen temperature (77°K) a stable intermediate, bathorhodopsin, is formed. This intermediate is characterized by an absorption maximum at 543 nm (compared with 500 nm for the parent bovine pigment) and is stable at temperature below 133°K.[1] At room temperature bathorhodopsin is formed from rhodopsin in less than 6 psec[2] and has a lifetime of about 50 nsec.[3] This intermediate has received much attention because the process of bathorhodopsin formation is seen as the "trigger" for visual transduction.

The chromophore in rhodopsin is retinal, in its 11-*cis* conformation, bound to opsin through a protonated Schiff base linkage. After bleaching rhodopsin thermally, the 11-*cis*-retinal can be recovered. Photobleaching, however, yields all-*trans*-retinal. Only the first step in the bleaching process requires light. These facts, along with spectral and photochemical properties of bathorhodopsin,[4] have led to the widespread belief that the chromophore in bathrohodopsin has a distorted all-*trans* conformation.

The fact that bathorhodopsin formation is a primary step in visual transduction and probably involves isomerization of the chromophore has led to many studies of the photoisomerization of retinal and related compounds. These studies have involved both quantum yield measurements of retinal isomer formations and direct kinetic measurements of photoisomerizations.

Quantum Yield Measurements

One of the earliest studies of the photoisomerization of retinal was that of Kropf and Hubbard.[5] They first obtained the absorption spectra of all-*trans*-, 9-*cis*-, 11-*cis*-, and 13-*cis*- retinals. Then, after irradiating solutions of these isomers to various degrees, they measured the spectra of the resultant isomer mixtures. The mixture spectra could then be analyzed in terms of the spectra of the component isomers. In this way quantum yields for the photoisomerization of the various isomers were determined.

[1] T. Yoshizawa and Y. Kito, *Nature (London)* **182**, 1604 (1958).
[2] G. E. Busch, M. L. Applebury, A. A. Lamola, and P. M. Rentzepis, *Proc. Natl. Acad. Sci. U.S.A.* **69**, 2802 (1972).
[3] T. Rosenfeld, A. Alchalal, and M. Ottolenghi, *Nature (London)* **220**, 482 (1972).
[4] A. Kropf, *Proc. Int. Sch. Phys. "Enrico Fermi"* **43**, 28 (1969).
[5] A. Kropf and R. Hubbard, *Photochem. Photobiol.* **12**, 249 (1970).

METHODS IN ENZYMOLOGY, VOL. 88

In order to make the analysis tractable it was necessary to assume that only one bond isomerizes at a time and that *cis* isomers preferentially convert to *trans* isomers and do not form di*cis* isomers. One interesting finding of this study was that the photoisomerization yield for 11-*cis* to all-*trans* retinal is 0.2 whereas the photobleaching yield of rhodopsin is 0.6.

Photoisomerization yield studies were extended by Raubach and Guzzo[6] to determine the yield of isomerization from the triplet state of retinal. They excited biacetyl in the presence of retinal. Biacetyl triplets are formed and energy is then transferred to retinal triplets. In this way triplets are formed (sensitized) without the production of retinal excited singlet states. Irradiation of the biacetyl/retinal solutions yielded triplet sensitized isomerization yields similar to photoisomerization yields found by Kropf and Hubbard for the all-*trans* to 13-*cis* to all-*trans* isomerizations (0.2 and 0.4, respectively). The triplet sensitized yield of 11-*cis* to all-*trans* retinal was, however, 0.75 compared to a direct excitation yield of 0.2.

Later studies of the role of the triplet state in retinal photoisomerization included laser excitation as well as conventional excitation for the photoisomerization experiments. Rosenfeld, Alchalel, and Ottolenghi[7] used kinetic spectroscopy to determine triplet state properties of retinal and combined these measurements with conventional measurements of isomerization yields. Kinetic spectroscopic techniques will be discussed in the next section. They are useful in obtaining spectra of transient species or following the kinetics of a photochemical reaction subsequent to initiation of the reaction with a short-lived light pulse. Rosenfeld used the technique to measure the triplet–triplet absorption spectrum of retinal. The lowest triplet state in retinal has a lifetime in hydrocarbon solvents of about 12 μsec in degassed and about 100 nsec in aerated samples.

Triplet absorption was used to obtain the quantum yield for triplet state formaton of retinal. This was done by first determining the extinction coefficient, ϵ_{TT}, for the $T_n \leftarrow T_1$ transition. A known concentration of retinal triplets was produced through energy transfer from biphenyl triplets and the optical density of the resultant retinal absorption determined ϵ_{TT}. Then, by comparing the triplet absorption of retinal with that of anthracene and pyrene, whose triplet quantum yields are known, the triplet quantum yield for retinal was determined to be 0.6 in *n*-hexane.

Once the triplet quantum yield was determined, photoisomerization yields were found for both directly excited and triplet sensitized processes. Direct yields were determined by measuring the optical densities

[6] R. A. Raubach and A. V. Guzzo, *J. Phys. Chem.* **77**, 889 (1973).
[7] T. Rosenfeld, A. Alchalel, and M. Ottolenghi, *J. Phys. Chem.* **78**, 336 (1974).

of samples at 254, 280, and 365 nm as a function of irradiation time. Quantum yields of 0.12 and 0.14 were found, respectively, for the 11-*cis* to all-*trans* and all-*trans* to *cis* retinal reactions. Triplet sensitization was carried out by exciting mixtures of biphenyl with tetramethyl-*p*-phenylenediamine (TMPD) in the presence of retinal. TMPD is excited and forms a TMPD–biphenyl exciplex. Triplet exciplex is then formed and energy transfer results in triplet retinal. This procedure eliminates production of retinal excited singlet states produced by fluorescence of TMPD or biphenyl alone (since the exciplex fluorescence is red shifted from that of the parent molecules). This method yielded quantum yields of 0.17 (11-*cis* to all-*trans*) and less than 0.002 (all-*trans* to *cis*) for triplet sensitized isomerization. Furthermore, it was determined that although oxygen quenches retinal triplet states, it does not affect the isomerization yields.

From these quantum yield measurements the authors made several conclusions about retinal photoisomerization. First, from isomerization yields and triplet yield they found that 80% of the 11-*cis* to all-*trans* isomerization occurs through the triplet (T_1) state. For the all-*trans* to *cis* reaction, however, no isomerization occurs through the thermalized T_1 state. Further, since *trans*- retinal undergoes no isomerization in the thermalized T_1 state, the *cis* and *trans* isomers must not be in thermal equilibrium in the T_1 state. Finally, they concluded from the lack of an oxygen effect on isomerization yield that isomerization must not occur in the thermalized T_1 state; it must occur in a vibrationally excited T_1 state.

Although studies of retinal isomerization yielded interesting.*results, the chromophore in the pigments is a protonated Schiff base rather than the aldehyde. Work thus proceeded on other analog of retinal. Work on the triplet sensitized photoisomerization of a retinal Schiff base[8] and protonated Schiff base[9] used techniques similar to the retinal photoisomerization study cited earlier. It is not clear that the triplet states of these molecules could contribute significantly to isomerization in visual pigments, since the triplet yields in the Schiff base (RSB) and protonated Schiff base (PSB) are 0.008 and less than 0.001, respectively.[10] It is conceivable, however, that protein interactions could enhance triplet formation and the effect of substitution on triplet isomerization properties is interesting in its own right. Triplet sensitized photoisomerization of PSB was thus measured by the technique of sensitization from a TMPD–biphenyl exciplex. Isomerization quantum yields were found to be 0.45 for 11-*cis* RSB, 0.06 for 9-*cis* RSB, 0.08 for 13-*cis* RSB, and 0.02–0.05 for

[8] T. Rosenfeld, A. Alchalel, and M. Ottolenghi, *Photochem. Photobiol.* **20**, 121 (1974).
[9] A. Alchalel, B. Honig, M. Ottolenghi, and T. Rosenfeld, *J. Am. Chem. Soc.* **97**, 2161 (1975).
[10] M. M. Fisher and K. Weiss, *Photochem. Photobiol.* **20**, 423 (1974).

all-*trans* RSB. Direct excitation resulted in a photoisomerization yield of 11-*cis* RSB of 0.005 in either degassed or aerated samples. These quantum yields are too small to determine whether direct excitation results in isomerization through the singlet or triplet manifolds.

Sensitization of PSB was accomplished with energy transfer from phenanthrene. The presence of O_2 quenched phenanthrene triplets very efficiently so no isomerization was observed in aerated samples. In degassed samples triplet isomerization yields were measured to be 1.0 for 11-*cis* PSB, 0.5 for 9-*cis* PSB, 0.2 for 13-*cis* PSB, and less than 0.05 for all-*trans* PSB.

With the advent of high-performance liquid chromatography (HPLC), studies like those cited earlier could be carried out with increased sensitivity. The first study of retinal photoisomerization to use HPLC was carried out in 1976 by Waddell *et al.*[11] They were able to determine not only the quantum yields for direct excitation and triplet sensitized photoisomerization but the product isomer distribution subsequent to photoisomerization. This was possible since in HPLC the product isomers are separated so the amounts of each isomer present can be determined.

Waddell *et al.* found that under similar conditions the photoisomerization yields of 9-*cis*, 11-*cis*, 13-*cis*, or 9,13-di*cis* retinals are similar. The yields in polar solvents, however, are 4 to 5 times smaller than those in nonpolar solvents. They also found that the photoisomerization yield of 11-*cis* retinal decreases with increasing solvent viscosity. In addition, by comparing direct and triplet sensitized (with biacetyl) excitation studies they concluded that photoisomerization occurs through both singlet and triplet routes.

Using similar procedures, they were also able to study the isomerization yields in various visual pigments. That is, they would irradiate rhodopsin (chromophore is 11-*cis* retinal), isorhodopsin (9-*cis* retinal), or 9,13-isorhodopsin (9, 13-*cis*-retinal) and observe the yield and product ratios of retinal isomers produced on pigment bleaching. The obtained yields of 0.67 for rhodopsin and 0.33 for the two isorhodopsins with direct excitation. They also irradiated the pigments in the presence of the triplet sensitizer 1,2-dioxetane and found complete isomerization. From these results they concluded that triplet sensitization does enhance pigment bleaching. One cannot conclude from this, however, that the triplet state is involved in the direct excitation of visual pigment bleaching.

Later studies of retinal photoisomerization used HPLC to gain more detailed information about conditions affecting isomerization. Photoproducts of all-*trans*- retinal were found to include 7-, 9-, 11-, and 13-*cis* iso-

---

[11] W. H. Waddell, R. Crouch, K, Nakanishi, and N. J. Turro, *J. Am. Chem. Soc.* **98,** 4189 (1976).

mers. The ratios of the different isomers produced was found to be wavelength independent from 430 nm to 310 nm. Irradiation at 270 nm produced a different product distribution, however. Product ratios were also different in polar versus nonpolar solvents.[12]

Waddell and West also found that 11-*cis* retinal in its 12-s-*trans* conformation had a higher photoisomerization yield than 11-*cis*- 12-s-*cis* retinal.[13] At room temperature 11-*cis* retinal had been shown to have a photoisomerization yield of 0.2, whereas at −65° the yield increases to 0.6.[5] This has been explained as an increase in the amount of 12-s-*trans* relative to 12-s-*cis* isomer at lower temperature.[14,15] Waddell and West confirmed this hypothesis by looking at the photoisomerization yields of 11-*cis*-13-demethylretinal and 11-*cis*-14-methylretinal. The 13-demethyl compound is expected to have a planar 12-s-*trans* geometry and the 14-methyl compound is thought to have a highly twisted 12-s-*trans* geometry. The photoisomerization yields were found to be 0.66 and 0.79 for the 13-demethyl and 14-methyl compounds, respectively. These yields are similar to the low-temperature 11-*cis* retinal yields as expected.

The use of HPLC to analyze products of photoisomerization greatly increased the sensitivity of quantum yield measurements and made it possible to gain much more information about photoisomerization products. As with many such advances, this provided much information that is difficult to understand in detail. Although some straightforward principles could be shown, like the fact that one carbon–carbon bond is isomerized per unit photon absorbed,[16] other features seem more complex. For example, isomerization quantum yields and compositions of product mixtures were shown to depend on wavelength, on solvent, and on irradiation time.[16]

Direct Kinetic Measurements

Though much mechanistic information of retinal isomerization had been obtained through quantum yield measurements, there was still a drive to measure the isomerization process more directly. Much of the incentive for this direct kinetic measurement came from a 1972 study of the buildup kinetics of bathorhodopsin using picosecond spectroscopy.[17] It

[12] W. H. Waddell and D. L. Hopkins, *J. Am. Chem. Soc.* **99**, 6457 (1977).

[13] W. H. Waddell and J. L. West, *Chem. Phys. Lett.* **62**, 431 (1979).

[14] A. M. Schaffer, W. H. Waddell, and R. S. Becker, *J. Am. Chem. Soc.* **96**, 2063 (1974).

[15] R. R. Birge, M. J. Sullivan, and B. E. Kohler, *J. Am. Chem. Soc.* **98**, 358 (1976).

[16] W. H. Waddell and J. L. West, *J. Phys. Chem.* **84**, 134 (1980).

[17] G. Busch, M. Applebury, A. Lamola, and P. Rentzepis, *Proc. Natl. Acad. Sci. U.S.A.* **69**, 2802 (1972).

was found, in this study, that bathorhodopsin is formed from rhodopsin within 6 psec of the absorption of a photon. This very rapid transformation prompted some to question whether retinal isomerization could occur so quickly and be responsible for the rhodopsin to bathorhodopsin transition. Microsecond, nanosecond, and picosecond kinetic spectroscopic measurements have since been used to study the isomerization of retinal and retinal analogs.

Kinetic spectroscopic measurements are generally made with the use of two light sources. The first is a pump source, which produces a new state to be studied or initiates a reaction to be followed. The second is a probe source with which one can monitor spectral changes resulting from the reaction being studied.

In microsecond kinetic spectroscopy[18] the pump source is usually a xenon flashlamp or a flashlamp-pumped dye laser. The probe source is usually a CW light source. One can monitor the change in intensity of the probe beam as a function of time and wavelength after the pump beam enters the sample. The change in beam intensity can be analyzed in terms of optical density. The wavelength dependence of the signal at a given time then yields the difference spectrum between the reactant and the product species. The time dependence of the signal at a given wavelength yields the buildup or decay kinetics of species absorbing light at that wavelength.

Nanosecond kinetic spectroscopy uses the same techniques used in a microsecond system.[19] The pump source is generally a laser with a pulse width of several nanoseconds. The probe source may be a CW light source, but usually a pulsed light source is preferable. This may be a xenon flashlamp or a CW lamp whose intensity is intermittently enhanced by discharging a capacitor across the lamp to produce high-intensity pulses. These pulsed probe sources are used because on a nanosecond time scale there is not time to average out photon noise in the light source. Since the signal-to-noise ratio (S/N) in light sources varies as the square root of the source power the pulsed sources, with their higher powers, give better S/N data.

Measurements in nanosecond and microsecond kinetic spectroscopy are made by converting the light of the probe source into an electrical signal proportional to the probe intensity. The electrical signal can be followed with an oscilloscope, an analog-to-digital converter, or some signal-averaging device. Although the principle behind picosecond spectroscopy is the same, present technology precludes following the

---

[18] S. Claesson, in "Spectroscopy of the Excited State" (B. Di Bartolo, ed.), p. 95. Plenum, New York, 1976.

[19] E. L. Russell, A. J. Twarowski, D. S. Kliger, and E. Switkes, Chem. Phys. 22, 167 (1977).

probe intensity in real time with electronics. Time measurements are made in a different way[20] (*vide infra*).

Picosecond pump sources are generally mode-locked lasers. Nd^{2+} glass lasers are most often used, though ruby lasers can also be mode-locked. The neodymium lasers emit pulses, of several picoseconds duration, at 1.06 μm that can be frequency doubled (530 nm), tripled (353 nm), or quadrupled (265-nm). There is enough power in these pulses that when they are focused into some liquids, resulting nonlinear effects produce picosecond pulses that are spectrally very broad. These picosecond continuum pulses are used to probe the reaction being studied.

Time measurements for picosecond kinetic studies are made by using the light source itself as a clock. Light travels 0.3 mm in 1 psec in a vacuum but smaller distances through materials of higher refractive index. By passing two beams through materials of different thicknesses they can thus be separated by known time intervals. Picosecond spectroscopy uses this principle by passing the continuum beam through an echelon. This is a device that separates the beam into several beams that pass through high refractive index materials of various known thicknesses. This creates a train of pulses separated by a known time span. With this pulse train one can follow kinetics by measuring the energy of each pulse in the probe beam following excitation of the sample by the pump beam. Only integrated intensities of each pulse need be determined, so ultrafast electronics is not necessary. Temporal information is obtained from knowledge of the time interval between pulses in the probe beam.

With any of these kinetic spectroscopic techniques one can study the kinetics of photoisomerization of retinal analogs by taking advantage of spectral differences between the various isomers. For example, in hydrocarbon solvents 11-*cis*- retinal has an extinction coefficient ϵ of ~ 25,000 at the peak of the α-transition (365 nm) and an ϵ of ~ 15,000 at the peak of the β-transition (250 nm). At these wavelengths all-*trans*- retinal has extinction coefficients of ~ 50,000 (~ 365 nm) and ~ 7,000 (250 nm). Thus, one could follow the kinetics of the reaction 11-*cis*-retinal → all-*trans*-retinal by exciting the 11-*cis* isomer with a short pulse of light and subsequently monitoring the change in optical density of the sample. At 365 nm the optical density will increase as the *cis* isomer is converted to *trans* isomer, whereas at 250 nm the optical density will decrease. An exact analysis of the kinetics is actually a little more complicated, since other isomers could be formed on photoisomerization.[21] Since, however, the predominant product of 11-*cis*-retinal isomerization is all-*trans*-retinal, the complications can be ignored.

[20] H. J. Kaufmann and P. M. Rentzepis, *Acc. Chem. Res.* **118**, 407 (1975).
[21] E. L. Menger and D. S. Kliger, *J. Am. Chem. Soc.* **98**, 3975 (1976).

When 11-*cis*-retinal was studied as described earlier[21] it was found that the optical density increased with a rate constant of about 100 nsec, the rate of triplet decay in aerated solution. This was interpreted as evidence for involvement of a thermalized triplet state in the isomerization process, though it has been pointed out[22] that this evidence is very weak because ground-state depletion effects would yield the same kinetic data even if isomerization occurred instantaneously.

Quantum yield measurements show that a triplet pathway must contribute to isomerization, but the question of thermalized versus nonthermalized triplet state involvement is not clear. Harriman and Liu[23] find identical spectra and decay rates of triplet states of several retinal isomers. This is consistent with equilibration of all retinal triplets. Veyret *et al.*[22] disagree with the conclusion, however, that the triplet spectra of the different intermediates are identical. Triplet absorption studies of retinal at low temperatures[24] reveal that different isomers do indeed have distinct spectra. The triplet spectra were interpreted as triplets of the ground-state isomers. Thus at low temperature, isomerization is assumed to take place from the thermalized triplet state.

The evidence for implicating a nonthermalized triplet comes primarily from O_2 quenching effects on retinal triplets and photoisomerization yields.[7,22] Quenching of the triplets has also been carried out with a nitroxyl radical, with which triplet quenching can be made to dominate the decay mechanism.[25] The results are consistent with either of two mechanisms. First, a nonthermalized triplet could be involved in isomerization. Product distributions would then be fixed in the thermalized triplet and triplet quenching would not alter the triplet yields. A second possibility is that isomerization does occur in the thermalized triplet state and that decay rates and quenching rates for triplets of all isomers are identical.

Since retinal is bound to the protein in visual pigments through a protonated Schiff base linkage, SB and PSB isomerizations were also investigated.[21] The quantum yield of SB isomerization is only 0.008 and is difficult to detect directly, as described earlier. Isomerization of PSB can be readily studied, however, and was found to occur very rapidly. With a nanosecond spectroscopic apparatus it could only be determined that isomerization occurs in less than 10 nsec and does not go through a triplet mechanism.

Investigation of PSB isomerization with picosecond spectroscopic

[22] B. Veyret, S. G. Davis, M. Yoshida, and K. Weiss, *J. Am. Chem. Soc.* **100**, 3283 (1978).
[23] A. Harriman and R. S. H. Liu, *Photochem. Photobiol.* **26**, 29 (1977).
[24] M. Grodowski, R. S. H. Liu, and W. G. Herkstroeter, *Chem. Phys. Lett.* **65**, 42 (1979).
[25] V. A. Kuzmin, D. S. Kliger, and G. S. Kliger, *Photochem. Photobiol.* **31**, 607 (1980).

techniques[26] yielded surprising results. Although some short-lived (20 psec) transient absorptions were observed, the total isomerization process was shown to take about 10 nsec. The total process involves exciting the 11-*cis* isomer, isomerization, and return to the ground state of all-*trans* PSB. That this process occurred in 10 nsec while bathorhodopsin formed within 6 psec of excitation of rhodopsin was surprising. It meant either that the formation of bathorhodopsin did not involve a simple isomerization or that the presence of the protein enhanced the isomerization rate by more than three orders of magnitude.

Several picosecond spectroscopic studies of bathorhodopsin formation have addressed the question of whether or not isomerization is occurring within a few picoseconds. Conclusions drawn by different workers have conflicted. Peters *et al.*[27] studied the temperature dependence of bathorhodopsin down to liquid helium temperatures. At low temperature they observed a short-lived transient that decayed to bathorhodopsin. The temperature dependence of the decay did not follow Arrhenius behavior. Instead, it becomes temperature independent at low temperatures. This asymptotic behavior is characteristic of tunneling phenomena.

In addition to the observed temperature dependence, Peters *et al.* observed a large deuterium effect on the kinetics of bathorhodopsin formation. They interpreted these effects to mean that a proton movement, rather than an isomerization, was responsible for bathorhodopsin formation. It should be pointed out that although the data are consistent with a mechanism of proton motion, they are also consistent with an isomerization process that could have a very small activation barrier. Deuterium could effect protein relaxation around an isomerized chromophore yielding the observed spectral effects.

Other picosecond spectroscopic experiments do, in fact, point to a rapid isomerization of the chromophore.[28,29] Bathorhodopsin formation was studied for both rhodopsin and isorhodopsin at room temperature. Bathorhodopsin in both pigments was formed in less than 3 psec. Rhodopsin contains an 11-*cis* chromophore and isorhodopsin contains a 9-*cis* chromophore. The bathorhodopsin of both species is spectrally the same and one can back-react the bathorhodopsin of each to form both rhodopsin and isorhodopsin. One could conclude that the same bathorhodopsin

[26] D. Huppert, P. M. Rentzepis, and D. S. Kliger, *Photochem. Photobiol.* **25**, 193 (1977).

[27] K. Peters, M. L. Applebury, and P. M. Rentzepis, *Proc. Natl. Acad. Sci. U.S.A.* **74**, 3119 (1977).

[28] B. H. Green, T. G. Monger, R. R. Alfano, B. Aton, and R. H. Callender, *Nature (London)* **269**, 179 (1977).

[29] T. G. Monger, R. R. Alfano, and R. H. Callender, *Biophys. J.* **27**, 105 (1979).

is formed from each species and that the chromophore must then be in a transoid conformation. If so, one must say that isomerization of the chromophore occurs within 3 psec.

There is much photochemical evidence[30] showing that bathorhodopsin must contain a chromophore that is in either an all-*trans* conformation or at least in a transoid conformation. It is still possible, of course, that other processes, such as proton translocations,[27] could be involved in bathorhodopsin formation. Such processes could, in fact, act to catalyze isomerization. In any event, it does seem clear that an isomerization does occur within 3 psec in visual pigments.

Conclusions

The preceding studies use either visible–ultraviolet spectroscopic absorptions or chromatographic techniques to study retinal isomerization. Through these methods a great deal has been learned about the isomerization of retinal analogs. Other tools could certainly add to our knowledge. Time-dependent Raman or infrared spectroscopies, for example, should prove useful, as would further studies of different isomers in more varied enviroments. These studies will undoubtedly follow two paths: (1) The study of retinal analogs themselves to work out details of the isomerization mechanism; and (2) the study of natural and synthetic visual pigments to determine the role of isomerization and other processes in the bleaching mechanism.

[30] T. Rosenfeld, B. Honig, M. Ottolenghi, J. Hurley, and T. G. Ebrey, *Pure Appl. Chem.* **49**, 341 (1977).

[69] Isolation and Purification of Retinals from Purple Membranes for Mass Spectral Analysis[1]

By STANLEY SELTZER and MOW LIN

Recent studies concerned with the action of the purple membrane proton pump of halobacteria have led to proposals concerning the mechanism of this process.[2] One mechanism suggests that upon photon absorption

[1] Research carried out at Brookhaven National Laboratory under contract with the U.S. Department of Energy and supported by its Office of Basic Energy Sciences.
[2] For recent reviews, see S. R. Caplan and M. Ginzburg, eds., "Energetics and Structure of Halophilic Microorganisms." Elsevier/North-Holland Biomedical Press, Amsterdam, 1978; W. Stoeckenius, R. H. Lozier, and R. A. Bogomolni, *Biochim. Biophys. Acta* **505**, 215 (1978); M. Eisenbach and S. R. Caplan, *Curr. Top. Membr. Transp.* **12**, 165 (1979).

the excited retinal leads to a charge-separated species that can induce proton hopping along the polyene chain to provide part of the route for the translocated proton. Other mechanisms suggest reversible proton or hydrogen atom transfer from the cyclohexenyl ring or from the 5-methyl of the cyclohexenyl ring of the bound retinal to a nearby acceptor amino acid residue of the protein. Similar proposals have been made for the rhodopsin system.[3]

Were reactions of this sort occurring one might expect some ultimate exchange of the hydrogen atoms of retinal with the migrating proton (i.e., those originally from the solvent) or exchange with the labile protons of the nearby acceptor amino acid side chains of the protein. Such processes can be tested by light pumping the membrane in a deuterated environment, detaching and isolating the chromophore from the pigment, purifying it by HPLC, and determining the deuterium content of the chromophore by mass spectrometry.[4] Other experiments such as those designed to examine kinetic and equilibrium isotope effects might also be carried out in similar fashion.

Removal of the chromophore can be accomplished by one of two general methods: (a) irradiation of a suspension of purple membrane with orange light in a neutral hydroxylamine solution,[5] or (b) extraction of the chromophore from the membrane with chloroform–cetyltrimethylammonium bromide (CTAB)[6] or with dichloromethane.[7] While the first method leads to intact apomembrane that can be used to regenerate purple membrane or an analog of the native membrane, residual absorption at 360 nm, even after washing, suggests that some retinal oxime remains in the membrane. In principle it should be possible to recover retinal by this method, but some thermal or photocatalyzed *cis-trans* isomerization of the free retinal or retinal oxime may occur unless precaution is taken to carry out reaction at low temperature[7] and with long wavelength light. This method will not be discussed further since it is treated more fully in another section.

Extraction of retinals from the purple membrane by methylene chloride or chloroform is preferred since it can be accomplished in a shorter period of time in the absence of light so that there is little chance of post-isomerization. Several extractions from a water suspension of the purple

[3] K. van der Meer, J. J. C. Mulder, and J. Lugtenburg, *Photochem. Photobiol.* **24**, 363 (1976).

[4] M. Lin and S. Seltzer, *FEBS Lett.* **106**, 135 (1979).

[5] D. Oesterhelt and L. Schuhmann, *FEBS Lett.* **44**, 262 (1974).

[6] L. Y. Jan, *Vision Res.* **15**, 1081 (1975).

[7] M. J. Pettei, A. P. Yudd, K. Nakanishi, R. Henselman, and W. Stoeckenius, *Biochemistry* **16**, 1955 (1977).

membrane have to be carried out to remove the major amount of the retinal present.

Since methylene chloride at the same time extracts some lipids, lipids must be removed from the retinal before subjecting it to mass spectral analysis. Purification and separation of retinal isomers is ultimately achieved by HPLC on a microparticulate silica gel column with the retinals eluted by a solvent of less than 1% ether in cyclohexane.[8] Silica gel, however, binds lipids much more strongly than retinals and the solvent system used to separate retinal isomers does not effectively elute lipids from the column. Thus repeated application of lipid–retinal mixtures would gradually inactivate the HPLC column. This is avoided by simply performing a gross separation of lipids and retinals on a coarser silica gel and eluting retinals and lipids sequentially by a more polar solvent. Alternatively, a microparticulate silica gel–alkyl nitrile column has been used to separate retinal and lipid components that have been extracted from rhodopsin by methylene chloride. Simultaneous purification of the retinals and then elution of the lipids is achieved with two separate solvent systems.[9] This method is discussed in another section and will not be discussed further here.

The retinal is introduced into the mass spectrometer as a solid. It is important to protect retinal samples against oxidation and therefore to carry out mass spectral analysis as soon after HPLC purification as possible. If samples have to be stored, it should be as a solid under a nitrogen atmosphere in the freezer.

The parent ion of retinal is 284. A smaller peak at 285, about 20% the height of the 284 peak, is mainly due to the approximately 1% natural abundance ^{13}C randomly distributed throughout the molecule. A 285/284 peak intensity ratio in excess of that measured for a control sample of retinal is taken as a measure of the degree of replacement of a light isotope with one that is one unit higher in mass, e.g., owing to the incorporation of an atom of deuterium. A similar increase in the 286/284 peak height ratio over the natural abundance ratio found in a control sample of retinal (i.e., ~0.03) can be used to detect the incorporation of two atoms of deuterium.

The presence of retinoic acid presents a problem. Retinoic acid exhibits a parent ion at mass 300. A relatively intense peak is generated at mass 285 (285/300 ≃ 0.3; 286/285 ≃ 0.22)[10] presumably because of the loss of a methyl group. Thus contamination by retinoic acid will lead to an

[8] R. Sack and S. Seltzer, *Vision Res.* **18**, 423 (1978).
[9] K. Nakanishi, *Pure Appl. Chem.* **49**, 333 (1977).
[10] R. L. Lin, G. R. Waller, E. D. Mitchell, K. S. Yang, and E. C. Nelson, *Anal. Biochem.* **35**, 435 (1970).

incorrectly determined degree of isotopic enrichment. Depending on the degree of precision desired, it may be reasonable to correct the 285 peak intensity for the presence of a small amount of retinoic acid (i.e., $\leqslant 5\%$) by measuring 285/300 for retinoic acid under the same conditions and correcting the 285 peak as determined from the size of the peak at 300.

Methods

Material
Reagent grade dichloromethane, hexanes, and ether
Spectrograde cyclohexane (Eastman Kodak)
Potassium phosphate buffer, 0.01 M, pH 7.4
Silica gel for column chromatography (60–200 mesh)
Silica gel HPLC column (60 × 0.2 cm, particle size, 10 μm);
Partisil 10 (Whatman) or equivalent is used
Procedure. If retinal isomer integrity is to be maintained, all operations are to be carried out under dim red light at 0°.

An aliquot of purple membrane containing the equivalent of about 50 μg of retinal is generally sufficient. Smaller quantities may be used if a more sensitive mass spectrometer than the one used here is available. Membranes which are initially in a basal salt solution are transferred to a 0.01 M phosphate buffer, pH 7.4, by centrifuging the basal salt suspension at 45,000 g for 20 min and then resuspending the pelleted material in about 10 ml of 0.01 M phosphate buffer, pH 7.4, as a homogeneous suspension. An equal volume of methylene chloride is added and an emulsion is formed by repeatedly sucking up and ejecting the mixture through a syringe needle (16–20 gauge).[7] The emulsion is then centrifuged at 35,000 g for 30 min at 0° and the layers are separated. Dichloromethane extractions of the residual aqueous phase and of the protein zone, located at the interface, are repeated four to five more times. The combined dichloromethane extracts should contain at this point about 80% of the total available retinal. The dichloromethane is removed with a rotary evaporator at ambient temperature.

Contaminating lipids are removed by column chromatography on silica gel. A column (23 × 0.9 cm) is slurry packed with silica gel (60–200 mesh) in hexane–ether (4:1, v/v). The effluent of the column is passed through a UV flow detector operating at 254 or 340 nm (Altex Model 153). The retinal sample is applied and eluted with the hexane–ether solvent mixture. The fraction between 16 and 32 ml generally contains all the retinal. The solution is concentrated in a stream of nitrogen to a small volume (~ 20 μl) for subsequent purification by HPLC.

HPLC is carried out on a column (60 × 0.2 cm) packed with microparticulate silica gel (10 μm). The eluate is passed through a UV flow detector operating at 254 or 340 nm. Ether (0.3–0.7%, v/v) in cyclohexane is used as the eluting solvent at a flow rate of 1–3 ml/min. The conditions of flow rate and ether content of the solvent will vary with the column but are adjusted to give efficient separation with a standard sample of retinals. Typical elution volumes for retinal isomers using 0.5% ether are as follows: 11-*cis*, 61.5 ml; 13-*cis*, 72.6 ml; 9-*cis*, 96.6 ml; all-*trans*, 112.5 ml. The desired isomers are collected separately and the solutions are concentrated with a rotary evaporator and then transferred to small conical tubes. Further concentration to a smaller volume ($\sim 50 \mu$l) is achieved with a stream of nitrogen. At this point an aliquot is transferred to the device used to introduce solid samples into the direct inlet of a mass spectrometer. If the sample holder has been previously cleaned in acid, it is imperative that it be washed with considerable distilled water, ammonia, and then distilled water before drying. Traces of residual acid in contact with retinal at elevated temperature lead to substantial decomposition and a considerably altered 285/284 ratio. After transfer, the solvent is evaporated off in a stream of nitrogen. The process is repeated until the whole solution is transferred to the sample holder and all the solvent is removed. The sample is introduced into the direct inlet system and the sample is analyzed. Typical conditions for the Hitachi-Perkin Elmer RMU-7 mass spectrometer are as follows: sample source chamber temperature, 60°; ionization chamber, 250°; photomultiplier voltage, 1.25 kV; ionizing voltage, 80 eV. The source pressure is generally 1×10^{-6} mm during measurement.

Prior to mass spectral analysis of the unknown, pure control samples of all-*trans*-retinal and/or 13-*cis*-retinal are analyzed to verify that the system is operational and to measure the 285/284 ratio in the absence of isotopic enrichment.

[70] Identification of cis/trans Isomers of Retinal Analogs by High-Performance Proton NMR Method

By PAUL TOWNER and WOLFGANG GÄRTNER

The first absolute identification of the isomers of retinal relied on their stereospecific synthesis; this afforded the assignment of an isomeric configuration to the chromatographic bands of retinal isomers when separated by thin-layer chromatography (TLC).[1] Further developments led to

[1] C. von Planta, U. Schwieter, L. Chopard-dit-Jean, R. Rüegg, and M. Kofler, *Helv. Chim. Acta* **45**, 548 (1962).

the identification of structure by using NMR methods, particularly in the lucid works of Patel[2] and Rowan et al.,[3] who accomplished the complete characterization of retinal using 220-MHz machines. In this article we describe an unambiguous method of identification of the isomers of retinal and retinal analogs using a high-performance NMR technique.

The essential role retinal plays in bacteriorhodopsin and rhodopsin is undisputed; however the mechanisms by which these proteins fulfill their function is still not clearly understood. Consequently many scientists interested in retinal protein interaction use retinal analogs as probes of the retinal binding sites, which in the case of bacteriorhodopsin should give valuable clues as to the mechanism of light-induced proton transport. In such experiments it is imperative to know the isomeric configuration of any analog used, and in some cases this has been done by inference to the properties of retinal. The chromatographic order of retinal isomers is known to be all-*trans*, 7-*cis*, 9-*cis*, 11-*cis*, and 13-*cis* with increasing R_f value; furthermore the apo-protein of rhodopsin (opsin) forms chromoproteins with 7-, 9- and 11-*cis*-retinal,[4,5] whereas bacterio-opsin reacts readily with all-*trans*- and 13-*cis*-retinal,[6] and to a lesser extent with 7-*cis*.[7]

Thus a pure but unknown isomer of an analog can be isolated from a TLC plate, mixed with opsin and bacterio-opsin and the λ_{max} value of any new chromophores observed, giving an inference of the isomeric type. However, as a means of identification this is clearly unsatisfactory since some retinal analogs are found to behave differently than retinal. A case in point is 13-demethylretinal where the chromatographic order of the 11-*cis* and 13-*cis* isomers is reversed, and where the 11-*cis* isomer readily reacts with bacterio-opsin.

To obtain unequivocal identification of retinal analog isomers the use of high-resolution NMR machines with magnetic fields of 400 and 500 MHz was found to yield excellent results with as little as 90 μg material, which required 40 min for accumulation of a Fourier transform spectrum. As an example the 400-MHz proton NMR spectrum of 11-*cis*-13-demethylretinal is shown in Fig. 1.

The signals in this spectrum fall into three categories: (1) the aldehyde proton at 9.5 ppm, (2) the olefinic protons of the chain (6.0–7.6 ppm), and

[2] D. J. Patel, *Nature (London)* **221**, 825 (1969).

[3] R. Rowan, III, A. Warshel, B. D. Sykes, and M. Karplus, *Biochemistry* **13**, 970 (1974).

[4] W. J. DeGrip, R. S. H. Liu, V. Ramamurthy, and A. Asato, *Nature (London)* **262**, 416 (1976).

[5] G. Wald, *Nature (London)* **219**, 800 (1968).

[6] D. Oesterhelt and L. Schuhmann, *FEBS Lett.* **44**, 262 (1974).

[7] K.-D. Kohl and W. Sperling, *Annu. Meet. Dtsch. Ges. Biophys.*, *1979* (1979).

[8] W. Gärtner, H. Hopf, W. E. Hull, D. Oesterhelt, D. Scheutzow, and P. Towner, *Tetrahedron Lett.* **21**, 347 (1980).

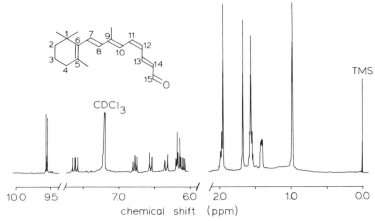

FIG. 1. 400-MHz [¹H]NMR spectrum of 11-*cis*-13-demethylretinal in deuteriochloroform. The chemical shifts of some of the protons of this spectrum differ from those presented in the table because of changes in concentration and temperature.

(3) the aliphatic protons of the ring and the methyl groups (1.0–2.0 ppm). It is the olefinic group of signals that reveals the cis/trans nature of the four double bonds of 13-demethylretinal; these signals are shown in expanded form in Fig. 2. Their interpretation first requires a short description of two basic NMR principles. First, it must be understood how the signal of a proton is influenced by neighboring protons on adjacent carbon atoms by spin–spin coupling, which effects splitting of the signal of the proton. Second, we explain how beginning with an unambiguously identified signal, an assignment of the residual signals to the other protons, is

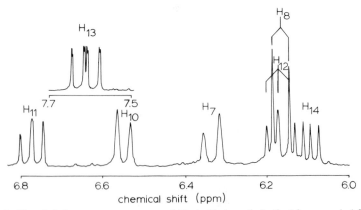

FIG. 2. The olefinic proton resonances of 11-*cis*-13-demethylretinal in expanded form.

determined by the double-resonance technique. These features can be best explained by reference to the partial olefinic structure I of 13-demethylretinal shown below.

STRUCTURE I

The effect of spin–spin coupling observed for any proton depends on (1) the number of protons bound to the adjacent carbon atoms, and (2) the type of carbon–carbon bond. The coupling between protons over three bonds is the most meaningful in interpretation of ^{1}H-NMR spectra; however coupling over four or more bonds can occur and yield useful information, but since this coupling is in the range of 1 Hz it is not totally resolved by machines of less than 250 MHz and leads only to a broadening of the signals.

If the coupling constants between the protons are similar, the splitting follows the rule $m = n + 1$, where m is the number of peaks in the signal group of a proton and n the number of proton neighbors on adjacent carbons. The coupling constant gives information on the type of bond between the carbon atoms. Thus the proton at position 2 undergoes splitting first to a doublet via coupling to proton at C-1 and is further split, with a different coupling constant, to a double doublet by simultaneous interaction with the proton at C-3. This is illustrated schematically in Fig. 3.

The symbol $^iJ_{k,1}$ designates the apparent coupling constant where i is the number of bonds over which coupling takes place and $k,1$ signify the coupled protons. In structure I, $^3J_{2,3}$, which is over a single carbon–carbon bond, has a value of 11–12.5 Hz; $^3J_{1,2}$ is over a trans double bond and

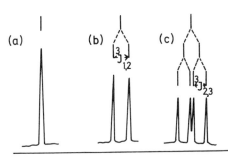

FIG. 3. Schematic representation of proton coupling. (a) Single resonance peak of the uncoupled proton at C-2 in structure I. (b) A doublet is formed by coupling with proton at C-1. (c) A multiplet is formed by simultaneously coupling to proton at C-3.

equals 14–16 Hz, and if the double bond was *cis* substituted, then $^3J_{1,2}$ is 11–13 Hz. Thus a knowledge of the coupling constants of protons across double bonds gives valuable information on the cis/trans nature of a polyene chain. The coupling constants in hertz can be directly measured from the spectrum as the distance between two peaks in a signal group, since 1 ppm (part per million) is by definition 400 Hz in the case of a 400-MHz machine.

The assignment of signals to the distinct protons is found by decoupling experiments using the double-resonance technique. This is dependent on the fact that a split proton signal, due to coupling with another proton, will collapse to a single peak when the resonance frequency of the effector proton is disturbed by irradiation.

The signal pattern of the protons at C-1, C-2, and C-3 in structure I are drawn in Fig. 4a. Upon irradiation at the frequency of the resonance of proton at C-3, only the residual coupling between protons 1 and 2 is found; the signal pattern of proton 1 remains largely uninfluenced (Fig. 4b).

The signals of the protons in a polyene chain could be anticipated to overlap, causing a spectrum of higher order to arise, which requires computer-assisted interpretation. NMR machines with large magnetic fields have the advantage that they produce well-defined first-order spectra from which valuable information of the proton interactions can be obtained at a glance. As a general rule, a spectrum can be classified as first order when the ratio of the difference in the chemical shift, in hertz, to the coupling constant between two protons is greater than or about 10. By reference to Fig. 2, protons 10 and 11, which have a coupling constant of

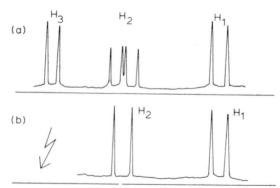

FIG. 4. Double-resonance experiment. (a) In structure I proton at C-2 couples with both H-1 and H-3, with different coupling constants. (b) Decoupling by irradiation at H-3 leaves the residual coupling between H-2 and H-1.

12 Hz, have a difference in chemical shift of 0.2 ppm; the ratio then equals about 7.

Using a knowledge of the structure of 13-demethylretinal, the signals of the olefinic protons can be constructed. Protons 7 and 8 are each observed as doublets, as is proton 10, whereas protons 11, 12, 13, and 14 show double doublets. The aldehyde proton, via coupling with proton 14, would also be present as a doublet at very low field at 9.5 ppm. Double resonance at this recognized site would change the signal form of proton 14, and if this proton in turn were subjected to double resonance it would lead to the identification of proton 13. This process can be repeated until protons 12, 11, and 10 are located. At position 9 the methyl group disrupts a link to protons 7 and 8; thus these two isolated protons can be recognized only by inference.

Once the signal pattern has been correlated to specific protons and their coupling constants measured, the isomeric configuration of the 7/8, 11/12, and 13/14 double bonds can be determined.

The assignment of cis or trans stereochemistry to the 9/10 double bond is impossible using the preceding method since no vicinal coupling is present; however there is a remarkable downfield shift of proton 8 for the *cis* configuration, as is found with retinal.[2]

Using these methods the chemical shifts of the olefinic protons of all-*trans* and all four mono-*cis* isomers of 13-demethylretinal can be measured. The coupling constants enabled the cis/trans geometry of the double bonds to be evaluated. These results are presented in the accompanying table.

Using these methods we have recorded the ^{1}H-NMR spectra of α-retinal, 5-demethylretinal, and 9-demethylretinal. The interaction of these analogs with opsin and bacterio-opsin was then studied and their biological activity determined.[9] As a practical guide we have usually used about 10 mg material, which substantially reduces the time required for accumulation of a Fourier transform spectrum. When loading samples into the machine, we were always careful to use dim light conditions. With small quantities of material ($\sim$ 100 μg) the relatively high temperature (30°) of the sample holder did not cause destruction or isomerization of the sample during the measuring time; thus we have not found it necessary to connect a cooling unit during experiments.

The main advantage of the sensitive 400- and 500-MHz machines is their ability to provide highly resolved spectra from small quantities of sensitive material. It is thus possible to extract retinal from a chromoprotein sample and use the extract directly for NMR analysis.

[9] P. Towner, W. Gärtner, B. Walckhoff, D. Oesterhelt, and H. Hopf, *FEBS Lett.* **117**, 363 (1980).

[¹H]NMR DATA OF 13-DEMETHYLRETINAL MONO-CIS ISOMERS IN DEUTERIOCHLOROFORM

Chemical shift (ppm)	all-*trans*	7-*cis*	9-*cis*	11-*cis*	13-*cis*
7-H	6.38	6.03	6.37	6.40	6.37
8-H	6.17	6.15	6.67	6.24	6.18
10-H	6.19	6.26	5.99	6.62	6.22
11-H	7.07	7.01	7.16	6.84	6.96
12-H	6.46	6.43	6.40	6.24	7.20
13-H	7.21	7.20	7.20	7.68	7.04
14-H	6.14	6.16	6.04	6.17	5.83
15-H	9.56	9.57	9.54	9.63	10.19
1-CH₃	1.04	1.04	1.05	1.05	1.05
5-CH₃	1.72	1.51	1.75	1.74	1.74
9-CH₃	2.02	1.92	2.03	2.02	2.04
Coupling constants (Hz)					
$J(7,8)$	16	12.5	16	16	16
$J(10,11)$	11.5	11.5	12	12	12
$J(11,12)$	14.5	14.5	14.5	11	14
$J(12,13)$	11.5	11.5	11.5	11.5	12
$J(13,14)$	15	15	15	15	11
$J(14,15)$	8	8	8	8	8

Acknowledgments

The authors are grateful for the NMR facilities provided by Bruker Messtechnik, Karlsruhe, and the encouragement offered by Dr. W. E. Hull. We thank Professor H. Hopf for helpful discussions and Professor D. Oesterhelt for his interest and provision of facilities. This work was supported by the Deutsche Forschungsgemeinschaft.

[71] Methods for Extraction of Pigment Chromophore

By MOTOYUKI TSUDA

I. Introduction

Visual pigments, retinochrome, and the purple membrane protein of the halobacteria (bacteriorhodopsin, BR) are chromoproteins consisting of a retinal molecule covalently bound to the protein. The isomeric conformation of retinal varies from one pigment to another; rhodopsin

(11-*cis*-retinal),[1] retinochrome (all-*trans*-retinal),[2] light-adapted bacteriorhodopsin; BR^L (all-*trans*-retinal), dark-adapted bacteriorhodopsin; BR^D,
(all-*trans*-retinal/13-*cis*-retinal).[3-5]

A method of extracting retinal from its pigment is described. An extension of this method that allows similar extractions at low temperatures
will also be described. The low-temperature extraction procedure enables
one to examine the isomeric composition of several of the photolytic intermediates of these pigments since these intermediates can be stabilized
at low temperatures.[6]

II. Methods of Extraction

The extraction procedure consists of four basic steps:

1. Preparation of the pigment solution or its intermediate at given conditions (temperature and pH of the sample, composition of the extracting
solution, with or without irradiation).
2. Denaturation of the apoprotein under proper conditions.
3. Extraction of the chromophore from the apoprotein.
4. Estimation of the yield of extraction and analysis of the isomeric
composition of extracted retinal.

A. Extraction of Chromophores from Bacteriorhodopsin at T > 0°

1. Purple membrane must be isolated from other membranes. The carotenoids of the red membrane of *Halobacterium halobium* can interfere
with assays of retinals in the purple membrane. Since the absorption spectrum of carotenoids, which is a major contaminant in PM preparation,
falls in the same region as that of retinals, the presence of carotenoids
may cause errors in estimating the yield of extraction. Though the retention times of carotenoids are longer than those of retinal isomers, their
elution peaks in chromatograms may overlap with retinal peaks of later
injections. Thus, for extraction studies of bacteriorhodopsin, a carote-
noidless mutant of *Halobacterium halobium*[7] is recommended instead of
the wild-type strain.

[1] R. Hubbard and A. Kropf, *Proc. Natl. Acad. Sci. U.S.A.* **44,** 130 (1958).
[2] T. Hara and R. Hara, *Nature (London)* **219,** 450 (1968).
[3] D. Oesterhelt and B. Hess, *Eur. J. Biochem.* **37,** 316 (1973).
[4] M. J. Pettei, A. P. Yudd, K. Nakanishi, R. Henselman, and W. Stoeckenius, *Biochemistry*
16, 1955 (1977).
[5] A. Maeda, T. Iwasa, and T. Yoshizawa, *J. Biochem. (Tokyo)* **82,** 1599 (1977).
[6] M. Tsuda, M. Glaccum, B. Nelson, and T. G. Ebrey, *Nature (London)* **287,** 351 (1980).
[7] W. Stoeckenius, R. H. Lozier, and R. Bogomolni, *Biochim. Biophys. Acta* **505,** 215 (1979).

2. An equal volume of cold (0°) dichloromethane is added to the aqueous solution or suspension of pigment (1 ml of OD_λ max~ 1) at 0°. Two or three times rapid mixing by a pipet usually causes complete denaturation of the apoprotein.

3. Two methods of extracting the chromophore are described below:

(a) EXTRACTION BY SYRINGE.[8,9] The mixture of denatured pigment and dichloromethane is transferred to a heavy-walled test tube (2 cm I.D. and 4 cm high) that has been precooled to 0°; it is then emulsified with a 5-ml precooled syringe fitted with a number 19 needle using 10 rapid extrusion–suction movements. The emulsion is next placed in a 15-ml Corex glass centrifuge tube and centrifuged at 5k rpm for 10 min at 0° to separate the organic (dichloromethane) layer from the aqueous phase. After centrifugation the membrane is sandwiched between the aqueous phase (on top) and the organic phase (underneath). The organic phase containing the extracted retinals is removed with a syringe fitted with a flat-tip needle and transferred to a precooled (0°) test tube. Two milliliters of fresh dichloromethane is mixed with the residual membrane mixture. A small amount of Na_2SO_4 (~1 g) is added to the dichloromethane sample and agitated to absorb any remaining H_2O. Particulate materials and membrane fragments in the solution are filtered with a Millipore Swinny stainless filter assembly attached to a syringe, which minimizes solution loss. The filtrate is then placed in a 5 ml pear-shaped flask and dried under a continuous stream of dry N_2. In humid atmosphere the evaporation of dichloromethane cools the flask to the point that the H_2O vapor from the air condenses on the inside walls. This condensation can be avoided by running N_2 into the vial through a stopper with small inlet and outlet holes, thus eliminating any influx of air into the interior of the flask.

(b) EXTRACTION BY SONICATION[10]. Precool the test tube and prepare dichloromethane and the sample as in Section A,2. The bottom half of the test tube is then placed in an ice–water bath and the tip of the sonicator probe is positioned ~4 mm into the sample. Sonication for t sec with power = 5 (Sonifier Cell Disruption Model W 140D, Branson) suffices to emulsify the mixture. Transfer emulsion to a 15-ml Corex centrifuge tube, centrifuge, and separate the organic phase. Add Na_2SO_4, filter, and dry samples as in (a). Sonication time of longer than 1 min should be avoided as it raises the temperature of the emulsion above 0°, which can cause thermal isomerization of the chromophore. Allow a 2-min cooling period

[8] R. Crouch, V. Purvin, K. Nakanishi, and T. G. Ebrey, *Proc. Natl. Acad. Sci. U.S.A.* **72,** 1538 (1975).

[9] F. G. Pilkiewicz, M. J. Pettei, A. P. Yudd, and K. Nakanishi, *Exp. Eye Res.* **24,** 421 (1977).

[10] M. Tsuda and T. G. Ebrey, *Biophys. J.* **30,** 149 (1980).

between each 1-min sonication for those samples that require longer sonication.

4. It is necessary to estimate the yield of extracted chromophore after each cycle of the extraction process because a greater yield gives a more reliable result. The total number of molecules of the original pigment to be extracted can be estimated from its absorption spectrum. Suppose that the optical density of pigment at its maximum absorption wavelength is A_p and total volume is V_p. The number of pigment molecules, N_p, can be calculated using the extinction coefficient, E_p, as follows:

$$N_p = \frac{A_p}{E_p V_p}$$

After measuring the optical density (A_r) of the extract in hexane at a given wavelength for a given volume, the number (N_r) of extracted chromophore molecules of the retinal can be estimated using the extinction coefficient ϵ_r^i for each isomer (i) and the percent composition C^i of isomers in the extract as follows:

$$N_r = \frac{A_r}{\sum_i C^i \epsilon_r^i V_r}$$

The yield of extraction (Y) can be calculated using these two numbers (N_p, N_r)

$$Y = \frac{N_r}{N_p}$$

High-pressure liquid chromatography is employed to separate and analyze the retinal isomers in the extract. This method is described in detail elsewhere in this volume (see Article [63], this volume). Dried retinal is solubilized in 50 μl of hexane. A 10-μl aliquot of this sample is then injected into an HPLC equipped with μ-Porasil column. The sample is eluted with a constant flow of 4% ether–hexane (v/v) at a rate of 2 ml/min. Peaks are detected by a spectrophotometer ($\lambda = 365$ nm); these peaks are identified by comparing their retention times with those of purified standard isomers. The percentage of each isomer is obtained by dividing the peak areas by their respective extinction coefficients at 365 nm: 4.58×10^4 for the all-*trans* and 3.84×10^4 for the 13-*cis* isomer.[11]

[11] R. Hubbard, *J. Am. Chem. Soc.* **78**, 4662 (1956).

B. Extraction of Chromophores from Bacteriorhodopsin at Low Temperature[6]

To extract chromophores for pigments at low temperatures the extraction solvent should not freeze at low temperatures (−74°). It should denature the pigments at low temperatures and should not isomerize the extracted chromophore. The following procedure meets the preceding requirements.

1. A mixture of 3 ml of solution A (buffer: glycerol = 33%:67%) and 1 ml of solution B (methylene chloride–hexane, 85%:15%) is first emulsified in the heavy-walled test tube by sonication (power = 5) for 30 sec at room temperature. Then a syringe cylinder (1 in Fig. 1) with the end sawed off and plugged by a Teflon piece (2 in Fig. 1), containing 0.5 ml of the pigment sample (OD 2.0) in 67% glycerol (3 in Fig. 1) is inserted into a test tube (4 in Fig. 1) containing the emulsion. This in turn is immersed into a dry-ice–ethanol bath at −74° (6,7 in Fig. 1) and, if necessary, the sample is illuminated in the syringe with an optical light guide (8 in Fig. 1).

2. A plunger is inserted into the syringe cylinder and the sample in-

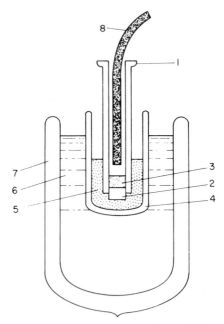

Fɪɢ. 1. Low-temperature extraction system: 1, syringe cylinder; 2, Teflon piece; 3, pigment sample in 67% glycerol; 4, heavy-walled test tube; 5, emersion for extraction; 6, ethanol–dry ice; 7, dewar; 8, optical guide.

jected into the extraction solvent. Next, sample denaturation is effected by thoroughly mixing the sample with the extraction solvent with a precooled ($-74°$) spatula. The temperature of the mixture can be monitored constantly with a thermocouple.

3. The mixture is then warmed up to $0°$ and the chromophore extracted by the sonication method as outlined in A,3.

4. Estimation of the yield of extracted chromophore and the analysis of its isomeric composition are done as outlined in A,4.

C. Extraction of Chromophores from Rhodopsin[12]

Rhodopsin and its chromophore, 11-*cis*-retinal, are much more labile than bacteriorhodopsin and its chromophores. The extraction procedure used with bacteriorhodopsin causes thermal isomerization of retinal if used on rhodopsin. Thus an alternative method is recommended for the extraction of chromophores from rhodopsin.

1. The membranes that contain the retinal pigment must be isolated from other membranes. In cephalopod retina the visual pigment and retinochrome can cause mutual contamination[2] because each contains a different isomeric form of retinal. Even if a single component of the photoreceptor membranes can be isolated, some fraction of the purified rhodopsin may already have been exposed to light during the initial isolation of retinas. Thus extraction of the chromophore from this preparation may give 10–40% all-*trans*-retinal due to the photolyzed rhodopsin. In order to reduce the amount of all-*trans*-retinal due to photolyzed rhodopsin in the preparation, purified membranes are incubated with hydroxylamine to form the resulting all-*trans*-retinal oxime. This is then removed by extraction with hexane as described elsewhere in this book (see Article [66], this volume).

2. 0.5 ml of pigment membranes (OD_λ max $\simeq$ 1.0) is mixed with 0.5 ml of hydroxylamine to form retinal oxime, followed by 3 ml methanol. A minimum 1000-fold molar excess with respect to the pigment is necessary to prevent thermal isomerization.[12]

3. 3 ml of dichloromethane is added to this mixture followed by 2 ml of buffer solution. The resulting suspension is thoroughly vortexed for 15 sec and centrifuged at 5k rpm for 10 min. The lower organic layer is removed with a syringe. Three milliliters of fresh dichloromethane is then mixed with the residual membrane mixture and the chromophore is once more extracted as above. The organic solutions from both extractions are combined and dried as shown in A,3.

4. Estimation of the yield of extracted chromophore and the analysis

---

[12] G. W. T. Groenendijk, W. J. De Grip, and F. J. M. Daemen, *Biochim. Biophys. Acta* **617,** 430 (1980).

of its isomeric composition is essentially the same as A,4. However, the extracted chromophore is retinal oxime, not retinal. The extinction coefficient and HPLC chromatogram of retinal oximes are different from those of retinals. Absorbance coefficients of retinal oximes at 360 nm are as follows: anti-all-*trans*, 51600; *syn*-all-*trans*, 54900; anti-11-cis, 29000; syn-11-cis, 3500.[12] Recovery of retinal oxime from rhodopsin originally present is about 90%.[12] A representative HPLC chromatogram of extracted rhodopsin from treated and untreated ROS is shown in Fig. 3. The sample is eluted with a constant flow of 8% ether-hexane (vol/vol) at a rate of 3 ml/min.

III. Application

A crucial problem is to determine whether or not the isomeric composition of the extracted chromophore exactly corresponds to the composition in the pigment before extraction. To check this, the extraction method should be tested with pigments whose isomeric composition of retinals has been determined by other well-established methods. In the present work, bovine rhodopsin (11-*cis*-retinal) and bacteriorhodopsin (all-*trans*-retinal for BR[L] form and an equimolar mixture of all-*trans* and 13-*cis*-retinal for BR[D] form at neutral pH) were chosen as the standard pigments. Isomeric stability of authentic retinal isomers should also be studied under the same conditions as those employed in the extractions.

A. Free Chromophore

Out of the 16 possible geometric isomers, 12 have been prepared. All-*trans*-retinal is the most stable isomer, but some of the *cis* isomers are unstable and can thermally isomerize to all-*trans* or other *cis* forms. 11-*cis*-Retinal is one of the less stable *cis* forms. When 11-*cis*-retinal is kept in dichloromethane at 4° overnight, about 10% of 11-*cis*-retinal isomerizes to all-*trans*, 13-*cis* and 9-*cis* forms. However, there is no apparent isomerization of 11-*cis*-retinal in dichloromethane at 4° within 2 hr. There is no apparent isomerization of 13-*cis*-retinal in dichloromethane at 4° overnight. In any case, retinals should be kept in solution for as short a period as possible. Thus it is recommended that the solvent be evaporated under a stream of N_2 as soon as possible and the dried sample of retinal stored at below −60°.

Sonication of the 11-*cis*-retinal in solution increases the thermal isomerization. For sonication times of up to 4 min (the solutions are allowed to cool at 0° for 2 min after every 1 min sonication), there is no apparent isomerization. However, sonication for 10 min results in about 5% isomerization. For all-*trans*-retinal and 13-*cis*-retinal there is no apparent isomerization after 10 minutes of sonication.

B. *Bacteriorhodopsin*

Retinals from purple membrane have been extracted using the syringe method. The disadvantage of this method is that it gives very low yield.[13] After 100 strokes of suction-extrusion movements the yield is at best 10%. The sonication method greatly increases the yield. Successive 2-min soni-

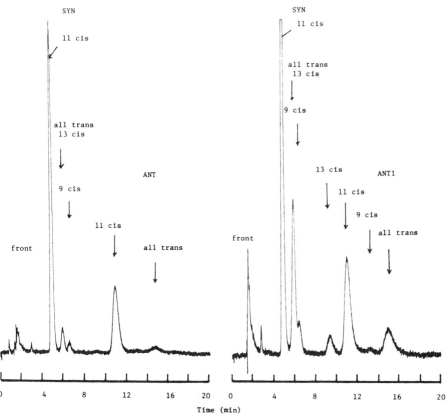

Time (min)

FIG. 2. HPLC chromatogram of retinals. Chromophore is extracted from treated bovine ROS by the method of A,3,a: A, the first cycle of extraction; β, the third cycle of extraction. Column, μ-Porasil column, 30 cm × 4 mm; elute, 4% ether in hexane; flow rate, 2.0 ml/min.

[13] P. C. Mowery, R. H. Lozier, Q. Chae, Y.-W. Tseng, M. Taylor, and W. Stoeckenius, *Biochemistry* **18**, 4100 (1979).

cations give yields of about 40, 20, 10, and 5% respectively. During the sonication, the isomeric composition of the extracted retinal does not change. The isomeric composition of extract after 2 min sonication is almost the same as that after 20 min sonication. pH, glycerol, and temperature affect the isomeric composition of retinal in the dark-adapted purple membrane since the equilibrium of $BR^{all-trans}$ and BR^{13-cis} in BR^D is affected by these conditions.

The isomeric composition of retinals in the photochemical intermediates L and M of purple membrane has been successfully determined by the low-temperature extraction method.[10]

C. Rhodopsin

Extraction of the chromophore from bovine rhodopsin has been tried under different conditions. The sonication method, which works very well for chromophore extraction from PM, was applied to rhodopsin. Al-

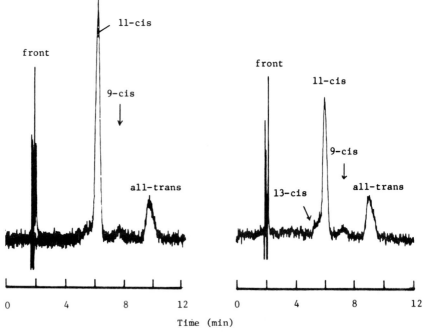

FIG. 3. HPLC chromatogram of retinal oximes. Chromophore is extracted by the method of C,3. a, Untreated bovine ROS. b, treated bovine ROS (same preparation used in Fig. 1). Column, μ-Porasil column; 30 cm × 4 mm; elute, 8% ether in hexane; flow rate, 2.0 ml/min.

though the yield of extraction was very high, this method induced isomerization even with only 5 sec sonication. Using the same preparation, the chromophore was extracted with a syringe method (10 suction–extrusion movements constitute one cycle of extraction.). Figure 2 shows the typical HPLC chromatograms of the first and third cycles of extraction from the treated ROS (ROS treated with NH_2OH and extracted with hexane). The content of 11-*cis*-retinal in the first extract was 87%, while that of the third was 76%. Figure 3 shows that these ROS contain at least 93% 11-*cis*-retinal. Thus the first 10 suction–extrusion movements caused thermal isomerization of 11-*cis*-retinal. Moreover, each cycle of extraction resulted in more isomerization (several percent for each), while the yield of each remained relatively constant (a few percent).

Although some groups[8,14] have utilized this method to determine isomeric composition of retinal of rhodopsin, I cannot recommend it. Extraction method C converts the original isomeric configuration of retinal from rhodopsin into corresponding retinal oxime with complete retention of geometric configuration. Another advantage of this method is its high yield of extraction (more than 80%). Figure 3 shows a typical HPLC chromatogram of extract from treated and untreated ROS.

Acknowledgments

    This manuscript was written while the author was a visiting Associate Professor of Biophysics at the University of Illinois. The author would like to thank Prof. T. G. Ebrey and his colleagues for many helpful discussions.

[14] A. Maeda, Y. Shichida, and T. Yoshizawa, *Biochemistry* **18**, 1449 (1979).

[72] Resonance Raman Spectroscopy of Rhodopsin and Bacteriorhodopsin: An Overview

By AARON LEWIS

I. Introduction

One of the most important physical techniques used to investigate rhodopsin and bacteriorhodopsin (BR) during the last decade has been resonance Raman spectroscopy. With this technique a sample is illuminated with a laser within the absorption of the retinylidene chromophore and the resulting scattered light is focused into a monochromator. The monochromator or spectrograph disperses the light so that the elastically scattered radiation can be separated from that of inelastically scattered photons, which represent the vibrational frequencies of the absorbing retinylidene chromophore, responsible for the scattering. A detector is then used to monitor the photon flux at predetermined wavelength intervals away from the laser frequency. Usually only lower energies relative to the laser energy are scanned, since the highest flux of inelastically scattered photons are expected in this region. The photon flux is then recorded from the detector as a function of wavelength using the laser line as a starting point.

Theoretically, in a classical vein, this process can be considered as a photon from the laser exchanging its momentum with particular vibrational energy levels of the molecule, sending the molecule into a higher vibrational state. The resulting photon is scattered with less momentum and energy, lower frequency and longer wavelength. These photons, detected as a function of wavelength, are the vibrational frequencies of the molecule in question. From a classical discussion it is not at all obvious why the process should be selective to the vibrational modes of the absorbing entity, the retinylidene chromophore. To appreciate fully this selective increase in intensity when a laser is in resonance (coincident or near coincident) with the absorption of a molecular entity, one has to resort to a quantum mechanical discussion,[1,2] which is beyond the purpose of this overview. However, it should be noted that when the exciting laser is in resonance, the vibrational properties of simply the atoms participating in the vibronic (vibrational and electronic) transition are observed. In the case of the retinal I species,

$$X = O, \qquad N-, \qquad \overset{\oplus}{\underset{\underset{H}{|}}{N}}-$$

it was shown by Cookingham et al.[3] that essentially only atoms ~6 − X (see drawing above), the isoprenoid chain, are observed in the resonance Raman spectrum and, thus, these are the atoms over which electron den-

[1] A. C. Albrecht, J. Chem. Phys. 34, 1476 (1961).
[2] J. Tang and A. C. Albrecht, in "Raman Spectroscopy" (H. A. Szymanski, ed.), Vol. 2, p. 33. Plenum, New York, 1970.
[3] R. E. Cookingham, A. Lewis, and A. T. Lemley, Biochemistry 17, 4699 (1978).

sity changes must be occurring in the electronic transition. Furthermore, the vibrational modes of these selectively observed atoms are those that exhibit the largest Franck–Condon factors.[4] Qualitatively, this corresponds to vibrational modes that would undergo the largest nuclear excursions as a result of the altered electronic distribution induced in the molecule in the transition from the ground to the excited-state. These, of course, are the very vibrational modes that give intensity to the allowed electronic transition for retinal chromophores, and practically speaking these are precisely the vibrational modes of critical importance in the photochemistry of the system.

II. An Overview of the Vibrational Spectrum

In Table I a list of the vibrational modes observed for the different double-bond isomeric forms of retinal and retinal Schiff bases is reproduced from the work of Cookingham et al.[3] The descriptions of these modes, which were gleaned from a variety of chemically modified retinals, have been sucessfully used by several subsequent workers in interpreting the spectra of the triplet state of retinal,[5] and of rhodopsin, BR, and their thermal intermediates.[6-11] Principally, several major groups of bands are observed, beginning with the C—H bending vibration of chain vinyl hydrogens around ~960 cm^{-1} (Δ cm^{-1} from laser). This is followed by C—CH$_3$ modes at ~1010 cm^{-1} arising from the C-9 and C-13 positions and the region between ~1100–~1200 cm^{-1} known as the fingerprint region. The fingerprint region is composed of C—C/C=C stretching modes, which are very sensitive to the configuration of the isoprenoid chain both in terms of the relative intensity and frequency of the vibrational modes. In Schiff bases (i.e., X = N) there is an additional mode between ~1220 and ~1250 cm^{-1} that is dependent on the state of protonation of the nitrogen and exhibits frequency alteration in concert with the C=N vibrational frequency.[3,6] The next higher-frequency mode occurs at ~1272 cm^{-1}. This mode occurs with varying intensity in most retinal-

[4] A. Y. Hirakawa and M. Tsuboi, Science 188, 359 (1975).
[5] R. Wilbrandt and N.-H. Jensen, J. Am. Chem. Soc. 103, 1037 (1981).
[6] M. A. Marcus and A. Lewis, Biochemistry 17, 4722 (1978).
[7] M. Sulkes, A. Lewis, and M. A. Marcus, Biochemistry 17, 4712 (1978).
[8] M. Stockburger, W. Klausmann, H. Gatterman, G. Massig, and R. Peters, Biochemistry 18, 4886 (1979).
[9] G. Eyring, B. Curry, R. Mathies, R. Fransen, I. Palings, and J. Lugtenberg, Biochemistry 19, 2410 (1980).
[10] T. Kitagawa and M. Tsuda, Biochim. Biophys. Acta 624, 211 (1980).
[11] R. E. Cookingham and A. Lewis, J. Mol. Biol. 119, 569 (1978).

TABLE I
Suggested Vibrational Assignments Frequencies (CM⁻¹)

Description of the Vibration	all-trans C=X: C=O	all-trans C=N	all-trans C=N—H⊕	13-cis C=O	13-cis C=N	13-cis C=N—H⊕	11-cis C=O	11-cis C=N	11-cis C=N—H⊕	9-cis C=O	9-cis C=N	9-cis C=N—H⊕
C=X stretch[a]	1656	1622	1655	1659	1622	1654	1658	1618	1655	1656	1623	1658 1606
C=C stretch[a]	1577 1568	1582	1555	1584 1573	1585	1565	1576	1579	1557	1586	1588	1567 1557
C-9 and C-13 assymetric methyl deformation[a]	1448[CCl₄]	[1]	[3]	1448[CCl₄]	[3] 1314	[3]	1448[CCl₄]	[1]	1438	1446[CCl₄]	[1]	[3]
C-9 symmetric methyl deformation[a]	1388[CCl₄]	[1]		1399[CCl₄]			1387[CCl₄]	[1]		1402[CCl₄] 1374[CCl₄]	[1]	[3]
C_1 gem-dimethyl[a]	1387[IR] 1362[IR]	[2] [2]	[2,4] [2]	1380[IR] 1362[IR]	[2] [2]	[2] [2]	1375[IR] 1358[IR]	[2] [2]	[2] [2]	1379[IR] 1360[IR]	[1] [1]	[3] [2]
C-14-H rock[b]	1337	1327		1352 1316	1331 1309	1323	1345	[3]	[3]	1337 1329	1327	1327
C—C—H bend + C=C stretch[a]; (C-11=C-12—H)[b]	1282	1284	1282	1282	[3]					1295	[3]	1294
C—C—H bend + C—C stretch[a]; (C-10—C-11—H)[b]	1272	1272	1272	1274	1274	1274	1271	1272	1272	1280		1273
C-14—C-15 stretch[b]			1237			1238			1234			
C—C stretch;[a] (C-12—C-13)[b]	1222	1222	1222	1222	1226	1220	1216	1220	1220	1216	1224	1238

	C-8—C-9 etc.											
C—C stretch;[a] (C-8—C-9)[b]	1198	1198	1193	1193	1198	1203	1206	1210	1212 1200 1190	1201	1203	1206
C—C (C-9—C-13) stretch[a]; (C-10—C-11)[b]	1163	1178 1170 1155	1159	1163	1163	1168	1143		1190	1187 1147	1140	1191 1140
C-14—C-15[a]				1118			1128			1117	1125	
C-9—CH$_3$[a]	1009	1011	1007	1012	1012	1011	1017	1020	1012	1009	1007	1008
C-13—CH$_3$[a]		[3]			[3]		997	1010	998		[3]	
C—H bend, out-of-plane[a]	970	965	965	969	972	972	970	968	965	963	965	965

[a] From Cookingham et al.[3]: Assignments made principally from chemically modified retinals in a variety of isomeric states.
[b] From B. Curry, A. Broek, J. Lugtenberg, and R. Mathies, "Vibrational Analysis of all-Trans Retinal" (to be published).
[1] Experiments performed in acetonitrile with interfering bands in this region. [2] No Infrared (IR) data available. [3] Weak bands or weak/broad bands detectable but not assigned.

type molecules. It appears, however, that this mode exhibits its largest intensity when the chromophore is in an 11-*cis* configuration. Between ~ 1300 and 1475 cm^{-1} there are a whole series of weak bands. Little use has been made of these bands, which derive in part from various deformations involving the methyl groups. However, in a series of investigations Cookingham *et al.*[3,11] have shown that the band in 11-*cis*-retinal at 1345 cm^{-1} is altered by temperature and chemical modification in a way that indicates it is sensitive to the presence of 12-s-*cis* and 12-s-*trans* conformers that this isomer is capable of attaining. The most prominent band in the spectrum follows this group of weak vibrational structure. The investigations of Cookingham *et al.*[3] clearly indicate that this mode is due to the C=C stretching vibration. In a number of cases the mode is split into several components. However, in none of the natural membrane-bound retinylidene chromophores has a splitting in this mode been confirmed. One possible splitting is in the spectra of intermediates of BR. In these spectra all major C=C stretching modes can be related to specific, previously identified species except for one mode that occurs at ~ 1550 cm^{-1}. There could be two explanations for this observation. Either the L species has a split C=C stretch or an additional species is present among these thermal intermediates previously unidentified by absorption spectroscopy. In addition to the preceding, the C=C stretching frequency generally seems to follow the absorption maximum of the chromophore, and thus protonated membrane-bound chromophores have frequencies ranging from ~ 1525 to ~ 1545 cm^{-1}, protonated isolated (unbound) chromophores have frequencies ~ 1555 cm^{-1}, unprotonated bound chromophores have frequencies ranging from possibly as low as ~ 1550 to 1570 cm^{-1}, and the frequencies of free retinals and unbound unprotonated chromophores range between ~ 1570 and 1580 cm^{-1}. Finally, one observes one of the weakest vibrational modes in the spectrum. This mode, which has provided a wealth of information, occurs between ~ 1620 and ~ 1655 cm^{-1}

$$\overset{\displaystyle H}{\underset{\displaystyle |}{}}$$

and corresponds to the $-\overset{|}{C}{=}X$ stretching mode, where X is O or N, a protonated or unprotonated Schiff base. For these species the vibrational frequencies are, respectively, ~ 1656 cm^{-1}, 1642–1655 cm^{-1}, and ~ 1620 cm^{-1}.

III. A Brief Experimental Perspective

Since this chapter is intended as a guide to our present knowledge in this area, the experimental aspects dealt with in the subsequent chapters of this section will not be reviewed again. However, as a guide to those chapters[12-14] and the topics not considered by their authors, a brief experimental perspective follows. The earliest work in this field was carried out by three groups in the early 1970s: Rimai, Gill, and co-workers[15-17]; Lewis and co-workers[18]; and Mendelsohn.[19] Rimai, Gill, and co-workers obtained the first resonance Raman spectra of retinals and Schiff bases. They even extended their studies to the retinylidene chromophore in bovine rod outer segments,[20] but they were barely able to see one band in the spectrum (the intense C=C stretch). This probably arose because a large fluoresence background was detected that could not be subtracted because of a lack of computer sophistication in the early laser Raman laboratories. Lewis and co-workers[18] used pulsed excitation from a mode-locked laser that emitted picosecond pulses every nanosecond to study a digitonin suspension of bovine rhodopsin. In this study the C=N—H⊕ mode could be clearly observed and was so interpreted. Mendelsohn used continuous wave (cw) laser excitation within the absorption of BR and observed two bands, one at 1620 cm^{-1} and one at ~ 1645 cm^{-1}. The 1645-cm^{-1} band was interpreted[19] as scattering from the water solvent, as was then customary in nonresonant protein Raman spectroscopy, and thus it was suggested that the 1620 cm^{-1} band indicated that BR had an unprotonated chromophore. This misinterpretation resulted from a lack of realization of sample photoability, and this criticism can be made in varying degrees on all these investigations reported before 1974. Even the lower-temperature bovine rhodopsin study of Rimai et al.[20] was performed in a temperature regime where photochemical back reactions produced a lumirhodopsin mixture with other species. Furthermore, the pulsed laser experiments of Lewis and co-workers,[18] which in principle should have produced a photostationary concentration of rhodopsin, could have been contaminated with other intermediates because of laser hot spots in the sample. Although, it is possible that, the observed differences in these spectra to more recent resonance Raman data on bovine rhodopsin could also be related to the detergent, digitonin, used in these early investigations which subsequently was shown to cause anomalous CD behavior in bovine rhodopsin.[21]

[12] R. Mathies, Chapters 75, this volume.
[13] R. H. Callender, Chapter 74, this volume.
[14] M. A. El-Sayed, Chapter 73, this volume.
[15] L. Rimai, R. Kilponen, and D. Gill, J. Am. Chem. Soc. **92**, 2824 (1970).
[16] L. Rimai, D. Gill, and J. Parsons, J. Am. Chem. Soc. **93**, 1343 (1971).
[17] D. Gill, M. Heyde, and L. Rimai, J. Am. Chem. Soc. **93**, 6288 (1971).
[18] A. Lewis, R. S. Fager, and E. W. Abrahamson, J. Raman Spectrosc. **1**, 465 (1973).
[19] R. Mendelsohn, Nature (London), **243**, 22 (1973).
[20] L. Rimai, R. G. Kilponen, and D. Gill, Biochem. Biophys. Res. Commun. **41**, 492 (1970).
[21] W. H. Waddell, A. P. Yudd, and K. Nakanishi, J. Am. Chem. Soc. **98**, 238 (1979).

In 1974 two investigations published simultaneously attacked the problems successfully. Lewis *et al.*,[22] investigating BR, and Oseroff and Callender,[23] investigating bovine rhodopsin, used low-temperature techniques to control sample lability successfully. In addition, both groups realized that resuspension of such membrane proteins in D_2O may result in exchange of the Schiff base proton by a deuterium if the Schiff base was protonated. If this occurred, then the frequency of the end group with a deuterium on the nitrogen would be shifted by a well-defined frequency reduction, ~ 20 cm^{-1}, because of a one mass unit increase on the nitrogen atom involved in the C=N—H⊕ stretch. The experiment was a complete success, and although there are still one or two dissenters,[24] it is nearly universally accepted that these two experiments established that a proton was in contact with the Schiff base nitrogen in both these retinylidene membrane proteins. Furthermore, by accurately controlling the temperature and the mixture of intermediates in the sample, Lewis *et al.*[22] were able to demonstrate that the unprotonated C=N stretch detected by Mendelsohn[19] arose from an intermediate produced after light absorption. With these Raman data as evidence, it is now universal dogma that light eventually causes the deprotonation of the Schiff base as the proton is released on one side of the membrane and that the protonated species is thermally regenerated when a proton is taken up on the other side of the membrane.

The rhodopsin experiment of Oseroff and Callender[23] also produced very significant information in addition to the definitive proof on the state of protonation. They demonstrated with the aid of a two-beam technique (described by Callender in this section) that the primary photochemical product (batho) in rhodopsin has intense low-frequency motions, and this has been a feature common to the resonance Raman spectrum of all batho-intermediates studied thus far.[7]

The next technical development was the application of continuous flow methods to the photoability problem. The sample was made to flow past a ~ 30 μm laser beam at a speed that was sufficient to renew the sample every ~ 10 μsec. The laser power was adjusted so that only one photon could interact with each molecule before the photolyzed molecules were replaced with previously unphotolyzed species. The technique took advantage of the rapidity of the Raman process ($\sim 10^{-13}$ sec), which is

[22] A. Lewis, J. Spoonhower, R. A. Bogomolni, R. H. Lozier, and W. Stoeckenius, *Proc. Natl. Acad. Sci. U.S.A.* **71**, 4462 (1974).

[23] A. R. Oseroff and R. H. Callender, *Biochemistry* **13**, 4243 (1974).

[24] J. Favrot, J. M. Leclercq, R. Roberge, C. Sandorfy, and D. Vocelle, *Photochem. Photobiol.* **29**, 99 (1979).

generated before a photochemical transformation in the molecule is effected. Since the photon density is low, it is improbable that photochemically altered species are being analyzed. However, it should be cautioned, especially with regard to the interpretation of intensities, that, in terms of yield, the Raman process is similar to a very weak fluorescence. Thus the question is, "Are we observing a special photochemically ineffective group of molecules, since we know that the molecules that emit their photons in a Raman event are not going on to photochemistry?" However, it is generally believed that the molecules that do participate in a Raman scattering event are a fair distribution of the molecules in the excited state, and this belief is implicitly accepted in all such Raman scattering investigations and by those[25,26] who have applied the flow technique which is described by Callendar in Chapter 74, this volume.[13]

The first kinetic data on retinylidene proteins were obtained by Marcus and Lewis.[27] In fact, this was the first microsecond kinetic data using vibrational spectroscopy. Principally this method *altered* the residence time of the molecule in the laser beam. As the residence time was lengthened, additional photons hit molecules that previously had been photochemically transformed and additional intermediates were observed. There have been several experimental manifestations of this method. In the original approach of Marcus and Lewis,[27] the residence time was altered by having a sample flow past a laser beam with a well-defined diameter at a variety of speeds. This method is still the method of choice for rhodopsin and BR samples where the material is readily available. However, in cases where material is not readily available, the method of choice will probably be a micro-spinning cell of the type recently described in the literature by Rodgers and Strommen.[28] The disadvantage of the spinning cell for such kinetic measurements is that the sample reexposure to the laser beam is very rapid, especially for short illumination times; in addition, critical alignment is necessary in order to minimize scattered light effects in the cell, causing molecules not in the laser beam to be exposed to scattered radiation. A full discussion of the theory behind these kinetic measurements and their application to spinning cells is given in this section by Lewis.[29]

---

[25] R. H. Callender, A. G. Doukas, R. Crouch, and K. Nakaniski, *Biochemistry* **15**, 1621 (1976).

[26] R. Mathies, A. R. Oseroff, and L. Stryer, *Proc. Natl. Acad. Sci. U.S.A.* **73**, 1 (1976).

[27] M. A. Marcus and A. Lewis, *Science* **195**, 1330 (1977).

[28] E. G. Rodgers and D. P. Strommen, *Appl. Spectrosc.* **35**, 215 (1981).

[29] A. Lewis, Chapter 78, this volume.

IV. Discussion

In this section we will focus on specific vibrational modes or on regions of the vibrational spectra of the retinylidene chromophore in order to present the current status of our resonance Raman knowledge on rhodopsin/BR.

A. Carbon-Nitrogen Stretching Vibration

1. Arguments on the Primary Event. As noted earlier, the C=N stretching vibration is probably the most studied mode in both the rhodopsin and BR systems. It has been used principally to determine the state of protonation of the Schiff base. Protonation state data obtained from several systems using this technique are summarized in Table II. Besides the major conclusion that the initial species of bacteriorhodopsin

TABLE II

A COMPARISON OF C=N & C=N—H⊕ STRETCHING FREQUENCIES IN
RHODOPSIN AND BACTERIORHODOPSIN[a]

Intermediate	Squid	Octopus	Bovine	Bacterio-rhodopsin	Intermediate
Rhodopsin	P[b] 1653[7]		P 1655[23]	P 1642[22]	BR$_{570}$
Isorhodopsin	P 1653[7]		P 1655[23]		
Bathorhodopsin	P 1653[7]		P 1655[7,31-33]	P[6,35-37,d]	K$_{610}$
Acid metarhodopsin or	P 1652[30]	P 1655[10]		P[6,8,38]	L$_{550}$
metarhodopsin I			P 1657[34]	U[6]	X?
Alkaline metarhodopsin or	U[c30]	U 1624[10]		U 1620[22]	M$_{412}$
metarhodopsin II			U[34]	P[39]	O$_{640}$

[a] Protonated and unprotonated retinal Schiff bases of butylamine have CN stretching frequencies at 1655 cm^{-1} and 1622 cm^{-1} independent of isomeric configuration.[3]

[b] P = protonated.

[c] Unprotonated.

[d] As a good working hypothesis although not definitely proved.

[30] M. Sulkes, A. Lewis, A. T. Lemley, and R. E. Cookingham, *Proc. Natl. Acad. Sci. U.S.A.* **73**, 4266 (1976).

[31] M. Marcus and A. Lewis, *Photochem. Photobiol.* **29**, 699 (1979).

[32] G. Eyring and R. Mathies, *Proc. Natl. Acad. Sci. U.S.A.* **76**, 33 (1979).

[33] B. Aton, A. Doukas, D. Narva, R. H. Callender, U. Dinur, and B. Honig, *Biophys. J.* **29**, 79 (1980).

[34] A. G. Doukas, B. Aton, R. H. Callender, and T. G. Ebrey, *Biochemistry* **17**, 2430 (1978).

[35] J. Spoonhover, Ph.D. Dissertation, Cornell University, Ithaca, New York (1977).

[36] J. Terner, C.-L. Hsieh, A. R. Burns, and M. A. El-Sayed, *Proc. Natl. Acad. Sci. U.S.A.* **76**, 3046 (1979).

[37] M. Braiman and R. Mathies, *Biophys. Soc. Abstr.* **33**, 216a (1981).

[38] J. Terner, C.-L. Hsieh, and M. A. El-Sayed, *Biophys. J.* **26**, 527 (1979).

[39] J. Terner, C.-L. Hsieh, A. R. Burns, and M. A. El-Sayed, *Biochemistry* **18**, 3629 (1979).

(BR_{570}) is protonated and the M_{412} intermediate is unprotonated, there are several other conclusions that can be deduced from these data.

Lewis et al.[7] and others[32,33] have reported that in rhodopsin there is no alteration in the frequency of the $C\!=\!N\!-\!H\oplus$ mode with respect to a free protonated Schiff base.[32,33] This in fact seems to be the rule in all protonated bovine, squid and octopus rhodopsin intermediates. In fact, the frequencies of all these protonated species are within ± 2 cm^{-1} of the position of this mode for retinal Schiff bases in solution. In addition, the unpublished material by Schulten and Lindner, which is often quoted when investigators in this field gather, indicates that, even if the counterion is altered, the solution $C\!=\!N\!-\!H\oplus$ stretching frequency is insensitive to environmental perturbations. They note the possibility that, as the double-bond nature of the $C\!=\!N$ is altered as a result of environmental perturbations, changes in hybridization may occur that could keep the vibrational frequency of the CN bond essentially constant. Such a point of view is principally held by those workers[40] who agree with Wald's original hypothesis[41] that the "*only* action of light is to isomerize retinal from the 11-cis to the all-*trans* configuration." Other workers, including this author,[42] have argued that in fact the $C\!=\!N\!-\!H\oplus$ stretch is sensitive to environmental perturbations. Why are these arguments important and what are the different conclusions they lead to?

Workers who argue against the sensitivity of the CN vibrational mode are able to disregard the invariance of the $C\!=\!N\!-\!H\oplus$ stretch in their models of the photochemistry. This has become particularly important in view of the need to adapt simple isomerization[40,41] to the more recent ideas that emphasize the fact that the protein as well as the retinal is altered in the primary photochemical step rhodopsin/BR $\rightsquigarrow$ batho. The first paper to focus on the protein *as well as* the retinal was by Lewis.[43] The principal contribution of this work was to indicate that the protein conformation was altered in going from rhodopsin to bathorhodopsin. Experimental verification for the hypothesis was obtained[43,44] by a combination of the resonance Raman observation of the CN frequency invariance discussed earlier with the experiments at Bell Laboratories of Peters, Applebury, and Rentzepis[45] of a large deuterium isotope effect in the rhodop-

[40] T. Rosenfeld, B. Honig, M. Ottolenghi, J. Hurley, and T. G. Ebrey, *Pure Appl. Chem.* **49**, 341 (1976).

[41] G. Wald, *Science* **162**, 230 (1968).

[42] A. Lewis, M. A. Marcus, B. Ehrenberg, and H. L. Crespi, *Proc. Natl. Acad. Sci. U.S.A.* **75**, 4642 (1978).

[43] A. Lewis, *Proc. Natl. Acad. Sci. U.S.A.* **75**, 549 (1978).

[44] A. Lewis, *Philos. trans. R. Soc. London, ser. A* **293**, 315 (1979).

[45] K. Peters, M. L. Applebury, and P. M. Rentzepis, *Proc. Natl. Acad. Sci. U.S.A.* **74**, 3119 (1977).

sin ~→ bathorhodopsin transition. Assuming that the CN frequency invariance is structurally meaningful, it clearly shows that the proton (or protons) that results in a 7 times deuterium isotope effect cannot be the exchangeable proton on the Schiff base as originally suggested by the Bell group[45] but must be some other proton (or protons) in the protein matrix.

In order to accommodate this large deuterium isotope effect the early advocates of simple isomerization[40] have now included protein conformation changes.[46] Thus both the mechanisms of Lewis[43] and Honig *et al.*[46] *have* ground state protein conformational changes; the latter mechanism,[46] however, limits protein conformational change to this state alone. The presence of ground-state protein structural alteration in the model by Lewis[43] should be stressed in view of the misimpression in the literature on this point.[46,47]

In addition to this ground state protein structural alteration, Lewis[43] has suggested that the protein also undergoes conformational rearrangement in the excited state. Conversely, Honig *et al.*[46] advocate that protein conformational change does not occur in the excited state. However, a simple consideration of their model, within the limitations they impose[46] will illustrate that excited-state protein conformational change must occur.

In Fig. 1 potential energy curves drawn in solid lines are those at zero protein coordinate change with only energy and retinal coordinate changes taking place. Both Lewis[43] and Honig *et al.*[46] agree that the final batho ground-state minimum has alterations in both the retinal and protein when compared to rhodopsin. Furthermore, both groups have imposed the constraint of a common rhodopsin/batho excited-state minimum, and this has received additional theoretical support[48] in spite of recent quantum yield measurements.[49] Therefore, it is evident that if a batho-structure with altered protein and retinal conformations must attain a minimum in the excited-state surface at zero on the protein coordinate, there must be excited-state surface at zero on the protein coordinate, there must be excited-state protein conformational alteration at least in the batho ~→ rhodopsin transition. Thus if the batho ~→ rhodopsin transition includes excited-state protein structural alteration, how can such structural alteration be excluded from the rhodopsin ~→ batho reaction? In fact, it cannot, and the BR, 6 psec resolution, deuteration data used by Honig *et al.*[46] cannot rule out excited-state protein alteration but can only support the ground-state protein structural alteration initially suggested by Lewis[43] and subsequently incorporated by Honig *et al.*[46]

[46] B. Honig, T. G. Ebrey, R. H. Callender, U. Dinur, and M. Ottolenghi, *Proc. Natl. Acad. Sci. U.S.A.* **76**, 2503 (1979).

[47] M. Ottolenghi, *Adv. Photochem.* **2**, 97 (1980).

[48] R. R. Birge and L. M. Hubbard, *Biophys. J.* **34**, 517 (1981).

[49] T. Suzuki and R. H. Callender, *Biophys. J.* **34**, 261 (1981).

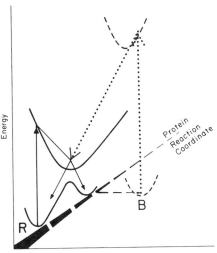

Fig. 1. A three-dimensional representation of protein and retinal alterations in going from rhodopsin (R) to the higher-energy bathorhodopsin state (B). Potential energy curves drawn in solid lines are at zero on the protein coordinate. Potential energy curves drawn with dashed lines are at some altered protein structural state. Arrows represent the alterations in the protein–retinal system in going from vertical rhodopsin excitation to form bathorhodopsin and then from vertical bathorhodopsin excitation to form rhodopsin.

Therefore, the most probable description of the primary events in BR and visual transduction is that light not only alters the retinal conformation but also causes excited- *and* ground-state protein conformational changes. A pictorial description of the suggested sequence of events for rhodopsin and bacteriorhodopsin is reproduced from a paper by this author.[44] Nearly universal agreement has been achieved on ground-state protein structural alteration and excited-state retinal structural alteration. Arguments for excited-state protein structural alteration have been made previously. Arguments for ground-state retinal structural alteration[43] can readily be made for BR, where the elegant experiments of Kalisky *et al.*[50] have shown that the L thermal intermediate will directly and thermally revert to BR_{570} at the appropriate temperature.

Finally, it seems that an important test of the specific molecular aspects of these two models would be the resonance Raman spectrum of hypsorhodopsin. The Honig *et al.*[46] model, which disregards the unaltered $C{=}N{-}H^{\otimes}$ stretching frequency in going from rhodopsin to bathorhodopsin and centers all protein structural alteration at the Schiff base, suggests that hypsorhodopsin is an unprotonated Schiff base. On the other hand, our model[43] emphasizes protein structural alteration removed

[50] O. Kalisky, M. Ottolenghi, B. Honig, and R. Korenstein, *Biochemistry* **20**, 649 (1981).

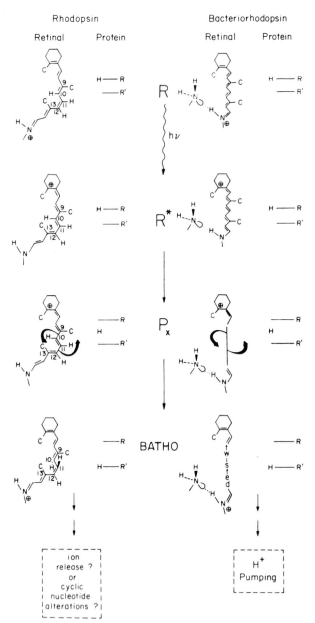

FIG. 2. Protein and retinal structural alteration in rhodopsin and bacteriorhodopsin on light absorption. Bacteriorhodopsin may have interactions at the Schiff base that rhodopsin may lack (see Section IV,A,2). The states R, R*, P_x, BATHO represent, respectively, rhodopsin or BR, vertically excited rhodopsin and BR, excited rhodopsin and BR in the *excited-state* minimum, and the primary photochemical product bathorhodopsin which stores >12 kcal of the light energy in an altered and separated R–R' salt linkage.[42] Thus electron motion in the isoprenoid molecular wire causes pK alterations in R and R' resulting in proton movement in the protein, and at least in terms of bacteriorhodopsin such a mechanism fits well into the overall view of metabolically and photosynthetically driven electron transport-

from the Schiff base. We would suggest that hypsorhodopsin is simply generated by incomplete protein relaxation at low temperatures, and this accounts for its altered spectral characteristics relative to bathorhodopsin. In fact, there may be several such strained states as the proton moves from one position in the protein to another, giving rise to additional hypsolike intermediates with protonated Schiff bases. These intermediates would then readily relax to bathorhodopsin as the temperature is raised. Thus this presents a concrete molecular difference between the two models that could be tested with resonance Raman spectroscopy.

2. *Bacteriorhodopsin—Active Site Architecture.* In addition to the central role played the CN stretch in discussions of the primary event and the state of Schiff base protonation, this vibrational mode has also begun to play a crucial role in resolving questions on BR active site architecture.

When a detailed analysis was made of the spectral features of bacteriorhodopsin BR_{570} it rapidly became obvious that the resonance Raman spectrum of the retinylidene chromophore of this species was very different from the resonance Raman spectrum of an all-*trans*-retinylidene Schiff base, even though all extractions of the chromophore clearly indicated that the chromophore was all-*trans*. Significantly, this was in marked contrast to the resonance Raman spectra of visual pigments,[30,51] which showed a striking correspondence between free and protein complexed chromophores. The lack of correspondence in the bR system was initially not given much emphasis and was principally explained as some bond rotations in the chromophore imposed by the protein-active site. The papers of Marcus and Lewis[6] and Lewis *et al.*[42] were the first to emphasize this lack of correspondence in the BR system and to note the strikingly different behavior in visual pigments. Furthermore, Lewis *et al.*[6,42] noted that the CN stretching frequency behaved in a way similar to the rest of the BR_{570} spectrum, namely, that, unlike visual pigments and model chromophores, the CN stretch in BR_{570} was significantly altered. They further argued[6] that there is no evidence that isomerizations or conformational changes of the polyene chain could have an effect on the CN stretching frequency.[3] For example, consider that the CN stretching frequency is the same in rhodopsin, isorhodopsin, metarhodopsin, and free protonated Schiff base models without regard to conformation (Table II). In BR, however, the protonated CN frequency moves toward a value intermediate between those normally observed for protonated (1655 cm^{-1}) and uprotonated (1620 cm^{-1}) Schiff bases. If bond rotations were the cause of

[51] R. Mathies, T. B. Freedman, and L. Stryer, *J. Mol. Biol.* **109**, 367 (1977).

generated proton gradients. Retinal structural alteration keeps R and R' separated, and for the rhodopsin system a definite structural alteration is suggested with double-bond twists around the 9–10 and 11–12 double bonds (see Section IV,H,3 for further discussion). (This figure is from Ref. 44.)

alterations in the BR resonance Raman spectrum, then such rotations should exhibit little effect on the CN stretch. Therefore, it was unlikely that only bond rotations could account for the CN alteration. Furthermore, earlier work[52] with protonated retinal in bacterior-opsin extracted from bacteria grown in fully deuterated media showed that the altered positions of the major bands in bR were not the result of the appearance of protein residue vibrations in the spectra. Thus, this led to a detailed investigation of the CN vibrational mode to determine the origin of its alteration in BR.

Bacteriorhodopsin is an ideal system in which one can elucidate the complex nature of even a single vibrational mode in a protein. Not only can the chromophore in BR be replaced with chemically modified structures, but chemically altered bacterio-opsin can be extracted from bacteria grown in different media. Thus we initiated a series of investigations the results of which are outlined in Table III.

We have interpreted this evidence to indicate that there is a protein interaction at the Schiff base linkage, which could alter the entire resonance Raman spectrum by protecting the Schiff base proton and causing a movement of this proton from its normal equilibrium position on the Schiff base nitrogen. Besides the obvious alterations in the resonance Raman spectrum of the BR_{570} chromophore discussed earlier, further evidence for this interaction is most consistently seen in our data on the C=N—H⊕ in H_2O for deuterated retinal in protonated and deuterated opsin. This mode clearly reflects the alteration in the opsin matrix by shifting down in wavenumber. The same experiment with H-retinal also shows this effect, but these samples reveal an excitation wavelength dependence of the frequency of this mode that is at present unexplained. There are other inequivalencies that lend further support to this hypothesis. For example, consider the difference between fully protonated ^{15}N BR and fully deuterated ^{15}N BR both in D_2O. The difference between both these systems is 33 cm^{-1}. However, we know that replacing protonated retinal with deuterated retinal in a deuterated matrix causes only a 25-cm^{-1} frequency shift; thus deuteration of the matrix may account for the remainder of the 33 cm^{-1} shift, although alteration in mode coupling could also be involved. There are also inconsistencies in protonated and deuterated Schiff base data that at present cannot be explained. For example, ^{15}N enrichment or deuteration of carbon 15 adjacent to the Schiff base on the retinal causes of 15- or 14-cm^{-1} shift in the protonated Schiff base but

[52] M. A. Marcus, A. Lewis, E. Racker, and H. L. Crespi, *Biochem. Biophys. Res. Commun.* **78**, 669 (1977).

[53] M. A. Marcus, A. T. Lemley, and A. Lewis, *J. Raman Spectrosc.* **8**, 22 (1979).

[54] B. Ehrenberg, A. T. Lemley, A. Lewis, M. von Zastrow, and H. L. Crespi, *Biochim. Biophys. Acta* **593**, 441 (1980).

TABLE III

VARIATIONS IN THE C=N, C=N—H⊕ AND C=N—D⊕ STRETCHING FREQUENCIES
AS A FUNCTION OF ISOTOPIC SUBSTITUTIONS

	—C=N— in H$_2$O or D$_2$O suspension (cm^{-1})	$\begin{array}{c}\text{H}\\ \vert\\ \text{C=N}\\ \oplus\end{array}$ in H$_2$O suspension (cm^{-1})	$\begin{array}{c}\text{D}\\ \vert\\ \text{C=N}\\ \oplus\end{array}$ in D$_2$O suspension (cm^{-1})
H-retinal + H-butylamine (all isomers)	1622[a]	1655[a,b]	1630[a]
H-retinal + H-bacterio-opsin	1620	1642	1620
H-retinal + D-bacterio-opsin	1618	1635	1617[c]
D-retinal + H-bacterio-opsin	1595	1625	1595[c]
D-retinal + D-bacterio-opsin	1594	1620	1592
H-retinal + H-bacterio-opsin (^{15}N)	1614	1627	1613
D-retinal + D-bacterio-opsin (^{15}N)	1590[c]	1600[c]	1580[c]
C-15-D-retinal + H-bacterio-opsin	1617 or 1601	1628	1612

[a] Butylamine, butylamine-d_9, propylamine, and hexylamine all exhibit C=N—H⊕ stretches at 1655 cm^{-1} whereas methylamine and ^{15}N methylamine have similar 15-cm^{-1} shifts due to isotopic enrichment in the C=N and C=N—H⊕ stretching vibrations.

[b] Data on model unprotonated and protonated Schiff bases were obtained by using ethanol as the solvent. Deuterated Schiff base data were obtained by using deuterated ethanol as the solvent. These model compound data are from Ref. 53.

[c] Previously unreported; all other data have been reported in Refs. 42, 44, 53, and/or 54.

only a 7- or 8-cm^{-1} shift in the deuterated Schiff base. In addition, deuteration of the Schiff base reduces the effect on the C=N—D⊕ stretch of deuterating the opsin matrix.

In spite of these anomalies, the weight of the evidence, which includes (1) the altered BR$_{570}$ CN stretching frequency as compared to visual pigments and free Schiff bases,[6,42] (2) the effect on the protonated and deuterated CN frequencies of deuterating the opsin matrix,[6,42] and (3) the inability to deprotonate the Schiff base in the dark even at a pH of 12,[55] in spite of the accessibility of this site to exchange on a millisecond time scale[56,57], strongly suggests that an interacting protecting amino acid residue is present when the Schiff base is protonated. On the other hand, the unprotonated CN stretch is unaffected by the preceding alterations and in fact is quite similar (in native membranes) to the frequency observed in free unprotonated Schiff bases of retinal. Thus the evidence indicates that

[55] B. Ehrenberg and A. Lewis, *Biochem. Biophys. Res. Commun.* **82,** 1154 (1978).
[56] B. Ehrenberg, A. Lewis, T. K. Porta, J. F. Nagle, and W. Stoeckenius, *Proc. Natl. Acad. Sci. U.S.A.* **77,** 6571 (1980).
[57] A. G. Doukas, A. Pande, T. Suzuki, R. H. Callender, B. Honig, and M. Ottolenghi, *Biophys. J.* **33,** 274 (1981).

the interaction that is present in the protonated species is absent in the unprotonated species. Nonetheless, even in the unprotonated data, questions remain. The most intriguing is the ^{15}N frequency shift of the unprotonated Schiff base. Unlike the protonated species, which has a ^{15}N shift of 15 cm^{-1} similar to model compounds, the unprotonated species has a ^{15}N shift of 5 cm^{-1} completely unlike model unprotonated Schiff bases, which also exhibit a 15-cm^{-1} ^{15}N shift.[53] In fact, the unprotonated ^{15}N shift appears to be comparable to the anomolous 7-cm^{-1} ^{15}N shift observed for the deuterated Schiff base BR. However, no similarity is implied by this comparison, which probably reflects the fact that deuteration not only deuterates the Schiff base but also deuterates any indirectly coupled protein species, which also affects the CN frequency.

The difference in the ^{15}N shift of protonated and unprotonated Schiff bases in the membrane when taken together with certain pK data in Section IV,A,5 initially suggested to us that the interacting residue could possibly[42] contain nitrogen in direct contact with the Schiff base proton. The pK data to be discussed in Section IV,A,5 further suggested the possibility[42] of the residue being lysine. However, the suggestions of the ^{15}N effect and the pK data have both been recently tested in two separate experiments and neither experiment bears out the possibility that these data are reflecting on the nature of the interacting residue. The ^{15}N effect has recently been tested by Argade et al.[58] These investigators grew Halobacterium halobium on a 50% lysine ^{14}N/50% lysine ^{15}N mixture. They demonstrate incorporation of the lysine and then argue that if ^{15}N from an interacting residue was affecting the protonated Schiff base frequency, then three vibrational modes should be observed—one at 1642 cm^{-1}, where both nitrogens are ^{14}N; one at 1627 cm^{-1}, the frequency we observed for fully ^{15}N[6,42]; and one intermediate in frequency resulting from the 50% ^{14}N/50% ^{15}N species. In fact, this intermediate species should have twice the intensity of either band at 1627 cm^{-1} or 1642 cm^{-1}. The data obtained clearly eliminate the possibility that the nitrogen from an interacting lysine residue is causing any shift on the Schiff base nitrogen. In addition, the pK argument (as discussed later) has recently been tested by Gogel and Lewis[59] using reagents to modify bR chemically. These data indicate that pK probably reflects a tyrosine in close proximity to the chromophore but unrelated to the interacting amino acid.

[58] P. V. Argade, K. J. Rothschild, A. H. Kawamoto, J. Herzfeld, and W. C. Herlihy, Proc. Natl. Acad. Sci. U.S.A. 78, 1643 (1981).

[59] G. Gogel and A. Lewis, Biophys. Soc. Abstr. 33, 216a (1981); Biochem. Biophys. Res. Commun. 103, 175 (1981).

In spite of these data the interacting lysine residue hypothesis has been slow to die. In fact, in addition to all the biochemical data that have been accumulating recently that the chromophore is probably very close to a second lysine residue,[60] additional resonance Raman evidence that was originally presented several years ago[61] has now been put on a firmer experimental basis. The recent experimental improvement is the development of an NMR assay that can be performed on the very samples used for the Raman measurements. Previously only similar mixtures of radioactive amino acids could be used to test incorporation. The Raman evidence that was obtained by Lemley *et al.*[61] is quite similar in character to the data presented by Argade *et al.*[58] Lemley *et al.*[61] grew *Halobacterium halobium* on defined mixtures of amino acids. Specifically, the bacteria were grown on fully deuterated amino acids except for a specific amino acid in the mixture that was contained in the growth medium as 50% in the fully deuterated form and 50% in the fully protonated form. Four different cultures were grown, with each culture having a different target amino acid. The amino acids chosen were tyrosine, tryptophan, lysine and phenylalanine. When the purple membrane was extracted from these cultures and analyzed in H_2O, no alterations in the protonated CN stretch were observed for any of the amino acids except lysine. The data for lysine are shown in Fig. 3. The shifts are very small but they are consistent, are above the noise level, and occur only for the protonated species. This latter point clearly can be seen by comparing the C=C stretch of BR_{570} and M_{412} under the three conditions shown. Only the bands of the protonated species bR_{570} are altered. The M_{412} vibrational modes are completely unperturbed. Of greatest interest, however, are the alterations in the protonated CN stretch. The 50% H-lysine/50% D-lysine sample has its main frequency between 1620 cm^{-1} (the frequency for the protonated CN stretch of fully deuterated bR in H_2O) and 1625 cm^{-1} (the protonated CN stretch of fully deuterated retinal in fully protonated protein in H_2O). These frequency shifts certainly are consistent with a lysine interaction hypothesis, but in view of the very small shifts involved, these data can only be taken as suggestive until appropriate difference Raman measurements are performed.

If indeed there are two lysines that only interact in the protonated state, the most reasonable assumption is that they are interacting through their nitrogens with the Schiff base proton between them.[42] Thus the immediate concern that must be expressed is how deuteration of a lysine side chain in the interacting residue can cause a frequency shift,

[60] Yu. A. Ovchinnikov, N. G. Abdulaev, V. I. Tsetlin, A. V. Kiselev, and V. I. Zakis, *Bioorg. Chem.* **9**, 1427 (1980).

[61] A. T. Lemley, A. Lewis, and H. L. Crespi, *Biophys. Soc. Abstr.* **25**, 78a (1979).

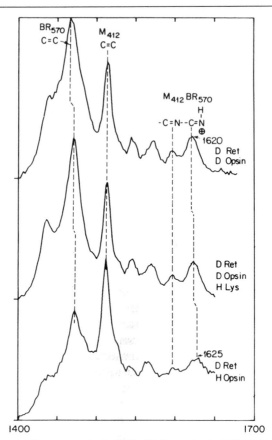

FIG. 3. Resonance Raman spectra in H_2O of fully deuterated BR (D Ret D Opsin), fully deuterated BR with 50% protonated lysine (D Ret D Opsin H Lys), and fully deuterated retinal in fully protonated opsin (D Ret H Opsin). Spectra obtained with 20 mW of 488.0-nm excitation; 2-cm^{-1} steps and 20 sec/step.

whereas ^{15}N on this same lysine, which is *directly* interacting with the proton, causes no shift.[58] The answer is far from clear, since all the experimental results obtained thus far indicate that this Schiff base mode is an extremely complicated admixture of vibrational modes. However, the following can be said qualitatively. First, all workers in the field agree that the N—H bending vibration plays an extremely important role in this mode and that the Schiff base mode appears not to extend into the *covalently* attached lysine side chain.[33,53] With this as a starting point it is conceivable that the interacting lysine, when its eight nonexchangeable protons are replaced with deuterons, could have a larger effect than a single mass unit change directly in line with the N—H axis. As an analogy, one could imagine the N—H . . . N as an axis from which extends the butyl side chain of the interacting lysine. In such an analogy it is quite conceivable that a one mass unit change on the interacting N axis will have a different effect on the N–H bend than an eight mass unit change on the butyl chain extending from the axis. Thus, in summary, there is no definitive proof that an interacting lysine residue is present in BR, but the data are extremely suggestive. In any event, the model certainly presents a useful framework for formulating experiments and analyzing results on an extremely complicated vibrational mode, which acts very differently from its solution–model counterparts.

3. Environmental Sensitivity. The principal lesson to be learned from the preceding data on the protonated CN vibrational mode is that it is very sensitive to environmental perturbations, even though this sensitivity does not extend to the protonated CN stretch in model protonated Schiff bases of retinal in solution (K. Schulten and M. Lindner, unpublished, see Section IV,A,1). This may arise because the freedom of the solution environment allows the Schiff base counterion to adjust to a minimum energy configuration. Such a configuration might result in a protonated CN stretching frequency independent of the size of the counterion present.

If this is indeed the case, then crystals of protonated Schiff bases of retinal with counterions of different sizes may result in altered CN stretching frequencies. In fact, such data have been obtained and the results are shown in Fig. 4. The crystalline spectra were obtained with 647.1-nm excitation and high-pressure liquid chromatography was performed to check for isomerization. Although only small percentages of isomerization (10%) were detected after the crystals were dissolved, the actual percentage isomerization of a sample that is not rotated in the laser beam could be considerably higher, since the molecules are prevented from diffusion. However, isomerization has no effect on the CN stretching frequency, as noted earlier. Only the fingerprint region could be affected, and even that effect is questionable in view of the fact that the fingerprint frequencies observed do not match previously observed spectral features for protonated Schiff bases.[3] In any event, the effect of the CN frequency, which is unlikely to be due to isomerization, must result from the altered nature of the counterion. These data then reflect the sensitivity of the CN frequency in condensed media to environmental effects and further confirm the interpretation of the sensitivity of the protonated CN, stretch based on the BR data.

Thus the preceding comparisons further highlight the lack of alteration of the protonated CN frequency in the rhodopsin-to-bathorhodopsin transition even though this transition is accompanied by a 45-nm red shift.

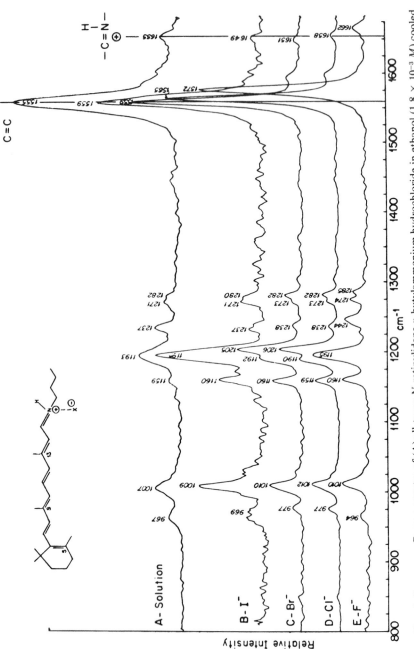

FIG. 4. Resonance Raman spectra of (A) all-*trans*-N-retinylidene-*n*-butylammonium hydrochloride in ethanol (1.8×10^{-3} M) cooled to $-20°$, flowed at 4 ml/sec, excited with 476.2-nm laser line. Resolution = 4.5 cm^{-1}. (B) Crystalline all-*trans* N⊕RB HI excited with 647.1-nm laser line. Resolution = 2 cm^{-1}. (C) Crystalline all-*trans* N⊕RB HBr excited with 647.1-nm laser line. Resolution = 2 cm^{-1}. (D) Crystalline all-*trans* N⊕RB HCl excited with 647.1 nm laser line. Resolution = 2 cm^{-1}. (E) Crystalline all-*trans* N⊕RB HF excited

This is a similar red shift to the one observed for the $BR_{570} \rightsquigarrow K_{610}$ transition in BR. Therefore, one must wonder if this indicates that the similarity in absorption red shift on similar time scales in both these systems[45,62,63] and the differences in the protonated CN frequency suggest that the photochemically induced absorption red shift has little to do with Schiff base alterations, but rather is controlled by protein–retinal interactions at some point removed from the Schiff base.

4. Deprotonation Kinetics of the Schiff base and Identification of the Unprotonated CN Frequency. No data are as yet available on the Schiff base deprotonation kinetics of visual pigments. However, the development and first application of the techniques of kinetic resonance Raman spectroscopy were to resolve the temporal characteristics of Schiff base deprotonation in bR.[27,64] The data obtained proved to be of interest, and there is a consensus supporting the interpretation of this original measurement.[8,65] The principle observation in this kinetic measurement was the development of a band at 1620 cm^{-1} between 2 and 8 μsec; long before appreciable scattering was observed at 1567 cm^{-1}, the $C=C$ stretching frequency of the M_{412} species. This early detection of a band at a frequency that corresponded to the unprotonated CN stretch suggested that deprotonation of the Schiff base preceded M_{412} production. Previous to this measurement it was generally assumed, based on the Lewis *et al.*[22] steady-state data, that the M_{412} state was the first unprotonated species. If Schiff base deprotonation precedes M_{412} formation, then there must be an unprotonated species that precedes M_{412}.[27] There was, however, no additional evidence for the existence of this species.

The preceding data also raised the possibility that, in spite of the good steady-state[22] evidence demonstrating that an increase in the intensity of the unprotonated CN stretching frequency occurs with a decrease in the scattering of the protonated species (which was definitely assigned[22] by deuteration), there was no unequivocal evidence, steady state or kinetic, demonstrating that the 1620-cm^{-1} band was associated with the CN stretch. This problem was addressed in a series of papers by Ehrenberg *et al.*[54,66] In the first paper,[54] data from various chemically modified BR species were analyzed under steady-state conditions. These included selective deuteration at carbon 15, which is adjacent to the Schiff base, ^{15}N enrichment, 3-dehydrobacteriorhodopsin, deuteration of the protein and

[62] K. J. Kaufmann, P. M. Rentzepis, W. Stoeckenius, and A. Lewis, *Biochem. Biophys. Res. Commun.* **68**, 1009 (1976).

[63] E. P. Ippen, C. V. Shank, A. Lewis, and M. A. Marcus, *Science* **200**, 1279 (1978).

[64] A. Lewis, *Spex Speaker* **21**, 1 (1976).

[65] A. Campion, M. A. El-Sayed, and J. Terner, *Biophys. J.* **20**, 369 (1977).

[66] B. Ehrenberg, A. Lewis, and H. L. Crespi, *Biochim. Biophys. Acta* **593**, 454 (1980).

lipid components only, and deuteration of the entire membrane (including the retinal). The most convincing data come from a measurement of the ^{15}N-labeled sample and 3-dehydrobacteriorhodopsin. An analysis of the entire spectrum of the ^{15}N-labeled sample clearly reveals that the 1620-cm^{-1} band arises from a vibrational mode that includes the motion of the nitrogen atom. The 3-dehydrobacteriorhodopsin measurements further support this conclusion by testing the applicability of the important results of Auerbach et al.[67] This group demonstrated that an anhydrovitamin A has a vibrational frequency at 1625 cm^{-1}, even though it does not have a nitrogen-containing end group. In this molecule the nature of the 1625 cm^{-1} vibrational mode is not clear, though it probably arises from an additional C═C stretching frequency. Our 3-dehydrobacteriorhodopsin measurements help to rule out the possibility that the 1620 cm^{-1} band in BR also has a similar origin. The data demonstrate that even though neighboring bands with significant C═C character are affected by the extension of retinal conjugation in 3-dehydrobacteriorhodopsin, the unprotonated CN stretching frequency is completely unperturbed.

Thus all the data appear to be consistent with the hypothesis that the 1620-cm^{-1} vibrational mode is indeed associated with the unprotonated carbon nitrogen stretch. The question still remains, however, whether the 1620-cm^{-1} vibrational mode that is detected before M_{412} production is also related to this CN stretching frequency that has been established on the basis of steady-state measurements. The second paper in the Ehrenberg et al. series[55] was aimed at trying to answer this question. Ehrenberg et al.[55] applied kinetic resonance Raman techniques[27] to a fully deuterated sample of purple membrane in H_2O. The object of this experiment was to determine whether the unprotonated CN frequency, which is now shifted in frequency by the isotopic labeling, also appeared at a rate that preceded M_{412} production. Although the data support this suggestion, it is not completely unambiguous because M_{412} appears to be generated in this system at an extremely rapid rate. Thus the time resolution was not sufficient to give a range in which the CN stretch was observed without any complication from M_{412} scattering. It is still not clear, however, why, in a protein where all the nonexchangeable protons are deuterons, the rate should be increased. Both in terms of its physiological and spectroscopic interest this result should be checked with an absorption measurement. If it is not verified by absorption flash photolysis, then it is possible that Raman intensities of the different intermediates are being altered in different ways by such deuteration.

In summary, although it certainly appears to be the case that the 1620-

[67] R. A. Auerbach, M. F. Granville, and B. E. Kohler, Biophys. J. 25, 443 (1979).

cm^{-1} band is a monitor of the unprotonated Schiff base and is detected before M_{412} production, there still remain unexplained anomalies in the unprotonated CN stretch. These include (1) the very small ^{15}N shift relative to protonated Schiff bases, even though model system data suggest the shift should be the same; (2) the data of BR labeled selectively at the C-15 position displays two vibrational modes, one that is close to the shift caused by ^{15}N enrichment of the unprotonated frequency and the other close to the effect of ^{15}N and C-15–D on the protonated frequency (Ehrenberg et al.[54] admit to not being able to resolve this C-15–D issue in terms of which is the unprotonated CN frequency, whereas Braiman and Mathies[68] do not address the question), (3) full deuteration of the retinal appears to have a larger effect on the unprotonated C=N mode relative to the protonated C=N—H⊕ mode. Conversely, it can be shown that, for the C=N—H⊕ mode in protonated Schiff base BR species deuterating the protein affects the CN frequency but deuterations on the retinal beyond C-15 have little effect. On the other hand, in the *unprotonated* species, deuterating the protein has no effect on the CN mode, but deuterations on the retinal beyond C-15 have a considerable effect. This is the case even though C=C stretching modes do not seem to play a significant role in this mode (see earlier), and thus it is possible that the preceding data indicate some coupling mechanism of C—H bends removed from the unprotonated Schiff base. In any event, the electronic and/or structural bases of these differences undoubtedly will be probed more fully in the future.

    5. *The pK of the Schiff Base Deprotonation.* With the development and application of kinetic resonance Raman spectroscopy to Schiff base deprotonation kinetics,[27] the way was opened to determine the pH dependence of this microsecond deprotonation process. This was reported by Ehrenberg and Lewis,[55] who showed that the process occurred much more rapidly at higher pH and displayed a characteristic titration-like pH dependence with a pK = 10.2 ± 0.3. These data indicated that a group of high pK was controlling the rate of Schiff base deprotonation. Thus, microseconds after light absorption the Schiff base pK is required to be $< 10.2 \pm 0.3$. This is in contrast to the pK of the Schiff base proton in the BR_{570} state, which was determined by the Ehrenberg and Lewis study[55] to be > 12. Therefore, two important results are directly derived from these data[42]: (1) The pK of the Schiff base is altered from > 12 before light absorption to $< 10.2 \pm 0.3$ within microseconds after light absorption. (2) The rate of Schiff base deprotonation is controlled by a high-pK group with pK = 10.2 ± 0.3. This latter result may turn out to be a crucial clue to the function of this proton pump. It clearly demonstrates that even

[68] M. Braiman and R. Mathies, *Biochemistry* **19**, 5421 (1980).

though BR can pump protons to pH 3, deprotonation of a high-pK group (pK = 10.2 ± 0.3) is essential in order for proton pumping to occur. Therefore, one is driven to the conclusion that the photochemical act must produce a species that either directly or indirectly leads to the deprotonation of the 10.2 ± 0.3 group in microseconds. The evidence in terms of time scales supports the production first of a group with a pK > 10.2 that then deprotonates the pK = 10.2 residue. (See Section IV,A,6, d.)

At room temperature light-driven deprotonation of the pK > 10.2 ± 0.3 group can readily occur; however, Kalisky et al.[50] have demonstrated that at − 90°C this deprotonation is blocked. This is indicated by warming in the dark the L species generated at − 90 to − 60°. The result of such an experiment[50] is the direct and ineffectual *regeneration* of the initial BR$_{570}$ species. On the other hand, if external pH is used to initially deprotonate the pK = 10.2 ± 0.3 group chemically and then the sample is irradiated at − 90°C, warming in the dark to − 60° forms directly the normal deprotonated Schiff base species M$_{412}$. Although Kalisky et al.[50] have not reported a careful temperature dependence, such an investigation will be valuable in determining the activation energy required for the formation of the suggested group of pK > 10.2 ± 0.3, which in all probability must be created, to deprotonate thermally (after L production) the pK 10.2 group at pHs < 10.2. In addition, the Kalisky et al. data clearly illustrate the ability of retinal in the appropriate light-activated protein structure to alter its isomeric structure *thermally* in the dark to the initial all-trans BR$_{570}$ species. This further verified those aspects of the Lewis excitation mechanism,[43] that include ground-state *retinal* structural alteration and that are not treated in the Honig et al. theory.[46]

A remaining nagging question is the character of the pK = 10.2 ± 0.3 group whose pK has been accurately determined and confirmed by two quite different methods.[55,50] Lewis et al.[42] initially suggested that the pK 10.2 group could be the lysine that was postulated to be interacting with the Schiff base proton. Although, as noted earlier, there is now some further proof for the hypothesis that a lysine may indeed be in such a complex, recent data obtained by us[59] cast doubt that the pK 10.2 group is this lysine.

Our experiments involved iodination of the purple membrane using the procedures of Hunter and Greenwood.[69,70] Using this procedure 4.5 out of the 11 tyrosines in BR were iodinated. Even though iodination of the purple membrane produces a change in the absorption maximum from

[69] W. M. Hunter and F. C. Greenwood, *Nature (London)* **194**, 495 (1962).
[70] J. Hermans and L. W. Lu, *Archiv. Biochem. Biophys.* **122**, 331 (1967).

570 nm to 550 nm, which is essentially independent of pH in the region we have considered (pH 7–11), the resonance Raman results demonstrated no alterations in any vibrational frequencies. This can be seen clearly in Fig. 5, which compares kinetic resonance Raman spectroscopy of native (A) and iodinated purple membrane (B). Notice the unaltered frequency of the BR_{570} (C=C stretch) at 1526 cm^{-1} and even the C=C stretch of the kinetic intermediate M_{412} at 1564 cm^{-1}. This is also the case for the important C=N—H⊕ and C=N vibrations at 1640 cm^{-1} and 1619 cm^{-1}, respectively, and other vibrational modes displayed in this spectrum and in other regions of the resonance Rama spectra of these membranes. This is most significant for two reasons. First, this highly sensitive structural data indicate that the center of photochemical and molecular activity in this proton pump is unaltered by iodination. Second, in

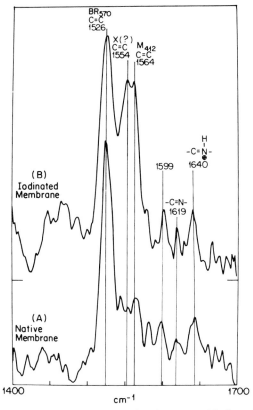

FIG. 5. Kinetic resonance Raman spectra of native (A) and iodinated (B) purple membrane with a 10 μsec transit time at pH 7. (For further details see Ref. 59, from which this figure is reproduced.)

view of the altered absorption spectrum, the unaltered Raman data indicate that the absorption changes are the result of a through space effect rather than the effect of some group directly interacting with the chromophore. Furthermore, the lack of any change in the unusually low-frequency $C{=}N{-}H\oplus$ stretch at 1640 cm^{-1} indicates that iodination does not directly affect the group or groups responsible for this lowered frequency.

In spite of the preceding there is one significant alteration in the spectra compared in Fig. 5. The contribution at short times ($\sim 10 \mu$sec) of the M_{412} $C{=}C$ stretch is significantly increased in the iodinated membranes. The $C{=}C$ stretch of other intermediate species that precede M_{412} production is also present with considerably enhanced intensity. The peak areas were used to determine the percentage of the 1564 cm^{-1} $C{=}C$ stretch area to the total $C{=}C$ stretch area. A plot of this percentage as a function of pH for the initial time of 10 μsec was used as an approximation of the initial slope of M_{412} formation. The results (shown in Fig. 6) indicate that the previously observed increase in M_{412} production between pH 9 and 10 is now observed[55] between pH 7 and 9. This is close to the pK of monoiodinated tyrosine, which occurs at 8.2. From the preceding data the following conclusions can be drawn: (1) Alteration in a predictable fashion of the pK of Schiff base deprotonation with iodination strongly suggests that tyrosine is the pK = 10.2 amino acid controlling deprotonation; (2) this group is not the amino acid responsible for the reduction in the protonated

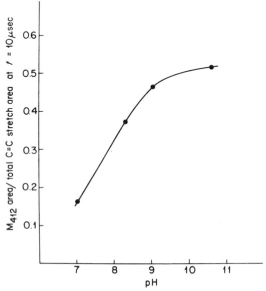

FIG. 6. A plot of M_{412} area per total $C{=}C$ stretch area at 10 μsec versus pH. (From Ref. 59.)

CN frequency in BR, since iodination causes no perturbation of this frequency. Therefore, we can conclude that there is another group responsible for the protonated CN frequency reduction, and lysine is certainly a good candidate for this group.

6. *Integration of Bacteriorhodopsin Schiff Base Data into a Proton Pumping Scheme.* Lewis *et al.*[42] have integrated the data on BR into a plausible scheme for proton pumping. This scheme was the first to introduce a framework of amino acid residues that were in a specific sequence necessary for proton relay. The proton relay was initiated in this framework by light and conformationally induced pK and structural changes between amino acids and in the retinal chromophore. This framework has been used successfully to interpret recently obtained results—for example, those of Hess and Kuschmitz[71] and Kalisky *et al.*[50] Since our framework of amino acids was introduced, several experiments have been to test specific aspects of the model, and we will incorporate in this section the results of these experiments.

(a) BR_{570}. The BR_{570} species is shown in Fig. 7 as having an interacting lysine residue in view of the Raman results on selectively labeled lysine discussed in Section IV,A,2. This lysine, however, cannot be the neighboring residue in the sequence, since it has been conclusively shown that retinal is not bound to lysine 41 as previously thought[72] but to lysine 216.[73,74] Even so, Bayley *et al.*[73] have pointed out that, given the constraints of the purple membrane structural data, the Schiff base of lysine-216 should still be very close to the amino group of a second lysine residue. In fact, they suggest that to be completely open-minded one might want to invoke that the retinal is complexed to two lysines as an aldamine. Although there is no evidence for this, it should be pointed out from discussions earlier that the protonated CN frequency in BR is certainly anomalous and not fully understood. In any event, we have chosen the more conventional structure depicted in Fig. 7 based on arguments discussed fully in Lewis *et al.*[42] In addition, included in Fig. 7 is a previously postulated salt linkage between a high and low pK group AB and a tyrosine residue thought to be involved in Schiff base deprotonation.[50,59] Finally, there is the tyrosine (to the right of the Schiff base in Fig. 7), which could reprotonate the Schiff base, and this is strongly indicated by the results of Konishi and Packer.[77]

[71] B. Hess and D. Kuschmitz, *FEBS Lett.* **100**, 334 (1979).

[72] J. Bridgen and I. D. Walker, *Biochemistry* **15**, 792 (1976).

[73] H. Bayley, K.-S. Huang, R. Radhakrishnan, A. H. Ross, Y. Takagaki, and H. G. Khorana, *Proc. Natl. Acad. Sci. U.S.A.* **78**, 2225 (1981).

[74] H. D. Lemke and D. Oesterhelt, *FEBS Lett.* **128**, 255 (1981).

[75] R. Renthal and B. Wallace, *Biochim. Biophys. Acta* **592**, 621 (1980).

[76] S. Tristram-Nagle and L. Packer, *Biochem. Int.* **3**, 621 (1981).

[77] T. Konishi and L. Packer, *FEBS Lett.* **92**, 1 (1978).

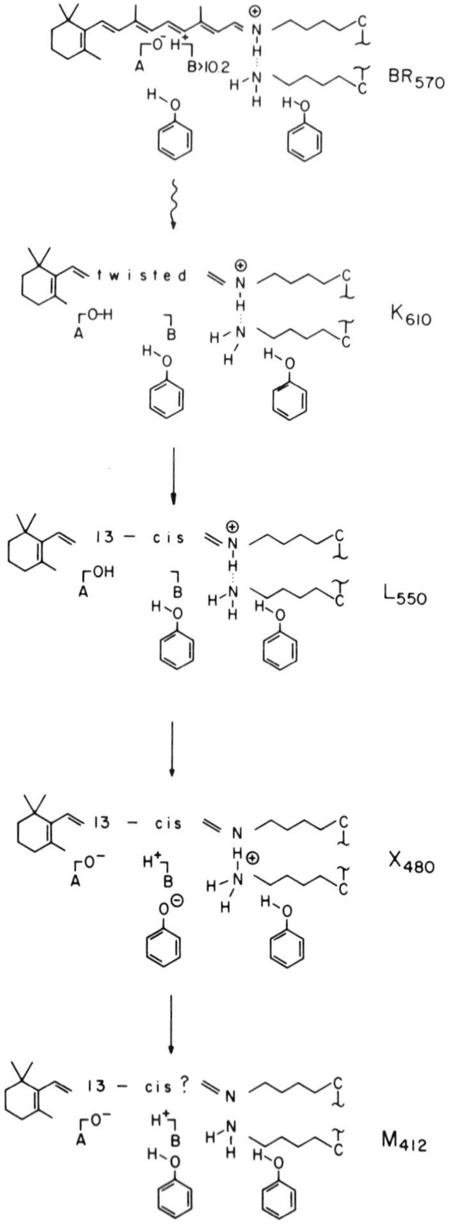

FIG. 7. Plausible molecular transformations in the light-activated transport of protons through the purple membrane. See Section IV,A,6 for further description.

(b) K_{610}. The structural transition to K_{610} in view of the discussion in Section IV,A,1 is probably a combination of protein and retinal structural alteration. The retinal structural alteration gives rise to increased intensity in the C—H out-of-plane bend[6,36,37] region. Similar alterations in the spectra of visual pigments have been thought to arise from out-of-plane twists in the isoprenoid chain,[7,9,43] and these alterations in BR probably also arise from twists in the isoprenoid chain.[37]

Protein structural alteration must involve proton movement.[43] The question is "Which proton?" Assuming sensitivity of the CN stretch to environmental perturbations, which the Raman evidence certainly bears out (see discussion in Sections IV,A,2,3), Lewis[7,43,44] demonstrated that the proton in visual pigments must be unrelated to the Schiff base proton. The case in BR is less clear-cut. Marcus and Lewis[6] based on dual-beam experiments could not detect alteration in the CN frequency in going from $BR_{570} \rightsquigarrow K_{610}$. Terner $et\ al.$[36] using a flow method reported a change in the CN frequency and also reported increased intensity in the C—H bending region, as has been reported by Marcus and Lewis[6] and Braiman and Mathies.[37] However, the Terner $et\ al.$[36] conclusions depend on significant amounts of computer subtraction, since the concentration of K_{610} in the sample was very low. In principle, Braiman and Mathies,[37] using a low-temperature rotating cell, are able to overcome the problem of low concentration of K_{610}, but were unable to detect the CN stretch of this species. In our laboratory we have successfully applied dual beam coherent antistokes Raman techniques[78,79] to bacteriorhodopsin, and these techniques should in principle yield in the near future a room temperature K spectrum. Thus, until more information on the CN stretch in K_{610} is available from Raman spectroscopy, the only evidence for deprotonation of B comes from the requirement that a $pK > 10.2$ group must be deprotonated by L in order to deprotonate the $pK = 10.2$ group (tyrosine) after L. Such a deprotonation appears not to be able to occur after K because the K → L transition time is independent of deuteration,[80] and therefore this process probably occurs between $BR \rightsquigarrow K$ in which the kinetics are deuteration dependent.[81] For a mechanism that suggests how such proton movement across an AB salt bridge the reader is referred to Ref. 42 and Fig. 2.

(c) L_{550}. The K → L transition appears to involve no proton movements based on the effects of deuteration on the kinetics,[80] but K → L probably does involve complete relaxation of the chromophore from a twisted structure to a 13-cis geometry. This is based on the appearance of

[78] E. T. Nelson, A. Lewis, and M. MacFarlane, $Biophys.\ Soc.\ Abstr.$ **25,** 79a (1979).

[79] E. T. Nelson, Dissertation, Cornell University, Ithaca, New York, (1979).

[80] R. Korenstein, W. V. Sherman, and S. R. Caplan, $Biophys.\ Struct.\ Mech.$ **2,** 267 (1976).

[81] M. L. Applebury, K. S. Peters, and P. M. Rentzepis, $Biophys.\ J.$ **23,** 375 (1978).

a strong band in the fingerprint region at 1185 cm^{-1} in L, which is close to an 1187 cm^{-1} band of 13-*cis* bR$_{560}^{DA}$, as was first pointed out by Marcus and Lewis.[6] This retinal structural alteration must also cause protein conformational changes that alter the interaction of retinal with specific charged groups and thus alters the absorption of the chromophore from 610 to 550 nm.

Movement of the CN stretching frequency band profile to lower frequencies has also been detected,[6] indicating that the CN stretching frequency may be altered in L$_{550}$, but this may be complicated by the possible presence of another unprotonated intermediate before M$_{412}$.

(d) X (?). Raman evidence for the existence of another unprotonated intermediate before M$_{412}$ was discussed in Section IV,A,4. The presence of such an intermediate called X (?)[6] is also supported by the data of Hess and Kuschmitz.[71] These workers detect the deprotonation of a tyrosine after L production but before M generation. The time scales they report are completely consistent with the time scales for X based on the Raman data. The lack of absorption spectral verification for the existence of X is certainly troubling. However, the resolution of intermediates with overlapping absorptions is a difficult task at best, as was the case in squid visual pigments, where an intermediate (mesorhodopsin λ_{max} = 475 nm) remained undetected by absorption spectroscopy for a decade.[82]

The Raman spectra clearly argue for X to be an unprotonated intermediate. The proton from the Schiff base would most logically move to the secondary lysine residue, which we have assumed has protected it thus far. The tyrosine simultaneously is deprotonated; this we determine from the large effect of iodination on the kinetic resonance Raman spectra[59] (see Fig. 5) and the time scale for tyrosine deprotonation as determined by kinetic absorption spectroscopy.[71] The pK > 10.2 residue accepts the proton from tyrosine. This is a requirement if the pK = 10.2 lysine residue is to deprotonate spontaneously in microseconds even at pH = 3. There is no evidence as to when A loses its proton. All we know is that it must take place after L based on the deuteration-independent kinetics of the K → L transition. We have put it between L → X, although it could just as well be between X → M or later. Raman evidence argues that X may still be 13-*cis*. This assignment, however, is not as definitive as the 13-*cis* conformation for L.[66]

(e) M$_{412}$. The X$_{480}$ → M$_{412}$ transition could simply be accounted for by a movement of a proton from the protecting lysine residue to the tyrosine. Such a proton movement would eliminate a positive charge at the Schiff base, causing a blue shift. If this is all that occurs, then the chromophore structure at M is 13-*cis* in view of our preceding discussion. However,

[82] Y. Ebina, N. Nagasawa, and Y. Tsukahara, *Jpn. J. Physiol.* **25**, 217 (1975).

there have been differences of opinion on the conformation of the chromophore in M. Chromophore extraction experiments clearly support the view that the conformation is 13-*cis*.[83] The resonance Raman results of Marcus and Lewis[6] questioned this assignment based on comparisons with model unprotonated Schiff bases of retinal and 3-dehydroretinal. On the other hand, the resonance Raman results of Braiman and Mathies[68] based on a comparison of 15-H- and 15-D-retinal Schiff bases question the assignment of Marcus and Lewis.[6] Now the results of Porta *et al.*[84] on 11-D-bacteriorhodopsin (to be discussed in more detail in Section IV,E,2,b) question the approach of Braiman and Mathies.[68] Finally, Mowrey and Stoeckenius,[85] using extraction procedures, claim that the 13-*cis* species disappears before M decays. These data have further complicated an already murky picture. Therefore, in Fig. 7 the retinal configuration is left in question.

Little structural detail is available on the decay of M. Two facts, though, have definitely been established. Resonance Raman spectroscopy has demonstrated (1) that as M returns to BR the Schiff base is reprotonated[22] and (2) that increasing the pH of the medium[86] to pH > 10 and iodination of the membrane[77] increases the lifetime of M. Both these results suggest that reprotonation of the Schiff base occurs either directly or through the secondary lysine by way of a tyrosine residue. Furthermore, the time scale of this reprotonation has been fairly definitively established by kinetic resonance Raman spectroscopy to be after 1.4 msec and apparently in parallel with the decay of M_{412}.[87] In addition, it has been shown[88] recently using glucose oxidase-lactoperoxidase that iodination of a surface tyrosine on the cytoplasmic surface of the membrane is responsible for the slowdown in M decay but that this tyrosine is not responsible for the acceleration of M production reported by Gogel and Lewis.[59] This then supports the placement of two different tyrosines in the framework of amino acids suggested by Lewis *et al.*[42] Furthermore, if the chromophore is not isomerized to *trans* in M, it will have to isomerize before BR and the postulated salt bridge (between a carboxylic acid group, A, and an amino acid, B, with pK > 10.2) will have to be reestablished to regenerate the starting BR_{570} species. Finally, it should be noted that this framework of amino acids, suggested by Lewis *et al.*[42] as the molecular switch, has

[83] M. J. Pettei, A. P. Yudd, K. Nakanishi, R. Henselman, and W. Stoeckenius, *Biochemistry* **16**, 1955 (1977).
[84] T. K. Porta, J. Aldern, G. Gogel, A. Lewis, and H. L. Crespi, *Biophys. Soc. Abstr.* **33**, 216a (1981).
[85] P. C. Mowrey and W. Stoeckenius, *Biochemistry* **20**, 2302 (1981).
[86] B. Becker and T. G. Ebrey, *Biophys. J.* **17**, 195 (1977).
[87] J. Terner, A. Campion, and M. A. El-Sayed, *Proc. Natl. Acad. Sci. U.S.A.* **74**, 5212 (1977).
[88] P. Scherrer L. Packer, and S. Selzter, *Arch. Biochem. Biophys.* **212**, 589 (1981).

been recently extended by Kalisky *et al.*[50] to demonstrate the pumping of more than one proton per photon. The scheme outlined in Fig. 7, which is identical to the previous Lewis *et al.*[42] suggestion except for the addition of a second tyrosine will also fit into the Kalisky *et al.*[50] two protons per photon extension. In fact, the existence of amino acid B (pK > 10.2) in our mechanism would be in better agreement with the two protons per photon results of Ort and Parson[89] than the Kalisky *et al.* scheme,[50] since it allows for a second high-pK group to deprotonate before Schiff base deprotonation.

G. Exchange Kinetics of the Schiff Base Proton

Kinetic resonance Raman spectroscopy has been recently applied to measure the exchange kinetics of the Schiff base in bacteriorhodopsin.[56,90] It was found[56,90] using the 8-jet rapid-mixing Durrum mixer with a mixing time of 0.9 ± 0.3 msec that the exchange time for the Schiff base is 4.7 msec. These workers[56] also reported the exchange time for all-*trans*-retinal Schiff bases of butylamine dissolved in various mixtures of C_2H_5 OD/D_2O. They found that as the relative D_2O concentration was increased the exchange time for the Schiff base decreased to less than the 0.9-msec time resolution of their instrument. This suggested that the bacteriorhodopsin Schiff base exhibited an exchange time that was at least five times longer than Schiff bases in solution. Such a result is certainly in agreement with the altered CN stretching frequency, which has been interpreted as indicating protection by an interacting amino acid.

The preceding results of Ehrenberg *et al.*[56,90] have been questioned within the last few months by Doukas *et al.*[57] These authors essentially repeated the Ehrenberg *et al.*[56,90] experiment. They claim that the Schiff base proton in BR exchanges in <3.0 msec, but they did not repeat the model Schiff base experiment. In addition, they[57] used a 2-jet mixer instead of the more conventional 8-jet mixer,[56] and this caused a significant difference in the time resolution of the Doukas *et al.*[57] measurement, which was 3.0 ± 0.5 msec, as compared to the 0.9 ± 0.3 msec resolution of Ehrenberg *et al.*[56,90] Furthermore, Doukas *et al.*[57] mixed 50 times excess of H_2O to 1 part of D_2O in which the bacteriorhodopsin was originally suspended. The comparable numbers for the Ehrenberg *et al.* experiments[56,90] are 12:1. These numbers are significant because H_2O has a band at 1650, which is close to the 1642 cm^{-1} of the protonated CN vibrational mode that has to be time-resolved. Thus a 50-fold excess of H_2O could obscure the time resolution of the very mode that has to be de-

[89] D. R. Ort and W. W. Parson, *Biophys. J.* **25**, 341 (1979).
[90] B. Ehrenberg and A. Lewis, *Biophys. Soc. Abstr.* **25**, 78a (1979).

tected. In spite of these differences, the claim by Doukas *et al.*[57] of an exchange time of <3.0 msec is difficult to understand on the basis of their published data. Specifically, we reproduce the pH 7.0 spectra of Doukas *et al.*[57] in Fig. 8. What is being shown in Fig. 8 is the disappearance of the deuterated CN stretching frequency (C—N—D⊕) at 1621 cm^{-1} and the appearance of the (C=N—H⊕) 1654 cm^{-1} vibrational mode. The times shown I and II are 3 msec and 15 msec. As can be seen by the dashed line, drawn by the author for comparison on the Doukas *et al.*[57] figure, there is significant scattering in I at the unexchanged 1621-cm^{-1} frequency, whereas at 15 msec this scattering has completely disappeared. Ehrenberg *et al.*,[56] in fact, do publish a pH 7 spectrum at 3.0 msec and the data are quite similar. One possible additional problem in all of this is that Ehrenberg *et al.*[56,90] used dark-adapted purple membrane, whereas in the Doukas *et al.*[57] paper light-adapted membrane was used in I and dark-adapted membrane was used for II. However, the similarity of the Doukas *et al.*[57] and Ehrenberg *et al.*[56] data at 3 msec and pH 7 strongly suggest that there is no difference between the light- and dark-adapted forms. Therefore, a comparison of the Doukas *et al.*[57] and Ehrenberg *et al.*[57] experiments suggests that the exchange time is 4.7 msec.

One region of agreement between the two groups is the pH indepen-

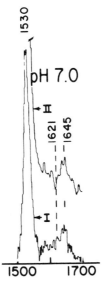

FIG. 8. Resonance Raman spectra showing hydrogen–deuterium exchange in bacteriorhodopsin at pH 7 (resolution 6 cm^{-1}). I, Deuterated sample (light-adapted) 3 msec after (1:50) mixing with H_2O. II, Deuterated sample (dark-adapted) 15 msec after mixing (1:50) with H_2O. (From Ref. 57 except for the dashed line, which was inserted by the author.)

dence of the exchange time between pH 2 and pH 10, which implies that the exchange is determined by the diffusion of a neutral water molecule to the Schiff base. Doukas et al.[57] make the further claim that acid–base catalysis cannot be the exchange mechanism and propose a concerted exchange mechanism with the direct participation of the solvent water molecules. However, Nagle[91] has recently pointed out that this assumption (i.e.; that acid–base catalysis is not the exchange mechanism) may not be valid in a hydrogen bonding environment buffered with ionic groups and thus having a pH relatively independent of the solution pH.

One disconcerting aspect of the Doukas et al.[57] mechanism is its prediction that both Schiff bases free in solution and the Schiff base in the membrane would undergo exchange controlled by a similar mechanism and thus should have similar time scales. Our experimental comparison[56] of the exchange time of model Schiff bases and the Schiff base in BR does not bear this out. It is clear, however, that the presence of an interacting amino acid would affect the rate, and the Doukas et al.[57] mechanism could certainly be extended to incorporate the presence of such an interacting amino acid.

Finally, however, one has to face up to the bothering question of how the Schiff base could exhibit an exchange time even as fast as 4.7 msec if indeed the Schiff base has a $pK > 12$ in an environment that is in all probability highly protected. The answer is not at all clear in view of the exchange times of hours and days seen for certain groups in such molecules as tRNA,[92] and this makes the Nagle suggestion[91] discussed above even more attractive. In any event it is good to end this section on a note of uncertainty in order to illustrate the level of our lack of understanding of even this most important vibrational mode after seven years and countless papers. It is also a warning against the overinterpretation in terms of bond angles and bond lengths of even more complicated vibrations in such regions as the fingerprint region. A lot more work is left and a lot more surprises await us as this sensitive selective probe of retinal structure and function is carried to new levels of sophistication. The previous seven sections simply show the possible definition that can be achieved with this powerful structural probe.

B. C=C Stretching Vibration

1. Linear Correlation of C=C Stretching Frequency with Absorption Maximum. The next set of vibrational modes in the spectrum is related to stretching vibrations of carbon–carbon double bonds. The strongest vi-

---

[91] John Nagle, private communication.
[92] P. R. Schimmel and A. G. Redfield, *Annu. Rev. Biophys. Bioeng.* **9**, 181 (1980).

brational mode in the spectrum is found in this region, as noted earlier in this chapter. This vibrational mode has been shown to be sensitive to the absorption maximum of the retinylidene chromophore. The first report of this sensitivity was by Heyde et al.,[93] and these workers[16,93] also demonstrated the linear dependence of the frequency of this mode to the absorption maximum of the retinal chromophore in λ (nm). Aton et al.[94] gathered together all the available data on protonated and unprotonated retinal Schiff bases free in solution and complexed to protein matrices and they demonstrated, with this expanded data set, that the linear correlation of absorption maximum with C=C vibrational frequency still appeared to hold. Marcus and Lewis[6] took this one step further. They studied the intermediates of fully deuterated and fully protonated bacteriorhodopsin and demonstrated that even though a comparison of the plots of absorption maxima versus C=C stretching frequency for the intermediates of these two systems showed a linear dependence, the deuterated system reflected an increased slope. This increase in slope would be predicted from the M (M = reduced mass) dependence on the slope for such proposed linear correlations. In spite of these data it is still unclear whether all retinal proteins can be placed on one correlation. There are some glaring examples of a lack of observed correlation. For instance, Sulkes et al.[7] noted that squid bacteriorhodopsin with a 550-nm absorption maximum has a C=C stretching frequency at 1530 cm^{-1}, whereas BR_{570} with an absorption maximum at 570 nm has a C=C stretching frequency also at 1530 cm^{-1}. In addition, squid isorhodopsin does not appear to fit even a linear correlation of squid intermediates. Therefore, it is clear that it cannot be generally assumed that a linear correlation exists, but most of the data certainly supports this suggestion with certain qualifications, as noted above.

It was in fact such a correlation of the intermediates of bacteriorhodopsin that led us to question the Terner et al.[87] assignment of the 1554-cm^{-1} C=C stretching mode as arising from the L intermediate with an absorption maximum of 550 nm. A linear correlation for the bacteriorhodopsin intermediates suggested such a species should have a C=C stretch that was close to 1537 cm^{-1}. The detailed investigations of Marcus and Lewis[6] in fact revealed such a vibrational mode and several workers[8,95] have since confirmed the Marcus and Lewis assignment.[6] According to Marcus and Lewis[6] the 1554-cm^{-1} band could arise from a species with an absorption maximum at 480 nm and could be associated with

[93] M. Heyde, D. Gill, R. Kilponen, and L. Rimai, J. Am. Chem. Soc. **93**, 6776 (1971).

[94] B. Aton, A. G. Doukas, R. H. Callender, B. Becher, and T. G. Ebrey, Biochemistry **16**, 2995 (1977).

[95] D. L. Narva, R. H. Callender, and T. G. Ebrey, Photochem. Photobiol. **33**, 567 (1981).

the unprotonated intermediate, X(?), that they proposed[27] may precede M_{412} formation. Whether or not this is the case, there is unanimous agreement that an extra C=C stretching mode at 1554 cm^{-1} is present in the spectra of the photocycle. Besides the obvious association of this C=C stretch with the X(?) intermediate there is the additional possibility that the L_{550} intermediate could have a second intense C=C stretch. There is good evidence for such additional intense C=C stretches in spectra of 3-dehydroretinal[3] and even some evidence in spectra of 3-dehydrobacteriorhodopsin.[54] The agreement between groups on this two C=C stretch issue is not unanimous. Low-temperature spectra of Marcus and Lewis[6] and Narva et al.[95] show no evidence of a second C=C stretch, but these spectra contain low concentrations of L. Stockburger et al.[8] clearly see the 1554-cm^{-1} band and also suggest that it may arise from X(?). Argade and Rothschild[96,97] are of two minds. They have agreed with the existence of X(?) and its deprotonated nature,[96] but they have also disagreed with this assignment.[97] Terner et al.[38] have partially retracted from their earlier 1554 assignment to L. A considered judgment of this author after all the evidence has been reviewed is that X(?) probably exists and that the 1554-cm^{-1} frequency is associated with this intermediate. In any event, until confirmed directly by absorption spectroscopy a question mark after X is still justified.

2. *Excitation Frequency-Dependent Alterations in the C=C Stretching Mode.* One of the more interesting observations on the C=C stretch is the apparent but small excitation frequency dependence that has been reported in the C=C stretch of BR. With 568. 2 nm excitation the C=C stretching frequency occurs at 1527 cm^{-1}, whereas at 457.9 nm excitation the C=C stretch is detected at 1530 cm^{-1}. In order to study this effect in a simpler polyene we have investigated the resonance Raman spectrum of all-*trans*-retinal with a variety of laser excitation frequencies. The spectra obtained are shown in Fig. 9. Although the general features of the spectra are unchanged in going from off resonance at 19,200 cm^{-1} to strict resonance at 27,488 cm^{-1}, there are some subtle alterations that can be noted.

As the region of strict resonance is approached, the intensity of the 1577-cm^{-1} and 1656-cm^{-1} vibrational modes increases more than the spectral features between 950 cm^{-1} and 1400 cm^{-1} (e.g., see the spectra obtained with 27,488-cm^{-1} and 24,587-cm^{-1} excitation frequencies). Thus these vibrations appear weaker in resonance. A second general observation is that the intensity of the 1163-cm^{-1} band, which is more intense than the 1198-cm^{-1} vibration in the preresonance region (Fig. 9, 19,200-cm^{-1}

[96] P. V. Argade and K. J. Rothschild, *Fed. Proc., Fed. Am. Soc. Exp. Biol.* **39,** 1846 (1980).
[97] P. V. Argade and K. J. Rothschild, *Biophys. Soc. Abstr.* **33,** 174a (1981).

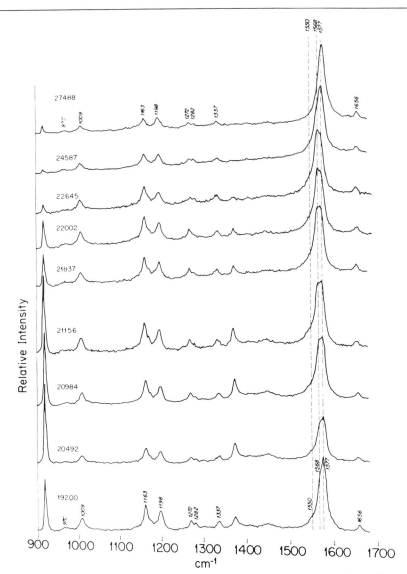

FIG. 9. Resonance Raman spectra of all-*trans* retinal obtained at selected laser frequencies. The acetonitrile intensity standard can be seen at 918 cm^{-1}. Other weak acetonitrile vibrational bands occur at 1375 cm^{-1}, 1412 cm^{-1}, and 1447 cm^{-1}. All acetonitrile bands are unlabeled. A 0.001 M solution was used to obtain all the spectra shown except for the 19,200 cm^{-1} spectrum, which was obtained with a 0.005 M solution.

excitation), becomes approximately equal in resonance (Fig. 9 for 27,488-cm^{-1} and 24,587-cm^{-1} excitation). Another general trend is the appearance and growth of two modes at 1568 cm^{-1} and 1550 cm^{-1} in the preresonance region and the disappearance of these vibrations in the strict resonance regime. These effects are probably caused by excited state phenomena and their interpretation should play an important role in understanding the resonance Raman spectrum of rhodopsin chromophores.

The changes in the intensity of the 1550-cm^{-1} and 1568-cm^{-1} vibrations are the most obvious of these effects and can be seen in Fig. 9. Visual inspection of Fig. 9 shows a shoulder at 1568 cm^{-1} in the 19,200-cm^{-1} spectrum. This shoulder becomes the dominant feature between 1500 cm^{-1} and 1600 cm^{-1} in the 22,645 cm^{-1} spectrum and then is undetectable in spectra obtained with laser lines above 24,587 cm^{-1}. Attempts to quantify this effect required the deconvolution of band profiles with a nonlinear least squares fitting routine to fit Voigt profiles to the observed data. The integrated areas of the deconvoluted bands were normalized to the preresonance value of the respective band detected with 19,200-cm^{-1} laser illumination. The relative intensities of each band[98] were then analyzed to determine the nature of the frequency dependence in the preresonance region and to locate the state or states giving rise to the preresonance behavior. It has been shown that the resonance Raman scattering from retinal[99] and similar chromophores[100] can be described by the Albrecht A-term[2]. Therefore, the functional form chosen for our analysis was

$$I^{1/2} \alpha \sum_i A_i \frac{v_{ei}^2 + v_0^2}{(v_{ei}^2 - v_0^2)^2},$$

where I is the relative intensity of a particular vibrational mode observed with a specific laser frequency v_0. A least squares fitting routine was used to determine the values of both the floating parameters v_{ei} (the frequency of the electronic state) and A_i (the relative oscillator strength of the state). In this fitting procedure, we used only the preresonance values below 21,000 cm^{-1}, since the preceding expression is only valid in the preresonance region. The results of this analysis suggest that the 1577 cm^{-1} C=C stretch and the 1656 cm^{-1} C=O stretch were enhanced by electronic transitions with a maximum near the 26,455 cm^{-1} absorption band of all-*trans*-retinal. This would be expected on the basis of the known position of the strongly allowed singlet–singlet transition of all-*trans*-retinal. The form of the observed enhancement profile for the 1577-cm^{-1} vibration can be seen

[98] R. E. Cookingham, Dissertation, Cornell University, Ithaca, New York (1978).
[99] A. Warshel and P. Dauber, *J. Chem. Phys.* **66**, 5477 (1977).
[100] F. Inagaki, M. Tasumi, and T. Miyazawa, *J. Mol. Spectrosc.* **50**, 2861 (1974).

in Fig. 10. A maximum in this enhancement profile probably occurs at about 25,500 cm^{-1}, which is red-shifted by approximately 1000 cm^{-1} from the maximum in the absorption of all-*trans*-retinal at 26,455 cm^{-1}. The enhancement profile of the 1656-cm^{-1} C=O stretching vibration also appears to have a maximum near this same frequency at 25,500 cm^{-1}. This can be visually verified in Fig. 9, especially in the spectra obtained with 27,488-cm^{-1} and 24,587-cm^{-1} laser excitation. Notice that the 1577-cm^{-1} and 1656-cm^{-1} bands have increased intensity relative to the other vibrational bands in the spectra obtained with these laser frequencies. Thus, both the complete enhancement profile and the preresonance behavior of the C=O and this C=C stretching vibration are very similar.

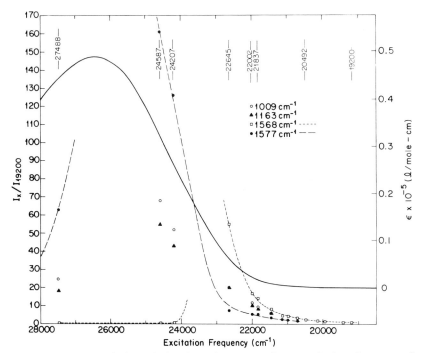

FIG. 10. A plot of the relative intensity versus laser excitation frequency for all-*trans*-retinal vibrations at 1009 cm^{-1}, 1163 cm^{-1}, 1568 cm^{-1}, and 1577 cm^{-1}. Below 21,200 cm^{-1} the points for the 1009 cm^{-1} and 1163 cm^{-1} vibrations falls between the 1568 cm^{-1} and 1577 cm^{-1} values and were not plotted to avoid confusing the presentation. The absorption spectrum of all-*trans*-retinal is plotted as a solid curve (with an extinction scale on the right ordinate) for comparison to the observed enhancement of the intensities of the vibrational bands. Smooth broken curves have been drawn to connect the points in the enhancement profiles of the 1568-cm^{-1} and 1577-cm^{-1} features. These profiles are not intended to imply any functional dependence but are included only as an aid to the reader. Positions of the laser lines referred to in the text are indicated in the figure.

The same analysis procedure described earlier was also used to locate the origin of the electronic state responsible for the enhancement of the 1568-cm^{-1} and 1550-cm^{-1} bands. In contrast to the previously described vibrational modes, the preresonance intensity values of the 1568-cm^{-1} and 1550-cm^{-1} bands indicated that a single electronic state in the vicinity of 23,000 cm^{-1} was responsible for the observed resonance enhancement. The enhancement profile of the 1568-cm^{-1} band is plotted in Fig. 10 and appears to reach a maximum in the region between 23,000 cm^{-1} and 23,500 cm^{-1}.

An analysis of the remaining vibrational modes in the spectrum (at 970, 1009, 1163, 1198, 1272, 1292, and 1337 cm^{-1}) supports the preceding conclusions. In addition, as was reported by Doukas et al.,[101] we find no evidence for a maximum in the excitation profile at 21,000 cm^{-1} that was detected by Rimai et al.[102] and Huong et al.[103] and ascribed to the triplet state. These reports probably arose from photochemical alterations in the sample. Our detection of this excitation profile anomaly at low energies probably does not arise from a triplet state, since Friedman and Hochstrasser[104] have argued that such a state would be unobservable. It also probably does not arise from an $n\pi^*$ transition, since $n\pi^*$ states exhibit weak vibronic coupling and since the enhancement profile of the C=O mode shows no evidence of a peak at 23,000 cm^{-1}. Rather, in light of several recent reports of low-lying dipole forbidden electronic states,[47,105,106] the preceding analysis and data suggest that we may be observing such a low-lying forbidden state in all-trans-retinal at approximately 23,000 cm^{-1}. Many of our observations about the suspected 23,000-cm^{-1} state support the preceding suggestion. The frequency of the 23,000-cm^{-1} state is approximately 3500 cm^{-1} below the strongly allowed transition at 26,455 cm^{-1}, which is very close to a value of 3000 cm^{-1} ± 2000 cm^{-1} [107] predicted for a polyene of retinals length. The results of our calculations, which show that the ratio of relative oscillator strengths are 1:35 for vibrational bands enhanced by both states, are remarkably similar to the 1:30[108] ratio quoted in reports on certain polyenes where such a low-lying forbidden state has been detected. As previously noted, the maximum of all the excitation profiles except those of the 1550 cm^{-1} and 1568 cm^{-1}

[101] A. G. Doukas, B. Aton, R. H. Callender, and B. Honig, Chem. Phys. Lett. **56**, 248 (1978).
[102] L. Rimai, M. Heyde, H. Eller, and D. Gill, Chem. Phys. Lett. **16**, 207 (1971).
[103] P. Huong, R. Cavagnat, and F. Cruege, in "Lasers in Physical Chemistry and Biophysics" (J. Joussot-Durben, ed.), p. 425. Am. Elsevier, New York, 1975.
[104] J. Friedman and R. Hochstrasser, Chem. Phys. Lett. **32**, 414 (1975).
[105] B. Hudson and B. Kohler, Annu. Rev. Phys. Chem. **25**, 437 (1974).
[106] R. R. Birge, Annu. Rev. Biophys. Bioeng. **10**, 315 (1981).
[107] R. Christensen and B. Kohler, Photochem. Photobiol. **19**, 401 (1974).
[108] B. Hudson and B. Kohler, J. Chem. Phys. **59**, 4984 (1973).

bands appears to be red shifted by approximately 1000 cm^{-1} from the observed maximum in the absorption spectrum. As has recently been observed in benzene,[109] this may be due to interference effects between the two states that lead to local minima in the enhancement profile even in regions of high absorption resulting from electronic transitions. This could explain the low relative intensities observed for the spectra taken at 27,488 cm^{-1} as compared to those with laser excitation at 24,587 cm^{-1}. Finally, recent theoretical and experimental work by Albrecht and co-workers has shown that dipole forbidden transitions can cause enhancement of vibrational modes in benzene.[110] On the basis of these observations we suggest that we may be observing a dipole forbidden state in all-*trans*-retinal at 23,000 cm^{-1}.

The relationship of these findings to the effect observed in BR of altered C=C stretching frequencies as a function of excitation frequency is not clear. The retinal results must be considered as one possibility in trying to understand the excitation-frequency-dependent alterations in BR. Other possibilities, such as the effect of excition interactions, that exist in BR[47] and that have been observed to cause excitation-frequency-dependent alterations in carotenoids[111] must also be considered in interpreting the BR observations.

C. The 1300 to 1500-cm^{-1} Region

Very little work has been focused on this region of the spectrum and the investigations that do consider the region in detail have mostly limited their attention to retinals rather than the retinal Schiff base. Cookingham and Lewis[11] discovered, using chemically modified retinals with various combinations of butyl groups replacing the methyl groups at C-9 and C-13, that there is a band in 11-*cis* retinal at 1345 cm^{-1} that is absent when the C-13 methyl is replaced by a butyl group. Cookingham *et al.*[3,11] hypothesized based on these data that the 1345-cm^{-1} band could be reflecting a well-established 12-s-*cis* and 12-s-*trans* conformational equilibrium around the 12–13 single bond in this isomer.[112,113] Cookingham *et al.*[3] further hypothesized that the presence of a butyl group at the C-13 position restricted this conformational equilibrium. To prove this deduction they obtained the spectrum of 11-*cis* retinal as a function of temperature and showed that the intensity of the band increased as the temperature was lowered. This demonstrated that the 1345-cm^{-1} band is associated with

[109] L. D. Zeigler and A. C. Albrecht, *J. Chem. Phys.* **67**, 2753 (1977).

[110] G. M. Korenowski, L. D. Ziegler, and A. C. Albrecht, *J. Chem. Phys.* **68**, 1248 (1978).

[111] P. R. Carey, *Q. Rev. Biophys.* **11**, 309 (1978).

[112] R. Gilardi, I. Karle, J. Karle, and W. Sperling, *Nature (London)* **332**, 187 (1971).

[113] B. Honig and M. Karplus, *Nature (London)* **229**, 558 (1971).

the 11-*cis* 12-s-*trans* conformation, since nuclear magnetic resonance studies had indicated that at lower temperatures the 12-s-*trans* conformer was stabilized.[114] Although such a mode would be extremely useful to probe the conformational state of the 11-*cis* chromophore of rhodopsin, the studies of Cookingham et al.[3,11,98] also demonstrated that this mode is extremely sensitive to the nature of the end group and in fact the 1345 cm^{-1} mode appears to be absent in 11-*cis* retinal Schiff bases. Thus it is unlikely that this vibrational mode will be capable of predicting the chromophore conformational state *in situ*.

D. The 1300 to 1240-cm^{-1} Region

There is one vibrational mode in this region that shows promise as a monitor of the 11-*cis* 12-s-*cis*/12-s-*trans* conformational equilibrium. If the 11-*cis* chromophore is in the 12-s-*trans* conformation, then a strong band is observed at ~ 1271 cm^{-1} in retinal. Any structural alteration that perturbs this state weakens the band. Thus this vibrational mode and the 1345-cm^{-1} band are the only vibrations in the resonance Raman spectrum of retinal that are sensitive to temperature alterations. In fact, Cookingham et al.[3] noted that this 1271 cm^{-1} vibrational mode is also relatively insensitive to the type of end group, and that can clearly be seen in Table I. Sulkes et al.[7] noted that in rhodopsin a strong band existed at ~ 1270 cm^{-1}, indicating that the chromophore initially is in an 11-*cis* 12-s-*trans* conformation. As the chromophore isomerizes from this conformation to all-*trans* the strong 1275-cm^{-1} band should become weaker and split into two vibrational modes of equal intensity, one at 1272 cm^{-1} and one at 1282 cm^{-1}. In going to bathorhodopsin the ~ 1270 cm^{-1} band has become weaker relative to rhodopsin, but there is no evidence for any splitting.[7,31,32,33] Although it should be noted that the importance of the splitting if not clear, since in BR$_{570}$, which is known to be all-*trans*, a weak band occurs at 1275 cm^{-1} without any splitting. However, BR$_{570}$ has a seriously perturbed spectrum (see Section IV,A,2) that does not resemble an all-*trans* chromophore in other ways, and thus it may not be valid to expect normative behavior in this mode.

E. The Fingerprint Region (1150–1240 cm^{-1})

The fingerprint region is principally composed of C—C stretching vibrations from C-9–C-13[3] and thus is very sensitive to the structural state of the chromophore. Each isomer of the retinals or the Schiff base has a distinctive signature, and this has been extensively catalogued in the work of Cookingham et al.[3] As is shown in Fig. 11 the substitution of a single

[114] R. Rowan, A. Warshel, B. Sykes, and M. Karplus, *Biochemistry* **13**, 970 (1974).

deuteron at the C-11 position causes significant alterations in the finger-print region of the isomers of retinal relative to the frequencies listed in Table I.

1. 1220–1240 cm⁻¹. A discussion of this region must initially focus on a band that occurs in unprotonated Schiff bases at ~1220 cm⁻¹ and in protonated Schiff bases at ~1237 cm⁻¹ and that is relatively unaffected by the isomerization state of the chromophore. Cookingham *et al.*[3] have assigned this band to a C—C stretch, CH₃ rock. Marcus and Lewis[6] have noted that this vibrational mode alters its position in protonated model

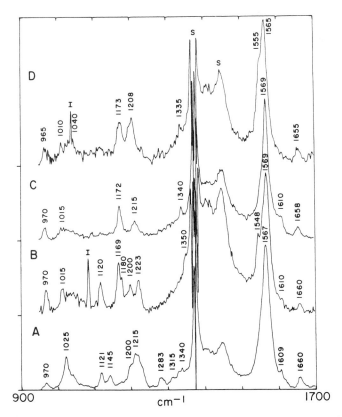

FIG. 11. The resonance Raman spectra of 11-D-retinal. (A) 11-*cis*, (B) 13-*cis*, (C) 9-*cis*, and (D) all-*trans*. Data were obtained with 45 mW of 647.1-nm excitation with a 2-cm⁻¹ step size and 20 sec/step. The bands marked S are those of the acetonitrile solvent. The bands marked I represent channels inadvertently dropped by the counting electronics. They are left uncorrected in this raw data to give the uninitiated reader a feel for a 2 cm⁻¹ step size on this scale. High-pressure liquid chromatography was performed on all samples to check for isomerization. Nuclear magnetic resonance spectroscopy verified 95% deuteration of the C-11 position.

systems as the saturated chain, which is coupled to the Schiff base, is altered in length. This suggested that the preceding vibrational mode may be sensitive to alterations in the region of the chromophore closest to the Schiff base. The same effect, though, was not observed for the ~ 1220-cm^{-1} band.[53] Marcus and Lewis[6] also noted that the position of this band in BR$_{570}$ was altered to 1255 cm^{-1}, and they reasoned that this alteration could be due to the perturbation that causes Schiff base alterations in this species. Stockburger et al.[8] claim that the 1255 cm^{-1} band disappears with deuteration, which would be consistent with the sensitivity of this mode to alterations in the Schiff base region, but other workers who have studied differences in purple membrane spectra in H$_2$O and D$_2$O[6,87] have not reported this difference. A further indication of the sensitivity of the 1255 cm^{-1} band in BR$_{570}$ to alterations in the Schiff base region is the effect of a single deuteron at the C-15 position. This modification eliminates the 1255 cm^{-1} mode.[54,68] This same modification in unprotonated all-trans Schiff base eliminates the band at ~ 1222 cm^{-1},[68] but the claim is made[68] that the somewhat more intense band at this frequency in the 13-cis chromophore arises from some other mode, since it increases in intensity on C-15 deuteration. In bovine bathorhodopsin two weak bands have been reported in 1220 to 1240-cm^{-1} region, although there is not universal agreement on this observation. Sulkes et al.[7] and Eyring and Mathies[32] have detected a band that is present only in batho at ~ 1277 cm^{-1}. Eyring and Mathies[32] detect an additional weak band in this region at 1240 cm^{-1} after subtraction. Aton et al.[33] do not report the 1277-cm^{-1} band but do detect the band at ~ 1240 cm^{-1}. It is the author's opinion that there probably are two peaks in this region, one at ~ 1277 cm^{-1} that is characteristic of bathorhodopsin, and one at ~ 1240 cm^{-1} that is in all three species—rhodopsin, isorhodopsin, and bathorhodopsin. The significance of the additional 1227-cm^{-1} peak in a chromophore that certainly appears to be a protonated Schiff base[7,32,33] is not clear, although it probably arises from a vibrational mode unrelated to the unprotonated 1222-cm^{-1} band discussed in this section.

2. *1150–1220 cm^{-1}*. This fingerprint region of the spectrum has been the focus of two sets of discussions, the first concerning the structure of the chromophore in bovine bathorhodopsin and the second concerning the structure of the chromophore in the M$_{412}$ intermediate of BR.

(a) THE STRUCTURE OF THE BATHORHODOPSIN CHROMOPHORE. The earliest work that accurately compared the spectra of bathorhodopsin in this region to a protonated all-trans chromophore was that of Sulkes et al.[30] They compared a spectrum of a squid rhodopsin, isorhodopsin, and bathorhodopsin mixture at low temperatures to the spectrum of squid metarhodopsin. The spectra obtained are shown in Fig. 12. The assign-

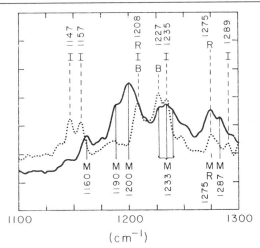

FIG. 12. A comparison of steady-state squid rhodopsin spectra at 77°K (dotted line) and 273°K (solid line). Spectral features are assigned for the dotted spectrum at the top of the figure in terms of rhodopsin, R; bathorhodopsin, B; and isorhodopsin, I. The assignments for the solid spectrum are at the bottom of the figure in terms of rhodopsin, R, and metarhodopsin, M. Assignments are based on Ref. 7 and 31.

ments are based on the work of Sulkes *et al.*[7,30] Both squid and bovine bathorhodopsin have bands at 1208 and 1227 cm^{-1}, whereas metarhodopsin has all its major fingerprint bands at 1200 cm^{-1} and below, a characteristic of the vibrational spectra of *all* the relaxed all-*trans* chromophores studied to date. Eyring and Mathies,[32] however, claim to detect at 1166 cm^{-1} a weak band in bathorhodopsin, which would be very important, since such a frequency compares well with a band in all-*trans* protonated Schiff bases. The band at 1166 cm^{-1} is only seen after considerable subtraction of high fluorescent backgrounds and pure rhodopsin and isorhodopsin spectra not necessarily obtained with the same background scattering. This author feels such subtraction procedures can only be used for the most intense features and are prone to errors otherwise. An example of such errors can be found by comparing the data of Aton *et al.*[33] with those of Eyring *et al.*[9,32] in the region of the suggested 1166-cm^{-1} mode. Once again in the Aton *et al.* data[33] subtraction does produce a feature in this region, but now it is at 1145 cm^{-1}, 21 cm^{-1} removed from the Eyring *et al.*[32] band at 1166 cm^{-1}, which is not detected by Aton *et al.*[33] In fact, the original data of Sulkes *et al.*[7,30] and Marcus and Lewis[31] show a valley in the 1166-cm^{-1} region rather than a peak and so it is considered opinion of this author that it is unlikely that bathorhodopsin exhibits significant scattering in this region.

In summary, therefore, the spectra of bathorhodopsin are probably composed of three bands in this region: an intense peak at 1208 cm^{-1} and a weak feature at 1227 cm^{-1}, which is common to squid and bovine batho, and possibly a weak band at 1240 cm^{-1}, which may also be common to both species. These features clearly are very different to those of a relaxed all-*trans* chromophore as is quite evident from Fig. 12. It should be noted that these alterations in the fingerprint bands of bathorhodopsin from those of a relaxed all-*trans* probably do not arise from single bond twists. This is deduced from our observations[3] on 11-*cis*-retinal, which can be thermally induced to undergo single-bond rotation around the C-12–C-13 bond and shows no change in its Raman spectrum as a function of temperature in this region of the spectrum (see Section IV,C,D). In addition, the resonance Raman spectrum of all-*trans*-retinal with the 5 methyl group removed (5-desmethylretinal) relieves a 59° twist around the 6–7 single bond and also leaves this region of the spectrum *completely* unaltered.[3]

(b) THE STRUCTURE OF M$_{412}$. Marcus and Lewis,[6] using a two-beam kinetic technique, obtained the spectrum of M$_{412}$ at physiological temperatures and pH. They concluded, based on a comparison of their M$_{412}$ spectrum and the M$_{412}$ spectrum of 3-dehydrobacteriorhodopsin with model compounds, that M$_{412}$ was probably all-*trans*. Braiman and Mathies,[68] using the Marcus and Lewis[6] technique, have now obtained a similar M$_{412}$ spectrum and they came to the conclusion that M$_{412}$ is 13-*cis* in agreement with extraction data.[83] The Braiman and Mathies[68] conclusion was reached by the effect on the M$_{412}$ spectrum of a deuterium substitution at C-15 and a comparison of this same effect on all-*trans* and 13-*cis* unprotonated Schiff bases. The logic of their argument is that the conformation of M$_{412}$ can be chosen by comparing the alterations induced by C15-D in M$_{412}$ with alterations induced by this same substitution in the all-*trans* and 13-*cis* Schiff bases. The data are indeed compelling. However, recently the spectrum of 11-D bacteriorhodopsin has been obtained under several illumination condtions.[84] The spectra obtained are shown in Fig. 13. The spectrum excited with 457.9 nm excitation (Fig. 13C) is principally composed of 11-D M$_{412}$. When the intensity analysis of Braiman and Mathies[68] is applied for this 11-D substition, one concludes that M$_{412}$ is all-*trans*. In fact, what one concludes is that several additional substitutions and comparisons will be needed before a final conclusion can be drawn.

Interestingly enough, though, our 11-D data do support the notion that isomerization back to all-*trans* does not occur until the O$_{640}$ to BR$_{570}$ transition. When spectra of 11-D BR at 514.5-nm excitation (principally composed of BR$_{570}$) and 568.2-nm excitation are compared (Figs. 13B and A) it is evident that there are two prominent bands and several other weaker

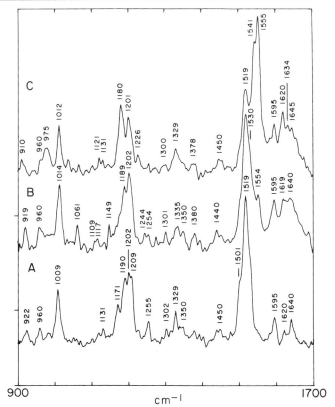

FIG. 13. Resonance Raman spectroscopy of 11-D-bacteriorhodopsin obtained with 568.2-nm (A), 514.5-nm (B) and 457.9-nm (C) excitation under steady-state conditions. The laser power was 20 mW and the step size was 2 cm⁻¹ with 20-sec residence time per step.

features with 568.2 nm excitation that are either less intense or absent at 514.5-nm excitation. The excitation frequency dependence of these features and the frequencies at which they occur assure us that they originate from the 11-D chromophore of a red-absorbing intermediate. The only such intermediate present under these conditions is O_{640}. This is the first time that any unique O_{640} vibrational frequencies have been observed in the fingerprint region without resort to subtraction and its associated problems. The 11-D O_{640} bands especially in the fingerprint region are clearly different from those of BR_{570} obtained with 514.5 nm excitation. This is internal proof that the structure of O_{640} is altered from BR_{570}, which is all-*trans*. The exact nature of the structural difference is not known but single-bond rotations cause no alterations in this region of the spectrum as the data on 5-desmethylretinal and the 11-*cis* 12-s-*trans* to

11-*cis* 12-s-*cis* conformational equilibrium have clearly indicated.[3] Therefore, in view of the implication of only 13-*cis* and all-*trans* in the BR cycle and because BR is all-*trans*, O_{640} is likely to be in a 13-*cis* configuration.

F. The C—CH₃ Region

The vibrational modes between 1000 and ~ 1030 cm⁻¹ can clearly be assigned to C—CH₃ vibrations involving only the C-9 and C-13 carbons. The C-5—CH₃ does not participate in this vibrational mode. All of the preceding was clearly demonstrated using a variety of chemically modified retinals with butyl substitutions at the C-9 and C-13 positions.[11] These studies also demonstrated that the C—CH₃ vibrational mode of 11-*cis* retinal in either the 12-s-*cis* or 12 s-*trans* conformers was unique from all other isomers. In essence this vibrational mode was split in both conformers of the 11-*cis* configuration with one band at 997 cm⁻¹ from the C-13—CH₃ and one band at 1017 cm⁻¹ from the C-9—CH₃. This effect is independent of the end group (see Table I) and in all cases disappears as the 11-*cis* isomeric configuration is lost. As was first noted by Sulkes *et al.*,[7] the effect is clearly lost in going from rhodopsin ⤳ bathorhodopsin in both the squid and bovine systems. Furthermore, the loss of the C-13—CH₃ vibration below 1000 cm⁻¹ is a consistent feature of all the bathorhodopsin assignments to date.[7,31–33] Therefore, this certainly demonstrates alteration of the 11-*cis* configuration in going to bathorhodopsin.

G. The C—H Bend Region

The C—H bend region is normally between 950 and 990 cm⁻¹ and a weak band or series of bands is always detected within these frequencies.

1. Bacteriorhodopsin. The most interesting variations in C—H bend vibrational modes have been found in BR. In reviewing the data presented in several papers[6,8,87] there appear to be inconsistencies as to the peak positions of the C—H bend modes in BR. After a consideration of all the data, which are outlined in Table IV, it appears that the variations result from whether the membranes were resuspended in H_2O or D_2O and/or whether the sample was flowed or was stationary. The H_2O–D_2O effect is that the major band in this region when excited between 514.5 nm and 568.2 nm moves from 958 cm⁻¹ in H_2O to 986 cm⁻¹ in stationary D_2O samples and 976 cm⁻¹ when the D_2O sample is flowed. In addition, there are intensity variations as a function of excitation between 514.5 nm and 568.2 nm in H_2O when the sample is flowed. The 959-cm⁻¹ band is always strong, but a shoulder at 969 cm⁻¹ that is seen at 568.2- and 514.5-nm excitation becomes approximately equal in intensity at 530.9 nm. Stationary and flow samples in D_2O show different intensities at ~ 959 cm⁻¹. The sta-

TABLE IV
VIBRATIONAL MODES IN BACTERIORHODOPSIN BETWEEN 950 AND 1000 CM^{-1}
OBSERVED UNDER A VARIETY OF CONDITIONS

Excitation (nm)	H$_2$O		D$_2$O	
	Stationary	Flow	Stationary	Flow
568.2	958 vs[a]	959 vs	945 vw	
	967 vwsh[b]	969 vwsh	960	958 s
			to	968 sh
			975 dsh[b]	976 vs
			986 s	985 sh
			945 w	
530.9	958 vs	959 s	959 s	958 s
	967 wsh	969 s	970 w	968 sh
			976 w	976 vs
			986 s	985 sh
			945 vw	
514.5	958 vs	959 s	960 w	958 s
	967 w	967 sh	970 w	968 sh
			974 w	976 vs
			986 s	985 sh
457.9	959 w		959 vvw	
	975 vs		975 s	
			986 sh	

[a] Bands are labeled as very strong, vs; strong, s; weak, w; very weak, vw; and very very weak, vvw. These assignments refer only to the relative intensities of these bands in this region.
[b] Shoulders are labeled as shoulder, sh; weak shoulder, wsh; very weak shoulders, vwsh; and diffuse shoulders, dsh.

tionary samples have much weaker 959 cm^{-1} scattering at 568. 2 and 514.5 nm and are more sensitive to excitation frequency alterations. Furthermore, different samples show different intensities at 959 cm^{-1} in D$_2$O. In some samples there is a clear band, whereas in others there is more of a shoulder. Nonetheless, the 976-cm^{-1} band in D$_2$O flow is always stronger than the 958-cm^{-1} band and the intensity differential appears greatest at 568.2 nm. The same is the case in stationary state D$_2$O samples when the strong 986-cm^{-1} band is compared to 959-cm^{-1} scattering. A factor complicating the analysis of vibrational modes in this region is their relatively weak intensities, which make it difficult to determine relative intensities accurately. Similar problems plague the ^{15}N data[6,54,58] in this region with the added problem of fewer available samples reducing the data pool. One definitive conclusion is that for BR the band profile shifts from a major peak at 959 cm^{-1} in H$_2$O to a major peak in D$_2$O at 976 cm^{-1} in flow and 986 cm^{-1} in stationary samples.

Basically two alternative explanations have been advanced for the H_2O-D_2O flow effect. First, Marcus and Lewis[6] have used the data of Craig and Overend[115] on simpler systems to suggest that deuteration of the Schiff base could alter coupling of this mode with the C—H bend or bends on the isoprenoid chain and cause a shift of the C—H bend to higher frequencies. Second, Stockburger et al.[8] have suggested that the 1255-cm^{-1} band in H_2O disappears in D_2O and reappears at 976 cm^{-1}. This would indeed be an exciting development if this can be proved; however, as was pointed out previously, two earlier studies[6,87] were not able to detect the disappearance of the 1255-cm^{-1} band.

2. *Rhodopsin.* In visual pigments the most interesting observation is the intense C—H bend frequency in rhodopsin at 970 cm^{-1} and in isorhodopsin at 960 cm^{-1}.[116] The intensity of vibrational modes in a resonance Raman spectrum is related to whether the atoms participating in a mode undergo large nuclear excursions on vertical excitation. Presumably, carbon atoms 9, 10, 11, and 12 are undergoing torisonal oscillation in the excited state in order to alter the isorhodopsin and rhodopsin chromophore structure to all-*trans*. Thus it may be expected that the hydrogen on these carbon atoms would be undergoing large oscillations and would therefore be responsible for the intense C—H bends observed. However, deuterium substitution of carbons 10, 11, and 12 suggests an alternate explanation. In isorhodopsin, for example, the C-10 hydrogen should experience large torsional oscillations in isorhodopsin as part of the photochemical conversion of this chromophore to the all-*trans* configuration. One may therefore predict that the C-10 carbon atom may be responsible for the 960-cm^{-1} isorhodopsin mode. However, a 10-deutero substitution does not cause a major alteration in the 960-cm^{-1} band but substitutions at the 11- and 12-carbon atoms eliminate this band.[117] In view of these results it is not at all clear that the intense C—H bends in visual pigments are solely a result of the large nuclear oscillations of the atoms participating in a photochemical isomerization.

An alternative hypothesis is that the hydrogens at atoms 10, 11, and 12 are interacting strongly with the protein environment because of the dipolar excited state of the chromophore. The interaction could then affect different atoms as the structure of the chromophore is altered from 11-*cis* to 9-*cis*, resulting, as observed,[117] in different carbon atoms contributing to the C-10, C-11, and C-12 isotope effect in the two configurations. Furthermore, protein-free 11-*cis* and 9-*cis* protonated Schiff bases that undergo photochemically driven isomerization do not show these effects, which only occur when the chromophores are complexed to the protein. Thus

[115] N. C. Craig and J. Overend, *J. Chem. Phys.* **51**, 1127 (1969).
[116] R. Mathies, T. B. Freedman, and L. Stryer, *J. Mol. Biol.* **109**, 367 (1977).
[117] G. Eyring, B. Curry, R. Mathies, A. Brock, and J. Lugtenburg, *J. Am. Chem. Soc.* **102**, 5390 (1980).

this plausible interpretation of the intensity of the C—H bend in rhodopsin and isorhodopsin must be taken as support for the hypothesis of a strong interaction between the protein and the excited state charge distribution in the chromophore. Such an interaction is an essential element of the postulated alteration of the protein by the excited state of the chromophore which was suggested by Lewis[43] and is discussed in Section IV,A,1.

H. The Region below 950 cm^{-1}

1. *Very-Low-Frequency Vibrations.* Cookingham et al.[3] were able to detect very-low-frequency vibrations down to a few wavenumbers from the exciting line in model compounds dissolved in acetonitrile. These frequencies are potentially sensitive monitors of structural changes but are obscured in H_2O and thus are difficult to observe in the natural system. The greatest potential use of these observations is to refine calculations on the potential energy surfaces of these chromophores. In fact, the calculations of Warshel and Karplus[118] accurately predicted many of the vibrational modes observed in the Cookingham et al.[3] study.

2. *700–950 cm^{-1}.* The investigation of Oseroff and Callender[23] set the stage for future experiments in this region of the spectrum with the important observation of intense low frequency motions which have been observed in all bathorhodopsin species studies to date.[7,37] These vibrational modes do not occur in the spectrum of rhodopsin, isorhodopsin, or all-*trans* protonated Schiff bases; thus, as Oseroff and Callender[23] correctly pointed out, this is a challenge to the structural chemist to find a structure that is not in one of the preceding conformations but can relax to all-*trans* and can be inconverted by light to the rhodopsin and isorhodopsin states.

There have been basically three plausible explanations for these vibrational modes. First, Lewis[43] proposed that, in addition to all-*trans,* there was another possible retinal structure that fit the preceding criteria and could produce such low-frequency vibrations. This structure could be generated from either the 11-*cis* rhodopsin or the 9-*cis* isorhodopsin chromophores by a *similar* motion, a simultaneous rotation of carbon atoms 10 and 11 out of plane to cause a chromophore twisted around the C-9–C-10 and 11–12 double bonds. Second, Honig et al.[46] have suggested that the double bonds achieve a planar configuration and that the low-frequency modes originate from single-bond twists throughout the isoprenoid chain. Third, Warshel[119] has suggested that ring distortions could give rise to such low-frequency modes.

The experiments of Lewis et al.[120] and Eyring et al.[9] on 5 desmethyl bathorhodopsin have definitely eliminated any participation of the ring in

[118] A. Warshel and M. Karplus, *J. Am. Chem. Soc.* **96**, 5677 (1974).

[119] A. Warshel, *Annu. Rev. Biophys. Bioeng.* **6**, 273 (1977).

[120] A. Lewis, A. Hochberg, M. von Zastrow, and A. T. Lemley, *Fed. Proc., Fed. Am. Soc. Exp. Biol.* **39**, 1847 (1980).

the bathorhodopsin low-frequency modes. Furthermore, Eyring et al.[9] used selective deuterium substitution to assign these lines definitely to the hydrogens on carbon atoms 10, 11, and 12, which are precisely the atoms that Lewis[43] predicted are involved in these vibrational modes based on molecular model building with space-filling models. Nonetheless, Erying et al.[9] have opted for the single-bond twist mechanism of Honig et al.,[46] although, they admit that both single- and double-bond twists in the 10, 11, and 12 carbon region could account for these intense low-frequency vibrations. Eyring et al.[9] base their single-bond twist conclusion not on their elegant low-frequency assignments but on the fingerprint region of bathorhodopsin, which they claim is very close to that of an all-trans protonated Schiff base. Therefore, they conclude that the double bonds must be trans in the batho chromophore. However, as we have pointed out in Section IV,E,2,a and Fig. 12 there is little or no correspondence between the spectrum of bathorhodopsin and the all-trans protonated Schiff base. The one correspondence that Eyring et al.[9] propose, of a very weak 1166-cm^{-1} band, is detected only after subtraction procedures to remove large backgrounds and major components. In addition, as noted (Section IV.E.2a), this band is not detected by a second study by Aton et al.[33] using similar substraction procedures. Thus such data cannot be used to suggest that the double bonds are trans in bathorhodopsin, especially in view of the data shown in Fig. 12, which clearly show little correspondence and do not depend on any subtraction.

   3. Interpretation of the Low-Frequency Bathorhodopsin Modes: The Double- versus Single-Bond Twist Hypothesis. The preceding clearly points out that there is a major region of agreement between Lewis et al.[7,43] and Eyring et al.[9] The low-frequency bathorhodopsin modes originate from carbon atoms 10, 11, and 12 of a twisted isoprenoid chain that is altered from 9-cis and 11-cis. However, in view of the preceding, what can an open-minded review of the data reveal about the nature of the twist—double bond or single bond?

   The best case for a single-bond twist mechanism is the intuition that has developed on retinals. This intuition clearly says that double-bond twists are very costly in terms of energy, whereas single-bond twists can and do occur fairly readily with little energy loss. In terms of the Raman data, however, one has to ask if there are previous experimental examples where single-bond or double-bond twists have caused drastic alterations in the low-frequency and fingerprint regions.

   There are three examples of alterations and possible alterations in single- and double-bond twisting of retinal chromophores that have been studied using resonance Raman spectroscopy. The Raman investigations on these three cases come up with fairly negative views as to the ability of single-bond twists to cause alterations in retinal chromophore low-frequency and fingerprint modes.

The first study that bears on this question is the spectra of 11-*cis* retinal as a function of temperature. In this retinal isomer there are two possible conformers 11-*cis*-12-s-*cis* and 11-*cis*-12-s-*trans*. The difference between these conformers is the twist around the 12–13 single bond. As the temperature is lowered the 12-s-*trans* conformer begins to predominate and only two vibrational modes in the entire spectrum change their intensity.[3] These vibrational modes occur at 1271 cm^{-1} and 1345 cm^{-1}.[3] There are *no changes* in the fingerprint or the low-frequency region. Therefore, this investigation does not lend support to the single-bond twist hypothesis.

The second set of results that reflects on this question includes the data obtained on 5-desmethylretinal[3] and 5-desmethyl rhodopsin.[9,120] These data are important because, as a result of the presence of the 5-desmethyl-group, the β-ionone ring is twisted 59° out of plane around the 6–7 single bond. Thus when the 5-desmethyl group is removed, the twist around the 6–7 bond should be relaxed. The effect on the spectra is clear; there are no effects on the low-frequency modes in any of the isomers studied by Cookingham *et al.*[3] and only minor intensity changes in the 5-desmethyl rhodopsin, isorhodopsin, bathorhodopsin mixtures.[9,120] Similarly, in the Cookingham *et al.*[3] study, the fingerprint regions of all 5-desmethyl isomers studied were identical to the parent (natural) isomer in both intensity and frequency except for the 13-*cis* isomer where all vibrational modes are identical except for the addition of an 1145-cm^{-1} mode in the 5-desmethyl molecule. The presence of this mode in this isomer is most interesting, but in view of all the other data on the preceding modification it must again be concluded that alteration in single-bond twists do not cause the characteristic features present in the spectrum of bathorhodopsin.

Finally, the third study that pertains to this question is the comparison of the resonance Raman spectra of 11-*cis* retinal in the solution[3,25] and crystalline state.[3,121] The comparison is important because the 11-*cis* crystalline state is the only other retinal system that has low-frequency vibrations of comparable intensity to bathorhodopsin. Fortunately, a crystal structure is available on this system and this crystal structure demonstrates that between carbon atoms 9 and 12 the only rotations out of plane are at the 9–10 and 11–12 double bonds and that the 10–11 single bond is completely planar.[112] The location of these out-of-plane double-bond twists are strongly reminiscent of similar twists in the structure proposed by Lewis[43]; thus, based on the available data, one is forced to conclude that twists around the 9–10 and 11–12 double bonds are the only way that the low-frequency bathorhodopsin spectrum can be reconciled with avail-

[121] R. E. Cookingham, A. Lewis, D. W. Collins, and M. A. Marcus, *J. Am. Chem. Soc.* **98,** 2759 (1976).

able Raman data. To test this hypothesis further it is certainly important to study model systems with controlled twists around both these double and single bonds. This will be a tremendous experimental achievement in terms of the difficult syntheses involved. However, the results will probably be the only definitive way of testing the single-bond versus double-bond twist hypotheses and of determining the degree of twist from the all-*trans* planar configuration.

V. Future Prospects

The future prospects for applying resonance Raman spectroscopy to rhodopsin and bacteriorhodopsin are extremely bright. It is almost certain that significant attempts will be made to obtain the resonance Raman spectrum of the chromophore in the rhodopsin and bacteriorhodopsin excited states. Already the resonance Raman spectrum of the triplet state of retinal has been reported[5,125] and a 30-psec study of bovine rhodopsin has recently appeared.[122] The 30-psec investigation was too slow either to investigate the excited state species or to probe the questions of ground-state protein and retinal structural alterations in bathorhodopsin that are produced in < 6 psec[45] but ring dye lasers with significantly better time resolution and pulse to pulse repeatability are now available in several laboratories and this should open new avenues for kinetic resonance Raman spectroscopy. In addition, little use has been made of coherent anti-Stokes Raman spectroscopy (CARS) in studying rhodopsin and bacteriorhodopsin.[78,79] This technique should be applicable to many problems such as the detection of the K species in bacteriorhodopsin that have been intractable becasue of high fluorescent backgrounds. Furthermore, UV resonance CARS and resonance Raman spectroscopy of protein amino acids will begin to take on new and important positions in rhodopsin and bacteriorhodopsin investigations. Finally, applications to cone systems and the bacteriorhodopsin Na+ pump[123] will soon take place and single-cell data will be recorded as resonance Raman spectrometers are coupled to microscopes. These data, along with recent applications of the reso-

[122] G. Hayward, W. Carlsen, A. Siegman, and L. Stryer, *Science* **211**, 942 (1981).
[123] R. MacDonald, V. R. Green, D. R. Clark, and E. Lindley, *J. Biol. Chem.* **254**, 11831 (1979).
[124] J. J. Johnson, A. Lewis, and G. Gogel, *Biochem. Biophys. Res. Commun.* **103**, 182 (1981).
[125] G. H. Atkinson, J. B. Phallix, T. B. Freedman, D. A. Gilmore, and R. Wilbrandt, *J. Am. Chem. Soc.* **103**, 5069 (1981).

nance Raman spectrum of carotenoids to detect bacteriorhodopsin induced membrane potentials,[124] should lead this powerful technique into
new vistas in the study of rhodopsin and BR-mediated cellular processes.

Acknowledgements

The author acknowledges support from National Eye Institute, the Army, National
Aeronautics and Space Administration and Naval Air Systems Command at various times
during the course of the investigations reported. A. L. was a Guggenheim Fellow when this
article was written.

[73] Time-Resolved Chromophore Resonance Raman and Protein Fluorescence of the Intermediates of the Proton Pump Photocycle of Bacteriorhodopsin*

By M. A. EL-SAYED

Different resonance Raman techniques are described in this paper that
are useful in studying intermediates of photolabile systems in the millisecond, microsecond, nanosecond, and picosecond time domains. The summary of the results of applying these techniques to study the retinal system in bacteriorhodopsin (BR) is given and discussed in terms of what is
known about its photochemical proton pump cycle. The main results are
as follows: (1) the largest retinal configurational changes, involving twisting and mixing of the σ and π electronic motion, occur in the first step (the
absorption step) and (2) the $-\overset{+}{C}=NH$ proton ionizes in 40 μsec in the
$BL_{550} \to BM_{412}$ step.

Two lasers, one acting as a photolysis source and the other exciting
the protein fluorescence, are pulsed with variable time delays, and are
thus used to study the level of the protein fluorescence intensity of the
different intermediates formed at different delay times after the photolysis
pulse in the BR photocycle. The results suggest that during the first step,
when the retinal suffers the largest configurational changes, the protein
fluorescence intensity does not change. However, as the first intermediate
decays into the subsequent ones, a decrease in the protein fluorescence is
observed until the M intermediate is formed, after which a recovery of the
fluorescence intensity is observed. The interplay between the retinal
changes, the protein fluorescence intensity changes, and the proton pump

* Modified versions of this paper were submitted to two different invited publications: (a)
The Proceedings of the Sergio Porto Laser Memorial Meeting, Rio de Janeiro, Brazil,
June 29–July 3, 1980; (b) The Proceedings of the VIIIth International Congress on Photo-
biology, Strasbourg, France, July 21–25, 1980.

process is qualitatively discussed in the light of the theoretical framework present in the literature.

Introduction

Although picosecond lasers have been used previously to obtain the resonance Raman spectra of stable molecules[1] (in particular, to reduce interference from fluorescence radiation), the first resonance Raman spectra of transients formed in the picosecond time domain have just been reported this year by our group[2] and others.[3] Of course, optical absorption and emission spectroscopy of picosecond transients has been useful during the past decade in determining the number, rise time, and decay time of picosecond intermediates in photolabile systems. However, because of the broad nature of the optical absorption of the system studied, these spectra did not yield the kind of structural information one would like to have in order to identify the exact structural changes taking place in these processes. Vibration spectra, as obtained from the Raman scattering process, are expected to give more structural information. For this reason, as well as the development of cavity dumped picosecond lasers and the imaging detection systems, time-resolved Raman spectroscopy is expected to be an active field of research over the coming decade or two.

In this chapter, we discuss the different techniques developed in our laboratory to obtain the resonance Raman spectra of transients of photolabile systems appearing in the millisecond, microsecond, nanosecond, and picosecond time domain. In Section II we summarize the results of our studying the bacteriorhodopsin system. In Section III we briefly describe a system used to record the protein fluorescence intensity of the intermediates of a photolabile system and summarize the results on the bacteriorhodopsin system. In the last paragraph we combine the protein fluorescence results with the retinal resonance Raman results to give a possible unified picture that may describe the proton pumping process in the primary process of the photosynthetic system of bacteriorhodopsin.

[1] M. Bridoux, *C. R. Hebd. Seances Acad. Sci.* **258,** 620 (1964); M. Bridoux, A. Chapput, M. Crunelle, and M. Delhaye, *Adv. Raman Spectrosc.* **1,** 65 (1973); M. Delhaye, *Proc. Int. Conf. Raman Spectrosc., 5th, 1976* p. 747 (1976); M. Bridoux, A. Deffontaine, and C. Reiss, *C. R. Hebd. Seances Acad. Sci.* **282,** 771 (1976); M. Bridoux and M. Delhaye, *Adv. Infrared Raman Spectrosc.* **2,** 140 (1976); P. P. Yaney, *J. Opt. Soc. Am.* **62,** 1297 (1972); R. P. Van Duyne, D. L. Jeanmaire, and D. F. Shriver, *Anal. Chem.* **46,** 213 (1974); F. E. Lyttle and M. S. Kelsey, *ibid.* p. 855; M. Nicol, J. Wiget, and C. K. Wu, *Proc. Int. Conf. Raman Spectrosc., 5th, 1976* p. 504 (1976).

[2] J. Terner, T. G. Spiro, M. Nagumo, M. F. Nicol, and M. A. El-Sayed, *J. Am. Chem. Soc.* **102,** 3238 (1980).

[3] M. Coppey, H. Tourbez, P. Valat, and B. Alpert, *Nature (London)* **284,** 568 (1980).

I. Time-Resolved Resonance Raman (TRRR) Techniques[4]

A. The Millisecond, Microsecond, and Nanosecond Transients*

Different time-resolved resonance Raman techniques have been developed[4] for determining the resonance Raman spectra of the intermediates formed from photolabile systems, e.g., the proton pump system of bacteriorhodopsin[5-8] and CO hemoglobin.[9,10] The method used varied and depends on the time scale in which the photointermediate builds up in concentration. All the methods that we have used have the following features in common:

1. One laser is used that acts as both the photolytic and the Raman probe light source. This is especially applicable to systems of broad absorption bands where one expects an overlap of the absorption band of the photolabile parent compound and that for the intermediate whose resonance Raman is being examined.

2. The experiment is carried out in a time scale appropriate for the rise and decay times of the intermediate being studied.

3. The laser used is adjusted at a wavelength that gives high photolytic probability and large Raman enhancement for the intermediate examined. Furthermore, the scattered Raman radiation should have minimum overlap with any fluorescence present.

4. For obtaining the spectrum of a certain intermediate, chemical or physical perturbations are used, if possible, to maximize its concentration.

5. Satisfying the preceding conditions, two Raman spectra are then detected and recorded by using the optical multichannel analyzer; one spectrum is collected at very low powers (to obtain the spectrum of the unphotolyzed parent compound at minimum photolysis), and the second spectrum is obtained at high powers (to maximize the concentration of the photoproduct).

*Written with James Terner, Alan Campion, Chung-Lu Hsieh, and Alan Burns.
[4] For a previous review, see Time-Resolved Resonance Raman Spectroscopy in Photochemistry and Photobiology. *ACS Symp. Series.* **102**, 215 (1979).
[5] A. Campion, J. Terner, and M. A. El-Sayed, *Nature (London)* **265**, 659 (1977).
[6] A. Campion, M. A. El-Sayed, and J. Terner, *Biophys. J.* **20**, 369 (1977).
[7] J. Terner, A. Campion, and M. A. El-Sayed, *Proc. Natl. Acad. Sci. U.S.A.* **74**, 5212 (1977).
[8] J. Terner, C.-L. Hsieh, A. R. Burns, and M. A. El-Sayed, *Proc. Natl. Acad. Sci. U.S.A.* **76**, 3046 (1979).
[9] W. H. Woodruff and S. Farquharson, *Science* **201**, 831 (1978).
[10] K. B. Lyons, J. M. Friedman, and P. A. Fleury, *Nature (London)* **275**, 565 (1978); R. F. Dallinger, J. R. Nestor, and T. G. Spiro, *J. Am. Chem. Soc.* **100**, 6251 (1978).

6. Computer subtraction techniques are then used to subtract the low-power spectrum from the high-power spectrum to obtain the Raman spectrum of the intermediate having maximum concentration in the time scale of the experiment used and having maximum enhancement at the wavelength of the laser used.

To satisfy condition 2, i.e., to adjust the time scale of the experiment to maximize the concentration of a certain intermediate, one of the following techniques is used:

i. Pulsed Lasers[5,9,10]. Only intermediates appearing in a time equal to or shorter than the pulse width can be detected by Raman spectroscopy if the wavelength is adjusted for maximum enhancement. Intermediates with rise times shorter, and decay times longer, than a few nanoseconds can be studied[5,9,10] by using the N_2-pumped (e.g., Molectron) or Nd-pumped (e.g., Quanta-Ray) dye lasers, which have few nanoseconds pulse width. Flashlamp-pumped dye lasers (e.g., Chromatix) could be used for intermediates with decay times in the microsecond time scale. In principle, picosecond intermediates could be detected by using the high-power pulsed picosecond lasers. However, the low-duty cycle of these lasers, as well as destruction from multiphoton processes, could hamper the observation of good S/N signals in these experiments.

ii. Modulation of cw Lasers[6,7]. Electric or mechanical modulation of cw lasers can produce pulses with different pulse width and at a given modulation frequency. Electric modulation could give short pulses with high-duty cycles. Mechanical modulation could give longer pulses with lower-duty cycles. We have used mechanical choppers (rotating disks) fitted with variable-size slits. Thus a cw laser could function as a pulsed laser with variable pulse width. The fact that the laser beam can be brought to a small focus in the micron range makes a slit width of a few microns usable in these experiments. With the available practical motors usable in this experiment, intermediates in the $10-200$ μsec time scale have been detected.[6] Of course, with slow motors and large slit width, millisecond intermediates can be easily observed. Two slits, one for the photolysis and one for the probe laser, can also be used[7] with variable time delays (i.e., separation between the slits). The duty cycle in these experiments is determined by the number of slits in the rotating disk as well as the motor speed. In general, they are much better than in the pulsed laser experiment.

iii. Flow Techniques[4,8]. Instead of pulsing the laser, the sample can be "pulsed" by flowing it across a focused cw laser beam. The flow technique was first used by Mathies *et al.*[11] to determine the Raman spectrum

[11] R. Mathies, T. B. Freedman, and L. Stryer, *J. Mol. Biol.* **109**, 367 (1977).

of the unphotolyzed rhodopsin. Marcus and Lewis[12] were the first to use it for kinetic studies in bacteriorhodopsin by varying the flow rate of the sample (which could be changed by a factor of 10) to obtain different time scales for different intermediates. We have extended the time scale of this method by realizing[4,8] that, for the same flow rate, the time scale of the experiment can be varied by changing the laser focus itself. By using a microscope objective and a flow rate of 10–40 m/sec, the experimental time resolution (determined by the time it takes the flowing sample to cross the focused laser beam) could be in the 50-nanosec time scale! More importantly, the scattered Raman relation is being collected continuously in this experiment.

B. Resonance Raman of Picosecond Intermediates[2*]

By replacing the cw laser in experiment A, iii with a mode-locked-cavity dumped argon or krypton ion-pumped dye laser, the resonance Raman spectrum of intermediates in the picosecond time scale could be recorded. In this case the time resolution of the experiment is no longer determined by the sample residence time in the laser beam as in A, iii, but rather by the pulse width of the picosecond laser used. This is true only if the time between the laser pulses (~ 1 μsec) is longer than the residence time of the sample in the beam (~ 0.1 μsec).

II. Resonance Raman Results on the Proton Pump System of Bacteriorhodopsin[6–8,13*]

Bacteriorhodopsin absorbs visible light ($\lambda_{max} = 570$ nm) and passes through a number of intermediates before the BR returns to its initial form[14]: $BR_{570} \xrightarrow{h\nu(psec)} BK_{590(610)} \xrightarrow{2\,\mu sec} BL_{550} \xrightarrow{40\,\mu sec} BM_{412} \xrightarrow{msec}$ $BO_{640} \rightsquigarrow BR_{570}$. As a result of this cycle, (one or two) protons are pumped out of the cell, thus creating proton gradients across the BR cell membrane.[15] It is this electric free energy that is believed to be used in the synthesis of the high-energy molecules (ATP).

We have attempted to use the TRRR technique to examine two problems concerning the preceding photocycle. The first one is concerned with the retinal conformation changes. Based on chemical reconstitution

*Written with J. Terner, T. G. Spiro, M. Nagumo, and M. F. Nicol.
*Written with James Terner, Chung-Lu Hsieh, and Alan Burns.
[12] M. A. Marcus and A. Lewis, *Science* **195**, 1328 (1977).
[13] M. A. El-Sayed and J. Terner, *J. Photochem. Photobiol.* **30**, 125 (1979).
[14] R. H. Lozier, R. A. Bogomolni, and W. Stoeckenius, *Biophys. J.* **15**, 955 (1975).
[15] D. Oesterhelt, *Angew Chem., Int. Ed. Engl.* **15**, 17 (1976).

studies,[16] it is believed that the retinal in BR_{570} is all-*trans*-retinal. In rhodopsin, resonance Raman studies showed that[11] the retinal is in the 11-*cis* configuration and it has been the common belief that the absorption process leads to isomerization to the all-*trans* form in the first step. This change in the retinal configuration leads to changes in the retinal–protein interaction energy as well as entropy. It is the change in the free energy upon the absorption that leads to the storage of the free energy necessary to drive the system through the latter process. The question then arises: If the retinal in the BR_{570} already contains an all-*trans*-retinal, what is the mechanism of converting solar energy into free energy in the absorption process of the BR system? Could it be that even if retinal in BR_{570} is in the all-*trans* configuration, further configuration changes take place in the first step of the cycle? (It should be pointed out that the all-*trans*-retinal *inside the protein* might not necessarily have the lowest value of the free energy of the system.) Another possibility is that when the all-*trans*-retinal combines with the bacteriorhodopsin to form the retinal–protein complex, it does not necessarily retain its all-*trans* configuration. In support of this is the difficulty in comparing the fingerprint region of the retinal in BR_{570} with that of all-*trans* model compounds,[13] unlike the rhodopsin system when good agreement between the 11-*cis* and the rhodopsin system is obtained.[11] In any case one would like to investigate whether or not the first step in the cycle involves a change in the retinal configuration. By using TRRR one can examine the fingerprint region (1000–1400 cm^{-1}) (sensitive to retinal configuration changes) for the $BK_{590(610)}$ intermediate and compare it to that for BR_{570}. The $BK_{590(610)}$ intermediate is formed in the picosecond time domain; however, it lasts a few microseconds. Thus the TRRR experiment can be carried out in the 0.1–1-μsec time scale.

The second problem concerns the origin of the protons pumped during the photochemical cycle. The retinal in BR_{570} is bonded to the protein via

$$a-\overset{|}{\underset{\underset{H^+}{|}}{C}}=N-\text{group.}$$

Although not definitely proved, it has been assumed[15] that one of the protons pumped out in the $BR_{570} \rightarrow BM_{412}$ transformation

comes from this group. We have followed the $-\overset{|}{\underset{\underset{H^+}{|}}{C}}=N-$ vibration (at

~ 1640 cm^{-1}) by TRRR techniques to determine the step at which this

band disappears (due to the formation of the $-\overset{|}{C}=N-$ group).

[16] M. J. Pettei, A. P. Yudd, K. Nakanishi, R. Henselman, and W. Stoeckenius, *Biochemistry* **16**, 1955 (1977).

The important results in the fingerprint region (which is sensitive to the retinal configuration) and the C=N stretching region (to follow the deprotonation of the Schiff base) can be summarized as follows:

1. Similar to rhodopsin,[17] large changes in the fingerprint region are observed during the first $BR_{570} \xrightarrow{h\nu} BK_{590}$ transformation.

2. The Raman bands in the frequency range just below 1000 cm^{-1} seem to be enhanced in a number of intermediates and *in particular for the K intermediate*. Furthermore, the position and relative intensity of these bands seem to be sensitive to changing H_2O to D_2O. This region might be assigned to out-of-plane (o-o-p) C—H vibrations[18] and their enhancement might suggest a distortion of the planar configuration of the polyene electronic system as proposed for bathorhodopsin.[18] The mixing between the σ and the π electrons upon distortion allows an enhancement of the out-of-plane C—H vibrations since the $\sigma-\pi$ coupling strength should be sensitive to their normal coordinate. The sensitivity to D_2O could arise from the coupling between these vibrations and C—H o-o-p vibration of the $C=\overset{+}{N}\overset{\diagup H}{\underset{|}{}}$ group.

3. If the assignment made in result 2 is correct, the fact that these vibrations (960–990 cm^{-1}) occur at higher frequency than those for bathorhodopsin[17] (850–925 cm^{-1}) might result for one or both of the following reasons: (a) The vibrations in bathorhodopsin have lower frequency than normal,[18] perhaps due to the presence of the countercharge nearby.[19] Since in bacteriorhodopsin, the countercharge is believed[20] to be near the ring, the o-o-p vibrations in K_{590} would be expected to have the normal value. (b) The enhanced o-o-p vibrations in the twisted intermediates of BR could involve different bonds (e.g., C_{14}—H and C_{15}—H) since it is a result of isomerization from all-*trans* to probably[19] 13-*cis* (contrasting the rhodopsin system, where the isomerization is from 11-*cis* to all-*trans*, these vibrations[19] are on C_{10}, C_{11}, C_{12}). Isotopic substitution[19] studies on the K intermediate should differentiate between these two possibilities.

4. Unlike in rhodopsin,[11] a complete identification of the isomeric form of the different intermediates is difficult to achieve from a comparison of our spectra with those of model compounds in solution. This could result for one or more of the following reasons: (a) Some distortion of the spectra obtained by subtraction techniques. (b) Larger perturbation of the

[17] G. Eyring and R. Mathies, *Proc. Natl. Acad. Sci. U.S.A.* **75**, 4642 (1979).

[18] G. Eyring, B. Curry, R. Mathies, R. Fransen, I. Palings, and J. Lugtenburg, *Biochemistry* **19**, 2410 (1980).

[19] Mathies, this volume [75].

[20] B. Honig, V. Dinur, K. Nakanishi, V. Balogh-Nair, M. A. Gawinowicz, M. Arnaboldi, and M. G. Motto, *J. Am. Chem. Soc.* **101**, 7084 (1979).

spectra of retinal by the retinal–opsin interaction in bacteriorhodopsin than that present in rhodopsin. (c) The small difference between the all-*trans* and 13-*cis* fingerprint spectra in solution.

5. In the $BK_{590} \rightarrow BL_{550}$ transformation, the 980-cm^{-1} region becomes normal. This might suggest a relaxation of the twisted polyene structure in L_{550} from the one that K_{590} has.

6. Small changes in the fingerprint region take place in the $BL_{550} \rightarrow BM_{412}$ process, the process accompanied by the largest change in the position of the optical absorption maximum in the cycle. It also involves the deprotonation of Schiff base nitrogen. This might suggest[13] that the theories based on ionic interactions[21] rather than retinal configurational changes[22] might be the correct ones in explaining the origin of the red shift in the retinal absorption upon combining with the opsin.

7. The CN stretching vibration is reduced in frequency during the first step ($BR_{570} \rightarrow BK_{590}$). However, it is found that it is reduced further in D_2O solvent. This might suggest that in the first step of the cycle the interaction between the Schiff base nitrogen and the proton has been reduced. This is unlike the results in the rhodopsin[17] system, in which the CN frequency in the batho form has remained similar to that in the parent compound.

8. The CN frequency in the BL_{550} form is found to be similar to the parent compound (and is thus protonated). This suggests that deprotonation of the nitrogen Schiff base takes place during the $BL_{550} \rightarrow BM_{412}$ process, i.e., in 40 μsec.

III. Time-Resolved Protein Fluorescence Results[23]*

One interesting question that immediately arises concerning the cycle in BR is: When does the protein learn about the absorption act by the retinal in the bacteriorhodopsin photocycle? By using two pulsed lasers, the first to initiate the BR cycle and the second (delayed by different times) to excite the protein fluorescence, the intensity of the fluorescence emission of the protein in the different intermediates has been monitored. The important results can be summarized as follows:

1. Although the largest configuration is observed in the first step, no change in the protein fluorescence is observed in this step.

2. The fluorescence intensity decreases during the $BK_{550} \rightarrow BM_{412}$

*Written with Joseph Fukumoto, William Hopewell, and Bela Karvaly.

[21] B. Honig, A. D. Greenburg, V. Dinur, and T. Ebrey, *Biochemistry* **15**, 4593 (1976).

[22] R. Korenstein, K. Muszkat, and S. Sharafy-Ozeri, *J. Am. Chem. Soc.* **95**, 6177 (1973).

[23] J. M. Fukumoto, W. B. Hopewell, B. Karvaly, and M. A. El-Sayed, *Proc. Natl. Acad. Sci. U.S.A.* **78**, 252 (1981).

transformation, with the BM_{412} having the lowest fluorescence intensity.

3. The quenching of the BM_{412} fluorescence could be explained by deprotonation of 1,2-tyrosine molecules or charge perturbation of 1,2-tryptophan molecules, as was recently suggested to explain the changes in the protein absorption by Hess and Kuschmitz.[24]

The Raman and fluorescence results on the BR system suggest that the absorption act ($BR_{570} \xrightarrow{h\nu} BK_{590}$) causes a change in the configuration of the retinal with some change in the nitrogen-proton binding of the Schiff base. These changes could be accompanied by changes in the energetics of the retinal–protein pocket (e.g., the electrostatic interaction between the Schiff base positive charge and the counterion(s)[25,26]). The protein then responds to this change during the $BK_{590} \rightsquigarrow BL_{550} \rightsquigarrow BM_{412}$ transformation in the microsecond region. During this transformation, two protons are probably released, the Schiff base proton is ionized, and there is a *possibility* that 1,2-tyrosine protons might also be ionized during this half of the photocycle. Whether these protons are the ones that are actually pumped out of the cell or are at least responsible for the pumping process remains to be proved.

Acknowledgments

The author wishes to thank Drs. Stoeckenius and Bogomolni for supplying some of the bacteriorhodopsin cultures. The financial support of the U.S. Department of Energy (Office of Basic Energy Sciences) is gratefully acknowledged.

[24] B. Hess and D. Kuschmitz, *FEBS Lett.* **100**, 334 (1979).
[25] A. Warshel, *Nature* (*London*) **260**, 679 (1976).
[26] B. Honig, T. Ebrey, R. H. Callender, V. Dinur, and M. Ottolenghi, *Proc. Natl. Acad. Sci. U.S.A.* **76**, 2503 (1979).

[74] Resonance Raman Techniques for Photolabile Samples: Pump–Probe and Flow

By ROBERT CALLENDER

Visual pigments and purple membranes, as well as other systems, offer an experimental problem to Raman spectroscopists. Although these materials are extremely photolabile, Raman cross sections, even in resonance-enhanced cases, are extremely small. For example, the resonance-enhanced Raman cross section of visual pigments is some eight orders of magnitude less than the absorption cross section.[1] Thus Raman measure-

[1] R. H. Callender, A. Doukas, R. Crouch, and K. Nakanishi, *Biochemistry* **15**, 1621 (1976).

METHODS IN ENZYMOLOGY, VOL. 88

ments cannot be performed fast enough before severe sample degradation occurs. Special techniques must be used to permit Raman work.

We discuss here two techniques that have had success in coping with photosensitive samples. The first can be used where the system is driven by light and perhaps also temperature to two or more species. The resulting Raman spectra can be assigned to a particular species by suitably controlling sample composition using a second irradiating laser beam. This is the "pump–probe" method. The second technique, which is useful for the broad class of photoreactive molecular systems that can be dissolved in solution, is the imposition of a sample molecular velocity transverse to the Raman exciting laser beam. In this method the flow speed is set high enough so that any given molecule has very little chance of absorbing a photon. Other techniques and an overview of Raman results are discussed elsewhere in this volume.

Pump–Probe Technique

Consider a system A that photochemically transforms to B, and vice versa, so that A $\rightleftarrows$ B. Eventually the system will reach a photostationary equilibrium composition. Thus the Raman measurement will yield a spectrum that is a composite of the two species. The intensity of a given line of one of the components will depend on the fractional sample concentration of that component as well as the Raman cross section of that mode. At equilibrium, the ratio of concentrations is given by

$$A/B = \sigma^B Y_B / \sigma^A Y_A$$

where σ is the absorption cross section of A and B, respectively, at the irradiating light frequency and Y_A is the quantum yield of A to B and vice versa for Y_B. It is easy to see that the species absorbing least will have the highest fractional population. Thus the laser frequency can be used to control the sample population by driving the sample toward one component or another.

This single laser beam procedure is useful but has complications. The laser frequency is now not in resonance with the highest concentration sample component so that the resonance-enhanced cross section is much weaker. On the other hand, it is very likely that the laser is in resonance with the second component. Thus although the concentration of, say, A is relatively high compared to B, the *reverse* is true of the relative Raman cross sections. Thus it may be quite difficult to ascribe a particular Raman band to species. Additionally, Raman bands may scale differently as a function of irradiating light frequency even for a given species. Thus, performing a series of Raman runs using different irradiating laser frequencies so as to control sample composition may still not allow identification to species of a particular Raman band.

These complications can be controlled by the use of a double-beam technique in which a relatively weak "probe" beam is used to excite the Raman scattering while an additional "pump" beam at a different wavelength applied at the sample is used to modify relative sample composition.[2] If the power level of the pump beam is sufficiently higher than that of the probe beam, the composition of the sample is determined by the pump beam wavelength solely. If the relative power levels are closer, then the sample concentration is a function of the wavelengths and power levels of the two beams. In either case the fixed probe beam wavelength holds the resonance Raman enhancement factors constant so that pump beam induced changes in the Raman spectrum are a direct representation of changes in sample composition. It is thus simply necessary to determine sample composition as a function of pump beam changes and relate this to changes in the Raman spectra. This approach can be clearly generalized to more than two sample species.

For example, this technique has been used recently to obtain the bathorhodopsin spectrum.[3,4] In one study[5] a 476.2-nm probe beam was used with (a) no pump beam (producing a relative composition of bathorhodopsin–rhodopsin–isorhodopsin of 56/29/15%), (b) a 568.2-nm pump bean with pump–probe power ratio of 25/1 (0.0/0.0/>95), and (c) a 590-nm pump beam of power ratio 22/1 (12/58/30). By suitably subtracting and adding the data sets,[3] the bathorhodopsin (Fig. 1a) and rhodopsin (Fig. 1b) are obtained from the composite spectra. It is pleasant to note that the rhodopsin spectrum obtained by this method and by the flow technique (next section) are the same even though sample photolability is controlled in vastly different ways. The key results from this work are that the Schiff bases of rhodopsin and bathorhodopsin are fully protonated, being unaffected by the rhodopsin–bathorhodopsin transformations,[3-5] and that a chromophore isomerization has occurred in this photochemical transition.[3,5]

Flow Technique

This method eliminates problems associated with sample photolability for any sample that can be dissolved in solution. Figure 2 gives a block

[2] A. R. Oseroff and R. H. Callender, *Biochemistry* **13**, 4243 (1974).
[3] B. Aton, A. G. Doukas, D. Narva, R. H. Callender, U. Dinur, and B. Honig, *Biophys. J.* **29**, 79 (1980).
[4] G. Eyring and R. Mathies, *Proc. Natl. Acad. Sci. U.S.A.* **76**, 33 (1979).
[5] D. Narva and R. H. Callender, *Photochem. Photobiol.* **32**, 273 (1980).

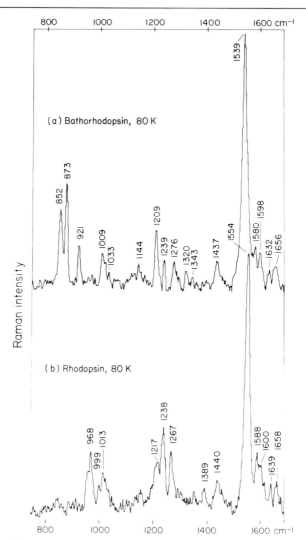

FIG. 1. Bovine bathorhodopsin and rhodopsin spectra isolated from composite spectra as described in the text. The resolution is 7.3 cm⁻¹.

diagram of the essential experimental features as first developed.[1,6] The sample is contained in a reservoir (how much is needed is discussed below) and, using a suitable pump, is made to flow transversely through the laser beam with a velocity sufficient to ensure that any given molecule

[6] R. Mathies, A. R. Oseroff, and L. Stryer, *Proc. Natl. Acad. Sci. U.S.A.* **73**, 1 (1976).

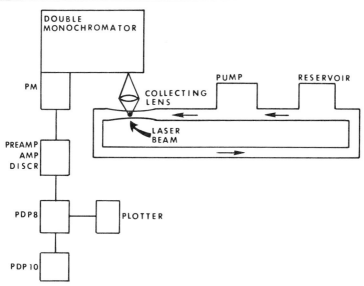

FIG. 2. A block diagram of a possible experimental arrangement using the flow technique. A pump flows sample through the laser beam (coming out of the paper) fast enough to ensure no significant building of molecules that have absorbed a photon. See Ref. 1 and 6 for details.

has a rare probability of absorbing a photon. In this way the sample in the laser beam is constantly replenished with a new sample. Although the sample in the laser beam is absorbing photons, those molecules that do absorb are not allowed to reach any appreciable percentage, being constantly removed and replenished by new sample. Since Raman scattering is an independent physical effect from absorption (a molecule can either absorb or Raman scatter an incident photon, not both at the same time), the Raman scattered photons arise from essentially pure starting material.

The numerics can be easily understood. Figure 3 shows a typical interaction area where the sample is contained in, say, a capillary tube. A typical molecule moves across a square laser beam (for simplicity) in a time $2r_0/v_b$, where r_0 is the beam radius and v_b is the molecular velocity transverse to the laser beam. While in the laser beam, the molecule is absorbing photons at a rate (photons per second) given by $I_0 \sigma Y/4r_0^2$, where I_0 is the laser flux (photons per second), σ the absorption cross section (square centimeters per molecule), and Y the quantum yield. Thus this molecule absorbs a total number of photons given by the product of these two numbers, i.e., $I_0 \sigma Y/2r_0 v_b$. The experimental parameters can be chosen so that this number is much less than 1; thus our molecule has a rare chance of absorbing a photon, which implies that nearly all the sample in the laser

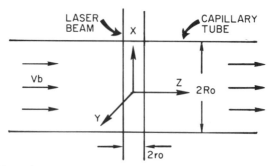

FIG. 3. A schematic representation of the flow experimental arrangement in the laser beam interaction area. The arrows point along the sample flow and the y axis points toward the spectrometer. (Reprinted with permission from Ref. 1.)

beam is the pure starting material. The effects of a nonuniform laser and flow pattern (due to turbulence) are considered in detail in Ref. 1 and 6. They have no major effect but should be considered.

In use this technique has typically been performed with a recirculating sample flow system as given in Fig. 2. The reason for this is that most of the sample does not pass through the laser beam in a single pass since the laser beam is focused and generally smaller than the capillary tube containing the sample. Thus, to minimize sample quantities, it is convenient to collect the sample and flow it through the laser beam many times. In addition, samples have been contained in both jet streams[6] and capillary tubes[1] in the laser beam interaction area. They are basically equivalent with the capillary tube arrangement, having the advantage of larger possible flow velocities (high velocity implies a high degree of turbulence, which breaks up jet stream flow) and the jet stream arrangement having the advantage of no glass walls, which may possibly give spurious Raman results.

There is one further important aspect of this technique and that is the necessary amount of starting material. Briefly, this quantity is determined by the ratio of absorption to Raman cross sections (it is clearly necessary to start with more material, as this parameter increases), the total Raman scattered photons counted (a better signal-to-noise ratio requires more sample), and apparatus efficiency. A detailed derivation can be found in Ref. 1 but it turns out that only in rare samples is an excessive amount needed. For example, using conventional Raman equipment and a noise-to-signal ratio of about 3%, it was necessary to start with about 200 ml of a 10^{-5} M solution of rhodopsin.[1] This visual pigment is an extreme example of photolability, and sample quantities of this size are typically readily obtainable.

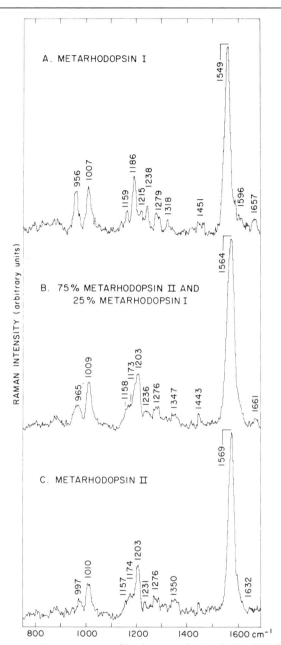

FIG. 4. Resonance Raman spectra of bovine metarhodopsin I and II. The resolution is 6.0 cm⁻¹ and sample temperature was 3°. See Ref. 9 for details. (Reprinted with permission from Ref. 9.)

Figure 4a shows the data[7] obtained by this method in our laboratory on metarhodopsin I and II. At this point in time many of the visual pigments have given independent simultaneous results, bovine rhodopsin,[1,6] isorhodopsin,[6] metarhodopsins I and II,[7] sqid rhodopsin and bathorhodopsin,[8] and membrane chromophores including the light and dark-adapted primary pigments,[9,10] and an important purple membrane photoproduct pigment M_{412}.[9] One of the most pleasing results, from the standpoint of technique, is the essentially identical Raman results on rhodopsin as taken by the pump-probe technique[5] described above (Fig. 1b) and as obtained from the flow method.[1,6]

Variations of the flow technique, essentially involving large laser fluxes and controlled fluid velocities that yield spectra from controlled quasi-photostationary state mixtures of photointermediates, are discussed elsewhere in this volume (see also Refs. 11, 12, and 13, and references therein).

The pump-probe technique and flow methods have yielded fundamental chromophore information. Very briefly, for visual pigments they have verified that the chromophore of rhodopsin is an 11-*cis* isomer,[14] that an isomerization occurs in the rhodopsin-to-bathorhodopsin transformation[3] (11-*cis* to probably distorted *trans*), that and metarhodopsin I and II contain an essentially *trans* chromophore.[7] The Schiff base is protonated until metarhodopsin II where it is deprotonated.[1–7] For the purple membrane system the primary pigment is a protonated Schiff base, whereas M_{412} is not[9] (see also Ref. 15); and there is likely a chromophore isomerization from the primary pigment to M_{412}.[9] See the original sources and this volume for more details and discussion.

The primary photochemical event, the formation of the batho chromophore from the primary pigment, thus appears to involve an isomerization, *cis* to *trans* for visual pigments and *trans* to *cis* for purple membrane systems, with no change in protonation of Schiff base.[3,5] These con-

[7] A. G. Doukas, B. Aton, R. H. Callender, and T. Ebrey, *Biochemistry* **17**, 2430 (1978).

[8] M. Sulkes, A. Lewis, and A. Marcus, *Biochemistry* **17**, 4712 (1978).

[9] B. Aton, A. G. Doukas, R. H. Callender, B. Becher, and T. G. Ebrey, *Biochemistry* **16**, 2995 (1977).

[10] B. Aton, A. G. Doukas, R. H. Callender, B. Becher, and T. G. Ebrey, *Biochim. Biophys. Acta* **576**, 424 (1979).

[11] M. Stockburger, W. Klusmann, H. Gattermann, G. Massig, and R. Peters, *Biochemistry* **18**, 4886 (1979).

[12] M. A. Marcus and A. Lewis, *Biochemistry* **17**, 4722 (1978).

[13] J. Terner, C.-L. Hsieh, A. R. Burns, and M. A. El-Sayed, *Proc. Natl. Acad. Sci. U.S.A.* **76**, 3046 (1979).

[14] R. Mathies, T. B. Freedman, and L. Stryer, *J. Mol. Biol.* **109**, 367 (1977).

[15] A. Lewis, J. Spoonhower, R. A. Bogomolni, R. H. Lozier, and W. Stoeckenius, *Proc. Natl. Acad. Sci. U.S.A.* **71**, 4462 (1974).

straints provided by the resonance Raman data must be taken into account in any model. We have recently proposed such a model[16] which makes use of the likelihood that a negatively charged amino group forms a salt bridge with the positively charged nitrogen of the retinyllic chromophore. The isomerization in either pigment system would cleave the salt bridge and thus separate charge in the interior of the protein. Photon energy conversion to chemical energy needed to drive subsequent thermal steps and the spectra red shift of the batho-photoproduct are natural consequences of the separation of charge.

Acknowledgments

This work was supported in part by the National Science Foundation (PCM 79-02683 and PCM82-02840), by the National Institutes of Health (EY03142), and by a City University PSC-BHE faculty award.

[16] B. Honig, T. Ebrey, R. H. Callender, U. Dinur, and M. Ottolenghi, *Proc. Natl. Acad. Sci. U.S.A.* **76,** 2503 (1979)

[75] Resonance Raman Spectroscopy of Rhodopsin and Bacteriorhodopsin Isotopic Analogs

By RICHARD MATHIES

Resonance Raman (RR) spectroscopy is a powerful technique for studying the photochemistry of retinal (I) in rhodopsin and bacteriorhodopsin because Raman scattering from the retinal chromophore in a photointermediate can be *selectively* enhanced, and because Raman spectra provide very precise vibrational information about the structure of the chromophore. Comprehensive reviews are available on the application of Raman to many biological systems,[1-5] including visual pigments and purple membrane.[6-9] Since the first RR spectrum of a visual pigment was ob-

[1] H. E. VanWart and H. A. Scheraga, this series, Vol. 49, p. 67.
[2] P. R. Carey, *Q. Rev. Biophys.* **II** (3), 309 (1978).
[3] T. G. Spiro and P. Stein, *Annu. Rev. Phys. Chem.* **28,** 501 (1977).
[4] T. G. Spiro and B. B. Gaber, *Annu. Rev. Biochem.* **46,** 553 (1977).
[5] N.-T. Yu, *CRC Crit. Rev. Biochem.* **4,** 229 (1977).
[6] R. Mathies, *in* "Chemical and Biochemical Applications of Lasers" (C. Bradley Moore, ed.), Vol. 4, p. 55. Academic Press, New York, 1979.
[7] R. Callender and B. Honig, *Annu. Rev. Biophys. Bioeng.* **6,** 33 (1977).
[8] A. Warshel, *Annu. Rev. Biophys. Bioeng.* **6,** 273 (1977).
[9] A. Lewis, *in* "Advances in Infrared and Raman Spectroscopy (R. J. H. Clark and R. E. Hester, eds.), Vol. 10, Heyden, London, 1982.

tained in 1970,[10] most work has involved the development of techniques to obtain well-characterized spectra of rhodopsin, bacteriorhodopsin, and their photointermediates. Low-temperature,[11] rapid-flow[12,13] and a variety of time-resolved techniques[14-16] are now available. With suitable combinations or modifications of these techniques, it is possible to obtain RR spectra of nearly all the photointermediates in these systems. However, the analysis of these Raman data has not progressed as rapidly.

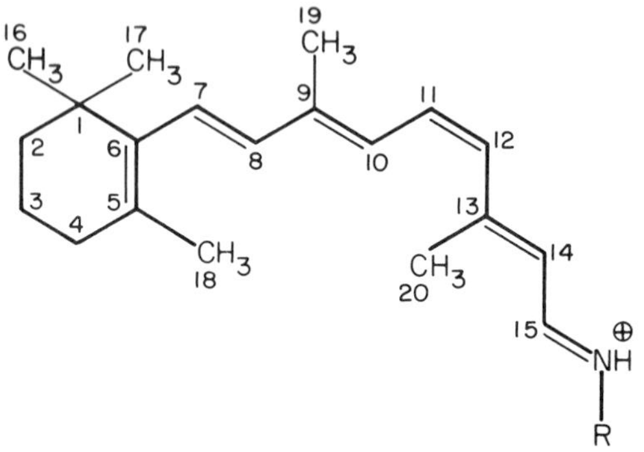

Scheme I

To transform the vibrational data in a RR spectrum into precise structural information, it is necessary to perform a normal mode analysis of the retinal chromophore *in the protein*. The vibrational analysis of small molecules has traditionally been accomplished through the use of specific isotopic derivatives. Isotopic substitution is an ideal modification because, unlike chemical modification, it introduces no ambiguities due to changes in chromophore structure or chromophore–protein interactions. We are now employing this technique to interpret the RR spectra of retinal pigments. This approach is difficult because the size of retinal (49 atoms) requires, in general, a large number of isotopic modifications, and because the modified retinal must be incorporated in the apoprotein for Raman analysis. Since only small amounts (~ 1 mg) of modified retinals are typically available, regeneration procedures must be efficient, purification

[10] L. Rimai, R. G. Kilponen, and D. Gill, *Biochem. Biophys. Res. Commun.* **41**, 492 (1970).
[11] A. R. Oseroff and R. H. Callender, *Biochemistry* **13**, 4243 (1974).
[12] R. Mathies, A. R. Oseroff, and L. Stryer, *Proc. Natl. Acad. Sci. U.S.A.* **73**, 1 (1976).
[13] R. H. Callender, A. Doukas, R. Crouch, and K. Nakanishi, *Biochemistry* **15**, 1621 (1976).
[14] M. A. Marcus and A. Lewis, *Biochemistry* **17**, 4722 (1978).
[15] J. Terner, C.-L. Hsieh, A. R. Burns, and M. A. El-Sayed, *Biochemistry* **18**, 3629 (1979), and references cited therein.
[16] M. Stockburger, W. Klusmann, H. Gattermann, G. Massig, and R. Peters, *Biochemistry* **18**, 4886 (1979).

should produce samples with low fluorescence and Raman backgrounds, and detection of the scattering should be as efficient as possible. The purpose of this article is to describe procedures for studying rhodopsin and bacteriorhodopsin analogs and to give rationales for the selection of isotopic derivatives and for the interpretation of these Raman results.

Rhodopsin Regeneration. Disk membranes, isolated from 100 frozen bovine retinas following Hong and Hubbell,[17] are suspended in 25 ml of pH 7 buffer containing 100 mM phosphate, 10 mM hydroxylamine, and 5 mM 2-mercaptoethanol. When assayed spectrophotometrically in Ammonyx-LO (see below), this suspension should have a net optical density of ~ 1.7 cm^{-1} at 500 nm, corresponding to a concentration of ~ 40 μM (using $\epsilon = 42,700$[17]). The suspension is bleached with ambient light until no absorbance remains at 500 nm. The bleached membranes are then pelleted by spinning the suspension at 25,000 g for 20 min. The pellet is resuspended twice in 25-ml aliquots of 100 mM phosphate buffer containing 5 mM 2-mercaptoethanol but no hydroxylamine and centrifuged at 25,000 g for 20 min each time. The final pellet is suspended in a third 25-ml aliquot of the 100 mM phosphate buffer containing no hydroxylamine. To this suspension a ~ 2-fold molar excess (600 μg) of the retinal analog is added in no more than 0.5 ml ethanol. The vortexed suspension is allowed to regenerate at room temperature for 90 min. The regeneration yield is assayed by comparing the absorbance of a 200-μl aliquot (diluted with 200 μl of 3% Ammonyx-LO) with an analogous sample of the native solubilized membranes. Typical yields are 70% based on opsin. A 3- to 4-fold excess of retinal will give up to 100% regeneration with more inefficient utilization of the retinal derivative. The regenerated membranes are centrifuged at 25,000 g for 20 min and the pellet is dissolved in 5–10 ml of dissolving buffer (3% Ammonyx-LO, 10 mM imidazole, 5 mM 2-mercaptoethanol) by stirring for 20 min at 4° with a magnetic stirrer.

Column Purification. The hydroxylapatite column procedure described below is a variation of the methods of Hong and Hubbell[17] and Applebury *et al.*[18] The hydroxylapatite (9 gm, DNA grade Bio-Gel HTP) is suspended in 60 ml start buffer (1% Ammonyx-LO, 10 mM imidazole, 5 mM 2-mercaptoethanol, pH 7) and swirled gently. After settling for ~ 10 min the supernatant is carefully decanted and the procedure is repeated twice. Finally the hydroxylapatite is suspended in 50 ml of buffer, swirled, and poured into the 1.5-cm-diameter column containing 20 ml of start buffer. The column is allowed to settle for 2–3 hr before flowing. The final column height should be ~ 10 cm. Prior to loading the sample, a

[17] K. Hong and W. L. Hubbell, *Biochemistry* **12,** 4517 (1973).
[18] M. L. Applebury, D. M. Zuckerman, A. A. Lamola, and T. M. Jovin, *Biochemistry* **13,** 3448 (1974).

volume of buffer equivalent to twice the column bed volume is flowed through the column. Typical flow rates are ~0.3 ml/min. To load the sample, the buffer level is permitted to drop to the hydroxylapatite. The dissolved pigment solution is carefully added to the column and allowed to flow completely into the column. About 10 ml of start buffer are placed on the column bed and a linear gradient of phosphate buffer eluent is initiated [low-phosphate buffer (100 ml): 25 mM phosphate, 1% Ammonyx-LO, 5 mM 2-mercaptoethanol, pH 7; high-phosphate buffer (100 ml): 150 mM phosphate, 1% Ammonyx-LO, 5 mM 2-mercaptoethanol]. Fractions are collected at 15-min intervals with ~4 ml per fraction. Excess retinal from the regeneration appears in fractions 5–15, while the pigment is eluted in fractions 25–35. Fractions with 280/500 ratios ≤1.85 (normally those with an optical density >0.4 at 500 nm) are pooled. Typical yield of purified rhodopsin is 20 ml with an absorbance of 0.9 cm^{-1} or about 40 retinas.

Raman Spectroscopy. The pooled pigment sample can be used directly to obtain room temperature Raman spectra of analog rhodopsins and isorhodopsins. The theory of rapid-flow resonance Raman spectroscopy and the experimental conditions and apparatus have been thoroughly described.[12,13,19] Because only small amounts of retinal analogs and hence visual pigment analogs are available, very efficient detection of the Raman scattering is necessary for the essentially destructive room temperature, rapid-flow experiments. With conventional single-channel photon-counting apparatus,[12] these experiments on pigment analogs would produce very poor signal-to-noise ratios. We have used an image intensified vidicon detector that provides a 50-fold improvement in the efficiency of these Raman experiments.[20] Optical multichannel analyzer (OMA) detection has proven to be very advantageous in resonance Raman experiments on visual pigments[21] and on purple membrane.[15,22] For our visual pigment analog experiments,[21] the 20-ml sample was recirculated with a stainless steel–Teflon gear pump (Micropump 40-33) through a 1-mm-diameter capillary at a velocity of 400 cm/sec. The laser was focused with a 32-mm lens and good spectra were obtained in 5 min with 200–400 μW of laser power at 488 nm. Typically, two to four separate 5-min vidicon integrations were averaged to improve the signal-to-noise ratio.

For low-temperature photostationary steady-state experiments[11] it was necessary to concentrate the purified pigment.[23] We used Schleicher

[19] See also, R. H. Callender, Article [74], this volume.
[20] R. Mathies and N.-T. Yu, *J. Raman Spectrosc.* **7,** 349 (1978).
[21] G. Eyring, B. Curry, R. Mathies, R. Fransen, I. Palings, and J. Lugtenburg, *Biochemistry* **19,** 2410 (1980).
[22] M. Braiman and R. Mathies, *Biochemistry* **19,** 5421 (1980).
[23] G. Eyring and R. Mathies, *Proc. Natl. Acad. Sci. U.S.A.* **76,** 33 (1979).

and Schuell collodion vacuum dialysis bags (no. 100, molecular weight cutoff 25,000). Concentration in a 4° cold room with a small dry-ice trapped vacuum pump or a water aspirator required ~16 hr. The final sample volume was ~1 ml with an optical density of ~18 at 500 nm. An aliquot of this sample was frozen on a 77°K cold finger for low-temperature photostationary steady-state experiments. When only one sample of a pigment analog could be prepared, the more concentrated low-temperature samples were prepared first. The remaining pigment was then rediluted for rapid-flow experiments. A detailed description of the low-temperature spectroscopy has been given by Oseroff and Callender[11] and by Eyring and Mathies.[23]

Purple Membrane Purification. Purple membrane was prepared from *Halobacterium halobium* using essentially the procedure of Oesterhelt and Stoeckenius.[24] The carotenoid-deficient strain R1-S9 was used to minimize carotenoid contributions to the resonance Raman spectra. Bacteria were grown in 4-liter flasks each holding 2 liters of growth medium containing 500 g NaCl, 40 g $MgSO_4 \cdot 7H_2O$, 4 g KCl, 6 g Na_3citrate$\cdot 2H_2O$, 0.2 g $CaCl_2$, trace amounts of $MnCl_2$ (10^{-6} M) and $FeCl_2$ (2×10^{-5} M), and 20 g of peptone (Inolex 49-111-03, Inolex Corp., Glenwood, Ill.). Flasks were maintained at 35–40° on a rotary shaker for 7–10 days and illuminated by several 150-W flood lamps. At this point the flasks should be noticeably purple if the culture has not been contaminated by a carotenoid-containing mutant. Cells were harvested by centrifugation and lysed by dialysis against 0.1 M NaCl for 12 hr with DNAse added to reduce viscosity. The purple portion of the lysate pellet was washed with distilled water, homogenized in a Thomas C-799 tissue grinder, and repelleted at 25,000 g (90 min, 4°). This wash cycle was repeated three to four times until no red carotenoid color could be detected in the supernatant.

At this stage a modification of the usual procedure provided efficient removal of the remaining carotenoids.[22] Two more differential centrifugations were performed, employing sonication rather than a tissue grinder to disrupt the membranes. Membranes from two flasks were resuspended in 20 ml of basal salt solution (growth medium without peptone) and sonicated at 4° using 50-W for 30 sec (Heat Systems-Ultrasonics W-220F). The membranes were pelleted and the sonication–washing procedure repeated once more. The resulting purple membrane fragments were stored as a pellet from basal salt solution at 4°. This sonication–differential centrifugation procedure removed carotenoid contamination[25] as completely

---

[24] D. Oesterhelt and W. Stoeckenius, this series, Vol. 31, p. 667.

[25] Carotenoid contamination was assayed by observing the intensities of the 1150 and 1510 cm^{-1} carotenoid Raman bands in the 514.5-nm room temperature steady-state purple membrane spectrum.

and consistently as centrifugation on a sucrose density gradient.[24,26] Therefore, density gradient purification was unnecessary for preparation of samples for Raman spectroscopy. Sonicated light-adapted samples, prepared as above, exhibited a 280/568 absorbance ratio of 1.8–2.0. These preparations gave a single band when subjected to SDS gel electrophoresis or density gradient ultracentrifugation.

Bleaching and Regeneration of Purple Membrane. The bleaching and regeneration procedure of Oesterhelt and Schuhmann[27] was modified to produce samples more suitable for Raman spectroscopy. Purple membrane was suspended in buffer (0.025 M HEPES, 1.0 M hydroxylamine, pH 7) at a concentration of ~1 mg/ml (4×10^{-5} M). Light from a 500-W quartz–halogen projector lamp was passed through a flowing water filter containing heat-absorbing glass and focused through a yellow filter (Corning 3-70) onto the magnetically stirred sample, maintained at 25–28°. The rate of bleaching was strongly dependent on temperature: Above 35° it required only a few hours, but at 4° several days are necessary. Our filters passed light from 500–800 nm and the area of the focused beam was ~1 cm² with an intensity of 1.2 W/cm². After 24 hr the bleaching was complete and the apomembrane was pelleted (25,000 g, 90 min) and resuspended in H₂O by gentle vortexing. This washing procedure was repeated three more times. Typically 30–40% of the apomembrane was lost in the supernatant after four washings.

For regeneration of purple membrane analogs, the apomembrane was suspended in 10 ml pH 7 HEPES buffer at a concentration of ~2 mg/ml. A 1.5 to 2-fold excess of the 13-*cis* or all-*trans* retinal analog was added in 0.1 ml of ethanol. Over 90% of the apomembrane regenerates in the first 5 min. If regeneration is carried out at 4° in the dark, it is possible to prepare either 13-*cis* or all-*trans* reconstituted purple membrane for rapid-flow resonance Raman experiments.

For Raman experiments with blue excitation it is necessary to remove residual retinal and retinaloxime from the regenerated purple membrane to reduce their contributions to the pigment spectra. In studies of the M_{412} intermediate with 413 nm excitation, we have removed these contaminants with chemical extraction or photobleaching[22]: (a) Extraction of excess retinals with organic solvent was performed according to Tokunaga and Ebrey.[28] Water-washed purple membrane was lyophilized and the pellet was pulverized in petroleum ether and vigorously mixed. After centrifugation and removal of the supernatant, the remaining petroleum ether

[26] B. M. Becher and J. Y. Cassim, *Prep. Biochem.* **5,** 161 (1975).
[27] D. Oesterhelt and L. Schuhmann, *FEBS Lett.* **44,** 262 (1974).
[28] F. Tokunaga and T. G. Ebrey, *Biochemistry* **17,** 1915 (1978).

was evaporated and the purple membrane resuspended in H_2O. Several extractions with petroleum ether may be necessary to completely remove the retinal or retinaloxime. (b) Alternatively, the retinyl contaminants may be photobleached by illuminating the regenerated sample at 4° with intense white light (preceding apparatus with no yellow filter). The bacteriorhodopsin chromophore is much less susceptible to photodestruction than free retinals.

If the preceding procedures are repeated until no detectable retinal or retinaloxime absorbance at ~360 nm is observed, then even with 413-nm excitation, M_{412} Raman spectra with virtually no retinyl contamination will be observed. As assayed by Raman spectroscopy, bacteriorhodopsin is not significantly perturbed by these purifications. Raman spectra of the M_{412} and the light- and dark-adapted forms of purple membrane regenerated with unmodified retinal are identical to those of native purple membrane.

Rationale for Isotopic Modifications. Two different types of isotopically modified retinal pigment experiments have been performed. First, for vibrations that are highly localized within the retinal chromophore, specific isotopic substitutions can be used to assign Raman bands to the motions of these atoms. When coupled with normal mode analyses, these assignments provide detailed information about the structure of the retinal chromophore. Second, even if motions of the isotopically substituted atom(s) contribute to several different vibrational modes, the analog spectra provide an "isotopic fingerprint" for the *cis-trans* configuration of the chromophore. The configuration of the chromophore is usually assigned in these Raman experiments by comparing the 1100–1400 cm^{-1} "fingerprint" region of the pigment spectrum with the fingerprint of model retinals of known configuration.[29] In some cases, an unambiguous assignment of pigment configuration is not possible with this method because the model compound spectra are not sufficiently distinctive. By comparing analog pigment and analog model compound spectra, an additional dimension of fingerprint information is provided that can facilitate configurational assignment. Examples of these two experimental approaches are given below to illustrate their utility. The isotopic derivatives we employed were synthesized by Prof. Johan Lugtenburg and co-workers.[21,30,31]

[29] R. Mathies, T. B. Freedman, and L. Stryer, *J. Mol. Biol.* **109**, 367 (1977).
[30] A. D. Broek and J. Lugtenburg, *Recl. Trav. Chim. Pays-Bas* **99**, 363 (1980); M. R. Fransen, I. Palings, J. Lugtenburg, P. A. A. Jansen, and G. W. T. Groenendijk, *Recl. Trav. Chim. Pays-Bas* **99**, 384 (1980).
[31] G. Eyring, B. Curry, R. Mathies, A. Broek, and J. Lugtenburg, *J. Am. Chem. Soc.* **102**, 5390 (1980).

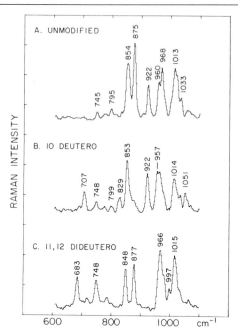

FIG. 1. Resonance Raman spectra of 77°K steady-state mixtures of native visual pigments and pigments regenerated with isotopically modified retinals are used to assign the 854, 875 and 922-cm^{-1} bathorhodopsin lines to hydrogen out-of-plane (HOOP) wags. Comparison of the unmodified (A), 10-monodeuterio (B), and 11,12-dideuterio spectra (C) demonstrate that the 875-cm^{-1} line is due to the $C_{10}H$ HOOP and the 922-cm^{-1} line is due to a combination of the $C_{11}H$ and $C_{12}H$ HOOPs. (From Eyring *et al.*[31].)

Assignment of Visual Pigment Vibrations. The resonance Raman spectrum of bathorhodopsin, the primary photochemical intermediate in vision, is dominated by three intense lines near 854, 875, and 922 cm^{-1} that are unique to this intermediate (Fig. 1A). Previous work with 18-trideuterio and 19-trideuteriobathorhodopsin analogs suggested that these lines are due to hydrogen out-of-plane (HOOP) wagging vibrations of chain vinyl hydrogens.[21] We have recently synthesized 10-monodeuterioretinal and 11,12-dideuterioretinal and regenerated the corresponding visual pigment analogs to confirm this HOOP assignment.[31] In Fig. 1B, deuteration on position 10 shifts the 875-cm^{-1} bathorhodopsin line to 707 cm^{-1}. Deuteration on the 11 and 12 positions (Fig. 1C) shifts the 922 cm^{-1} bathorhodopsin line to 683 (748) cm^{-1}. The nearly $\sqrt{2}$ deuterium-induced shifts confirm that these modes involve predominantly hydrogen motion. The 875 cm^{-1} line is due to the $C_{10}H$ hydrogen out-of-plane wag and the 922 cm^{-1} line is due to the coupled (A_u in C_{2h})

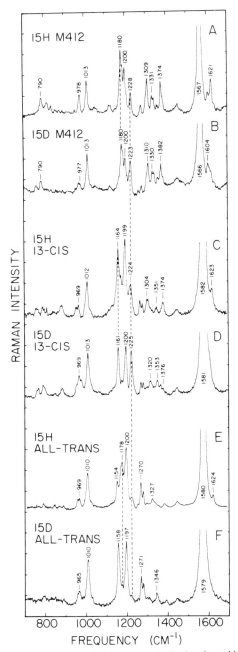

FIG. 2. Resonance Raman spectra of native bacteriorhodopsin and bacteriorhodopsin regenerated with 15-deuterioretinal (15D) are compared with the appropriate model compounds to assign the *cis-trans* configuration of retinal in the M_{412} intermediate. (A) Native 15H M_{412} dual-beam flow spectra. (B) M_{412} regenerated using 15D-retinal. (C) Spectrum of the 13-*cis*-retinal Schiff base in CCl_4 with solvent lines subtracted. (D) 13-*cis*-15D-retinal Schiff base. (E) All-*trans*-15H-retinal Schiff base. (F) All-*trans*-15D-retinal Schiff base. (From Braiman and Mathies[22].)

HC_{11}=$C_{12}H$ wag. The assignment of the unique bathorhodopsin vibra-tions to *specific,* uncoupled vibrational modes is an important step in the determination of the structure of the retinal chromophore in bathorhodop-sin. First, the 922-cm^{-1} HC_{11}=$C_{12}H$ HOOP is unusual because this fre-quency is substantially below the expected range for this vibration (960–970 cm^{-1}).[21] Calculations show that this frequency can be substantially lowered by placing a negatively charged residue near C_{11}=C_{12}. Thus these Raman experiments on bathorhodopsin provide evidence for a po-tentially catalytic charge in the retinal binding pocket. A similar perturba-tion was recently suggested by Honig *et al.*[32] to explain the absorption spectra of rhodopsin analogs. Second, the intensities of the unique bath-orhodopsin lines provide information about conformational distortion of the retinal chromophore. The observed enhancement of the 10 and the 11,-12 HOOP modes results from a relative displacement of these hydrogen equilibrium positions between the ground and excited electronic state. Because the HOOP modes are highly localized and uncoupled, their in-tensities provide very specific information about chain twists near C_{10} and C_{11}=C_{12}. The bathorhodopsin HOOP intensities provide evidence for chain single-bond twists of 10–30° at C_{10}–C_{11}, C_{12}–C_{13} and C_{14}–C_{15} in the ground state.[21,31] Thus once a vibrational mode is definitively assigned, very detailed information about charge and steric protein–retinal interac-tions can be derived from the subsequent vibrational analysis.

Configuration Assignment in Bacteriorhodopsin. The utility of the "isotopic fingerprint" method is demonstrated in studies of 15-deuterio-bacteriorhodopsin analogs. In the proton pumping photocycle, absorption of a photon by light-adapted bacteriorhodopsin (BR_{LA}) causes the pig-ment to cycle through a series of intermediates ($BR_{LA} \xrightarrow{h\nu} K \rightarrow L \rightarrow M$-412 → O → BR_{LA}) returning to BR_{LA} in ~ 10 msec.[33] BR_{LA} is known from extraction experiments to contain an all-*trans* chromophore.[34] Although several arguments have been given for a *trans* → 13-*cis* isomerization in the conversion to M-412,[35] direct measurements of the configuration of M-412 by extraction[34] and resonance Raman[14,16,36] have not unambi-

[32] B. Honig, U. Dinur, K. Nakanishi, V. Balogh-Nair, M. A. Gawinowicz, M. Arnaboldi, and M. G. Motto, *J. Am. Chem. Soc.* **101,** 7084 (1979).

[33] R. H. Lozier, R. A. Bogomolni, and W. Stoeckenius, *Biophys. J.* **15,** 955 (1975).

[34] M. J. Pettei, A. P. Yudd, K. Nakanishi, R. Henselman, and W. Stoeckenius, *Biochemis-try* **16,** 1955 (1977).

[35] J. B. Hurley, B. Becher, and T. G. Ebrey, *Nature (London)* **272,** 87 (1978); B. Honig, T. Ebrey, R. H. Callender, U. Dinur, and M. Ottolenghi, *Proc. Natl. Acad. Sci. U.S.A.* **76,** 2503 (1979).

[36] J. Terner, A. Campion, and M. A. El-Sayed, *Proc. Natl. Acad. Sci. U.S.A.* **74,** 5212 (1977); B. Aton, A. G. Doukas, R. H. Callender, B. Becher, and T. G. Ebrey, *Biochemis-try* **16,** 2995 (1977).

guously supported this conclusion. Because Raman provides an *in situ* probe of the retinal configuration, it is preferable to chemical extraction. However, the interpretation of the M_{412} Raman spectrum has been inconclusive because the 13-*cis* and all-*trans* Schiff base spectra alone are not sufficiently distinctive. Braiman and Mathies[22] have recently resolved this problem by obtaining resonance Raman spectra of 15-deuterio (15D) M_{412} analogs. 15D and 15H M-412 spectra, taken using a dual-beam rapid-flow technique, are compared in Fig. 2A,B. A distinctive 2-fold increase in scattering at ~ 1225 cm^{-1} is observed in 15D M_{412}. The pattern of deuterium-induced changes in the 13-*cis* Schiff base spectra (Fig. 2C,D) is very similar to that observed upon deuteration of M_{412}. A line at 1225 cm^{-1} in 15H 13-*cis* nearly doubles in intensity upon deuteration. In Fig. 2E,F, spectra of the 15H and 15D all-*trans* Schiff bases are compared. Deuteration causes a dramatically different pattern of changes. The line at ~ 1225 cm^{-1} in 15H all-*trans* *drops* in intensity and the 1178-cm^{-1} line shifts to 1158 cm^{-1}. The unique pattern of isotopically induced changes in these spectra demonstrates that the chromophore in M_{412} is 13-*cis*. Together with the fact that BR_{LA} contains an all-*trans* chromophore,[22,34] these data confirm that *trans* → *cis* isomerization is a key event in proton pumping. Further vibrational analysis should provide even more precise information about the structure of the retinal chromophore in bacteriorhodopsin. However, it is not necessary to wait until all the chromophore–protein interactions are accurately modeled before it is possible to specify the *cis-trans* configuration of the chromophore in bacteriorhodopsin's photointermediates.

Acknowledgments

This research was performed by G. Eyring, M. Braiman and B. Curry in collaboration with A. Broek and Prof. J. Lugtenburg at the University of Leiden, The Netherlands. Cultures of *H. halobium* were obtained from Prof. W. Stoeckenius. This work was supported by the National Institutes of Health (EY 02051) and the Alfred P. Sloan Foundation.

[76] Kinetic Resonance Raman Spectroscopy of Purple Membrane Using Rotating Sample

By Pramod V. Argade and Kenneth J. Rothschild

In order to characterize the spectral intermediates in the photocycle of bacteriorhodopsin (BR) from purple membrane (PM) of *Halobacterium halobium*, it is necessary to record resonance Raman spectra (RRS) of the pure spectral species. A rotating cell has proved particularly useful in these studies. Kiefer and Bernstein[1] originally suggested use of such a cell and it was used by Stockburger et al.[2] to record kinetic RRS of PM. Here at Boston University we have extensively used the rotating cell to record kinetic RRS of PM.[3,4] We give here a design of the cell that is simple and has proved effective in our studies.

The experimental arrangement is shown in Fig. 1. The cell is rotated with frequency f about the axis. A laser beam with waist d is incident a distance r from the axis of rotation of the cell. In this case the time t spent by PM fragment in the beam is

$$t = d/2\pi fr$$

and the time between successive exposures T is

$$T = 1/f$$

In our case the cell is rotated at 40 rev/sec, $d = 40\ \mu m$, and $r = 2.32$ cm. Then

$$t = 7\ \mu sec$$

and

$$T = 25\ msec$$

Thus T is about five times the time for completion of the BR photocycle. This time is sufficient for BR exposed to a laser to complete the photocycle and return back to the native BR_{570} state before the next exposure. In this case the Raman spectrometer records the signal from BR_{570} as well as the intermediates formed in the 7 μsec after the exposure to the laser*. In order to record a spectrum of BR_{570} species, it is necessary to keep the contribution of the other intermediates low. This can be achieved by choosing a suitable exciting laser line to resonance enhance BR_{570} and

---

* A magnified (5×) image of the scattered laser is focused on the entrance slit (s) of the double monochromator. If the slit-width, s, is narrower than 5d, the spectrometer records intermediates formed in time given by

$$t_e = \frac{d}{2\pi fr} \frac{(5d/2 + s/2)}{5d}$$

[1] W. Kiefer and H. J. Bernstein, Appl. Spectrosc. 25, 500 (1971).
[2] M. Stockburger, W. Klusmann, H. Gattermann, G. Massig, and R. Peters, Biochemistry 18, 4886 (1979).
[3] P. V. Argade, K. J. Rothschild, A. H. Kawamoto, J. Herzfeld, and W. C. Herlihy, Proc. Natl. Acad. Sci. U.S.A. 78, 1643 (1981).
[4] P. V. Argade Ph.D Dissertation, Boston University (1982).

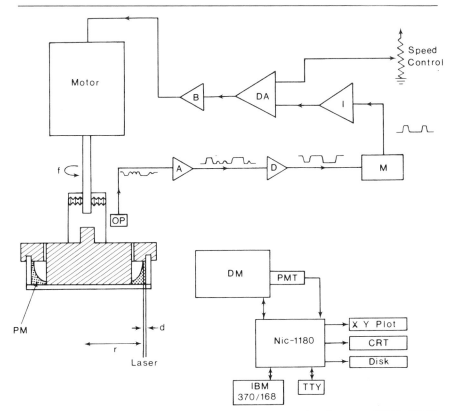

FIG. 1. The experimental arrangement using a rotating sample cell to record RRS of PM is shown. The cell is rotated with frequency f. A laser beam with waist d is incident a distance r from the axis of rotation of the cell. Light scattered at 90° is incident on the entrance slit of the double monochromator (DM). A block diagram of the photoelectric feedback system used to regulate the speed of rotation of the cell is also shown. The voltage pulses from the optical pickup (OP) are amplified (A) and then passed through a discriminator (D). They are then passed through a monostable circuit (M) and resulting pulses are integrated (I). The output voltage is compared (DA) to the reference voltage adjusted through the speed control. The resulting voltage is applied to the dc motor (Spex) through a buffer (B). The action of some of the circuit elements is shown by waveforms. A block diagram of our data acquisition system is also shown. The arrows indicate the flow of control/data.

also by decreasing the power of the incident laser (~ 5 mW), thereby decreasing the photoreaction rate constant[2] and hence the photoalteration parameter F.[5] Using this approach, we have recorded spectra where the contribution of BR_{570} exceeds 95%.

The rotating cell can also be used to record spectra of intermediates other than BR_{570}. By using a defocused "pump" beam at 514.5 nm and a coaxial but focused "probe" beam at 457.9 nm, Stockburger et al.[2] recorded a spectrum of M_{412}. In general, because of closely overlapping absorption bands of various intermediates and similar lifetimes of intermediates, it is difficult to obtain spectra of the other intermediates. We have combined the method of curve fitting and difference spectroscopy to obtain a spectrum of L_{550}.[4]

Figure 2 shows various parts of the cell. For a 90° scattering geometry, the Raman cell needs two transparent surfaces, one for the incident laser (E) and one for the scattered radiation (D). A Plexiglas piece (C) is machined such that it has a groove in which a Pyrex glass cylinder (D) can fit. A flat glass disk (E) (Oriel) can then rest touching both the lower face of D and bottom face of C. There are four symmetric holes (B) in C through which the sample can be inserted or removed using a syringe with a nee-

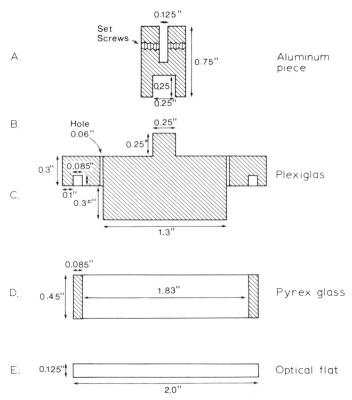

FIG. 2. Various parts of the rotating cell are shown. The Plexiglas piece (C) has a circular groove in which the Pyrex cylindrical piece (D) fits. A flat glass disk (E) is glued to C and D. The holes in C (B) are used to insert or remove the sample. The aluminum cylindrical piece (A) is glued to C and it has a hole in which the shaft of the motor can be fitted using set screws.

dle. The pyrex cylinder (D) is glued into the groove of C. The flat glass piece (E) is then glued to C and D. A cylindrical aluminum piece (A) is made such that it fits on the top of C and is glued there. This piece has a hole in which the shaft of the motor (Spex) can be fitted using set screws. Extra precaution must be taken while gluing all the pieces, since a small misfit will make the cell wobble and it may crack when it is rotated.

In order to regulate the speed of rotation of the cell, a power supply consisting of a photoelectric feedback system was built (Fig. 1). The aluminum piece (E) is covered with a black tape except for a small rectangular slot. The optical pickup (OP) consists of a light-emitting diode and a phototransistor. When the slot passes across the pickup, the light from LED is reflected by the exposed aluminum on the phototransistor, which then produces a voltage pulse. The pulses are amplified (A) and are passed through a discriminator (D). The accepted pulses are passed through a monostable circuit (M) and then integrated (I). The resulting voltage is compared to the reference voltage (determined by the speed control) through a differential amplifier (DA). The control output then goes to the motor through a buffer (B). This power supply regulates the speed of rotation of the motor within 3% error.

When the cell is rotated, all the sample is spun to the wall owing to centrifugal force. For this reason it is not necessary to fill the cell com-

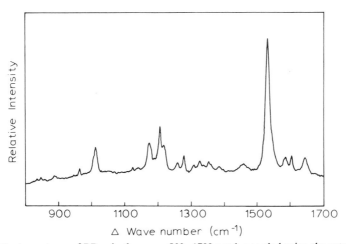

FIG. 3. A spectrum of BR_{570} in the range 800–1700 cm^{-1} recorded using the rotating cell as shown. The cell was rotated at 40 rev/sec. Thus $t = 7$ μsec (since the entrance slit of the monochromator was 100 μm, $t_e = 5.25$ μsec), and $T = 25$ msec. The spectrum was averaged for 10 scans. The resolution was 2 cm^{-1}. The exciting laser was at 514.5 nm, incident power ~5 mW (photoreaction rate constant $I_0 = 4.15 \times 10^4$ sec^{-1} and photoalteration parameter $F = 0.22$).

pletely with PM. About 1 ml sample forms a thickness of about 2 mm near the wall when the cell is rotated. The laser is focused through this portion of the sample and as close as possible to the wall in order to avoid self-absorption.

The rotating cell enables recording a spectrum of BR_{570} using only 1 ml of sample, as opposed to the flow method,[5,6] which requires a more than 100-ml sample. Rotating the sample also minimizes the heating effect encountered with a fixed sample.

In Fig. 3 we show a spectrum of BR_{570} recorded using the rotating cell discussed here. The contribution of BR_{570} to the spectrum as determined by the band at 1530 cm is more than 95%.

Acknowledgments

We thank Mr. James Taaffe for building the power supply for the rotating cell. This work was supported by grants from the NIH and NFF. K.J.R. is an established investigator of the AHA.

[5] R. Mathies, A. R. Oseroff, and L. Stryer, *Proc. Natl. Acad. Sci. U.S.A.* **73**, 1 (1976).
[6] R. H. Callender, A. Doukas, R. Crouch, and K. Nakanishi, *Biochemistry* **15**, 1621 (1976).

[77] Spinning Sample Raman Spectroscopy at 77°K: Bacteriorhodopsin's Primary Photoproduct

By Mark Braiman *and* Richard Mathies

Resonance Raman spectroscopy has recently begun to find wide application as an *in situ* structural probe of biological chromophores.[1] For example, resonance Raman spectra of bacteriorhodopsin have helped to determine the isomeric configuration of the retinal chromophore,[2] the state of protonation of its Schiff base linkage to the protein,[3] and the conformational distortion it undergoes as a result of chromophore–protein interactions.[4] An understanding of these structural features in each of the bacteriorhodopsin photointermediates is crucial for explaining the mechanism of light-induced proton pumping by purple membrane. The structure of bacteriorhodopsin's primary photoproduct, K, is of special importance because it is the earliest detected intermediate that has

[1] R. Mathies, in "Chemical and Biochemical Applications of Lasers" (C. B. Moore, ed.), Vol. 4, p. 55. Academic Press, New York, 1979.
[2] M. Braiman and R. Mathies, *Biochemistry* **19**, 5421 (1980).
[3] A. Lewis, J. Spoonhower, R. A. Bogomolni, R. H. Lozier, and W. Stoeckenius, *Proc. Natl. Acad. Sci. U.S.A.* **71**, 4462 (1974).
[4] M. Braiman and R. Mathies, *Proc. Natl. Acad. Sci. U.S.A.* **79**, 403 (1982).

trapped the energy of the absorbed photon. We have developed a novel liquid N_2–temperature spinning sample Raman cell that has enabled us to obtain excellent spectra of K. This apparatus should also be generally useful for observing Raman spectra of other photosensitive molecules at low temperatures.

Our technique begins with the pump-and-probe method used earlier to obtain the Raman spectrum of bathorhodopsin, the primary photoproduct in visual excitation.[5,6] For bacteriorhodopsin as well as rhodopsin, cooling to 77°K blocks the thermal decay of the primary photoproduct, so that illumination causes only the rapid interconversion of the parent molecule (BR) with its primary photoproduct (K). The composition of BR and K in this steady-state mixture is a function of the excitation wavelength. Under red illumination ($\lambda \geq 620$ nm), almost no K is present, because K ($\lambda_{max} = 627$ nm) strongly absorbs red light and rapidly undergoes photolysis back to BR. On the other hand, by using green (514-nm) illumination, it is possible to generate up to ~30% K in the photostationary steady state. However, green illumination gives a poor resonance enhancement of the K Raman lines, since it lies significantly to the blue of K's absorption band. An obvious solution is to use two coincident laser beams: a 514-nm "pump" beam to convert BR in the sample to K and a coincident 676-nm "probe" beam to give a favorable resonance enhancement for the K that is produced. The advantages of such a red probe–green pump configuration have been demonstrated in analogous 77°K Raman experiments on bathorhodopsin.[6] However, in the case of bacteriorhodopsin, we have not been able to observe the K spectrum using the coincident red probe– green pump technique because bacteriorhodopsin, unlike rhodopsin, has significant intrinsic fluorescence at 77°K that falls in the same wavelength region as the desired red probe Raman spectrum.

Our solution to this problem has been to *spin* the sample and to illuminate it *successively* with spatially separated pump and probe beams (see Fig. 1). The BR sample is converted to K each time it passes through the 514-nm pump beam. The K that is produced then spins to the red probe beam, where its Raman spectrum can be observed, unobscured by the pump beam fluorescence. Elimination of the pump beam fluorescence is the major reason for spinning the sample in this experiment.

Besides its usefulness for this dual-beam pump–probe Raman technique, our spinning cell has the additional advantage that it allows us to overcome thermal or photochemical degradation[7] of the sample caused by the high laser intensities needed to obtain satisfactory signal levels. For

[5] A. R. Oseroff and R. H. Callender, *Biochemistry* **13**, 4243 (1974).
[6] G. Eyring and R. Mathies, *Proc. Natl. Acad. Sci. U.S.A.* **76**, 33 (1979).
[7] T. Gillbro, A. N. Kriebel, and U. P. Wild, *FEBS Lett.* **78**, 57 (1977).

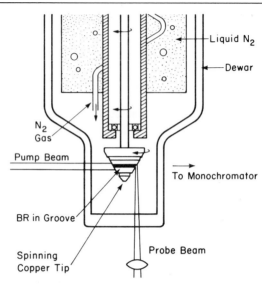

FIG. 1. Schematic representation of spinning sample apparatus, showing illumination geometry of pump and probe beams (from Braiman and Mathies[4]).

example, typical laser powers in a bacteriorhodopsin or rhodopsin experiment result in the absorption of $\geq 10^6$ photons per molecule per second. Our Raman sample spinner greatly mitigates this problem, since it allows us constantly to replace the sample in the illuminated volume. A variety of Raman flow cells and sample spinners designed to do the same thing at higher temperatures have been described.[8-12] The spinning sample apparatus we have designed now permits Raman experiments down to 77°K to be performed using ~1000 times less input energy per molecule than the corresponding stationary experiments.

Raman Sample Spinning Apparatus. The 77°K Raman sample spinner and dewar is depicted, roughly to scale, in Fig. 2. A dc motor is connected to a conical copper tip inside a glass dewar with an unsilvered pyrex tail. The sample is cooled by a stream of dry N_2 gas, which flows through a copper coil in the liquid N_2-filled canister. We run the shaft through a narrow core in the center of the canister, because direct contact with liquid

[8] R. Mathies, A. R. Oseroff, and L. Stryer, *Proc. Natl. Acad. Sci. U.S.A.* **73,** 1 (1976).
[9] M. Stockburger, W. Klusmann, H. Gattermann, G. Massig, and R. Peters, *Biochemistry* **18,** 4886 (1979).
[10] See Articles [72–78], this volume.
[11] W. H. Woodruff and T. G. Spiro, *Appl. Spectrosc.* **28,** 74 (1974).
[12] W. Kiefer and H. J. Bernstein, *Appl. Spectrosc.* **25,** 500 (1971).

nitrogen would require a mechanical seal on the shaft to prevent leaks. A thermocouple mounted next to the copper tip monitors the temperature during experiments and shows that it can easily be maintained near 77°K over the course of a 15-min Raman scan, with a moderate flow (several liters/min) of precooled N_2 gas.

We have found that to obtain satisfactory Raman spectra, it is necessary to ensure that the sample spins concentrically; otherwise, the Raman scattering is poorly imaged on the monochromator slit. To this end, we have used a stiff $\frac{1}{4}$-in.-diameter stainless steel shaft that spins on two tightly held bearings. The lubricant in the bearings is washed out and replaced with a very light coat of spray silicone lubricant (Dayton Demkote 2X988A). To further guarantee a concentric surface, the final machining of the copper tip is accomplished with it mounted on the shaft, using our sample spinner as a miniature lathe. To reduce vibration the shaft is connected to the motor via a flexible coupling made of surgical tubing. Before and during use, the assembled apparatus is thoroughly purged with dry N_2 gas. Positive N_2 pressure maintained inside the dewar and the Plexiglas hood protects the bearings from condensation and subsequent corrosion.

Sample Preparation. Best results are achieved by using a tightly packed purple membrane pellet centrifuged (30,000 g, 1 hr) from a basal salt solution. The sample is spread around a groove (1 mm deep $\times$ 2 mm wide $\times$ 1.5-cm diameter) cut into the conical copper tip until the surface is flat and even. The sample is light-adapted at 20° by illuminating the slowly rotating copper tip ($\sim$1 rev sec^{-1}) with the 514-nm pump beam ($\sim$5 min, 1 W cm^{-2}). The sample is then kept in complete darkness for 1 min prior to and during cooling. Until the sample is frozen, the N_2 purge is kept at a minimum to prevent dehydration of the sample, and the spinning speed is kept low enough to prevent displacing the sample from the groove.

Once the sample is spinning at 77°K, the probe beam is focused to a $\sim$75-μm spot in the center of the sample groove. The pump beam is directed onto the sample at a different point, generally without focusing, in order to ensure uniform photolysis of the sample (see Fig. 1). The details of the laser illumination, monochromator control, and detection of Raman scattering have been described previously.[2,4]

Choice of Excitation Wavelengths. The relative concentrations of BR and K in the 77°K photostationary steady state are predicted to be given by

$$\frac{K}{BR} = \frac{\phi_{BR \to K} \, \epsilon_{BR}(\lambda)}{\phi_{K \to BR} \, \epsilon_{K}(\lambda)} \tag{1}$$

The quantity on the right is just the ratio of the quantum yields (ϕ) for the

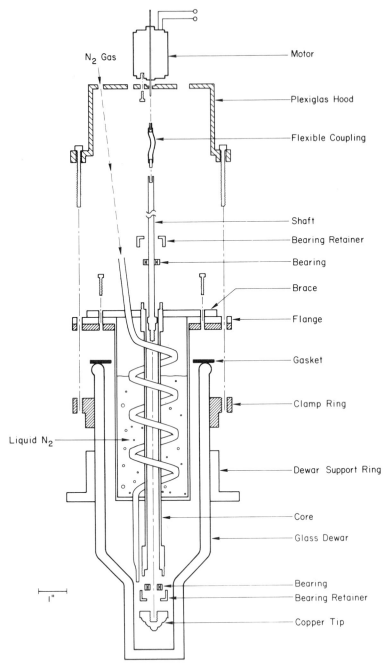

Motor

N₂ Gas

Plexiglas Hood

Flexible Coupling

Shaft

Bearing Retainer

Bearing

Brace

Flange

Gasket

Clamp Ring

Liquid N₂

Dewar Support Ring

Core

Glass Dewar

Bearing

Bearing Retainer

Copper Tip

1"

FIG. 2. Raman sample spinner and dewar. The motor is a 12 V dc, 1700 rpm model train engine. The Plexiglas hood has a hinged window (not shown) to permit access to the flexible coupling. The shaft is a ¼-in. stainless steel rod, which is threaded into the copper tip. The stainless steel core with bearing holders built into both ends acts as a rigid bracket for the shaft. The bearings (Delco R4) are fixed in their holders by brass bearing retainers. The bottom end of the core is welded to a stainless steel canister, which has a double-layer flange

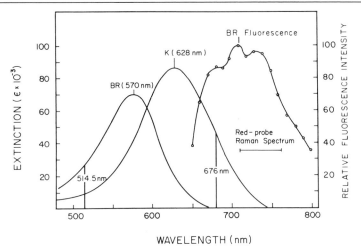

WAVELENGTH (nm)

FIG. 3. Choice of pump and probe wavelengths is dictated by the BR and K absorption spectra. The 514-nm pump beam is preferentially absorbed by BR, and will produce up to ~30% K in the photosteady state. The 676-nm probe beam lies inside K's absorption band but outside BR's, and thus generates a favorable resonance enhancement of K over BR. However, Raman scattering from the 676-nm probe beam lies in the same wavelength region as the BR fluorescence excited by the pump beam, which must therefore be excluded from the detector by our spinning technique. (Spectra adapted from Hurley and Ebrey[13] and Govindjee et al.[14].)

forward and reverse photoreactions times the ratio of the extinction coefficients (ϵ) of BR and K at the chosen wavelength (λ) of illumination. One would expect to produce the most K by illuminating at a wavelength for which ϵ_{BR}/ϵ_K is a maximum. Figure 3 shows that this is expected to occur somewhere below 570 nm. Actual experimental determination of BR and K concentrations at different wavelengths has shown that the maximum K yield of 28% is achieved with 500- to 520-nm illumination.[13] The 514-nm argon ion laser line is thus an optimal choice for a pump beam.

Since there is always more BR than K in the sample, it is important to choose a Raman probe beam that will selectively enhance the K Raman spectrum. Qualitatively, the best relative enhancement for K is expected to be obtained with a probe beam that is within K's absorption band, but outside that of BR. As can be seen in Fig. 3, obvious choices for the probe wavelength are 647, 676, or 752 nm from a krypton ion laser. We have in

[13] J. B. Hurley and T. G. Ebrey, Biophys. J. 22, 49 (1978).

(stainless steel on top, linen-based phenolic below) at its upper end. The entire assembly is held together by bolting the hood and flange to the clamp ring (linen-based phenolic), which catches the upper lip of the dewar. The aluminum dewar support ring is attached with silicone rubber cement. N_2 gas, precooled to 77°K with a separate dewar, is blown through a copper coil running down through the liquid N_2-filled canister.

fact found that 676 and 752 nm give a somewhat better relative enhancement of K over BR than 647 nm.

Another important consideration when using red excitation is fluorescence. Figure 3 shows that the intrinsic BR fluorescence at 77°K[14] is near its maximum in the wavelength region where a 647- or 676-nm Raman spectrum falls. Without the spatial separation of pump and probe beams, the BR fluorescence excited by 514-nm light would swamp the photodetector, and the background level would be too high to obtain a Raman spectrum. The probe beam itself is also capable of generating BR fluorescence, especially at 647 nm where BR still has an appreciable absorbance. Even using 676-nm excitation, which is absorbed much less than 647-nm light by BR, the fluorescence background produced is ~ 10 times as intense as the strongest Raman lines. We have been able to completely circumvent the intrinsic BR fluorescence only by using 752-nm probe excitation.

Choice of Powers. For stationary samples with coincident pump and probe beams, the sample composition is constant over time and is determined only by the ratio of pump and probe powers. For our spinning sample, however, the sample composition changes as it passes successively through the pump and probe beams. We must therefore take into account the *kinetics* of the photolysis of the sample. With a few minor modifications, we can use the photoalteration parameter F, originally used to describe the effects of laser photolysis in rapid-flow resonance Raman spectra,[8] to describe the effects of the pump and probe beams on the 77°K Raman sample.

We first consider both the forward and reverse photolysis of a sample irradiated with a single laser beam of uniform intensity I (photons cm^{-2} sec^{-1}) and wavelength λ. The rate of the photolysis reaction

$$\text{BR} \underset{k_r}{\overset{k_f}{\rightleftharpoons}} \text{K}$$

is given by

$$\frac{d(\text{BR})}{dt} = -\frac{d(\text{K})}{dt} = -k_f(\text{BR}) + k_r(\text{K}) \tag{2}$$

where

$$k_f = I\epsilon_{\text{BR}}(\lambda)\phi_{\text{BR}\rightarrow\text{K}} \cdot 3.824 \times 10^{-21}$$
$$k_r = I\epsilon_{\text{K}}(\lambda)\phi_{\text{K}\rightarrow\text{BR}} \cdot 3.824 \times 10^{-21}$$

---

[14] R. Govindjee, B. Becher, and T. G. Ebrey, *Biophys. J.* **22**, 67 (1978).

The steady-state sample composition (Eq. 1) is obtained by setting $d(K)/dt = 0$. The approach to the steady state is a first-order exponential decay, characterized by a total "light-dependent photochemical relaxation rate," $k_{total} = k_f + k_r$, that is analogous to the familiar chemical relaxation rate. The concentration of K in the sample after a given time t spent in the laser beam is

$$K\ (t) = K_{ss} + (K_0 - K_{ss})\ \exp\ (-k_{total}t) \tag{3}$$

where K_{ss} is the steady-state concentration and K_0 is the concentration prior to entering the laser beam. The term in the exponential, $k_{total}t$, is the "total photoalteration parameter," F_{total}, which is just the sum of the photoalteration parameters calculated for the forward and reverse photoreactions (see Eq. 10 in Ref. 8).

When two spatially separated laser beams at different wavelengths are used, the sample successively approaches each of two different steady-state compositions. For our experiment these compositions are 28% K for the 514-nm beam, and 0% K for the 676-nm beam. How close each beam comes to achieving its particular steady state depends on how high the total photoalteration parameter for that beam is. To maximize the K contribution to the spectrum we want to keep F_{pump} high and F_{probe} low. At 1700 rpm, an unfocused (3-mm dia.) 514-nm laser beam with a power of 400 mW yields an F_{pump} of about 1.5, which is sufficient to generate a near-steady-state amount of K in a single pass through the beam. The lower limit on the probe power is, of course, determined only by the minimum signal level that is needed to obtain a satisfactory Raman spectrum in a given period of time. We have found that ~ 3 mW of focused (75-μm dia.) 676-nm probe power (corresponding to $F_{probe} \approx 1.2$) gives excellent spectra after several hours of signal averaging, and the fraction of scattering from K is only $\sim 40\%$ less than the maximum that can be obtained by using a substantially weaker probe beam ($F_{probe} \approx 0.1$) (see Fig. 2 in Ref. 8). When using a 752-nm probe beam, it is necessary to employ more power than with 676 nm, since 752 nm is farther off resonance. However, the K extinction is sufficiently small at 752 nm that powers up to ~ 250 mW do not cause excessive photoalteration.

Effect of Self-absorption on Subtraction Parameters. With the optimal pump and probe wavelengths and powers, the K in the photostationary steady state contributes about half of the Raman scattering in the pump-and-probe Raman spectrum; the remainder comes from BR (see Fig. 4B). Thus, most of the K Raman features can be seen in the raw spectrum, making it possible to perform an unambiguous subtraction of the residual BR scattering to produce a pure K spectrum. The spectrum of pure BR is obtained by simply turning off the 514-nm pump beam (see Fig. 4A). In

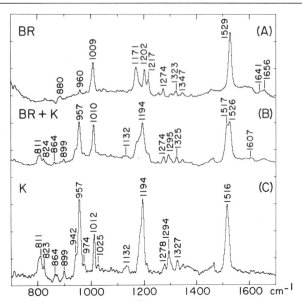

FIG. 4. 77°K spinning sample Raman spectra. The BR spectrum in **A**, obtained with a 3-mW 676-nm probe, is essentially identical to room temperature BR spectra. When we add the pump beam (514 nm, 400 mW) in **B**, the BR lines drop in intensity, and new lines due to K appear at 1517, 1194, and 957 cm⁻¹. The BR contribution to this pump-and-probe spectrum can be completely removed by subtracting 55% of spectrum **A**. The resulting K spectrum, corrected for detector sensitivity, is given in **C**. (Adapted from Braiman and Mathies[4].)

fact, if the probe power and data accumulation parameters are identical for both scans, then one would nominally expect the relative amounts of BR scattering in the probe-only and pump-and-probe spectra to reflect directly the relative amounts of BR in the sample under the two sets of conditions. That is, since the 676-nm probe-only sample is 100% BR, and the pump-and-probe sample is 72% BR, it should be possible to eliminate completely the BR contribution by subtracting the probe-only spectrum, weighted by a factor of 0.72, from the pump-and-probe spectrum.

This nominally "correct" subtraction parameter actually results in a substantially oversubtracted spectrum, with large inverted peaks at all the positions of strong BR Raman lines. Experimentation with a variety of subtraction parameters shows clearly that the BR Raman scattering in the pump-and-probe spectrum is only 55 ± 5% as intense as in the probe-only spectrum; the proper subtraction parameter is thus 0.55, not 0.72. The reason for the discrepancy between composition and subtraction parameter occurs because in these samples, Raman scattering intensity is *not* a direct measure of concentration. Turning on the pump beam effectively introduces a colored filter into the path of the probe laser beam: the K that is produced absorbs 676-nm light strongly, and so significantly attenuates the probe power "seen" by the average sample molecule. This "inner filter" effect is observed because we generate a K concentration that is high enough to attenuate the beam when it has passed through just a fraction of the total sample thickness. This can be seen by a simple calculation of the relative intensities, I_1 and I_0, of the probe beam at a depth x into the sample, in the presence and absence of the K produced by the pump beam:

$$\log(I_1/I_0) = -\Delta\epsilon(\lambda)\,\Delta C\,x \tag{4}$$

where $\Delta\epsilon$ is the difference in the extinction coefficients of K and BR, i.e. $\Delta\epsilon(\lambda) = \epsilon_K(\lambda) - \epsilon_{BR}(\lambda)$, and ΔC is the change in K's molar concentration when the pump beam is turned on (approximately 10^{-4} M for our pelleted samples). A plot of $\Delta\epsilon(\lambda)$ is given in Fig. 5. For example, using a value of 30,000 M^{-1} cm^{-1} for $\Delta\epsilon(676$ nm$)$, one calculates that at a depth of only 0.2 mm into the sample, the laser beam intensity is reduced by a factor of 0.87 when the pump beam is turned on.

The effective average attenuation, i.e., the correction factor for the subtraction parameter, is easier to determine empirically than to calculate, since there are many factors, some of them difficult to measure precisely, that weight the contributions to the spectrum from different regions of the sample. We can say that for a 676-nm probe, the effective average attenuation is 0.55/0.72, or 75%. For other wavelengths, we ex-

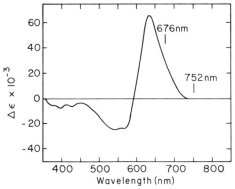

FIG. 5. A plot of the difference in extinction coefficients of K and BR as a function of wavelength (adapted from Lozier *et al.*[15]) indicates that self-absorption is insignificant with a 752-nm probe, but must be taken into account when using shorter probe wavelengths.

[15] R. H. Lozier, R. A. Bogomolni, and W. Stoeckenius, *Biophys. J.* **15**, 955 (1975).

pect the attenuation qualitatively to go as $\Delta\epsilon(\lambda)$. When we use a probe beam at 752 nm, where $\Delta\epsilon$ is very small, the appropriate subtraction parameters do in fact turn out to be in the range 0.75 ± 0.05, the same as predicted in the absence of self-absorption. With a 647-nm probe beam, on the other hand, the effective average attenuation is about 50%. In fact, there is a further complication with a 647-nm probe, because the Raman spectrum now falls within K's absorption band, and it too is affected by the "inner filter." Not only is the Raman scattering attenuated as it passes out through the sample, it is attenuated *differently* in different frequency regions of the spectrum. In this case a single subtraction parameter cannot be used over the whole 700–1700 cm^{-1} Raman spectrum, but rather only over portions of it in which the absorption can be approximated as constant. This wavelength dependence of the inner filter is insignificant for the portion of the 676-nm Raman spectrum above 800 cm^{-1}, but can be observed below this frequency as a change in the shape of the BR fluorescence background when the pump beam is turned on.

Intensity Correction of the Spectrum. It is important to take into account the wavelength dependence of the detector sensitivity when using red excitation. A relative quantum sensitivity curve for the detection apparatus can be generated by obtaining a spectrum of white light from a tungsten lamp and comparing it to the emissivity curve predicted for a black-body source of known filament temperature. In the region of the 676-nm or 752-nm Raman spectrum, the sensitivity of our spectrometer (RCA 31034 photomultiplier) drops by a factor of 1.5 between 700 and 1700 cm^{-1}. Also, the sensitivity correction curve provides an important explanation for the appearance in our raw spectra of a rather sharp "band" at 1650 cm^{-1} (Fig. 4A,B). This results from a $\sim 2\%$ variation in the detector sensitivity at this particular wavelength and is removed by intensity correction. This detection anomaly only shows up in spectra with large fluorescence backgrounds, where it results in a feature that could be mistaken for a Raman line.

Results. Figure 4C shows our Raman spectrum of K with 676-nm excitation. We have obtained almost identical Raman spectra by using 752- and 647-nm probe beams. K's Raman spectrum is distinctively different from that of any of the other bacteriorhodopsin photointermediates. There is a single intense C=C stretching vibration at 1516 cm^{-1}, indicating that we are observing a single species. There are also distinctive fingerprint lines (e.g., 1295, 1276, and 1194 cm^{-1}), as well as several very intense hydrogen-out-of-plane (HOOP) vibrations (e.g., 957 cm^{-1}, 811 cm^{-1}).

We have developed an "isotopic fingerprint" method that has allowed us to determine the configuration of the retinal chromophore in BR and M even in the presence of protein perturbations.[2] We have recently applied this method to K and its isotopic analogs.[4] Spectra of 15-deuterio-K analogs exhibit vibrational shifts that are very similar to those seen in the 13-*cis* model compounds, but are distinctively different from those seen in all-*trans* models. These data argue that K's chromophore has a 13-*cis* configuration and that the primary event in bacteriorhodopsin is a trans → cis photoisomerization. However, the strong enhancement of the HOOP vibrations in K indicates that K's conformation is significantly distorted by out-of-plane twists of its polyene chain.

Our 77°K Raman sample spinner has enabled us to observe a chromophore structure that exhibits specific protein-induced distortions and that is prevented from relaxing because the low temperature freezes its protein-binding pocket. This apparatus should also be useful for obtaining Raman spectra of intermediates in other chemical and biological systems that are constrained by protein or solvent environments as a result of cooling to 77°K. There is also the as yet unexplored possibility of performing kinetic resonance Raman on these trapped structural intermediates by altering the time between pump and probe beams, the rotational velocity of the tip, and the temperature in the spinning cell.

Acknowledgments

This work was supported by a grant from the National Institutes of Health (EY 02051). R.M. is an Alfred P. Sloan Research Fellow. M.B. was supported by a University of California Regents Fellowship. We are grateful to Thomas Ebrey and Richard Lozier for permission to adapt their data. The Raman spectrometer intensity correction procedures were developed by Anne Myers.

[78] Kinetic Resonance Raman Spectroscopy with Microsampling Rotating Cells

By AARON LEWIS

One of the technical achievements of resonance Raman spectroscopy of visual pigments and bacteriorhodopsin has been the introduction of a whole variety of kinetic techniques to the field of Raman spectroscopy. In this paper we will focus specifically on one of these techniques: the use of rotating cells. However, the principles that apply to this technique are in fact identical to all aspects of kinetic resonance Raman spectroscopy as they have been applied to rhodopsin and bacteriorhodopsin. In fact these same principles can be applied generally to other forms of low-light-level kinetic spectroscopy of light-driven chemical and biological systems.

METHODS IN ENZYMOLOGY, VOL. 88

In principle all low-light-level forms of spectroscopy suffer from the same problem: Not enough signal can be accumulated in a short period of time to obtain kinetic information. There are two possible solutions to this problem; either significant effort can be expended to develop rapid detection capabilities or innovative sample-handling procedures and temporal modulation of the excitation source can be developed to work around limitations in detection. The latter route has been the principal approach taken by Raman spectroscopists interested in bacteriorhodopsin and visual pigments. Basically, three different methods have been employed to modulate the sample exposure time to the laser beam. They are as follows: (1) using a cw (continuous wave) laser source and varying the flow rate of the sample in a well-defined manner, (2) using a stationary sample while modulating the temporal duration of the laser on the sample, or (3) using a cw or modulated laser beam and rotating the sample at various speeds. The most generally applicable of the methods is (1). Using this method even samples that do not regenerate the starting materials can be investigated. However, the biggest disadvantage of flow methods is the amount of material that is needed for most flow systems. To circumvent this problem rotating cell techniques have been employed, but they suffer from their lack of applicability to nonregenerable samples. In other words, in terms of the photochemically active retinylidene proteins, they are limited to bacteriorhodopsin.

The first to apply rotating cell techniques to resonance Raman spectroscopy were Kiefer and Bernstein.[1] Their aim was not to obtain kinetic data but simply to reduce the effects of heating on samples that were being exposed to lasers at wavelengths within their absorption. The cells of Kiefer and Bernstein[1] were large and thus did little to solve the problem of sample concentration. More recently, however, Rodgers and Strommen[2] have developed a spinning cell for obtaining Raman spectra of microsamples, and their cell could be quite useful when applied to bacteriorhodopsin.

Spinning cells are simple devices. The Rodgers and Strommen version is shown in Fig. 1. As the sample is spun in the cell, it is driven to the cell perimeter where a laser beam illuminates an ~40-μm spot. The transit time of that 40-μm region is inversely related to the angular velocity of the cell, which depends on the rotation frequency. Thus the faster one rotates the cell, the smaller is the residence time of the sample in the laser beam and the more rapidly the same spot in the sample will be illuminated by the laser beam. Therefore in the example of the Rodgers and Strommen microcell with a sample volume of 150 μl and a 2.2 cm diameter the fas-

[1] W. Kiefer and H. J. Bernstein, *Appl. Spectrosc.* **25**, 500 (1971).
[2] E. G. Rodgers and D. P. Strommen, *Appl. Spectrosc.* **35**, 215 (1981).

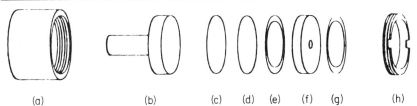

(a)   (b)   (c)   (d)   (e)   (f)   (g)   (h)

FIG. 1. A micro-Raman spinning cell (adapted from Rodgers and Strommen):[2] a, housing; b, aluminum backing shaft; c, e, g, Teflon cushioning disks and rings; d, mirror for signal improvement; f, quartz cell with filling leveled hole for inserting samples even when cell is spinning; h, tightening ring. The cell is illuminated off axis with a laser beam and the scattering is collected and separated as to wavelength in the monochromator. The photons at each wavelength are then detected by a photocathode surface and the counts are then plotted by the computer as the monochromator is stepped from the incident laser frequency.

test speed that would be feasible with a 25-msec return time (which is required for the bacteriorhodopsin system to regenerate thermally the starting species) is 14 μsec in a 40-μm laser beam. This could be improved considerably by focusing the laser beam or pulsing the laser but neither one of these improvements has been reported.

The residence time of the sample in the laser beam is the kinetic parameter that is the basis of all kinetic resonance Raman measurements on photolabile retinylidene proteins. In order to appreciate the preceding fact let us recall that the absorption of a photon by rhodopsin and bacteriorhodopsin creates a sequence of intermediates. The retinylidene chromophore in each of these intermediates can also absorb photons, and this photochemically regenerates the initial pigment state. Thus the transit time of the molecules in the laser beam determines whether there is sufficient time to produce a particular intermediate state, and determines the complex admixture of pigment states present in the laser beam. For such a system of multiple photolabile species the number of detected counts S at a photomultiplier for a given scattered frequency (ν) is given by the general expression

$$s(\nu) = \sum_i E(\nu)\sigma_{R_i}(\nu) \int_v I(\mathbf{r})\, C_i(\mathbf{r})\, dV \qquad (1)$$

where $E(\nu)$ is the overall detection efficiency of the Raman apparatus at frequency ν, σ_{R_i} is the resonance Raman cross section (cm^2/molecule) of the ith species, $I(\mathbf{r})$ is the incident laser light intensity, and $C_i(r)$ is the concentration (molecule/cm^2) of the ith species. This equation by Marcus and Lewis[3] is the starting point of the theoretical description of their ki-

[3] M. A. Marcus and A. Lewis, *Biochemistry* 17, 4722 (1978).

netic experiments[3,4] and is a generalization of the equation by Mathies *et al.*[5] and Callender *et al.*[6] for a single photolabile species. Specifically, Mathies *et al.*[5] and Callender *et al.*[6] were not considering the kinetic evolution of the Raman spectrum, but they did apply their equation to determine the rate of flow needed to eliminate all but the initial pigment state in the bovine visual pigment rhodopsin. They showed that with transit times of ~ 10 μsec through a cw, 25-mW, 600-nm laser beam only 5% of the rhodopsin molecules were photoaltered, and they recorded the resonance Raman spectrum of this pigment in rod outer segments and the detergent Ammonyx-LO using these experimental conditions.

To use Eq. (1) effectively for the analysis of the kinetic measurements performed by Lewis and co-workers[3,4] and other investigators, such as El Sayed[7] and co-workers and more recently Eyring and Mathies and Stockburger and co-workers,[9] it is convenient to determine the average concentration $C_i(r)$ of the species i in the laser beam. In a flowing sample with uniform or turbulent flow and neglecting depth effects on laser intensity (justified in dilute solutions of effective OD < 0.1) the average concentration $C_i(\mathbf{r})$ of species i is given by

$$\overline{C_i(\mathbf{r})} = \overline{C_i(t_f - t_i)} = \int_t^{t_f = 2r_0/v_b + t_i} \frac{c_i(t)\ dt}{t_f - t_i} \tag{2}$$

where r is the laser beam radius, v_B is the bulk flow velocity, and $t_f - t_i$ is the transit time through the laser beam. In simple laser beam experiment t_i is considered zero. As will be discussed below, certain kinetic experiments have been performed with two laser beams, and in these experiments the value of t_i has to be considered in terms of the particular experiment.

For the single-beam experimental situation, Marcus and Lewis[3] have combined Eqs. (1) and (2) and written the result in terms of species observed in bacteriorhodopsin by absorption spectral studies at times < 100 μsec. The equation they get is

$$S(\nu, t_f) = E(\nu)I_0[\sigma_R(\nu)\ \overline{R(t_f)} + \sigma_K(\nu)\ \overline{K(t_f)} + \sigma_L(\nu)\ \overline{L(t_f)} + \sigma_M(\nu)\ \overline{M(t_f)}] \tag{3}$$

where $\overline{R(t_f)}$ is the average concentration of bacteriorhodopsin (BR_{570}) and $\overline{K(t_f)}$, $\overline{L(t_f)}$, and $\overline{M(t_f)}$ are the average concentrations of the K, L, and M

[4] M. A. Marcus and A. Lewis, *Science* **195**, 1330 (1977).
[5] R. Mathies, A. R. Oseroff, and L. Stryer, *Proc. Natl. Acad. Sci. U.S.A.* **73**, 1 (1976).
[6] R. H. Callender, A. Doukas, R. Crouch, and K. Nakanishi, *Biochemistry* **15**, 1621 (1976).
[7] M. A. El-Sayed, this volume, Chapter 73.
[8] G. Eyring and R. Mathies, *Proc. Natl. Acad. Sci. U.S.A.* **76**, 33 (1979).
[9] M. Stockburger, W. Klusmann, H. Gattermann, G. Massig, and R. Peters, *Biochemistry* **18**, 4886 (1979).

intermediates, respectively. In addition, Marcus and Lewis[3] have used these equations to calculate theoretical curves for average concentration of the preceding species versus transit time in the laser beam. These are seen in Fig. 2 for a cw, 27-μm, 30-mW, and 530.9-nm laser beam. This graph clearly shows that variations in the laser beam transit time will vary the average composition of species detected. Thus the time evolution of vibrational modes characteristic of these species can be readily followed.

All kinetic spectra obtained by the preceding single-beam method are necessarily composed of a superposition of vibrational modes of various intermediates with, of course, whatever specificity, differences in the absorption of the various intermediates impart to the measurements. However, in certain instances the specificity of resonance enhancement and the techniques of conventional flash photolysis can be combined to provide spectra of an intermediate without complications from other species. An example of this is an experiment performed by Marcus and Lewis[3] to record kinetically the spectrum of the M species in bacteriorhodopsin at physiological pH and temperature. Their experiment uses an intense photolyzing laser beam at the peak of bacteriorhodopsin's (BR_{570}) visible ab-

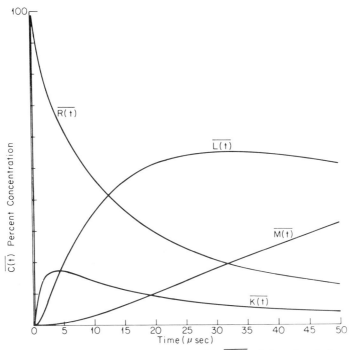

FIG. 2. Theoretical curves for average concentration $[\overline{C(t)}]$ of photochemical species as a function of transit time for 30 mW of 530.9-nm laser light focused to 27 μm.

sorption and a weak probe beam within the blue absorption of the M species. Figure 3 shows a moving sample with two such laser beams incident upon it. In the region between the two laser beams the light-dependent rate constants for the conversion of BR_{570} and K and vice versa, k_1 and k_2, respectively, are both zero. Thus the concentration of BR_{570} when it enters the probe beam is the same as its concentration upon leaving the pump beam. If it is experimentally possible to make $R(t_1)$ and $K(t_1)$ (concentrations of BR_{570} and K at t_1) equal to zero, then expressions for $L(t)$ and $M(t)$ for $t_1 \leq t \leq t_2$ (where no laser beam is present) become

$$L(t) = L(t_1)e^{-K_4(t-t_1)}$$
$$M(t) = 1 - L(t_1)e^{-K_4(t-t_1)}$$

(4)

where K_4 is the rate constant for M_{412} formation. Figure 4 shows the concentration as a function of time for the various species in the photochemical cycle of bacteriorhodopsin for a photolyzing pump beam excitation of 300 mW at 573.0 nm focused to 100 μm with a cylindrical lens. It is seen that after ~35 μsec of excitation the concentrations of BR_{570} and K are effectively zero. This occurs because not all K molecules produced can absorb a second photon to return to BR_{570} before K decays to L and M, which have progressively smaller extinctions at 573.0 nm. Thus, if the blue probe laser beam is directed at a point that is sufficiently downstream from the photolyzing laser, the resonance Raman spectrum of the M species will be preferentially recorded. This will be the case so long as the extinction of the M species at the blue laser wavelength is high enough to give significant resonance enhancement but low enough to prevent photoalteration of the M species at the flow velocities required to assure the

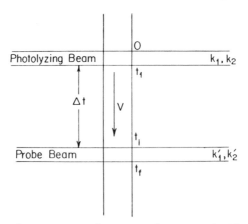

FIG. 3. Dual laser beam geometry for a nonstationary sample. The photolyzing beam pumps the photochemistry, and the probe beam is utilized to obtain a resonance-enhanced Raman spectrum of some intermediate with an optimal concentration corresponding to Δt.

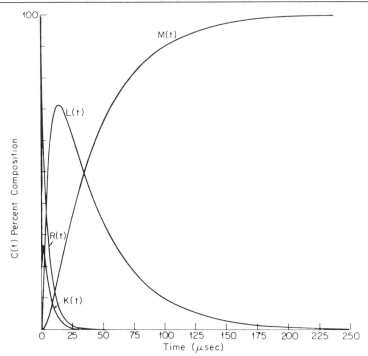

FIG. 4. Theoretical curves of concentration versus transit time for various photochemical species with 300 mW of 573.0-nm laser light focused to 100 μm. (From Ref. 3.)

validity of Eq. 4. This condition is readily met with the 457.9-nm emission of an argon ion laser, and a description of an experimental arrangement that fits this and the preceding requirements is schematically illustrated in Fig. 5 of Marcus and Lewis[3] and described in the figure caption. Stockburger et al.[9] recently adapted this experimental arrangement for similar measurements with a rotating cell. They also applied the technique to get pure spectra of M_{412}.

In summary, there is little doubt that the kinetic resonance Raman techniques outlined in this chapter will play a significant role in the future applications of resonance Raman spectroscopy to rhodopsin and bacteriorhodopsin. Specifically, kinetic resonance Raman spectra of rhodopsin and bacteriorhodopsin with selective chemical modifications will be essential for further advances in this field. These molecules can often be synthesized in only limited quantities and thus microsampling rotating cell techniques will be invaluable in the next stage of applying the powerful spectroscopic techniques of kinetic resonance Raman spectroscopy to the further elucidation of the mechanism of action of rhodopsin and bacteriorhodopsin.

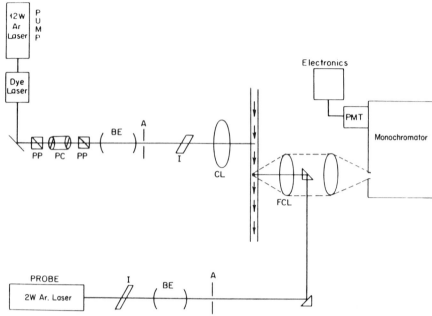

FIG. 5. Schematic of the dual laser beam apparatus utilized in our experiments. The sample is moved as indicated by the vertical arrows. A Coherent Radiation 12-W argon ion laser is used to pump a Coherent Radiation Model 490 rhodamine 6G dye laser. The frequency of this pump is 570–580 nm. The probe comes from a 2-W Coherent Radiation argon ion laser or a Model 52 krypton laser. The laser beams are made nearly uniform by passing them through 10× beam expanders (BE) and then through 1-mm apertures (A). Interference filters (I) are used in both beams to block out laser emission lines and dye laser fluorescence. The probe beam is incident in a backscattering arrangement. In some experiments the focusing-collection lens (FCL) is cylindrical to ensure a uniform line image on the sample. This allows direct correlation of the experimental data with the theoretically calculated curves such as those shown in Fig. 2, since these calculations assumed that the laser beam is a uniform line image on the sample. In all cases the pump beam is focused on the sample with a 50-mm focal length cylindrical lens (CL) for the reason cited above and to ensure illumination of the entire sample volume in all dual-beam experiments. For some experiments a Pockels cell (PC) is inserted between two Glan–Thomson polarizing prisms (PP). By switching the half-wave voltage through the Pockels cell on and off, the pump beam can be effectively switched on and off. In this manner single-beam and double-beam experiments can be performed simultaneously, and the data subsequently separated by computer. The monochromator is stepped only after data are obtained in two channels, one with and the other without the pump beam present. This is possible since the thermal relaxation time of bacteriorhodopsin molecules activated by the pump beam is on the order of msec, whereas the time per channel is 5-10 sec. (From Ref. 3.)

Acknowledgements

The author acknowledges support from the National Eye Institute, the Army, the National Aeronautics and Space Administration, and the Naval Air Systems Command at various times during the course of these investigations. A. L. was a John Simon Guggenheim Fellow when this review was written.

[79] Calorimetric Measurements of Light-Induced Processes

By Alan Cooper

Direct calorimetric methods have been used to determine energy (enthalpy) changes during the process of rhodopsin photolysis.[1–3] The techniques are applicable, in principle, to any photochemical process of high quantum efficiency and can also be used as an analytical tool for the determination of other light-induced changes (e.g., H^+ uptake or release, Ca^{2+} efflux, etc.). Sample requirements are modest, typically 50–100 nmol per determination (i.e., of order 3 mg rhodopsin), and, within limits, turbid suspensions can be used. Accessible temperature range is $-196°$ (liquid nitrogen) to ambient, or higher, so that various intermediate steps in a photoreaction may be resolved.

Basic Principles

Photocalorimetry utilizes the principle of isothermal heat flow microcalorimetry used widely in other areas of biological thermochemistry.[4] Unlike classic calorimetry, in which thermal energy changes are measured directly as changes in temperature, heat energy changes in the sample are allowed to dissipate rapidly across sensitive thermopiles, producing an output voltage that is proportional to the rate of energy dissipation in the sample. Integration over time yields the total heat energy change in the sample during reaction.

To initiate photochemical reactions we must shine light into the calorimeter. In the absence of any photoreaction the absorbed light energy is rapidly converted into heat and produces a measurable heat flux, which constitutes the baseline for any measurement. If physical or chemical changes are induced in the sample by illumination, then some of this light energy may be diverted and not appear as heat (endothermic reaction). Alternatively, photon absorption might stimulate exothermic processes in the sample and give rise to an enhanced heat flux. Measurement of the molar enthalpy change for a photon-induced reaction, therefore, requires the determination of two quantities: (1) the change in heat energy content

---

[1] A. Cooper and C. A. Converse, *Biochemistry* **15**, 2970 (1976).
[2] A. Cooper, *FEBS Lett.* **100**, 382 (1979).
[3] A. Cooper, *Nature (London)* **282**, 531 (1979).
[4] N. Langerman and R. L. Biltonen, this series, Vol. 61, p. 261.

METHODS IN ENZYMOLOGY, VOL. 88

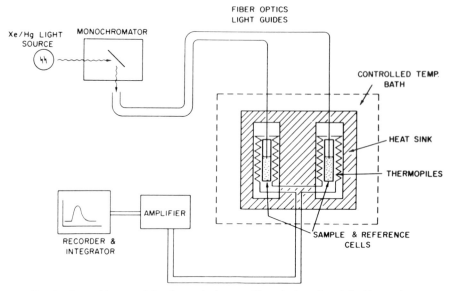

FIG. 1. General layout of the photocalorimeter. The calorimetric unit in this version was a modification of an LKB batch microcalorimeter. (Reprinted with permission from Cooper and Converse.[1]) Copyright 1976, American Chemical Society.

of the sample brought about by the isothermal photochemical transformation, and (2) the amount of material, in moles, transformed in the reaction.

The basic plan of the instrument is shown in Fig. 1 and consists of a microcalorimetric unit, with associated temperature control and data collection equipment, together with a variable wavelength light source and fiber optics light guides for irradiation of samples within the calorimeter. Construction details, operation, and performance are described below.

Construction

Calorimetric Unit. Commercial microcalorimeters, for example the LKB Batch Microcalorimeter, can be adapted readily for use as photocalorimeters above $0°$,[1] but are not easily modified to operate at much lower temperatures. A more versatile and cheaper calorimetric unit[3] may be constructed as shown in Fig. 2, and subsequent discussion will be limited to this version.

The central components of the calorimeter are sealed within a cylindrical brass canister, fitted with open-ended stainless-steel access tubes. A separate tube (not shown in Fig. 2) provides for external electrical connections. This arrangement allows sample insertion and removal while the main body of the calorimeter remains submerged in a temperature-con-

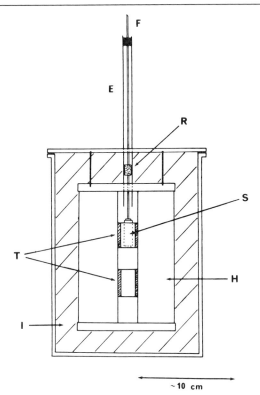

Fig. 2. Photocalorimeter module suitable for use down to $-196°$. Key: E, sample and reference access tubes; F, flexible fiber optics light guides; R, thermal radiation shield; S, sample cell in place in the calorimeter block (the reference cell lies in the same block, directly behind the sample as viewed in this sketch); H, heatsink; T, thermoelectric sensors; I, insulation. (Reprinted with permission from Cooper.[3]) Copyright 1979, Macmillan Journals Ltd.

trolled liquid. Samples are held in cylindrical cells of 2.5-ml capacity, made of thin-walled stainless steel and gold plated for chemical stability. The top of each cell is sealed with a screw-fitting Teflon stopper into which is fixed one end of the fiber optics light guide. When properly aligned, each sample cell can be lowered into the calorimeter block via the access tubes so as to rest in matching cylindrical cavities in an aluminum block mounted between solid-state thermoelectric detectors (Borg-Warner 950-71) and the massive brass heat sink. The flat faces of the thermopiles are coated with thermally conducting grease (Cambion/Cambridge Thermionic Corp.) for good thermal contact. Two such cells are mounted side by side in the calorimeter block, each with a separate access tube, to hold either the sample or an inert reference mixture, respectively.

Small holes bored in the base of the aluminum block accommodate 30-ohm calibration heaters, wound on glass formers with a low thermal coefficient heater tape (Gilby-Brunton, 54 ohm m^{-1}), and electrically insulated with epoxy resin. A second pair of thermoelectric detectors with a dummy block is mounted below the actual sample block. All the thermopiles are connected together in series, the dummy detectors in reverse polarity to the sample thermopiles such that the voltage response due to global thermal fluctuations over the entire unit tend to cancel. Electrical connections involving the thermopile circuits are made using a low thermal emf cadmium solder (Keithley Instr. Inc.) and low-noise coaxial cable. The entire heat sink and thermoelectric assembly is suspended from the roof of the canister by thin stainless steel rods, and intervening spaces are filled with rigid foam insulation. A single copper–constantan thermocouple mounted between the heat sink and the inner canister wall provides a useful diagnostic probe of thermal equilibration and stability during use.

Thermal radiation down the access tubes, which can cause major instabilities in the calorimeter performance, is reduced by attachment of reflecting screens on each of the fiber optic guides. These consist simply of light plastic formers attached to the light guides and coated with reflective aluminum foil to give a loose sliding fit within the access tube.

Temperature Control. Detection of the very small heat energy changes in biological samples requires good thermal uniformity and stability within the calorimeter, with long-term temperature drifts in the thermostat bath less than 0.001° hr^{-1}, or better. Short-term fluctuations of order 0.001° can be tolerated provided they are cyclic and of short period (about 1 min or less) since they are damped out by the insulation and large thermal capacity of the calorimeter heat sink.

Constant-boiling liquid nitrogen provides sufficient temperature stability for experiments at $-196°$ (77°K). The calorimeter is suspended in a large open-necked Dewar vessel filled to constant level with liquid nitrogen. The open ends of the access tubes project from the top of the Dewar into a dry-gas-flushed chamber, to minimize frosting, and the entire system is further lagged and protected from drafts. The gradual boil-off of nitrogen during use results in small baseline drifts, but these are normally uniform and linear and can be corrected for.

For control at higher temperatures (down to $-80°$) the Dewar vessel is filled with methanol and a copper cooling coil is attached around the calorimeter canister. Constant temperature methanol is circulated around the coil from a low-temperature bath (Forma Scientific Inc., Model 2762/2103), modified for proportional temperature control with a platinum resistance probe and precision temperature controller (Oxford Instr.

Ltd.). The temperature sensor is mounted directly on the cooling coil, with control heaters situated at the circulation pump intake, so as to provide a tight response. Auxiliary heaters, running at constant power, are immersed in the main body of the Forma bath and adjusted to offset most of the cooling power.

Above 0° rather simpler (refrigerated) water baths may be used together with a similar form of proportional temperature control. We have not explored control of temperatures in the region between − 196 and − 80°.

All temperature control equipment is run from a stabilized power supply for isolation from main voltage fluctuations. Considerable patience and perseverance are required at this stage to eliminate all sources of thermal instability. Installation in a temperature-controlled darkroom helps.

Optics. Only relatively modest light intensities are required. A 200 to 250-W xenon or mercury–xenon arc lamp, with stabilized power supply (Oriel Scientific or Applied Photphysics Ltd.), is adequate. Light is focused through a grating (Bausch & Lomb) or filter monochromator (Oriel) via a simple shutter onto the entry faces of the PVC-clad, flexible glass fiber optics light guides (Corning No. 5010, 120-cm length), which are mounted on a precision sliding support so that either the sample or the reference cell may be identically illuminated at any one time. Slight differences in optical transmission properties of pairs of light guides may be compensated by use of neutral density filters.

This arrangment gives a light intensity in the calorimeter cell equivalent to about 0.1 mW, or of order 3×10^{14} photons per second at 500 nm. The glass fiber optics guides do not transmit light below about 400 nm.

Data Acquisition. The output voltage of the calorimeter thermopiles, usually of the order of microvolts, is amplified by a chopper-stabilized dc microvoltmeter (Keithley 150-B) and recorded on a standard linear chart recorder fitted with an integrator. We also use a simple electronic integrator, with printer output, connected in parallel to the recorder and giving integrated signal output at 1-min intervals.

Calibration. Absolute calibration requires a constant current source of a few milliamperes applied to the heaters of known resistance in the calorimeter block. Secondary calibration of the entire system is possible using a known photochemical reaction of high quantum yield, such as the potassium ferrioxalate photoredox process,[1,5] and this also gives an indication of the photoefficiency of the equipment. About 70% of the light entering the calorimeter cell is normally available for photochemistry, the remain-

[5] C. G. Hatchard and C. A. Parker, *Proc. R. Soc. London, Ser. A* **235**, 518 (1956).

der being absorbed by the cell walls, etc. This photoefficiency may be somewhat reduced with turbid samples due to light scattering.

The thermal response of the current instrument is in the region of 0.1 μV μW^{-1} at 25°, falling to about 0.07 and 0.02 μV μW^{-1} at -75 and $-196°$, respectively. Energy changes in the sample as low as 5×10^{-4} J can be detected, with standard deviations down to 10% of this for a series of determinations.

Use

Energy Measurements. Sample solutions or suspensions, 2–2.5 ml, are loaded in the dark and the stoppered cell is lowered into the calorimeter block on the end of the fiber optic light guide. Care should be taken to ensure that the cell is well sealed, since even a trace of sample evaporation will cause erratic and spurious heat effects. A suitable reference mixture, preferably a fully bleached sample, is similarly loaded into the adjacent cell position. After thermal equilibration, which normally takes several hours, the total light intensity is determined by irradiation of the reference for a specific length of time, typically 0.5–5 min, depending on how much photoreaction is required. The sample is then irradiated for the same period under identical conditions, and additional reference irradiations may then follow. The difference in integrated energy flux between sample and reference gives the total light-induced heat energy change in the sample. The sample is removed in the dark and the extent of reaction determined spectroscopically.

Differential measurements may be performed by continuous illumination of the reference cell to give a steady-state baseline equivalent to the total light energy flux. Once this baseline is established, the illumination is switched briefly to the sample cell and then back to the reference. This gives a transient response whose integrated area measures directly the energy change in the sample. Blank experiments are done to correct for any slight imbalance in the fiber optics to the two cells.

For rhodopsin, with sample concentrations about 3×10^{-5} M ($A_{500} \simeq 1.5$), a typical 2-min irradiation will bleach about 15 nmol of material and produce a differential heat effect of up to 2×10^{-3} J, depending on the bleaching product.

Although optically clear samples are not required, some artefacts have been seen at low temperatures because of freezing and extensive cracking of the sample or reference solutions. Use of suitable glycerol–water mixtures has eliminated this.

Analytical Uses. In addition to absolute energy measurements, various chemical tricks may be devised to detect other significant changes during a photoreaction. For example, the enthalpy differences observed when using buffers of identical pH, but different heats of proton ionization, have been used to determine H^+ uptake and release by rhodopsin during some of the intermediate bleaching processes.[1] Similarly, additional heat effects upon bleaching in the presence of sulfhydryl reagents or calcium chelating agents may be used to detect light-induced exposure of $-SH$ groups and the (lack of) release of Ca^{2+} ions, respectively.[6] Additional large heat effects have been seen on bleaching rhodopsin at room temperature in detergents such as CTAB (cetyltrimethylammonium bromide), presumably because of the thermal instability of opsin under these conditions.[6]

Limitations. The large thermal inertia of the sample and calorimeter block result in a relatively slow response time for the system ($t_{1/2}$ approx. 1 min). Although this is good for thermal stability and detectability, it implies that except for very slow processes the kinetics of photoreactions cannot be followed by this technique. More important, it means that we can measure accurately only those reactions that are essentially irreversible over several minutes following irradiation. A further limitation is that we have no means of measuring the extent of reaction *in situ* in the calorimeter. Rather, we must remove the sample for spectral examination at room temperature, or thereabouts. Although this is generally no problem with visual pigments, it has so far restricted direct measurements on bacteriorhodopsin since, even though it is possible to produce metastable photoproducts in the calorimeter at low temperatures, these revert to the parent pigment on warming, and the amount of reaction cannot be estimated with confidence. An alternative microphonic technique used by Ort and Parson[7] overcomes this problem.

Acknowledgment

Development of the instrument described here was supported by a grant from the U.K. Science Research Council.

[6] A. Cooper, unpublished observations (1976).
[7] D. R. Ort and W. W. Parson, *Biophys. J.* **25**, 355 (1979).

[80] Tritium–Hydrogen Exchange Kinetics

By NANCY W. DOWNER and JOAN J. ENGLANDER

Tritium exchange methods for disk membranes, purple membrane, and purified rhodopsin in detergent solutions allow direct, accurate determination of the number of hydrogens per rhodopsin remaining unexchanged at a given time. Thus one of the technique's major advantages over most physical or spectroscopic means of structure analysis is that it is not subject to ambiguities of interpretation introduced by rhodopsin's membrane or detergent environment. Measurement of ^{3}H–^{1}H exchange kinetics is a method of choice for comparing rhodopsin's structure in a given detergent with the membrane-bound state,[1] because exchange kinetics reflect both amount and stability of structure. ^{3}H–^{1}H exchange also has much potential for exploring the issue of whether membrane proteins such as rhodopsin[2] and bacteriorhodopsin[3] may differ significantly in some structural aspects from general patterns observed in soluble proteins. A previous article in this series[4] describes ^{3}H–^{1}H exchange techniques and should be consulted for the basic procedures. Here we will discuss only the modifications or new techniques developed for handling rhodopsin and bacteriorhodopsin. Some issues specific to ^{3}H–^{1}H exchange studies of these membrane proteins will be emphasized.

Exchange-in

^{3}H–^{1}H exchange studies are usually performed by monitoring exchange-out kinetics after the protein has been prelabeled with ^{3}H by incubation in ^{3}H^{1}HO. The conditions chosen for exchange-in determine which hydrogens become labeled and therefore what can be observed in the subsequent exchange-out. Temperature, pH, and denaturing solvents are the primary variables that can be manipulated to alter the exchange rate during exchange-in.

For some purposes it is desirable to label the peptide hydrogens fully during exchange-in. The most slowly exchanging peptide hydrogens of soluble proteins have rates 10^9 slower than the free peptide rate.[5] The time required to equilibrate peptides this slow can be estimated from Eq. (1), which gives the exchange half-time (minutes) for amides of PDLA as a function of temperature (T, °C) and pH.

$$t_{1/2} = 200/(10^{pH-3} + 10^{3-pH})(10^{0.05T}) \tag{1}$$

For example, at pH 10.5 and 4° the expected half-time for peptides slowed nine orders of magnitude is close to 3 days. After labeling purple membrane under the same conditions for 5 days, only $\sim$ 180 exchangeable hydrogens, roughly 70% of the expected amide number, were observed.[3]

[1] H. B. Osborne and E. Nabedryk-Viala, *Eur. J. Biochem.* **89**, 81 (1978).
[2] N. W. Downer and S. W. Englander, *J. Biol. Chem.* **252**, 8092 (1977).
[3] J. J. Englander and S. W. Englander, *Nature (London)* **265**, 658 (1977).
[4] S. W. Englander and J. J. Englander, this series, Vol. 26 Part C, p. 406.
[5] S. W. Englander, N. W. Downer, and H. Teitelbaum, *Annu. Rev. Biochem.* **41**, 903 (1972).

Therefore, more rigorous exchange conditions are required for equilibration of the peptide hydrogens of bacteriorhodopsin. In the case of vertebrate disk membranes labeled under rather moderate conditions, the number of hydrogens observable matches the number of expected amides.[2] However, infrared-deuterium exchange measurements indicate that labeling of the amides was incomplete under the conditions used; incubation of disk membranes at 60° for several hours[1] or denaturation in SDS[6] are required to equilibrate 100% of amide hydrogens. The implication that nonamide hydrogens contribute to disk membrane $^3H-^1H$ exchange curves is discussed in a following article.[7]

In summary, because of problems presented by hydrogens resistant to labeling and the possible role of protein side chains and lipids in membrane studies, the exchange-in conditions used in studies on rhodopsin and bacteriorhodopsin need to be given particular consideration. If extreme conditions are used, the reversibility of exchange should be evaluated. For many purposes, including studies of structure change, complete labeling is not required. However, the role of exchange-in conditions in influencing the relative contributions of amide and nonamide exchange needs to be considered.

Exchange-out

To initiate exchange-out and determine the number of hydrogens remaining unexchanged at a given time, the sample handling procedures and column techniques already described are generally applicable for rhodopsin and bacteriorhodopsin. Disk membranes, purified rhodopsin in detergent, or purple membranes appear in the excluded volume fraction of columns prepared with Sephadex G-25 or Sephadex ion exchange resins. Plastic or glass columns can be used. It is desirable to keep the dead volume and tubing volume below the gel bed at a minimum. Since aggregation or binding to resins can present a problem with membrane samples, any given combination of buffer and column material should be checked for adequate sample recovery from the column run. For purple membrane samples, which tend to aggregate under pressure, the use of medium-grade (particle size 50–150 μm) Sephadex G-25 and low hydrostatic pressure is recommended. In the case of disk membranes, charged resins appear to give better separation between the excluded volume and tritiated water peaks; DEAE- or SP-Sephadex can be used, depending on the pH of the experiment.

To obtain long time points for disk membrane exchange, a centrifuge

[6] H. B. Osborne and E. Nabedryk-Viala, *FEBS Lett.* **84,** 217 (1977).
[7] N. W. Downer and H. B. Osborne, this volume [82].

method is more convenient than either dialysis or a second column. Small polyethylene centrifuge tubes (4 × 49 mm, containing ~0.4 ml) are prefilled with buffer and overlayed with about 50 µl silicone oil. At a given time, a 20- to 50-µl sample from the exchanging disk membrane pool is layered onto the oil and spun for a few seconds in a small centrifuge (e.g., Beckman Microfuge). The tip of the centrifuge tube is cut off to recover the pellet, which is then solubilized in the appropriate SDS buffer for Lowry protein determination. A sample of the buffer is collected to determine background counts.

Analytical Methods

Bound tritium is measured as described.[4] It is most useful to express the $^3H-^1H$ exchange results as hydrogens remaining unexchanged per molecule of rhodopsin or bacteriorhodopsin in the sample. The concentration of bacteriorhodopsin is determined using the optical density of the collected samples and the appropriate extinction coefficient. For rhodopsin and disk membranes, a modification of the Lowry method provides a relatively sensitive and linear assay for rhodopsin.[8] A standard curve is generated for each preparation of disk membranes by determining the Lowry optical density for several samples of disk membranes. The rhodopsin concentration of the disk membrane stock is determined from the optical density at 500 nm. The exchange experiments should be arranged so that final samples (or replicates) from column dialysis or centrifuge runs contain 20–40 µg protein.

[8] H. H. Hess, M. B. Lees, and J. E. Derr, *Anal. Biochem.* **85**, 301 (1978).

[81] Infrared Measurement of Peptide Hydrogen Exchange in Rhodopsin

By H. Beverley Osborne and Eliane Nabedryk-Viala

Infrared spectroscopy can measure directly and specifically the $^1H-^2H$ substitution in peptide groups of proteins, according to

$$-CON^1H- + {}^2H_2O \rightarrow -CON^2H- + {}^1H^2HO$$

Figure 1 shows the infrared absorption spectrum from 1800 to 1300 cm^{-1} of membrane-bound rhodopsin in ^2H_2O buffer, p^2H 7, 26°. The amide I band, centered at 1655 cm^{-1}, is due to the C=O stretching vibration of the peptide bond, and its absorbance is proportional to the total number of peptide groups present. The amide II band, centered at 1546

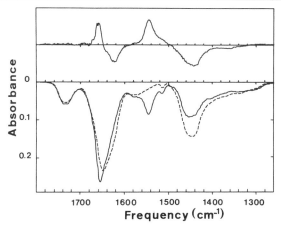

FIG. 1. The amide region of the infrared spectrum of rod outer segment membranes in ^1H_2O buffer p^{2}H 7. Spectrum 1 (————): rod outer segment membranes after 24 hr incubation at 26° in 10 mM imidazole, 100 mM sodium phosphate p^{2}H 7, ^2H_2O buffer. Spectrum 2 (----------): the same sample after heating for 6 hr at 60° in the spectrophotometer cell and then cooling to 26°. The computed difference between these two spectra is presented in the upper graph: This difference spectra shows that the maximum of the amide I and amide II bands are centered at 1655 and 1546 cm^{-1}, respectively; it also demonstrates (at 1455 cm^{-1}) that no leakage of ^1H_2O from the atmosphere into the cell has occurred during the heating procedure. (From Osborne and Nabedryk-Viala.[12])

cm^{-1}, is due to the coupled N^{1}H bending and CN stretching vibrations; it decreases as peptide hydrogens become deuterated and is then replaced by another band at about 1455 cm^{-1}. Thus the ratio of the peak intensities of the amide II to the amide I bands is proportional to the number of still unexchanged N^{1}H peptide hydrogens.[1]

It must be noted that the frequency of the amide I band is sensitive to protein conformation.[2] In the present case it indicates that rhodopsin contains α-helical structures. Indeed, infrared dichroism measurements of highly oriented rod outer segments (ROS) have recently shown that the α-helices are preferentially oriented perpendicular to the membrane plane.[3–5]

For other aspects of the infrared exchange method, the reader is re-

[1] E. R. Blout, C. De Loze, and A. Asadourian, *J. Am. Chem. Soc.* **83,** 1895 (1961).
[2] H. Susi, S. N. Timasheff, and L. Stevens, *J. Biol. Chem.* **242,** 5460 (1967).
[3] M. Michel-Villaz, H. Saibil, and M. Chabre, *Proc. Natl. Acad. Sci. U.S.A.* **76,** 4405 (1979).
[4] K. J. Rothschild, N. A. Clark, K. M. Rosen, R. Sanches, and T. L. Hsiao, *Biochem. Biophys. Res. Commun.* **92,** 1266 (1980).
[5] K. J. Rothschild, W. J. DeGrip, and R. Sancho, *Biochim. Biophys Acta* **596,** 338 (1980).

ferred to available review articles.[6-8] We will only discuss here the specific application of this technique to rhodopsin. Infrared peptide exchange and tritium exchange techniques are compared in the following paper.

ROS and Rhodopsin Preparation

ROS membranes and rhodopsin, purified and delipidated, are prepared by standard techniques.[9,10] When necessary, the Ammonyx-LO associated with the purified rhodopsin can be exchanged for another detergent by chromatography on Sepharose-concanavalin A.[10]

Preparation of Samples for ^{1}H–^{2}H Exchange

^{1}H–^{2}H exchange is initiated by transferring the sample from a ^{1}H$_2$O buffer into a ^{2}H$_2$O (99.7% enrichment) buffer. In samples of soluble proteins and peptides this is normally achieved by lyophilizing the sample and then dissolving it in ^{2}H$_2$O exchange-in buffer (buffer in which the exchange is to be studied). This approach cannot be used with membrane samples, since the relatively slow hydration of lyophilized membranes will cause a large error in the definition of the zero time. In addition, purified rhodopsin in detergent is denatured by lyophilization. Hence, for the ^{1}H–^{2}H exchange of ROS and purified rhodopsin samples the following method was used.[11]

At all times samples and ^{2}H$_2$O buffers are manipulated in a glove box under dim red light, through which a flow of dry nitrogen (3 liters/min) is maintained. In addition, the presence of 100 mM phosphate in the exchange-in buffer will increase by 6% the number of peptide hydrogens of rhodopsin that exchange at 26°C.[12]

ROS samples were first thoroughly washed in ^{1}H$_2$O exchange-in buffer. Then exchange was initiated by diluting a 0.5-ml sample 25 times with ^{2}H$_2$O exchange-in buffer. The membranes are collected by centrifugation (20 min, 100,000 g) and the pellet is resuspended in the appropriate volume of supernatant. Final rhodopsin concentrations of approximately 10 mg/ml are readily attainable. Higher concentrations are not easily manipulated because of the increased viscosity that causes air bubbles when loading the infrared cell. The final isotopic enrichment of ^{2}H$_2$O in samples

[6] A. Hvidt, and S. O. Nielsen, *Adv. Protein Chem.* **21**, 287 (1966).

[7] L. Willumsen, *C. R. Trav. Lab. Carlsberg* **38**, 223 (1971).

[8] E. Nabedryk-Viala, Thesis Doctorat d'Etat, University of Paris VII (1978).

[9] H. B. Osborne, C. Sardet, and A. Helenius, *Eur. J. Biochem.* **44**, 383 (1974).

[10] H. B. Osborne, C. Sardet, M. Michel-Villaz, and M. Chabre, *J. Mol. Biol.* **123**, 177 (1978).

[11] H. B. Osborne and E. Nabedryk-Viala, *Eur. J. Biochem.* **89**, 81 (1978).

[12] H. B. Osborne and E. Nabedryk-Viala, *FEBS Lett.* **84**, 217 (1977).

so prepared is 96%. This means that the measured values for the number of exchanged peptide hydrogens must be increased by 4%.

Samples of purified rhodopsin (3–4 mg/ml) were first extensively dialyzed against ^1H_2O exchange-in buffer containing the appropriate detergent. The exchange is then initiated by dialyzing (60–90 min) a 0.5-ml sample against two changes (15 ml) of ^2H_2O exchange-in buffer. Equilibration (to 99%) of hydrogen ions across the dialysis membrane ($\phi = 9$ mm) is attained in less than 20 min (measured with ^3H_2O). Therefore, the isotopic enrichment of ^2H_2O in these samples is 99%. If larger samples are to be prepared, the surface-to-volume ratio can be increased by inserting a small glass rod into the dialysis bag.

With these methods, exchange-in times of less than 60 min are not possible. In principle, measurements at shorter exchange-in times could be obtained by coupling a gel filtration column to the spectrophotometer cell, in a way similar to that described by Johansen.[13] However, lower rhodopsin concentrations would be obtained that would increase the error in the measurements. It would also necessitate the construction of a dark room around the infrared spectrophotometer compartment, which could be a practical limitation.

It is important to note that the exchange rate of solvent-exposed peptide hydrogen increases by a factor of 10 for an increment of 1 in the pH.[6] Hence the pH of the exchange-in buffer must be carefully controlled. For deuterated buffers the equivalent concentration parameter, p^2H, is given by the formula $p^2H = pH$ (meter reading) + 0.4.[14]

Infrared Spectroscopy Measurement

The ^1H–^2H exchange of peptide groups is monitored by standard infrared spectroscopy method.[1,6–8] Infrared spectra are recorded with a double-beam spectrophotometer equipped with matched thermostated cells (CaF_2, 0.1-mm pathlength). The temperature inside the sample should be directly measured with a chromel–alumel thermocouple. The reference cell contains either supernatant from the ^2H_2O buffer wash step or the dialyzate from the last dialysis, as appropriate. The spectrophotometer is purged with dry nitrogen obtained by evaporating liquid nitrogen. Analysis of the infrared spectra can be performed with a PDP 12 computer after digitalizing the data. Difference absorption spectra and integrated band intensities can then easily be obtained.

The fraction of unexchanged peptide hydrogens (N^1H) is estimated as

$$N^1H = A_{\text{amide II}} / (w A_{\text{amide I}})$$

[13] J. T. Johansen, *Biochim. Biophys. Acta* **214**, 551 (1970).
[14] P. K. Glasoe and F. A. Long, *J. Phys. Chem.* **64**, 188 (1960).

where $A_{\text{amide II}}$ and $A_{\text{amide I}}$ are the absorbances at the maximum of the amide II and amide I bands, respectively (after background correction). No attempt was made to estimate w, the ratio $A_{\text{amide II}}/A_{\text{amide I}}$ at time zero, i.e., prior to any deuterium substitution in ROS. For soluble proteins, $w = 0.45 \pm 0.05$[1,6–8]; the w value is usually determined either by extrapolating to zero time the exchange kinetics at $10°$, pH 3 (where the exchange rate constant is minimal), or by recording infrared spectra of protein films. For ROS membranes and purified rhodopsin, w was taken as 0.45, which is comparable to the 0.50 value recently reported by Rothschild et al.[4] for suspensions of ROS and disoriented films of rhodopsin.

The amide II band background, arising from side chains and carboxyl groups, is determined after deuteration of all the peptide groups. Complete peptide exchange is obtained either by heating samples to $60°$ for 6 hr, or by treatment with 5% SDS at $60°$. For samples of rhodopsin in detergent micelles, heating to $60°$ for 1 hr was sufficient to obtain complete exchange of the peptide groups.[11] These preparations give similar values for the ratio $A_{\text{amide II}}/A_{\text{amide I}}$ after total exchange (0.04 ± 0.02).

Errors

Since the concentration of purified rhodopsin is about one-fifth that generally used in standard infrared peptide exchange measurements, the accuracy of these measurements is less ($\pm 5\%$). For membrane-bound rhodopsin, the higher concentrations permit more reproducible measurements ($\pm 2\%$). The major source of error in these measurements is the determination of the proportionality factor (w). The uncertainty in the value of w introduces an error of up to $\pm 10\%$ in the determination of the percentage of unexchanged peptide hydrogens, whose total error will therefore be $\pm 15\%$ of its value. This is particularly important when the peptide hydrogen exchange data are compared with other data obtained by different techniques. When changes, caused by an external stimulus, in the percentage of unexchanged peptide hydrogens are to be measured, the associated errors are much smaller. For infrared spectra recorded from the same sample, before and after illumination, differences as small as 0.5% are clearly and reproducibly observable.

[82] A Comparison of Hydrogen Exchange Methods Applied to Rhodopsin

By NANCY W. DOWNER and H. BEVERLEY OSBORNE

Complementary kinds of information are obtained from the two different isotope methods that are available for studying the hydrogen exchange kinetics of membrane-bound rhodopsin. The use of both methods turns out to be particularly important, since the analysis of tritium exchange data for membrane systems is complicated by the potential contribution to the observed exchange curves of components other than peptide hydrogens.

The infrared deuterium method provides a convenient and relatively continuous monitor of the exchange kinetics.[1] Its most significant feature for the study of rhodopsin and other membrane proteins is that it is specific for the amide hydrogens. The percentage of amides deuterated is determined directly and quantitatively. The most serious drawback of the method is that the manipulations involved in introducing disk membranes or purified rhodopsin into ^2H_2O require at least 30 min. Even under conditions of slowest peptide exchange the kinetics of the free peptide class cannot be observed and hence the fraction of free peptides cannot be estimated.

Tritium exchange methods determine directly and with high accuracy the total number of hydrogens remaining unexchanged at a given time.[2] In contrast to the former method, tritium exchange measurements can be made within 10 sec. Therefore, exchange of the fastest amide hydrogens is seen as well as that of the much slower internally bonded peptide hydrogens. The fraction of peptides in this latter class gives an estimate of the amount of secondary structure in a protein. This is in principle one of the most straightforward and useful pieces of information to be extracted from hydrogen exchange studies on membrane proteins.

The interpretation of the tritium exchange data for rhodopsin is complicated by the fact that side chain and/or lipid hydrogens appear to make a significant contribution to the total number of hydrogens observed in these experiments.[3,4] This was not a critical issue in hydrogen exchange studies on soluble proteins, since it has been well established that their tritium exchange data contain virtually no detectable contribution from

---

[1] H. B. Osborne and E. Nabedryk-Viala, this volume [81].

[2] N. W. Downer and H. B. Osborne, this volume [80].

[3] H. B. Osborne and E. Nabedryk-Viala, *FEBS Lett.* **84,** 217 (1977).

[4] J. J. Englander, N. W. Downer, and S. W. Downer, *J. Biol. Chem.* **257,** 7982–7986 (1982).

METHODS IN ENZYMOLOGY, VOL. 88

protons other than amide protons. One consequence of the presence of slowly exchanging nonpeptide hydrogens for the tritium exchange of rhodopsin and other membrane proteins is that equilibriation may not be complete when the total number of exchanged hydrogens is equal to the number of peptide hydrogens. Infrared spectroscopy allows the conditions for complete labeling of the peptide hydrogens to be defined. Furthermore, by comparing the data of peptide exchange measured by infrared spectroscopy with the total proton exchange measured by the tritium method, the contribution of side chain and/or lipid hydrogens can be evaluated. In order to do this, it is necessary to convert the percentage of amides deuterated into the number of exchanged amides/rhodopsin chromophore. This requires the determination of the total number of amino acid residues per rhodopsin chromophore in the membrane samples. Given the errors involved in the necessary measurements, it should be possible to estimate with an accuracy of ± 20% the number of side chain and/or lipid hydrogens that are contributing to the overall membrane exchange.

Both deuterium and tritium methods can be used to detect structural changes that would be reflected in modifications of the protein exchange kinetics. Again the advantage of the infrared method derives from its specificity for peptide hydrogens and the relative convenience of the measurements. Tritium exchange gives a more comprehensive view of the possible structural changes, since the entire kinetic range of peptide exchange is observable. Changes in secondary structure related to perturbations in the portions of the exchange curve inaccessible to the infrared deuterium method can be detected. However, with improved sample handling and instrumentation it may be possible to obtain better time resolution with the infrared method. For the investigation of structure changes, the complementary information from both methods will be required to analyze the respective roles of the peptide backbone and protein side chains in the observed changes.

[83] Spin-Label Probes of Light-Induced Electrical Potentials in Rhodopsin and Bacteriorhodopsin

By DAVID S. CAFISO, WAYNE L. HUBBELL, and ALEXANDRE
QUINTANILHA

Two well-studied retinal containing proteins involved in photochemical conversion are rhodopsin in the rod outer segment (ROS) disk membrane and bacteriorhodopsin found in the purple membrane of *Halobac-*

terium halobium. Even though the chemical intermediates of the photocycle in both systems have been well characterized, the electrical events accompanying the cycle and their physiological significance still remain to be clarified.

Here we describe the use of a number of derivatives of spin-labeled molecules to

a. Measure and distinguish between transmembrane ($\Delta\psi$) and interfacial boundary potentials in the ROS disk membrane of the vertebrate photoreceptor in order to elucidate the nature of an early electrical event that occurs upon the photolysis of rhodopsin (contributed by D. Cafiso and W. Hubbell).

b. Measure the kinetics and stoichiometry of surface potential changes ($\Delta\psi_s$) in bacteriorhodopsin purple membrane sheets in an attempt to understand the mechanism of H^+ release and uptake that accompanies the photocycle (contributed by A. Quintanilha).

The use of molecular probes for the quantitation of electrical events is essential when such measurements are inaccessible to microelectrodes. The advantages of using spin probes have been discussed by several authors,[1-6] and the rapid determination of the membrane-bound and aqueous signals from the electron paramagnetic resonance (EPR) spectra is an important one.

Estimation of Transmembrane and Boundary Potentials in ROS Membranes

Properties of Hydrophobic Ions. Hydrophobic ions, such as tetraphenylphosphonium (ϕ_4P^+), unlike inorganic ions (Na^+, K^+, etc.), are membrane-permeable ions. These ions also associate with the membrane in two interfacial boundary regions. The boundary region lies beneath the membrane–solution interface in a region that is inaccessible to salts in solution. This behavior is predicted on theoretical grounds[7] and results from a summation of electrical and chemical energy terms for the transfer of an ion and hydrophobic groups into the membrane, respectively. The resulting free energy profile for these ions is shown in Fig. 1. Despite the sim-

[1] B. J. Gaffney, this series, Vol. 32, p. 161.
[2] J. D. Castle and W. L. Hubbell, *Biochemistry* **15,** 4818 (1976).
[3] D. S. Cafiso and W. L. Hubbell, *Biochemistry* **17,** 187 (1978).
[4] D. S. Cafiso, Ph.D. Thesis, University of California, Berkeley (1979).
[5] R. J. Mehlhorn and L. Packer, this series, Vol. 56, p. 515.
[6] A. T. Quintanilha, *in* "Frontiers of Biological Energetics" (P. L. Dutton, J. S. Leigh, and A. Scarpa, eds.), p. 257. Academic Press, New York, 1978.
[7] B. Ketterer, B. Neumeke, and P. Läuger, *J. Membr. Biol.* **5,** 225 (1971).

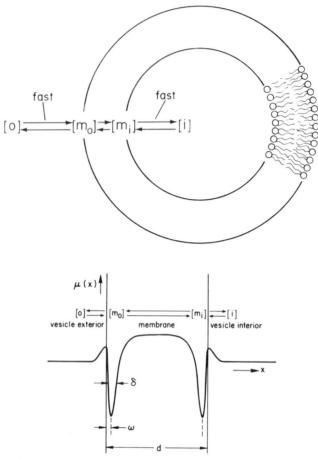

Fig. 1. Multiple equilibria for spin-labeled hydrophobic ions in a phospholipid vesicle. The symbols o, m_o, m_i, and i represent the four regions of space occupied by the ion: the exterior solution, the exterior interface, the interior interface, and the vesicle interior, respectively. Shown below is the potential energy profile across the bilayer for a spin-labeled hydrophobic ion in the absence of a transmembrane potential. δ is the width of the potential minima, ω is the displacement of the minima from the membrane surface, d is the bilayer thickness, and $\mu(x)$ is the potential energy of the ion.[7]

plicity of this model, there is presently strong evidence that the general features of this binding are correct. Recent ^{13}C-NMR studies of the binding of TPP$^+$ indicate that this type of ion resides near the C_2 and carbonyl carbons of the phospholipid.[8]

Spin-Labeled Hydrophobic Ions: Equilibria. Spin-labeled derivatives

[8] W. L. Hubbell, D. S. Cafiso, and M. F. Brown, *Fed. Proc., Fed. Am. Soc. Exp. Biol.* **39**, 1983 (1980).

of phosphonium ions have been synthesized.[3,4] These ions are membrane permeable, and in the presence of small bilayer vesicles, the EPR spectra indicate that there are two distinct populations of probe. One population arises from spin label in aqueous solution; the other arises from spin label associated with the bilayer in the boundary region. (The equilibria and binding for the phosphonium are shown in Fig. 1.) Thus the spin label resides in four regions of a vesicle system: the interior and exterior aqueous spaces that give rise to a *free* signal and the internal and external interfacial boundary regions that give rise to a *bound* signal. The partitioning of the label, that is, the ratio of bound-to-free probe λ, may be varied by changing either the vesicle concentration or the length of the alkyl chain on the probe molecule.

Phosphonium Ions: Potential Dependence. Since the spin label is charged and membrane permeable, its distribution throughout the vesicle system will depend on the relative electrostatic potentials in the regions where it resides, and thus the partitioning of the label is dependent on the transmembrane and interfacial boundary potentials (see Fig. 2). In the presence of an inside negative transmembrane potential, the ratio of the bound-to-free spin label increases. This potential dependence is shown in Fig. 3. Physically, the bound-to-free ratio changes because the ratio of aqueous volume to membrane surface is much less on the vesicle interior than on the vesicle exterior. When the label is on the interior, as in the

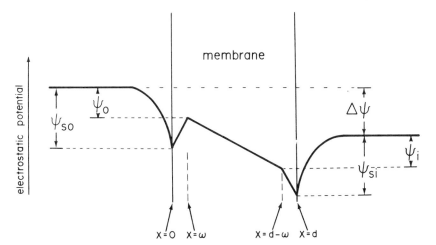

FIG. 2. Simplified electrostatic profile across a membrane. The bilayer thickness is d and the boundary regions are at $x = \omega$ and at $x = d - \omega$. ψ_o and ψ_i are the external and internal potentials of the boundary regions with respect to their corresponding aqueous phases. ψ_{so} and ψ_{si} are the corresponding surface potentials and $\Delta\psi$ is the transmembrane potential. The boundary potential is defined here as the sum of a surface potential ψ_{so} and ψ_{si} and an interfacial potential drop (i.e., between $x = 0$ and $x = \omega$ or $x = d$ and $x = d - \omega$).

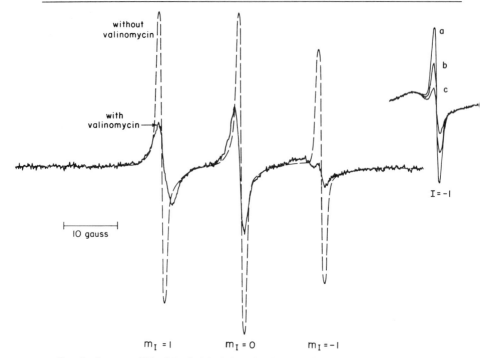

FIG. 3. Spectra of 20 μM spin-label phosphonium I(3)[3] in the presence of sonicated egg PC vesicles 2% (w/v) with (—) and without (----) an inside negative transmembrane potential. The potential was established by the addition of 1 μM valinomycin in the presence of a K$^+$ gradient (450 mM K$^+$ inside, 450 mM Na$^+$ outside, SO$_4^{2-}$ anion). (Inset) The high field resonance at three K$^+_{in}$/K$^+_{out}$ ratios in the presence of 1 μM valinomycin (a) 3/1, (b) 10/1, and (c) 45/1.

case of a large inside-negative transmembrane potential, the probe is essentially all bound.

Quantitation. In practice, the quantity that is required from the EPR spectrum is the ratio of bound-to-free probe λ. This quantity is easily obtained from the spectrum, and procedures for extracting this information have been given elsewhere; consequently, they will not be discussed here.[2]

A thermodynamic description of the changes in partitioning as a function of the potentials $\Delta\psi$, ψ_i, and ψ_o has been given for the phosphonium spin labels[3]; the result is given in the following equation.

$$\lambda = \frac{V_{m_i}}{V_i} \cdot \frac{K_{m_i}e^{-\psi_i\beta} + K_{m_o}\dfrac{V_{m_o}}{V_{m_i}}e^{-\psi_o\beta}e^{\Delta\psi\beta}}{1 + \dfrac{V_o}{V_i}e^{\Delta\psi\beta}} \tag{1}$$

where

$$\beta = ZF/RT$$

V_o, V_i = the volumes of the external and internal vesicle aqueous phases, respectively.

V_{m_o}, V_{m_i} = the "volumes" of the external and internal membrane phases, respectively. These volumes are the effective volumes of the interfacial phases occupied by the phosphonium.

K_{m_o}, K_{m_i} = binding constants for the label to the external and internal sides of the bilayer, respectively.

$\psi_i, \psi_o, \Delta\psi$ = potentials defined in Fig. 2.

Z, F, R, T = charge on the label, the Faraday constant, the gas constant, and the absolute temperature, respectively.

In most systems $V_{m_o}/V_{m_i} \simeq 1$ and V_o/V_i is determined from the vesicle geometry and concentration. This equation has been tested experimentally for the case where membranes are symmetric ($K_{m_i}e^{-\psi_i\beta} = K_{m_o}e^{-\psi_o\beta} \equiv K$). In this case the equation given above reduces to

$$\lambda = \frac{KV_{m_i}}{V_i} \frac{1 + V_{m_o}/V_{m_i}e^{\Delta\psi\beta}}{1 + V_o/V_ie^{\Delta\psi\beta}} \tag{2}$$

and KV_{m_i}/V_i may be measured directly from a study of the binding of the spin label and knowledge of the vesicle geometry. There is an excellent agreement between predicted and measured values for ψ in model systems.[3]

For asymmetric systems in which $K_{m_i}e^{-\psi_i\beta} \neq K_{m_o}e^{-\psi_o\beta}$, the dependence of λ on $\Delta\psi$ may be dramatically altered. If the binding constant on the inside of the vesicle becomes significantly less than the outside, the dependence of λ on $\Delta\psi$ will decrease. In native membranes the effect may be significant and a check by an empirical calibration will be necessary. At the same time, this dependence on asymmetry may offer a unique way of determining the sidedness of vesicle systems. For example, by establishing known diffusion potentials with known values of V_o/V_i, $K_{m_i}e^{-\psi_i\beta}$ and $K_{m_o}e^{-\psi_o\beta}$ may be obtained. By varying the ionic strength it should be possible to establish whether the asymmetry resides in the binding constants or in the surface charge density. Thus direct measurements of potential using the phosphonium labels are best suited to reconstituted or other symmetric systems. However, their careful and sophisticated application in native membranes probably will reveal valuable information regarding charge distributions and membrane asymmetry.

In practice, a change seen in λ, the partitioning of the phosphonium, may be complex. Changes in transmembrane, surface, or boundary potential as well as nonelectrostatic changes resulting in a change in K_{m_i} and K_{m_o} may all give rise to a change in λ. Fortunately, it is possible to distinguish between these possibilities.[9] Transmembrane potential components may be dissected out by their sensitivity to ionophores under conditions that short-circuit the membrane. Surface potentials may be characterized by their salt dependence and may also be directly measured using alkylammonium labels described later. Boundary potentials remain under conditions in which $\Delta\psi$ is short-circuited and surface potentials have been titrated by high salt. Nonelectrical changes may be tested by examining the partitioning or mobility of a number of uncharged hydrophobic probes or by the use of oppositely charged probes. Because the mode of response for the phosphonium is well defined, it is possible in a number of electrostatic cases to predict in a quantitative manner the effect on the partitioning of the label.

Electrical Changes in ROS Disk Membranes. The spin-labeled phosphonium ions have been used to study the electrostatic properties of the vertebrate photoreceptor membranes.[9,10] In Fig. 4(a), the response of the spin-labeled phosphonium in isolated ROS disk membranes on flash illumination is shown. The recording made here is one of the high field, $m_I = -1$, resonance at low time resolution. In practice, λ (and hence the potential) is calculated from a measure of this resonance amplitude.[6] This response is clearly a composite of two components, a rapidly and a much more slowly decaying component.

Evidence has been presented indicating that these phenomena are electrical in origin.[9,10] The most convincing evidence is that spin-label probes of opposite charge give opposite response. Using the phosphonium probes, the complex changes seen in Fig. 4(a) were shown to be due to a change in *transmembrane potential,* $\Delta\psi$, outer *surface potential* $\Delta\psi_{so}$, and outer *boundary potential,* $\Delta\psi_o$. In the presence of ionophores or in other cases where the disk is made "leaky," the rapidly decaying component disappears. The response of the phosphonium in the disk membrane with 10 μM valinomycin present in a K^+ medium is shown in Fig. 4(b). This rapidly decaying component is due to the formation and decay of a transmembrane potential. The surface potential change was identified by alkylammonium spin labels. The boundary potential component was identified by measurements under conditions where the surface potential was titrated with high salt and the membranes were made leaky with valinomycin.

All these potential changes may be accounted for by a single event, the translocation of charge (approximately 1 per photolyzed rhodopsin) from

[9] D. S. Cafiso and W. L. Hubbell, *Biophys. J.* **30,** 243 (1980).
[10] D. S. Cafiso and W. L. Hubbell, *Photochem. Photobiol.* **32,** 461 (1980).

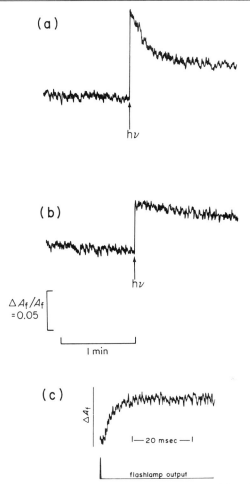

FIG. 4. Recordings of the high-field resonance amplitude, A_f, of the spin-label phosphonium, 35 μM^9 in ROS disk membranes (4 mg/ml). ROS are suspended in 125 mM K$_2$SO$_4$, 10 mM MES, pH = 6.15. Recording (a) is without valinomycin; (b) and (c) contain 1 mM valinomycin. The samples are bleached through a radiation window in the EPR cavity using a 600-μsec, 100-J flash.

the aqueous solution, across the external disk interface, and into the disk membrane interior. Calculations of the salt dependence of the phosphonium response based on this model agree well with the measured dependence. The charge transferred in this event is believed to be H$^+$and the rise time (see Fig. 4) is similar to that for the formation of the MII spectral intermediate of the visual pigment.

The charge transfer appears to be the molecular event that results in the macroscopic potential known as the early receptor potential (ERP) where rhodopsin in the plasma membrane (not the disk membrane) is the source of this potential.[11] The charge-transfer event is linear in the amount of rhodopsin photolyzed and is, therefore, not directly involved in the later stages of visual transduction. It could, however, be an event that is important for electrostatic interactions and coupling within the disk membranes between rhodopsin molecules or other disk membrane proteins. In summary, measurements using the technique described here reveal an interfacial charge transfer occurring across the external disk membrane interface. This charge transfer results in a salt-dependent surface potential, transmembrane potential, and boundary potential change. The boundary potential change represents a change in potential within the low dielectric of the disk membrane, a potential that is not screened under high-salt conditions.

Surface Potential Changes in Bacteriorhodopsin Purple Membranes

Bacteriorhodopsin functions as an electrogenic light-activated proton pump.[12] The detailed molecular mechanism of H^+ translocation may involve the reversible protonation of the retinal Schiff's base[13] and charge displacements measured in other amino acids.[14-16]

Purified purple membranes appear as membrane sheets that do not form closed membrane vesicles[17]; during steady-state illumination they produce pH changes[18] in their suspension medium that are strongly pH and temperature dependent.[19] As expected, the H^+ release can be correlated with the accumulation of the deprotonated intermediate M_{412} in the bacteriorhodopsin photoreaction cycle. However, the light-induced proton release and its stoichiometric relation to M_{412} vary with salt concentration in the aqueous medium. This effect suggests the involvement of surface potentials in the mechanism of the proton pump. As will be shown later, it is indeed possible that, when the highly negative surfaces of the

[11] H. Ruppell and W. A. Hagins, in "Biochemistry and Physiology of Visual Pigments" (H. Langer, ed.), p. 257. Springer-Verlag, Berlin and New York, 1973.
[12] W. Stoeckenius, R. H. Lozier, and R. A. Bogomolni, Biochim. Biophys. Acta 505, 215 (1979).
[13] A. Lewis, S. Spoonhower, R. A. Bogomolni, R. H. Lozier, and W. Stoeckenius, Proc. Natl. Acad. Sci. U.S.A. 71, 4462 (1974).
[14] T. Konishi and L. Packer, FEBS Lett. 92, 1 (1978).
[15] B. Hess and D. Kuschmitz, FEBS Lett. 100, 334 (1979).
[16] R. A. Bogomolni, L. Stubbs, and J. K. Lanyi, Biochemistry 17, 1037 (1978).
[17] K. A. Fisher and W. Stoeckenius, Science 197, 72 (1977).
[18] D. Oesterhelt and B. Hess, Eur. J. Biochem. 37, 316 (1973).
[19] H. Garty, G. Klemperer, M. Eisenbach, and S. R. Caplan, FEBS Lett. 81, 238 (1977).

purple membrane[12] are not screened by ions in the aqueous phase, the protons released from the Schiff's base or other protonatable groups may never leave the membrane surface.

Surface Potential Measurements. The estimation of surface potentials (and charge density) of membranes from the partitioning of alkylammonium spin labels has been surveyed by several authors.[1,2,5] The method has been applied in the case of several biological membranes.[6,9,20,21] Since the probes used—e.g., 4-(dodecyldimethylammonium)-1-oxyl-2,2,6,6-tetramethylpiperidine bromide (CAT_{12})—are essentially detergents, it is important that they do not modify the activity parameters studied.

The partitioning of the label between both sides of purple membrane sheets and the aqueous environment will depend on the membrane concentration, the ionic aqueous concentration, and the pH of the medium. Mathematically, the distribution of the probe will depend on the hydrophobic interaction and the electrical surface potential of the membrane. As pointed out previously, activity-dependent changes in surface potential can be distinguished from changes in hydrophobic interactions by the simple fact that the former are ionic strength dependent and should be abolished at high salt concentrations.

The surface potential of any membrane (ψ_s) may be determined directly by measurement of the spin-label partition coefficient (λ) if the value in the absence of surface charge (λ_0) is known. Changes in surface potential ($\Delta\psi_s$) will result in changes in the partition of the probe, which can be calculated from the Boltzmann relation[2,5,6]

$$\Delta\psi_s = \frac{RT}{ZF} \ln \frac{\lambda_1}{\lambda_2} \tag{3}$$

where λ_1 and λ_2 stand for the partitioning of the probe at two different states.

The surface charge density σ and the surface potential can be related by the following equation:

$$\sinh \frac{ZF\psi_s}{2RT} = \frac{500\pi}{DRT} \frac{\sigma}{\sqrt{c}} \tag{4}$$

where c is the ionic concentration (moles/liter) and D is the aqueous dielectric constant.

In the case of purple membranes we use Eq. (3) and assume that the binding constants for the label to both sides are the same and uniform.

[20] A. T. Quintanilha and L. Packer, *FEBS Lett.* **78**, 161 (1977).
[21] A. T. Quintanilha and L. Packer, *Arch. Biochem. Biophys.* **190**, 206 (1978).

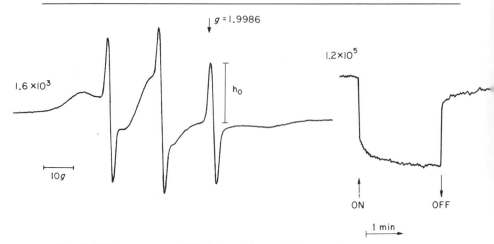

FIG. 5. EPR spectrum and light-induced changes in the amplitude of the high field aqueous CAT_{12} signal (h_o) in a purple membrane suspension. A 50-μl sample contained 0.256 mM bacteriorhodopsin, 0.256 mM valinomycin, 0.256 mM beauvericin, 100 mM KCl, 2 mM CAT_{12}, pH 7.2 at 22°C. Time course of the light- (40.5 mW/cm²) induced changes in h_o were measured in a flat cell (0.254 mm light path length).

Light-Induced Surface Potential Changes in Purple Membranes. At room temperature, light of moderate intensity converts only a very small fraction of bacteriorhodopsin into its M_{412} intermediate; to increase the steady-state concentration of M_{412} one can use higher light intensities or decrease the rate of its decay. Continuous illumination at high light intensities damages the system, so we have used the antibiotics valinomycin and beauvericin at a 1:1 molar ratio with the protein, to slow down the decay of M_{412}. In the presence of the antibiotics, two distinct phases in the decay of M_{412} were observed.[22] The kinetics of the flash-induced rise of M_{412} did not change. Under these conditions the spectral changes in the high-field aqueous EPR line of CAT_{12} during the continuous illumination are shown in Fig. 5.[23] Control studies with an uncharged spin probe suggested that there was very little change in the hydrophobicity of the purple membrane during illumination and that these changes were ionic strength independent.

The change in surface potential seems to be independent of ionic strength up to about 100 mM (KCl or NaCl), decreasing only at higher ionic strengths. It is clear from Eq. (4) that the same change in ψ_s at two

[22] Y. Avi-Dor, R. Rott, and R. Schnaiderman, *Biochim. Biophys. Acta* **545**, 15 (1979).
[23] C. Carmeli, A. T. Quintanilha, and L. Packer, *Proc. Natl. Acad. Sci. U.S.A.* **77**, 4707 (1980).

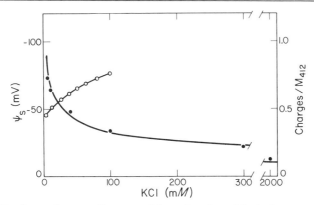

FIG. 6. Purple membrane surface potential as a function of the ionic strength of the medium. Purple membranes (0.128 mM bacteriorhodopsin) in the presence of 1 mM CAT$_{12}$, 0.128 mM valinomycin, and 0.128 mM beauvercin, pH 6.1 were titrated with KCl to give the indicated final concentration. The line represents a theoretical curve calculated according to Eq. (4) (assuming $\sigma = 0.00174$ negative charges/Å^2), whereas the points (●) were calculated experimentally from Eq. (3). The contribution to the surface charge due to the binding of CAT$_{12}$ to the membrane was corrected for; the error for each point is of the order of 10 mV. The ratio of light-induced charge changes per M$_{412}$ (○) were calculated by measuring the steady-state M$_{412}$ in the same sample using an Aminco DW 2 spectrophotometer (flat cell at 45° to measuring beam and to the illuminating beam), assuming $\Delta\epsilon^{412} = 23,000$ M^{-1} cm^{-1}.

different ionic strengths will correspond to a higher change in surface charge density at the higher ionic strength.

The charge density changes under steady-state illumination were related to the steady-state level of M$_{412}$ photointermediate for increasing ionic strength. As can be seen from Fig. 6, the number of negative charges per M$_{412}$ induced by light increased from less than 0.5 at 2 mM KCl to values close to 1.0 at higher ionic strengths (>100 mM KCl). A similar response was observed[22] for the ratio of light-induced proton release/M$_{412}$ under the same conditions.

The kinetics of surface potential changes could also be studied. However, since the generation of the $\Delta\psi_s$ was much faster at room temperature than the limit of sensitivity of our EPR instrumentation, we studied the laser-flash-induced decay of M$_{412}$ and the corresponding decay of the surface potential change (see Fig. 7). Both decays could be resolved into a fast and a slow component. The large number of CAT$_{12}$ molecules bound per bacteriorhodopsin did not change the kinetic constants of the decay of M$_{412}$, but increased by approximately 30% the contribution of the fast phase to the overall decay of M$_{412}$. Over a wide range of pH values, the kinetics of both the fast and slow components of the surface potential changes were slower than the kinetics of M$_{412}$ decay. At pH 7, the half-

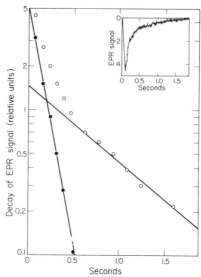

FIG. 7. Kinetics of the laser-flash-induced decay of CAT_{12} binding to the purple membranes. A phase-R dye laser with rhodamine 575 (0.2 J/flash, 150 nsec flash rise time) was used. Other conditions are as in Fig. 5. The inset is a trace average of 50 flashes. The EPR time constant was 8 msec. To analyze the kinetics, the decay of the EPR signal (○) was plotted on a semilog plot, and the "curve peeling" technique was used to distinguish the slow ($\tau_{1/2}$ = 574 msec) from the fast (●) ($\tau_{1/2}$ = 103 msec) first-order kinetic components.

times for the two phases in surface potential decay were about 110 and 370 msec, whereas the corresponding constants for the M_{412} decay were about 50 and 250 msec. This cannot be due to a limitation of the response time of the spin probe, since changes in the EPR signal can be generated with half-lives as short as 7 msec.

Low-Temperature Studies. It was important to provide independent experimental proof that the surface potential changes observed can also be observed in the absence of antibiotics.[24] Instrumentational limitations required that the reaction be slowed down. At − 10°, in the presence of 25% ethylene glycol, the flash-induced decay of $\Delta\psi_s$ showed biphasic kinetics; the fast and slow components had half-times of 200 and 750 msec, respectively. Because of the relatively high noise contribution from the instrumentation, the rise time for $\Delta\psi_s$ could not be accurately measured. Nevertheless, the data obtained gave a correlation coefficient of 0.77 for a half-time of 9.3 msec.[23] The half-times for the decay and the rise of M_{412} at − 10° are approximately 70 msec and 1.5 msec.[25,26] It seems therefore that

[24] S. Tokutomi, T. Iwasa, T. Yoshizawa, and S. Ohnishi, *FEBS Lett.* **114**, 145 (1980).
[25] W. V. Sherman and S. R. Caplan, *Nature (London)* **258**, 766 (1975).
[26] M. Chu-Kung, D. DeVault, B. Hess, and D. Oesterhelt, *Biophys. J.* **15**, 907 (1975).

both the rise and the decay of the $\Delta\psi_s$ lag slightly behind the kinetics of M_{412}.

Discussion. The stoichiometry of proton release per M_{412} has been shown by other workers to be ionic strength dependent.[27,28] These results were interpreted to indicate a change in proton release due to surface potential changes in the protein. The spin probe technique made it possible to measure directly the surface potential and the changes thereof during steady-state illumination. If light polarizes the membrane, this could mean that the stoichiometric values given are underestimated, as the charge changes were assumed to take place homogeneously on both sides of the purple membrane.

Furthermore, under all conditions studied, both the rise and the decay of the surface potential changes were slightly slower than the M_{412} changes, indicating that the dissociation of a proton from the Schiff's base precedes the appearance of charge changes on the surface of the membrane fragments. Measurements of flash-induced proton absorbance changes in pH indicators in purple membrane suspensions, which were interpreted to measure proton release, were slower than the formation of M_{412}.[29] Flash-induced volume changes, interpreted to measure proton changes in the medium, also showed that the release and uptake of protons is slower than the formation and decay of the M_{412}.[30]

We have suggested that the charge changes measured on the surface of the purple membrane may reflect the dissociation of amino acid residues that are on the path of transfer of protons across the membrane.

Comments

The spin-labeled phosphonium ions discussed here change their partitioning as a function of the transmembrane and boundary potentials in a vesicle system. The alkylammonium probes change their partitioning as a function of surface potential. For each of these techniques, the quantitative theory has been carefully tested under controlled conditions in model systems, where the interpretation of results is relatively straightforward. The success of these methods in model membrane systems makes them ideal tools for investigations in reconstituted or other symmetric systems. For native membranes that are typically not symmetric and of variable

[27] B. Hess and D. Kuschmitz, in "Frontiers of Biological Energetics" (P. L. Dutton, J. S. Leigh, and A. Scarpa, eds.), p. 257. Academic Press, New York, 1978.

[28] D. Ort and W. W. Parson, *Biophys. J.* **25**, 341 (1979).

[29] R. D. Lozier, W. Niederberger, R. A. Bogomoloni, S. B. Hwang, and W. Stoeckenius, *Biochim. Biophys. Acta* **440**, 545 (1976).

[30] D. R. Ort and W. W. Parson, *J. Biol. Chem.* **253**, 6158 (1978).

morphology, one needs to be more cautious in the interpretation of partitioning changes. However, through the proper studies, an entirely new dimension of information becomes accessible.

Acknowledgments

Work cited by the authors was supported by grants from the National Institutes of Health (EY 00729 to W.L.H.), the Jane Coffin Childs Fund for Medical Research to D.S.C., and by the Department of Energy under Contract W-7405-ENG-48 to A.T.Q.

[84] Infrared Absorption of Photoreceptor and Purple Membranes

By K. J. ROTHSCHILD, R. SANCHES, and NOEL A. CLARK

I. Overall Perspective

Although extensive visible and ultraviolet absorption studies (200–700 nm) have been made on photoreceptor and purple membrane,[1-3] only recently have measurements been extended to the mid-infrared region (200–4000 cm^{-1}).[4-7] Such infrared measurements probe structural groups of the membrane and are sensitive to the amplitude, frequency, and polarization of their vibrational modes.[8] As will be discussed in this section, information about the state of the membrane components such as lipid fluidity,[9,10] membrane protein secondary structure,[8,11-13] specific bond orien-

[1] C. D. B. Bridges, in "Biochemistry of the Eye" (C. N. Graymore, ed.), Chapter 9. Academic Press, New York, 1970.
[2] T. B. Ebrey and B. Honig, Q. Rev. Biophys. **8**, 129 (1975).
[3] W. Stoeckenius, R. H. Lozier, and R. A. Bogomolni, Biochim. Biophys. Acta **505**, 215 (1979).
[4] H. B. Osborne and E. Nabedryk-Viala, FEBS Lett. **84**, 217 (1977).
[5] H. B. Osborne and E. Nabedryk-Viala, Eur. J. Biochem. **89**, 81 (1978).
[6] K. J. Rothschild and N. A. Clark, Biophys. J. **25**, 473 (1979).
[7] K. J. Rothschild, W. J. DeGrip, and R. Sanches, Biochim. Biophys. Acta **596**, 338 (1980).
[8] F. Parker, "Applications of Infrared Spectroscopy in Biochemistry Biology and Medicine." Plenum, New York, 1971.
[9] D. F. H. Wallach, S. P. Verma, and J. Fookson, Biochim. Biophys. Acta **559**, 153 (1979).
[10] D. Chapman, "The Structure of Lipids by Spectroscopy and X-ray Techniques". Wiley, New York, 1965.
[11] S. N. Timasheff, H. Susi, and L. Stevens, J. Biol. Chem. **23**, 5467 (1967).
[12] H. Susi, S. N. Timasheff, and L. Stevens, J. Biol. Chem. **23**, 5460 (1967).
[13] R. D. B. Fraser and E. Suzuki, in "Physical Principles and Techniques of Protein Chemistry" (S. J. Leach, ed.) Part B, p. 213. Academic Press, New York, 1970.

tation,[6,8,9,14] α-helix tilt,[6] deuteration exchange rates,[4,5,8] conformational changes,[15,16] and effects of hydration[17] can be obtained with infrared absorption spectroscopy.

Advantageously, the problems normally encountered with infrared spectroscopy of biological samples such as high background water absorption and relatively long sampling time can be partially alleviated with the Fourier transform technique.[18] This method has been increasingly applied to biological membranes,[19-21] and is particularly valuable in the detection of small spectral changes.[15] Hence infrared absorption spectroscopy presently offers a promising tool in photoreceptor and purple membrane research and an attractive complement to other methods.

In contrast to resonance Raman studies, which measure only the chromophore vibrations of rhodopsin and bacteriorhodopsin,[22,23] infrared absorption will sample all membrane components. (By removing the chromophore, as in the case of opsin, Raman spectroscopy also reveals nonchromophoric vibrations.[24]) Consequently, the infrared spectrum of membranes is complex, often consisting of many overlapping peaks. It is possible, however, to separate vibrations due to protein and lipids by careful delipidation and reconstitution procedures as will be described. Further, comparison with model systems are extremely useful for peak assignment.

II. Basic Principles

We shall very briefly discuss the fundamentals of absorption spectroscopy with particular emphasis on the structure of membrane proteins. Several excellent texts are available that deal in greater detail with this subject.[8,10,13] In addition, the recent review on infrared and Raman spec-

[14] B. Jasse and J. L. Koenig, *J. Macromol. Sci., Rev. Macromol. Chem.* **C17**, 61 (1979).
[15] K. J. Rothschild, M. Zagaeski, and W. A. Cantore, *Biochem. Biophys. Res. Comm.* **103**, 483 (1981).
[16] K. J. Rothschild and H. Marrero, *Proc. Natl. Acad. Sci. U.S.A.* **79** (1982), in press.
[17] A. S. Schneider, C. R. Middaugh, and M. D. Oldewurtel, *J. Supramol. Struct.* **10**, 265 (1979).
[18] J. B. Bates, *Science* **191**, 29 (1976).
[19] K. J. Rothschild and N. A. Clark, *Science,* **204**, 311 (1979).
[20] H. L. Casal, I. C. P. Smith, D. G. Cameron, and H. H. Mantsch, *Biochim. Biophys. Acta* **550**, 145 (1979).
[21] D. G. Cameron, H. L. Casal, and H. H. Mantsch, *J. Biochem. Biophys. Methods* **1**, 21 (1979).
[22] A. Lewis, J. Spoonhower, R. A. Bogomolni, R. Lozier, and W. Stoeckenius, *Proc. Natl. Acad. Sci. U.S.A.* **71**, 4462 (1974).
[23] A. R. Oseroff and R. H. Callender, *Biochemistry* **13**, 4243 (1974).
[24] K. J. Rothschild, J. R. Andrew, W. J. DeGrip, and H. E. Stanley, *Science* **191**, 1176 (1976).

troscopy as applied to biomembranes by Wallach et al.[9] is highly recommended.

Amide carbonyl groups display a stretching frequency (amide I mode) that can vary between 1625 and 1680 cm^{-1}.[8,25] The actual vibrational frequency of a protein carbonyl group will depend on its specific environment and conformation, which includes such factors as hydrogen bonding, peptide bond angles, and neighboring atoms. We can consider this molecular environment the "fine tuning" of the carbonyl vibrational frequency.

In the case of proteins and polypeptides, the amide carbonyl stretching frequency can be used as an indicator of protein secondary structure.[8,11-13] Most α-helical proteins and polypeptides have a carbonyl stretching frequency, referred to as the amide I band, between 1650 and 1655 cm^{-1}.[11] Antiparallel β structure, in contrast, absorbs strongly near 1630 cm^{-1} with a weaker band at 1680 cm^{-1}.[8,11-13] Random-coil structure produces a band near 1655 cm^{-1}.[11] Hence, additional methods must be used to distinguish between random and α-helical structure on the basis of the amide I frequency. Exchange of amide hydrogens with deuterium is often used, since the amide I vibration of random-coil structure shifts to 1640 cm^{-1} upon deuteration, whereas the amide I α-helix frequency is only affected by a 2–4 cm^{-1} shift.[8,11]

The absorption intensity of a carbonyl stretching vibration depends on the relative angle between its transition moment and the electric field direction (maximum absorption occurs at 0°, minimum absorption at 90°). It is known from polarized absorption studies on peptide-containing crystals that the direction of the transition moment for the amide I mode is almost parallel to the carbonyl bond.[26-28] In contrast, the amide II vibration, which consists predominantly of in-plane N—H bending, makes an angle between 72 and 78° with the carbonyl bond. The amide A mode, consisting primarily of N—H stretching, has its transition moment also approximately parallel to the carbonyl group (cf. Fig. 1c).

In the case of an α helix, it can easily be seen from its structure that the maximum C=O and N—H stretching absorptions will occur when light is polarized parallel to the α-helix axis. In contrast, the N—H bending mode absorbs maximally when the electric vector is oriented predominantly perpendicular to the α-helix axis. Hence the overall α-helix orientation of a membrane protein may be probed, as shall be discussed in the

[25] L. J. Bellamy, "The Infrared Spectra of Complex Molecules." Wiley, New York, 1958.
[26] E. J. Ambrose and A. Elliot, Proc. R. Soc. London Ser. A 205, 47 (1951).
[27] M. Tsuboi, J. Polym. Sci. 58, 139 (1962).
[28] I. Sandeman, Proc. R. Soc. London, Ser. A 232, 105 (1955).

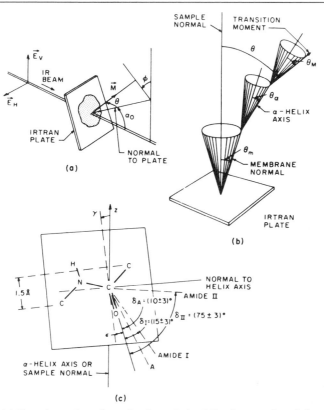

FIG. 1. (a) Experimental configuration for polarized Fourier transform infrared absorption measurements. (b) Geometry of the axially symmetric distributions of the transition moment M, the α-helix axis, and the membrane normal. (c) Peptide group plane geometry. (From Rothschild and Clark.[6])

case of rhodopsin and bacteriorhodopsin, by measuring the infrared dichroism of the amide I, II, and A vibrations.

III. Infrared Absorption of Photoreceptor Membrane

The infrared absorption of intact photoreceptor membrane can be measured using a variety of methods. In order to reduce water absorption, it is sometimes desirable to work with partially dried films.[7,29,30] Highly uniform films of photoreceptor membrane or reconstituted rho-

[29] K. J. Rothschild, N. A. Clark, K. M. Rosen, R. Sanches, and T. L. Hsiao, *Biochem. Biophys. Res. Commun.* **92**, 1266 (1980).
[30] K. J. Rothschild, R. Sanches, T. L. Hsiao, and N. A. Clark, *Biophys. J.* **31**, 53 (1980).

dopsin membranes up to 50 μm thick can be deposited on infrared transmitting materials such as KRS-5 or AgCl.[31] These films display optical linear dichroism of the 500-nm α band of rhodopsin, indicating that the membranes are oriented.[32] Furthermore, the rhodopsin in these films appears to be fully regenerable with 11-cis-retinal subsequent to bleaching, an important indication of the structural integrity of rhodopsin. Intact rod outer segments can also be studied in ^2H_2O.[4,5] This method has the advantage of allowing measurements of membranes in an environment similar to their native state. In this case magnetic fields can be used to orient the rods for polarization studies.[33] In both of these methods care must be taken to demonstrate that the results are not affected by dehydration or, in the case of ^2H_2O suspended rod outer segments, by deuteration, which shifts peaks in the infrared spectrum.

Assignment of Protein and Lipid Vibrations. Many of the peaks in the infrared spectrum of photoreceptor membrane from 400 to 4000 cm^{-1} can be assigned on the basis of a comparison of native, delipidated, and reconstituted rhodopsin membrane spectra and from infrared lipid and model compound studies.[7] As is shown in Fig. 2, many of the peaks due to rhodopsin can be clearly distinguished from lipid vibrations. In addition to the intense amide I and II peaks, weaker vibrations appear, such as at 1520 cm^{-1} (tyrosine ring vibration) and 1500 cm^{-1} (lysine, N—H deformation stretch), these being more prominent in the spectrum of delipidated rhodopsin. In some cases, such as the 1045 cm^{-1} (C—N stretch) peak, only a shoulder appears in the spectra of the native membrane, whereas a separate peak is found in delipidated rhodopsin. Another example, is the small 1735 cm^{-1} vibration due to glutamate and aspartate, which is hidden by the more dominant ester C=O stretch peak due to lipids.

Major lipid vibrations appear at 1237 cm^{-1} (P=O stretch), 1070 cm^{-1} (P—O—C stretch) and 1172 cm^{-1} (C—O—C stretch) in the reconstituted membrane. A complete list of the major peaks in the absorption spectrum of photoreceptor membrane and tentative assignments is shown in Table I.

The Amide I and II Region. The infrared spectrum of photoreceptor membrane has been measured in ^2H_2O suspended rod outer segments[4,5] and films.[7] In both cases two large peaks are found near 1655 and 1545 cm^{-1} that are predominantly due to the amide I and amide II modes, respectively, of rhodopsin. These frequencies are in the range found for

[31] N. A. Clark, K. J. Rothschild, B. A. Simon, and D. A. Luippold, Biophys. J. 31, 65 (1980).
[32] K. J. Rothschild, K. M. Rosen and N. A. Clark, Biophys. J. 31, 45 (1980).
[33] M. Michel-Villaz, H. R. Saibil, and M. Chabre, Proc. Natl. Acad. Sci. U.S.A. 76, 4405 (1979).

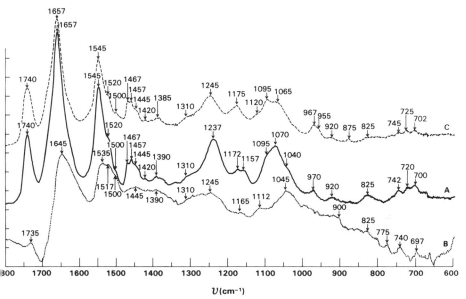

FIG. 2. Fourier transform infrared absorption spectra from 600 to 1800 cm⁻¹ of (A) cattle photoreceptor membrane, (B) delipidated rhodopsin, and (C) reconstituted rhodopsin. Samples were prepared on AgCl window using the isopotential spin-dry method (described in Clark et al.[31]). Spectra were obtained with 100 scans of the sample and 50 scans of the reference beam at resolution 4 cm⁻¹. (From Rothschild et al.[7])

highly α-helical proteins, although the amide I mode is at a somewhat higher than average frequency. Since random coil can also contribute near 1655 cm⁻¹, this assignment is not considered conclusive. However, deuteration of both rod outer segments and photoreceptor membrane film leads to only small shifts from 1655 cm⁻¹, in contrast to random-coil polypeptides, which show a shift to 1640 cm⁻¹.[11] This is a strong indication that rhodopsin contains a large fraction of alpha-helix structure. The absence of a peak or shoulder at 1630 cm⁻¹ also indicates that there exists little β structure in rhodopsin, in agreement with Raman spectroscopy of fully bleached rod outer segments.[24]

The C—H Stretch Region. The C—H stretch modes arise mainly from the long hydrocarbon chains of the membrane lipids and the shorter chains of the rhodopsin residues such as valine and leucine and also from the protein skeletal C—H.[8] As listed in Table I, the 2925 and 2855 cm⁻¹ peaks are due to the antisymmetric and symmetric vibrations of CH_2 groups while the 2960 and 2872 cm⁻¹ are due to the antisymmetric and symmetric vibrations, respectively, of CH_3 groups. Since the ratio of methyl to methylene groups is greater in the short hydrocarbon chains of

TABLE I

TENTATIVE ASSIGNMENT OF MAIN PEAKS IN THE INFRARED SPECTRUM OF
PHOTORECEPTOR MEMBRANE[a]

Wave number (cm^{-1})	Assignment	Wave number (cm^{-1})	Assignment
3295	N—H stretch associated (CONH amide A)	1390	CO$_2^-$ stretch (symmetric) C—H bend (CH$_3$ symmetric) Asp
3200	N—H stretch associated (CONH amide B)	1310	C—H wag (in CH$_2$)
3015	C—H stretch (in —C=C—)		C—H in plane deformation (in trans—C=C—)
2960	C—H stretch (CH$_3$ antisymmetric)	1245	N—H bend (CONH amide III)
2925	C—H stretch (CH$_2$ antisymmetric)	1237	P=O stretch (PO$_2^-$ antisymmetric) C—O—C stretch (antisymmetric)
2872	C—H stretch (CH$_3$ symmetric)	1172	C—O—C stretch (symmetric)
2855	C—H stretch (CH$_2$ symmetric)	1165	(CH$_3$)$_2$—C skeletal vibration Val, Leu
1740	C=O stretch (in esters) Glu, Asp	1157	C—C skeletal stretch
1680	Arg (guanidinium modes)	1112	N—H bend (?)
1657	C=O stretch (CONH amide I) C=C stretch (cis only)	1095	P=O stretch (PO$_2^-$ symmetric)
		1070	P—O—C stretch
1545	N—H bend (CONH amide II)	1045	C—N stretch
1520	Tyr (modes of p-disubstituted benzene ring)	1040	C—O stretch (in hydroxyl groups)
1500	Lys (NH$_3^+$ deformation modes) Phe (mode of monosubstituted benzene ring)	970	C—H bend (in trans—C=C—) C—C—N$^+$ stretch
1467	C—H scissor (in CH$_2$)	955 920 875 825 775 742 720	C—H rock (in CH$_2$)
1457	C—H bend (CH$_3$ antisymmetric)		
1445	C—H scissor (in CH$_2$) C—H bend (CH$_3$ antisymmetric) Ala		
1420	C—H scissor (in CH$_2$ next to C=O)	700	N—H bend (out of plane—amide V) C—H bend (in cis—C=C—)

[a] From Rothschild et al.[7]

rhodopsin residues relative to the long hydrocarbon chains of lipids, we expect the 2960 and 2872 cm^{-1} peaks to be more dominant in the delipidated rhodopsin spectrum, as confirmed in Fig. 3.

The C—H stretch region is sensitive to lipid fluidity. For example, Raman studies reveal a decrease in the ratio of the antisymmetric to symmetric methylene absorptions at the gel–liquid crystal-phase transition of

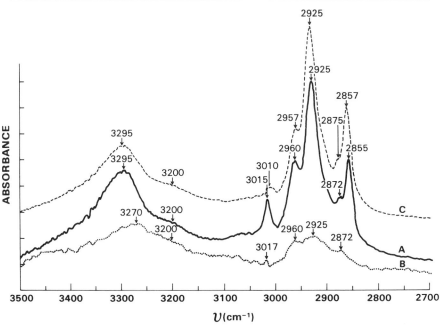

FIG. 3. Fourier transform infrared absorption spectra; same as Fig. 2. Spectra from 2700–3500 cm⁻¹. (From Rothschild *et al.*[7])

the lipids.[9,34,35] Infrared absorption changes in this ratio in the case of photoreceptor membrane film have been measured as a function of hydration.[36] It is found that increasing hydration results in a decrease in the ratio, a result consistent with an increase in the fluidity of the lipid chains. Changes in lipid fluidity may also be detected by using the 720 cm⁻¹ (CH_2 rock) and 1467 cm⁻¹ (CH_2 scissor) infrared peaks.[9,37] Recently, a phase change in phosphatidylethanolamine from lamellar to hexagonal II was observed by monitoring these peaks.[38]

Effects of Delipidation and Reconstitution of Rhodopsin. Rhodopsin can be fully delipidated and reconstituted with different lipids using stan-

[34] K. G. Brown, W. L. Peticolas, and E. Brown, *Biochem. Biophys. Res. Commun.* **54,** 538 (1973).
[35] R. Mendelsohn, S. Sunder, and H. J. Bernstein, *Biochim. Biophys. Acta* **419,** 563 (1976).
[36] W. J. DeGrip, R. Sanches, and K. J. Rothschild, unpublished results.
[37] D. G. Cameron, H. L. Casal, and H. H. Mantsch, *Biochemistry* **19,** 3365 (1980).
[38] H. H. Mantsch, H. Martin, H. L. Casal, J. Unemura, and D. G. Cameron, *Fed. Proc., Fed. Am. Soc. Exp. Biol.* **39,** 2190 (Abstr. Mo. 3076) (1980).

dard procedures.[39,40] Infrared measurements of these preparations can be useful for determining the extent of rhodopsin conformational changes. For example, a substantial shift is observed in the amide I and II peaks of delipidated rhodopsin relative to native and reconstituted membranes (cf. Fig. 2), suggesting a more disordered conformation of rhodopsin upon complete delipidation. The addition of ^2H_2O is found partially to restore the native structure. Upon reconstitution with different lipids, including dioleyl phosphatidylcholine, egg phosphatidylcholine, or phosphatidylethanolamine,[36] a complete restoration of the native rhodopsin structure can be inferred from the normal position of the protein peaks (cf. Figs. 2 and 3) relative to the native membrane. In fact, almost all differences in the infrared spectrum of native and reconstituted membranes can be attributed to the presence of different lipids.

Effects of Detergents. Almost no changes are observed in the infrared spectrum of rhodopsin when it is solubilized in the nonionic detergents Ammonyx-LO or Cemulsol LA 90.[5] However, bleaching of rhodopsin in detergent does cause a change in the shape and frequency maximum of the amide I band, in contrast to bleaching of native rod outer segments, which produces no observable changes. This indicates that rhodopsin is less stable in detergent, in agreement with circular dichroism studies.[41,42]

Hydrogen–Deuterium Isotope Exchange of Rhodopsin. Infrared spectroscopy can be utilized in order to determine the extent of deuterium–hydrogen exchange of the peptide groups in rhodopsin. This method has advantages over radioactive measurements since only peptide groups and not the total labile hydrogen pool can be sampled. Since the amide II peak shifts from 1545 to 1455 cm^{-1} owing to the replacement of N^{1}H with N^{2}H, the amide II/amide I ratio can be used as an indicator of the extent of deuteration as a function of time.

^{1}H–^{2}H exchange in rod outer segments was performed by Osborne and Nabedryk-Viala[5] by diluting the sample 25 times in ^2H_2O buffer, centrifuging for 20 min at 100,000 g, and then resuspending to a final concentration of 10 mg/ml. The changes in the infrared spectrum could only be monitored after 60–90 min using this method, thus excluding observation of the rapid-exchange phase. Figure 4 shows a comparison of the spectra of membrane-bound rhodopsin and rhodopsin/Cemulsol LA 90 micelles after 24 hr at 26° in ^2H_2O buffer. It is concluded from these measurements that there is less than 9% increase in deuterium–hydrogen exchange of the detergent-solubilized rhodopsin relative to the membrane-bound rho-

[39] M. Montal, *Biochim. Biophys. Acta* **559**, 231 (1979).
[40] W. J. DeGrip, F. J. M. Daemen, and S. L. Bonting, this series, Vol. 67, p. 301.
[41] C. N. Rafferty, J. Y. Cassim, and D. G. McConnell, *Biophys. Struct. Mech.* **2**, 277 (1977).
[42] H. Schichi and E. Shelton, *J. Supramol. Struct.* **2**, 7 (1974).

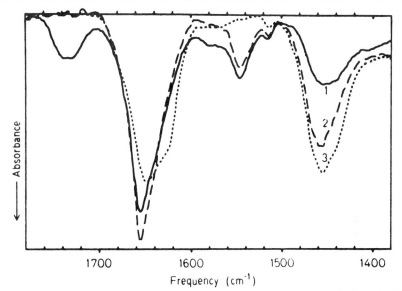

FIG. 4. Infrared spectra of (1) membrane-bound rhodopsin, (2) rhodopsin Cemulsol LA 90 micelles in ^2H_2O buffer, and (3) same as (2) after heating for 1 h. (From Osborne and Nabedryk-Viala.[5])

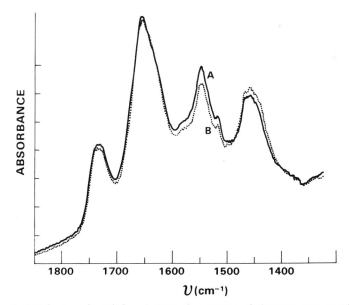

FIG. 5. Fourier transform infrared absorption spectra of photoreceptor membrane (A) after 6 min and (B) after 31 min of deuteration. (From Rothschild et al.[7])

dopsin. In contrast, a much larger difference was found in the case of bleached samples, indicating again a conformational change upon bleaching in the detergent-solubilized rhodopsin that increases solvent accessibility.

Faster exchange kinetics can be measured by utilization of Fourier transform infrared spectroscopy.[7] The sample is partially dried by lyophilization using the isopotential spin-dry method[30–32] and then resuspended in ^2H_2O inside a dry-box to minimize reexchange with H_2O. Using this method, an infrared spectrum of the sample can be recorded after only 6 min of deuteration (Fig. 5). It shows the appearance of a shoulder in the amide I peak that corresponds to the rapidly deuterating random structure of rhodopsin. Using this method, differences in the deuteration rate of delipidated and Na-borohydride-reduced rhodopsin have been found, again indicating increased solvent accessibility.[36]

IV. Infrared Absorption of Purple Membrane

Infrared studies of purple membrane have thus far concentrated on utilization of the amide I, II, and A vibrations to reveal information about the structure of bacteriorhodopsin and the orientation of bacteriorhodopsin α helices.[6,19] As described below, these studies have largely confirmed the evidence based on X-ray scattering[43,44] and electron microscopy[45] that bacteriorhodopsin consists to a large extent of a bundle of α helices arranged predominantly perpendicular to the purple membrane plane. The infrared spectrum shows also that the amide carbonyl stretching frequency of the bacteriorhodopsin α helices is 5 cm^{-1} higher than that normally observed in α-helical proteins. (cf. Section VI for more recent progress)

Anomalous Amide I Infrared Absorption of Purple Membrane. Although circular dichroism,[46] X-ray,[43,44] and electron microscopy[45] indicate a predominance of α-helical structure in bacteriorhodopsin, the amide I frequency (1660 cm^{-1}) and amide A frequency (3310 cm^{-1}) fall outside the normal range for an alpha helix (i.e., 1650–1655 cm^{-1} and 3290–3300 cm^{-1}, respectively).[8] This rise in frequency is also found, but to a lesser extent, in rhodopsin (1657 cm^{-1}) and cytochrome oxidase (1654 cm^{-1}).[47] Since the Pimental–Sederholm relation[48] predicts that only a 0.05 Å increase in hydrogen bonding distance in the C=O . . . H—N bond could

[43] R. Henderson, *J. Mol. Biol.* **93**, 123 (1975).
[44] A. Blaurock, *J. Mol. Biol.* **93**, 139 (1975).
[45] R. Henderson and P. N. T. Unwin, *Nature (London)* **257**, 28 (1975).
[46] J. A. Reynolds and W. Stoeckenius, *Proc. Natl. Acad. Sci. U.S.A.* **74**, 2803 (1977).
[47] R. A. Capaldi, *Biochim. Biophys. Acta* **303**, 237 (1973).
[48] G. C. Pimental and C. H. Sederholm, *J. Am. Chem. Soc.* **24**, 639 (1956).

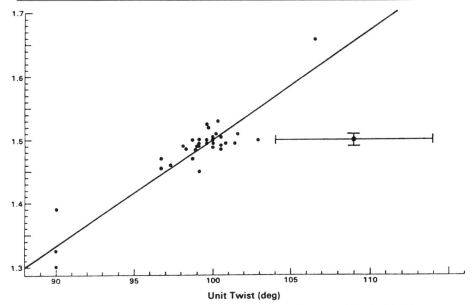

FIG. 6. Plot of h, unit height per residue, versus t, unit twist per residue for 34 different α-helical polypeptides and protein segments (adapted from Fraser and MacRae[49]) including h and t measured for bacteriorhodopsin (from Henderson[43]) as shown with error bars.

produce a 10-cm^{-1} upshift in the amide I peak, it is possible that the α helices of bacteriorhodopsin are in a somewhat distorted conformation from the normal Pauling–Corey structure. This is supported by a measured pitch of the bacteriorhodopsin α helix of 5.05 Å[43] as compared to 5.4 Å measured in most α helices.[49] A plot of the pitch versus twist for a variety of α-helical polypeptides and proteins emphasizes the deviation of the bacteriorhodopsin α helices (cf. Fig. 6). Krimm and Dwivedi recently proposed[50] on the basis of the IR data[6,19] and normal mode analysis that bacteriorhodopsin helices are of the α_{II} type.

The Orientation of Bacteriorhodopsin Alpha Helices as Measured by Polarized Infrared Spectroscopy

If purple membrane is oriented so that the membrane sheets are arranged parallel to each other, the sample displays linear dichroism of the amide I, II, and A vibrations which can be used to measure the average

---

[49] R. D. B. Fraser and T. P. MacRae, "Conformation in Fibrous Proteins and Related Synthetic Polypeptides." Academic Press, New York, 1973.

[50] S. Krimm and A. M. Dwivedi, *Science* **216**, 407 (1982).

tilt of the bacteriorhodopsin α helices. Orientation can be accomplished by drying purple membrane on a clean Irtran-4 window. Linear dichroism of the 570-nm absorption indicates that this method results in a high degree of orientation of the purple membrane fragments.[52] Furthermore, the proton–pump cycle of bacteriorhodopsin appears to be functional, although the kinetics is altered in this partially dehydrated state.[53]

Infrared linear dichroism is measured by directing a horizontally or vertically polarized beam through the sample that is tilted around a horizontal axis at some angle α_0 (cf. Fig. 1a). As expected, the infrared spectrum measured with the horizontal polarization does not vary, but the vertical does as the tilt angle changes (cf. Figs. 7 and 8).

The molecular basis for the observed dichroism changes of the amide I, II, and A bands can be understood by considering the direction of the bacteriorhodopsin α helices in the oriented sample, as shown in Fig. 9. In this orientation the helix C=O . . . H—N groups will be predominantly perpendicular to the membrane plane. Hence both the amide I and A transition moments will also be roughly perpendicular to this plane. In contrast, the amide II vibration should be predominantly in-plane. When $\alpha_0 = 0°$, a reduction in the absorption of the amide I and A peaks and an increase in the amide II band will occur, since the electric polarization will be entirely in the film plane. Hence the amide II/amide I ratio of the purple membrane film is 1 as compared to 0.45 found for most unoriented proteins.[51] As the sample is tilted, the angle the carbonyl groups make with the vertical polarization becomes smaller and hence the amide I and A absorptions increase while the amide II decreases (cf. Fig. 7).

A complete analysis of the infrared dichroism of the amide I, II, and A peaks is possible, allowing one to determine the average tilt of the α helices. The equation that describes the dependence of the dichroism on α (as discussed by Rothschild and Clark[6]) is

$$R = 1 + \sin^2 \alpha \frac{3p}{1 - p} \tag{1}$$

where R is the dichroic ratio, α is the sample tilt angle corrected for the refraction in the film plane, and $p = p_M p_T p_\alpha$ with p_M, p_T, and p_α the order parameters for the mosaic spread of the membranes, the transition moment relative to the α-helix axis, and the α-helix axis relative to the membrane plane, respectively (cf. Fig. 1b). Since the membrane orientation is high $p_M \cong 1$, and p_T is known from the α-helix structure and the angle the transition moments make with the peptide group,[26-28] p_α, which is related

[51] E. R. Blout, C. DeLoze, and A. J. Asadourian, *J. Am. Chem. Soc.* **83**, 1895 (1961).
[52] M. P. Heyn, R. J. Cherry, and U. Muller, *J. Mol. Biol.* **117**, 607 (1977).
[53] R. Korenstein and B. Hess, *Nature (London)* **270**, 184 (1977).

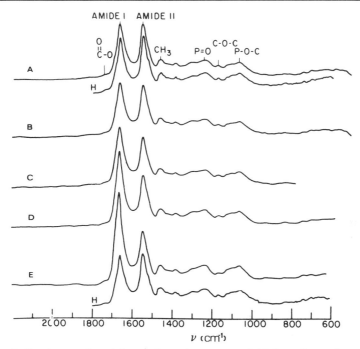

FIG. 7. Fourier transform infrared absorption spectra of dried purple membrane on Ir-tran-4 from 600 to 2200 cm⁻¹. Spectra were recorded with 100 scans of the sample and 50 scans of the reference beam at resolution 4 cm⁻¹. Spectra denoted H were obtained with horizontally polarized light. Spectra A to E were obtained with vertically polarized light and with the sample tilted at angles of α_0 = 0, 15, 30, 45 and 60°. (From Rothschild and Clark.[6])

to the α-helix tilt angle θ_α by

$$p_\alpha = (3 \cos^2 \theta_\alpha - 1)/2 \qquad (2)$$

can then be determined independently from the slope of the linear dichroism curves for the amide I, II, or A absorptions. An upper limit of $\theta_\alpha = 26°$ is predicted for the average α-helix tilt, in good agreement with the electron microscope determined model.[45]

V. Orientation of Rhodopsin α Helices as Determined by Polarized Infrared Spectroscopy

Two polarized infrared absorption studies have been made of the rhodopsin α-helix orientation. Michel-Villaz et al.[33] examined magnetically oriented rod outer segments in ^2H_2O solution. Samples were prepared by orientation in a 15-kG field followed by sedimentation onto a CaF_2 win-

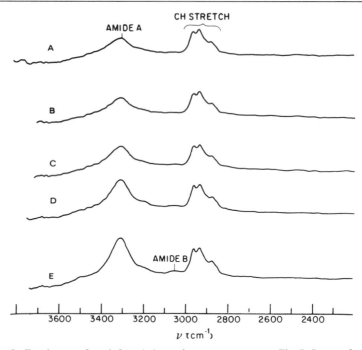

FIG. 8. Fourier transform infrared absorption spectra; same as Fig. 7. Spectra from 2200 to 3800 cm⁻¹. (From Rothschild and Clark.[6])

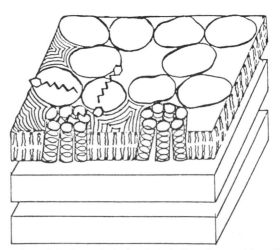

FIG. 9. Orientation of the bacteriorhodopsin α helices in a multilamellar array. (With permission of L. Duong.)

dow. Sedimentation was accomplished either with an air-driven centrifuge arranged axially between the poles of an electromagnet or by allowing samples to sediment for 30 min in the presence of 2.5% glutaraldehyde. Samples were then immersed in ^2H_2O Ringer's solution and measured in a cell with a 25–50 μm path length. The resulting orientation of the rods was studied microphotographically and it was determined that the average angular deviation from the magnetic field direction was 31°.

The infrared spectrum of the oriented rod outer segments preparation, as shown in Fig. 10, exhibited a difference in absorption depending on whether the polarization of the incident light was perpendicular or parallel to the rod outer segment symmetry axis. This effect can be directly related to the α-helix orientation in rhodopsin using the analysis described in Section IV. The most probable estimate of θ_α was 38° based on dichroism of the amide I band. Dichroism of the residual amide II band, which corresponds to unexchanged NH groups, also gave a similar estimate of θ_α. The authors note that since these unexchanged groups are likely to be part of the hydrophobic interior of the protein, it can be deduced that the oriented α helices are hydrophobic. In contrast, no dichroism was detected in the exchanged amide II band corresponding to N^2H stretching. This may indicate a more unoriented structure for the hydrophilic regions of rhodopsin.

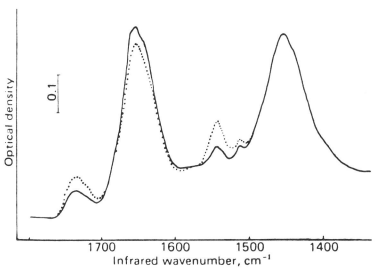

FIG. 10. Infrared spectra of rod outer segments oriented in a magnetic field in the presence of gluteraldehyde. Polarization (—) parallel and (----) perpendicular to the rod axis. (From Michel-Villaz et al.[33])

Rothschild *et al.*[29,30] have measured the infrared dichroism of oriented photoreceptor membrane films using methods previously applied to oriented purple membrane.[6] It has been shown on the basis of the visible dichroism of the α band that the rhodopsin chromophore in these films has an orientation of approximately 17° from the film plane.[32] Hence the film consists of a multilamellar stacking of photoreceptor membranes with an orientation as shown in Fig. 9.

One noticeable feature of the infrared spectra of photoreceptor membrane films is the relatively high amide II/amide I ratio, which approaches 1, in contrast to disordered photoreceptor preparations with an amide II/amide I ratio of 0.50.[36] The cause of this high ratio, as in the case of purple membrane, is the preferential orientation of rhodopsin α helices

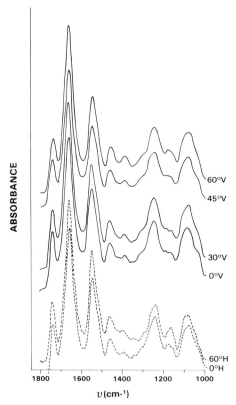

FIG. 11. Fourier transform infrared spectra of cattle photoreceptor membrane film from 1000–1800 cm⁻¹. The sample was prepared on AgCl windows using the isopotential spin-dry method. Spectra were obtained with 200 scans of the sample and 100 scans of the reference beam at 8-cm⁻¹ resolution. The tilt angle of the sample and the polarization of the incident beam are indicated in the spectra. (From Rothschild *et al.*[30])

perpendicular to the membrane. This can be verified by measuring the linear dichroism of the amide I, II, and A vibrations as a function of α (cf. Figs. 11 and 12). The dichroism measurements can be analyzed as described in Section IV and yield α-helix tilt estimates shown in Table II. These estimates must be treated as upper limits. In addition to the factors discussed in Section IV, which lead to an overestimate of θ_α, the nondichroic contribution to the bands, which may be as high as 40% in rhodopsin,[42] will also significantly raise θ_α. This contribution can be accounted for in Eq. (1) with an additional f factor[30] that is the percentage of α-helical contribution to the peak intensity for a nonoriented sample. The equation becomes

$$R = 1 + \sin^2 \alpha \, \frac{3fp}{1 - fp} \qquad (3)$$

Using Eq. (3) and an $f = 0.6$ results in much lower estimates of θ_α, as shown in Table II.

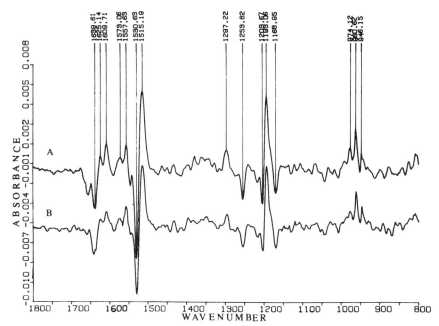

FIG. 12. Fourier transform difference spectra of purple membrane film deposited on AgCl at 77°K and fully humidified. (A) Difference spectrum obtained by subtracting a reference spectrum recorded in the dark from the same sample recorded during illumination for 3 min with 500 nm light. (B) Difference spectrum obtained by subtracting a reference spectrum in the dark obtained after 500 nm illumination from a spectrum recorded during 3-min illumination with light of wavelengths greater than 690 nm. For additional details see reference 16.

TABLE II
ESTIMATION OF RHODOPSIN α-HELIX TILT (θ_α) FROM
DICHROISM DATA[a]

	p_α	θ_α (using $f = 1$)	θ_α (using $f = 0.6$)
Amide A	0.39	40	29
Amide I	0.36	41	31
Amide II	0.35	41	32

[a] From Rothschild et al.[30]

VI. Recent Progress

The method of FTIR difference spectroscopy has recently been used to study the conformational changes occurring in purple membrane[15,16,54] and photoreceptor membrane.[55] Vibrational changes of both the chromophore and protein can be detected providing an important new tool for examining in detail the molecular alterations that occur during the bacteriorhodopsin photocycle and rhodopsin bleaching. Specific findings thus far include evidence for the carboxylate group protonation which occurs during the formation of the M_{412} intermediate of bacteriorhodopsin.[15] Low temperature measurements reveal that the local environment of the Schiff base proton in BR_{570} is altered in the formation of the first photoproduct, K.[16] Such studies combined with biochemical and genetic methods of altering proteins should be a powerful tool to help elucidate the molecular mechanisms active in these membranes.

[54] G. Dollinger, K. Bagley, S. F. Bowne, L. Eisenstein, W. Mantele, E. Shyamsunder, and F. Siebert Biophys. J. 37, 382a (1982).
[55] K. J. Rothschild, W. Cantore, and H. Marrero (unpublished data).

[85] Effect of Pressure on Visual Pigment and Purple Membrane

By MOTOYUKI TSUDA

Some sort of functional interaction between rhodopsin and other components in the membrane of the photoreceptor cells (proteins, lipids, ions) is implied in most models for signal transduction in photoreception. A similar interaction has been proposed for bacteriorhodopsin in which it is responsible for proton pumping across the purple membrane of *Halobacterium halobium*. Changes of these interactions in the membrane are expected to take place during the steps in the photolysis of the retinal containing pigments. To study these interactions, the molar free volume change (ΔV) can be a useful parameter, because large ΔV and $\Delta V\ddagger$ are expected for processes in which association between the pigment and other components in the membranes is altered. Starting with the pioneering work of Bridgman,[1] a number of studies have shown that most biomolecules are denatured at pressures above 3 kbar. Rhodopsin (bovine rhodopsin[2] and squid rhodopsin[3]) is an exception and the effects of pressure on rhodopsin and its intermediates can be studied up to 6 kbar; however, most interesting pressure phenomena are observed within 3 kbar, so no attempt will be made to review ultra-high-pressure techniques. Moreover, construction and operation of high-pressure equipment below 3 kbar is quite easy.

High-Pressure Optical Cell

For biological samples the following features in a high-pressure optical cell are desirable: (1) the sample should have a small chamber, (2) the sample should be easily replaceable, (3) the sample should not be in contact with the metal surface of the optical cell, (4) the cell should transmit a wide wavelength range from UV to near IR, (5) the temperature of the cell should be controlled from 0 to 60°, (6) the optical pressure cell should be small enough to be interfaced easily with most types of optical instruments, (7) the apparatus should be easy to operate.

Figure 1 shows a cross section of a reasonably small, portable high-pressure apparatus that can be used with many spectrophotometers and other optical setups. The principle of this high-pressure equipment follows after Fishman and Drickamer,[4] but modifications have been made in order to satisfy the conditions described earlier. The low-pressure end can be attached to a portable hydraulic pump and a good Bourdon gauge, such as a Heise gauge. Pressure on the high-pressure side can be read using a manganin gauge.[5] The solution to be studied can be introduced directly on the high-pressure side (4 ml) from the top. For the study

---

[1] P. W. Bridgman, *J. Biol. Chem.* **19**, 511 (1974).
[2] A. A. Lamola, T. Yamane, and A. Zipp, *Biochemistry* **13**, 738 (1974).
[3] M. Tsuda, I. Shirotani, S. Minomura, and Y. Terayama, *Biochem. Biophys. Res. Commun.* **76**, 989 (1977).
[4] E. Fishman and H. G. Drickamer, *Anal. Chem.* **28**, 804 (1956).
[5] K. Suzuki, this series, Vol. 26, p. 424.

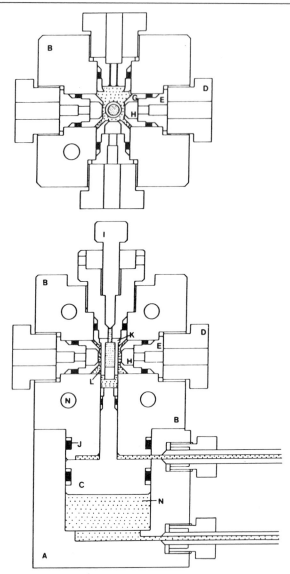

FIG. 1. High-pressure apparatus (stainless steel) with optical windows. A, Cylinder of the low-pressure side; B, cylinder of the high-pressure side; C, piston head of the Bridgman type; D, nut; E, plug with optical window; F, plug with electrode for a thermocouple and manganing gauge; G, cap to support optical window; H. optical window (8 mmp × 6 mm, sapphire); I, stopper; J, rubber packing; K, cylindrical internal cell of fused quartz; L, pressure transmitting in the high-pressure side (water); M, pressure transmitting in the low pressure side (machine oil); N, thermostated fluid.

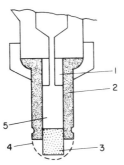

FIG. 2. Cylindrical internal cell with piston of quartz. 1, Spacer of Teflon; 2, cylindrical cell of quartz; 3, cylindrical piston of quartz; 4, stopper of Teflon; 5, sample.

of biological molecules, a cylindrical internal cell of fused quartz (Fig. 2) is very useful.

Materials used for optical windows of high-pressure cells are glass, quartz, sapphire, and/or diamond. For the present purpose, synthetic sapphire was selected for its transmittance (150 nm to 6 μm with selected material), strength (up to 10 kbar), and availability in large sizes. Windows are mounted without gaskets in a manner described by Poulter.[6] The mode of dealing is a simple application of the unsupported area principle of Bridgman.[7] The window is supported from the low-pressure side on a surface with a contact area smaller than the window by the amount of the aperture. If the two surfaces are optically flat, the sapphires will usually seal as purchased; otherwise epoxy resin can be used to cement the windows. A brass cap is useful for holding the window in place during assembly.

As a pressure fluid for up to 10 kbar, water, glycerin, silicon oil, machine oil kerosene, hexane, and alcohols can be used. The selection of the fluid is determined by the experiment and the material for high-pressure cells. For the cells made of stainless steel, water is convenient and relatively inert. Moreover, water has a wide transmittance range and large specific heat (a temperature rise follows compression) and is the solvent for most samples (thus, no problem of contamination). To interface the high-pressure optical cell with the sample compartment of some spectrophotometers, optical guides (UV grade) are convenient.

[6] T. C. Poulter and R. O. Wilson, *Phys. Rev.* **40,** 877 (1932).
[7] P. W. Bridgman, "The Physics of High Pressure." Dover, New York, 1970.

Temperature Control of High-Pressure Cell

The temperature of the high-pressure optical cell should be regulated. If temperature changes during the course of a measurement, the pressure of the solution will change because the solution is at constant volume. A second reason to control temperature is to use it as a variable thermodynamical parameter. The sample temperature can be controlled using a thermostated jacket that fits snugly over the entire length of the high-pressure vessel. In the high-pressure cell of Fig. 1, the thermostated fluid is circulated directly in the optical cell. The temperature can be monitored using a thermocouple on the high-pressure side and controlled to within $\pm 0.1°$.

Operation

The procedures for the operation of an optical high-pressure cell are basically the same regardless of the specific system. Here I will discuss some of these with reference to the high-pressure system shown in Fig. 1.

1. After water is filled in both the outer and inner cells (K), the pistons of the intensifier (C) and inner cell (K) should be placed in the bottom. The inner cell is filled with buffer and the inlet (I) closed. The spectrum of the reference (solvent) should be measured as a function of pressure up to maximum pressure to be used and kept for more than 30 min at maximum pressure. Sometimes the baseline of the spectrum changes at different pressures. This is because of differential solubility of dissolved gases with pressure. The high-pressure fluid can be degassed to avoid this.

2. After the stopcock of the low-pressure side is closed at the maximum pressure, it is desirable to make sure the pressure is maintained. A decrease in pressure indicates a fluid leak. If the outer surface of the high-pressure cell is dried before the application of pressure, any leak can be easily detected as a wet spot.

3. After exchanging the solvent with the sample, the spectra are recorded at various pressures up to the maximum pressure employed and finally at atmospheric pressure after releasing the pressure. If the spectra at 1 bar are different before and after pressing the sample, it is possible the sample has been denatured at the high pressures used. Pressure range should be carefully chosen to avoid denaturing the samples.

4. When the sample solution is pressed, the temperature of the solution will rise. Consequently, care should be taken to change the pressure slowly during experiments so that the temperature control system can

maintain a constant sample temperature. It is best to monitor the temperature of the pressure fluid at the high-pressure side with a thermocouple.

5. After the measurement at a given pressure is finished, the pressure should be released slowly to prevent sample cooling. Pressure equilibration takes longer when the pressure is released from a high to a low value. Thus, when making measurements at various pressures starting from high pressure going to a low value, care should be taken to allow sufficient equilibration. The true pressure exerted is higher than the calculated pressure due to friction of the piston of the intensifier. To measure the correct value, pressure at the high-pressure side should be measured directly using a pressure gauge, such as a manganin gauge.[5]

Correction for Solvent Compression

The absorbance of any solution in the optical cell will increase with pressure owing to compression. In order to obtain the absolute value of absorbance at a given pressure, the experimentally determined value must be corrected by multiplying it by the relative volume $V(P,T)/V$ of the solvent. (For water see Vedam and Holton[8] and Neuman et al.[9])

Effect of Pressure on the pH of Buffers

Biochemical systems are best studied under buffered conditions. However, the pH values of many commonly used buffers can exhibit very strong pressure dependences.[9,10] (It is best to choose a buffer with a negligibly small pressure coefficient.) It was found that conjugated acids of neutral bases such as Tris-HCl[9] and imidazole-HCl[10] change their pH by no more than 0.1 unit on increasing the pressure from 1 bar to 6 kbar. If it is necessary to use other buffers, the pH of the solution should be calibrated using the data summarized in Ref. 11.

Pressure Effect on Absorption Spectra of Rhodopsin, Bacteriorhodopsin, and Their Intermediates

Pressure-Induced Intermediate in the Photolysis of Pigment. If thermal transformation of an intermediate has a positive value of $\Delta V\ddagger$, the lifetime of the intermediate will be increased with pressure. Thus it is pos-

[8] R. Vedam and G. Holton, *J. Acoust. Soc. Am.* **43**, 108 (1968).
[9] R. C. Neuman, Jr., W. Kauzman, and A. Zipp, *J. Phys. Chem.* **77**, 2687 (1973).
[10] M. Tsuda, I. Shirotani, S. Minomura, and Y. Terayama, *Bull. Chem. Soc. Jpn.* **49**, 2952 (1976).
[11] T. Asano and W. J. LeNoble, Chem. Rev. **78**, 407 (1978).

sible that an intermediate, which cannot be observed at 1 bar, can be measured under high pressure using a regular spectrophotometer. New intermediates seen at high pressures have been found in squid rhodopsin.[3] For both bovine rhodopsin[2] and bacteriorhodopsin (Tsuda) no pressure-induced intermediates have yet been found. To calculate the spectrum of the pressure-induced intermediate in the photolysis of squid rhodopsin, the absorption spectrum in alkaline solution is measured before and after pressing. Then the preparation under pressure (6 kbar) is irradiated with light. If the spectrum after irradiation is different from that when the pigment is irradiated at 1 bar, these changes may be due to a pressure-induced intermediate. The absorption spectra of transformations of the high-pressure intermediate to a final photoproduct must be observed at several pressures because the spectrum of the intermediate is affected by pressure. For example, the difference spectrum per unit of the intermediate and final photoproduct (alkaline metarhodopsin) was calculated from (a) the photosteady-state data at 6 kbar, (b) the spectrum at 6 kbar after releasing pressure to 1 bar, and (c) the content of irradiated rhodopsin, which is estimated with complete bleaching of the preparation in alkaline solution. The addition of this difference spectrum to alkaline metarhodopsin of an equivalent concentration yields the spectrum of the pressure-induced intermediate. If more than two intermediates exist in the photosteady-state mixture under high pressure and have different stability in pressure, a more complex spectral analysis (see Ref. 3) must be used to give the absorption spectrum of each intermediate.

Pressure Effect on Equilibrium of Intermediates

Most of the photochemical reactions and subsequent thermal reactions in the photolysis of rhodopsin and bacteriorhodopsin are essentially irreversible, but some are equilibrium reactions—for example, metarhodopsin I $\rightleftharpoons$ metarhodopsin II in bovine rhodopsin and $BR^{13-cis} \rightleftharpoons BR^{all-trans}$ in dark-adapted bacteriorhodopsin (BR^D). The effect of pressure P on the equilibrium constant K at constant temperature T is related to the molar free volume change ΔV for the reaction by

$$\left(\frac{\partial \ln K}{\partial P}\right)_T = -\frac{\Delta V}{RT} \tag{1}$$

where R is the gas constant.

The equilibrium constant K in the equilibrium reaction $A \rightleftharpoons B$ can be estimated from their concentration and $K = (A)/(B)$. The concentration of each component can be estimated from their absorption spectra if the absorption spectrum of each component is sufficiently different, for example, metarhodopsin I and metarhodopsin II.[2]

To calculate the ratio $(A)/(B) = K$ from the absorption spectrum of an irradiated sample of rhodopsin or dark-adapted purple membrane preparation at high pressure, several manipulations must be made. One must correct for sample compression (discussed earlier), scattering, and the absorption due to other residual retinal pigments. In addition, one must correct for pressure-induced changes in the spectrum of these materials. Absorption spectra of the sample solution should be the same before and after pressing as far as the reaction is reversible. However, sometimes these spectra are not the same. In the case of the reaction metarhodopsin I $\rightleftharpoons$ metarhodopsin II, metarhodopsin III can appear even at 5°. Thus this contribution to absorption spectra must be corrected for. As examples, the dark equilibrium in bacteriorhodopsin[12] and the metarhodopsin I $\rightleftharpoons$ metarhodopsin II equilibrium[2] will be discussed.

The absorption spectra of BR^D solution at 1 bar before and after pressing are different if one measures the absorption spectrum immediately after releasing the pressure. This is because the reaction $BR^{13-cis} \rightarrow BR^{all-trans}$ is slow and its lifetime is temperature dependent. If one waits long enough at 1 bar after releasing the pressure, one can measure the same spectrum as before pressing. On the other hand, the slow reverse reaction from $BR^{13-cis} \rightarrow BR^{all-trans}$ allows one to determine the equilibrium constant in terms of the content of isomeric composition of the chromophore after equilibration at high pressures by extraction methods. Details are discussed in another section of this volume (see Article [71], this volume).

The pressure effect on the equilibrium constant of metarhodopsin I $\rightleftharpoons$ metarhodopsin II is measured spectroscopically as follows. First, absorption spectrum of rhodopsin and irradiated rhodopsin at 0°, 1 bar, and pH = 6.5 is measured. This pH is slightly lower than the pK_a of the metarhodopsin I $\rightleftharpoons$ metarhodopsin II equilibrium. After absorption spectra are measured at several pressures, pressure is released to 1 bar, and the absorption spectrum is measured again. Two μl of hydroxylamine (1 M) are added to the solution (200 μl) and the absorption spectra are measured before and after irradiating the sample with yellow light ($\lambda > 440$ nm). From these results, one can estimate the content of photolyzed rhodopsin in the photosteady mixture under high pressure. Absorption spectra of metarhodopsin I and metarhodopsin II can be calculated assuming 100% conversion of photolyzed rhodopsin to these intermediates. The equilibrium constant K at a given pressure can be determined from the absorp-

[12] M. Tsuda and T. G. Ebrey, *Biophys. J.* **30**, 149 (1980).

tion spectra at a given pressure and the pure metarhodopsin I and meta-rhodopsin II spectra.

Pressure Effect on the Decay Rates of Intermediates

The pressure dependency of a rate constant gives the molar volume change of activation as follows

$$\left(\frac{\partial \ln k}{\partial P} \right)_T = - \frac{\Delta V\ddagger}{RT}$$

Flash photolysis studies at high pressures allow the measurement of the pressure dependence of the rate constant of each intermediate process in rhodopsin[13,14] or bacteriorhodopsin.[15] Most of the operating procedures for the high-pressure system are the same for normal absorption measurements. Again, we have to correct for the contraction of the solvent under high pressure if we need the absolute value of the absorption change with time. Since bovine rhodopsin is bleached with a single flash, the sample must be changed after each observation. The high-pressure apparatus in Fig. 1 is especially useful because sample exchange is easy.

The lifetime of a single photochemical cycle of purple membrane is less than 100 msec and purple membrane is stable up to 3 kbar over several days. Thus, the pressure effect on the rate of bacteriorhodopsin can be studied without changing the sample. However, a purple membrane solution under high pressure must be frequently irradiated in order to keep the sample in the light-adapted form, because pressure favors the transformation of $BR^{all-trans}$ to BR^{13-cis}, and only $BR^{all-trans}$ pigment has a photochemical cycle, having intermediates K, L, M, N and O.

Acknowledgments

This manuscript was written while the author was a visiting Associate Professor of Biophysics at the University of Illinois. The author would like to thank Prof. T. G. Ebrey and his colleagues for many helpful discussions.

[13] M. Tsuda, *Biophys. J.* **25**, 315 (1979).
[14] M. Tsuda, T. G. Ebrey, unpublished observations (1980).
[15] M. Tsuda, R. Govindjee, and T. Ebrey, submitted (1982).

[86] Some Approaches to Determining the Primary Structure of Membrane Proteins

By N. G. ABDULAEV and YU. A. OVCHINNIKOV

As interest continues in general aspects and theories of transport one may clearly discern a tendency in this field toward detailed investigation of the proteins directly involved in translocation of ions and molecules across biological membranes. One of the most interesting and extensively explored membrane proteins is bacteriorhodopsin—the only protein of purple membranes of halophilic bacteria that thrive under the extreme conditions and effectively utilize light energy in the performance of their vital activity. Bacteriorhodopsin has been found to function as a light-driven proton pump, establishing a considerable pH gradient in the membrane utilized in the synthesis of ATP and other vital functions.[1]

The investigation of bacteriorhodopsin seems very important not only for the elucidation of the mode of action of proton pumps but also because the close resemblance of bacteriorhodopsin to the visual rhodopsin can be exploited in the analysis of the early events in visual excitation.

Much can be learned of the molecular mechanism of bacteriorhodopsin functioning by studying the spatial structure of the protein and the topography of its functionally important groups which, in turn, can be firmly established only on the basis of its complete amino acid sequence.[2,3]

Bacteriorhodopsin is an interesting object for structural analysis, not only because it is the most simple of all the known membrane systems capable of the active proton transport, but also because it belongs to the class of integral membrane proteins of a highly hydrophobic nature and consequently of a low solubility in aqueous media. Therefore, we had reason to suspect that bacteriorhodopsin structural study could hardly be based on the standard arsenal of methods of protein chemistry. The experience accumulated when studying the bacteriorhodopsin primary structure shows the way for the development of the general methodology for the structural analysis of membrane proteins.

[1] W. Stoeckenius, R. H. Lozier, and R. A. Bogomolni, *Biochim. Biophys. Acta* **505**, 215 (1979).

[2] Yu. A. Ovchinnikov, N. G. Abdulaev, M. Yu. Feigina, A. V. Kiselev, N. A. Lobanov, and I. V. Nazimov, *Bioorg. Khim.* **4**, 979 (1978).

[3] Yu. A. Ovchinnikov, N. G. Abdulaev, M. Yu. Feigina, A. V. Kiselev, and N. A. Lobanov, *FEBS Lett.* **100**, 219 (1979).

The progress in the elucidation of the primary structures of membrane proteins is determined by several circumstances, namely:

1. The most complete delipidation of the investigated protein.
2. Quantitative and specific cleavage of the polypeptide chain by chemical or enzymatic methods.
3. The availability of effective and reproducible methods for separation and purification of peptide fragments.

Delipidation

Either gel chromatography in detergent solutions or extraction with various mixtures of water–organic solvents is the method employed for delipidation of membrane proteins. The following delipidation procedures were used in the case of visual- and bacteriorhodopsin:

a. Ten milligrams of lyophilized purple membranes or disk membranes were dissolved in 1 ml of chloroform–methanol mixture (2 : 1); it was added in drops to 20 ml of the preliminarily cooled acetone–ammonium hydroxide solution (5 : 1) under vigorous stirring. The precipitate was stirred for 15–20 min, centrifuged, and washed several times with cold acetone–25% ammonium hydroxide mixture and, finally, with distilled water.

b. Lyophilized membranes were dissolved in a 4% solution of sodium dodecyl sulfate and kept overnight at 35–40°. The protein was then precipitated with ethanol (the final alcohol concentration, 80%) and the pellet was washed several times with 80% ethanol and water. The protein material so obtained was dissolved in a minimal volume of concentrated formic acid; the protein solution was diluted with distilled water (the final acid concentration, 10%) and ammonium hydroxide was added. The resultant precipitate was extensively washed with distilled water, and 80% ethanol and kept as water suspension at −20°.

The preceding methods have their own advantages and disadvantages. Both these methods not only separate lipids but also retinal from the protein; the removal of the latter sometimes occurs with much difficulty. Undoubtedly a disadvantage is that delipidation with organic solvents results in a high degree of aggregation of the protein material and its complete inaccessibility to the action of proteolytic enzymes. On the contrary, the delipidation procedure based on the use of sodium dodecyl sulfate decreases considerably the amount of protein aggregation and thus makes possible the digestion of the polypeptide chain. This method was successfully used for delipidation not only of bacteriorhodopsin but also of visual rhodopsin as well as DCCD-binding protein from ATPase complex of chloroplast membranes.

Preparation of Membrane Proteins for the Chemical and Enzymatic
  Digestion

The important step in determination of the protein primary structure is
the fragmentation of their polypeptide chains by various chemical or enzymatic methods. The high specificity, as well as the easiness of removal
from the reaction medium, makes cyanogen bromide one of the powerful
tools in the quantitative fragmentation of proteins and peptides. The interest in this reagent has increased rapidly, since under controlled conditions
it can quantitatively cleave proteins at both the methionine and tryptophan residues.[4]

We regularly employed cyanogen bromide cleavage (with some modifications) for fragmentation of bacteriorhodopsin and visual rhodopsin.
An aqueous suspension of the protein (20 mg) delipidated with sodium dodecyl sulfate was centrifuged and dry guanidinium hydrochloride (1 g)
was added to the resulting pellet. The mixture was carefully triturated
with a glass rod until a fine suspension was obtained, and distilled water
was added gradually in drops until the precipitate was dissolved completely. The concentrated formic acid (final concentration, 70%) and 500-
fold excess of cyanogen bromide were added to the protein solution in
guanidinium hydrochloride. The reaction was carried out for 24 hr at
room temperature in the dark. Then the peptide mixture was evaporated
on a rotor evaporator and vacuum-dried over sodium hydroxide.

The following analysis showed that this method allowed not only carrying out of the quantitative cleavage of the protein but also avoidance of
the aggregation characteristic of cyanogen bromide peptides, thus considerably facilitating their subsequent separation. It is noteworthy that the
peptide material obtained after cyanogen bromide cleavage of the protein,
delipidated with organic solvents, manifested an evident tendency toward
aggregation, and for its separation we used the method based on the stepwise extraction of peptides with different solvents and the subsequent
separation on different types of Sephadex.

The experience accumulated shows that the determination of the complete amino acid sequence of the membrane proteins cannot be based
solely on the analysis of fragments obtained after chemical cleavage.
The hydrophobic character and low solubility of these proteins in aqueous
media make difficult the use of enzymatic digestion methods. In this respect the development of the methods of solubilization and enzymatic
digestion of the membrane proteins is of considerable interest. To solubilize bacteriorhodopsin the protein solution in guanidinium hydrochloride

---

[4] J. Ozols and C. Gerard, *J. Biol. Chem.* **252,** 5987 (1977).

was subjected to extensive modification at the ε-amino groups of lysine with succinic or maleic anhydride.[5,6] After the modification the protein solution was slowly dialyzed against small volumes of 0.1 M ammonium bicarbonate buffer, pH 8.3. Preparations of protein solubilized in such a way underwent enzymatic hydrolysis with little difficulty.

Proteolysis of the bacteriorhodopsin suspension obtained after the removal of guanidinium hydrochloride as a result of slow dialysis of the nonmodified protein solution against 0.1 M ammonium bicarbonate buffer, pH 8.3 also produced satisfactory results.

In the case of bacteriorhodopsin and visual rhodopsin we succeeded in the effective application of another procedure for obtaining structural information. It entails the limited proteolysis of the protein integrated in the membrane, i.e., the treatment of the native purple membranes with aqueous solutions of proteolytic enzymes.[3]

Separation of Hydrophobic Peptides

Isolation of pure peptides from the mixtures obtained after the cleavage of the protein by chemical or enzymatic methods is laborious. In this respect the most difficult steps are the isolation and purification of the most hydrophobic peptides formed during cleavage of the membrane proteins. For separation of large hydrophobic peptides, the procedures based on the step-by-step extraction of the peptides by various concentrations of organic acids or by their chromatography in high concentrations of guanidinium hydrochloride on different carriers appeared to be most successful. As an example we present here two procedures for separation of cyanogen bromide peptides of bacteriorhodopsin.

According to the first one the mixture of cyanogen bromide peptides (cleavage in 70% formic acid) was first divided into two fractions—soluble (group I) and insoluble (group II) in 50% formic acid. To dissolve the peptides of the second group 6 M urea solution in 50% formic acid was used. The peptides insoluble in this system (group III) were solubilized in the concentrated solution of guanidinium hydrochloride (see dissolving of the protein in guanidinium hydrochloride). Such stepwise dissolving of the peptides permitted not only facilitating and accelerating their subsequent fractionation but also achieving good separation of peptides of similar molecular weights and charges.

Separation of the fragments of the first group in 50% formic acid by gel filtration is represented in Fig. 1. The analysis of fractions obtained after

[5] F. S. Chu, E. Crary, and M. S. Bergdoll, *Biochemistry* **8**, 2890 (1969).
[6] P. J. G. Butler, J. I. Harris, B. S. Hartley, and R. Liberman, *Biochem. J.* **112**, 679 (1969).

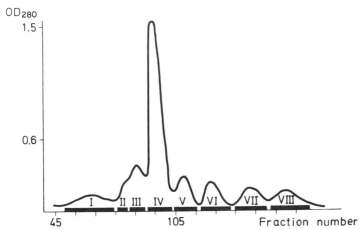

Fig. 1. Separation of CNBr fragments of bacteriorhodopsin soluble in 50% formic acid on Sephadex G-50 (Superfine). Column size 1.5 × 100 cm. Peptides were eluted with 50% formic acid. Flow rate 3.5 ml/hr.

the separation showed that four of them (I, V, VI, and VII) are the individual peptides. Rechromatography of the fractions corresponding to peaks III and IV on the same column resulted in two more pure peptides. In such a way one succeeded in obtaining 6 of 10 cyanogen bromide peptides of bacteriorhodopsin quite easily and with high yields.

Figures 2 and 3 demonstrate the results of separation of peptides of groups II and III. The stepwise fractionation and separation allowed us to obtain 8 of 10 cyanogen bromide peptides of bacteriorhodopsin.

The use of HPLC for separation of peptides is becoming increasingly popular. But in spite of the definite successes achieved in separation of

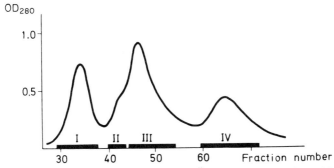

Fig. 2. Separation of CNBr fragments of bacteriorhodopsin soluble in 50% formic acid and 6 M urea on Sephadex G-75 (Superfine). Column size 1.5 × 100 cm. Peptides were eluted with 6 M urea in 50% formic acid. Flow rate 3.5 ml/hr.

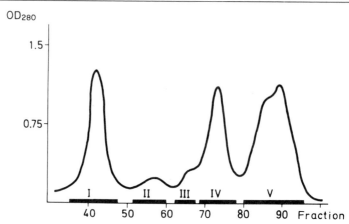

FIG. 3. Separation of CNBr peptides of bacteriorhodopsin soluble in 8 M guanidinium hydrochloride on Sephadex G-75 (Superfine). Column size and flow rate are the same as in Figs. 1 and 2.

native proteins and short peptides obtained after exhaustive enzymatic hydrolysis, separation of large hydrophobic peptides is still a complicated problem.[7-9] Nevertheless, we used this method for the separation of fraction I.

The high solubility of the delipidated bacteriorhodopsin and the peptides obtained after the cleavage of the protein with cyanogen bromide in guanidinium hydrochloride prompts use of concentrated solutions of this denaturating agent as an eluent for the separation of hydrophobic peptides.

The application of guanidinium hydrochloride solutions arose from the considerable destruction of tryptophan during peptide chromatography employing high concentrations of formic acid. Moreover, this method makes it possible to obtain in one step a considerable number of pure peptides, and separation of unresolved peaks is not difficult. The separation of cyanogen bromide hydrolysate on the column with Sephadex G-75 equilibrated with 8 M solution of guanidinium hydrochloride enabled us to obtain 6 of 10 peptides in one step (Fig. 4). Fractions I and II were separated by their rechromatography on the same column under the same conditions.

One should note that these approaches were widely used for separation of the peptides obtained after the cleavage of bacteriorhodopsin with

[7] M. Rubinstein, *Anal. Biochem.* **98,** 1 (1979).
[8] F. Naider, R. Sipzer, A. S. Steinfeld, and J. M. Becker, *J. Chromatogr.* **176,** 264 (1979).
[9] W. C. Mahoney and M. A. Hermodson, *J. Biol. Chem.* **355,** 11199 (1980).

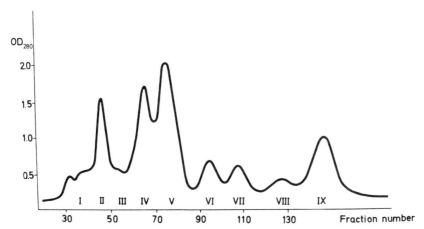

FIG. 4. Separation of CNBr peptides of bacteriorhodopsin in 8 *M* guanidinium hydrochloride. Column size, 1.5 × 100 cm. Flow rate 3 ml/hr.

BNPS-skatole,[10-12] N-bromosuccinimide,[13] and cyanogen bromide at methionine and tryptophan residues, as well as for separation of large tryptic peptides and can be useful when the primary structure of integral membrane proteins is analyzed.

[10] A. Fontana, this series, Vol. 25, p. 419.
[11] E. H. Eylar, J. J. Jackson, C. D. Bennet, P. J. Knisken, and S. W. Brostoff, *J. Biol. Chem.* **249**, 3710 (1974).
[12] B. Debuire, K. K. Han, M. Dautrevaux, G. Biserte, F. Reghout, and R. Kassab, *J. Biochem. (Tokyo)* **81**, 611 (1977).
[13] A. Fontana, N. M. Green, and B. Witkop, *J. Am. Chem. Soc.* **86**, 1846 (1964).

[87] Kinetic Properties of Rhodopsin and Bacteriorhodopsin Measured by Kinetic Infrared Spectroscopy (KIS)

By WERNER MÄNTELE, FRITZ SIEBERT, and W. KREUTZ

In rhodopsin as well as in bacteriorhodopsin, photochemical reactions are coupled to intramolecular reactions, leading to increased enzyme activity and proton pumping, respectively.[1-4]

[1] P. A. Liebman and E. N. Pugh, *Nature (London)* **287**, 734 (1980).
[2] B. K. K. Fung, J. B. Hurley, and L. Stryer, *Proc. Natl. Acad. Sci. U.S.A.* **78**, 152 (1981).
[3] H. Kühn, *Nature (London)* **283**, 587 (1980).
[4] D. Oesterhelt and W. Stoeckenius, *Proc. Natl. Acad. Sci. U.S.A.* **70**, 2853 (1973).

The nature of this coupling of the photochemistry to the functional properties still presents an unsolved problem. Flash photolysis and spectroscopy in the visible spectral range[5,6] have well characterized the photochemical reactions of the chromophore, and resonance Raman spectroscopy, favored by its great selectivity, has yielded abundant information on the molecular structure of the chromophore in rhodopsin, bacteriorhodopsin, and their intermediates.[7-9] A review is given by Mathies.[10]

The part of a chromoprotein that is accessible by means of resonance Raman spectroscopy is, however, restricted to the chromophore and its direct environment in the protein. In addition to this restriction, kinetic parameters of the intermediates can only be deduced indirectly.[11] Thus, considerable lack of information exists on the molecular changes in the protein coupled to chromophore reactions.

Supplementary information on the molecular changes of a chromoprotein can be obtained by infrared spectroscopy.[12] The normal infrared spectrum of a protein, however, is not detailed enough to allow the investigation of molecular events in a small part of a protein because of the overlap of many absorption bands. This nonselectivity imposes a major problem and makes the interpretation of static infrared spectra of proteins very troublesome. Infrared difference spectroscopy, again achieved by static measurements, is hard to perform within the lifetime of the intermediates of the reaction sequence of rhodopsin and bacteriorhodopsin, respectively.

Kinetic infrared spectroscopy, however, monitoring transmission changes only, renders infrared spectroscopy selective. In addition, information on the dynamics of the reaction and the existence and nature of the intermediates is obtained from the kinetics of the transmission changes. The difference spectrum generated from kinetic infrared signals then contains useful information on the molecular processes in the chromophore as well as in the protein.

One major advantage is connected with kinetic infrared spectroscopy: The measuring light intensity is not limited by causing photoreactions or

[5] E. W. Abrahamson, in "Biochemistry and Physiology of Visual Pigments" (H. Langer, ed.), p. 47. Springer-Verlag, Berlin and New York, 1973.

[6] R. H. Lozier, R. A. Bogomolni, and W. Stoeckenius, Biophys. J. 15, 955 (1975).

[7] A. R. Oseroff and R. H. Callender, Biochemistry 13, 4243 (1974).

[8] R. H. Callender, A. Doukas, R. Crouch, and K. Nakanishi, Biochemistry 15, 1621 (1976).

[9] M. Stockburger, W. Klusmann, H. Gattermann, G. Massig, and R. Peters, Biochemistry 18, 4886 (1979).

[10] R. Mathies, in "Chemical and Biochemical Application of Lasers" (C. B. Moore, ed.), Vol. 4, p. 55. Academic Press, New York, 1979.

[11] M. A. Marcus and A. Lewis, Science 195, 1328 (1977).

[12] D. F. Hoelzl Wallach and A. R. Oseroff, this series, Vol. 32, Part B, p. 247.

sample decomposition as is the case with a measuring beam in the visible spectral range. The difficulty that is caused in infrared spectroscopy and kinetic infrared spectroscopy by the strong absorbance of water can be overcome by the use of very concentrated samples or suspensions in D_2O as a solvent.

In the following a flash photolysis apparatus with an infrared monitoring beam will be described briefly. This instrument is sensitive enough to detect the very small changes that occur in a small part of the total complex molecule. The preparation of samples for the investigation of the rhodopsin reaction sequence and the bacteriorhodopsin reaction cycle with kinetic infrared spectroscopy will be reported, and the results obtained with our method will be discussed in connection with other experimental techniques, mainly resonance Raman spectroscopy.

Description of the Apparatus

Figure 1 shows a schematic diagram of our apparatus. Its basic features are those of a conventional flash photolysis instrument. The globar

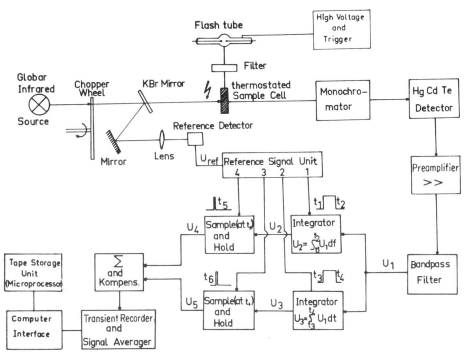

FIG. 1. Schematic diagram of the flash photolysis apparatus with infrared monitoring beam. The reference signal unit together with integrators and sample-and-hold unit operates as a boxcar integrator.

infrared source and the monochromator are part of a Perkin–Elmer infrared spectrophotometer, model PE 180. An Hg–Cd–Te detector is used instead of the standard thermopile detector, providing high sensitivity and high time resolution. The $1/f$ noise exhibited by the detector, which dominates at frequencies below about 2 kHz, requires chopping of the infrared beam for measurements with time constants larger than 1 msec. Chopping is performed with a variable frequency motor at a rate of up to 10 kHz.

For measurements at shorter time constants, the $1/f$ noise is filtered by a high-pass filter, eliminating the need of a chopper. The thin-film samples that will be described are excited by xenon flashlamps almost coaxial with the measuring beam. The excitation energy is variable; in the case of rhodopsin, up to 20% can be bleached. Kinetic signals are stored in a transient recorder, where signal averaging can be performed to improve the signal-to-noise ratio (SNR). Signals stored on a tape unit can be combined with a computer to form difference spectra at various time intervals. With this instrument, relative transmission changes of the order of 10^{-3} at a spectral resolution of about 6 cm^{-1}, depending on the intensity at the detector, can be measured at a time resolution of 1 msec with a SNR of 1. If the sample under investigation allows averaging, as in the case of bacteriorhodopsin, time resolution under similar conditions can be up to a few microseconds.

Sample Preparation for Kinetic Infrared Spectroscopy

Depending on the biological system to be investigated, several types of infrared samples were prepared: Soluble proteins (e.g., myoglobin) were prepared in concentrated solutions at a thickness of 25 μm. The preparation of these samples and the investigation of the photodissociation of CO-myoglobin by means of kinetic infrared spectroscopy are reported elsewhere.[13] In the case of rhodopsin and bacteriorhodopsin, we were constrained to get concentrated protein samples at a thickness of a few micrometers with an area density of 0.2–1 mg/cm^2. This value was calculated for a reasonable absorbance of the thin-film samples in the visible part of the spectrum (for excitation) as well as in the infrared.

Suspensions of rod outer segments (ROS) or purple membrane (PM) fragments cannot be used at such a concentration because of their high viscosity. In addition, aqueous samples of rod outer segments or purple membranes at the preceding concentration still absorb strongly at the O—H-bending vibration of water around 1650 cm^{-1}, which makes kinetic measurements impossible. For the investigtion of ROS samples and PM samples at a defined ionic strength and pH, however, we have used such concentrated aqueous samples prepared by a rather crude but successful technique: Suspensions were sedimented in small polyethylene tubes at

[13] F. Siebert, W. Mäntele, and W. Kreutz, *Biophys. Struct. Mech.* **6**, 139 (1980).

50,000 g, frozen in liquid nitrogen, and cut slightly below the interface between sediment and supernatant. The sediment was then squeezed out of the bottom of the tube onto a CaF_2 window and pressed by a second window to a spacing of 10–20 μm. This method can be performed with suspensions in H_2O or D_2O, and at buffer concentrations corresponding to the final protein concentrations. Its disadvantage is the inhomogeneity of the sample in its final form. The large diameter of the infrared measuring beam, which at the moment is about 1 cm, partially compensates for this. To get samples of higher homogeneity and to reduce water absorption further, we developed a technique to prepare thin ROS or PM films that can be hydrated or deuterated to the desired degree. Similar films have been used for the investigation of the BR photocycle by Korenstein and Hess.[14] The preparation of thin rhodopsin films for the spectroscopic study of α-helix orientation and of purple membrane films was also described by Rothschild et al.[15,16] Purple membrane or rod outer segments, respectively, are centrifuged from a suspension in a vessel consisting of a CaF_2 window with a frame sealed to it. The supernatant is soaked with a syringe and the sample is moderately dried by a water-jet pump. The vessel is then sealed with a second CaF_2 window, thus forming a cuvette with a thin protein film on the inner side of one window. The frame inside contains a small reservoir, providing a constant relative humidity inside the sealed cuvette when filled with saturated salt solutions, water, or D_2O if necessary for ^1H–^2H exchange. Filling of the reservoir is achieved with a syringe through a septum-sealed hole in the frame. Rehydration of the sample is monitored by following the absorbance at 3400 cm^{-1} (O—H stretching vibration of water). This process generally took about 1 hr. Deuteration effects were investigated using the same sample by simply removing H_2O from the reservoir, drying the sample, and filling the reservoir with D_2O. The time course and the degree of deuteration could be monitored by measuring the decrease of the amide II band.[17] This hydrogen–deuterium exchange via the vapor phase took about 2 hr at room temperature.

This technique yields very homogeneous films on the CaF_2 windows. They allow static and kinetic spectroscopic measurements from the near UV to about 10 μm. To extend this range to longer wavelengths a similar cuvette was constructed with polyethylene films of 10 μm thickness as windows. The sampling procedure is carried out as with CaF_2 windows.

One disadvantage is connected with the use of thin hydrated films: Although ionic strength in the films can be approximately controlled by

[14] R. Korenstein and B. Hess, FEBS Lett. **82,** 7 (1977).
[15] K. J. Rothschild, K. M. Rosen, and N. A. Clark, Biophys. J. **31,** 45 (1980).
[16] N. A. Clark, K. J. Rothschild, D. A. Luippold, and B. A. Simon, Biophys. J. **31,** 65 (1980).
[17] A. R. Osborne and E. Nabedryk-Viala, FEBS Lett. **84,** 217 (1977).

starting with suspensions of low ionic strength, pH cannot be adjusted by the use of standard buffers.

Hence, only samples prepared from suspensions of purple membranes in distilled water or of rod outer segments in low ionic concentration were used. The pH in such films can be estimated from kinetics of known photoreactions and was found to be approximately pH 7.

All infrared samples were tested for correct functioning by conventional UV–visible spectroscopy and flash photolysis. No deviations from spectra or kinetic signals of aqueous suspensions at low protein concentration were found.

Results

Kinetic Infrared Spectroscopy of Bacteriorhodopsin

Figure 2a shows the time and wavelength dependence of the IR-absorbance changes of a small section of the fingerprint region from 1500 to

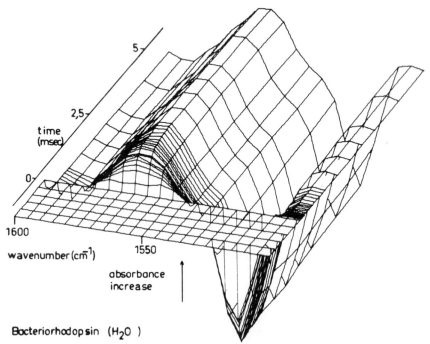

FIG. 2a. Time and wavelength dependence of the IR absorbance changes in the wavelength region from 1500 to 1600 cm^{-1}. In the time scale shown, almost no reformation of photolyzed bacteriorhodopsin has occurred. The absorbance changes shown are of the order of $\Delta A/A \approx 10^{-2}$.

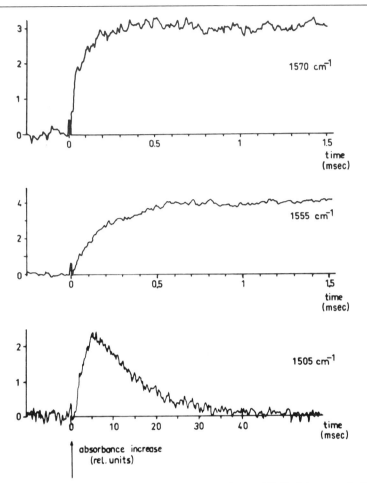

FIG. 2b. Kinetic signals at 1570, 1555, and 1505 cm^{-1}. $T = 20°$, flash is at $t = 0$; 100 signals were averaged to improve SNR. The signal at 1570 cm^{-1} represents the occurrence of the C=C vibration of M_{412} with a characteristic $t_{1/2}$ of ~ 50 μsec. It clearly correlates with the absorbance changes observed at 412 nm. The signal at 1555 cm^{-1}, with a half-time of its rise of about 200 μsec, has no corresponding absorbance change in the visible spectral range. The signal at 1505 cm^{-1} represents the rise and decay of the O_{640}-intermediate as monitored by the C=C vibration of the chromophore. (Note change in time scale.) This absorbance change correlates with the signals observed at 700 nm.

1600 cm^{-1}. The time scale extends to 5 msec; the time-dependent difference spectrum therefore reflects intermediates that have only partially decayed back to bacteriorhodopsin. The wavelength region displayed in Fig. 2a mostly contains absorbance changes due to the C=C vibration of the chromophore. The fast absorbance decrease at about 1525 cm^{-1} with a corresponding absorbance increase at 1550 cm^{-1} represents the shift of the C=C vibration corresponding to the shift of λ_{max} from 570 to 550 nm at the L-intermediate. The C=C vibration is further shifted to around 1565 cm^{-1}, where an absorbance increase, corresponding to the C=C vibration of M_{412}, is observed.

The absorbance increase at about 1560 cm^{-1} is, however, not homogeneous. In addition to the well-known risetime of the M_{412}-intermediate, a slow component is superimposed. It can be separated from M_{412}, although in some cases, where its contribution is small, separation may be incorrect. Two signals that represent almost pure kinetics of M_{412} and of the slow component at 1570 cm^{-1} and at 1555 cm^{-1}, respectively, are shown in Fig. 2b. Figure 3 shows the difference spectrum constructed from the separated amplitudes. The difference spectra obtained from deuterated samples are indicated by the dashed line. Between 1505 and 1515 cm^{-1} a pronounced effect occurs (Fig. 2b). Its rise time resembles that of the O_{640} form of the chromophore and its amplitude increases with deuteration and increasing temperature, as does O_{640}.[18] This absorbance increase can be assigned to the C=C vibration of O_{640}, which, from its λ_{max}, should be expected around 1510 cm^{-1}. The amplitude of the slow components observed at other wavenumbers also increase with deuteration and increasing temperature. Their kinetics, significantly slower than the risetime but faster than the decay time of the intermediate M_{412} seem to correlate with O_{640}. It is generally assumed that O_{640} originates from M_{412}. This has recently been questioned by us on the basis of the influence of deuteration and Triton X-100 on O_{640} and its precursors.[19] The kinetic infrared signals observed around 1560 cm^{-1} provide additional evidence that M_{412} does not decay to O_{640}, since this would imply that a slow transmission increase is observed around 1560 cm^{-1}. Instead, we detect a slow transmission decrease. No slowly increasing spectroscopic intermediate with a λ_{max} of about 400 nm, corresponding to a C=C vibration at 1560 cm^{-1}, is known, which could be assigned to the slow component observed. We would, therefore, rather suggest that this slow component is not caused by the chromophore. Since a certain correlation between O_{640} and the extrusion of protons from *Halobacterium halobium* ghosts has been reported,[20] these nonchromophore signal components deserve further attention. Measurements are in progress to clarify the meaning of these

[18] W. Mäntele and F. Siebert, *Biophys. Struct. Mech.* **6,** Suppl., 124 (1980).

[19] W. Mäntele, F. Siebert, and W. Kreutz, submitted for publication.

[20] R. H. Lozier, W. Niederberger, M. Ottolenghi, G. Sivorinosky, and W. Stoeckenius, *in* "Energetics and Structure of Halophilic Microorganisms" (S. R. Caplan and M. Ginzburg, eds.), p. 123. Elsevier/North-Holland Biomedical Press, Amsterdam, 1978.

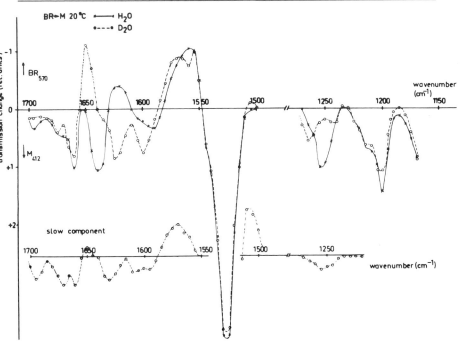

FIG. 3. Bacteriorhodopsin-M_{412} difference spectrum generated from kinetic signals. Dashed line shows difference spectrum from deuterated sample. For each data point 100 signals were averaged. Spectral resolution varied between 4 and 6 cm^{-1}. The dashed-dotted line around 1650 cm^{-1} represents a fast component of the deuterated sample. The lower insert shows the difference spectrum of the slow component of the deuterated sample.

components for the coupling of the photochemical cycle of bacteriorhodopsin to its proton-pumping function.

The wave number region from 1150 cm^{-1} to 1250 cm^{-1} mostly contains C—C-stretching and C—C—H-bending vibrations and is, therefore, sensitive to changes due to isomerization. A comparison with model compounds has, however, not been performed yet. The decreasing absorbance observed in this wavelength region can be explained by the deprotonation of the Schiff base.[21] The band at 1250 cm^{-1}, showing a strong deuteration effect, has tentatively been assigned to the bending vibration of N$^+$ —H of the Schiff base.[9] Its C=N-stretching vibration should appear around 1650 cm^{-1}. In this wavelength region also several slow components are superimposed on the absorbance changes due to BR_{570}-L_{550}-M_{412}-reaction, which complicates the separation of the signals. The large spectral changes, however, which are caused by deuteration,

[21] F. Siebert and W. Mäntele, *Biophys. Struct. Mech.* **6**, 147 (1980).

are in agreement with a normal protonated Schiff base. The strong transmission decrease observed in deuterated samples at 1650 cm^{-1} seems to indicate a further nonchromophore component. Its risetime is between that of L_{550} and M_{412}, which rules out a chromophore band but suggests that it might reflect a molecular change in the protein leading to the proton ejection.

In recent experiments we have investigated the wavelength region from 1500 to 1250 cm^{-1} that was omitted in the difference spectra of Fig. 3. Around 1400 cm^{-1} absorbance changes were found that contain slow components as observed at 1555 cm^{-1}. In addition, a fast absorbance increase with subsequent decay in a time range slower than the M_{412} formation is observed. In this wavelength region the spectra of retinal model compounds[21] show bands that are attributed to the CH$_2$- and CH$_3$-deformation vibrations. These bands are not expected to depend strongly on the isomeric form of the chromophore or on its state of protonation. The large changes in this wavenumber region, together with the kinetic information, again suggest that these signals are not due to the chromophore.

Kinetic Infrared Spectroscopy of Rhodopsin

As mentioned in the description of the apparatus, the time resolution for a single signal is of the order of 1 msec. The reactions of the rhodopsin reaction sequence that can be monitored at that time resolution are the rhodopsin MI and MII reaction. The MI reaction cannot be time-resolved, but its absorbance difference can be evaluated. The difference spectra are much more difficult to obtain than in the case of bacteriorhodopsin, since for one data point, corresponding to one signal, about 20% of the rhodopsin has to be bleached.

Figure 4 shows the rhodopsin MII difference spectrum generated in this way. A new sample was used for each data point. The second flash at the sample was used to obtain a signal at 1235 cm^{-1}. This strong transmission increase that reflects changes due to the rhodopsin MI transition, independent of pH and temperature, was therefore used for normalization.

Three main groups of bands in the difference spectrum reflect the state of protonation, the isomeric state of the chromophore, and, presumably, changes in the protein that follow chromophore reactions. A detailed discussion is given in Siebert and Mäntele,[21] where the absorbance changes are interpreted on the basis of model compounds.

The 1100 to 1300-cm^{-1}-wavelength region mostly reflects changes caused by the 11-*cis* to all-*trans* isomerization of the chromophore. The dominating negative components are due to the fact that a markedly structured positive rhodopsin MI difference spectrum as reported in Siebert and Mäntele,[21] (caused by isomerization) is superimposed by a strong but

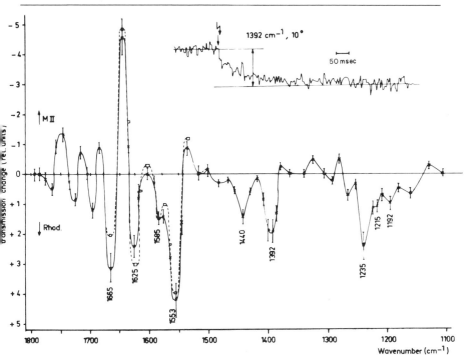

FIG. 4. Kinetic rhodopsin–metarhodopsin II difference spectrum. A new sample was used for each data point. Each flash bleached approximately 20% of the rhodopsin. Temperature was 20°. Spectral resolution varied between 10 and 6 cm^{-1}, depending on the intensity at the detector. Ordinate scale unit corresponds to a relative transmission change of $\Delta I/I = 3 \times 10^{-3}$. Insert shows kinetic signal at 1392 cm^{-1} and 10°. Dashed line: deuterated sample.

less structured negative difference spectrum of the MI-MII reaction (mostly caused by deprotonation). In this wavelength region there is good agreement with resonance Raman measurements.[22,23]

The spectral region between 1500 and 1600 cm^{-1} reflects the changes of the C=C vibration of the retinal polyene chain. They too can be understood in terms of isomerization and shift of the C=C-stretching frequency, corresponding to the shift of λ_{max} and followed by an absorbance decrease due to deprotonation.

In the 1600- to 1700-cm^{-1} wave number region, typical for the C=N-stretching vibration of the Schiff base, the difference spectrum appears very complex. For a normal protonated Schiff base in model compounds a strong band at 1650 cm^{-1} is seen, shifting to 1620 cm^{-1} and decreasing in

[22] R. Mathies, A. R. Oseroff, and L. Stryer, *Proc. Natl. Acad. Sci. U.S.A.* **73**, 1 (1976).
[23] A. Doukas, B. Aton, R. H. Callender, and T. Ebrey, *Biochemistry* **17**, 2430 (1978).

intensity on deprotonation. Thus the rhodopsin MII difference spectrum should show an increase in transmission at 1650 cm^{-1} and a decrease around 1620 cm^{-1}. Obviously, the difference spectrum is not in accordance with a simple protonated Schiff base. This discrepancy is further discussed in Siebert and Mäntele.[21]

The bands in the difference spectrum at 1392 cm^{-1} and at 1440 cm^{-1} deserve further attention. For the reasons discussed for bacteriorhodopsin, no major changes can be expected between 1300 cm^{-1} and 1500 cm^{-1} from the chromophore. In addition, preliminary measurements of their temperature dependence indicate that their amplitude does not follow the MI/MII equilibrium as observed at other wavenumbers or in the visible. Hence we suppose that these signals are not caused by the chromophore. Whether they reflect molecular processes in the opsin or in the disk membrane cannot be decided yet and demands further investigations.

General Discussion

These examples have demonstrated the feasibility of kinetic infrared spectroscopy (KIS) as a method of vibrational spectroscopy complementary to Raman spectroscopy. The results partially confirm results with resonance Raman spectroscopy, although in some cases, as, for instance, in the Schiff base region of the rhodopsin kinetic difference spectrum, discrepancies are seen.

Kinetic infrared spectroscopy appears to be especially useful for the investigation of biological systems with a chromophore-protein link. In such systems, in addition to information on the chromophore that may be obtained by other methods, further parts of the chromoprotein are accessible by kinetic infrared spectroscopy.

It should, however, be pointed out that the apparatus described actually represents a specialized realization of kinetic infrared spectroscopy for the investigation of chromoproteins. A further application of KIS would be the temperature jump or field jump method. Both could be easily performed with the samples described earlier and would be generally applicable to biological systems with or without a chromophore.

[88] Photodichroism and the Rotational Motions of Rhodopsin and Bacteriorhodopsin

By PATRICK L. AHL and RICHARD A. CONE

Introduction

The rotational motions of the two membrane proteins rhodopsin and bacteriorhodopsin can be readily detected using polarized light since (1) the chromophore (retinal) is tightly bound to each protein and (2) the photoproducts of both of these molecules persist throughout the time scale of their rotational motions.[1,2] A linearly polarized actinic flash can photoselect chromophores aligned with the electric vector of the flash, rendering an initially random sample photodichroic. If either the photoactivated and/or the remaining unactivated chromophores undergo rotational displacements following the actinic flash, the photoinduced dichroism will change, and such changes in dichroism can be easily detected with polarized measuring light. Using this method, rhodopsin has been shown to undergo rapid rotational diffusion in photoreceptor disk membranes,[3] and recently rotational mobility of bacteriorhodopsin has been observed in the purple membrane.[4] Moreover, studies of the rotational mobility of bacteriorhodopsin monomers in artificial membrane systems aid in our understanding of lipid–protein interactions.[5-7] The rotational and lateral diffusion of membrane proteins in general has been recently reviewed.[8]

Here we will discuss the problems involved in measuring the rotational motions of rhodopsin and bacteriorhodopsin in various preparations. First we examine the theory of photodichroism relaxation for samples such as rhodopsin and bacteriorhodopsin. Although the theory is similar to the familiar theory for fluorescence relaxation,[9] there is a key difference that must be recognized: In fluorescence relaxation only the excited chromophores contribute to the signal, whereas in the relaxation of photodichroism both the photoactivated and the unactivated chromo-

[1] W. Stoeckenius, R. H. Lozier, and R. A. Bogomolni, *Biochim. Biophys. Acta* **505**, 215 (1979).

[2] T. G. Ebrey and B. Honig, *Q. Rev. Biophys.* **8**, 129 (1975).

[3] R. A. Cone, *Nature (London), New Biol.* **236**, 39 (1972).

[4] P. L. Ahl and R. A. Cone, *Fed. Proc., Fed. Am. Soc. Exp. Biol.* **39**, 1847 (1980); P. Ahl and R. Cone, *Biophys. J.* **37**, 229a (1982).

[5] R. J. Cherry, U. Muller, R. Henderson, and M. P. Heyn, *J. Mol. Biol.* **121**, 283 (1978).

[6] R. J. Cherry, U. Muller, and G. Schneider, *FEBS Lett.* **80**, 465 (1977).

[7] C. L. Wey, P. L. Ahl, R. A. Cone, and B. J. Gaffney, *Biophys. J.* **25**, 169a (1979).

[8] R. J. Cherry, *Biochim. Biophys. Acta* **559**, 289 (1979).

[9] R. Rigler and M. Ehrenberg, *Q. Rev. Biophys.* **6**, 139 (1973).

METHODS IN ENZYMOLOGY, VOL. 88

phores contribute to the signal since both populations of chromophores absorb the measuring light. Thus chromophore rotations in either or both populations produce changes in the observed dichroism. In the last section we briefly discuss some technical details for measuring the rotational motions of rhodopsin and bacteriorhodopsin.

Photodichroism Relaxation Theory

The process of photoselection has been previously reviewed[10,11] and here we want only to stress that both the activated and unactivated populations of chromophores are photoselected, rendering both populations partially oriented. Subsequent motions of these chromophore populations can be monitored with linearly polarized measuring light oriented with its electric vector parallel or perpendicular to that of the actinic flash. After the actinic flash the absorbance of the sample is the sum of the absorbancies of the new photoproducts and the remaining unactivated molecules. The observed absorbance change will be positive or negative depending on the relative contribution of these two populations of molecules. If there is more than one photoproduct, the final absorbance change will be

$$\Delta A = \Delta A_d + \sum \Delta A_{pj} \tag{1}$$

where ΔA_d is the change in absorbance of the unactivated molecules and ΔA_{pj} is the change in absorbance due to photoproduct j. Since the actinic flash depletes the number of unactivated molecules, ΔA_d is always negative, whereas ΔA_{pj} is always positive.

Dichroism can be expressed several ways; two of the most useful are polarization, p, or anisotropy, r, as defined in Table I. These expressions are useful in analyzing the time dependence of chromophores undergoing rotational diffusion: If the conditions are appropriate, the decay of p will be a single exponential and the decay of r will be the sum of exponentials.[8] Polarization is used for analyzing dichroism in samples such as retinas or oriented multilayers where the membranes containing the chromophores are highly ordered and the chromophores are free to undergo rotational motions only around the axis parallel to the direction of the measuring light. Anisotropy is useful for analyzing situations in which the membranes are randomly oriented, such as suspensions of lipid vesicles containing bacteriorhodopsin or suspensions of purple membrane fragments. The

[10] A. C. Albrecht, *Prog. React. Kinet.* **5**, 301 (1970).
[11] A. C. Albrecht, *J. Mol. Spectrosc.* **6**, 84 (1961).

TABLE I
DICHROISM EXPRESSIONS

	Anisotropy	Polarization
Definition	$r \equiv \dfrac{\Delta A_{\parallel} - \Delta A_{\perp}}{\Delta A_{\parallel} + 2\Delta A_{\perp}}$	$p \equiv \dfrac{\Delta A_{\parallel} - \Delta A_{\perp}}{\Delta A_{\parallel} + \Delta A_{\perp}}$
Single population	$r = \frac{1}{2}\{3 \overline{\cos^2 \theta} - 1\}$	$p = 2 \overline{\cos^2 \theta} - 1$

main absorption band of the chromophores of both rhodopsin and bacteriorhodopsin can be adequately treated as arising from electric-dipole transitions[12]; thus the probability of absorbing a photon is proportional to $\cos^2 \theta$, where θ is the angle between the transition dipole and the electric vector of the polarized light. As shown in Table I, for a single population of dipoles the dichroism depends on $\overline{\cos^2 \theta}$, the average $\cos^2 \theta$ of the dipoles produced or depleted by the actinic flash. These expressions can be calculated by applying the methods of Ref. 13 to photodichroism.

For a multipopulation sample the observed polarization or anisotropy will be the sum of p's or r's for each individual population of dipoles weighted by that population's share of the total absorbance change. Thus

$$P = \sum_{d,pj} f(\Delta A_k)p_k \tag{2}$$

and

$$r = \sum_{d,pj} f(\Delta A_k)r_k \tag{3}$$

where

$$f(\Delta A_k) = \frac{\Delta A_k}{\Sigma_{d,pj} \, \Delta A_k} \tag{4}$$

Both ΔA_k and $\overline{\cos^2 \theta}$ are defined differently for the two expressions. For p, $\Delta A_k = \Delta A_{k\parallel} + \Delta A_{k\perp}$, whereas for r, $\Delta A_k = \Delta A_{k\parallel} + 2\Delta A_{k\perp}$. For p

$$\overline{\cos^2 \theta} = \frac{\int_0^{\pi/2} n_k(\theta) \cos^2 \theta \, d\theta}{\int_0^{\pi/2} n_k(\theta) \, d\theta} \tag{5}$$

and for r

---

[12] J. Hofrichter and W. A. Eaton, *Annu. Rev. Biophy. Bioeng.* **5**, 511 (1976).
[13] A. J. Pesce, C. Rosen, and T. L. Pasby, "Fluorescence Spectroscopy." Dekker, New York.

$$\overline{\cos^2 \theta} = \frac{\int_0^{\pi/2} n_k(\theta) \cos^2 \theta \sin \theta \, d\theta}{\int_0^{\pi/2} n_k(\theta) \sin \theta \, d\theta} \tag{6}$$

$n(\theta)_k$ is the angular distribution of dipoles as a functions of θ. Therefore, $n_{pj}(\theta)$ is the angular distribution of the dipoles for photoproduct j. Rather than using the angular distribution of the remaining unactivated dipoles $n_u(\theta)$, we will use $n_d(\theta)$, which is the angular distribution of the depleted dipoles. If $n_o(\theta)$ is the dipole angular distribution for the pigment before the flash, then $n_d(\theta) = n_0(\theta) - n_u(\theta)$.

$\cos^2 \theta_k$ can be calculated from Eqs. (5) and (6) if $n_k(\theta)$ is known. The initial dipole distributions for the unactivated and photoactivated dipoles are determined by many factors, including the intensity and polarization of the actinic flash, light scattering by the sample, and possible energy transfer between chromophores. Initially $n_{pj}(\theta)$ for the first photoproduct will equal $n_d(\theta)$ unless the photoproduct dipoles have rotated with respect to the original unactivated dipole orientations before the earliest observation of dichroism is accomplished. After the photoselection process by the actinic flash, the dipoles of both the unactivated molecules and the photoproducts may independently undergo rotational motions, changing $\cos^2 \theta_d$ and $\cos^2 \theta_{pj}$ as they do. We will now calculate what $n_d(\theta)$ and $n_{p1}(\theta)$ will be for optimum photoselection. Then we will examine the effects of rotational motions of the dipoles on the observed dichroism.

If $no(\theta)$ is the initial θ distribution of the unactivated dipoles, then the distribution of unactivated dipoles after the actinic flash will be

$$n_u(\theta) = n_0(\theta)e^{-E\epsilon\gamma \cos^2 \theta} \tag{7}$$

where

E = energy of the actinic flash
ϵ = napierian extinction coefficient at the actinic flash wavelength
γ = quantum efficiency

If the dipoles of the first photoproduct have not rotated, then their initial distribution will be

$$n_{p1}(\theta) = n_0(\theta) (1 - e^{-E\epsilon\gamma \cos^2 \theta}) \tag{8}$$

For optimum photoselection, $E\epsilon\gamma$ is $\ll 1$, and in this case the distributions simplify to

$$n_d(\theta) = n_0(\theta) - n_u(\theta) = n_0(\theta)E\epsilon\gamma \cos^2 \theta \tag{9}$$
$$n_{p1}(\theta) = n_0(\theta)E\epsilon\gamma \cos^2 \theta \tag{10}$$

Table II shows $\overline{\cos^2 \theta}$ for these dipole distributions as well as for randomly oriented dipoles for both polarization and anisotropy. Thus $\overline{\cos^2 \theta}$ will

TABLE II
DIPOLE DISTRIBUTIONS

	Optimum photoselection	Randomly oriented
Anisotropy	$r = \frac{2}{5}, \overline{\cos^2 \theta} = \frac{3}{5}$	$r = 0, \overline{\cos^2 \theta} = \frac{1}{3}$
Polarization	$p = \frac{1}{2}, \overline{\cos^2 \theta} = \frac{3}{4}$	$p = 0, \overline{\cos^2 \theta} = \frac{1}{2}$

vary between the limits shown in Table II as the chromophores undergo randomizing rotations.

As shown in Table I, for a single population of chromophores the observed dichroism is completely determined by $\overline{\cos^2 \theta}$. Thus for a single population these measures of dichroism are independent of the magnitude of the absorption changes and directly reveal the angular distributions of the chromophores. However, if more than one population of chromophores contributes to the absorption change, then the angular distributions of the various dipoles cannot be determined from the observed polarization or anisotropy alone, as shown in Eqs. (2) and (3). The observed dichroism will no longer be a function of $\overline{\cos^2 \theta}$ alone, and the limits shown in Table II for the observed p and r no longer apply. To illustrate this we will discuss, using polarization, a simple system with two chromophore populations: unactivated molecules plus a single stable photoproduct. In Fig. 1 we show what would be observed if the photoproduct (activated) dipoles randomized but the unactivated dipoles remained fixed. Under optimum photoselection conditions both the depleted and photoproduct dipoles will have an initial $\overline{\cos^2 \theta}$ of $\frac{3}{4}$. As the photoproduct randomizes, $\overline{\cos^2 \theta_p}$ falls from $\frac{3}{4}$ to $\frac{1}{2}$, but both the total absorption and $\overline{\cos^2 \theta_d}$ remain constant. Figure 1a shows the changes in $\Delta A_{\parallel}$ and $\Delta A_{\perp}$ that would be observed if only the photoproduct dipoles were detected by the measuring light, and Fig. 1b shows $\Delta A_{\parallel}$ and $\Delta A_{\perp}$ if only the unactivated dipoles were detected. Experimentally, these two conditions can only be achieved if there is no spectral overlap between the unactivated molecules and the photoproducts. The changes in polarization for these cases are shown by the correspondingly labeled traces at the bottom of the figure. Often there is a significant spectral overlap between the two populations, in which case $\Delta A_{\parallel} = \Delta A_{d\parallel} + \Delta A_{p\parallel}$ and $\Delta A_{\perp} = \Delta A_{d\perp} + \Delta A_{p\perp}$. Figure 1c demonstrates the absorbance changes that would occur when the photoproduct absorbs somewhat less than the unactivated molecule, in this example $|\Delta A_p|/|\Delta A_d| = \frac{3}{4}$. Notice that as the photoproduct *randomizes* the observed polarization *increases*, as shown by the trace c in the bottom of the figure. However, if the photoproduct absorbs more than the unactivated

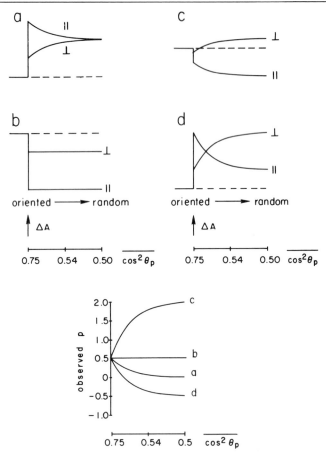

FIG. 1. Dichroism of a photoselected sample with two chromophore populations: unactivated pigment and a stable photoproduct. In this example the unactivated molecules are shown as stationary, and only the photoproduct chromophores are shown undergoing randomizing rotational motion. The lower scale gives $\overline{\cos^2 \theta}$ of the photoproduct as it randomizes. Parts (a) through (d) show what would be observed if the photoproduct and the unactivated molecules absorb the measuring light to different degrees. In (a) only the product absorbs, whereas in (b) only the unactivated molecules absorb. In (c) and (d) both populations absorb the measuring light with $|\Delta A_p|/|\Delta A_d| = \frac{3}{4}$ in (c) and 2 in (d). The observed polarizations (p) for these four cases are shown by the correspondingly labeled traces at the bottom of the figure.

molecule, then the observed polarization *decreases* as is shown in Fig. 1d for the case $|\Delta A_p|/|\Delta A_d| = 2$. Thus the polarization of a two-population system can greatly exceed the values attainable by a single-population system and that disordering of one population relative to another can produce both increases or decreases in the observed polarization.

It is obvious from the preceding discussion that, if experimentally possible, it is highly advantageous to use measuring wavelengths where each chromophore population can be separately observed. If this is not possible, then a full analysis requires solving simultaneous equations for p or r at several measuring wavelengths to separate out absorbance changes and dipole angular distributions, as is evident from Eqs. (2) and (3). When such multiple populations occur, diagrams such as Fig. 1 can aid in qualitative interpretation of the results.

Photodichroism Relaxation Techniques

A flash photometer with linearly polarized light can be used to measure the rotational motions of rhodopsin and bacteriorhodopsin. Such flash photometers tend to be unique instruments designed for particular applications. In this section we discuss special problems involved in using these techniques to measure the rotational motions of rhodopsin and bacteriorhodopsin.

For biological samples it is usually simpler to prepare relatively small samples such as isolated pieces of retina for rhodopsin or small oriented films for bacteriorhodopsin. Thus we have found it useful to mount our sample on a microscope slide and expose the sample to the measuring and actinic illumination through a microscope, as shown in Fig. 2. The microscope slide is placed in a temperature-regulated chamber mounted on the microscope stage. Our system has two available measuring light sources: a Coherent Radiation C500 krypton gas laser for high-intensity monochromic illumination and a 300-W (filtered dc) zirconium arc lamp for white light. The actinic light source is a dye laser pumped by an Avco C-950 nitrogen pulse laser. Both the actinic and measuring light travel the same path through the sample, i.e., "head-on configuration." This is the most sensitive configuration for measuring rotational motions around an axis parallel to the direction of the actinic light. An important problem with the head-on configuration is blocking the actinic flash from the photomultiplier tubes to avoid a large actinic flash artifact. This can usually be accomplished by protecting the photomultiplier tubes with a stack of two or more matched interference filters, a procedure that greatly enhances side-band blocking with little loss in transmitted light. Our instrument is a dual-beam system with a neutral-density wedge for balancing the sample and reference beams. This arrangement significantly reduces noise due to fluctuations in the intensity of the measuring light source. The signal is the voltage difference between the two photomultiplier tubes and is proportional to the transmission change in the sample produced by the actinic flash. The signal from the photomultiplier tubes is

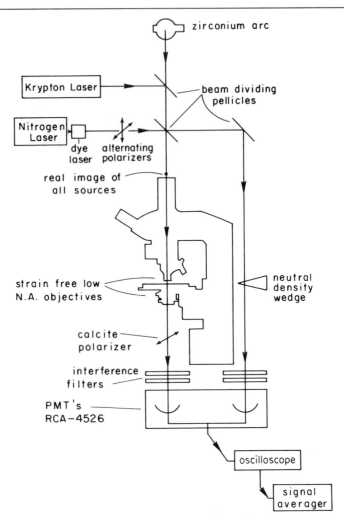

FIG. 2. Basic design of our dual-beam flash photometer.

sent into an oscilloscope for amplification and display. For photocycling pigments such as bacteriorhodopsin, signal averaging can be used to enhance the signal-to-noise ratio greatly and we do this with a Biomation 805 waveform recorder and a Tracor Northern NS 575A signal averager. However, for a photolabile pigment such as rhodopsin, signal averaging does not improve the signal-to-noise ratio and single-flash experiments provide the highest signal-to-noise ratios.

The actinic source should be a short flash of light with wavelength near the peak absorbance of the pigment, and should have enough energy to activate a significant fraction of the pigment molecule in a single flash. We use a coumarin dye for rhodospin, λ = 470 nm, and rhodamine dye for bacteriorhodopsin, λ = 590 nm. The duration of our actinic flash is around 10 nsec. Since the actinic flash is focused on the sample through a microscope objective, the light intensity at the sample will depend on the power of the objective and the focus position of the microscope. Too much actinic energy activates not only a significant fraction of the dipoles closely aligned with the electric vector of the actinic light, but also a large fraction that is far from alignment with the electric vector. This can reduce the dichroism [see Eqs. (7) and (8)].[14] The dichroism can therefore be maximized by reducing the actinic flash energy until the dichroism is independent of the actinic flash energy, thereby satisfying the condition $E\epsilon\gamma \ll 1$. The microscope objective should be of the strain-free type and the numerical aperture should be small enough to keep the actinic light highly polarized.[15] High polarization of the actinic flash is essential for optimum photoselection, and high polarization of the measuring light is necessary to observe small changes in dichroism. Different polarizers and polarizer positions are used depending on the light sources and experimental procedures and requirements. For experiments with bacteriorhodopsin the arrangement shown in Fig. 2 is used. Polaroid HN32 polarizers rotate the polarization of the actinic flash exactly 90° as detected at the sample while a stationary calcite polarizer is used below the sample to select that fraction of the measuring light that is polarized parallel or perpendicular to the actinic flash.

The rotational motions of rhodopsin in disk membranes and bacteriorhodopsin incorporated into lipid vesicles occur on microsecond time scales. To detect such rapid rotational motion high-intensity measuring illumination is needed since the signal-to-noise ratio is proportional to the square root of both the measuring light intensity and the observation interval. For these measurements we use the krypton gas laser. Although increasing the measuring light intensity improves the signal-to-noise ratio, two effects limit the intensity that can be used. First, the measuring light may deplete (bleach) a significant amount of pigment molecules in the sample. Second, intense measuring light can reduce the dichroism in the sample the same way that excessive actinic flash energy reduces dichroism. We have found that for measuring the rotation of rhodopsin in the photoreceptor disk membrane a reasonable signal-to-noise ratio is obtained when the measuring light and the actinic flash both bleach approximately one-fifth of the pigment. The rotational motions of bacteriorhodopsin in the purple membrane are much slower (millisecond time scale),

[14] S. Kawato and K. Kinosita Jr., *Biophys. J.* **36**, 277 (1981).
[15] F. I. Harosi and F. E. Malerba, *Vision Res.* **15**, 379 (1975).

permitting the use of much lower light intensities.[16] For these measurements we use the zirconium arc lamp. To ensure that the sample remains light adapted we leave the measuring light on continuously, but at an intensity that does not degrade the dichroism of the sample.

Selection of the measuring wavelength is very important because this determines the relative absorbances of the different populations of chromophores observed following the actinic flash. Because of this, the magnitude and/or the time dependence of the observed dichroism may vary as a function of measuring wavelength as discussed in the theory section. This phenomenon is particularly important with purple membrane samples since the photodichroism observed depends strongly on the measuring wavelength used.[4] For this preparation we illuminate the sample with "white" light from the zirconium arc lamp. The measuring light wavelength is then determined by the interference filters at the photomultiplier tubes. By changing these filters we can easily observe dichroism at several important wavelengths (400 nm, 570 nm, and 650 nm) without altering the illumination to which the sample is exposed. To make observations at all these wavelengths on one sample we have found that samples with an optical density of about 1.0–1.3 at 570 nm provide reasonable signal-to-noise ratios as well as high dichroic ratios.

For the microsecond rotational motions of rhodopsin in the disk membrane and bacteriorodopsin monomers in lipid vesicles we often use the 568-nm line from the krypton gas laser. For monomeric bacteriorhodopsin in lipid vesicles the absorption change at this wavelength during the rotational motions of the protein results primarily from the depletion of bacteriorhodopsin. Thus the dichroism at this wavelength is due primarily to unactivated bacteriorhodopsin molecules. In contrast, for rhodopsin at this measuring wavelength and observation time interval the photoproducts (bathorhodopsin and lumirhodopsin) absorb more strongly than rhodopsin. The dichroism in this case is primarily due to the photoproducts. By selecting the isosbestic wavelength between a pigment and its photoproduct, it is possible to make a sensitive comparison between the rotational motions of these two populations. For example, at the isosbestic point for rhodopsin and lumirhodopsin (approximately 510 nm) no dichroism was detected, implying both populations must undergo the same rate of rotational diffusion.[3]

[16] P. Ahl and R. Cone, in preparation.

[89] Measurements of Volumes and Electrochemical Gradients with Spin Probes in Membrane Vesicles

By Rolf J. Mehlhorn, Pedro Candau, and Lester Packer

Volumes and electrochemical gradients in cells and vesicles can be measured by several available methods. It is important to employ at least two independent techniques for measuring these parameters to avoid misinterpretation of data due to artifacts. Spin probe techniques complement traditional methods which focus upon extravesicular and extracellular water. By observing probe concentrations inside sealed membrane systems, these methods provide an important check on other techniques in terms of criteria of complementarity, i.e., whether inferred changes of electrochemical potentials on two sides of a membrane are opposite and equal after correcting for volume differences. Spin probes also give unique information: volume measurements are independent of assumptions of cell shape; probe binding is quantitated and corrected for; and with several probes, kinetic measurements appear to be feasible.

General Considerations

The technique of spin labeling[1] has recently been extended to include measurements of bioenergetics parameters such as volume changes, pH gradients, and electrical potentials.[2,3] To differentiate between the signal of spin probes on two sides of a membrane, the phenomenon of exchange broadening has been exploited. Exchange broadening of the paramagnetic signal of a spin label occurs when paramagnetic molecules collide with a high enough frequency to allow significant exchange of unpaired electrons between paramagnetic species, thus quenching the spin-label signal. Quenching can be used to observe either intracellular or extracellular spin signals. The intracellular signal is elicited by quenching the extracellular signal with impermeable transition metal complexes, whereas enhancement of the extracellular signal can be achieved indirectly by means of increased membrane binding of the probe resulting from uptake or at high spin-label concentrations, where increases in intracellular probe concentrations lead to self-quenching of the nitroxides. When intracellular signal changes are observed, these are linearly related to probe concentrations

[1] L. J. Berliner, ed., "Spin Labeling - Theory and Applications." Academic Press, New York, 1976.

[2] R. J. Mehlhorn and I. Probst, this volume.

[3] D. Cafiso and W. L. Hubbell, *Biochemistry* **17**, 187 (1978).

METHODS IN ENZYMOLOGY, VOL. 88

under the appropriate experimental conditions, whereas extracellular signal changes have a markedly nonlinear concentration dependence and require calibration techniques for quantitation. Therefore, accurate probe concentrations are more readily obtained by observing intracellular spin signals with impermeable quenching agents. The signal observed in the electron spin resonance spectrometer is proportional to the total number of unquenched spins in the sample, so considerable signal changes can occur during energization of biological membrane preparations. For example, in chloroplasts it was shown that the intrathylakoid signal of a spin-labeled amine increased more than a thousandfold during illumination.[4]

Probes useful for measuring bioenergetic parameters are shown below.

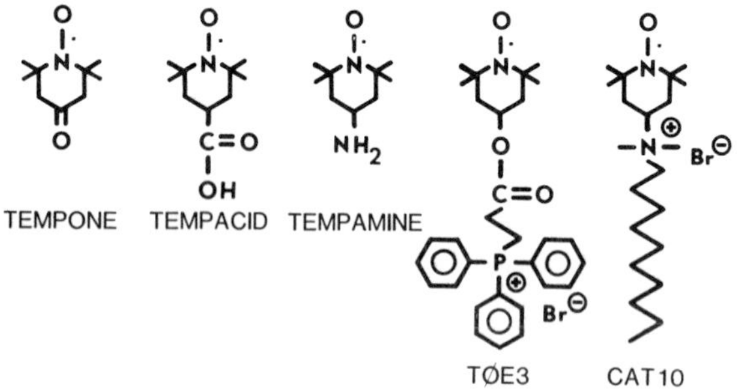

Volume Measurements

Bulk ion transport can be measured directly with radioactive isotopes in flow-dialysis schemes, or by performing specific-element analysis of isolated cells or membrane vesicles. A considerably more convenient albeit indirect method for monitoring coupled ion fluxes of cations and anions (symport) is to measure volume changes associated with these bulk ion movements. The spin probe method is ideally suited for this purpose because volumes can be measured accurately and easily.

A schematic diagram representing volume measurements is shown in Fig. 1. The probe of choice for measuring volumes should satisfy the following criteria: efficient quenching by paramagnetic agents to minimize osmotic effects of the quencher, rapid membrane permeability and inertness toward electrical or ionic gradients. Efficient quenching is realized with probes having narrow intrinsic line widths since a given increment of line broadening causes relatively more signal decrease than is observed

[4] A. T. Quintanilha and R. J. Mehlhorn, *FEBS Lett.* **91**, 104 (1978).

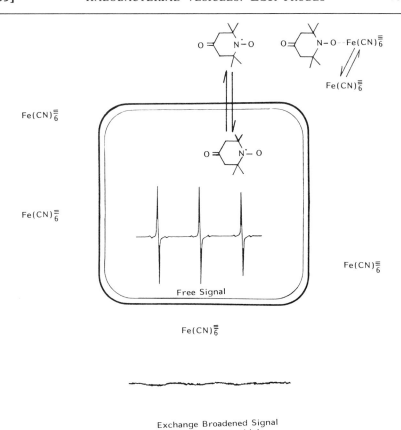

Exchange Broadened Signal
(100 mM Ferricyanide)

FIG. 1. Impermeable ferricyanide broadens the external spin-probe signal and the remaining unbroadened spin signal intensity is directly proportional to the cell or vesicle volume.

with broad intrinsic line widths. This requirement is met by ring nitroxides, having effectively planar structures over the lifetimes of the excited spin states. Rapid permeability requires that the probe have weak polarity, i.e., just sufficient for adequate water solubility. Inertness toward electrochemical gradients excludes probes with charge and titratable groups. All of these criteria are met by 2,2,6,6-tetramethyl-4-oxopiperidinooxy (TEMPONE). The ketone group on the ring confers enough polarity to give adequate water solubility and also maintains the ring in a time-averaged planar configuration that gives narrow lines. Permeability of this nitroxide is rapid (half-time for equilibration is less than 50 msec in human red cells). The procedure for absolute volume measurements consists of obtaining a TEMPONE spectrum in the presence of vesicles and a second spectrum with added quenching agent. The final aqueous signal is expressed as a fraction of the initial signal and generally converted to membrane concentration in terms of milligrams of protein. Relative changes can be measured in a single sample with quenching agent present during volume changes. Control experiments should be conducted without quenching agent present during the volume changes and by quantitating volumes with subsequent quenching agent additions.

Measurements of pH Gradients

Procedures for measuring pH gradients are shown schematically in Fig. 2. Spin-labeled amine and carboxylic acids have proved useful for following proton movements across membranes.

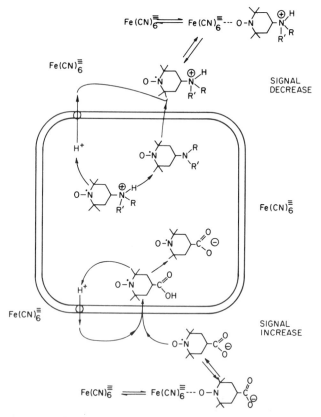

FIG. 2. Unprotonated amines and protonated carboxylate spin probes are freely membrane permeable whereas the charge species are not; hence, as protons are pumped across membranes, a redistribution of the amine or acid probes occurs and the equilibrium concentration gradient of the probes provides a direct measure of the pH gradient.

Amines with different substituents can serve several purposes: Dimethyl TEMPAMINE and other tertiary amines do not react with activated carboxyl groups and are useful for carbodiimide-treated samples where the potential for such reactions exists. With various substitutions on the nitrogen, partitioning differences of spin-labeled amines into lipids can be used to demonstrate whether the permeability of a probe is rate limiting.

Quaternary amines serve to determine whether a given membrane is leaky to the charged amines. Spin-labeled phosphates and sulfates serve a similar purpose for the carboxyl probe.

By determining pH gradients at different probe concentrations, thresholds for inhibition can be ascertained. Decreases of pH gradients become apparent at about 20 μM of a spin-labeled carboxylic acid in envelope vesicles of *Halobacterium halobium*.[2] These probes are accumulated inside the vesicles during illumination. On the other hand, the spin-labeled amine, which is extruded in envelope vesicles of *Halobacterium halobium*, does not inhibit pH gradient formation until its concentration exceeds 1 mM.

The high sensitivity of spin probes and availability of both weak acid and amine labels permits measurements of steady-state pH gradients of small magnitude (ΔpH < 0.1). The procedure consists of an "effective volume" determination of cells or vesicles with both acid and amine probes. In the absence of a pH gradient these should be equal to the true volume determined with TEMPONE after correcting for binding. A small difference in the measured apparent volumes is easily converted to a pH gradient, i.e.,

$$\text{pH} = \tfrac{1}{2} \log \left(\frac{h_{TC}}{h_{TA}} \right)$$

where h_{TC} and h_{TA} are the fractional intracellular signals of the acid and amine probes, respectively. As an additional control, it can be demonstrated that the difference in "effective volume" is abolished upon treating the cells with ion exchanging agents like tributyltin or nigericin.

Measurements of Electrical Gradients

Spin-labeled phosphonium derivatives first introduced by Cafiso and Hubbell[3] are useful for measuring equilibrium transmembrane electrical potentials although their permeability is insufficient to allow them to be exploited for kinetic measurements (see Fig. 3). Charge effects dictate that anionic quenching agents be used with these probes in media of low ionic strength. There is appreciable binding of these nitroxides to membranes. Because of this binding, membrane perturbation effects are a con-

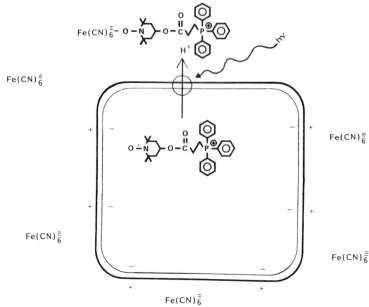

FIG. 3. Measurement of $\Delta\Psi$. The large phosphonium ion, whose charge center is surrounded by hydrophobic groups, is membrane permeable and thus responds to electrical transmembrane potentials by being accumulated within the more negative aqueous compartment.

cern and careful control experiments to demonstrate that probe uptake is directly proportional to the bulk probe concentration are required to assess inhibitory effects.

Addition of the permeable spin-labeled ion to vesicles will generate a transmembrane electrical potential that will oppose equilibration of the probe. Generally, the rate-limiting process in achieving equilibrium will be the movement of some other ion, e.g., an electrogenic proton counterflow through "leaks" in the membrane. Accordingly, equilibration may proceed slowly and the rate constant for uptake of the probe under nonenergized conditions should be established for any given membrane preparation prior to attempting measurements of energized electrical gradients.

Quantitation of probe concentrations under nonenergized and energized conditions must take into account the appreciable and variable fraction of probe molecules bound to the membrane. The magnitude of the total bound signal at both membrane interfaces in the energized state relative to that in the nonenergized state is readily determined by quantitation of aqueous signal intensities before and after energization. With a knowledge of the energy-dependent behavior of the bound population of the probe, the net change of probe concentration outside the vesicles can be determined by quantitating its intravesicular concentration with quenching agent present. The electrical potential in millivolts will be related to the concentration gradient of aqueous probes directly; i.e.,

$$\Delta \Psi = 60 \log \left(\frac{[T\emptyset E3]_{in}}{[T\emptyset E3]_{out}} \right)$$

where $[T\emptyset E3]$ refers to the aqueous concentration of the phosphonium probe.

The following is an example of a membrane potential calculation carried out for envelope vesicles of the S9 strain of *Halobacterium halobium:* Membranes at pH 6.7 were suspended at 7 mg/ml of protein and the cell volume was determined with TEMPONE as 3.1% of the total volume. An aliquot of the vesicles was spin-labeled with 30 μM T$\emptyset$E3 and treated with 2 mM sodium ferricyanide to ensure full oxidation of the probe without suffering any reduction in nitroxide line heights due to exchange broadening. The aqueous line heights in the presence of membranes after equilibration were 40% of the line heights observed in 4 M NaCl, implying that 17.7 μM of the probe was membrane-bound under these conditions. Under illumination there was a substantial decrease of the total aqueous signal. Careful analysis of the signal intensity of the bound component revealed that the aqueous line height reduction was due entirely to increased membrane binding of the probe and not to concentration-dependent self-quenching or chemical reduction. The final aqueous line height was equivalent to a 4.0 μM aqueous solution of T$\emptyset$E3. Another aliquot of vesicles labeled with T$\emptyset$E3 was treated with 100 mM sodium ferricyanide to quench completely the spin signal outside of the vesicles. Upon illumination the unquenched aqueous signal increased, reaching a final value equivalent to a 1.5 μM aqueous solution of T$\emptyset$E3. Subtracting this value from the total aqueous line height observed under illumination with 2 mM ferricyanide, the external concentration of probe is inferred to be 2.5 μM. The calculation of internal probe concentration requires that the vesicle volume be taken into account; i.e., 1.5 μM divided by 0.031 yields an intravesicular probe concentration of 49 μM. The resulting membrane potential is

$$\Delta \Psi = 60 \log \left(\frac{49}{2.5} \right) = 78 \text{ mV}$$

It is instructive to use the same data to perform a calculation that has been used in calculating membrane potentials with phosphonium ion electrode measurements corrected for the initial binding of the phosphonium ions.[5] One obtains a potential of 126 mV, i.e., a 48-mV overestimate. The

---

[5] M. Murastugu, N. Kamo, Y. Kobatake, and K. Kimura, *Bioelectrochem. Bioenerg.* **6,** 477 (1979).

overestimate results from neglecting the increased membrane binding of the probe during the energization process.

Structures of currently available phosphonium derivatives of nitroxide include the ester TØE3 and corresponding amides. The latter probes are resistant to hydrolysis but suffer from the disadvantage that their greater polarity renders them less permeable than ester labels. Work is in progress to synthesize other phosphonium nitroxides that may be more permeable and stable than TØE3.

Computer Methods

Increased accuracy of volume and potential determinations can be achieved with computer manipulations of the spin resonance spectra. The most useful aspect of computer analysis is correction for spectral features due to incomplete quenching of the probes. Reference spectra of partially quenched signals are obtained by performing experiments with probes and quencher either in the absence of membranes or by incorporating quenchers inside cells by sonication. Membrane-bound spectra are obtained directly by subtracting a spectrum of probe and quencher in buffer from a spectrum of membranes sonicated in the presence of the probes and quenching agent. Double integration of the derived spectrum yields quantitative data on how much of the probe is membrane-bound.[2]

Dual-Probe Methods

The availability of nitroxides highly enriched in the ^{15}N isotope makes it possible to conduct simultaneous measurements of two parameters since spectra of the two nitrogen isotopes in water are clearly resolved. By having dual-probe stock solutions one can avoid the difficulties associated with variability of EPR signals among different samples. These dual-probe methods have proved most useful for volume studies because volume changes associated with lytic events occur frequently when cells or vesicles are mixed with solutions of different osmotic activities.

Production of Sustained pH and Electrical Gradients across Vesicle Membranes in the Dark

Prior to attempting to understand light-induced ion movements across membranes, passive as well as coupled ion movements should be characterized in the dark. In halobacterial vesicles suspended in concentrated salt solutions, substantial and stable pH gradients can be induced easily and rapidly by adding either ammonium chloride or sodium acetate solutions to the vesicles. Only the uncharged ammonia or acetic acid species are membrane permeable; hence the interior vesicle compartments will be rendered alkaline or acidic, respectively. Spin-labeled acids or amines can be used to quantitate the resulting pH gradients, as well as the kinetics of their collapse, the latter providing an index of proton leakage.

For example, addition of a sodium acetate solution to vesicles equilibrated at pH 7 produces a pH gradient as freely permeable acetic acid carries protons into the vesicle interior with a concomitant alkalinization of the vesicle exterior. The magnitude of the resulting pH gradient is difficult to estimate mathematically because of the unknown concentrations of buffering substances at membrane interfaces and in the aqueous phases. However, these gradients can be determined experimentally. Thus with spin-labeled amines it was observed that addition of 400 mM sodium acetate at pH 7 to envelope vesicles of S9, a wild-type strain of *Halobacterium halobium* suspended in 4 M NaCl, generated a pH gradient of 0.8. The subsequent collapse of this pH gradient was slow—the time required for the gradient to diminish to half its initial maximum value was about 1 hr.

Rapid Determination of Membrane Surface Areas

Amphiphilic nitroxides partition between aqueous and membrane domains giving rise to easily resolved narrow aqueous and broad membrane-bound spectral components, respectively. Membrane translocation of quaternary amine probes is slow (several hours for 50% uptake in envelope vesicles of halobacteria) so their bound signal can be ascribed to the membrane surface facing the extravesicular aqueous phase. To determine the aqueous and membrane fractions of probes quantitatively, the aqueous signal is determined by direct measurement of the high field aqueous line in the presence of membranes. The membrane fraction is deduced by subtraction of this component from the measured line height of the probe in the absence of membranes. In concentrated salt solutions surface potential effects are negligible and partitioning is a direct function of the membrane concentration. This was confirmed for several envelope vesicles preparations derived from wild-type and mutant strains of *Halobacterium halobium* where serial dilutions of several sets of membranes were spin-labeled with the cationic nitroxide CAT10.[6]

This is reflected in the accompanying table which summarizes partitioning data obtained as a function of binary volume dilutions of different cell envelope vesicle preparations of several strains of halobacteria. The ratio of bound-to-free signal fractions was plotted against protein concen-

---

[6] R. J. Mehlhorn and L. Packer, this series, Vol. 56, p. 515.

	Cell strain[a]				
	L-33	L-33	L-33*	R_1mW	S9
Protein mg/ml	48	7.7	4	20	28.3
Slope	0.21	0.54	0.16	0.20	0.34
Coefficient of determination	0.999	0.996	0.998	0.988	0.997

[a] L-33* vesicles were prepared in 3.8 M choline chloride, 0.2 M NaCl; all others in 4 M NaCl.

trations of the samples diluted serially from the initial concentrations shown in the table. An excellent linear fit to the data was obtained by least squares as reflected in the coefficients of determination in the table and all lines passed through the origin within the experimental error.

Assuming that membrane surface areas are proportional to membrane protein concentrations, the slopes in the table should be equal (except for the choline chloride solution, which may solubilize CAT10 differently than sodium chloride solutions). However, slopes shown in the table vary significantly, perhaps owing to differing amounts of cytoplasmic proteins entrapped in the vesicles during the membrane isolation procedures, or, less likely, to variability in membrane composition.

Quenching Agents

Line broadening of nitroxides by transition metal ions and chelates is proportional to the concentration of the transition metal.[7] The observed linewidth w is the sum of the intrinsic linewidth w_0 and the line broadening Δw caused by the broadening agent. The line height of a broadened spectrum h, is related approximately to the intrinsic line height h_0 by the expression:

$$h = \left(\frac{w_0}{w}\right)^2 h_0$$

Therefore, at sufficiently high concentrations of a quenching agent, quenching effectiveness is proportional to the square of its concentration. Line-broadening data for several nitroxides and two useful quenching agents are presented in Fig. 4. Intrinsic linewidths for the probes shown are as follows: TEMPONE, $w_0 = 0.4$ G; TEMPACID, $w_0 = 1.6$ G; TEMPAMINE, $w_0 = 1.8$ G; TØE3, $w_0 = 1.5$ G.

Generally, 100 mM of quenching agent is required for efficient eradication of the extracellular concentration of nitroxides in the absence of

[7] A. D. Keith, W. Snipes, R. J. Mehlhorn, and T. Gunter, *Biophys. J.* **19**, 205 (1977).

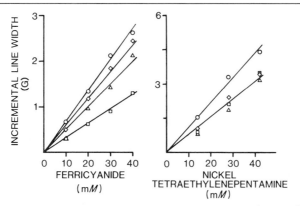

FIG. 4. Line broadening Δw of several nitroxides by paramagnetic quenching agents in 4 M NaCl. □, TEMPACID, △, TØE3, ○, TEMPONE, ◇, TEMPAMINE.

charge effects. However, suboptimal quencher concentrations can be used in conjunction with careful mathematical analysis, of quenched and unquenched spectral components preferably with a computer, to ascertain that the quenching agent does not exhibit inhibitory effects on activity at high concentrations.

The most important feature of useful quenching agents is impermeability in the membrane system being studied. Uptake of the quenching agent into cells should cause negligible broadening effects inside the cells on the time scale of a given experiment. This constraint is most difficult to meet for volume change studies because such phenomena generally occur over long time spans. Other important requirements of a quenching agent include: that no interfering resonances be produced by the agent, that quenching occur at the lowest possible concentrations so as to minimize osmotic effects, and for work with pH-sensitive probes, that the quenching agent have negligible buffering capacity in the pH range of interest. Alternatively, extracellular buffering can be made large so that observed pH changes can be ascribed to the intracellular compartment. Finally, the quenching agent should not inhibit or promote the reactions being assayed, e.g., by altering electron transport. In practice, not all of these criteria are met by a single quenching agent so a combination of agents or other control experiments must be employed for accurate work.

Many of the preceding criteria are met by ferricyanide and nickel chelates. These agents exhibit no ESR signals to interfere with the nitroxides. Ferricyanide is quite impermeable while a number of nickel chelates are only slightly permeable. However, most effective nickel complexes, unlike ferricyanide, suffer from the disadvantage that they have appreciable buffering capacity at physiological pH values.

Acknowledgments

Research supported by NIH (GM-24-273); the U.S. Department of Energy, Division of Biological Energy Research, Office of Basic Energy Sciences; and a fellowship to P.C. from the U.S.–Spain Joint Committee for Scientific and Technological Cooperation.

[90] Diffusible Spin Labels Used to Study Lipid–Protein Interactions with Rhodopsin and Bacteriorhodopsin

By DEREK MARSH and ANTHONY WATTS

Introduction

The electron spin resonance (ESR) spectra of spin-labeled lipid analogs introduced at probe concentrations in rhodopsin- or bacteriorhodopsin-containing membranes differ considerably from the spectra of the same spin-labeled lipids in bilayers of the extracted or reconstituting lipids. For reasons of spectral resolution, these effects of lipid–protein interactions are best seen with phospholipids labeled close to the terminal methyl end of the acyl chain (the 14-C atom being optimal) or with labeled steroids. The principal feature is the appearance of a second spectral component corresponding to labeled lipids, which are more motionally restricted than the normal fluid bilayer lipids and are in slow or intermediate exchange with them (exchange frequencies $\leqslant 10^7$ sec^{-1}).[1,2] This component is attributed to lipids interacting directly with the membrane proteins. General details of spin-label ESR spectroscopy are not given here but can be found in Refs. 3–6.

[1] D. Marsh and A. Watts, in "Lipid-Protein Interactions" (P. C. Jost and O. H. Griffith, eds.) Vol. 2, Chapter 2. Wiley (Interscience), New York, 1982.

[2] A. Watts, J. Davoust, D. Marsh, and P. F. Devaux, *Biochim. Biophys. Acta* **643**, 673 (1981).

[3] P. F. Knowles, D. Marsh, and H. W. E. Rattle, "Magnetic Resonance of Biomolecules." Wiley, New York, 1976.

[4] D. Marsh, in "Membrane Spectroscopy" (E. Grell, ed.), p. 51. Springer-Verlag, Berlin and New York, 1981.

[5] D. Marsh, in "Techniques in Lipid and Membrane Biochemistry" (J. C. Metcalfe and T. R. Hesketh, eds.), Vol. B4. Elsevier/North-Holland, Amsterdam, 1982.

[6] A. Watts, in "Progress in Retinal Research" (N. N. Osborne and G. J. Chader, eds.). Chapter 5. Pergamon, Oxford, 1982.

Lipid Spin-Label Synthesis

Doxyl fatty acids, $I(m,n)$, form the basic building block for producing phospholipid spin labels. The synthetic route for $I(m,n)$ is shown in Fig. 1.[1,7]

The amino fatty acid ester $IV(m,n)$ is produced by condensing an excess (5- to 10-fold in moles) of 2-methyl-2-amino-1-propanol (III) (Fluka, Buchs) with the keto fatty acid ester $II(m,n)$,[8] in refluxing toluene (10 ml/mM of II) and toluene-p-sulphonic acid · H_2O (1.5 g/mol of II). After 10 days, with the continuous removal of water, the colorless product $IV(m,n)$ is oxidized to the yellow nitroxide free radical $I(m,n)$ by slowly adding perchlorobenzoic acid (1.5 mol-eq of IV as estimated by thin-layer chromatography) in dry ether on ice. After 2 days at room temperature the product is purified by dry column chromatography (Woelm, Eschwege, Federal Republic of Germany) using high-activity silica gel with n-hexane:ether (7:3) ($R_f \sim 0.7$). The pure doxyl ester is hydrolyzed in ethanol (95%) with NaOH pellets. After acidification to pH 3, the fatty acid spin label $I(m,n)$ is extracted into ether to give a yellow solid or oil. Yields of $I(m,n)$ vary depending on the value of m. For I(3,12), the yield was 15.5% from II(3,12); ir(KBr) acid band 1705 cm^{-1}; elemental analysis calculated for $C_{23}H_{44}NO_4$ (for Me ester, $m + n = 15$), C, 69.34; H, 11.05; O, 16.08; N, 3.51. Found: C, 69.40; H, 11.05; O, 16.09; N, 3.40.

A limited number of spin labels $I(m,n)$ are available commercially (Syva, 3221 Porter Drive, Palo Alto, Calif. 94304; Molecular Probes, 849J Place, Suite B, Plano, Texas 75074).

The phosphatidylcholine spin labels, $VI(m,n)$, are produced by condensing the doxyl fatty acid, $I(m,n)$, onto sn-2-OH(lyso)phosphatidylcholine.[9] Other phospholipid spin labels are produced from the phosphatidylcholine labels by headgroup exchange catalyzed by phospholipase D.[1,10] The overall scheme is given in Fig. 2.

The imidazole derivative $V(m,n)$ of the doxyl fatty acid $I(m,n)$ is produced in dry benzene (~ 4 ml/100 mg of spin label) by reaction with N,N-carbonylimidazole in stoichiometric molar proportions to the fatty acid. After 30 min at room temperature, a 1.2 molar excess of lysophosphatidylcholine is added and stirred overnight. Removal of the solvent is fol-

---

[7] Reviews on nitroxide spin label chemistry are given by B. J. Gaffney, *in* "Spin Labelling: Theory and Application" (L. J. Berliner, ed.), p. 183. Academic Press, New York, 1976; J. F. W. Keana, *Chem. Rev.* **78**, 37 (1978).

[8] Keto fatty acids $II(m,n)$ are synthesized by a standard Grignard reaction for which precursors are readily available for making the $n = 1$–13 stearic acids ($m + n = 15$). Extensive details are given in Ref. 1.

[9] W. F. Boss, C. J. Kelley, and F. R. Landsberger, *Anal. Biochem.* **64**, 289 (1975).

[10] P. Comfurius and R. F. A. Zwaal, *Biochim. Biophys. Acta* **488**, 36 (1977).

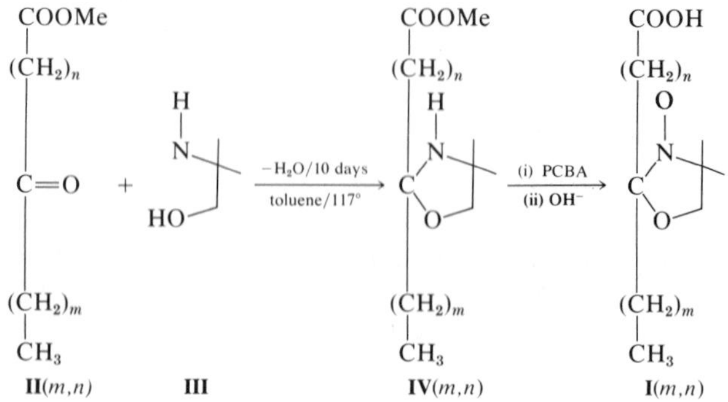

FIG. 1. Reaction scheme for synthesizing doxyl fatty acids, $I(m,n)$.

lowed by heating to $65-70°$ for $2-3$ days, after which time water $(1-2$ ml) is added and then removed by repeated rotary evaporation with CCl_4. The phosphatidylcholine spin label $VI(m,n)$ is purified on silica gel to give approximate yields of 35% from the lysolipid, and any unused fatty acid label is also recovered.

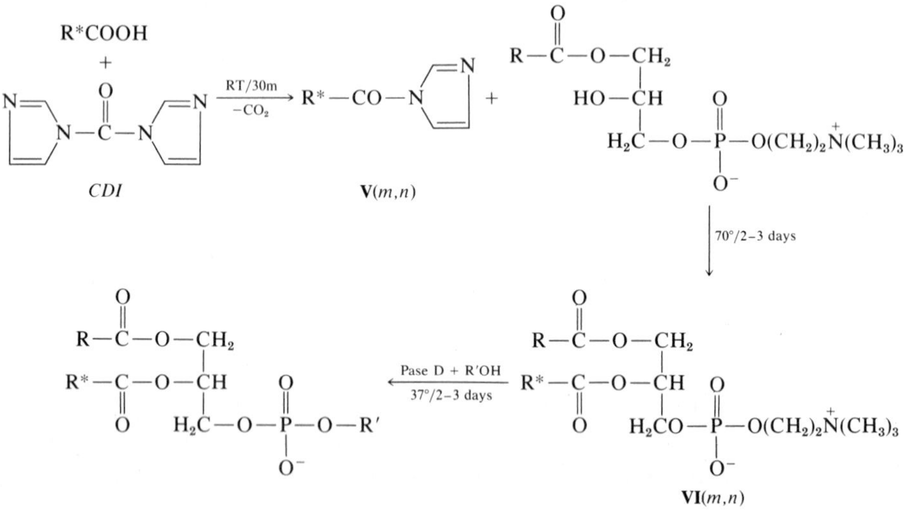

$R*$—COOH = doxyl fatty acid, $I(m,n)$.

R' = serine, —CH_2—$CH(\overset{+}{N}H_3)COO^-$; phosphatidylserine.

= ethanolamine, —$(CH_2)_2$—$\overset{+}{N}H_3$; phosphatidylethanolamine.

= glycerol, —CH_2—CHOH—CH_2OH; phosphatidylglycerol.

= water, —H; phosphatidic acid.

FIG. 2. Reaction scheme for making phospholipid spin labels. Pase D, phospholipase D; RT, room temperature.

Phospholipase D (Boehringer-Mannheim, from white cabbage) is added (1 IU/ml of reaction mixture) to a 1:1 aqueous (0.1 M sodium acetate/0.1 M calcium chloride, pH 5.6)–ether solution of VI(m,n) (20 mg/ml) containing a large excess of the appropriate headgroup alcohol, and incubated at 35–45° for 1–3 days. In the presence of the following alcohols: glycerol (50 v/v), serine (46% w/v), ethanolamine (22% v/v), or water, the corresponding phospholipid spin-label derivatives shown in Fig. 2 are produced. The phospholipids are extracted into chloroform–methanol (2:1) and purified on silica gel to separate from any phosphatidic acid spin label produced and from unreacted phosphatidylcholine spin label.

The androstanol spin label is available commercially, but can be synthesized from the corresponding keto derivative, 17-ol-3-androstanone,[1] by a similar method to that shown in Fig. 1.

Membrane Preparation

a. ROS Disk Membranes. Rod outer segments (ROS) are removed by shaking or mild homogenization of isolated retina (in 1 ml buffer[11] per two retina) from bovine or frogs eyes, which are excised and put on ice in the dark immediately *post mortem.*[12–14] Isolated rods, by virtue of their relatively low density (1.08–1.10), can be purified on either a discontinuous[12] or continuous[15] density gradient. The red band containing rods is recovered and either stored directly in sucrose ($< -20°$ under N_2), or immediately washed in low ionic strength buffer[11] to effect lysis. Centrifugation gives a pellet of ROS disk membranes suitable for ESR spin-labeling experiments. When all operations are carried out under dim red light, the A_{280}/A_{500} (total protein/rhodopsin) is 2.3–2.5 and the yield of protein is about 1 mg rhodopsin per two retina. Intact, isolated disks can be produced by the use of Ficoll gradients.[14]

b. Purple Membranes. *Halobacterium halobium* S-9 cells are grown and the purple membranes isolated from lysed cells on a linear sucrose gradient.[16] After several washes and differential centrifugations to minimize possible red membrane contamination, the purple membranes (at

[11] A wide range of buffers have been used, such as Tris, HEPES, and phosphate, usually at pH 7.3 and in 60–150 mM salt (no salt for lysis).

[12] D. S. Papermaster and W. J. Dreyer, *Biochemistry* **13**, 2438 (1974).

[13] H. G. Smith, Jr., G. W. Stubbs, and B. J. Litman, *Exp. Eye Res.* **20**, 211 (1975).

[14] H. G. Smith, Jr. and B. J. Litman, Vol. 81, this series, Article [10].

[15] W. J. DeGrip, F. J. M. Daemen, and S. L. Bonting, *Vision Res.* **12**, 1697 (1972).

[16] D. Oesterhelt and W. Stoeckenius, this series, Vol. 31, p. 667.

3 mg protein/ml in 0.1 M sodium acetate, 1 mM sodium azide, pH 5.0) can be used directly for spin-labeling or stored at 4°C.[17]

Partially delipidated (about 75% lipid removal) purple membranes can be prepared by incubation of 10 mg of purple membrane in 5 ml of 20 mM DTAB (dodecyltrimethylammonium bromide) followed by density gradient centrifugation (15 hr at 130,000 g at 4°) with steps of 10, 20, 25, 30, 35, 40, and 50% sucrose, the first two steps containing 20 mM DTAB. The membranes are recovered from the 50% cushion.[17]

c. Reconstituted Membranes. Purified rhodopsin is obtained by octyl glucoside (50 mM detergent in buffer) solubilization of ROS disk membranes followed by concanavalin A affinity chromatography with detergent.[18,19] Rhodopsin ($A_{280}/A_{500} \sim 1.7$–2.1) and lipid at the desired ratio are cosolubilized in detergent (octyl glucoside or cholate, 50–75 mM in buffer). Exhaustive dialysis against buffer (5 mM HEPES, 1 mM EDTA, pH 7.4) containing progressively decreasing concentrations of NaCl (200, 150, 100, 50, and 0 mM) produces recombinants that are recovered by centrifugation (16,000 g; 45 min; 4°). The pellet is washed (65 mM NaCl, 10 mM HEPES, pH 6.8) and checked for homogeneity by linear density gradient (15–60% w/w sucrose) ultracentrifugation (160,000 g; 15 hr) to give a single sharp band. Rhodopsin recovery after dialysis is 85 ± 5% from its optical purity.

Purple membranes can be solubilized in Triton X-100[20] (24 hr at room temperature in the dark with a 4-fold weight ratio of Triton to membrane protein). Lipid is then added with agitation and the detergent removed by dialysis (4° against 0.1 M sodium acetate–3 mM sodium azide at pH 5.0 for 5 days). Density gradient centrifugation (4.5 to 40% sucrose) gives predominantly one band when dimyristoyl phosphatidylcholine is used, but two bands with the dipalmitoyl derivative, each with a different lipid-to-protein ratio. After washing (40,000 g, 10 min, in 0.1 M acetate, pH 5) the recombinants can be used directly for ESR spin labeling.

d. Lipid Extraction. The lipids can be extracted from purple membrane and from rod outer segment disk membranes using simple chloroform–methanol extraction, see, e.g., Ref. 21. Extraction in the presence of cyanoborohydride,[22] or prior treatment of the membranes with hydrox-

[17] P. C. Jost, D. A. McMillen, W. D. Morgan, and W. Stoeckenius, *in* "Light Transducing Membranes," (D. Deamer, ed.) p. 141. Academic Press, New York, 1978.

[18] B. J. Litman, Vol. 81, this series, Article [23]. "Purification of Rhodopsin by Concanavalin A Chromatography."

[19] A Kusumi, T. Sakaki, T. Yoshizawa, and Ş.-I. Ohnishi, *J. Biochem. (Tokyo)* **88**, 1103 (1980).

[20] R. J. Cherry, U. Muller, R. Henderson, and M. P. Heyn, *J. Mol. Biol.* **121**, 283 (1978).

[21] S. C. Kushwaha, M. Kates, and W. G. Martin, *Can. J. Biochem.* **53**, 284 (1975).

[22] R. S. Fager, P. Sejnowski, and E. W. Abrahamson, *Biochem. Biophys. Res. Commun.* **47**, 1244 (1972).

ylamine,[23,24] yields a retinal-free lipid extract. Purple membrane lipids are chemically very stable, consisting entirely of dihydrophytol, ether-linked chains.[21] The rod outer segment disk lipids are highly unsaturated[24] and require extreme care to prevent oxidative degradation. Details for handling and extraction of these lipids, under argon and in the presence of butylated hydroxytoluene, are given in Ref. 25. The dispersed rod outer segment disk lipids appear to exist close to a lamellar-hexagonal phase boundary,[25,26] and only under carefully controlled conditions are the lipids obtained entirely in the lamellar phase. In our experience, small contaminations by hexagonal-phase lipid do not appreciably change the spectra of the lipid spin labels,[27] presumably because both phases have very similar fluidities.

Spin Labeling of Membranes

a. ROS Disk and Purple Membranes. Two different methods are used for incorporating freely diffusible lipid spin labels into biomembranes, depending on the label being used.[28]

Sterol and fatty acid spin labels $I(m,n)$ are deposited as a film on the inside of a glass vial from a chloroform stock solution of the label (typically 1 mg/ml). Solvent is removed under high vacuum for at least 6 hr before adding a cold membrane suspension to the vial (around $1-2$ μmol of membrane phospholipid in < 1 ml of buffer). Labeling levels are dictated by the amount of membrane lipid present, but are usually kept less than one label per $100-200$ endogenous lipids, assuming most of the label is taken up. Incorporation appears to be fairly efficient with the rather "fluid" ROS disk membranes,[28] but purple membrane fragments were briefly sonicated to improve uptake of the label, which was initially added at a level of one fatty acid per $60-70$ membrane lipids.[17]

A brief centrifugation ($3000-20,000$ g for $10-15$ min depending on the membrane density) concentrates the membrane, and any free label can be removed in the supernatant. The membrane pellet is then transferred to a sealed-off 100-μl capillary (I.D. ≤ 1 mm, Supracaps or a similar brand) with a drawn-out Pasteur pipette. If required, the capillary can be centrifuged up to 12,000 g using a solid plastic insert with a suitable hole (and

[23] G. W. Stubbs, B. J. Litman, and Y. Barenholz, *Biochemistry* **15**, 2766 (1976).
[24] G. P. Miljanich, L. A. Sklar, D. L. White, and E. A. Dratz, *Biochim. Biophys. Acta* **552**, 294 (1979).
[25] M. F. Brown, A. J. Deese, and E. A. Dratz, this volume.
[26] A. J. Deese, E. A. Dratz, and M. F. Brown, *FEBS Lett.* **124**, 93 (1981).
[27] D. Marsh, A. Watts, R. D. Pates, R. Uhl, P. F. Knowles, and M. Esmann, *Biophys. J.* **37**, 265 (1982).
[28] A. Watts, I. D. Volotovski, and D. Marsh, *Biochemistry* **18**, 5006 (1979).

balance for the rotor).[29] More dense membranes can be sedimented in a bench centrifuge without a supporting insert. All the membrane sample should ideally be within the lower 3 cm of the capillary. Excess aqueous phase should be removed to minimize dielectric loss, and also because fatty acid spin labels tend to partition into the free aqueous volume and complicate the membrane ESR spectrum with narrow isotropic lines.

Phospholipid spin labels form bilayers spontaneously in water, and therefore to minimize distortion of the ESR spectrum by an interaction-broadened signal, an ethanolic labeling procedure is used.[28] To a fairly dilute, cold membrane suspension (typically 1 μmol of membrane phospholipid in 10–12 ml of buffer) is added an ethanolic solution of phospholipid spin label ($\sim$5–10 μl of a 4 mg/ml stock solution). This gives a maximum labeling level of 4 mol % of the endogenous membrane lipids, although different labels are incorporated to varying degrees, and none completely. Centrifugation and washing (four or five times with 10 ml of buffer at 20,000 g for 15 min for ROS membranes) removes any spin-label vesicles, which, having no protein in their bilayers, tend not to sediment with the pellet, as judged by the ESR spectrum. Final transfer to the capillary for ESR measurements is as for the sterol and fatty acid labels described earlier.

Phosphatidylserine spin labels are insoluble in ethanol. Therefore labeling of ROS disk membranes is performed by adding membranes to a spin-label film as for the fatty acid labels, but diluting and washing the membranes many times as for the other phospholipid labels. Incubation of membranes with the film for 24 hr at 4° tends to improve incorporation.[28]

b. Reconstituted and Extracted Lipid Membranes. Lipid spin labels can be added directly to organic solvent solutions of extracted lipids before solvent removal under vacuum and dispersion in buffer.[17,28] Labeling levels are routinely maintained at <2 mol % of labeled lipids relative to unlabeled lipids. Dispersion is best performed above any lipid phase transition and with a minimal amount of buffer (0.10–0.15 ml for 1–2 μmol of lipid), since protein-free membranes are not as readily concentrated into the ESR capillary by centrifugation.

Reconstituted complexes can be labeled as detailed above for natural membranes, either by shaking with a spin-label film or by ethanolic injection. These methods are particularly useful if a reconstituted complex of defined lipid–protein ratio is to be labeled with a range of different labels.[30] Incorporation is more efficient above the phase transition of the lipid.

[29] R. Cammack and A. Watts, "ESR in Biochemistry." Academic Press, New York, 1982.
[30] P. F. Knowles, A. Watts, and D. Marsh, *Biochemistry* **20**, 5888 (1981).

Alternatively, the label can be cosolubilized with the lipid and protein in detergent, prior to reconstitution by dialysis.[31] This method is clearly only possible for one particular label per complex. During the detergent removal step it is possible that spin labels may be preferentially lost through the dialysis bag, because they are more soluble than their unlabeled analogs. This can make it difficult to control and predict the labeling levels. In addition spin-label intensity is lost due to chemical reduction of the spin label during dialysis. In principle, however, this is the more favorable method.

Spectral Subtraction

The spectra of rhodopsin–lipid systems appear as a fluid component (somewhat broadened with respect to the similar component from the lipid bilayers) with a second component, which is not present in the spectra of lipid bilayers alone, superimposed.[27,28] The second component has a larger spectral splitting, corresponding to labeled lipids with a reduced mobility, and is visible in the outer wings of the narrower fluid-bilayer lipid component. Computer-aided subtraction of the digitized spectra is necessary to obtain the spectra of the individual components, to determine their relative integrated intensities, and to check whether the spectra truly are composed of independent components or represent composite spectra indicative of intermediate exchange.

The spectral subtraction requires the choice both of a suitable single-component spectrum to subtract and of criteria for an acceptable subtraction endpoint. A recommended protocol for subtractions is the following: (1) A fluid lipid spectrum that corresponds most closely to the fluid component in the membrane spectrum is chosen; usually this is the spectrum from the lipids recorded at a similar or slightly different temperature from that of the membranes. (2) The lipid spectrum is subtracted from the membrane spectrum to give a motionally restricted difference spectrum that most closely corresponds to one of a series of trial spectra from motionally restricted labels. The trial spectra can be from simulations, from lipid-depleted samples at various temperatures, or from sonicated dimyristoylphosphatidylcholine vesicles at various temperatures in the gel phase. This motionally restricted spectrum is then used in assessing the subtraction endpoint, by comparing with the difference spectrum at considerably expanded vertical gain. (3) The selected motionally restricted spectrum is used to subtract from the membrane spectrum to give a fluid difference spectrum whose endpoint is determined by comparison with the fluid lipid spectrum selected in (1). Steps (2) and (3) yield the differ-

[31] P. F. Knowles, A. Watts, and D. Marsh, *Biochemistry* **18**, 4480 (1979).

ence spectra of the two individual components, and two independent values for the fraction f of motionally restricted lipid in the original spectrum. (It is assumed that the digitized spectra used in the subtractions are all normalized to the same double-integrated intensity.) In general, the two complementary methods of subtraction give comparable values for f, although the subtraction (3) is often the more reliable and can be extended to higher temperatures.

For its success the method relies on a correct choice of the single-component spectra to be subtracted. This has to be done by trial and error. It cannot necessarily be expected that the spectra of the extracted lipid at the same temperature, or the spectra from a delipidated membrane protein sample, will perfectly match the corresponding components in the membrane spectra. For this reason interactive computer graphics are almost indispensable, allowing one to check and optimize a large number of trial spectra in a relatively short time. It is also useful to be able to slightly expand or contract the spectral width digitally, in order to match exactly the splittings when the overall line shape already fits well. This can compensate for small differences in polarity, line widths, etc.

For further details of subtraction techniques, see Refs. 5 and 32.

Data Evaluation

Assuming that the spectra of the membranes can be satisfactorily decomposed into two components, the first important parameter is the fraction f of lipid labels that give rise to the more motionally restricted or immobilized component. If the spin label reflects the unlabeled lipid distribution in an exact 1:1 fashion, the effective number of lipid molecules, n_1, associated with each protein molecule is given by $n_1 = f \times n_t$, where n_t is the lipid–protein ratio of the sample. For multicomponent lipid systems, this expression holds for the individual lipid components, assuming no lipid specificity. In reconstituted systems it can be checked whether there is a discrimination between the spin-labeled lipid and the parent unlabeled lipid host by performing a lipid–protein titration. If K_r is the association constant of the spin-labeled lipid *relative* to that of the host lipid[1]:

$$n_f^* / n_b^* = n_t / (n_1 K_r) - 1/K_r \qquad (1)$$

where $n_f^* / n_b^* = (1 - f)/f$ is the ratio of the fluid to motionally restricted components in the spin-label spectrum. Values of n_1 and K_r are then obtained from a plot of n_f^* / n_b^* versus n_t (see Ref. 1). A value of $K_r = 1$ implies no discrimination between labeled and unlabeled lipid. Equation (1)

---

[32] P. C. Jost and O. H. Griffith, this series, Vol. 49, p. 369.

can also be used to analyze specificities of interaction between different phospholipids. In the case of different classes of binding sites one measures the average relative binding constant, K_r^{av}, and thus specificity could be due to a large increase in binding constant of a few specific sites or a proportionally smaller increase in binding constant of a larger number of sites.[33]

The other important properties of the lipid–protein interaction to be derived from the spectra are the degree of immobilization of the labeled lipids contributing to the motionally restricted component and the rate of exchange of the lipids between the motionally restricted and fluid components. The spectra of the motionally restricted component lie in the slow-motion regime of ESR spectroscopy for which lengthy and involved computer simulations are necessary to describe the motional properties completely.[4,34] Approximate, empirical methods are available based on linewidth and line-splitting measurements in the motionally restricted spectrum. The effective rotational correlation time for isotropic motion is given by $\tau_R = a_m(1 - A_{zz}/A_{zz}^R)^b$ or $\tau_R = a_m'(\Delta H_m/\Delta H_m^R - 1)^{b_m}$, where $\Delta H_{1_{(h)}}$ are the linewidths of the low (high) field outer peaks in the difference spectrum and $2A_{zz}$ is the separation between them.[4,5,34] A_{zz}^R and $\Delta H_{1(h)}^R$ are the corresponding values for a rigidly immobilized nitroxide spectrum. The empirical constants in these calibrations are as follows: $a = 5.4 \times 10^{-10}$ sec, $b = -1.36$; $a_1' = 1.15 \times 10^{-8}$ sec, $b_1' = -0.943$; $a_h' = 2.12 \times 10^{-8}$ sec, $b_h' = -0.778$ (see Ref. 34). Although these calibrations are for isotropic motion, in the case of anisotropic motion they will approximate the correlation time for the motion of the nitroxide z axis, provided this is of sufficiently large amplitude. Thus the calibrations are likely to give an upper limit for the correlation time of segmental or long-axis motion of the chain-labeled phospholipids. A treatment for rotation about the long axis of the steroid labels can be found in Ref. 35. The rigid limit values A_{zz}^R, ΔH_m^R can be determined from delipidated samples at low temperature. Care is required with A_{zz}^R since this is also sensitive to polarity, and in addition the spectra at low temperature may be distorted by spin-spin broadening due to partial segregation of the label.

The fluid and motionally restricted components may not be truly independent but may be coupled by exchange of the lipids between the two components, if this is sufficiently rapid relative to the ESR time scale ($\tau_{ex} \lesssim 10^{-7}$ sec). A check on this is whether the spectrum can be decom-

[33] J. R. Brotherus, O. H. Griffith, M. O. Brotherus, P. C. Jost, J. R. Silvius, and L. E. Hokin, *Biochemistry* **20**, 5261 (1981).

[34] J. H. Freed, *in* "Spin Labelling: Theory and Applications" (L. J. Berliner, ed.), Vol. 1, p. 53. Academic Press, New York, 1976.

[35] C. F. Polnaszek, D. Marsh, and I. C. P. Smith, *J. Magn. Reson.* **43**, 54 (1981).

posed into two independent components by subtraction. In general, the quality of the motionally restricted difference spectrum decreases with increasing temperature, and this could be due to increasing difficulty of exactly matching the narrow linewidths of the fluid component, but it could also be because the spectra are changing from a state of slow exchange to one of intermediate exchange. Spectral simulations for this latter situation are given in Ref. 36, from which values for the exchange rate can be estimated. The rotational correlation time τ_R, as estimated above, gives an upper estimate for the rate of exchange, assuming that this is the fastest motion of the motionally restricted lipids relative to the protein.[27] The increase in linewidth ($\delta H = \Delta H_{memb} - \Delta H_{lipid}$) of the fluid component relative to that of the extracted lipids also gives an approximate estimate for the exchange rate ($\nu_{ex} \sim (g\beta/h) \times \delta H$, see Ref. 27). For further details on spectral interpretation reference should be made to Refs. 3, 4, and 5.

[36] P. F. Devaux, J. Davoust, and A. Rousselet, *Biochem. Soc. Symp.* **46**, 207 (1981).

[91] A Computer Technique for Structural Studies of Bacteriorhodopsin

By HAYWARD ZWERLING, ROLF MEHLHORN,
LESTER PACKER, and ROBERT MACELROY

Introduction

Bacteriorhodopsin (BR) is an energy-transducing protein embedded in the membrane of *Halobacterium halobium*. Light initiates a photocycle which results in the transport of a proton across the cell's membrane.[1] To illuminate the proton pumping mechanism of this protein, many investigators have studied its structure. An underlying assumption in most of these structural investigations has been the hope that information derived from them may be useful in understanding the protein's mechanism of action.

This article discusses a technique which should prove helpful in defining the protein's secondary and tertiary structure. The premise of the iterative structure refinement technique is that structural similarities and differences between native BR and a hypothetical structure of BR can be highlighted by comparing the electron diffraction pattern of the actual bacteriorhodopsin molecule to the electron diffraction pattern that the hypothesized molecular structure would create. We suggest that currently

[1] D. Oesterhelt and W. Stoeckenius (1973). *Proc. Nat. Acad. Sci. U.S.* **70**, 2853.

METHODS IN ENZYMOLOGY, VOL. 88

available models of BR, although crude, are sufficiently detailed to be used as a first approximation in an iterative procedure that will produce information to allow the construction of a more accurate model.

Even when precise determinations of biopolymer structures have not, or cannot be made, it is possible to infer some information about probable conformations of both single molecules and the structure of molecular complexes. Using techniques of molecular description that have been developed during the last three decades, and the concurrent emergence of increasingly cheap computational power, it has been possible to use mathematical representations of molecular structures, of interactions within molecules, and of interactions between molecules to achieve descriptions of structure, and also to explore hypothesized mechanisms of action. When coupled with experimentally determined information of physical structure, the use of molecular models can provide a power tool for identifying structure.

We will first briefly describe the functions of one group of computer programs, the Ames Interactive Molecular Modeling System (AIMS),[2] which has been found to be useful in exploring molecular structure. We will then summarize the literature relevant to the study of the structure of bacteriorhodopsin. The derived information will provide a basis for the selection of the first-guess models that may be used at the beginning of the iterative structure refinement technique.

No effort has been made to discuss the theory supporting the use of this technique; rather, the description of this technique is intended as a guide for its practical implementation. We stress that this is only one of several refinement methods that rely on the comparison of experimental crystallographic data with theoretical information derived from a hypothesized structure; a summary of such techniques can be found in Blundell and Johnson.[3]

Classification of Molecular Modeling Techniques

ab initio Approaches

Two fundamentally different techniques, the *ab initio* approach and the empirical or "classical" method, based upon methods of energy calculation, are available for the exploration of molecular structure. The *ab initio* technique, in its most refined state, essentially considers the probable location of every electron around a specific atomic nu-

[2] Y. Coeckelenbersh, J. Hart, R. MacElroy, and R. Rein (1978) *Comp. Graph.* **3**, 9.

[3] T. Blundell and L. Johnson (1976). "Protein Crystallography." Academic Press, New York.

cleus while it calculates the interactions between all electrons associated with the same atom and with adjacent interacting atoms. The results of these computations are very accurate descriptions of the positions of atoms within a molecule, and of the interactions between atoms within a particular molecule. The technique has been used to explore pathways of molecular reaction, to identify reaction intermediates and their geometries, and to calculate enthalpies, entropies, and activation energies of reactions.[4] The advantage of this technique lies in its accuracy of predicting molecular characteristics. Its disadvantage is that its application, due to its use of large amounts of computational time and computer memory, is practically limited to molecules with a relatively small number of electrons.

In order to calculate the characteristics of rigid molecular systems various "semiempirical" quantum mechanical methods have also been developed. These methods use empirical parameters to approximate the calculation of large numbers of complex integrals. They were developed to address problems involving specific classes of molecules, and they differ primarily in their assumptions regarding the treatment of outer orbital electrons. Examples of some of these techniques are the Iterative Extended Hückel (IEH), the Complete Neglect of Differential Overlap (CNDO), and the Perturbative Configuration Interaction using Localized Orbitals (PCILO).

Empirical Approaches

The "classical" or empirical approach assumes that the fundamental characteristics of a molecule remain constant; more specifically, it assumes that both partial electronic charge and attractive and repulsive behavior do not change with the molecular environment. Approximate parameter data describing electronic charge and attractive and repulsive parameters can be obtained by experiment, for example, from crystal packing or from microwave data. The primary advantage of empirical calculations is high speed, which allows for the consideration of large numbers of atoms and large molecules. The disadvantages include the assumption of static atom behavior and a difficulty in adequately representing changes induced by the environment (for example, by solvents and ions).

[4] T. Oie, G. Loew, S. Burt, J. Binkley, and R. MacElroy (1982). *Intern. J. Quant. Chem.: Quant. Biol. Symp,* in press.
[5] J. Egan, J. Hart, and R. MacElroy (in preparation).

Molecular Modeling: The AIMS Program

The central feature of AIMS, the program for molecule construction and storage (MOLECULE), was designed to allow user interaction both with a variety of empirical and semiempirical energy calculation and minimization programs and with several graphical display devices.

The MOLECULE Program

MOLECULE is used to construct molecules, adjust conformations, and to establish the relative configuration of interacting molecules. The program consists of a large number of coordinated subroutines, each of which permits user interaction. Together they allow the assembly of mathematical representations of molecular structures. Such structures can be altered by the manipulation of bond length, bond angle, and torsional angle, to produce any desired molecular conformation. Molecules can be juxtaposed (bonded) to one another, or a collection of molecules can be assembled. Each molecule or collection of molecules can be stored in a library containing up to several dozen elements. Several hundred such libraries can be used at will. Structures can be moved both within a library or between libraries. Using specific libraries a user can, simply by defining its sequence, construct any of the usual biological polymers (proteins, ribonucleic acids, deoxyribose acids), in any large number of predefined geometries (such as α helices, A-DNA). Libraries can also contain specific molecules such as fatty acids, NADP, or phytochrome. Information which is assembled into a library element may be transferred to other AIMS programs.

Graphic Displays

The data produced by MOLECULE can be used to produce graphical representations of molecular structures. The method easiest to access uses a program designed to construct a ball-and-stick representation with a vector display system, such as the Tektronix microcomputer. This has been modified to also allow display on a raster screen, such as a Terak. The Terak display system has software allowing local manipulation of structures.[3]

Graphic displays can also be easily created for use on the Evans and Sutherland Computer System[2] which allows the user to interact with the stick figure and permits on-screen changes in conformation or spatial position. The displays are dynamic and can be used with a scripting program to make movie sequences. Single images can also be transferred to a Versatec printer.

The graphic data produced by MOLECULE can also be used to create displays on a Computer Output Microfiche (COM) device, which writes on photographic film. Displays can range from black and white stick structures to full color space-filling models.

Conformational and Spatial Minimization

Access to methods of molecular energy calculation permits the use of energy minimization techniques. An optimizing program, such as the quasi-Newton difference method (QNMDIF),[6] is used with a geometry manipulation program to alter molecular conformation, and/or the spatial relationships among molecules, until a minimum system energy is calculated. The program functions by calculating the energy of the system, arbitrarily moving the components recalculating energy, and repeating this procedure until a "direction" is found in which continued movement results in increasingly lower energies. Repetitions of this procedure allow the calculation of a minimum energy molecule conformation, a minimum energy spatial relationship among a set of molecules, or both.

Selection of a First-Guess Model of Bacteriorhodopsin

In order to implement the iterative structure refinement technique that is described in Section V, a reasonable first-guess model of BR must be constructed. What follows is a brief review of BR structural information.

The complete primary sequence of BR has been determined[7,8] and, although there are minor ambiguities in the amino acid sequence, for the purposes of this procedure, the disagreements are inconsequential.

The secondary structure of BR is composed largely of α-helix[9,10] and, possibly, β-sheet conformations.[10] Most models rely on data which concludes that 70–80% of the amino acids[9] are arranged in

---

[6] P. Gill, W. Murray, and R. Pitfield (1972). National Physics Laboratory Division and Numerical Analysis and Computation Department, No. 11.

[7] Y. A. Ovchinnikov, N. Abdulaev, M. Y. Feigina, A. V. Kiselev, and N. A. Lobanov (1979). FEBS Lett. **100**, 219.

[8] H. G. Khorana, G. Gerber, W. Herlihy, C. Gray, R. Anderegg, K. Nihei, and K. Biemann (1979). Proc. Nat. Acad. Sci. U.S. **76**, 5046.

[9] M. M. Long, D. W. Urry, and W. Stoeckenius (1977). Biochem. Biophys. Res. Commun. **75**, 725.

[10] B. Yap, M. Maestre, S. Hayward, and R. Glaeser. To be published.

α-helical conformation; however, one model relies more heavily on data which concludes that no more than 50% of the amino acids are arranged in α-helical conformation and at least 20% of the peptides are arranged in β-sheet conformation.[10] Recently, it was reported that the α-helical component is in a α_{II} configuration.[11]

Having selected the type(s) of secondary structure(s), the polypeptide sequence is partitioned into secondary structural domains. This partitioning has been based on energy calculations, chemical modifications, enzymatic cleavage, and/or protein crystallography. This has resulted in the construction of several models of the secondary structure of BR incorporating seven α-helical coils[7,12-14] and one model which consists of five α-helical and four β-sheet segments.[10]

After secondary structural domains have been assigned to the primary sequence, the domains are assembled into a reasonable tertiary structure. Visual inspection of the three-dimensional electron density map reveals several (seven[14,15]), transmembrane rods, the position and inclination of which may be measured directly from the three-dimensional density map. This information, combined with knowledge about the extracellular/intracellular orientation of the three-dimensional electron density map,[16] provides a first approximation of the geometric orientation of the transmembrane rods within the context of the unit cell.

Each secondary structural domain is then assigned to one of the seven transmembrane positions whose coordinates have been determined. Obviously, the model builder should ensure that consecutive secondary structural domains are close enough to each other so that the intertransmembrane link will be long enough to accommodate the chosen placement of the intramembrane domains. One of the most important clues regarding the assignment of the secondary structural domains to the transmembrane position in the electron density map has been provided by the "visualization" of the extramembrane polypeptide segments which link the transmembrane rods.[14] Based on this linking information alone (and assuming the BR monomer is composed of seven α-helices), the total possible number of models (vis-à-vis the placement of the secondary structural domains) is reduced to, at most, sixteen models.

[11] S. Krimm and A. M. Dwivedi (1982). *Science* **216**, 407.

[12] D. M. Engelman, R. Henderson, A. McLachlan, and B. Wallace (1980). *Proc. Nat. Acad. Sci. U.S.* **77**, 2023.

[13] D. M. Engelman, A. Goldman, and T. A. Steitz (1982), this volume, Article [11].

[14] D. Agard and R. Stroud (1981). *Biophys. J.* **37**, 589.

[15] R. Henderson and P. N. T. Unwin (1975). *Nature (London)* **257**, 28.

[16] S. Hayward, D. Grano, R. Glaeser, and K. Fisher (1978). *Proc. Nat. Acad. Sci. U.S.* **75**, 4320.

By invoking additional constraints the total number of possible models can be reduced even further. Useful constraining information includes: (1) the position of the β-ionone ring and the Schiff base end of the retinal has been located in the electron density map;[17] (2) the retinal does not leave the binding site during the photocycle;[18] and (3) the retinal is (probably) covalently bonded to Lys-216 and not to Lys-41.[19-21] Collectively, this information precludes Lys-216 from being located in several of the transmembrane positions.

Having assigned the polypeptides to a secondary structure and arranged the secondary structure into a crude tertiary structure, the helices are oriented about their long axes. Useful information for this task includes evidence that the phenylalanine residues tend to be located in the central region of the monomer while the valine residues tend to be located near the monomer's periphery,[22] and the knowledge that Tyr-26 is probably located in the vicinity of the active site of the retinal.[23]

Taken as a whole, the above information will result in the construction of several "reasonable" models, which can be used as a first approximation of the structure of bacteriorhodopsin. The first-guess model is then used as the starting point in the iterative structure refinement technique that is explained below.

Mathematical Formulation of the Iterative Structure Refinement Technique

The first-guess model that was envisaged in accordance with the data summarized in Section IV can be constructed using MOLECULE. (The model to be used in this refinement technique must include all of the atoms that are contained in one unit cell, or in an equivalent structural unit. In the case of BR, an appropriate structural unit is the BR trimer.)

[17] G. King, P. Mowery, W. Stoeckenius, H. Crespi, and B. Schoenborn (1980). *Proc. Nat. Acad. Sci. U.S.* **77**, 4726.

[18] N. Katre, P. Wolber, W. Stoeckenius, and R. Stroud (1981). *Proc. Nat. Acad. Sci. U.S.* **78**, 4068.

[19] H. Bayley, K. Huang, R. Radhakrisnan, A. Ross, Y. Takagaki, and H. G. Khorana (1981). *Proc. Nat. Acad. Sci. U.S.* **78**, 2225.

[20] H. D. Lemke and D. Oesterhelt (1981). *FEBS Lett.* **128**, 255.

[21] E. Mullen, A. Johnson, and M. Akhtar (1981). *FEBS Lett.* **130**, 187.

[22] D. M. Engelman and G. Zaccai (1980). *Proc. Nat. Acad. Sci. U.S.* **77**, 5894.

[23] H. D. Lemke and D. Oesterhelt (1981). *Eur. J. Biochem.* **115**, 595.

The refinement technique requires four major steps:

A. The redefinition of the positions of the atoms are defined by their location within the unit cell.
B. The calculation of the model's diffraction pattern.
C. A comparison of the model's calculated diffraction pattern with the experimentally obtained diffraction pattern.
D. The refinement of the model based on a difference Fourier electron density map.

The iterative structure refinement technique is implemented as described below.[2]

Step A. Redefinition of the Atomic Coordinates

The molecular model that was constructed by MOLECULE was defined in a cartesian coordinate system. The first step in the implementation of this technique is to define the position of each atom based on its location within the unit cell. Therefore, the position of atom "i" will be defined by the vector, $\bar{R}_i$:

$$\bar{R}_i = x_i\bar{a} + y_i\bar{b} + z_i\bar{c} \tag{1}$$

where $(\bar{a},\bar{b},\bar{c})$ represent *unit* vectors and (x_i,y_i,z_i) are scalar quantities. The directions of the unit vectors are defined by the edges of the unit cell. The magnitude of the unit cell vector shall be defined such that each edge of the unit cell has a length of one. For example, BR forms a P3 hexagonal lattice that is one unit cell, approximately 45 Å deep. The membrane face of the BR unit cell has the shape of a parallelogram; the edges of the parallelogram measure 63 Å in length and enclose angles of 60° or 120°.[15] If $\bar{a}$ and $\bar{b}$ are defined by the edges of the parallelogram then the vector $\bar{R} = 0.5\bar{a} + 0.0\bar{b} + 0.0\bar{c}$ represents a point halfway (31.5 Å) down the $\bar{a}$ edge of the unit cell. The $\bar{c}$ unit vector is normal to the $\bar{a} - \bar{b}$ plane. One unit displacement in the $\bar{c}$ direction represents a movement 45 Å (one unit cell) normal to the plane of the membrane.

Step B. Calculation of the Model's Diffraction Pattern

The diffraction vector, which defines a point in the diffraction pattern (also called reciprocal space), is defined by the vector $\bar{S}$:

$$\bar{S} = h\bar{a}^* + k\bar{b}^* + l\bar{c}^* \tag{2}$$

where h, k, and l are integers. The relation between $(\bar{a},\bar{b},\bar{c})$ and $(\bar{a}^*,\bar{b}^*,\bar{c}^*)$ is defined by the shape of the unit cell. (For a table of such relations see reference).[24] Henceforth, $|\bar{S}|$ will represent the magnitude of the vector $\bar{S}$. In the case of BR, $\bar{a}^*$ and $\bar{b}^*$ are not orthogonal; therefore, caution should be exercised in calculating $|\bar{S}|$.

If $\bar{a}$ and $\bar{b}$ are defined in the unit cell so that they enclose an angle of 120°, $\bar{a}^*$ and $\bar{b}^*$ must be defined, in reciprocal space, so that they subtend an angle of 60°.[24] One unit displacement in the $\bar{a}^*$ (or $\bar{b}^*$) direction represents a displacement in reciprocal space of $2/[63(3)^{1/2}]$ Å^{-1}. $\bar{c}^*$ is defined as normal to the $\bar{a}^* - \bar{b}^*$ plane. One unit displacement in the $\bar{c}^*$ direction represents a movement in reciprocal space of $1/45$ Å^{-1}.

After having selected and appropriately defined a molecular model, the diffraction pattern of the model is calculated using the Structure Factor equation:[3]

$$\bar{F}(\bar{S}) = \sum_{\bar{S}} f_i(|\bar{S}|) \exp [2\pi i(x_i h + y_i k + z_i l)] \tag{3}$$

for $|\bar{S}| \leq$ (Resolution)$^{-1}$. The symbol $\sum_{\bar{S}}$ is an abbreviation for $\sum_h \sum_k \sum_l$. Henceforth, the symbol $|\bar{F}(\bar{S})|$ will represent the amplitude of the vector $\bar{F}(\bar{S})$.

The scalar quantity $f_i(|\bar{S}|)$ in the structure factor equation is called the atomic scattering factor. It represents the relative strength of the scattered incident radiation by an individual atom at a distance $|\bar{S}|$ in reciprocal (diffraction) space. Numerical values and approximations for the atomic scattering factor as a function of the parameter $|\bar{S}'|$ have been published;[24,25] in these sources $|\bar{S}'| = 0.5|\bar{S}|$.

The structure factor, which is a complex number, has a real and an imaginary component. After both components of the structure factor have been calculated, the structure factor is redefined in terms of its amplitude, $|\bar{F}(\bar{S})|$, and its phase, $\sigma(S)$. The structure factor is related to the experimentally obtained diffraction pattern such that $(|\bar{F}(\bar{S})|)^2$ is equal to the intensity of the diffraction pattern at $\bar{S}$.

To determine the electron density of the model at every point in real (x,y,z) space, the Fourier transform of the structure factor is calculated.[3] Specifically,

$$\text{Density } (x,y,z) = \frac{1}{\text{volume}} \sum_{\bar{S}} |\bar{F}(\bar{S})| \cos [\sigma(\bar{S}) - 2\pi(xh + yk + zl)] \tag{4}$$

An electron density contour map of the model may be constructed by using the values of the electron density matrix [Eq. (4)] as the input to

[24] International Table of X-Ray Crystallography (1962). Kynoch Press, Birmingham, England.
[25] P. Doyle and P. Turner (1968). Acta Crystallogr. A24, 390.

a contour plotting subroutine. If a two-dimensional electron density map is desired, then the "*l*" and "*z*" coordinates in Eqs. (1), (2), (3), and (4) are ignored.

Step C. Assessment of the Model's Diffraction Pattern

Having calculated the model's diffraction pattern (structure factor), an assessment is made to determine how accurately the model mimics the true molecular structure. This is accomplished by comparing the experimental (observed) structure factor amplitude, ($|\bar{F}o(\bar{S})|$), to the model's (calculated) structure factor amplitude, ($|\bar{F}c(\bar{S})|$).

In order to allow the direct comparison of $|\bar{F}c(\bar{S})|$ to $|\bar{F}o(\bar{S})|$, both of the structure factors are normalized using the relation

$$|\bar{E}(\bar{S})| = \frac{|\bar{F}(\bar{S})|}{\left(\sum_{\bar{S}} |\bar{F}(\bar{S})|^2\right)^{1/2}} \tag{5}$$

With this redefinition of the structure factors, the overall (global) similarity of the trial structure and the actual crystalline structure is specified by the value of the residual, R, where:

$$R = \frac{\sum_{\bar{S}} (\||\bar{E}o(\bar{S})| - (\bar{E}c(\bar{S})\|)}{\sum_{\bar{S}} |\bar{E}o(\bar{S})|} \tag{6}$$

As the value of the residual is reduced toward zero, the hypothesized structure more closely resembles the actual crystalline structure.

Step D. Structural Refinements Based on the Difference Fourier Density Map

Although the residual quantifies the global similarity between the trial structure and the true structure, it does not specify which regions of the model are in the best, and worst, agreement with the actual molecular structure. This type of information may be obtained from a difference Fourier density map.

To construct a difference Fourier density map, a structure factor amplitude is defined by

$$|\bar{F}(\bar{S})| = |\bar{E}o(\bar{S})| - |\bar{E}c(\bar{S})| \tag{7}$$

Using the new amplitude, and the phase information that was generated by the hypothetical structure in Step B [Eq. (3)], the difference Fourier density matrix can be calculated using Eq. (4). A difference Fourier contour map can be created by using the difference Fourier density matrix as the input to a contour plotting subroutine. The resultant difference Fourier density map will consist of negative and positive contours.

When the difference Fourier contour map is superimposed upon the electron density map of the model [that was obtained in Step B, Eq. (4)], the negative and positive peaks of the difference map indicate areas within the model which have, respectively, too much or too little mass as compared to the actual crystalline structure. The information thus obtained will enable the model builder to selectively alter those

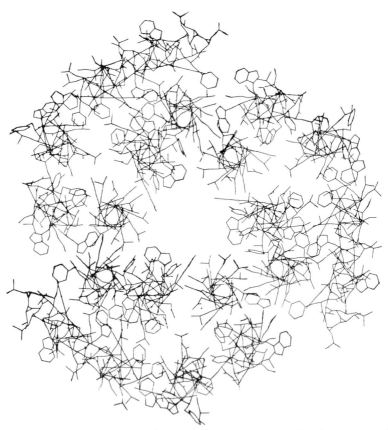

Fig. 1. A graphic representation of a first-guess bacteriorhodopsin trimer generated on the Evans and Sutherland Computer System. Details concerning the construction of this model are discussed in the text.

areas of the molecular model which are in the poorest agreement with the actual crystal.

The entire procedure could be repeated until a sufficiently small residual is obtained.

Graphic Demonstration of the Refinement Technique

Based on information that was presented in Section IV, a reasonable first-guess model of bacteriorhodopsin was assembled using MOLECULE. The model consists of seven α-helices. Its primary sequence is defined in Khorana et al.,[8] its secondary sequence is defined in Engelman et al.,[13] and its tertiary sequence is model No. 1 in Engelman et al.[12] The helices are oriented about their long axes so that they do not violate the data in Engelman and Zaccai[22] and Lemke and Oes-

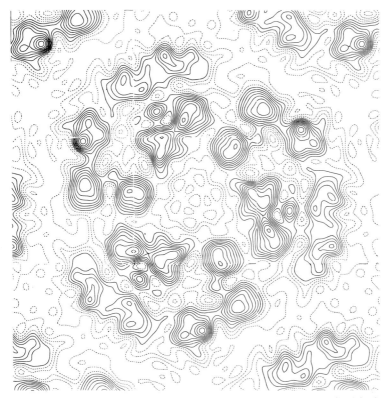

FIG. 2. The electron density map generated with the technique described in Section V,A and B. The hypothetical molecular structure used to generate this map is shown in Fig. 1.

terhelt.[23] The model, intended purely for demonstrative purposes, does not include water molecules, lipid molecules, the retinal or interhelical polypeptide links. The cytoplasmic face of this model, generated on an Evans and Sutherland picture system, is shown in Fig. 1.

Having first specified all atomic positions in a cartesian coordinate system, the coordinates of the atom are then redefined using the coordinate system of the unit cell [Step A, Eq. (1)]. The structure factor, at 3.7 Å resolution, is then calculated as was done in Step B [Eq. (3)]. Subsequently, the density matrix is calculated using Eq. (4). The values of the density matrix are used to create the contour map of the electron density shown in Fig. 2. The map shows a gross qualitative similarity to the 3.7 Å electron density map of BR.[26]

The residual, which quantifies the similarity between the trial structure and the true structure, is calculated using Eqs. (5) and (6). If the residual is not as small as desired, a difference Fourier density map should be created using Eqs. (5), (7), and (4). Regions within the model that have too much or too little mass can be highlighted by superimposing the difference Fourier density map on the model's electron density map. The peaks and valleys in the resultant difference Fourier map indicate regions of the model that contain, respectively, too little or too much mass as compared to the true molecular structure. The trial structure may then be selectively modified.

The entire procedure can be repeated as necessary until a sufficiently small residual is obtained.

Acknowledgments

This work was funded by NASA-AMES Interchange Agreement No. NCA2-ORO50-202. We thank S. Hayward and B. Yap for useful discussions and B. Yap for permission to quote unpublished results.

[26] S. Hayward and R. Stroud (1982). *J. Mol. Biol.* **151,** 491.

Author Index

Numbers in parentheses are reference numbers and indicate that an author's work is referred to although the name is not cited in the text.

Subject Index

A

B